THE PERIODIC TABLE OF THE ELEMENTS

	IA	IIA	IIIB	IVB	VB	VIB	VIIB	VIIIB			IB	IIB	IIIA	IVA	VA	VIA	VIIA	0
1	H 1																	He 2
2	Li 3	Be 4											B 5	C 6	N 7	O 8	F 9	Ne 10
3	Na 11	Mg 12											Al 13	Si 14	P 15	S 16	Cl 17	Ar 18
4	K 19	Ca 20	Sc 21	Ti 22	V 23	Cr 24	Mn 25	Fe 26	Co 27	Ni 28	Cu 29	Zn 30	Ga 31	Ge 32	As 33	Se 34	Br 35	Kr 36
5	Rb 37	Sr 38	Y 39	Zr 40	Nb 41	Mo 42	Tc 43	Ru 44	Rh 45	Pd 46	Ag 47	Cd 48	In 49	Sn 50	Sb 51	Te 52	I 53	Xe 54
6	Cs 55	Ba 56 *	Lu 71	Hf 72	Ta 73	W 74	Re 75	Os 76	Ir 77	Pt 78	Au 79	Hg 80	Tl 81	Pb 82	Bi 83	Po 84	At 85	Rn 86
7	Fr 87	Ra 88 †	Lw 103	— 104	— 105													

*Lanthanons	La 57	Ce 58	Pr 59	Nd 60	Pm 61	Sm 62	Eu 63	Gd 64	Tb 65	Dy 66	Ho 67	Er 68	Tm 69	Yb 70
†Actinons	Ac 89	Th 90	Pa 91	U 92	Np 93	Pu 94	Am 95	Cm 96	Bk 97	Cf 98	Es 99	Fm 100	Md 101	No 102

contemporary chemistry

contemporary chemistry

JOHN E. HEARST
University of California, Berkeley

JAMES B. IFFT
University of Redlands

W. H. FREEMAN AND COMPANY
San Francisco

Frontispiece courtesy of Professor Ignacio Tinoco, Department of Chemistry, University of California, Berkeley. See p. 326 for an explanation of the presence of the cloud at the Matterhorn's peak.

Library of Congress Cataloging in Publication Data

Hearst, John E. 1935–
Contemporary chemistry.

Bibliography: p.
Includes index.
1. Chemistry. I. Ifft, James B., 1935– joint author. II. Title.
QD31.2.H4 540 75-28230
ISBN 0-7167-0172-3

Printed in the United States of America

9 8 7 6 5 4 3 2 1

TO

Daniel
David
Evelyn
Jean
Joanna
Leslie
Stephen

contents

preface

This is a textbook for an introductory, one-year college or university chemistry course. It is designed to accompany 60 to 80 one-hour lectures. The student is expected to have taken high-school algebra, geometry, trigonometry, and chemistry.

We wrote the book because of the widespread opinion that most high-quality chemistry textbooks are written by professional chemists for future professional chemists, whereas most students enrolled in elementary college chemistry courses are premedical students and life-science majors. This book is therefore directed more to the needs of these students, without compromising the principles that must be a part of every good, basic chemistry course. We have done this primarily by using examples from organic chemistry and biochemistry and eliminating most of the descriptive inorganic chemistry found in many introductory texts. We feel that descriptive chemistry is best learned in the laboratory with the chemicals in one's hands, and not just from the printed page. Extensive tables of data are presented in both the text and the appendixes to further acquaint the student with this subject. In addition, references are given to other sources of information.

We believe that the material in this book is suitable for all science majors. We also believe that the order of presentation of the material is superior to the more common chronological order. The book begins with atoms and ends with the genetic code.

We are pleased to express our appreciation to our friend and expert typist, Eleanor Scott, for her skill and patience in typing the several drafts of our manuscript.

October 1975

JOHN E. HEARST
JAMES B. IFFT

1 introduction

This chapter is about science, particularly the science of chemistry—what it is, how it is practiced, and what it is "good for." Perhaps this material will help you formulate some questions for yourself about the nature of chemistry and why anyone should bother studying it. We hope you will be encouraged to seek the answers. Although we may never get to know you personally, we already know something about each other: you are beginning a course in chemistry and we have written a textbook about chemistry. Why are you taking chemistry? And why have we written this book? At least it can be said that we all have an interest in, a need for, or possibly an amazement at the subject of chemistry. Although we cannot get to know each other well through the pages of a book, we hope that, by reading this book, you will be able to learn something about us and why we are interested in chemistry.

1.1 WHAT IS CHEMISTRY?

Since the beginnings of history, science has been studied and practiced by only a small portion of humanity. The knowledge obtained by the practice of science has, however, played a dominant role in the history of civilization. The first "chemists" lived in the Stone Age and discovered how to control fire, then start it at will. To the Stone Age man, wood was changed to heat by fire. With the control of heat came the beginnings of the common chemistry of daily life, such as cooking, which altered the taste, color, and texture of food. The heating of clay made the production of bricks and pottery possible. The heating of the mineral kaolinite produced glazes and the heating of sand produced glass.

As early as 4000 B.C., man discovered that heating certain blue rocks produced droplets of copper. Thus began the science of metallurgy, which ushered in the Bronze Age and Iron Age. We see early in recorded history that those who made systematic observations of matter, its properties, and the changes it undergoes contributed enormously to the culture of man. The metal frying pan doubtless came into being simultaneously with metal spear heads and swords. Science served man's comfort and fostered his self-destruction. The amoral character of technical knowledge became apparent.

The origin of the word *chemistry* is obscure, but it comes to us via the Greek word *chēmeia,* meaning "the art of metalworking." In 900 B.C. the Egyptians were knowledgeable in metallurgy, in the production of pigments from minerals, and in the extraction of juices from plants. *Chēmeia* may be derived from the word *Kham,* the Egyptians' name for their land. It has also been suggested that *chēmeia* derives from the Greek word *chymos,* meaning "the juice of a plant," or from the Cantonese word *kem-mai,* meaning "gone astray in search of gold."

For those of us "going astray" today, chemistry is the observation and description of matter, its properties, and the changes it undergoes. Our knowledge is still amoral and is, unfortunately, still limited to a small portion of humanity. It remains a dominant influence upon human endeavors, however, and the quality of its use will ultimately determine how long the human race will survive on Mother Earth.

1.2 THE ADVANTAGES OF IGNORING HISTORY

People have been posing chemical questions—and answering some of them—for a very long time. In 2500 B.C. the Chinese postulated, in the *Shu-ching,* the Canonical Book of Records, that there were five elements. Progress in understanding matter was painfully slow from that time until about two centuries ago, when scientific knowledge began its present period of explosive growth.

The question raised by these thoughts is whether or not the arrangement of a general chemistry textbook should reflect the historical development of the subject. Apparently, many authors believe that it should. Most textbooks begin with a discussion of gases—one of the first areas of science to be treated quantitatively—and end with the most recent concepts in nuclear chemistry.

If you turn to the outline of this book, you will find a rather different approach. Our sequence of topics does not correspond to the historical development of science in general or chemistry in particular. We are not antihistory—not at all! The study of history provides invaluable insights into the development of man's understanding of everything that he has thought and written about: economics, government, philosophy, as well as science. However, two factors deter us from the historical approach. First, the progress of science has often been confusing at its best and illogical at its worst. To follow the actual, chronological progress of chemistry is an exercise in frustration. Second, we are writing a book about science, not the history of science. Many books have been written about the history of science and numerous courses in this fascinating field are available at our major universities. We encourage your pursuit of these studies as a supplement to the material presented here.

If we have rejected the traditional, historical approach to the presentation of general chemistry, what approach *have* we taken? One of our goals is that you learn enough basic chemistry from the earlier portions of this book to be able to understand some of the elementary aspects of molecular biology (that branch of biology in which the structure and development of biological systems are analyzed in terms of the physics and chemistry of their molecular constituents). The book builds up to and concludes with a discussion of the structure of proteins and nucleic acids and a description of the genetic code.

In order to achieve this goal, an elementary understanding of a number of areas of physical chemistry, such as thermodynamics, kinetics, and oxidation-reduction, is necessary. It is difficult to understand the structure and properties of a gigantic molecule without first understanding the structure and properties of the smallest molecules. It is impossible to understand a molecule without understanding the bonds that hold atoms together in a molecule. And finally, one cannot comprehend bonds between atoms without some appreciation of what atoms are really like.

And so our book begins—with atoms. The description of the atom is elementary but modern. Chapter 2 presents an introduction to the properties of waves and the associated concept of quantum states. In this way a rationale for atomic structure is provided. Chapter 3 presents a survey of some of the physical and chemical properties of the elements, as correlated by the periodic table of the elements. We believe that most of you are more interested in the biological implications of chemistry than in the descriptive inorganic chemistry that constitutes a large part of many general chemistry texts. Although the emphasis in this book is decidedly on such biological implications, Chapter 3 does give a brief introduction to the properties of inorganic compounds. If you wish to pursue inorganic chemistry further (and you should!), we refer you to such works as E. S. Gould's *Inorganic Reactions and Structure,* revised ed. (Holt, Rinehart and Winston, New York, 1962), R. T. Sanderson's *Inorganic Chemistry* (Reinhold, New York, 1967), or the monumental series by J. W. Mellor and G. D. Parkes, *A Comprehensive Treatise on Inorganic and Theoretical Chemistry* (Wiley, New York, 1922–), to which new volumes are added periodically.

Chapters 4 and 5 describe the bonds that hold atoms together to form molecules, and molecules to form liquids and solids. These intramolecular and intermolecular forces are very close to the heart of much of modern scientific research. Several theories of bond formation are included in Chapter 4, along with information about which theory does the better job of predicting bond energies or molecular geometries in various instances. A number of modern experimental techniques that are useful in the study of the properties of molecules are included in Chapter 5.

The chapter on the states of matter describes the properties of large collections of atoms or molecules in the gaseous, liquid, and solid states. It begins with the best understood state of matter, the gaseous, and presents and compares two descriptions, the experimental and the theoretical. Quantitative relations for the liquid and solid states are much more difficult to derive and understand, so the descriptions of these states are more qualitative.

Chapters 7 and 8 present introductions to two very important branches of physical chemistry, namely, thermodynamics and kinetics. These chapters are necessarily more mathematical than any others in the book. If this is a problem for you, we recommend that you spend extra time trying to understand the underlying concepts of these two disciplines rather than memorizing formulas. The *best* way to gain understanding is to practice solving problems, and then practice some more. After all, solving problems in order to understand nature is what science is all about.

The chapter on chemical equilibrium is one of the longest in the book. Many things chemists do concern systems at equilibrium; hence there is much information to assimilate. The mathematical level seldom ranges beyond algebra here. As is always true, an understanding of the underlying principles should be the primary goal of your study. So—solve problems! The chapter on oxidation-reduction reactions is concise, but it explains the methods for calculating electrochemical potentials of galvanic and electrolytic cells and the use of such potentials in determining equilibrium constants. The chapter ends with a description of the electrochemical processes in living cells.

The book concludes with an extensive discussion of biological molecules, both small and large. This chapter represents the climax of a year's study, in that nearly all the principles learned are applied in the description of the chemistry of life. The amino acid side chains are classified according to their acidity and polarity. The structures of monosaccharides and nucleotides are presented. The relation between molecular structure and optical properties, first introduced in Chapter 5, is a recurrent theme throughout Chapter 11. Polymerization and some general properties of polymers are then discussed, culminating in a description of the biopolymers: proteins, polysaccharides, and nucleic acids. The chapter concludes with a brief illustration of the central dogma of molecular biology, or molecular genetics. The final objective of the book is thus an explanation in molecular terms of the essential features of all life as we know it. It describes the molecular mechanism for the propagation and transfer of biological information.

1.3 THE SCIENTIFIC METHOD

What do you do when you are confronted with a refrigerator that doesn't work, a zipper that's stuck, or a jib sail that can't be raised? If you can stifle your anger or impatience for a few minutes, think about the probable source of the problem, postulate how the problem might be solved, and then try your solution, your life becomes somewhat less frustrating than it might otherwise be.

These are precisely the skills that the scientist must develop. If we replace the concept of a problem with the concept of a question, we have the first step in a scientific inquiry. A broken appliance is an obvious problem. "How can it be repaired?" is the obvious question. In scientific research it is easy to think of questions, but not all questions are likely to lead to productive lines of investigation. The formulation of a *good* question by a scientist is far more difficult. It requires a disciplined and highly trained mind as well as the crucial ingredient of curiosity. If people were not really interested in why nature behaves as it does, they would ask no questions and there would be no science.

Once a question is formulated, thereby defining a scientific problem, an experiment or set of experiments must be devised to provide data that will have a direct bearing on the question. Depending on the results of the initial experiments, further experiments may be indicated. If the question was perceived with sufficient insight and the initial experiments carried out with sufficient ingenuity and precision, the investigator may be able to proceed directly to the next step, which is the formulation of a hypothesis to explain his results, i.e., answer the question he originally asked. He can then devise further experiments to test this hypothesis and, if it seems valid, use it to provoke new, more sophisticated questions so that the cycle can start again.

Scientists seldom work within such a rigid framework of inquiry. The problems they work on are infinitely varied, and the kinds of insight and approach they use on these problems are often utterly different and highly personal in style. In one form or another, however, the elements of the scientific method described above enter every investigation.

1.4 WHY STUDY CHEMISTRY?

From our experience with several thousand students over the past decade, there are a number of answers to this question. Some of them are not very good: "Most people seem to take it," "My girlfriend is in the class," or "It's considered a snap course." Other answers are at least practical: "I've got to have it to get into medical school," "My engineering major requires it," or "I want to be a patent lawyer for a chemical company." The first group of students could benefit from some alternative answers, and the second group could also. In addition, there is a third group. These are the students who do not really know what chemistry is or what it is good for, but may suspect it has something to do with many of society's problems (e.g., problems of health, the environment, energy resources, and economics), and would like to find out. We are particularly interested in these students.

One reason for studying chemistry has already been provided in the preceding section: it gives experience in using the scientific method in solving problems or answering questions. This should considerably strengthen your ability to think carefully and analytically about the many problems you face daily.

If you are at all curious about why things behave as they do, you have undoubtedly asked many questions about the world around you. The course of study you are about to begin will provide some immediate answers and enough background information that you can begin digging out other answers for yourself. Some questions you will soon be able to answer are: Why does a cake rise? Why is blood red and grass green? What is smog? Why can I see better several minutes after entering a dark room? How do those glasses that become darker in the sunlight work? Why do children tend to look like their parents? How does antifreeze keep a car radiator from freezing? How do you make wine? Why does a warm bottle of cola froth when it is opened? And on and on.

A third and very important reason for studying chemistry has to do with the quality of the environment in which you will spend the next half-century. Mercury poisoning, smog, overpopulation, water pollution, lead contamination, DDT, food shortages, radioactive wastes, thermal pollution. If you have read a newsmagazine or a newspaper or talked with a friend during the past year, you have undoubtedly heard of all these problems, and many more. Who is going to solve them?

Chemists will certainly not do it alone. These are extremely complex problems that will require the coordinated efforts of governmental agencies, scientists, engineers, sociologists, economists, etc. But the chemist will certainly play a major role because most of these problems are, at least in part, chemical. Smog consists of a variety of gaseous compounds and particulate matter generated in large part by the internal combustion engine. Chemists are playing a major role in identifying the components of smog, studying the microscopic atmospheric particles on which many chemical reactions occur, and working with automotive engineers to design cleaner-burning engines.

DDT is a chemical. It has virtually destroyed the osprey population of Long Island, found its way into the penguins of Antarctica, even though it has never been sprayed there, and accumulated sufficiently in the fat of some fish that they are unfit for humans to eat. Chemists made DDT; they must now make a safe substitute.

Mercury poisoning has eliminated the West Coast swordfish fleet. The reality of mercury poisoning has been demonstrated by the death and grotesque deformation of many of the residents of the tragic town of Minamata, Japan. A chemical plant discharged mercury into the bay, where it entered the fish that were caught by local fishermen and eaten by the townspeople. Chemists must find substitutes for certain mercury compounds and advise legislators how to prevent the further release of mercury into the environment.

To increase your interest in chemistry and keep you conscious of the relations between your studies and the world around you, we have included an *Interlude* at the end of each chapter except this one. Each Interlude is related to the chemistry you have learned from the preceding chapter. The Interludes may be divided into three classes. Four come under the heading of Ecology; they deal with resources and pollution, smog, energy sources for society, and the supersonic transport. Three we classify as Biology; they describe the chemical bases for smell and night vision and the scientific aspects of winemaking. The remaining three pertain to Technology, as represented by transistors, water softening, and fuel cells.

Many more Interludes could be written. We hope that the ones in this book will stimulate your desire to find out more about the role of chemistry in human endeavors, and that some of them will demonstrate to you that, in order for society to deal intelligently with many of its problems, we, the people, need as much technical knowledge as we can get. Furthermore, we need professional chemists who are willing to devote their time and energy to help clean up our befouled environment and keep it clean, and provide us with better health and a better understanding of what we are.

2 the fundamental properties of matter

2.1 THE ATOMIC HYPOTHESIS

The hypothesis that all matter consists of atoms is central to a modern scientific description of matter. For reasons that will become apparent throughout this book, the evidence in favor of this hypothesis is overwhelming. It is also very recent (less than two centuries old), considering that the atomic hypothesis was first propounded by the Greek philosopher Demokritos about 2400 years ago.

If we could magnify a drop of water about 50 million times, we would see clusters of three atoms in constant, rapid motion. Figure 2-1 is a two-dimensional representation of what magnified water might look like. Each group of three atoms is called a *molecule*. We will discuss the definition of a molecule in more detail later, but for now, a molecule can be thought of as a discrete group of atoms having a fixed, unvarying composition.

There are several noteworthy features of Figure 2-1. Not all atoms are the same size. There are, in fact, 105 known kinds of atoms having distinct values of physical properties such as radius and mass. Furthermore, the clusters of atoms (molecules) have a definite geometry. Each of the larger atoms (oxygen) has two smaller atoms (hydrogen) connected to it. The angle between the two center lines from hydrogen atom to oxygen atom is 104.5° and the distance between the center of each hydrogen and the center

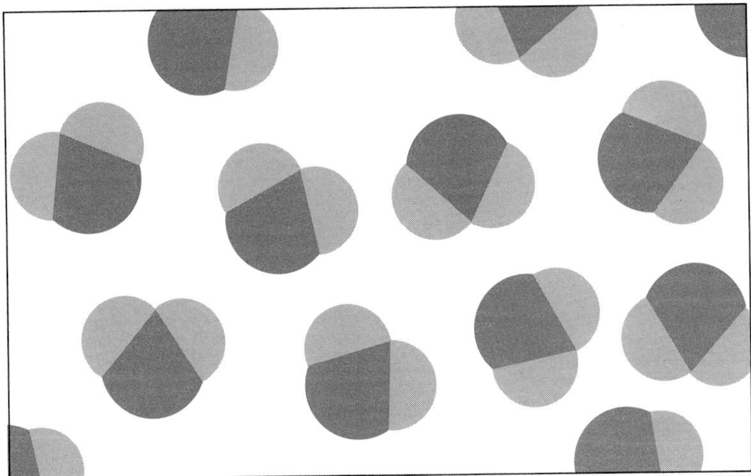

Figure 2-1. Water magnified about 50 million times.

of the oxygen is 0.958×10^{-8} centimeter (cm).* Since atomic radii are about 1 or 2×10^{-8} cm, it is convenient to express atomic and molecular dimensions in terms of the *angstrom* (Å), which is defined as 10^{-8} cm. Thus, 0.958×10^{-8} cm = 0.958 Å. The angstrom was named after the Swedish physicist Anders Ångström (1814–1874), who proposed it.

If we heat our drop of water with a flame, it evaporates; the resulting vapor (steam) is represented in Figure 2-2, again magnified about 50 million times. The molecules are now far apart, on the average, and moving about very rapidly. Since there is very little interaction between them, the steam will expand indefinitely unless it is confined in a container. This is a characteristic property of all gaseous substances.

If, instead of heating the water drop, we put it in a freezer, it solidifies; the resulting solid (ice) is represented in Figure 2-3. In ice the molecules do not move freely because the forces of attraction between them are very strong. Ice is therefore very rigid, a property of most solids. At the molecular level most solids are characterized by repetitive, highly symmetric arrays of molecules. Such symmetric arrays are called *crystals*.

Our simple description of water leaves us with a large number of questions that underlie the subject of chemistry. Just a few examples are: (1) Why do hydrogen and oxygen bind to each other? What determines the distance between their centers? What determines the angle between their center lines? (2) Why do molecules attract each other in a liquid? What

*See Appendix A for the definitions and abbreviations of units, and their conversion factors. See Appendix B for a review of exponentials and logarithms.

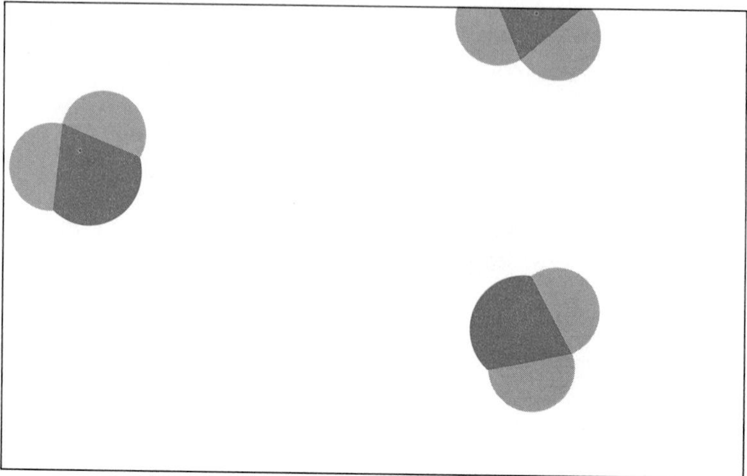

Figure 2-2. Steam magnified about 50 million times.

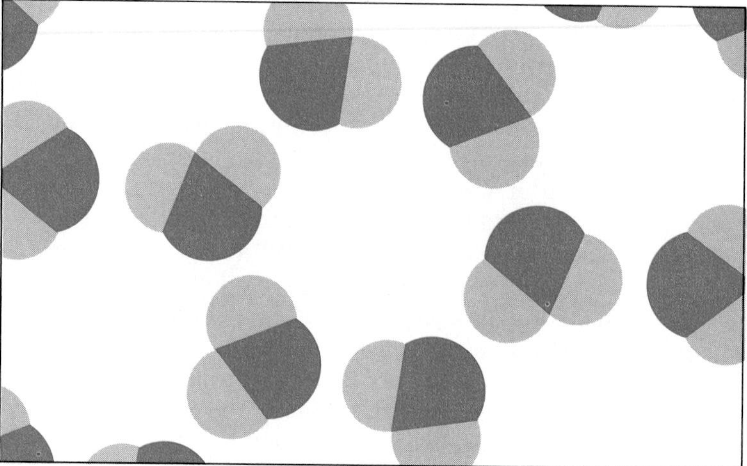

Figure 2-3. Ice magnified about 50 million times.

determines the temperature at which a liquid solidifies? What determines the structure of its crystal? (3) What is heat? How much heat is required to vaporize the water drop? How does the vapor respond to changes in temperature and pressure? (4) What other molecules exist? How can one group of molecules be changed into others? The questions are endless, but before attempting to answer them, we must ask what the fundamental components of matter are.

2.2 ATOMS CONSIST OF PROTONS, NEUTRONS, AND ELECTRONS

Atoms are the smallest units of matter that, when brought together in large numbers, constitute the chemical elements. But atoms are by no means the smallest units of matter. Every atom contains a tiny, central body called the *nucleus,* which has a positive electric charge and about 99.9% of the mass of the atom. The nucleus contains particles called *protons,* each of which has one positive charge* (in units called the *electron charge,* which can be positive or negative) and a mass of 1.6726×10^{-24} gram (g). The nucleus also contains *neutrons,* which have no charge (they are electrically neutral) and a mass of 1.6749×10^{-24} g. Note that the mass of a proton and the mass of a neutron are nearly identical (they differ by two parts in one thousand). The total number of protons in the nucleus, Z, plus the total number of neutrons, N, is called the *mass number, A*. Thus,

$$A = Z + N \tag{2-1}$$

The mass of any nucleus is approximately $A(1.67 \times 10^{-24}\text{ g})$. For this reason the helium nucleus, which contains two protons and two neutrons, has a mass about four times greater than that of a hydrogen nucleus, which is a single proton. The charge on the nucleus is equal to the number of protons, Z, which is also called the *atomic number* (do not confuse with mass number!). Thus, the hydrogen nucleus ($Z = 1$; $A = 1$) has a charge of +1 electron charge unit, and the helium nucleus ($Z = 2$; $A = 4$) has a charge of +2 electron charge units.

The actual volume of an atom is defined—roughly—by a diffuse distribution of very rapidly moving *electrons,* centered around the nucleus. An electron has one negative electron charge and a mass of 9.1096×10^{-28} g, or about 1/1836 the mass of a proton or neutron. Electrons therefore contribute very little to the mass of the atom. Figure 2-4 is an artist's interpretation of the hydrogen atom. The origin and meaning of this type of representation of atoms are discussed in detail in the later sections of this chapter.

An atomic nucleus is extremely small compared to the radius of the atom. A nuclear diameter is about 10^{-13} to 10^{-12} cm (10^{-5} to 10^{-4} Å). Since an atomic diameter is about 10^{-8} cm (1 Å), this means that the nucleus, with almost all the mass of the atom, occupies only about one part in 10^{14} (on the average) of its volume. If we could magnify a typical atomic nucleus

*The sign convention for electric charges is arbitrary and was set by Benjamin Franklin (1706–1790). Electric charges repel each other if they have the same sign and attract each other if they have opposite signs.

Figure 2-4. The electron cloud of the hydrogen atom.

to the size of a Ping-Pong ball, the diameter of the atom would be about 2 miles. Thus, even "solid" matter as we know it consists almost entirely of empty space! If we could eliminate this space by compressing atomic nuclei into a truly compact mass, a human being would be reduced to a microscopic speck. A lump of such nuclear matter the size of our Ping-Pong ball would weigh about 4 billion tons! As incredible as it may seem, nuclear matter does exist—in neutron stars, whose previously hypothetical existence was confirmed in 1968.

There are Z electrons in an atom, which means that an atom is electrically neutral (has no net charge): the number of negative charges (electrons) in the cloud around the nucleus equals the number of positive charges (protons) in the nucleus. If an electrical imbalance is created in the atom by the removal or addition of one or more electrons, the result is an *ion*. Thus, an ion is an electrically charged atom (or molecule). It can be either positive or negative, depending on whether its electron cloud contains fewer electrons or more electrons than does the neutral atom (or molecule). Some examples are the hydrogen ion, H^+, which is a free proton, and the fluoride ion, F^-, which contains an extra electron in its electron distribution.*

2.3 ISOTOPES

The mass of any atom can be approximated by adding the mass of its component parts. The hydrogen atom contains one proton and one electron, so its mass is

*See Appendix C for the origins of the symbols for the elements. You should always refer to Appendixes A and C for the meanings of physical units and inorganic chemical terms that may be unfamiliar to you.

$$(1.67 \times 10^{-24}\ \text{g}) + (9.1 \times 10^{-28}\ \text{g}) \cong 1.67 \times 10^{-24}\ \text{g}$$

(Note that the electron mass is negligibly small compared to the proton mass.) The sodium atom ($Z = 11$) contains 11 protons, 12 neutrons, and 11 electrons. Its mass number is $A = 23$ and its approximate mass is therefore

$$23(1.67 \times 10^{-24}\ \text{g}) = 38.4 \times 10^{-24}\ \text{g}$$

The chlorine atom ($Z = 17$) is found in two abundant forms in nature: 75.5% of chlorine atoms contain 18 neutrons, while 24.5% contain 20 neutrons. Atoms with the same nuclear charge, or atomic number (Z), but different numbers of neutrons (N) are called *isotopes*. The mass of the isotope ^{35}Cl is therefore

$$35(1.67 \times 10^{-24}\ \text{g}) = 58.5 \times 10^{-24}\ \text{g}$$

The mass of ^{37}Cl is

$$37(1.67 \times 10^{-24}\ \text{g}) = 61.8 \times 10^{-24}\ \text{g}$$

On the average, Cl atoms have a mass of

$$0.755(58.5 \times 10^{-24}\ \text{g}) + 0.245(61.8 \times 10^{-24}\ \text{g}) = 59.3 \times 10^{-24}\ \text{g}$$

although no individual Cl atom can have this mass. Note that the mass number A is given as a superscript in front of the symbol for the element.

The relative natural abundances of the isotopes of some elements on the Earth's surface, as well as in the universe as a whole, are variable and depend upon the history of the object containing the atoms. The reasons for these variations derive from the phenomenon of *natural radioactivity*. In this process an unstable nucleus spontaneously "decays," by the emission of a subatomic particle, to yield either a different isotope of the element or a nucleus of a different element. The rate at which radioactive decay occurs (called the *half-life*, i.e., the time required for one-half the initial amount of the species to disappear) is characteristic of the species and almost completely independent of external effects, such as temperature or pressure. These two facts have a very important consequence: they allow the time elapsed since the formation of an object containing a *known* concentration of a radioactive species to be determined. This is done by measuring the concentration that remains in the object after the *unknown* time interval since its formation. The half-life is then used to calculate the time that must have elapsed to produce this concentration—in other words, the age of the object.

The procedure described above is the basis of the *carbon dating* technique for determining the ages of ancient artifacts made of wood or other forms of vegetation (this includes the durable remains of animals, since the food cycle of all animals begins with plants). The normal, stable isotope of carbon is ^{12}C, but there is also a trace radioactive isotope, ^{14}C, that decays with a half-life of 5730 years. At the Earth's surface the steady loss of ^{14}C atoms due to radioactivity is counterbalanced, however, by the steady production of new ^{14}C atoms by the interaction of cosmic-ray neutrons with nitrogen molecules in the upper atmosphere. Thus, the overall level of ^{14}C in the biosphere (everywhere that life exists) remains relatively constant, even though marked differences exist between different ecosystems. Part of this ^{14}C is present in the form of atmospheric carbon dioxide, which is assimilated into the molecular components of plants via the process of *photosynthesis*. Thus, whenever and wherever plants grow, they absorb the same, constant ratio of ^{14}C to ^{12}C that exists in the atmosphere. From then on, however, the ratio in the plant drops steadily because of the radioactivity of the ^{14}C. By measuring this ratio in an ancient artifact of organic (i.e., living) origin, the age of the artifact can be calculated. This technique is of inestimable value to archaeologists and historians; it earned its inventor, the American chemist Willard Libby (b. 1908), a Nobel Prize in chemistry.

The hydrogen atom has mass number $A = 1$ and is the lightest atom. The sodium atom weighs 23 times as much as the hydrogen atom; on the average, chlorine atoms weigh 35.5 times as much as the hydrogen atom. Since the relative average masses of atoms are nearly integral numbers in most cases, it is convenient to define an *atomic weight* scale that specifies the weights of all atoms relative to an arbitrary standard.* We could choose this standard as the hydrogen atom, ^{1}H, and we could just as arbitrarily choose the atomic weight of this species to be the dimensionless number 1 (exactly). Suppose we now express this atomic weight in grams, which are the most convenient units of mass for most chemical purposes. The number of H atoms in 1.00 g of hydrogen is approximately

$$\frac{1.00 \text{ g}}{1.67 \times 10^{-24} \text{ g/atom}} = 5.99 \times 10^{23} \text{ atoms}$$

*Because the *weight* of an object at any given altitude (i.e., the gravitational force exerted on it by the Earth) is a measure of its *mass* (the amount of matter it contains), and because most human activities occur at or very near the surface of the Earth, where the gravitational field is essentially constant, it is convenient and common to use the term *weight* to denote mass. The "weight" of an object, when its mass is meant, is then properly expressed in units of mass, e.g., grams or kilograms. (True weights are expressed in units of force, e.g., dynes or newtons.) This rather loose usage can really be justified only when we deal with *relative* weights, since the relative weights of two objects are always the same at any given altitude, and are thus always equivalent to the relative masses.

Therefore, for a pure sample of *any* element, the mass corresponding to the atomic weight of the element, expressed in grams, would contain about 6×10^{23} atoms, provided that the atomic weight scale used were one in which hydrogen had a value very close to 1.

The number 6×10^{23} is called *Avogadro's number* (symbol: N_A) after the Italian physicist Amedeo Avogadro (1776–1856). It is one of the fundamental constants of nature, depending for its numerical value (as do all the fundamental constants) only on arbitrary definitions—here, of the gram and the numerical scale of atomic weights. Once these definitions are made, the value of the constant is immutable. All that remains is to measure it as accurately as possible.

The international atomic weight scale presently in use is based on the choice of the exact number 12 for the atomic weight of the carbon-12 isotope, ^{12}C. The table of atomic weights based on this standard can be found inside the back cover of this book. The average atomic weights of the ten lightest elements are

H	1.00797	C	12.01115
He	4.0026	N	14.0067
Li	6.939	O	15.9994
Be	9.0122	F	18.9984
B	10.811	Ne	20.183

Note that, although most of these numbers are nearly integers, the deviations from integral values are due largely to the existence of stable isotopes. For example, although ^{12}C has an atomic weight of 12.00000 and accounts for 98.89% of all carbon atoms on the Earth, the other 1.11% of all carbon atoms (ignoring trace isotopes such as ^{14}C) is ^{13}C; therefore the atomic weight of natural carbon is not 12.00000, but a weighted average of the two isotopes equal to 12.01115.

The precise value of Avogadro's number for this ^{12}C atomic weight scale is 6.022169×10^{23} atoms per gram atomic weight. One gram atomic weight of any element (i.e., the atomic weight of the element, expressed in grams) is called a *mole*. One mole of *any* pure substance—whether it is composed of atoms, molecules, ions, electrons, or any other kind of particle—contains Avogadro's number of particles. For this reason, Avogadro's number is given as

$$N_A = 6.022169 \times 10^{23} \text{ mole}^{-1}$$

with the term "particle" being understood. The mass of one mole of any substance is the atomic weight (or molecular weight, etc.) of the substance, expressed in grams. Thus, 1 mole of H atoms weighs precisely 1.00797 g;

1 mole of N atoms weighs 14.0067 g. Similarly, 1 mole of water molecules, H_2O, weighs

$$15.9994\ \text{g} + 2(1.00797\ \text{g}) = 18.0153\ \text{g}$$

2.4 ELECTRIC CHARGE IS QUANTIZED AND CONSERVED

The electric charge on any object, in units of electron charge, is equal to the number of protons in the object minus the number of electrons in the object. The electron charge $e = 4.80 \times 10^{-10}$ electrostatic unit (esu), a unit of electric charge. The electric charge on any object must therefore be an integral multiple of e, if the first sentence in this paragraph is true. It follows that electric charge cannot be subdivided into units smaller than e and that the charge on an object can have only a discrete set of values rather than a continuous range of them. A property having only a discrete set of values is said to be *quantized*.

In everyday life anything intangible that we count is quantized. For example, the score of a basketball game is always given in integral numbers, such as 78–56, and is therefore quantized. The weight of a human is not quantized, because it can have a continuous range of values, nor are humans regarded as quantized, because they are tangible.

Innumerable experiments have indicated that total charge is always conserved. It can have either a positive or negative sign, depending on whether there is an excess of protons or electrons in the object. Its total value cannot change unless protons or electrons (or other elementary charged particles, which are unimportant in chemistry) are exchanged with the surroundings. Total charge can be neither created nor destroyed by any process in nature. This does not, for example, say that a neutron cannot decay into a proton plus an electron, since the total charge of a neutron is 0 and the total charge of a proton plus an electron is 0. The *law of the conservation of charge* states that no process that changes the total charge of the universe can occur. Thus, chemical reactions (reactions in which at least one chemically distinct form of matter is created or destroyed) cannot create or destroy charge, a fact whose significance will become apparent when we study such reactions.

2.5 THE FUNDAMENTAL FORCE OF CHEMISTRY IS ELECTRIC

When two electric charges of the same sign (both positive or both negative) are brought close together, they repel each other. If the charges have opposite signs, they attract each other. The force F, which has both magnitude and direction, acts along the line between the two charges. Its magnitude is given by *Coulomb's law,* named after the French physicist Charles de Coulomb (1736–1806):

$$F = \frac{q_1 q_2}{Dr^2} \tag{2-2}$$

where q_1 and q_2 are the charges (always multiples of e), r is the distance between them, and D is a dimensionless proportionality factor called the *dielectric constant.* When the charges are expressed in electrostatic units and the distance in centimeters, the force is given in dynes. If the charges are in a vacuum, $D = 1$. If they are in a medium such as a gas, a liquid, or a solid, however, the protons and electrons in the medium interact with the charges in such a way that the absolute value of the force between the two charges of interest is smaller than it would be in a vacuum. Therefore, the dielectric constants of all substances are greater than one.

According to Equation 2-2, a force in the direction of increasing r (repulsion) is positive because the sign of the product of two like charges is positive. A force in the direction of decreasing r (attraction) is negative because the sign of the product in this case must be negative. Thus, the equation is consistent with the statement that charges of like sign repel and charges of opposite sign attract.

Coulomb's law is not a complete description of the interactions between charged particles, since magnetic interactions are also important between moving electric charges. The effects of magnetic forces are observable in the chemical and physical properties of molecules, and will be discussed shortly. The electric forces described by Coulomb's law are called *electrostatic forces* since their existence does not depend on the particles' being in motion. Electrostatic and magnetic interactions together constitute the *electromagnetic interaction,* one of the four fundamental interactions in nature.

The other three fundamental interactions play no important role in chemistry. Gravitational forces are completely insignificant relative to electromagnetic forces in any chemical system. The two fundamental nuclear interactions, called the *strong interaction* and the *weak interaction,* hold nuclei together and govern radioactive phenomena, respectively, but likewise play no role in chemical phenomena, which are concerned entirely with the interactions between electrons and nuclei, electrons and other electrons, and nuclei and other nuclei. Since it is electromagnetic forces alone that govern these interactions, all of chemistry—and biology—is ultimately governed by the laws of electromagnetism.

Example 2-1. Let us compare the Coulomb and gravitational interactions between a sodium ion and a chloride ion in crystalline NaCl.

The gravitational force between any two objects is*

$$F_{\text{grav}} = -G\frac{m_1 m_2}{r^2}$$

where m_1 and m_2 are the masses of the objects, r is the distance between them, and G is the *universal gravitational constant,* which has the value

$$G = 6.67 \times 10^{-8} \text{ dyn cm}^2/\text{g}^2$$

We can now calculate the gravitational force between a sodium ion and a chloride ion at a distance of 2.81×10^{-8} cm, the interionic spacing in the NaCl crystal at 25°C. The mass of the sodium atom (or ion) calculated earlier in this chapter was 38.4×10^{-24} g. A more precise value is

$$\frac{22.99 \text{ g}/\cancel{\text{mole}}}{6.022 \times 10^{23} \cancel{\text{mole}^{-1}}} = 38.18 \times 10^{-24} \text{ g}$$

Similarly, a more precise value for the chloride ion is 58.87×10^{-24} g. The gravitational force between the two ions is therefore

$$F_{\text{grav}} = (-6.67 \times 10^{-8} \text{ dyn } \cancel{\text{cm}^2}/\cancel{\text{g}^2}) \frac{(38.2 \times 10^{-24} \cancel{\text{g}})(58.9 \times 10^{-24} \cancel{\text{g}})}{(2.81 \times 10^{-8} \cancel{\text{cm}})^2}$$

$$= -1.90 \times 10^{-37} \text{ dyn}$$

The Coulomb force between these same ions in the NaCl crystal (at 25°C, at which $D = 5.9$) is

$$F_{\text{Coul}} = \frac{(4.8 \times 10^{-10} \text{ esu})(-4.8 \times 10^{-10} \text{ esu})}{5.9(2.8 \times 10^{-8} \text{ cm})^2}$$

$$= -5.0 \times 10^{-5} \text{ dyn}$$

We see that, in the NaCl crystal, the electrostatic force is 2.6×10^{32} times larger than the gravitational force; the latter is therefore totally insignificant for chemical properties.

*Note that gravitational force is always attractive, regardless of whether the objects are charged or uncharged. We therefore arbitrarily define it as being a negative force, in analogy with the attractive Coulomb force mentioned above.

2.6 VELOCITY, MOMENTUM, ACCELERATION

Velocity, momentum, and *acceleration* are concepts that pertain to objects in motion. When a car traverses a California desert road, it can easily cover 55 miles in 1 hour; we say it is moving at 55 miles per hour, or 55 mile/h. This is its *speed.* Its *velocity* is obtained by specifying both its speed *and* its direction. We see that a velocity v is obtained by dividing a distance by the time required to traverse that distance:

$$v = \frac{x_{\text{end}} - x_{\text{begin}}}{t_{\text{end}} - t_{\text{begin}}} = \frac{\Delta x}{\Delta t}$$

where x denotes position, t denotes time, and Δ (Greek capital *delta*) is a symbol that means "difference in" or "change in." As the distance and the corresponding time are reduced to infinitesimal numbers, we replace the Δ by a small d. The expression is now called a *derivative* and is written

$$v = \frac{dx}{dt} \tag{2-3}$$

Derivatives are mathematical expressions that denote the rate of change of one variable with respect to another, in infinitesimal increments. They are fundamental to the exact descriptions of many natural phenomena.*

Momentum p is defined as the product of the mass of an object and its velocity:

$$p = mv \tag{2-4}$$

Thus, momentum is also a directional quantity. Momentum is a familiar term: we use it intuitively to predict the result of a 125-lb end's attempting a head-on tackle of a 250-lb fullback running at full speed.

Acceleration a is defined as the rate of change of velocity with time. Thus, the derivative

$$a = \frac{dv}{dt} \tag{2-5}$$

represents the acceleration most accurately. The most common acceleration we encounter is the acceleration due to (not "of"!) gravity. This

*Note that the d's in derivatives, like the Δ's used above, are *not* multipliers (i.e., dx is not to be interpreted as "d times x," etc.); they are *operators*—symbols that denote a mathematical operation to be performed on the symbol(s) immediately following.

acceleration, given the symbol g, is approximately constant at sea level and has a "standard" value of 980.665 cm/s^2. Indeed, deviations from this value provide the geophysicist with valuable information about the shape and composition of the Earth.

2.7 ENERGY

Kinetic energy is energy that a body or system possesses owing to its motion with respect to some frame of reference. A pile driver in action and a surfer riding a wave are familiar macroscopic examples of objects possessing kinetic energy. Gas molecules undergoing translational motion and atoms vibrating with respect to each other in a molecule are microscopic examples of kinetic energy. The expression for the kinetic energy of a body is

$$KE = \tfrac{1}{2}mv^2 \tag{2-6}$$

It can be applied to any system, provided that the mass m and velocity v of each moving part of the system are known.

Potential energy is energy that is stored in a body or system by virtue of the position of the body with respect to a reference position, or the configuration of the system with respect to a reference configuration. In either case the energy thus stored has the *potential* of being converted to kinetic energy if the mass in question moves toward the reference state. For example, a ball at the top of a hill has potential energy with respect to the bottom of the hill by virtue of the gravitational force between Earth and ball. This gravitational potential energy is converted to kinetic energy when the ball rolls down the hill. Because of friction, most of the ball's kinetic energy is converted to kinetic energy of the molecules (i.e., heat) in the surface over which the ball rolls, while the remainder is converted to acoustic energy in the form of whatever noise the ball makes as it rolls. Thus, when the ball comes to rest at the bottom of the hill, it has zero kinetic energy, and zero potential energy with respect to the bottom of the hill. However, it still has potential energy with respect to even lower positions, such as the bottom of a nearby well, and so on.

The calculation of potential energy requires that a specially defined force that is a function only of position be exerted on the particle. Frictional forces are not in this category, since they depend on velocity. The two most common forces that depend only on position are the gravitational force and the electrostatic, or Coulomb, force. Which of these do you think would lead to the greater potential energy of a system consisting of any two oppositely charged ions separated by a small distance?

At the surface of the Earth, the work required to lift a mass m a distance h is the applied force times the distance. The force exerted on the mass by

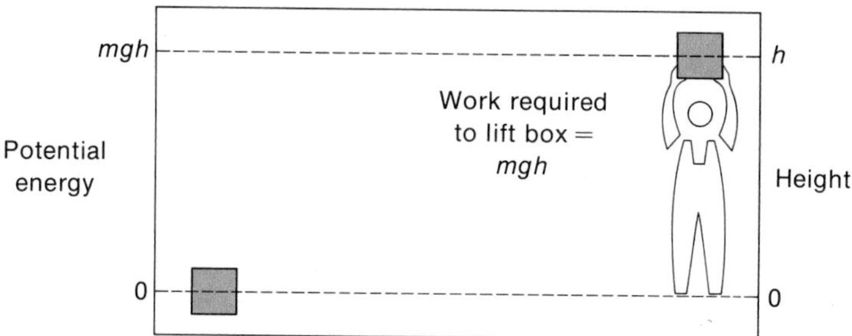

Figure 2-5. The relation between potential energy and work in a gravitational field.

the Earth's gravitational field is mg, where g is the acceleration due to gravity. Thus, the work required to lift the object is

$$\text{Work} = mgh$$

and this is defined as the potential energy of the object, once it is lifted (see Figure 2-5).

2.8 MASS, ENERGY, AND THE CONSERVATION OF ENERGY

There is an equivalence between mass and energy that was first expressed by Albert Einstein (1879–1955):

$$E = mc^2 \tag{2-7}$$

where m is the mass of an object, c is the speed of light (3.00×10^{10} cm/s), and E is the total energy of the object. This law, from the special theory of relativity, accounts for the difference between the measured mass of an atom and the sum of the masses of its Z electrons, Z protons, and $A - Z$ neutrons.

Example 2-2. An atom of the carbon-12 isotope contains six protons, six neutrons, and six electrons. The mass of this atom is $12.000/(6.0222 \times 10^{23}) = 1.9926 \times 10^{-23}$ g. Calculate the difference between this mass and the sum of the masses of the atom's component parts.

Mass of 1 proton:	1.6726×10^{-24} g
Mass of 1 neutron:	1.6749×10^{-24} g
Mass of 1 electron:	0.0009×10^{-24} g
Sum:	3.3484×10^{-24} g
	$\times 6$
Total:	20.090×10^{-24} g

Mass of 6 protons + 6 neutrons + 6 electrons − mass of ^{12}C atom = 0.164×10^{-24} g

The difference in mass calculated in the above example is called the *mass defect*. It arises from the enormous amount of energy released in the hypothetical process of fusing six protons and six neutrons in the carbon nucleus. (A much smaller amount of energy is released in adding the six electrons to form the neutral atom.) From the Einstein equation we see that the energy difference between the 18 free particles and the ^{12}C atom is

$$E = (0.164 \times 10^{-24}\ \text{g})\,(3.00 \times 10^{10}\ \text{cm/s})^2 = 1.48 \times 10^{-4}\ \text{erg}$$

which is about 100 million times greater than the energy differences observed (per atom) in chemical processes. This nuclear binding energy, or fusion energy, is the basis for the enormous destructiveness of the hydrogen bomb as compared to ordinary chemical bombs. It also explains why nuclei are stable despite the repulsive forces between protons: the strong interaction, as manifested by the nuclear binding energy, is more than sufficient to overcome the repulsive Coulomb interaction. The mass differences associated with the energies released or absorbed in chemical reactions are so small that they need generally not be considered. For all practical purposes, mass is conserved in chemical reactions. It is thus possible to add the atomic weights of the atoms in a molecule to obtain the molecular weight, even though some energy is released in forming the molecule from the atoms.

Energy is conserved. No process has ever been observed that changes the total energy of the universe. There are many examples of potential energy being converted to kinetic energy, e.g., a ball falling in the Earth's gravitational field. In this process all the gravitational potential energy of the ball is converted to kinetic energy of the ball as it falls. Some of this kinetic energy is transferred to the gas molecules of the atmosphere; the remainder is transferred to the ground upon impact. (If the ball fell in a vacuum, all of the initial potential energy would eventually become kinetic energy of the ground.) In fact, all *spontaneous* mechanical processes that a single object undergoes result in a decrease in the object's potential energy and an equivalent increase in the kinetic energy of the surroundings (i.e., the universe) through friction or impact. Thus, we say that *a system seeks its lowest possible potential energy*. This generalization is very useful when one is considering microscopic objects such as atoms or molecules. Note, however, that a reduction in the potential energy of any object, even an atom or molecule, requires that energy be transferred to another object by collision or that energy be converted to another form, such as light, because total energy must be conserved.

2.9 THE CONSERVATION OF MASS

In the previous section we saw that energy is conserved and that for chemical processes the energy lost to or gained from the surroundings is negligible compared to the mass energy of the atoms. Under these conditions the conservation of energy is equivalent to the conservation of mass. An example is the reaction of two hydrogen atoms to form a hydrogen molecule:

$$H + H \rightarrow H_2$$

This reaction releases 104 kcal of energy for each mole of H_2 gas formed, a negligible fraction of the mass energy of the H_2 molecule. Therefore the chemist knows that, if he starts with 1.000 g of H atoms in an inert container, he will form 1.000 g of H_2 molecules.

In chemical processes atoms are conserved and the mass of each atom is conserved. This is important for balancing chemical equations. For example, when propane gas, C_3H_8, is burned in air, yielding carbon dioxide and water, the unbalanced chemical equation is

$$C_3H_8 + O_2 \rightarrow CO_2 + H_2O$$

The species on the left are called *reactants*, and those on the right are *products*. To balance this equation we must have equal numbers of all atoms on each side of the arrow. Since we start with 3 carbon atoms on the left, we must produce 3 carbon dioxide molecules on the right. Similarly, there are 8 hydrogen atoms on the left, so there must be 4 water molecules on the right:

$$C_3H_8 + ?O_2 \rightarrow 3CO_2 + 4H_2O$$

Finally, we must balance the oxygen atoms without unbalancing any other atoms. There are $3 \times 2 + 4 \times 1 = 10$ oxygen atoms on the right, so we must have 5 oxygen molecules on the left. The completely balanced equation is

$$C_3H_8 + 5O_2 \rightarrow 3CO_2 + 4H_2O$$

This equation says that 1 molecule of propane reacts with 5 molecules of oxygen to produce 3 molecules of carbon dioxide and 4 molecules of water. Any multiple of this statement is correct. In particular, 1 mole of C_3H_8 reacts with 5 moles of O_2 to produce 3 moles of CO_2 and 4 moles of H_2O:

$$
\begin{aligned}
1 \text{ mole } C_3H_8 &= (3 \times 12) + (8 \times 1) &&= 44 \text{ g } C_3H_8 \\
5 \text{ moles } O_2 &= 5 \times 2 \times 16 &&= 160 \text{ g } O_2 \\
3 \text{ moles } CO_2 &= 3 \times [12 + (2 \times 16)] &&= 132 \text{ g } CO_2 \\
4 \text{ moles } H_2O &= 4 \times [(2 \times 1) + 16] &&= 72 \text{ g } H_2O
\end{aligned}
$$

It follows that 44 g C_3H_8 reacts with 160 g O_2 to produce 132 g CO_2 and 72 g H_2O. Note that $44 + 160 = 204$ g of reactants produces $132 + 72 = 204$ g of products. Mass is conserved. The methodology of balancing chemical equations and calculating the masses of reactants and products is called *stoichiometry*.

Example 2-3. The gravimetric analysis (analysis by weight) for dihydrogen phosphates is performed by reacting dihydrogen phosphate ion, $H_2PO_4^-$, with ammonium and magnesium ions in water. The balanced chemical equation for this reaction is*

$$H_2PO_4^- + NH_4^+ + Mg^{2+} + 6H_2O \rightarrow MgNH_4PO_4 \cdot 6H_2O \downarrow + 2H^+$$

(Note that the sum of the charges on the left is +2, which equals the sum of the charges on the right. This is another example of the conservation of charge.) The product of the reaction, magnesium ammonium phosphate hexahydrate, $MgNH_4PO_4 \cdot 6H_2O$, is a *precipitate* (an insoluble solid). When heated it decomposes to yield magnesium pyrophosphate, $Mg_2P_2O_7$, which is then weighed. The equation is

$$2MgNH_4PO_4 \cdot 6H_2O \xrightarrow{\text{Heat}} Mg_2P_2O_7 + 2NH_3 \uparrow + 13H_2O \uparrow$$

Question: If 0.762 g $Mg_2P_2O_7$ is produced by these reactions, what was the weight of the $H_2PO_4^-$ in the original solution? *Answer:*

$$\text{Mol wt of } Mg_2P_2O_7 = 2 \times 24.3 + 2 \times 31.0 + 7 \times 16.0 = 222.6 \text{ g/mole}$$

$$\text{Moles of } Mg_2P_2O_7 \text{ product} = \frac{0.762 \text{ g}}{223 \text{ g/mole}} = 3.42 \times 10^{-3} \text{ mole}$$

From the two chemical equations shown above, we see that 1 mole of $H_2PO_4^-$ produces 1 mole of $MgNH_4PO_4 \cdot 6H_2O$ and that 2 moles of $MgNH_4PO_4 \cdot 6H_2O$ produce 1 mole of $Mg_2P_2O_7$. It therefore requires 2 moles of $H_2PO_4^-$ to produce 1 mole of $Mg_2P_2O_7$, so

*It is customary to show multiple ionic charges in chemical formulas as 2+, 2−, etc., rather than +2, −2, etc., in order to avoid possible confusion with mathematical exponents.

$$\text{Moles of } H_2PO_4^- \text{ reactant} = 2 \times \text{moles of } Mg_2P_2O_7 \text{ product}$$

$$= 6.84 \times 10^{-3} \text{ mole}$$

$$\text{Mol wt of } H_2PO_4^- = 2 \times 1.01 + 31.0 + 4 \times 16.0 = 97.0 \text{ g/mole}$$

$$\text{Weight of } H_2PO_4^- = (6.84 \times 10^{-3} \cancel{\text{mole}})(97.0 \text{ g/}\cancel{\text{mole}}) = 0.663 \text{ g}$$

2.10 THE WAVE NATURE OF ELECTROMAGNETIC RADIATION

Light is the term used to describe one narrow portion of the electromagnetic spectrum, and is defined by our ability to see it. Electromagnetic radiation has wave properties. Figure 2-6 is a representation of the electric component of an electromagnetic wave.

Two important properties of waves are evident in Figure 2-6. The maximum *amplitude* A is a measure of the maximum displacement (from the zero value) of the property exhibiting the wave behavior—in this instance, the electric field strength. The intensity (or brightness) of a light wave is proportional to the square of its maximum amplitude. The *wavelength* λ is the distance between any two repeating points on the wave. Depending on the region of the electromagnetic spectrum under consideration, it is expressed in whatever unit of length is most convenient, e.g., angstroms (Å) for x rays, micrometers (μm) for infrared radiation, or meters (m) or kilometers (km) for radio waves.

Figure 2-6 represents a wave at one instant. Waves have motion: for example, the waves on the surface of water. In Figure 2-6 we need only think of the curve's moving to the right as time progresses in order to

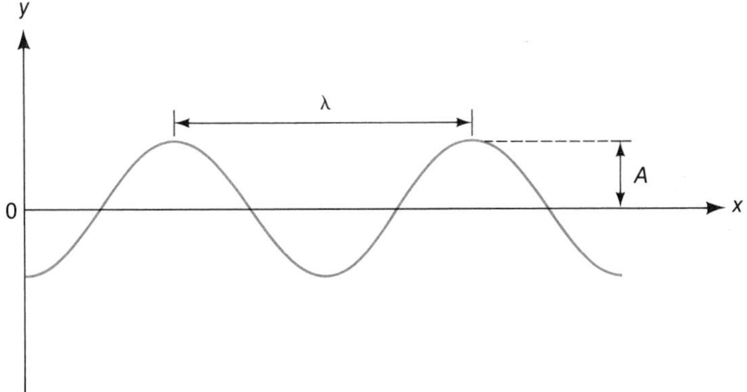

Figure 2-6. Representation of a sinusoidal wave; λ is the wavelength and A is the amplitude.

achieve the correct visualization of a moving wave. When any point x has increased by the amount λ, y has undergone one complete cycle and returned to its original value. The amplitude varies somewhat like a yo-yo on a string between $-A$ and $+A$.

The *frequency* ν of a wave is the number of complete oscillations that the property exhibiting the wave behavior undergoes in unit time. The unit of measure for frequency is therefore that of reciprocal time. Frequency is usually expressed as reciprocal seconds, s^{-1}.

It is easy to see that frequency and wavelength are related to the speed with which the wave is propagating. If ν waves of wavelength λ (expressed, e.g., in centimeters) pass a given point in one second, the speed of the waves is $\nu\lambda$ cm/s. For all electromagnetic radiation in a vacuum, this speed, called the speed of light, c, is a constant:

$$c = \nu\lambda = 3.00 \times 10^{10} \text{ cm/s} \tag{2-8}$$

It is one of the fundamental constants of nature.

The other component of electromagnetic radiation is the magnetic wave that accompanies the electric wave. The relation between these two components is shown in Figure 2-7, in which two features of all electromagnetic waves can be observed: (1) the electric and magnetic components are at right angles to each other—they are *orthogonal*; (2) the electric and magnetic field strengths reach their maxima and minima simultaneously—they are *in phase*.

It is fascinating to be aware of the tremendous variety of electromagnetic radiation, e.g., x rays, light, and radio waves, and to realize that all these

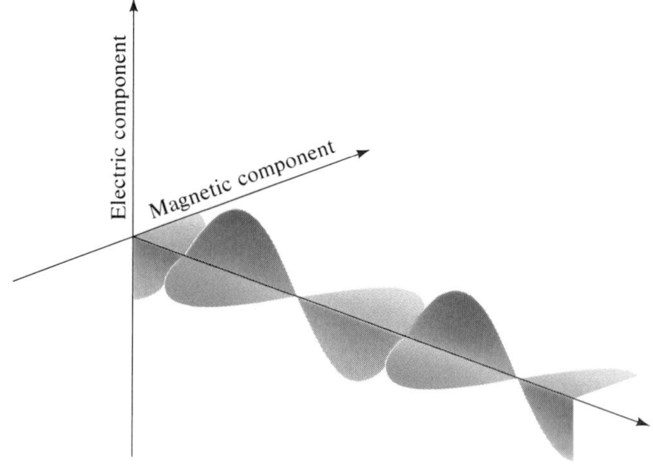

Figure 2-7. An electromagnetic wave.

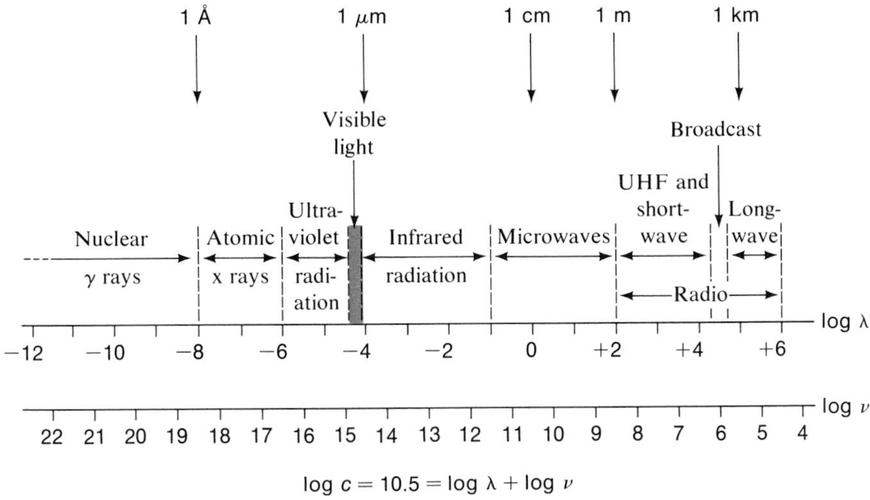

Figure 2-8. The electromagnetic spectrum. The units of λ and ν are cm and s^{-1}, respectively. Except for the visible-light range, the boundaries between the regions of the spectrum are approximate, and there are varying degrees of overlap of the regions.

types of radiation propagate at the same speed and have the properties represented in Figure 2-7. The x rays used to examine your teeth have a wavelength of about 0.00000001 cm, whereas the wavelength of the radio wave used to carry the FM signal to your tuner may be as long as the room you are sitting in.

Figure 2-8 illustrates several aspects of electromagnetic radiation. The common names for the various regions of the spectrum are given. The first row of numbers gives the logarithm of the wavelength; the second row gives the logarithm of the frequency. The sum of corresponding vertical points on these two scales equals the logarithm of 3.0×10^{10}.

Many experiments demonstrate the wave nature of electromagnetic radiation. Diffraction by a grating, resolution by a prism, and interference are some examples. The wave properties of radiation are exploited by chemists in the technique of x-ray diffraction, which has provided important information regarding the structure of molecules and crystals (this technique is described in Chapter 6). In the past few years, the determination of the three-dimensional structures of a number of protein molecules by x-ray diffraction has caused great excitement in the scientific community.

2.11 THE DIFFRACTION AND INTERFERENCE OF WAVES

An understanding of the properties of waves is crucial to the understanding of small (atomic) systems and is at the heart of most modern descriptions of

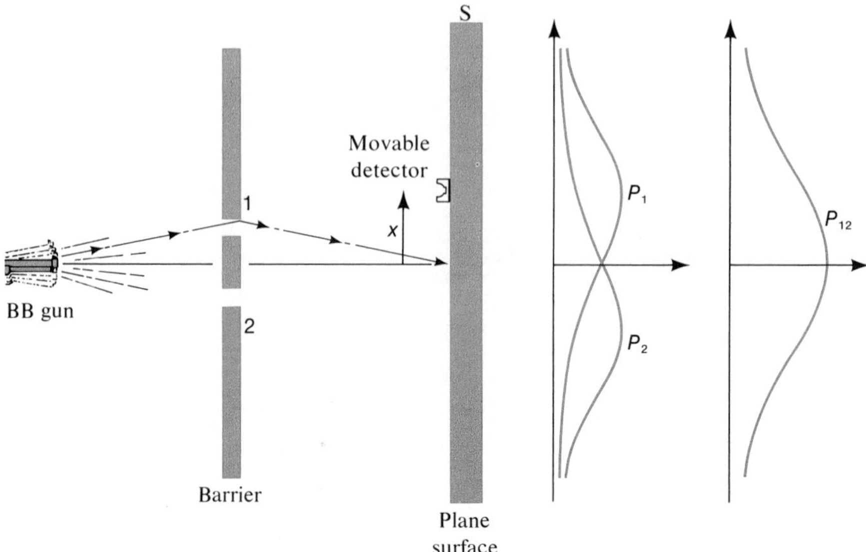

Figure 2-9. The passage of BBs through two slits; there is no diffraction. (Adapted from R. P. Feynman, R. B. Leighton, and M. Sands, *The Feynman Lectures on Physics*, Vol. I, Addison-Wesley, Reading, Mass., 1963.)

matter. Earlier we referred to an electron cloud in an atom. Such a concept can be misinterpreted unless some of the subtle properties of light are properly comprehended. We will use a description given by Richard Feynman in his elegant *Lectures on Physics** to assist us in arriving at this goal. A comparison between the properties of large particles, such as BBs, and those of light are an important feature of this description.

Let us first consider two slits in a barrier and a source of particles – BBs – on one side of the barrier (see Figure 2-9). The BBs fired at the barrier in a random pattern can only pass through the two slits, or be bounced back. Because the slits are narrow, many of the BBs are deflected by the walls as they pass through. Let us place a detector in a plane S some distance from the barrier. This detector can be moved up and down in plane S and can count the number of BBs that arrive at a given position. If a large number of BBs is fired at the barrier a sufficient number of times with the detector in different positions, we can determine a *probability distribution P* that tells us the probability of a BB's arriving at some position x in plane S. For example, the BB gun is fired 10,000 times and the number of BBs arriving at x is counted. The value of P at x is just the number counted divided by 10,000. The detector is then moved to a new position and the

*Richard P. Feynman, Robert B. Leighton, and Matthew Sands, *The Feynman Lectures on Physics*, Vol. I, Addison-Wesley, Reading, Mass., 1963.

number of BBs arriving at the new x is determined for another 10,000 shots. This process is repeated for many values of x in order to determine the complete probability distribution P.

Many BBs are deflected and scattered by the slits. The probability of a BB's arriving at a given position in plane S of Figure 2-9 has a certain value at each point in the plane. The probability distribution for BBs passing through Slit 1 with Slit 2 closed is P_1, and the probability distribution for BBs passing through Slit 2 with Slit 1 closed is P_2. The total probability of a BB's passing through the barrier with both slits open and arriving at a position x is just the sum of the probability of its passing through Slit 1 and arriving at x and the probability of its passing through Slit 2 and arriving at x. The probability distribution with both slits open, P_{12}, is just the sum of the two probability distributions with only one slit open at a time:

$$P_{12} = P_1 + P_2 \tag{2-9}$$

This result agrees with our intuition about large particles.

Waves, however, behave differently: they undergo *diffraction*. If we use a wave source, such as a lamp, we observe that new spherical waves originate from each slit. Like all electromagnetic waves, these new waves are sinusoidal and add in such a way that, if two of them have a maximum at a given point in space, the amplitude of the resultant wave is twice that of either wave alone. This is called *constructive interference*. Conversely, if one wave has a maximum where the other has a minimum, the amplitude

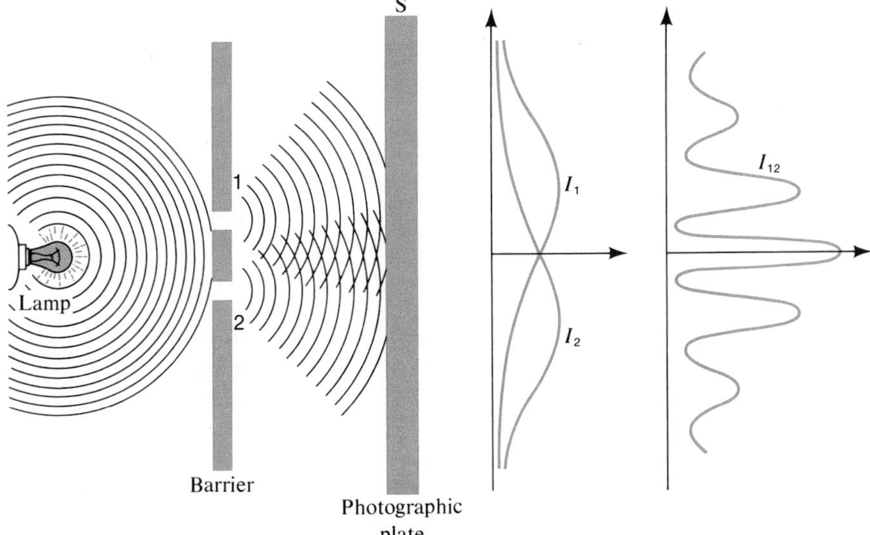

Figure 2-10. The diffraction of light by two slits. (Adapted from R. P. Feynman, R. B. Leighton, and M. Sands, *The Feynman Lectures on Physics*, Vol. I, Addison-Wesley, Reading, Mass., 1963.)

of the resultant wave is zero. This is called *destructive interference*. Figure 2-10 shows the *intensity* of the wave (its rate of energy transport) when only Slit 1 is open, I_1, when only Slit 2 is open, I_2, and when both are open and interference occurs, I_{12}. The intensity of the light is measured by a photographic plate in plane S. Just as with BBs, the intensity is a function of position in the plane. The intensity distribution is similar to the probability distribution for BBs, provided that only one slit is open and that the slit is narrow compared to the wavelength of the light.

The intensity distribution with *both* slits open, I_{12}, is entirely different from the corresponding probability distribution for the BBs, however, so we must conclude that, for waves,

$$I_{12} \neq I_1 + I_2$$

The intensity distribution for light passing through two slits (called a *diffraction pattern*) is *not* the sum of the intensity distributions for light passing through each slit alone.

Figure 2-11 presents some examples of wave diffraction. The first two diagrams show the properties of water waves, which are analogous to the light properties we have just discussed. The third shows the diffraction of helium-neon laser light by a double slit.

An important feature of waves is that the amplitudes of waves from all possible paths must be added at any given point before the total amplitude is squared to obtain the wave intensity at that point. Since, for any given point in a diffraction pattern,

$$I_{12} \neq I_1 + I_2$$

it follows that

$$(y_{12})^2 \neq (y_1)^2 + (y_2)^2$$

where y is the amplitude of the wave. The correct equation for waves is

$$I_{12} = (y_{12})^2 = (y_1 + y_2)^2 \tag{2-10}$$

Let us again consider the two-slit experiment with light, but replace the photographic plate with a detector that converts light energy to an electric current. At high light intensity the results remain unchanged: as we move the detector up and down in plane S, it measures light intensity as a function of position and verifies the results obtained with the plate. As the intensity of the lamp is decreased, however, something surprising happens: eventually, the detector no longer gives a steady signal. Instead, it produces electric pulses randomly distributed in time, as though it were counting BBs instead of measuring the intensity of a wave. If the detector is placed in each position for a long time and the pulses counted, the intensity distribution

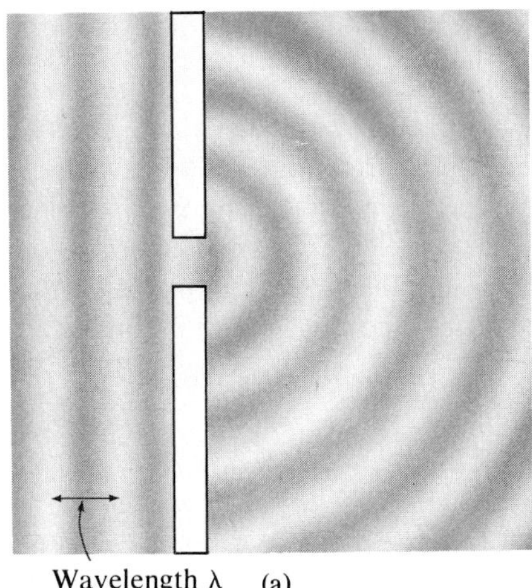

(a)

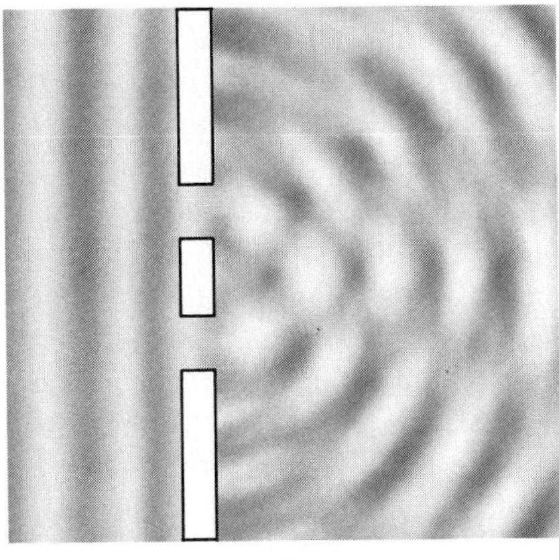

(b)

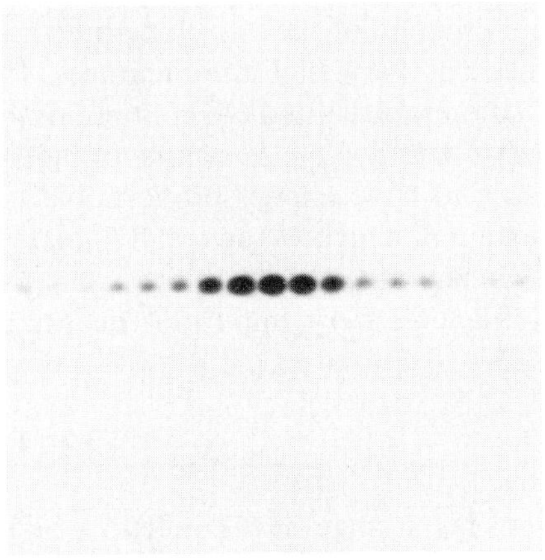

(c)

Figure 2-11. Some examples of wave diffraction. (a) Propagation of waves through one slit, generating circular waves by simple diffraction. (b) Propagation of waves through two slits, generating two sets of circular waves; constructive and destructive interference are evident. (c) Two-slit diffraction pattern of light from a helium-neon laser. [(a) and (b) from L. Pauling and P. Pauling, *Chemistry,* W. H. Freeman and Company. Copyright © 1975. (c) John E. Hearst with the assistance of David W. Appleby, Department of Chemistry, University of California, Berkeley.]

produced by the two slits is unaltered, indicating that wave interference (diffraction) is still occurring even though pulses of light arrive at plane s only one at a time! We will see in the next section that light has particulate properties that explain these pulses. Light particles are called *photons* and the amazing thing about the above observation is that it demonstrates that a photon interferes with itself. Two photons arriving simultaneously are not needed for interference—one is enough.

2.12 THE PARTICULATE NATURE OF ELECTROMAGNETIC RADIATION

Late in the nineteenth century it was found that light shining on certain metals, such as sodium, potassium, and calcium, caused the emission of electrons from the surface of the metal. This was called the *photoelectric effect.* It was found, for each metal that exhibited the effect, that only if light of a wavelength shorter than some characteristic value were used did electron emission occur. Careful experimentation revealed a most puzzling aspect of this phenomenon. It was shown that the number of electrons emitted was proportional to the intensity of light shining on the metal—this was to be expected. The unexplained observation was that the *energy* of the electrons emitted was *independent* of the intensity of the incident light. According to the wave theory of light, the energy of the electrons should be proportional to the intensity of the light. Instead, the energy of the electrons depended solely on the *frequency* ν of the light.

In 1905 Einstein, who was 26 at the time and working as a patent examiner because he could not get into graduate school, provided the answer. He utilized the work of the German physicist Max Planck (1858–1947), who had been wrestling with the troublesome problem of the wavelength distribution of radiation emitted by substances at very high temperatures. In 1900 Planck had had to propose that radiation was emitted not continuously, but rather in discrete packets, or *quanta,* of energy. These quanta of radiation, now called photons, were postulated to have energy proportional to their frequency. The proportionality constant h, which we now call *Planck's constant,* is one of the fundamental constants of nature; its value is $h = 6.63 \times 10^{-27}$ erg s (erg second). Planck's very important quantum relation for the energy E of a photon is

$$E = h\nu \tag{2-11}$$

Although even Planck was unhappy with his strange new concept, which flagrantly violated the principles of physics known at that time, Einstein realized that in it lay the explanation of the photoelectric-effect anomaly.

Einstein took Planck's quantum hypothesis a giant step further by proposing that photons ("particles" of electromagnetic radiation) are not just artifacts of the emission (or absorption) of radiation by substances, but a fundamental property of all radiation, everywhere. Thus, radiation—like electric charge and matter itself—is quantized in discrete units. Armed with this theory, Einstein went on to propose that a given light photon can impart its energy to one electron in an atom at the surface of the metal. If the energy of the photon is equal to or greater than a certain characteristic value, the electron overcomes the attractive force of the nucleus and is emitted. Thus, until the frequency of the incident light is sufficiently high to provide this energy, no electrons can be emitted. This minimum frequency is called the *threshold frequency* and is given the symbol ν_0. Light of higher frequency imparts greater energy to the electrons but does not liberate additional electrons.

These ideas are summarized in the following equations, in which E_{in} is the energy of the incident photon and E_{out} is the energy required to just remove the electron from the metal (the *work function* W) plus whatever kinetic energy (KE) the electron has as it leaves the surface:

$$E_{in} = h\nu$$

$$E_{out} = W + KE = h\nu_0 + \tfrac{1}{2}mv^2$$

$$E_{in} = E_{out} \qquad \text{(Conservation of energy)}$$

$$h\nu = h\nu_0 + \tfrac{1}{2}mv^2 \tag{2-12}$$

Equation 2-12 is the Einstein equation for the photoelectric effect. A plot of the kinetic energy of the emitted electrons as a function of the incident photon frequency should be a straight line. That this is true is shown in Figure 2-12.

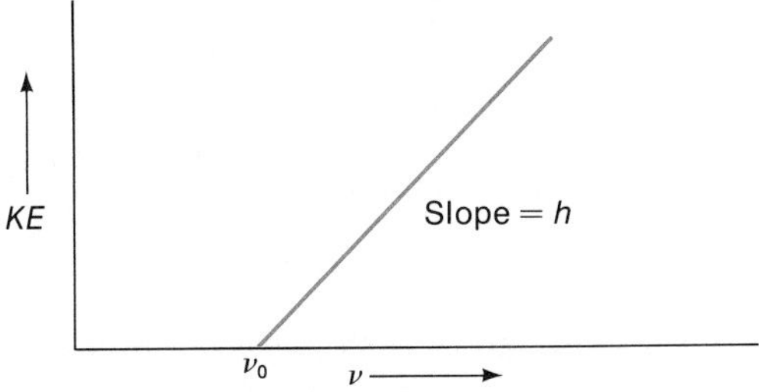

Figure 2-12. Kinetic energy of electrons emitted from a photoelectric surface, as a function of the frequency of incident radiation. The x-intercept is the threshold frequency ν_0 and the slope is Planck's constant h.

The quantum theory of radiation explains the pulses that were observed at low light intensity in the two-slit diffraction experiment. We are left with the following description of a photon: a photon is both a particle and a wave—or we could just as well say it is neither a particle nor a wave. It is a particle-wave. A particle-wave has no analogy in the macroscopic world of our daily observations. The best we can say is that light (or radiation in general) has some properties that we normally associate with particles and others that we associate with waves. Light is light, however, and the fact that we cannot force it into one or another descriptive picture that is consistent with our observations of the macroscopic world is clearly not light's problem or nature's problem, but ours.

2.13 THE WAVE PROPERTIES OF PARTICLES

We have discussed three elementary particles: the electron, the proton, and the neutron. A clear diffraction pattern can be observed when electrons, protons, or neutrons are scattered by crystals, indicating that even these particles of genuine matter have wave properties! Figure 2-13b shows an electron diffraction pattern obtained by scattering electrons from aluminum foil. Thus, these particles are apparently particle-waves with properties similar to those of light. If this is true it should be possible to relate the velocity, or energy, of the particle to a wavelength.

From Planck's quantum hypothesis we know that the energy of a particle, or photon, of light is $h\nu$. If we now assume that the photon has a mass m (why not, since it *is* a wave-*particle?*), the Einstein mass-energy relation allows us to equate the energy of a wave with the relativistic energy of a particle (and, of course, vice versa):

$$E = h\nu = mc^2$$

The momentum of this hypothetical photon is

$$p = mc = \frac{h\nu}{c} = \frac{h}{\lambda}$$

The French theoretical physicist Prince Louis Victor de Broglie (b. 1892) postulated in 1924 that this relationship between momentum and wavelength was, indeed, valid for *all* particles, not just photons. Thus, for particles that have velocities less than that of light (i.e., for any particle of matter),

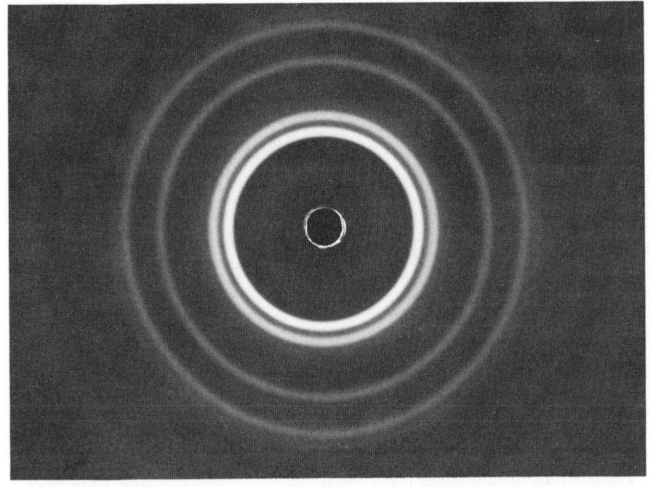

(a)

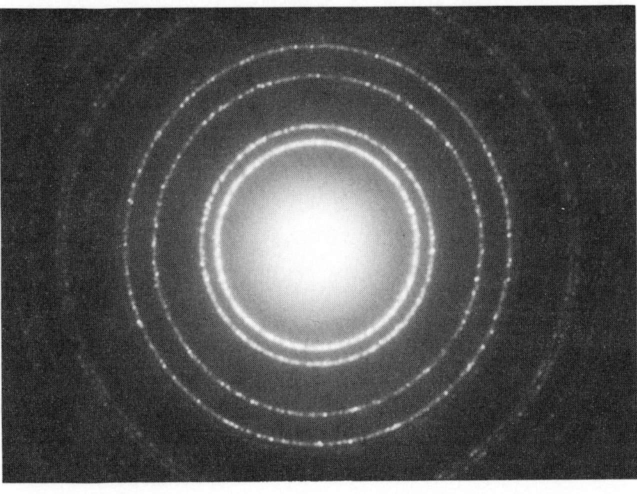

(b)

Figure 2-13. Diffraction of two kinds of waves by aluminum. (a) X rays of wavelength 0.71 Å. (b) Electrons of wavelength 0.50 Å. (Permission by Education Development Center, 39 Chapel St., Newton, Mass.)

we write

$$p = mv = \frac{h}{\lambda} \tag{2-13}$$

All particles in motion have a wavelength that depends exclusively on the particle's momentum. The larger the momentum, the smaller the wavelength. Experimental evidence has proved the validity of this brilliant theoretical discovery.

Example 2-4. Calculate the wavelength of an electron that has been accelerated in an electron microscope through 100,000 volts.

$$KE = \tfrac{1}{2}mv^2 = 10^5 \text{ electron·volts (eV)}$$

$$1 \text{ electron volt} = 1.6 \times 10^{-12} \text{ erg}$$

$$1 \text{ electron mass} = 9.1 \times 10^{-28} \text{ g}$$

$$v^2 = \frac{2(10^5 \text{ eV})(1.6 \times 10^{-12} \text{ erg/eV})}{9.1 \times 10^{-28} \text{ g}} = 3.5 \times 10^{20} \text{ cm}^2/\text{s}^2$$

$$v = 1.9 \times 10^{10} \text{ cm/s}$$

$$p = mv = (9.1 \times 10^{-28} \text{ g})(1.9 \times 10^{10} \text{ cm/s})$$

$$= 1.7 \times 10^{-17} \text{ g cm/s}$$

$$\lambda = \frac{h}{p} = \frac{6.6 \times 10^{-27} \text{ erg s}}{1.7 \times 10^{-17} \text{ g cm/s}} = 3.9 \times 10^{-10} \text{ cm}$$

$$\lambda = 0.039 \text{ Å}$$

(This result neglects a certain effect due to relativity: the increase in the mass of a particle at extremely high velocities. Since the calculated velocity of the electron is 63% of the velocity of light, the relativistic correction will be substantial.)

Example 2-5. Calculate the wavelength of a BB of mass 0.35 g and velocity 10^4 cm/s.

$$p = (0.35 \text{ g})(10^4 \text{ cm/s}) = 3.5 \times 10^3 \text{ g cm/s}$$

$$\lambda = \frac{6.6 \times 10^{-27} \text{ erg s}}{3.5 \times 10^3 \text{ g cm/s}} = 1.9 \times 10^{-30} \text{ cm}$$

In a two-slit diffraction experiment, it would be impossible to observe an interference pattern from a wave of such short wavelength because the peaks would be too close together. The interference pattern would thus look continuous.

2.14 THE HEISENBERG UNCERTAINTY PRINCIPLE

In 1927 Werner Heisenberg (b. 1901), a German theoretical physicist, discovered the *uncertainty principle*. This fundamentally important principle states that, if a position measurement is made with an uncertainty Δx and a momentum measurement is made simultaneously with an uncertainty Δp, then the product of the two uncertainties is equal to or greater than h:

$$\Delta x \Delta p \geq h \tag{2-14}$$

Exercise 2-1. Verify that the product of the units of position and momentum is equivalent to the units of Planck's constant h.

A two-slit diffraction experiment using electrons instead of BBs cannot be performed because the apparatus would have to be made on an impossibly small scale in order to reveal the effects of interest. However, it is known from many other studies of electrons that, if it *were* possible to perform such an experiment, one would obtain the same diffraction pattern as that obtained with light of the same wavelength (see Figure 2-10). This is a direct consequence of the wave nature of matter. In performing the experiment, either the uncertainty Δx in the position of an electron at a given instant or the uncertainty Δp_x in its momentum in the x-direction at that instant would be so large that it would be impossible to determine which slit the electron had passed through on its way to the detector.

If we attempted to use a light beam to observe the electron passing through one or the other of the slits, thereby pinpointing its position, the interaction of the photons with the electron would significantly change the motion of the electron, and hence its momentum, in a way that could not be precisely known. This change in the motion of the electrons would result in the destruction of the diffraction pattern, I_{12}. Under these conditions the distribution of electrons at plane S would become the same as that for BBs, given by P_{12} in Figure 2-9.

The uncertainty principle should help resolve in your mind the apparent logical contradiction between the properties of particles, which have position and momentum in space, and those of waves, which always exist at more than one point in space. Although the uncertainty principle is valid for *all* particles, only submicroscopic particles are small enough and have small enough momenta that uncertainties in these quantities have significant magnitudes relative to the quantities themselves. We have already discussed the fact that, in practice, only submicroscopic particles exhibit wave-like characteristics, and it is precisely these particles that exhibit the effects of the uncertainty principle. The more accurately we know the position of a particle-wave at a given instant, the less accurately we can know its momentum at that instant, i.e., the direction and speed at which it is moving. Conversely, the more accurately we know its momentum at a given instant, the less accurately we can know just where it is at that instant.

It is a fact of physics that all wave phenomena can be described by a certain class of equations called *wave equations*. Therefore, if particles are waves, an equation that describes the properties of such waves must exist. This equation was discovered by the Austrian theoretical physicist Erwin Schrödinger (1887–1961) in 1926.

2.15 WAVE EQUATIONS

The solution of a wave equation provides the detailed shape of a wave in space and time. For example, a rock thrown into water creates a surface wave pattern that depends upon both position and time. The solution of the appropriate wave equation provides the mathematical description of the concentric rings that spread out from the point of impact.

A still simpler example of wave motion is provided by a vibrating guitar string. Here the wave is one-dimensional instead of two-dimensional. This means that the displacement y of a point on the string depends only on the position of the point along the string (one dimension) and the time. The value of the *maximum* displacement (amplitude) depends only on the position along the string. If the string is clamped at both ends and plucked, its motion is generally complex, but it is always some combination of the simple patterns shown in Figure 2-14. We will call each of these patterns a *state*; with each state we associate a number n equal to one plus the number of points on the wave (not counting the ends) at which the displacement y is always zero. Such points are called *nodes*. Thus, the state $n = 1$ has no nodes, $n = 2$ has one node, and so forth.

There is a simple relation between the length l of the string, the number n that designates a given state, and the wavelength λ_n for that state (called the nth state):

$$\lambda_n = \frac{2l}{n} \qquad (2\text{-}15)$$

We will see shortly why this relation is important to us.

In quantum mechanics there is a problem analogous to that of the vibrating string. The problem is called the *particle in a one-dimensional box* and applies to an electron or any other small particle on a one-dimensional "track," such as a conducting wire. The word *box* refers to the shape of the potential energy well in which the particle is trapped. Figure 2-15 is a diagram of such a potential energy well. It shows that the potential energy V

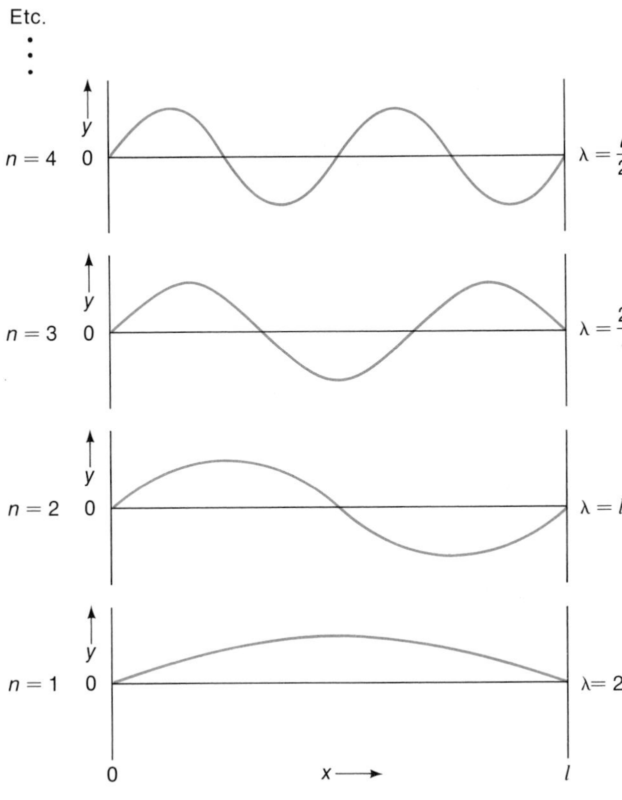

Figure 2-14. Allowed standing waves for a vibrating string of length l. The straight horizontal lines represent the zero point of amplitude.

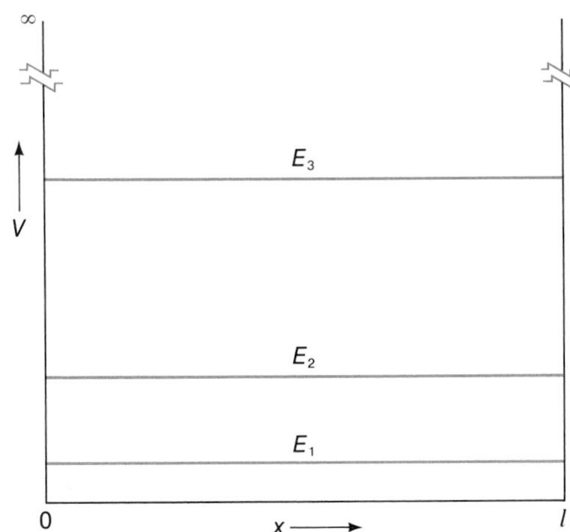

Figure 2-15. Potential energy well for a particle in a one-dimensional box.

of the particle is zero if x is between 0 and l and infinite everywhere else (i.e., outside the potential energy well). The particle will therefore never get "out of the box." The lines labeled E_1, E_2, and E_3 represent the first three of the many energy states allowed the electron in the potential well.

The first four solutions of the Schrödinger wave equation for the particle in the box are shown in Figure 2-16, where they are labeled ψ (Greek psi, pronounced psī). ψ_n is called the *wave function* for the nth state; it is a mathematical function denoting the amplitude of the wave that describes the particle in the box. The same figure contains the "code" number of the state, n. This is again equal to the number of nodes plus 1, and is called the *quantum number.* The column labeled E_n gives the energy of the particle when it is in the nth state, or, simply, the energy of the nth state. We will soon see where these energies come from, but for now, you should realize that the solutions of the Schrödinger equation are waves, or states, of definite shape and energy. The last column in Figure 2-16 shows the very important quantity ψ_n^2. Just as with electromagnetic waves, it is the *square* of the wave function (amplitude) that has physical meaning and represents "intensity." In this case, ψ^2 is called the *probability density function;* it represents the probability of finding the particle within a unit volume element centered on a given point.

Note that there is no state with zero energy, since a particle in such a state would not be moving; it would violate the uncertainty principle because its momentum would be zero. Also, a node in the wave function (or, more properly, the square of the wave function) represents a point at which the probability of finding the particle is zero. The particle will also not be found at the "walls" of the potential energy well or beyond, since that would require that it have infinite potential energy.

The energies of these states can be calculated from the de Broglie equation,

$$p = mv = \frac{h}{\lambda}$$

The velocity of a particle is therefore related to its wavelength by

$$v = \frac{h}{m\lambda} \tag{2-16}$$

The only kind of energy that the particle in the box has is kinetic energy. Thus,

$$E = \tfrac{1}{2}mv^2 = \frac{h^2}{2m\lambda^2} \tag{2-17}$$

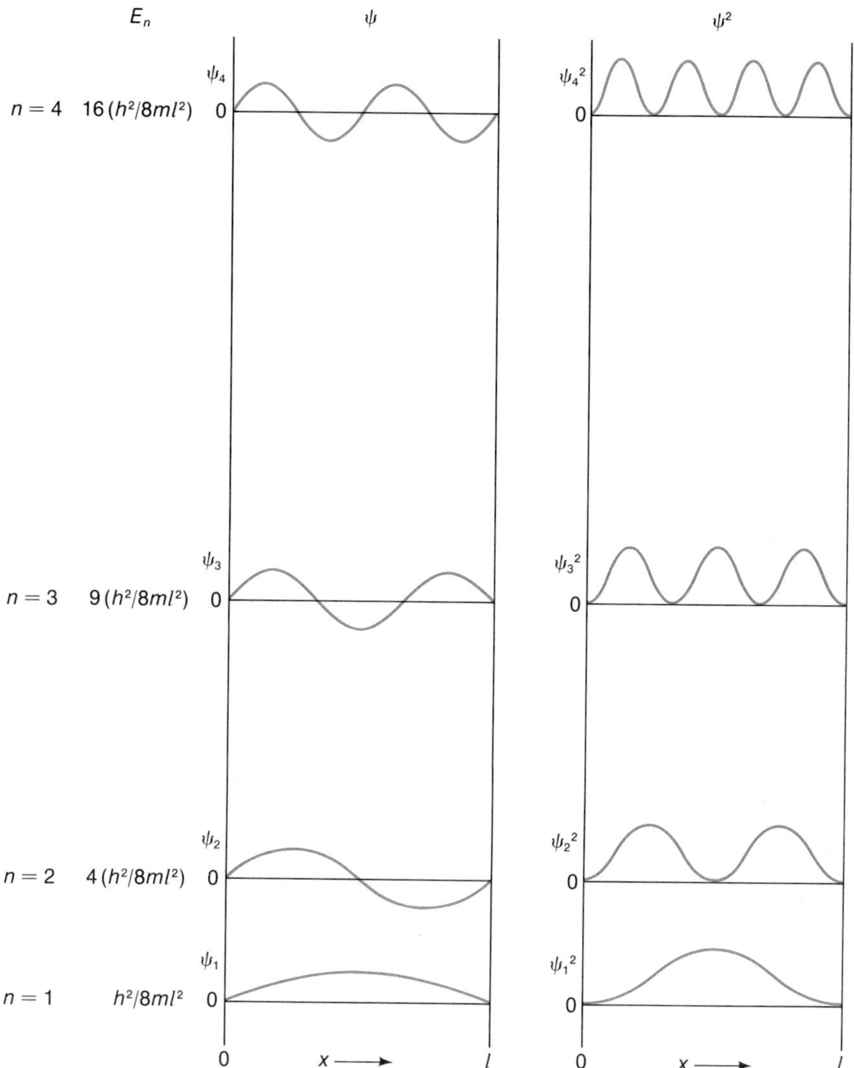

Figure 2-16. Energies, amplitudes, and probability density functions for the first four solutions of the wave equation for a particle in a one-dimensional box of length *l*.

We have already seen that

$$\lambda_n = \frac{2l}{n}$$

Therefore,

$$\frac{1}{\lambda_n} = \frac{n}{2l} \text{ and } \left(\frac{1}{\lambda_n}\right)^2 = \frac{n^2}{4l^2}$$

The energy of the *n*th state is then

$$E_n = \frac{h^2}{2m}\left(\frac{1}{\lambda_n}\right)^2 = \frac{h^2n^2}{8ml^2} \tag{2-18}$$

where m is the mass of the particle, l is the length of the box, n is the quantum number, and h is Planck's constant. This is the same result for the energy as that calculated from the wave equation and shown in Figure 2-16.

Application of the Schrödinger equation to the particle-in-a-box problem has yielded a set of allowed energy levels, which are quantized, and a set of probability density functions characterized by the presence of nodes. Both features of our answer contain a quantum number n that can assume any integral value greater than zero. These properties are characteristic of every system described by the Schrödinger equation. We turn now to our first atomic system.

2.16 THE HYDROGEN ATOM

The hydrogen atom consists of one proton and one electron. The force holding the electron in the proximity of the proton is an attractive Coulomb force between the two particles. Solving the wave equation for this electron in a coulombic potential well represents a more complicated wave problem, for two reasons. First, it is three-dimensional: three coordinates are required to specify the position of the electron relative to the proton. The spherical polar coordinate system shown in Figure 2-17 is mathematically the most convenient for this purpose.

The second complication is that the potential energy of the system depends on position, unlike the potential energy in the particle-in-a-box problem. The mathematical solutions are therefore more complex than a simple wave, but there are still a wave function ψ and a definite energy E for each state. The solution of the Schrödinger equation can be written as the product of a radial part (depending only on r) and an angular part (depending on θ and ϕ):

$$\Psi = (R)\,(\Theta\Phi) \tag{2-19}$$

The R part is called the *radial* wave function and the $\Theta\Phi$ part is called the *angular* wave function.

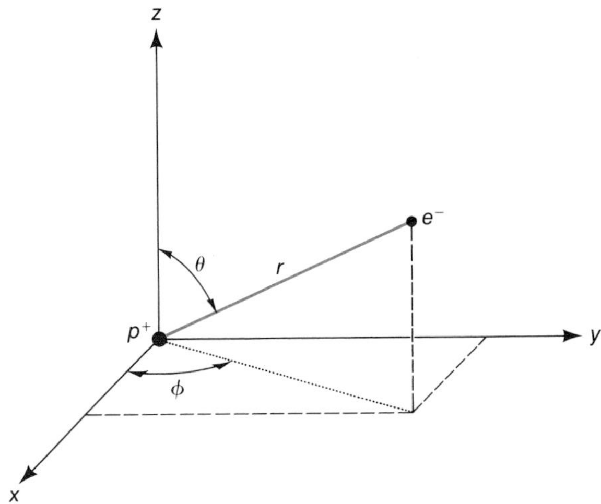

Figure 2-17. Spherical polar coordinates for the solution of the wave equation for the hydrogen atom.

There are three quantum numbers associated with the spatial distribution of the electron, because there are three coordinates. These quantum numbers are of great importance in describing the behavior of electrons in atoms. They are the *principal* quantum number n (equal to the number of nodes in the wave function, plus 1), the *azimuthal* quantum number l, and the *magnetic* quantum number m. The R function depends only on the values of n and l, whereas the $\Theta\Phi$ function depends on l and m. The allowed values of each quantum number and the physical characteristics it describes are summarized in Table 2-1. The allowed n values are obvious from the table. For each n value, the allowed l values are all the integers from 0 to $n - 1$, inclusive. For each l value, m can have any integral value from $-l$ to $+l$, inclusive. It is essential that you become proficient in handling these three quantities.

Table 2-1. **Three Quantum Numbers.**

Symbol	Name	What It Describes	Allowed Values
n	Principal	Size, energy, number of nodal surfaces	$1, 2, 3, 4, \ldots$
l	Azimuthal	Shape, total angular momentum	$0, 1, 2, \ldots, (n-2), (n-1)$
m	Magnetic	Direction, angular momentum along the z-axis	$0, \pm 1, \pm 2, \ldots, \pm l$

Reference to just n and l is often sufficient to give all the information needed about a given energy state of an electron (called an *electronic state*, or *energy level*). It is awkward to write "the $n = 4$, $l = 2$ state," and writing "the 4,2 state" could be confusing. The problem is solved by identifying each value of l with a letter. The following traditional scheme is used:

Value of l:	0	1	2	3
Symbol:	*s*	*p*	*d*	*f*

(For all practical purposes, l is never greater than 3.) Thus, the electronic state mentioned above is called the 4*d* state, and the electrons that it describes are called 4*d* electrons. The allowed combinations of the three quantum numbers for the first four principal energy levels are displayed in Table 2-2.

Example 2-6. Give all possible sets of the three quantum numbers for 3*s* and 4*f* electrons.

3*s*	4*f*
$n = 3$	$n = 4$
$l = 0$	$l = 3$
$m = 0$	$m = 0, \pm 1, \pm 2, \pm 3$

Only one set of quantum numbers is possible for 3*s* electrons. Seven different sets are possible for 4*f* electrons.

Example 2-7. Which of the following sets of quantum numbers are impossible? Why?

a. $n = 2, l = 2$	d. $n = 2, m = +2$	g. $n = 5, l = 4, m = 3$
b. $n = 4, l = 2$	e. $l = 2, m = +2$	h. $n = 2, l = -1, m = 0$
c. $n = 0, l = 0$	f. $l = 0, m = 0$	i. $n = 1, l = 0, m = 0$

Impossible combinations:

a. l can have values up to $n - 1$ but cannot equal n.
c. n cannot be zero.
d. m cannot be greater than $|l|$, which must be less than n.
h. l is either zero or positive; it cannot be negative.

Having learned the properties of the quantum numbers, we must now examine the shapes of the electron density distributions in the hydrogen

Table 2-2. Allowed Quantum Number Combinations for the $n = 1, 2, 3,$ and 4 States.

	$l = 0$ *s* electrons	$l = 1$ *p* electrons			$l = 2$ *d* electrons					$l = 3$ *f* electrons						
$n = 4$	$n=4$ $l=0$ $m=0$	$n=4$ $l=1$ $m=-1$	$n=4$ $l=1$ $m=0$	$n=4$ $l=1$ $m=1$	$n=4$ $l=2$ $m=-2$	$n=4$ $l=2$ $m=-1$	$n=4$ $l=2$ $m=0$	$n=4$ $l=2$ $m=1$	$n=4$ $l=2$ $m=2$	$n=4$ $l=3$ $m=-3$	$n=4$ $l=3$ $m=-2$	$n=4$ $l=3$ $m=-1$	$n=4$ $l=3$ $m=0$	$n=4$ $l=3$ $m=1$	$n=4$ $l=3$ $m=2$	$n=4$ $l=3$ $m=3$
$n = 3$	$n=3$ $l=0$ $m=0$	$n=3$ $l=1$ $m=-1$	$n=3$ $l=1$ $m=0$	$n=3$ $l=1$ $m=1$	$n=3$ $l=2$ $m=-2$	$n=3$ $l=2$ $m=-1$	$n=3$ $l=2$ $m=0$	$n=3$ $l=2$ $m=1$	$n=3$ $l=2$ $m=2$							
$n = 2$	$n=2$ $l=0$ $m=0$	$n=2$ $l=1$ $m=-1$	$n=2$ $l=1$ $m=0$	$n=2$ $l=1$ $m=1$												
$n = 1$	$n=1$ $l=0$ $m=0$															

atom—and other atoms. These distributions are called *orbitals* (not to be confused with "orbits," such as those of the planets). An orbital is actually a one-electron wave function representing a solution of the Schrödinger equation for a one-electron system. As such, it is not a "real" entity, but a mathematical equation for Ψ as a function of R and $\Theta\Phi$. However, because Ψ^2 defines a region of space in which the probability of finding the electron is high, and because this region has a characteristic shape that can be drawn, it is customary and very convenient to equate the orbital with the electron density distribution itself. Thus, we speak of electrons as being "in" orbitals, and of orbitals that are "filled" or "empty" and that "overlap." Strictly speaking, such usages are incorrect, but they are so valuable in visualizing atomic (and molecular) structure that it is well worth using them.

Let us now consider an *atomic orbital.* First we will look at the radial wave function R. It is important to see how the wave function and its square, the probability density function, vary with the distance from the nucleus. It is also important to see how the function $4\pi r^2R^2$ varies, because R^2 gives only the probability of finding the electron in a unit volume element at a distance r from the nucleus. A much more meaningful quantity is the probability of finding the electron in any arbitrarily thin spherical shell bounded by r and $r + dr$. When the probability density function at distance r is multiplied by the volume of this shell, $4\pi r^2dr$, one obtains the probability that the electron is in that shell. For all shells bounded by r and $r + dr$, this probability is proportional to the thickness dr of the shell. The relative probabilities of the electron's being in any two shells, both of thickness dr, is equal to the relative magnitudes of the function $4\pi r^2R^2$ for the two shells. A plot of $4\pi r^2R^2$ as a function of r therefore indicates the probability of the electron's being in a thin shell at a distance r, relative to the probability of its being in another thin shell at a different value of r but with the same thickness dr.

As a two-dimensional analogy, consider the problem of a baseball player who drops a contact lens out in center field. In Figure 2-18 an X marks the spot where he threw down his cap when the lens fell out. This represents the center of the area where he feels there is the greatest probability of finding the lens. There is a much smaller probability of finding the lens at third base, but he cannot entirely rule out that possibility. In reality the ballplayer has defined a probability diagram such as that shown in Figure 2-19.

Figure 2-19 shows the probability of finding the lens in a unit area—a square centimeter, say—at a distance r from X. Obviously, the more square centimeters the ballplayer has to look in, the more likely he is to find it. The area of the annulus bounded by r and $r + dr$ is $2\pi rdr$. For any small but constant thickness dr for the annuli, this area increases linearly with distance, as shown in Figure 2-20.

dr X r

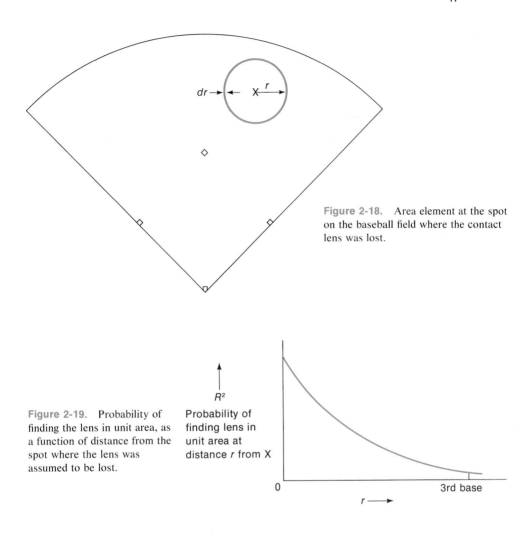

Figure 2-18. Area element at the spot on the baseball field where the contact lens was lost.

R^2
Probability of finding lens in unit area at distance r from X

0 3rd base

r

Figure 2-19. Probability of finding the lens in unit area, as a function of distance from the spot where the lens was assumed to be lost.

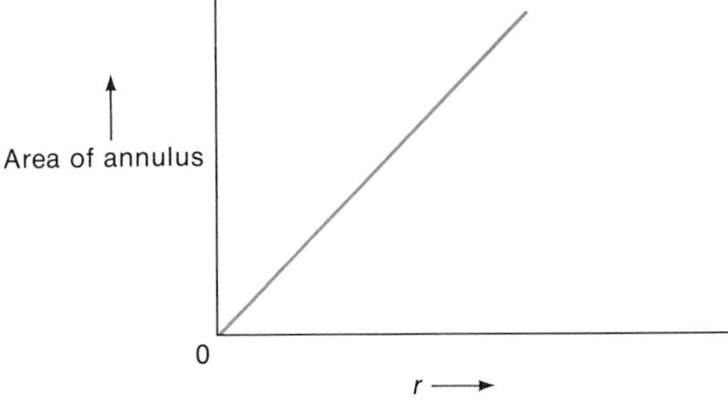

Figure 2-20. Variation of the area of an annulus with radial distance.

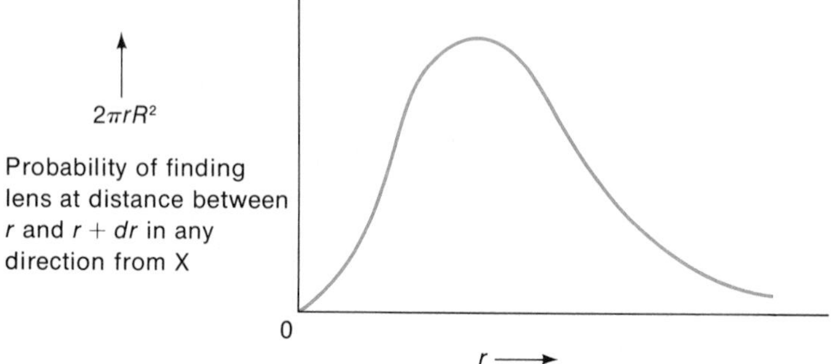

Figure 2-21. Probability of finding the lens in unit area at a distance between r and $r + dr$ in any direction from the spot where the lens was assumed to be lost.

The total probability of finding the lens in the annulus between r and $r + dr$ is the product of R^2 and the area of the annulus. The exponentially decreasing R^2 curve and the linearly increasing area curve combine to give a plot such as that shown in Figure 2-21. Again, because all the annuli are of equal thickness dr, we need not multiply by dr. The resulting curve shows the relative probabilities of finding the lens at different values of r.

Now let us look at some probability curves for electrons. The complete wave function for a 1*s* electron, Ψ_{1s}, is shown in Figure 2-22a. As in the two previous examples, we must square this function to obtain the probability density function Ψ^2 shown in Figure 2-22b. Figure 2-22c shows the relative probability of finding the electron at a distance between r and $r + dr$ from the nucleus in any direction.

Diagram c in Figure 2-22 has several interesting features. The maximum probability of finding the electron is no longer right at the nucleus (where $r = 0$), as it is in Diagrams a and b. Instead, it occurs at $r = 0.529$ Å. Note that the curve is asymptotic: it approaches zero only as r approaches infinity. The probability is infinitesimally small, but not zero, that the electron of a hydrogen atom in New York City may find itself momentarily in Los Angeles.

Plots of R^2 and $4\pi r^2 R^2$ are given in Figure 2-23 for a number of atomic orbitals of the hydrogen atom. Except for the orbital having the highest value of l for a given value of n, all the $4\pi r^2 R^2$ plots show values of r for which the probability of finding the electron is zero. These values define nodal surfaces (three-dimensional). The shapes of the $4\pi r^2 R^2$ plots should be studied carefully. The small peaks located near $r = 0$ for the higher *s* and *p* orbitals will have particular significance when we consider the energy levels of many-electron atoms later on.

The angular part ($\Theta\Phi$) of an *s* orbital is just a constant, so the shape of the orbital depends only on the radial part (R) and is therefore spherically symmetric. The shapes of the *p*, *d*, and *f* orbitals do depend on the angles θ and ϕ, however, so they are not spherically symmetric. The three 2*p* orbitals ($n = 2$, $l = 1$) are orthogonal and are called p_x, p_y, and p_z; by convention, they correspond to the *m* values ± 1 (for the *x*- or *y*-direction) and 0 (for the *z*-direction). Their figure-eight shapes are shown in Figure 2-24.

Both the radial and angular parts of the $2p_z$ orbital are positive for values of θ between 0° and 90°, making the entire wave function positive above the *x*-*y* plane. For this reason a plus sign is placed inside the corresponding lobe. Remember that the sign indicates only that the *wave function* is positive in that region. An identical lobe for values of θ between 90° and 180° exists below the *x*-*y* plane. Now, however, the wave function is negative, as indicated by the minus sign.

The shapes of the angular wave functions and their squares for the *s* and the three *p* orbitals are shown in Figure 2-24; the shapes of the five *d* or-

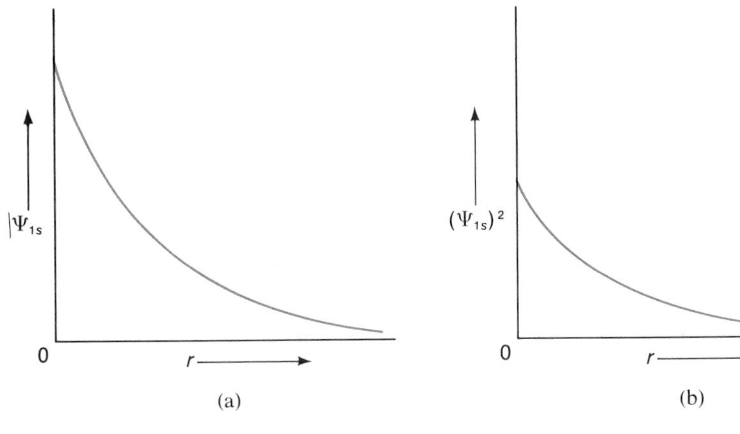

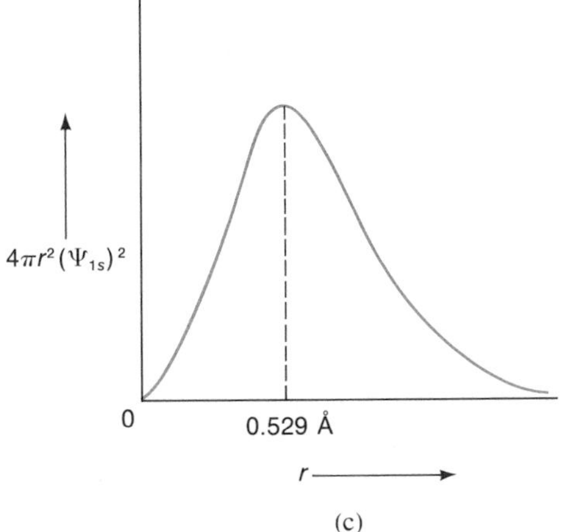

Figure 2-22. Representations of the 1*s* state of the hydrogen atom. (a) The amplitude of the 1*s* wave function (atomic orbital) as a function of distance *r* from the nucleus. (b) The probability density function (probability of finding the electron in unit volume) as a function of *r*. (c) The relative probability of finding the electron in a thin shell at distance *r*.

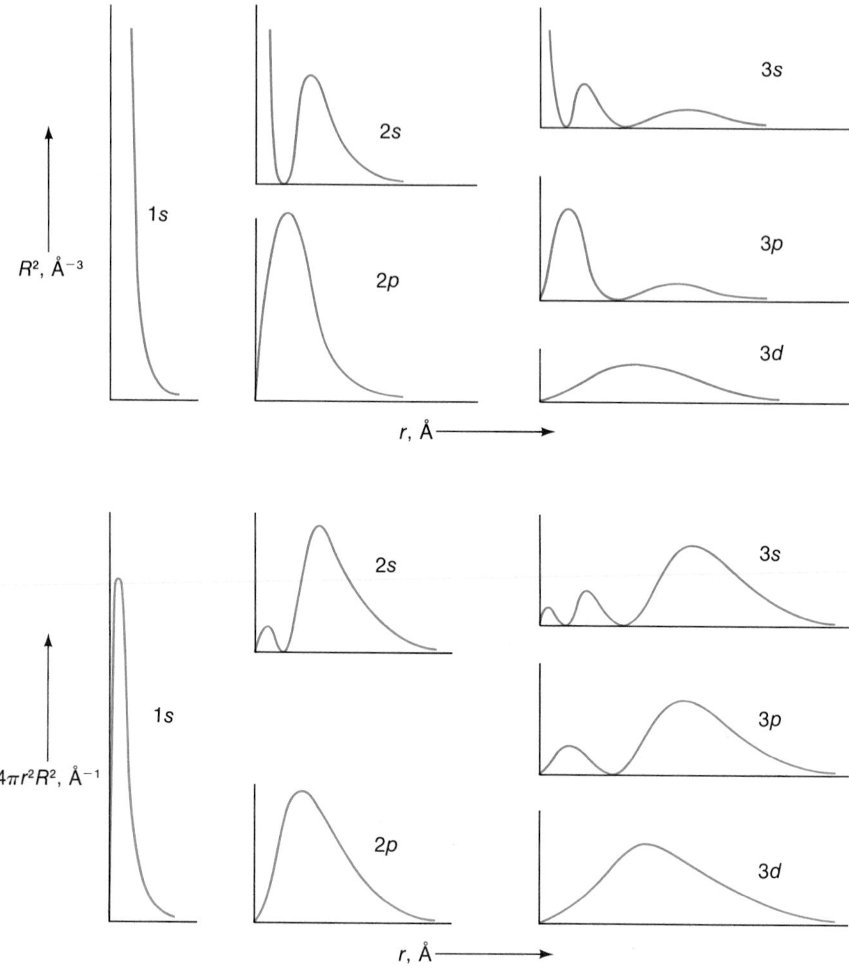

Figure 2-23. Radial probability density functions and the corresponding relative probabilities of finding the electron in a thin shell at distance r. Note that, apart from the differences in scale, the $4\pi r^2R^2$ plot for the $1s$ orbital is essentially identical to the $4\pi r^2\psi^2$ plot shown in Figure 2-22c. They differ only by the factor $(\Theta\Phi)^2$, which is a constant for s orbitals.

bitals are shown in Figure 2-25. It is important that you study these diagrams and learn the shapes and directional properties of these orbitals. They will be used repeatedly during our study of the geometry of molecules.

Another useful skill to cultivate is the visualization of the actual, three-dimensional electron density distributions for the s and p orbitals of the hydrogen atom. This is necessary because the *complete* probability density function is given by the product $R^2(\Theta\Phi)^2 = \Psi^2$. You must think of the

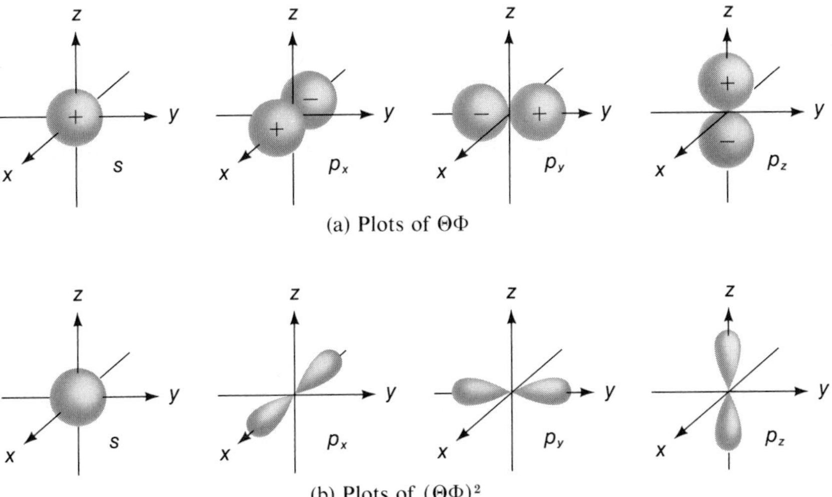

Figure 2-24. Angular wave functions and the corresponding probability density functions for s and p orbitals of the hydrogen atom.

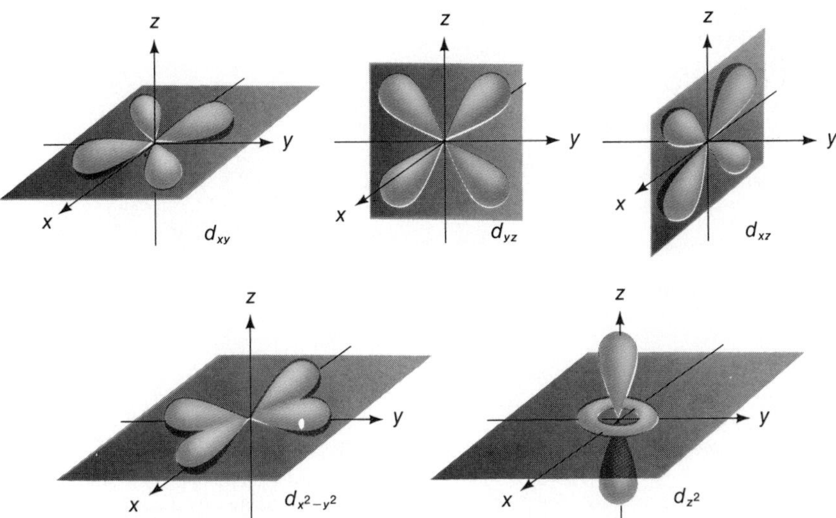

Figure 2-25. The shapes of the d orbitals.

variation in R^2 as r increases in any direction, but subject to the changing values of $(\Theta\Phi)^2$ as θ and ϕ vary. An attempt at 3-D visualization of both parts of the distribution simultaneously is shown in Figure 2-26. The maximum density of shading corresponds to the peak in the R^2 plot of the $2p$ orbital in Figure 2-23. The denser the shading, the greater the probability of finding the electron in that region of space.

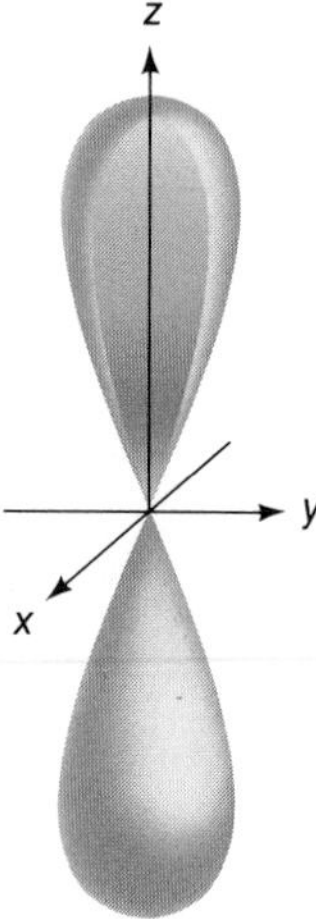

Figure 2-26. Representation of the electron density distribution "inside" the $2p_z$ orbital of the hydrogen atom.

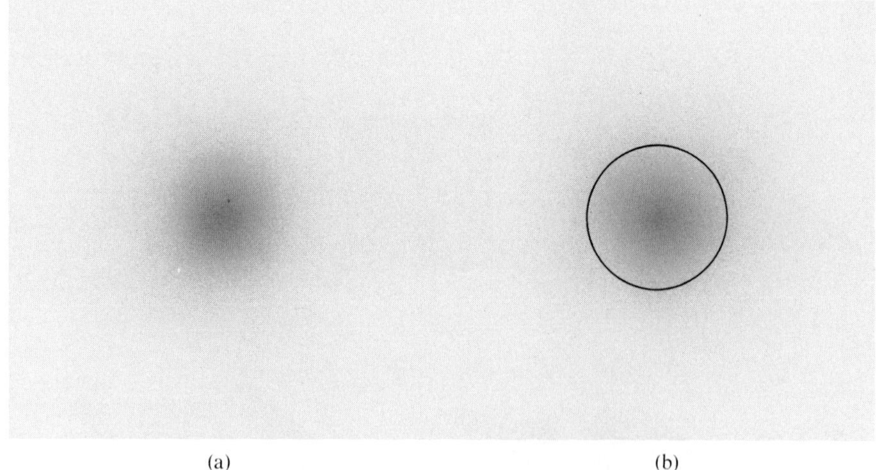

Figure 2-27. Representation of the electron density distribution in the $1s$ orbital of the hydrogen atom. (a) The probability of finding the electron is proportional to the density of shading within a unit area. (b) The $1s$ orbital is generally represented by a circle. As shown in this diagram, the circle implies that a majority (generally $\approx 90\%$) of the shading is contained within it.

This shading technique is useful to show what the surfaces of Figure 2-24 mean. The 1*s* orbital can be portrayed (Figure 2-27) as a circular shaded area extending to infinity but peaking in density at the nucleus in an R^2 plot. A circle drawn at $r = 1.4$ Å encompasses 90% of the shading and shows that there is a 90% chance that the electron will be found inside this region at any instant.

A further method of presenting both the shapes and the radial variations within these shapes is given in Figure 2-28, which shows three-dimensional representations of several atomic orbitals of the hydrogen atom. Each orbital is accompanied by a cross-sectional view of the electron density in the most revealing plane of symmetry for that orbital. These forms provide a useful way to visualize the nodes in the squared wave functions as well as the delicate and beautiful geometries of the orbitals with larger values of l.

Note, in Figure 2-28, that the 1*s* orbital contains no nodes. The 2*s* orbital contains one radial node, a spherical surface where there is no electron density. The 2*p* orbitals do not have a radial node but they do have one angular node, a plane where there is no electron density. The 3*s* orbital has two radial nodes and the 3*p* orbitals have one radial node and one angular node. The 3*d* orbitals have two angular nodes (two orthogonal planes in each orbital except the $3d_{z^2}$, in which they are two conical surfaces that meet apex-to-apex at the nucleus). This confirms the relationship, mentioned earlier, between the principal quantum number and the total number of nodes in the wave function. It also explains why 1*p*, 1*d*, and 2*d* states cannot exist: they require more nodes than the principal quantum number will allow.

2.17 SPECTROSCOPY

The above discussion of the nature of the electron density distributions in the hydrogen atom is important for an understanding of many of the properties of molecules. However, it represents only part of the quantum mechanical results obtained by solving the Schrödinger equation for the hydrogen atom. This solution also reveals the quantized values of the energy that the electron can possess. Before we discuss these energy levels, it will be helpful to backtrack for a moment and look at some experimental data on the hydrogen atom.

The electron in the hydrogen atom is electrically charged; it therefore has forces exerted upon it by both the electric and magnetic components of electromagnetic radiation. At certain wavelengths (frequencies) the energy $E = h\nu$ of this radiation is absorbed by the atom as a single photon.

Spectrometers measure the wavelengths at which absorption occurs;

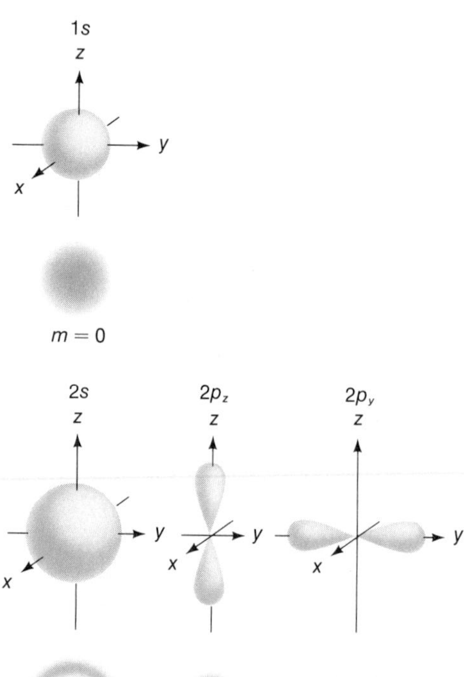

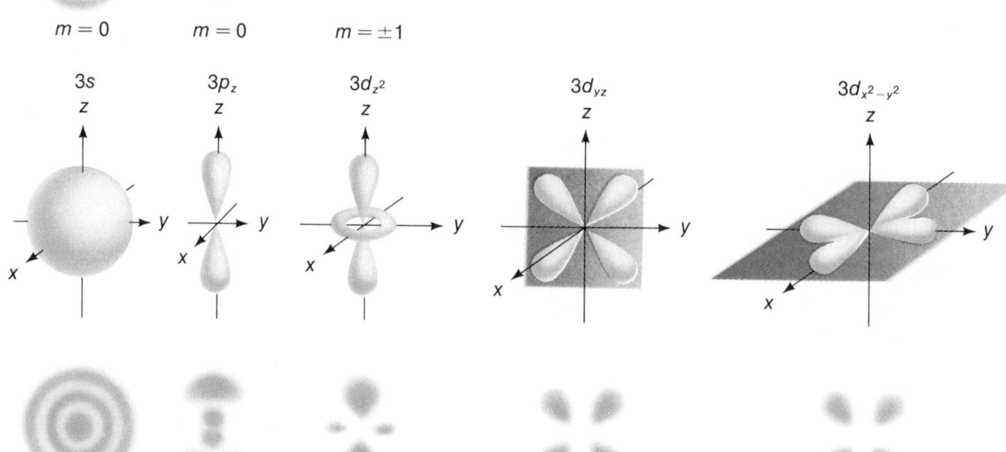

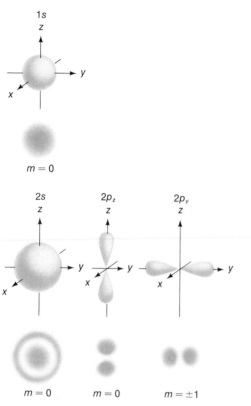

Figure 2-28. Electron distributions for several energy states of the hydrogen atom. The top member of each pair of diagrams is a three-dimensional representation of an atomic orbital. The bottom member is a cross section through the most revealing plane of symmetry in that orbital and represents the approximate electron density in that plane. The other $2p$ orbital (the $2p_x$, not shown) is directed along the x-axis; its cross-sectional representation is equivalent to those of the $2p_y$ and $2p_z$ orbitals. The other $3p$ orbitals (the $3p_x$ and $3p_y$) are directed along the x- and y-axes, respectively. As in the $2p$ orbitals, their cross sections are equivalent in shape but different in orientation. Two $3d$ orbitals are not shown: the $3d_{xy}$ orbital has four lobes directed along perpendicular axes oriented in the xy-plane 45° between x and y; similarly, the $3d_{xz}$ orbital has four lobes directed along perpendicular axes in the xz-plane 45° between x and z. Each indicated $\pm$ value of m corresponds to two orbitals with equivalent cross sections oriented differently in space. For example, the $3d_{xz}$, $3d_{yz}$, $3d_{xy}$, and $3d_{x^2-y^2}$ orbitals have identical cross sections. The former two are associated with the values $m = \pm 1$ and the latter two with the values $m = \pm 2$.

spectrophotometers measure these wavelengths and the amount of radiation absorbed as well. The basic components of both instruments are a source of radiation in the appropriate wavelength range, a scanning monochromator (an instrument that uses a prism or diffraction grating to isolate a very narrow range of wavelengths from the source spectrum) to vary the wavelength of radiation focused on the sample, a slit system, a container appropriate to the physical state of the sample, a detector to determine the amount of radiation transmitted by the sample, and a recorder to display the resultant *absorption spectrum*. These basic components are shown schematically in Figure 2-29.

If a source that emits a continuous spectrum of radiation is used and if this radiation is passed through a vapor consisting of hydrogen atoms, an absorption spectrum similar to that shown in Figure 2-30 is obtained. The white lines in this spectrum are characteristic of the H atom and represent "missing" radiation, i.e., radiation that was absorbed by the sample. In Figure 2-30 these lines are identified not by wavelength λ or frequency ν, but by *wave number* $\tilde{\nu} = 1/\lambda$. This useful quantity is derived from the Planck equation as follows:

$$E = h\nu = hc\left(\frac{1}{\lambda}\right) = hc\tilde{\nu} \quad (2\text{-}20)$$

Since h and c are constants, the wave number $\tilde{\nu}$ is directly proportional to the energy E. It is generally expressed in the units cm^{-1}, called *reciprocal centimeters*, although the less cumbersome term *kayser* (symbol K) is gaining favor; $1\ \text{cm}^{-1} \equiv 1\ \text{K}$. Thus, the lines in Figure 2-30 represent the absorption of photons of specific energies that increase with increasing wave number.

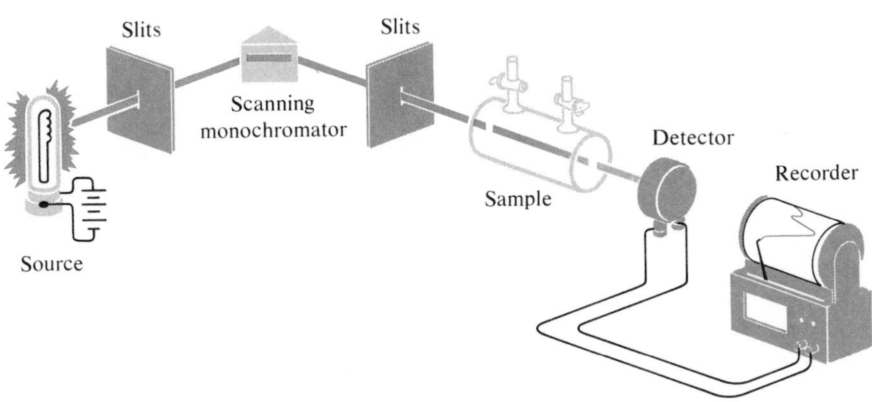

Figure 2-29. Schematic drawing of a simple spectrophotometer.

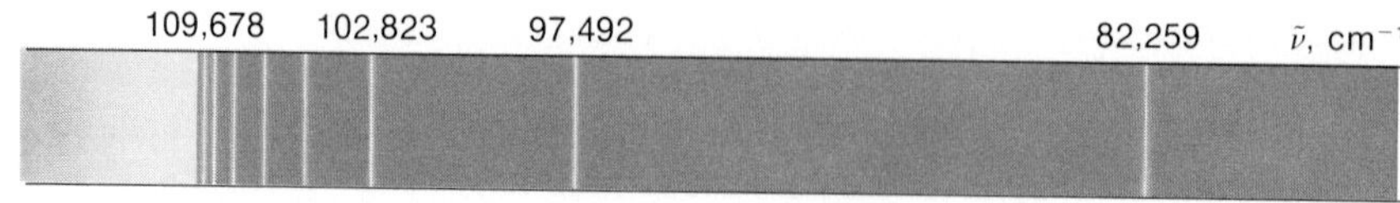

Figure 2-30. Part of the absorption spectrum of the hydrogen atom.

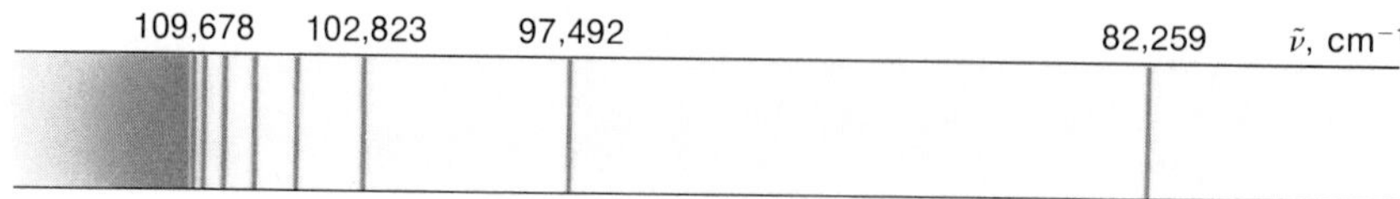

Figure 2-31. Part of the emission spectrum of the hydrogen atom.

The experiment can also be performed in a different manner, however. If the hydrogen atoms are heated to a very high temperature, they are found to *emit* radiation. Examination of this radiation with a spectrometer yields an *emission spectrum* similar to that shown in Figure 2-31. Compare the absorption and emission spectra of the hydrogen atom. Obviously there must be some unique property of this atom that allows it to absorb and emit radiation of only certain energies.

The lines shown in Figures 2-30 and 2-31 are not the only ones in the hydrogen-atom spectrum, there being several such series of lines in different regions of the electromagnetic spectrum. The series shown above is in the ultraviolet region. The first series discovered was observed in the visible region, and soon thereafter three more series were found in the infrared region. A Swiss schoolteacher named Johann Balmer (1825–1898) provided the first "explanation" of these lines. In 1885 he found empirically that the wave numbers of these series of lines could be obtained from an equation similar to the following equation:

$$\tilde{\nu}\ (\text{cm}^{-1}) = 109{,}678\left(\frac{1}{n_2^2} - \frac{1}{n_1^2}\right) \tag{2-21}$$

where n_1 and n_2 are integers, with $n_1 > n_2$. The equation in the general form shown here was subsequently discovered by the Swedish physicist Johannes Rydberg (1854–1919). It is called the *Rydberg equation* and the constant 109,678 cm^{-1} is called the *Rydberg constant.* The five spectral series of the hydrogen atom, named after their discoverers, are described in Table 2-3.

From Planck's hypothesis we know that the emission or absorption of radiation of specific frequencies corresponds to the emission or absorption of discrete amounts of energy. Thus, the data of Table 2-3 suggest that the hydrogen atom must possess a discrete set of energy levels and that it can change from one level to another by absorbing or emitting a photon of precisely the right energy. Figure 2-32, in which the origin of the energy scale is chosen as the energy of a free electron, i.e., one that is not trapped

Table 2-3. Spectral Series of the Hydrogen Atom.

Series	n_2	n_1	Region of Spectrum
Lyman	1	2, 3, 4, . . .	Ultraviolet
Balmer	2	3, 4, 5, . . .	Visible
Paschen	3	4, 5, 6, . . .	Infrared
Brackett	4	5, 6, 7, . . .	Infrared
Pfund	5	6, 7, 8, . . .	Infrared

in the potential well of an attracting proton, illustrates the observed data.

Both the absorption and emission spectra of the hydrogen atom (Figures 2-30 and 2-31) show that electromagnetic radiation with any wave number greater than 109,678 cm^{-1} is absorbed and emitted by hydrogen. This region of the spectrum is called the *continuum* because the absorption and emission lines are no longer discrete. The absorption of such photons by ground-state hydrogen atoms causes the formation of hydrogen ions (protons) and free electrons. Photon energy in excess of the $hc(109{,}678)$ erg required to remove the electron from the atom is imparted to the free electron and proton as kinetic energy. Similarly, the emission of radiation in the continuum results from the capture of a free electron with kinetic energy by a hydrogen ion (proton), yielding a hydrogen atom.

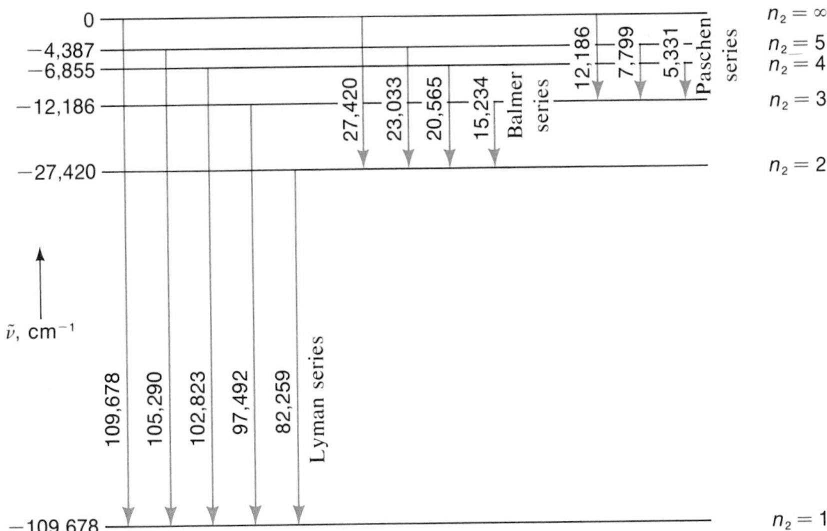

Figure 2-32. Energy-level diagram of the hydrogen atom. The transitions from upper levels to lower levels are shown. These differences in energy correspond to the lines of the emission spectrum. The right-hand column shows the values of n_2 from the Rydberg equation.

The origin of the observed energy levels, as well as their actual values, is given by the quantum theory of atomic structure. Solution of the Schrödinger equation for the hydrogen atom requires that the energy of the atom be quantized. As indicated in Table 2-1, the allowed energies are determined by the values of the principal quantum number n. The following equation provides the allowed energies of the H atom:

$$E_n = -\frac{2\pi^2 m_e e^4}{n^2 h^2} \qquad n = 1, 2, 3, \ldots \tag{2-22}$$

where m_e is the mass of the electron and e is its charge. A Z^2 is inserted in the numerator of Equation 2-22 for one-electron systems other than the hydrogen atom, e.g., the helium ion He^+. (Z represents the nuclear charge of such ions.)

Example 2-8. Verify that the Rydberg equation (2-21) can be derived from the quantum mechanical energy-level equation (2-22).

$$E_n = -\frac{1}{n^2}\left[\frac{2(3.142)^2(9.110\times10^{-28}\text{ g})(4.803\times10^{-10}\text{ cm}^{3/2}\text{ g}^{1/2}/\text{s})^4}{(6.626\times10^{-27}\text{ erg s})^2}\right]$$

$$= -\frac{1}{n^2}\left(\frac{2\times9.872\times9.110\times5.322\times10^{-66}}{4.390\times10^{-53}}\right)\text{ erg}$$

$$= -\frac{2.180\times10^{-11}}{n^2}\text{ erg}$$

In order to compare this answer with the Rydberg constant, we must convert from ergs to wave numbers, using the relation $\tilde{\nu} = E/hc$.

$$\tilde{\nu} = -\left(\frac{2.180\times10^{-11}\,\cancel{\text{erg}}}{n^2}\right)\left(\frac{1}{6.626\times10^{-27}\,\cancel{\text{erg}}\,\cancel{\text{s}}\times3.00\times10^{10}\text{ cm}/\cancel{\text{s}}}\right)$$

$$= -\frac{109{,}700}{n^2}\text{ cm}^{-1}$$

A transition between two energy levels n_1 and n_2 yields

$$\Delta E = E_2 - E_1 \propto \left(-\frac{109{,}700}{n_2^2}\right) - \left(-\frac{109{,}700}{n_1^2}\right)$$

$$= \tilde{\nu} = -109{,}700\left(\frac{1}{n_2^2} - \frac{1}{n_1^2}\right)$$

Figure 2-33 displays the energy levels of the hydrogen atom again. This time the fact that more than one type of orbital exists for all energy levels except the first (the *ground state*) is emphasized. Energy levels consisting

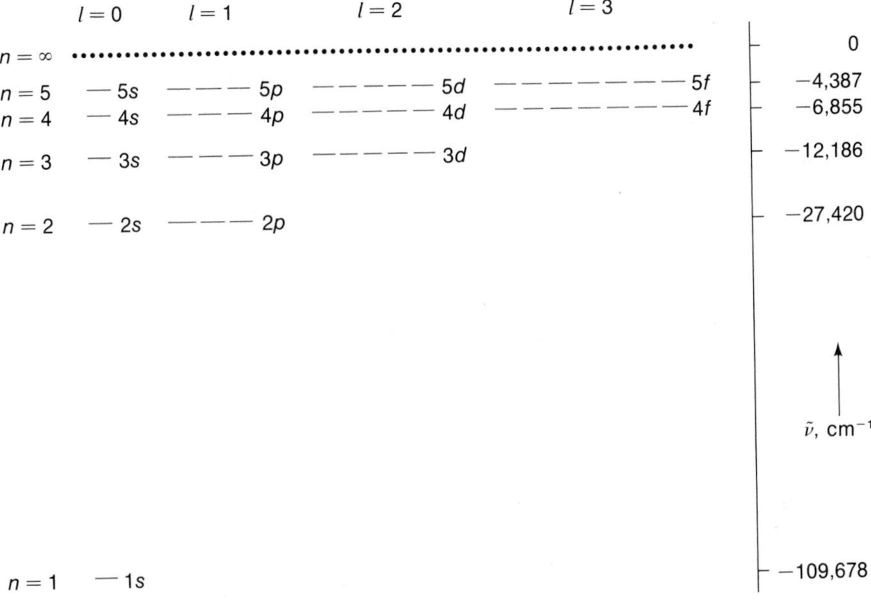

Figure 2-33. Energy-level diagram of the hydrogen atom. This diagram shows not only the energy levels for different values of the principal quantum number n, but also the degeneracies of these levels for different values of the azimuthal quantum number l.

of two or more identifiable states of identical energy are said to be *degenerate*. Before we can turn to the many-electron atom, a fourth quantum number must be introduced to help us determine how these energy levels (atomic orbitals) are to be "filled" with electrons.

2.18 ELECTRON SPIN AND THE PAULI EXCLUSION PRINCIPLE

In 1925 the Dutch physicists George Uhlenbeck (b. 1900) and Samuel Goudsmit (b. 1902) postulated that the electron has an intrinsic angular momentum called *spin,* because the electron behaves as though it were spinning about an axis. This spin has been experimentally verified by careful analysis of numerous atomic spectra. It is now known that not only electrons, but most elementary particles, have finite values of spin, which is quantized in units of $h/2\pi$. For the electron, the quantum number s can have the values $\pm\frac{1}{2}$. There is a magnetic field associated with a spinning charged particle, just as there is with any circulating electric current. Thus, the electron is also a small magnet.

The *Pauli exclusion principle,* discovered by the Austrian theoretical physicist Wolfgang Pauli (1900–1958), states that no two electrons in the same atom or molecule can have precisely the same wave function. Since the wave function for any electron is determined by the four quantum numbers n, l, m, and s, it follows that no two electrons in the same atom or molecule can have the same values of all four quantum numbers, i.e., they cannot be in precisely the same state. Each electron in an atom or molecule must therefore have a unique description in terms of the four quantum numbers. This discovery has proved to be of fundamental importance in the theory of the electronic structures of atoms and molecules.

If two electrons in the same atom have the same values of n, l, and m, they are said to be *paired.* This is because they have opposite spins—a necessity according to the exclusion principle—but are identical in every other respect. Since one electron has the spin quantum number $s = +\frac{1}{2}$ and the other has $s = -\frac{1}{2}$, the magnetic moments associated with these spins cancel. An atom or molecule with an *unpaired* electron is attracted by an external magnetic field because the magnetic field of the unpaired electron aligns itself in the direction of the external field, thereby reducing its energy and causing the attraction. Such a substance is called *paramagnetic.*

In *all* substances, whether they contain an unpaired electron or not, a much weaker magnetic force is induced by an external magnetic field; this force is opposite to the direction of the external field, causing a slight

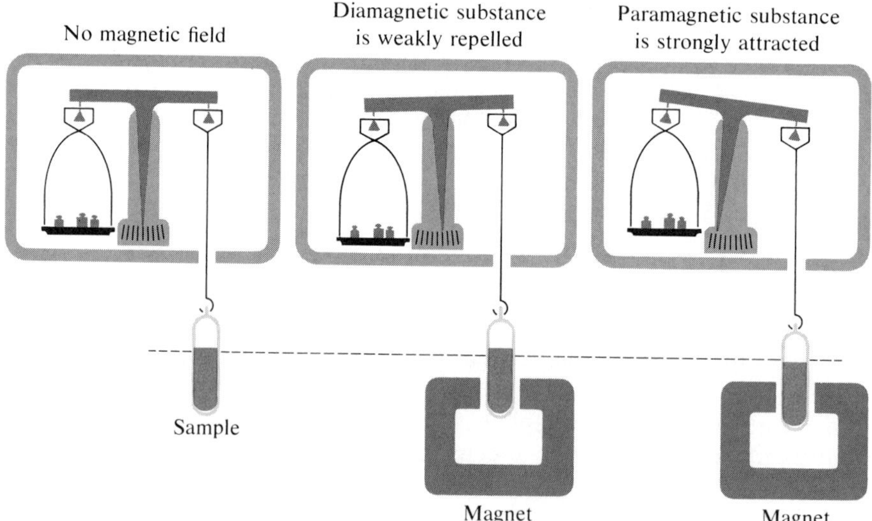

Figure 2-34. Experimental determination of paramagnetism and diamagnetism.

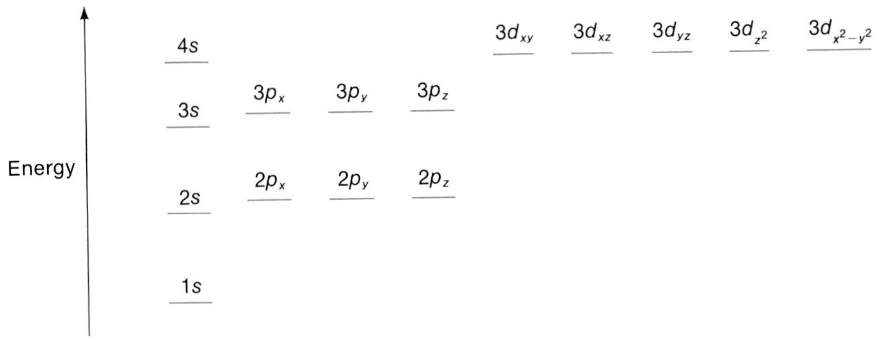

Figure 2-35. Energy-level diagram for the many-electron atom.

repulsion. If no other type of magnetism is present, such a substance is called *diamagnetic*. Diamagnetism is an extremely weak effect. Paramagnetism is several orders of magnitude stronger, so in those substances in which it occurs, it overwhelms the intrinsic diamagnetism. Figure 2-34 illustrates the method used for determining if a compound is paramagnetic or diamagnetic.

2.19 THE MANY-ELECTRON ATOM

With many electrons in an atom, solution of the Schrödinger equation is extremely difficult because the repulsive interactions between the electrons must be taken into account. One major effect of these interactions is the elimination of the degeneracy between the *s*, *p*, *d*, and *f* electronic states of the same principal quantum number (although the degeneracies *within* the *p*, *d*, and *f* states themselves remain unaffected). The reason for this effect is that the *p* electrons do not penetrate the volume elements nearest the nucleus as much as do the *s* electrons with the same value of *n*. The *p* electrons are thus "shielded" from the positive nuclear attraction by the *s* electrons, and therefore have slightly higher energies than do the corresponding *s* electrons. This effect is even greater for *d* electrons. The order of the first few energy levels in the many-electron atom is presented in Figure 2-35.

There are several things we must keep in mind. The exact energy values in a diagram such as Figure 2-35 change as the nuclear charge (and therefore the number of electrons in the atom) changes. In fact, the order of the energy levels sometimes changes at large values of *n*, where the levels are rather close together. Although we retain the labels *s*, *p*, *d*, and *f* from the hydrogen-atom wave functions, the exact wave functions for the electronic states in the many-electron atom cannot be the hydrogen wave functions; they are much more complex and not precisely known. The symmetry properties of the wave functions are retained, however, so the general shapes and nodal properties of the orbitals (the electron density distributions, actually) remain the same. Thus, our labels for the different energy levels refer mainly to the symmetry of the wave functions.

The impossibility of obtaining *exact* solutions of the Schrödinger equation for the many-electron problem is the major reason that chemistry is still primarily an experimental science. Nevertheless, powerful and sophisticated mathematical approximation methods are being used to solve the Schrödinger equation for complex systems. The resulting detailed predictions of atomic and molecular energy levels and geometries are very valuable and sometimes as accurate as experimental measurements.

2.20 HUND'S RULE AND THE ELECTRON CONFIGURATIONS OF THE ELEMENTS

The order of filling of the available energy levels as atomic number increases is governed by the general rule that the electrons always enter the available level with the lowest energy, just as large particles always seek their lowest potential energy. Remember that no more than two electrons can occupy any level and that these electrons must have opposite spins in order to satisfy the Pauli exclusion principle.*

If we consider a set of degenerate energy levels, such as those represented by the $2p_x$, $2p_y$, and $2p_z$ orbitals, *Hund's rule* [discovered by the German physicist Friedrich Hund (b. 1896)] states that, with increasing atomic number, these orbitals are occupied in succession by only one electron each, and that these electrons have parallel spins. Only after each orbital contains one electron does any of the orbitals acquire a second electron, of opposite (antiparallel) spin. Electrons with the same spin are farther apart, on the average, than electrons with opposite spins. Because of the Coulomb repulsion between electrons, the configuration in which they are more "spread out" is energetically favored. There are some minor exceptions to Hund's rule, but it is generally correct.

*Remember from our earlier discussion that the idea of "available" orbitals—as though orbitals were little boxes waiting to be filled—is, strictly speaking, nonsense (but *useful* nonsense). An orbital is a wave function that describes an electron; it "exists" only when the electron is "in" it.

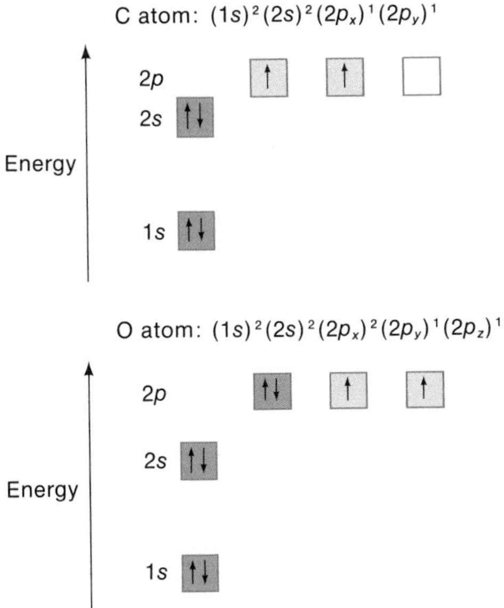

Figure 2-36. The electron configurations of carbon and oxygen atoms in their ground states. The electron population of each energy level is shown by means of arrows, the directions of which are arbitrary designations of the directions of the electron spins.

It is now a simple matter to write the electron configuration of an atom, using superscripts to denote the numbers of electrons in the various orbitals, e.g.,

Carbon, with 6 electrons, is $(1s)^2(2s)^2(2p_x)^1(2p_y)^1$

Oxygen, with 8 electrons, is $(1s)^2(2s)^2(2p_x)^2(2p_y)^1(2p_z)^1$

The meaning of this notation is illustrated more clearly in Figure 2-36. The electron configurations of the 105 known elements are shown in abbreviated form in Table 2-4.

The periodic properties of the elements, which are discussed in the next chapter, are a direct result of the order of filling of these energy levels and the spacing between the levels.

Table 2-4. **Electron Configurations of the Elements.**

		Principal Quantum Number *n*																		
		1	2		3			4				5				6				7
Z	Element	*s*	*s*	*p*	*s*	*p*	*d*	*s*	*p*	*d*	*f*	*s*	*p*	*d*	*f*	*s*	*p*	*d*	*f*	*s*
1	H	1																		
2	He	2																		
3	Li	2	1																	
4	Be	2	2																	
5	B	2	2	1																
6	C	2	2	2																
7	N	2	2	3																
8	O	2	2	4																
9	F	2	2	5																
10	Ne	2	2	6																
11	Na	2	2	6	1															
12	Mg	2	2	6	2															
13	Al	2	2	6	2	1														
14	Si	2	2	6	2	2														
15	P	2	2	6	2	3														
16	S	2	2	6	2	4														
17	Cl	2	2	6	2	5														
18	Ar	2	2	6	2	6														
19	K	2	2	6	2	6		1												
20	Ca	2	2	6	2	6		2												
21	Sc	2	2	6	2	6	1	2												
22	Ti	2	2	6	2	6	2	2												
23	V	2	2	6	2	6	3	2												
24	Cr	2	2	6	2	6	5	1												
25	Mn	2	2	6	2	6	5	2												
26	Fe	2	2	6	2	6	6	2												
27	Co	2	2	6	2	6	7	2												
28	Ni	2	2	6	2	6	8	2												
29	Cu	2	2	6	2	6	10	1												
30	Zn	2	2	6	2	6	10	2												
31	Ga	2	2	6	2	6	10	2	1											
32	Ge	2	2	6	2	6	10	2	2											
33	As	2	2	6	2	6	10	2	3											
34	Se	2	2	6	2	6	10	2	4											

Table 2-4. **(continued)**

		Principal Quantum Number *n*																		
		1	2		3			4				5				6				7
Z	Element	*s*	*s*	*p*	*s*	*p*	*d*	*s*	*p*	*d*	*f*	*s*	*p*	*d*	*f*	*s*	*p*	*d*	*f*	*s*
35	Br	2	2	6	2	6	10	2	5											
36	Kr	2	2	6	2	6	10	2	6											
37	Rb	2	2	6	2	6	10	2	6			1								
38	Sr	2	2	6	2	6	10	2	6			2								
39	Y	2	2	6	2	6	10	2	6	1		2								
40	Zr	2	2	6	2	6	10	2	6	2		2								
41	Nb	2	2	6	2	6	10	2	6	4		1								
42	Mo	2	2	6	2	6	10	2	6	5		1								
43	Tc	2	2	6	2	6	10	2	6	6		1?								
44	Ru	2	2	6	2	6	10	2	6	7		1								
45	Rh	2	2	6	2	6	10	2	6	8		1								
46	Pd	2	2	6	2	6	10	2	6	10										
47	Ag	2	2	6	2	6	10	2	6	10		1								
48	Cd	2	2	6	2	6	10	2	6	10		2								
49	In	2	2	6	2	6	10	2	6	10		2	1							
50	Sn	2	2	6	2	6	10	2	6	10		2	2							
51	Sb	2	2	6	2	6	10	2	6	10		2	3							
52	Te	2	2	6	2	6	10	2	6	10		2	4							
53	I	2	2	6	2	6	10	2	6	10		2	5							
54	Xe	2	2	6	2	6	10	2	6	10		2	6							
55	Cs	2	2	6	2	6	10	2	6	10		2	6			1				
56	Ba	2	2	6	2	6	10	2	6	10		2	6			2				
57	La	2	2	6	2	6	10	2	6	10		2	6	1		2				
58	Ce	2	2	6	2	6	10	2	6	10	2	2	6			2?				
59	Pr	2	2	6	2	6	10	2	6	10	3	2	6			2?				
60	Nd	2	2	6	2	6	10	2	6	10	4	2	6			2				
61	Pm	2	2	6	2	6	10	2	6	10	5	2	6			2?				
62	Sm	2	2	6	2	6	10	2	6	10	6	2	6			2				
63	Eu	2	2	6	2	6	10	2	6	10	7	2	6			2				
64	Gd	2	2	6	2	6	10	2	6	10	7	2	6	1		2				
65	Tb	2	2	6	2	6	10	2	6	10	9	2	6			2?				
66	Dy	2	2	6	2	6	10	2	6	10	10	2	6			2?				
67	Ho	2	2	6	2	6	10	2	6	10	11	2	6			2?				
68	Er	2	2	6	2	6	10	2	6	10	12	2	6			2?				
69	Tm	2	2	6	2	6	10	2	6	10	13	2	6			2				

Table 2-4. (continued)

		Principal Quantum Number *n*																		
		1	2		3			4				5				6				7
Z	Element	*s*	*s*	*p*	*s*	*p*	*d*	*s*	*p*	*d*	*f*	*s*	*p*	*d*	*f*	*s*	*p*	*d*	*f*	*s*
70	Yb	2	2	6	2	6	10	2	6	10	14	2	6			2				
71	Lu	2	2	6	2	6	10	2	6	10	14	2	6	1		2				
72	Hf	2	2	6	2	6	10	2	6	10	14	2	6	2		2				
73	Ta	2	2	6	2	6	10	2	6	10	14	2	6	3		2				
74	W	2	2	6	2	6	10	2	6	10	14	2	6	4		2				
75	Re	2	2	6	2	6	10	2	6	10	14	2	6	5		2				
76	Os	2	2	6	2	6	10	2	6	10	14	2	6	6		2				
77	Ir	2	2	6	2	6	10	2	6	10	14	2	6	7		2				
78	Pt	2	2	6	2	6	10	2	6	10	14	2	6	9		1				
79	Au	2	2	6	2	6	10	2	6	10	14	2	6	10		1				
80	Hg	2	2	6	2	6	10	2	6	10	14	2	6	10		2				
81	Tl	2	2	6	2	6	10	2	6	10	14	2	6	10		2	1			
82	Pb	2	2	6	2	6	10	2	6	10	14	2	6	10		2	2			
83	Bi	2	2	6	2	6	10	2	6	10	14	2	6	10		2	3			
84	Po	2	2	6	2	6	10	2	6	10	14	2	6	10		2	4?			
85	At	2	2	6	2	6	10	2	6	10	14	2	6	10		2	5?			
86	Rn	2	2	6	2	6	10	2	6	10	14	2	6	10		2	6			
87	Fr	2	2	6	2	6	10	2	6	10	14	2	6	10		2	6			1?
88	Ra	2	2	6	2	6	10	2	6	10	14	2	6	10		2	6			2
89	Ac	2	2	6	2	6	10	2	6	10	14	2	6	10		2	6	1		2?
90	Th	2	2	6	2	6	10	2	6	10	14	2	6	10		2	6	2		2
91	Pa	2	2	6	2	6	10	2	6	10	14	2	6	10	2	2	6	1		2?
92	U	2	2	6	2	6	10	2	6	10	14	2	6	10	3	2	6	1		2
93	Np	2	2	6	2	6	10	2	6	10	14	2	6	10	4	2	6	1		2?
94	Pu	2	2	6	2	6	10	2	6	10	14	2	6	10	6	2	6			2?
95	Am	2	2	6	2	6	10	2	6	10	14	2	6	10	7	2	6			2?
96	Cm	2	2	6	2	6	10	2	6	10	14	2	6	10	7	2	6	1		2?
97	Bk	2	2	6	2	6	10	2	6	10	14	2	6	10	9	2	6			2?
98	Cf	2	2	6	2	6	10	2	6	10	14	2	6	10	10	2	6			2?
99	Es	2	2	6	2	6	10	2	6	10	14	2	6	10	11	2	6			2?
100	Fm	2	2	6	2	6	10	2	6	10	14	2	6	10	12	2	6			2?
101	Md	2	2	6	2	6	10	2	6	10	14	2	6	10	13	2	6			2?
102	No	2	2	6	2	6	10	2	6	10	14	2	6	10	14	2	6			2?
103	Lw	2	2	6	2	6	10	2	6	10	14	2	6	10	14	2	6	1		2?
104	—	2	2	6	2	6	10	2	6	10	14	2	6	10	14	2	6	2		2?
105	—	2	2	6	2	6	10	2	6	10	14	2	6	10	14	2	6	3		2?

Bibliography

Robin M. Hochstrasser, **Behavior of Electrons in Atoms**, W. A. Benjamin, New York, 1964. The first half of this small text provides a rather thorough introduction to the electronic structure of atoms, paralleling in large part the discussion in the present chapter. Electromagnetic radiation and atomic spectra are used to introduce the principal concepts of modern quantum theory. The discussion is essentially nonmathematical. The latter half of the book deals with energy terms, collisions between excited atoms, and the behavior of atoms in magnetic fields.

Emil J. Margolis, **Formulation and Stoichiometry**, Appleton-Century-Crofts, New York, 1968. This is a problem book. The material in it covers a wide variety of chemical calculations. Thus, you may find it useful not only for this chapter but for a number of later chapters as well. Chapter 1 will provide you with some good practice in making stoichiometric calculations involving numbers of atoms, the mole concept, and Avogadro's number.

Problems

1. Radioisotopes are often used in medicine. For example, ^{131}I is used in the treatment of goiter and cancer of the thyroid, ^{32}P is used in the treatment of bone cancer, and ^{60}Co is used in the treatment of cancer. Give the atomic mass and the number of protons, neutrons, and electrons in each of these isotopes. (For a list of isotopes, consult the CRC *Handbook of Chemistry and Physics.*)

2. The isotopes of lithium found in natural abundance are ^{6}Li and ^{7}Li, with masses 6.015 and 7.016, respectively. Natural lithium has an atomic weight of 6.939. What are the relative abundances of ^{6}Li and ^{7}Li?

3. When an electric discharge is passed through neon gas confined in a glass tube, the gas gives off a bright orange-red light that is commonly seen in advertising. If a lamp contains 7.20 g of neon, how many Ne atoms are there in the lamp?

4. Given that the atomic weight of zinc is 65.37, what is the mass in grams of an average zinc atom?

5. Which of the following samples contains the largest number of atoms:
   a. 6.70 g copper
   b. 0.150 mole copper
   c. 7.80×10^{22} atoms of copper

6. Monomers are simple compounds that, when joined together, form large molecules called polymers. Calculate the molecular weight of each of the following monomers:
   a. In polyethylene: $CH_2{=}CH_2$
   b. In Orlon: $CH_2{=}CH{-}CN$
   c. In Saran: $CH_2{=}CCl_2$
   d. In Teflon: $CF_2{=}CF_2$

7. Carbonated beverages consist of flavored water that has been charged with CO_2 under pressure. Calculate the number of grams in one mole of CO_2. One gram of carbon dioxide contains how many molecules of carbon dioxide? What is the mass of one molecule of carbon dioxide?

8. Freon 12, CCl_2F_2, is an easily liquefiable gas used in commercial refrigerators and air conditioners. How many molecules are contained in 9.20 g of Freon 12?

9. The 1970 standard established by the U.S. Government for carbon monoxide emission from automobiles was 23 g CO per vehicle-mile. Assume that in a typical city there are 600,000 people with 75,000 automobiles, driven an average of 15 miles per day. How many tons of CO are emitted into the city's atmosphere daily, assuming that the standard is not being exceeded?

10. Nitrous oxide, N_2O, is commonly known as laughing gas. It is a colorless gas with a rather sweet, pleasant odor and is used by dentists as an anesthetic. How many atoms of nitrogen are contained in 0.150 mole N_2O? How many grams of oxygen are contained in this amount? How many moles of N_2O can we make from 4.00 g of nitrogen (N_2) and 2.00 g of oxygen (O_2)?

11. The air in Cambridge, Mass., was analyzed for vanadium [*Analytical Chemistry* **42**, 257 (1970)], since this element is important as a catalyst in the oxidation of SO_2 to SO_3. A maximum allowable concentration of 2 mg V_2O_5/m^3 of air has been established in the Soviet Union. Has this standard been exceeded in Cambridge, Mass., if, after 100 m^3 of air is filtered, the filter is found to contain 40 mg of vanadium? If so, by how much?

12. Calculate the force between two electrons 10^{-8} cm apart. Is the force an attractive force or a repulsive force?

13. Barium chromate ($BaCrO_4$) is very insoluble in water, whereas sodium chromate (Na_2CrO_4), barium chloride ($BaCl_2$), and sodium chloride (NaCl) are all very soluble in water. For this reason a solid precipitate of $BaCrO_4$ forms when $BaCl_2$ and Na_2CrO_4 are mixed in water. We can write this (unbalanced) chemical reaction as

$$BaCl_2 + Na_2CrO_4 \rightarrow BaCrO_4 \downarrow + NaCl$$

a. Balance the above chemical reaction.
b. How many molecules of $BaCl_2$ react with one molecule of Na_2CrO_4?
c. How many moles of $BaCrO_4$ precipitate if one mole of Na_2CrO_4 reacts?
d. How many grams of $BaCl_2$ are required to react completely with 10.0 g of Na_2CrO_4?
e. How many grams of $BaCl_2$ are required to form 40.0 g of $BaCrO_4$?
f. How many grams of $BaCl_2$ were added if 13.0 g of NaCl are left in solution?
g. How many kilograms of Na_2CrO_4 react to yield 253.3 kg of $BaCrO_4$?

14. The following reaction is used in making superphosphate fertilizer:

$$\underset{\text{Calcium phosphate}}{Ca_3(PO_4)_2} + \underset{\text{Sulfuric acid}}{2H_2SO_4} \rightarrow \underset{\text{Calcium dihydrogen phosphate}}{Ca(H_2PO_4)_2} + \underset{\text{Calcium sulfate}}{2CaSO_4}$$

The superphosphate fertilizer is the 1:2 mole mixture of the two products. (For each molecule of $Ca(H_2PO_4)_2$ there are two molecules of $CaSO_4$.) How many tons of fertilizer are made from one ton of calcium phosphate? How many tons of sulfuric acid must be used to react completely with one ton of calcium phosphate?

15. Iron is manufactured from iron oxide, Fe_2O_3, by reaction with coal (carbon, C) in a blast furnace in the presence of oxygen gas. The two reactions in the furnace are

$$2C + O_2 \rightarrow 2CO \uparrow$$

$$3CO + Fe_2O_3 \rightarrow 2Fe + 3CO_2 \uparrow$$

a. Starting with one metric ton of Fe_2O_3, how many grams of C must be added to the furnace to reduce all the Fe_2O_3 to Fe? (1 metric ton = 10^3 kg = 10^6 g.)
b. How many grams of product, Fe, are obtained?

16. When a sample of pure manganese dioxide, MnO_2, is heated, only oxygen gas is evolved. If a sample of MnO_2 weighing 5.23 g is heated until no more O_2 is evolved, the remaining solid weighs 4.59 g.
a. How many moles of O_2 were evolved?
b. What is the formula of the remaining pure oxide of manganese?
c. Write the chemical reaction.

17. An automobile uses an experimental fuel, octane, which has the chemical formula C_8H_{18}. The octane is burned in the engine (reacted with oxygen) to form carbon dioxide and water vapor. The unbalanced equation for this reaction is

$$C_8H_{18} + O_2 \rightarrow CO_2 \uparrow + H_2O \uparrow$$

a. Balance the equation.
b. If one gallon of this fuel weighs 2660 g, how many moles of CO_2 are produced when one gallon is consumed?
c. How many grams of O_2 must be sucked into the engine to burn the one gallon of fuel?

18. Much of this country's highway lighting is being converted from sodium-vapor lamps to mercury-vapor lamps. These lamps emit bluish-white radiation with a maximum intensity at about 4300 Å. What is the energy per photon of this light in ergs and in kcal/mole?

19. A tungsten-filament lamp emits continuous radiation in the region from 350 to 2500 nm. What wave numbers, frequencies, and energies (ergs) do these two limits correspond to? The tungsten-filament lamp is often used as a radiation source in scientific instruments. For what regions of the spectrum is this lamp an effective source?

20. Calculate the energy in kcal/mole of photons with a 5.00-Å wavelength. What do we call such radiation? Since a carbon-oxygen single bond has a bond energy of 85 kcal/mole, would you expect this radiation to produce a chemical reaction?

21. The work function for aluminum is 4.2 eV. Calculate the threshold frequency ν_0 for aluminum and the longest-wavelength electromagnetic radiation that can eject electrons from this metal.

22. When ultraviolet radiation of wavelength 2000 Å falls on an aluminum screen door whose work function is 4.2 eV, what is the speed in cm/s of the electrons emitted from the surface?

23. The winning car (M16B McLaren) of the 1972 Indianapolis 500 was driven at an average speed of 163.465 mph. The approximate mass of the car and driver was 770 kg. Calculate the de Broglie wavelength of the car and driver at their average speed.

24. The eight ball in pool weighs about 300 g and is moving toward the corner pocket at about 300 cm/s. Assume that we can locate the ball with an error equal in magnitude to the wavelength of light used, which is 5000 Å. Calculate the ratio of the uncertainty in the momentum of the eight ball to the momentum itself.

25. How many electrons can occupy the orbitals corresponding to the first, second, and third principal quantum levels? To obtain your answer, make a diagram of the allowed values of n, l, m, and s. How many electrons are needed to completely fill the s, p, and d orbitals of the third level?

26. Draw a picture of the angular part of the following orbitals on the x-, y-, and z-coordinates:

a. p_x c. p_z e. d_{xy}
b. d_{yz} d. s f. d_{z^2}

Show where the nodal planes are, if any, for each orbital. Also, show the sign of the wave function on each lobe of the orbital.

27. For the gas of atomic hydrogen, what wave numbers correspond to transitions from $n = 5$ to $n = 4$ and $n = 2$ to $n = 1$? What are the names of the two spectral series to which these lines belong?

28. For the hydrogen atom, what energy, wavelength, and frequency of radiation emitted correspond to a transition from $n = 4$ to $n = 2$? What is the name of the spectral series containing such an emission? What region of the spectrum is it in?

29. When the gas of atomic hydrogen is ionized and then recaptures an electron, the emission spectrum shows not only the Lyman series of lines but also the Balmer and Paschen series. Explain this.

30. An electron is confined to a square potential well, 3 Å wide, with infinitely high walls. What is the wavelength of the ground-state electron in this simple particle-in-a-box model?

31. Hexatriene has the structure $CH_2{=}CH{-}CH{=}CH{-}CH{=}CH_2$. Its length is about 7.3 Å and the longest-wavelength peak in its absorption spectrum is at 2580 Å. Compare this wavelength with that predicted by the simple particle-in-a-box model. (*Hint*: Hexatriene has six electrons that are free to travel the length of the molecule.)

32. Which of the following atoms are paramagnetic: Be, Ca, N, O, Al?

33. What, if any, principles or rules are violated in each of the following electron configurations:

a. Boron: $(1s)^2(2s)^3$

b. Nitrogen: $(1s)^2(2s)^2(2p_x)^2(2p_y)^1$

c. Beryllium: $(1s)^2(2p)^2$

Write the correct electron configuration for each of these elements.

34. Indicate which sets of the following species are isoelectronic (i.e., having the same number of electrons) and then write the electron configuration for each species: O^{2-}, Ar, Be^{3+}, Ne, Cl^-, He^+, Li.

"This is the man who ate the steak that came from the steer that nibbled the grass that grew in the field where roamed the cat that caught the bird that ate the fish that fed on the bug that floated around in the oil slick."

INTERLUDE

resources and pollution

The conservation of mass in chemical processes is a principle that governs much of our everyday lives. When an automobile is manufactured, the raw materials (resources) used in its production are reshaped to provide us with a vehicle that enriches our lives by providing us with great mobility and a vastly expanded set of experiences. However, when the automobile is no longer useful because of wear and tear, it becomes junk and adds to the undesirable waste (pollution) that we must somehow deal with. Even when it is operating properly, its exhaust pollutes our atmosphere, making us less comfortable and less healthy. Man is having a profound effect on his environment, both good and bad, by virtue of his interaction with it.

Our advanced technological society has reached its present, uneasy state by adhering to a wasteful *linear materials economy*, a schematic of which is shown in the upper portion of Figure 1. Very few of our raw materials other than some of the metals are effectively recycled, so most of what we consume becomes waste and pollution. Most of our energy sources produce huge amounts of pollution, which is very expensive to prevent—but it *can* be done.

The ideal economy would be a *closed materials economy* such as that shown schematically in the lower portion of Figure 1. In such an economy almost everything we use would be recycled, so the need for raw materials would be limited to providing whatever growth was desirable or necessary, compensating for the inevitable loss of a small amount of waste material, and providing the energy needed to keep things running. Only in this way is there hope of controlling the destructive effects of advanced technological society on the natural environment.

Nature has many such closed cycles by which the raw materials required for life are permanently maintained. Before we discuss two such cycles—the carbon dioxide cycle and the water cycle—we should point out that even in pristine nature there are linear processes at work. For example, the erosion of land results in the transfer of billions of tons of material (not

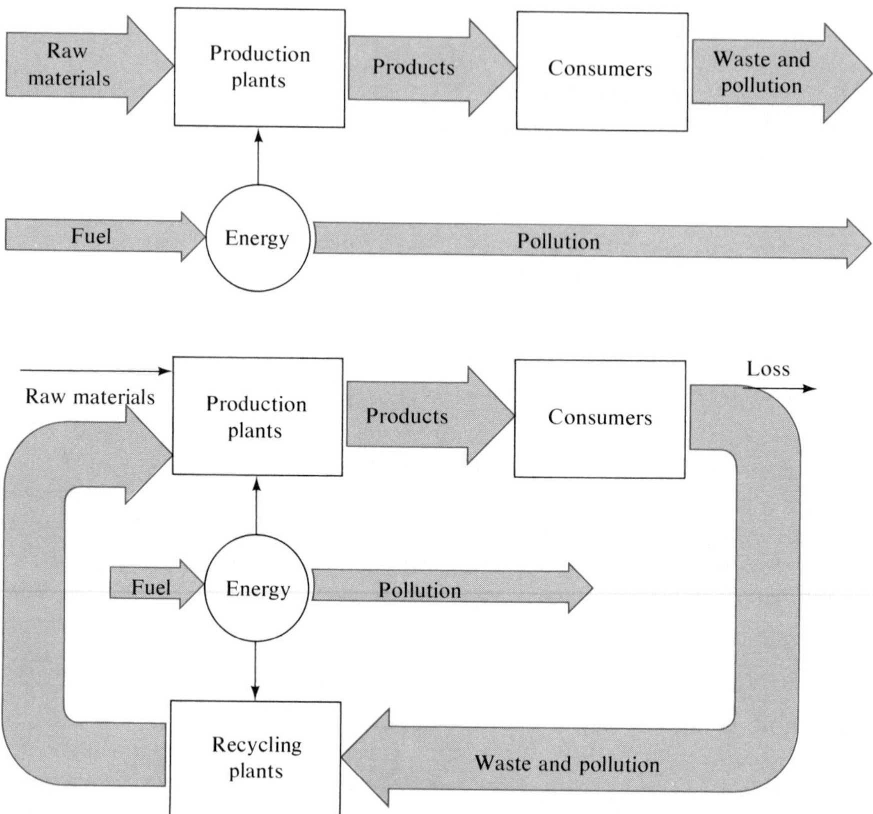

Figure 1. Linear materials economy (top) and closed materials economy (bottom). The former is inherently wasteful, whereas the latter recycles the material resources of the society with minimal raw-materials input and minimal net loss. The amounts of fuel and pollution associated with the energy source are highly variable and are much greater for energy derived from the combustion of fossil fuels than for nuclear, solar, or hydroelectric energy. (Adapted from W. C. Gough and B. J. Eastlund, "The Prospects of Fusion Power." Copyright © 1971 by Scientific American, Inc. All rights reserved.)

including the water) from land to the oceans each year. Figure 2 shows the annual loss of minerals and organic matter to the oceans. This is the source of many of the ions in seawater. There is no reason to be concerned about these inexorable linear processes because they are part of the natural history of the Earth's crust and represent no immediate danger to us—except insofar as we upset the natural order of things by adding excessive amounts of such substances as nitrates and phosphates (from fertilizers) to the runoff. The linear processes introduced by man, however, are cause for great concern because many of them introduce natural or synthetic chemicals into the environment in concentrations that are a threat to life. The fact that not *all* linear processes pose an immediate or even foreseeable threat means that informed, intelligent judgments are needed in determining which pollutants are benign and which are dangerous. Society has not yet dealt adequately with these decisions.

The carbon dioxide cycle is shown in Figure 3. We see that there are vast reservoirs of CO_2 in the atmosphere and dissolved in the oceans. These amounts are in equilibrium, about 100 billion tons leaving the oceans each year and the same amount being redissolved each year. In addition, the biological reservoir (plants and animals) is in equilibrium with the atmosphere, removing about 60 billion tons of CO_2 per year by photosynthesis and returning the same amount by respiration and decay. All other natural

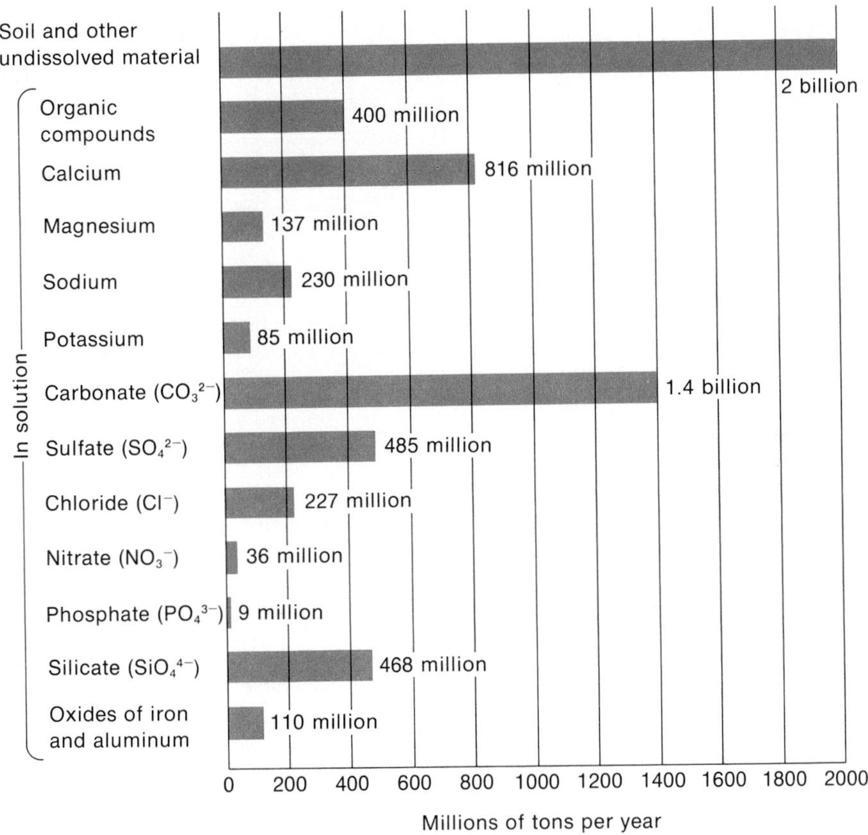

Figure 2. The annual loss of minerals and organic matter washed into the oceans amounts to billions of tons. Much nitrogen and carbon (as CO_2) eventually return to land via the atmosphere; the loss of phosphate is more serious because almost all of it remains in the oceans. (L. C. Cole, "The Ecosphere." Copyright © 1958 by Scientific American, Inc. All rights reserved.)

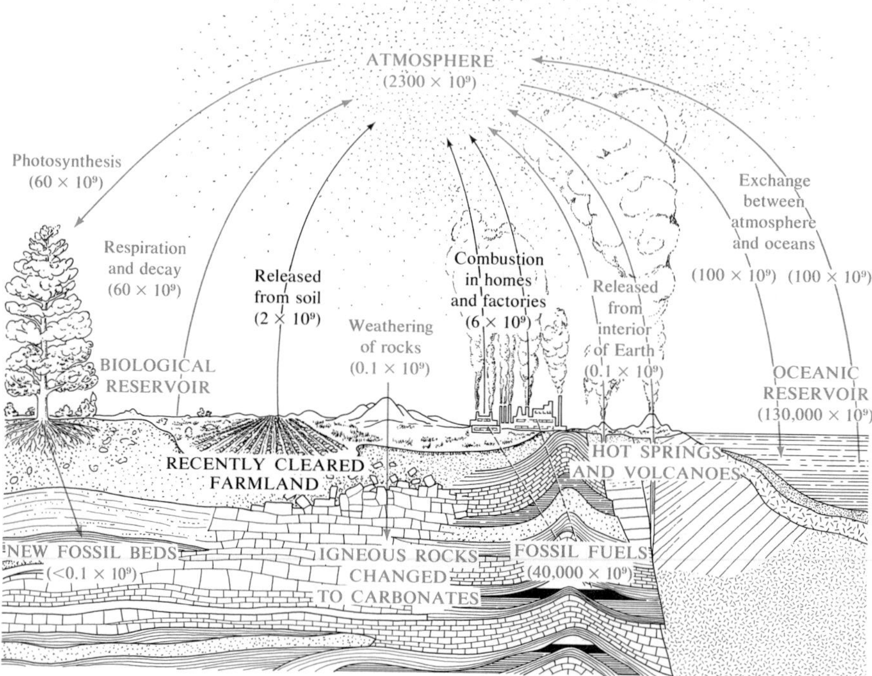

Figure 3. The carbon dioxide cycle, illustrating the processes that add and remove CO_2 to and from the atmosphere. The number given with each process indicates the tons of CO_2 transferred per year. (G. N. Plass, "Carbon Dioxide and Climate." Copyright © 1959 by Scientific American, Inc. All rights reserved.

sources are of little consequence by comparison. Human technology, however, increases the annual amount of CO_2 entering the atmosphere by about 8 billion tons, or 13% of the amount returning to the atmosphere from the biological reservoir.

Carbon dioxide is not generally regarded as a pollutant. Those who favor this position feel that most of the CO_2 added to the atmosphere by man will be dissolved in the oceans and that any net increase in atmospheric CO_2 will increase the efficiency of photosynthesis, thereby restoring the equilibrium. There is still reason for concern, however. An increase in atmospheric CO_2 will decrease the efficiency with which energy is lost by the Earth as infrared radiation, and will therefore upset the thermal-energy balance between the sun and the Earth. This could increase the average temperature of the Earth's surface and partially melt the polar ice caps. That, in turn, could result in the flooding of many of our largest cities. Such unresolved arguments provide a powerful incentive for adopting a closed materials economy. The idea is: when in doubt, play it safe.

The Earth's atmosphere is 21% oxygen, nearly all of which was produced from CO_2 by photosynthesis in plants. Since we are utterly dependent on a permanent supply of oxygen, our destiny is tied to that of the Earth's plants, not only for food but for the very air we breathe. Similarly, the destiny of technological society is tied to the availability of vast amounts of energy, but we are consuming fossil fuels at least 60 times faster than nature is producing them. Clearly, alternative sources are not only desirable but, in the long run, absolutely necessary.

Figure 4 shows the estimated lifetimes of the recoverable reserves of a number of our mineral resources, most of which will be seriously depleted within *your* lifetime. The figure represents a very pessimistic outlook, for it ignores technological advances that are sure to occur and that can change

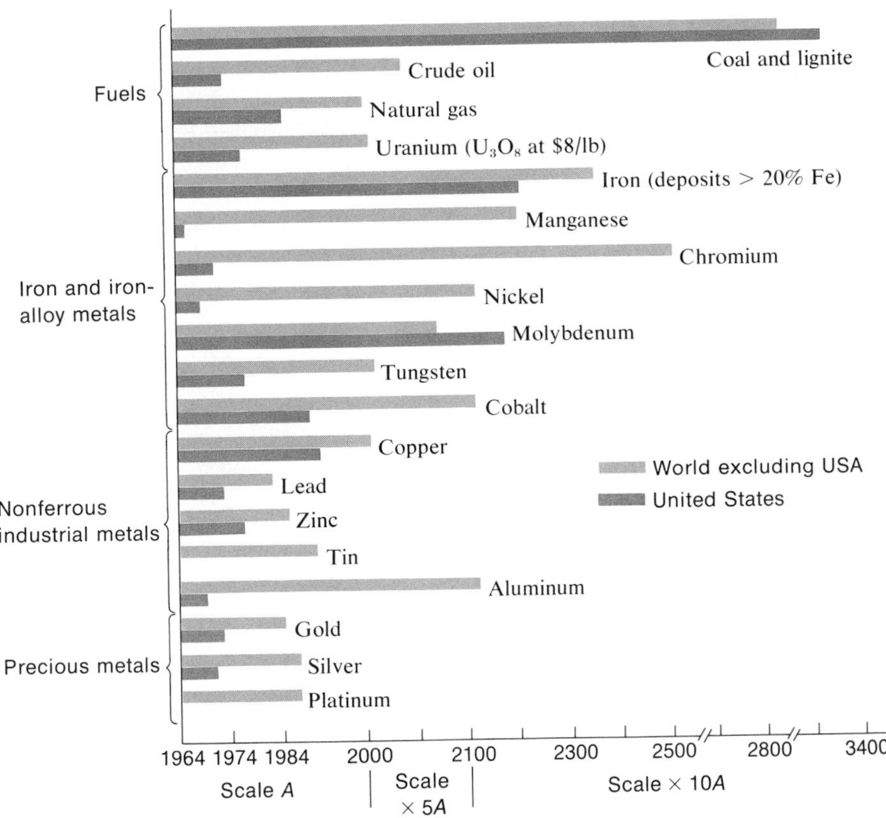

Figure 4. The lifetimes of estimated recoverable reserves of mineral resources. Reserves are those estimated to be of high enough grade to be mined by present techniques. Increasing population and consumption rates, unknown deposits, and future use of present submarginal ores are not considered. [After P. E. Cloud, "Realities of Mineral Distribution," *Texas Quarterly* **11,** 103 (1968).]

the picture entirely. There are, in fact, enough minerals in low-grade ores to provide a nearly limitless supply of minerals, but winning metals from such ores is very expensive and requires enormous amounts of energy. More energy means more pollution. You can see that the linear materials economy is an expressway to disaster.

The natural water cycle is shown in Figure 5. Since water evaporates as pure H_2O, leaving the dissolved minerals and organic matter behind, rain water is nearly pure. With nature providing such a marvelous closed cycle, we are guaranteed that fresh water will always be available to us – provided that we do not pollute our lakes and streams. But there will still be difficulties. Irrigation in natural flood plains has been a boon to farming for thousands of years. The Central Valley of California is a modern wonder of agricultural technology. The farmer enjoys warm temperatures and a large supply of water that he can control for maximum crop yield. The use of river water instead of direct rain is the problem, however. River water contains salts from the erosion of the mountains whence it came. If the farmer irrigates just enough to keep his crops thriving, the majority of the water evaporates from the surface of the soil and from transpiring plants, leaving the salts behind. The buildup of salts in the soil eventually makes it agriculturally sterile. The only solution is to irrigate with far more water than is actually needed by the plants, so that the salts are carried deep into the ground by seepage or run off into the rivers. The problem of salts in irrigated agri-

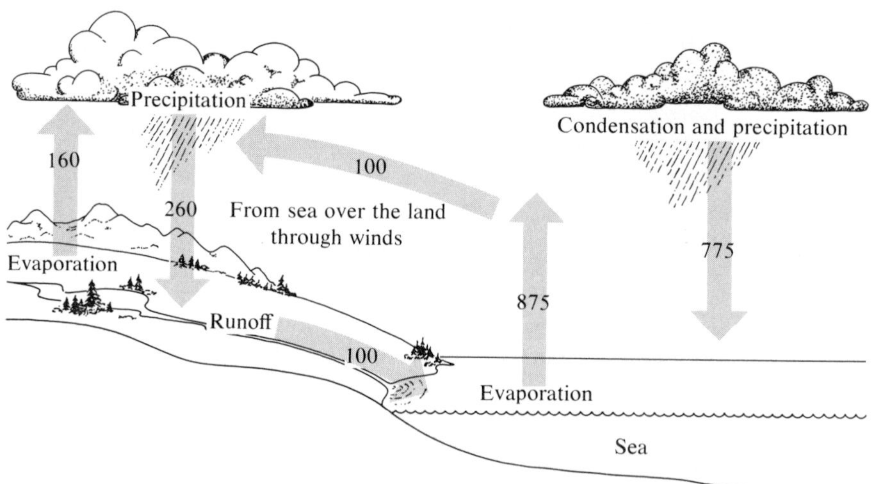

Figure 5. The water cycle. The numbers represent cubic kilometers per day. (P. R. Ehrlich and A. H. Ehrlich, *Population, Resources, Environment,* 2nd ed., W. H. Freeman and Company. Copyright © 1972.)

cultural lands has been part of man's history for some time. The downfall of the Babylonian civilization in the valley of the Tigris and Euphrates rivers may have been caused by the increased salinity (salt content) of the soil. The same problem exists in West Pakistan today and threatens to bring famine to millions.

Even with modern technology, unforeseen problems develop. The Central Valley is fertilized extensively to improve crop yields. Nitrogen in the form of inorganic nitrates is an important type of fertilizer. Because of the heavy irrigation to prevent excessive salinity of the soil, the nitrates have seeped into the well water in parts of the valley. The nitrates themselves are not dangerous, but both farm animals and human infants have bacteria in their intestinal tracts that convert nitrate ions to nitrite ions. Nitrite oxidizes the iron in hemoglobin to form methemoglobin, a form of hemoglobin that is unable to bind oxygen. The resultant disease, called *methemoglobinemia*, has caused infant deaths by suffocation in the Central Valley, and doctors now advise the use of bottled mountain water for infants. So, even with a natural cycle that assures the continual reuse of a vital resource, the local problems of pollution can be substantial.

Remember, atoms are conserved. When man extracts large numbers of atoms of a given kind from his environment, he must ultimately deal with the problems of recycling them. Failing this, he must dispose of them—but where? How much longer can such failure be tolerated?

3 the periodic properties of the elements

The quantum mechanical analysis of the electronic structure of the atom is a fascinating exercise in its own right. However, we may legitimately ask how it is related to the reality that we define as that which we can measure. In this chapter we will see how valence orbital structures are correlated with atomic and molecular properties. The emphasis will be on *periodic properties*—those properties that vary in a systematic way as the atomic number varies. We will compare the properties of several elements rather than treat the properties of one element in isolation.

That this effort is important can be appreciated from the fact that there are 105 elements now known. These elements can combine to form compounds in a staggering number of ways. The current edition of the important reference book, the *Handbook of Chemistry and Physics,* lists 4126 inorganic and 13,520 organic compounds. This is but a small fraction of the more than 3 million known chemical compounds. New ones are discovered or synthesized at the rate of many thousands per year. For the present we will deal primarily with inorganic compounds. Knowledge of the periodic properties of the elements, however, also contributes greatly to an understanding of the behavior of organic compounds. It is obviously preferable to commit to memory only a small number of descriptive facts about all of these elements and compounds and then rely on some general principles that can be used to estimate the properties of other elements and compounds.

Examples of the questions we might pose would arise from a comparison of the physical and chemical properties of, e.g., chromium and iodine atoms. How do they compare in size? In chemical reactivity toward oxygen? In the number of atoms they can bond to? What is the color of the simplest ion of each? How much energy is required to remove one electron from each atom? The following sections will demonstrate how the quantum mechanical description of the electron configuration of each atom provides the basis for these comparisons.

3.1 THE PERIODIC TABLE

Chemists have been arranging the elements in a variety of tables for a long time. In 1829 Johann Döbereiner (1780–1849), a German chemist, noted similarities in groups of three elements, such as Li, Na, and K, or S, Se, and Te. In these *Döbereiner triads,* as they have come to be called, the three elements have similar chemical properties and the middle element has an atomic weight that is approximately the average of the other two. This early effort by Döbereiner was important in spurring further efforts at classification, although the number three turned out to have no significance.

Then in 1864 John Newlands (1837–1898), an English chemist, proposed a listing of the elements in groups of seven in order of increasing atomic weight. He noted that every eighth element had similar properties. Newlands was moved to propose that these chemical "octaves" were perhaps related to musical octaves. Alas, this rapprochement between science and the arts was doomed to failure when a new class of elements called the *noble gases* was discovered in 1894 and the octaves became nonaves.

The German chemist Julius Lothar Meyer (1830–1895) and the Russian chemist Dmitri Ivanovich Mendeléev (1834–1907) independently proposed nearly identical *periodic tables* in 1869. The table shown in Figure 3-1 is Mendeléev's, as given in Volume I of his famous book, *The Principles of Chemistry,* 3rd English edition,* which was translated from the 7th Russian edition (1902). Mendeléev must have taken extraordinary pride in the fact that this table contained three elements—scandium, gallium, and germanium—that did not appear in his first table. He had had the brilliant insight to leave spaces where he felt new elements would be discovered. He predicted not only the existence of these three elements (and of three more that were discovered only many years later) but also their physical and chemical properties, by averaging the properties of neighboring elements in his periodic table.

---

*Longmans, Green, and Co., London, 1905.

PERIODIC SYSTEM OF THE ELEMENTS IN GROUPS AND SERIES.

Series	GROUPS OF ELEMENTS								
	0	**I**	**II**	**III**	**IV**	**V**	**VI**	**VII**	**VIII**
1	—	Hydrogen **H** 1·008	—	—	—	—	—	—	
2	Helium **He** 4·0	Lithium **Li** 7·03	Beryllium **Be** 9·1	Boron **B** 11·0	Carbon **C** 12·0	Nitrogen **N** 14·04	Oxygen **O** 16·00	Fluorine **F** 19·0	
3	Neon **Ne** 19·9	Sodium **Na** 23·05	Magnesium **Mg** 24·3	Aluminium **Al** 27·0	Silicon **Si** 28·4	Phosphorus **P** 31·0	Sulphur **S** 32·06	Chlorine **Cl** 35·45	
4	Argon **Ar** 38	Potassium **K** 39·1	Calcium **Ca** 40·1	Scandium **Sc** 44·1	Titanium **Ti** 48·1	Vanadium **V** 51·4	Chromium **Cr** 52·1	Manganese **Mn** 55·0	Iron Cobalt Nickel **Fe Co Ni (Cu)** 55·9 59 59
5		Copper **Cu** 63·6	Zinc **Zn** 65·4	Gallium **Ga** 70·0	Germanium **Ge** 72·3	Arsenic **As** 75	Selenium **Se** 79	Bromine **Br** 79·95	
6	Krypton **Kr** 81·8	Rubidium **Rb** 85·4	Strontium **Sr** 87·6	Yttrium **Y** 89·0	Zirconium **Zr** 90·6	Niobium **Nb** 94·0	Molybdenum **Mo** 96·0	—	Ruthenium Rhodium Palladium **Ru Rh Pd (Ag)** 101·7 103·0 106·5
7		Silver **Ag** 107·9	Cadmium **Cd** 112·4	Indium **In** 114·0	Tin **Sn** 119·0	Antimony **Sb** 120·0	Tellurium **Te** 127	Iodine **I** 127	
8	Xenon **Xe** 128	Cæsium **Cs** 132·9	Barium **Ba** 137·4	Lanthanum **La** 139	Cerium **Ce** 140	—	—	—	— — —
9		—	—	—	—	—	—	—	
10	—	—	—	Ytterbium **Yb** 173	—	Tantalum **Ta** 183	Tungsten **W** 184	—	Osmium Iridium Platinum **Os Ir Pt (Au)** 191 193 194·9
11		Gold **Au** 197·2	Mercury **Hg** 200·0	Thallium **Tl** 204·1	Lead **Pb** 206·9	Bismuth **Bi** 208	—	—	
12	—	—	Radium **Rd** 224	—	Thorium **Th** 232	—	Uranium **U** 239		
	HIGHER SALINE OXIDES								
	R	**R_2O**	**RO**	**R_2O_3**	**RO_2**	**R_2O_5**	**RO_3**	**R_2O_7**	**RO_4**
	HIGHER GASEOUS HYDROGEN COMPOUNDS								
					RH_4	**RH_3**	**RH_2**	**RH**	

Figure 3-1. Mendeléev's periodic system of the elements. (From Dmitri Mendeléev, *The Principles of Chemistry,* Vol. I, 3rd English ed., Longmans, Green, and Co., London, 1905; translated from the 7th Russian ed., 1902.)

One of the crucial tests of a theory lies in its ability to predict—can the theory predict natural phenomena that have not yet been observed? Table 3-1 gives Mendeléev's predictions for one of the three missing elements (which he called eka-aluminum) and the data measured a few years later by the Frenchman Paul Émile Lecoq de Boisbaudran (1838–1912). (Interestingly, all three elements predicted by Mendeléev were named after the homelands of the European scientists who discovered them.) The success of his predictions places Mendeléev in the company of men like Linus Pauling (b. 1901), who predicted the existence and formulas of a number of noble-gas compounds, and Murray Gell-Mann (b. 1929), who predicted the existence and properties of several new subatomic particles. All of these predictions eventually became reality, when the crucial experiments were performed.

Table 3-1. The Properties of a "Missing" Element.

Properties Predicted for Eka-aluminum (Ea) by Mendeléev	Properties Found for Gallium (Ga) by Lecoq de Boisbaudran
Atomic weight: about 68	*Atomic weight:* 69.9
Metal of specific gravity 5.9; melting point low; nonvolatile; unaffected by air; should decompose steam at red heat; should dissolve slowly in acids and alkalies.	*Metal* of specific gravity 5.94; melting point 30.15; nonvolatile at moderate temperature; not changed in air; action of steam unknown; dissolves slowly in acids and alkalies.
Oxide: formula Ea_2O_3; specific gravity 5.5; should dissolve in acids to form salts of the type EaX_3. The hydroxide should dissolve in acids and alkalies.	*Oxide:* Ga_2O_3; specific gravity unknown; dissolves in acids, forming salts of the type GaX_3. The hydroxide dissolves in acids and alkalies.
Salts should have tendency to form basic salts; the sulfate should form alums; the sulfide should be precipitated by H_2S or $(NH_4)_2S$. The anhydrous chloride should be more volatile than zinc chloride.	*Salts* hydrolyze readily and form basic salts; alums are known; the sulfide is precipitated by H_2S and by $(NH_4)_2S$ under special conditions. The anhydrous chloride is more volatile than zinc chloride.
The element will probably be discovered by spectroscopic analysis.	Gallium was discovered with the aid of the spectroscope.

SOURCE: M. E. Weeks, *Discovery of the Elements,* 7th ed., Chemical Education Publishing Company, Easton, Pa., 1968.

One flaw in all the nineteenth-century periodic tables was that the elements were arranged in order of increasing atomic *weight.* Several pairs of elements (e.g., Ar and K) were then found to be "out of order" in terms of their obvious physical and chemical properties. The difficulty was circumvented by assuming that one of the atomic weights in question was erroneous and reassigning the value to make it fit the required sequence. It was hoped that better measurements would eventually resolve the discrepancies.

The real solution lay elsewhere, however. In 1913 the young English physicist Henry G.-J. Moseley (1887–1915) made the fundamental discovery that each of the elements is characterized by an atomic number, i.e., the number of positive charges in its nucleus, or the number of orbital electrons. When the elements were arranged in order of increasing atomic *number,* the above-mentioned discrepancies vanished. The nearly simultaneous theory of isotopes, proposed by the English chemist Frederick Soddy (1877–1956), explained the atomic weight variability that had caused the discrepancies in the first place. The periodic table was thus placed on a secure and rational basis.

In the preceding chapter, we asserted that almost all the chemistry of atoms originates in the electron cloud surrounding the nucleus. Thus, we should expect that the periodic properties of the elements are determined by the number of these electrons and by their spatial configurations. We will construct our periodic table following two general principles: (1) the elements are to be arranged in order of increasing atomic number and (2) a break should occur in the table whenever the configuration ns^2np^6 is reached for the largest principal quantum number n in an atom (this configuration corresponds to the noble gases and is called an *octet*). Figure 3-2 was constructed following these principles. Except for the first row, each row begins with the filling of the ns orbital and ends with the filling of the np orbitals. The breaks are apparent in Table 2-4, which shows the electron configurations of the elements.

Figure 3-2 is interesting in that its symmetry is pleasing and that it does represent the variation in electron configurations rather well. Let us see whether similar elements are grouped in contiguous positions. We note that the noble gases of atomic numbers 2, 10, 18, 36, 54, and 86 are found by beginning at element number 2, moving down one row and over three columns, then down one, then down one and over five, then down one, and, finally, down one and over seven. This crossword-puzzle approach actually works when applied to other groups of elements that are similar to each other. Following the above procedure and beginning with "down one, down one and over five, etc.," a search for the elements with chemical properties similar to those of element number 5 indicates elements 13, 31, 49, and 81. If our procedure is extended to include moving in both directions, we find

l																																	*n*
s																1	2																1
s, *p*													3	4	5	6	7	8	9	10													2
													11	12	13	14	15	16	17	18													3
s, *p*, *d*								19	20	21	22	23	24	25	26	27	28	29	30	31	32	33	34	35	36								4
								37	38	39	40	41	42	43	44	45	46	47	48	49	50	51	52	53	54								5
s, *p*, *d*, *f*	55	56	57	58	59	60	61	62	63	64	65	66	67	68	69	70	71	72	73	74	75	76	77	78	79	80	81	82	83	84	85	86	6
	87	88	89	90	91	92	93	94	95	96	97	98	99	100	101	102	103	104	105														7

Figure 3-2. Possible periodic chart of the elements. The elements are arranged in horizontal rows according to the filling of orbitals of increasing *l* values.

that elements 5, 13, 21, 39, 57, and 89 might also have similar properties—and they do. With the exception of boron, atomic number 5, they are all metals and form ions of charge +3.

The obvious problem with this approach is that it is visually a little tricky, to say the least. A great number of similar and more sophisticated approaches have been attempted, including such configurations as spiral cylinders. One of the most useful forms, and the one we will use in this text, is the *long form.* The elements are arranged according to the order in which the *s, p, d,* and *f* orbitals of the various principal quantum number levels are filled. This is shown schematically in Figure 3-3.

The periodic table itself is shown in its present form in Figure 3-4. The power of this form is that it presents the elements in order of increasing atomic number, creating breaks where valence orbital structures are completed, and it also presents similar elements (called *congeners*) in columns. Thus, definite trends in chemical and physical properties can be found upon traversing a row (called a *period*) or by going down a column (called a *group*).

There are two classes of elements, one labeled A and one B; this labeling provides the clearest representation of the continuous variation of physical and chemical properties. The elements labeled B are called the *transition elements.* As will be shown in the succeeding pages, there are substantial similarities in their properties, leading to a distinctive chemistry.

The 28 elements shown at the bottom of the table are called the *lanthanons* and *actinons* (for lanthanum and actinium), or, collectively, the *rare earths.* They correspond to the filling of the 4*f* and 5*f* orbitals, respectively, and are

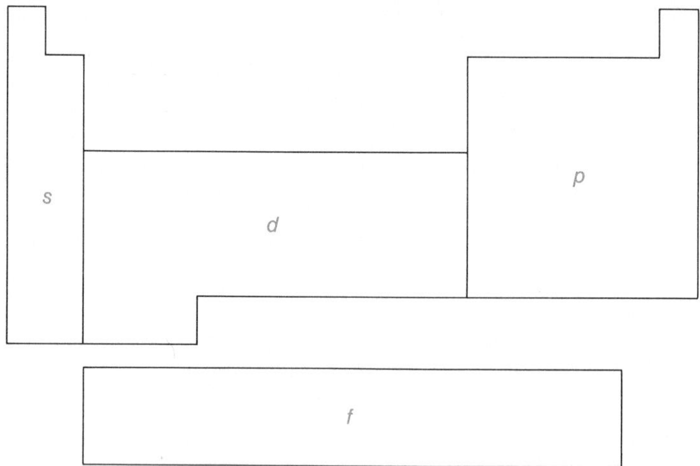

Figure 3-3. Electron configuration as the basis of the modern form of the periodic table. The letters in the diagram correspond to the values of *l* for the highest-energy electron in that portion of the periodic table.

	IA	IIA	IIIB	IVB	VB	VIB	VIIB	VIIIB			IB	IIB	IIIA	IVA	VA	VIA	VIIA	0
1	H 1																	He 2
2	Li 3	Be 4											B 5	C 6	N 7	O 8	F 9	Ne 10
3	Na 11	Mg 12											Al 13	Si 14	P 15	S 16	Cl 17	Ar 18
4	K 19	Ca 20	Sc 21	Ti 22	V 23	Cr 24	Mn 25	Fe 26	Co 27	Ni 28	Cu 29	Zn 30	Ga 31	Ge 32	As 33	Se 34	Br 35	Kr 36
5	Rb 37	Sr 38	Y 39	Zr 40	Nb 41	Mo 42	Tc 43	Ru 44	Rh 45	Pd 46	Ag 47	Cd 48	In 49	Sn 50	Sb 51	Te 52	I 53	Xe 54
6	Cs 55	Ba 56	* Lu 71	Hf 72	Ta 73	W 74	Re 75	Os 76	Ir 77	Pt 78	Au 79	Hg 80	Tl 81	Pb 82	Bi 83	Po 84	At 85	Rn 86
7	Fr 87	Ra 88	† Lw 103	— 104	— 105													

*Lanthanons	La 57	Ce 58	Pr 59	Nd 60	Pm 61	Sm 62	Eu 63	Gd 64	Tb 65	Dy 66	Ho 67	Er 68	Tm 69	Yb 70
†Actinons	Ac 89	Th 90	Pa 91	U 92	Np 93	Pu 94	Am 95	Cm 96	Bk 97	Cf 98	Es 99	Fm 100	Md 101	No 102

Figure 3-4. Modern long form of the periodic table. (Elements 104 and 105 have not yet been officially named.)

often found in various combinations in natural ores. Since most of their properties (and those of their compounds) are remarkably similar, they are rather difficult to separate by physical or chemical means.

Some periodic tables place H at the top of both groups IA and VIIA because it exhibits both +1 and −1 ionic states. Because of the predominance of the +1 state and for simplicity, we place it only at the top of group IA. In many respects, however, it can be considered to be "in a group by itself."

3.2 BINDING OF ELECTRONS BY ATOMS

Ionization Potentials

The *ionization potential* of a particle is the energy required to remove one of its electrons to an infinite distance, yielding a new particle having a charge larger by one unit. If the original particle is a neutral atom, the energy is labeled I_1 and called the *first ionization potential.* This is the energy required to produce the reaction

$$X(g) + \text{energy} \rightarrow X^+(g) + e^-(g) \qquad (3\text{-}1)$$

where X is the element and g denotes the gaseous state. Higher ionization potentials, labeled I_2, I_3, I_4, etc., correspond to the successive removal of additional electrons. The energy required to ionize the particle is generally measured in electron volts (eV) per particle. An *electron volt* is the energy acquired by an electron when it is accelerated through a potential difference of one volt (see Example 2-4). Another unit employed is the wave number (in units of cm^{-1}); 1 eV is equivalent to 8065 cm^{-1}.

The value 13.60 eV per particle is equivalent to about 109,700 cm^{-1}, which should be familiar: it is the same value shown (more accurately) in Figure 2-32 for the energy level of the 1s ground state of the hydrogen atom. Thus, the first—and only—ionization potential of H corresponds to the maximum value of $\tilde{\nu}$ in Figures 2-30 and 2-31, where the atomic line spectrum becomes continuous.

The measurement of I_1 must be made on neutral, monatomic atoms in the gaseous state. The noble gases are thus amenable to measurement in their natural state. Although some of the nonmetals in groups VA through VIIA exist as gases, they must first be dissociated into neutral atoms from their normal diatomic molecular state. The metals must be heated to very high temperatures and thus vaporized. The energy required to ionize the atom is frequently supplied by a bombarding stream of electrons whose energy is continuously raised by increasing the voltage between two accelerating plates. When the value corresponding to I_1 is reached, a large surge in the current occurs and the energy of the impinging electrons is recorded as I_1. As the accelerating potential is raised further, additional surges of current are noted and recorded as the higher ionization potentials. Figure 3-5 shows the basic components of the apparatus required to measure ionization potentials.

All the known ionization potentials of the elements are given in Table 3-2. Because the values of I_1 provide perhaps the greatest insight into the correlation between the electronic structure of atoms and the periodic table, these values are also plotted as a function of atomic number in Figure 3-6. The interesting, pinnacle-shaped figure displaying I_1 deserves our careful consideration. We will first note the gross features of the plot and then examine the fine structure. In each instance, we attempt to explain the observed behavior in terms of atomic electron configurations.

The first striking feature is that the values of I_1 for the noble gases are much higher than those for their neighboring elements. This means that

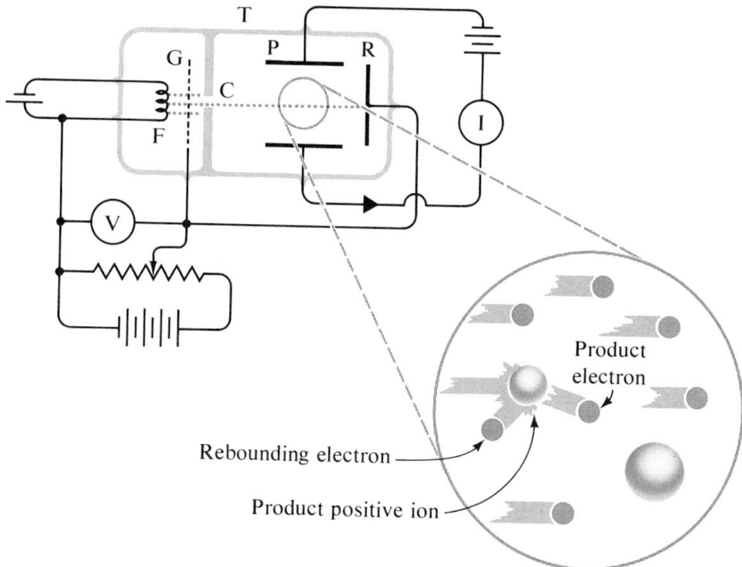

Figure 3-5. Determination of the ionization potential of an element in gaseous form. Electrons from a hot filament (F) are accelerated by a positively charged grid (G) through a collimating slit (C) into a gas-filled tube (T). There they bombard the gas under study. Any ionization of the gas is detected as a current (I) between the two parallel plates (P). The electrons are removed by the perpendicular plate (R), and their energy is measured as a voltage (V). The lowest voltage at which ionization is observed is the first ionization potential of the element. (Adapted from J. A. Campbell, *Chemical Systems: Energetics, Dynamics, Structure,* W. H. Freeman and Company. Copyright © 1970.)

Table 3-2. Ionization Potentials.
$X^n + \text{energy} \rightarrow X^{n+1} + e^-$
(n = charge on X = 0, +1, +2, . . . , +7).

Atomic Number	Symbol	I_1	I_2	I_3	I_4	I_5	I_6	I_7	I_8
1	H	13.60							
2	He	24.58	54.40						
3	Li	5.39	75.62	122.42					
4	Be	9.32	18.21	153.85	217.66				
5	B	8.30	25.15	37.92	259.30	340.13			
6	C	11.26	24.38	47.86	64.48	391.99	489.84		
7	N	14.54	29.61	47.43	77.45	97.86	551.93	666.83	
8	O	13.61	35.15	54.93	77.39	113.87	138.08	739.11	871.12
9	F	17.42	34.98	62.65	87.23	114.21	157.12	185.14	953.60
10	Ne	21.56	41.07	64	97.16	126.4	157.91		
11	Na	5.14	47.29	71.65	98.88	138.60	172.36	208.44	264.16
12	Mg	7.64	15.03	80.12	109.29	141.23	186.86	225.31	265.97
13	Al	5.98	18.82	28.44	119.96	153.77	190.42	241.93	285.13
14	Si	8.15	16.34	33.46	45.13	166.73	205.11	246.41	303.87
15	P	11.0	19.65	30.16	51.35	65.01	220.41	263.31	309.26
16	S	10.36	23.4	35.0	47.29	72.5	88.03	280.99	328.80
17	Cl	13.01	23.80	39.90	53.5	67.80	96.7	114.27	348.3
18	Ar	15.76	27.62	40.90	59.79	75.0	91.3	124.0	143.46
19	K	4.34	31.81	46	60.90		99.7	118	155
20	Ca	6.11	11.87	51.21	67	84.39		128	147
21	Sc	6.56	12.89	24.75	73.9	92	111.1		159
22	Ti	6.83	13.63	28.14	43.24	99.8	120	140.8	
23	V	6.74	14.2	29.7	48	65.2	128.9	151	173.7
24	Cr	6.76	16.6	31?	50.4?	72.8?			
25	Mn	7.43	15.70	32?	52?	75.7?			
26	Fe	7.90	16.16						
27	Co	7.86	17.3						
28	Ni	7.63	18.2						
29	Cu	7.72	20.34	29.5					
30	Zn	9.39	17.89	40.0					
31	Ga	6.00	20.43	30.6	63.8				
32	Ge	8.13	15.86	34.07	45.5	93.0			
33	As	10?	20.1	28.0	49.9	62.5			
34	Se	9.75	21.3	33.9	42.72	72.8	81.4		
35	Br	11.84	19.1	25.7	50?				
36	Kr	14.00	26.4	36.8	68?				
37	Rb	4.18	27.36	47?	80?				

Table 3-2. **(continued)**

Atomic Number	Symbol	I_1	I_2	I_3	I_4	I_5	I_6	I_7	I_8
38	Sr	5.69	10.98						
39	Y	6.6	12.3	20.4					
40	Zr	6.95	13.97	24.00	33.8				
41	Nb	6.77		24.2					
42	Mo	7.18							
43	Tc								
44	Ru	7.5							
45	Rh	7.7							
46	Pd	8.33	19.8						
47	Ag	7.57	21.4	35.9					
48	Cd	8.99	16.84	38.0					
49	In	5.79	18.79	27.9	57.8				
50	Sn	7.33	14.5	30.5	39.4	80.7			
51	Sb	8.64	18?	24.7	44.0	55.5			
52	Te	9.01		30.5	37.7	60.0	72?		
53	I	10.44	19.4						
54	Xe	12.13	21.1?	32.0	46?	76?			
55	Cs	3.89	23.4	35?	51?	58?			
56	Ba	5.21	9.95						
57	La	5.61	11.4	20.4?					
58	Ce	6.9?	14.8						
59	Pr	5.8?							
60	Nd	6.3?							
61	Pm								
62	Sm	5.6	11.4						
63	Eu	5.67	11.4						
64	Gd	6.16							
65	Tb	6.7?							
66	Dy	6.8?							
67	Ho								
68	Er								
69	Tm								
70	Yb	6.2							
71	Lu	5.0							
72	Hf	5.5?	14.8?						
73	Ta	6?							
74	W	7.98							
75	Re	7.87							
76	Os	8.7							
77	Ir	9.2							

Table 3-2. **(continued)**

Atomic Number	Symbol	I_1	I_2	I_3	I_4	I_5	I_6	I_7	I_8
78	Pt	8.96							
79	Au	9.22	19.95						
80	Hg	10.43	18.65	34.3	72?	82?			
81	Tl	6.11	20.32	29.7	50.5				
82	Pb	7.42	14.96	31.9?	42.11	69.4			
83	Bi	8?	16.6	25.42	45.1	55.7			
84	Po								
85	At								
86	Rn	10.75							
87	Fr								
88	Ra	5.28	10.10						
89	Ac								
90	Th			29.4					
91	Pa								
92	U	4?							

SOURCE: M. J. Sienko and R. A. Plane, *Physical Inorganic Chemistry*, W. A. Benjamin, New York, 1963.

a great deal more energy is required to remove an electron from these atoms than from other types of atoms. Another way of stating the same fact is to say that these electron configurations are the most stable ones that atoms can possess. With the exception of helium, all these configurations are ns^2np^6, where n is the highest principal quantum number. Although considerable variation occurs as the d and f orbitals begin filling, the maximum stability of this particular configuration (the octet) has no exceptions. This fact is of great importance in explaining the chemistry of the neighboring elements.

The next trend that stands out is that, relative to their neighboring elements, the alkali metals possess the lowest ionization potentials. All these elements have one 1*s* electron in their outermost orbital. The loss of this electron makes them *isoelectronic* (having an identical electron configuration) with the preceding noble gases, and is precisely what characterizes their chemistry most. This again points to the great stability of the ns^2np^6 configuration.

The last major trend to note is that, in general, the values of I_1 decrease as the atomic number increases. Although there is considerable fluctuation, this trend indicates that, as atomic radii increase, the removal of an electron becomes progressively easier.

Now let us examine the detailed features of Figure 3-6. The value of

I_1 for hydrogen, atomic number 1, is relatively high. Apparently it is not easy to remove the electron, leaving a bare proton behind. However, the ionization potential of helium, atomic number 2, is much higher: a second electron has been added to the 1*s* orbital. From Equation 2-22 the binding energy of an electron in a helium atom might be expected to be about four times as great as in a hydrogen atom (see the text immediately following the equation). The fact that I_1 for He is only 24.6 eV rather than $4 \times 13.6 = 54.4$ eV indicates that electron-electron repulsion is important. It is apparent from the much higher value of I_1 for He compared to the value for H that the first electron does not completely shield the second electron from the increased nuclear charge. Note, from Table 3-2, that the experimental value of I_2 for He is exactly 54.40 eV, as it must be, since this I_2 for He is the ionization potential of a one-electron atom with $Z = 2$.

Lithium, atomic number 3, has these same two 1*s* electrons plus an electron in the 2*s* orbital. The I_1 value is very much lower, indicating that the increased nuclear charge of +3 does not exert its full influence on this third electron. In fact, the inner two 1*s* electrons appear to shield it from the positive charge of the nucleus. This *screening* effect is important in determining the values of I_1 in this and subsequent atoms. Note that I_3 for Li equals $9 \times 13.60 = 122.4$ eV. In general, the outer, or valence, electrons

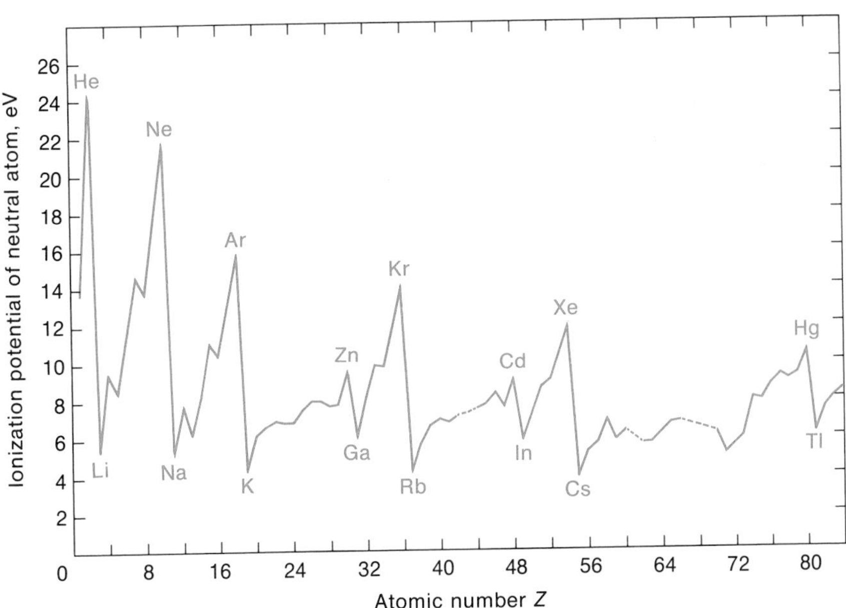

Figure 3-6. Variation of the first ionization potentials of the elements with atomic number.

are shielded from the nucleus by electrons in inner shells. (A *shell* is the collection of atomic orbitals having the same principal quantum number n.)

Beryllium has a higher first ionization potential than lithium because of its larger nuclear charge, as in the case of helium and hydrogen. Interestingly, a dip in the curve is noted at atomic number 5, boron. This corresponds to the addition of one electron to the $2p$ orbitals. The resulting electron configuration has less stability than one in which the outer shell contains only the completed $2s$ orbital.

The I_1 values increase linearly from atomic numbers 5 through 7 as a result of increasing nuclear charge and incomplete screening of the $2s$ and $2p$ electrons. However, another dip in the figure occurs at atomic number 8, oxygen. This indicates that atomic number 7, nitrogen, has special stability. Its electron configuration is $1s^2 2s^2 2p^3$, or, more explicitly, $1s^2 2s^2 2p_x 2p_y 2p_z$. An especial stability is associated with the half-filled p orbitals. The ionization potentials increase steadily from atomic numbers 8 through 10, with the maximum value being achieved by Ne, which has the $2s^2 2p^6$ configuration. Elements of atomic numbers 11 through 18 behave in a manner similar to that of the eight preceding elements, for the same reasons.

The first deviation from the trends observed so far occurs at element number 21, scandium. Here the plot continues upward as the first d electron is added, rather than downward, as occurred when the first p electron was added. The large decreases in I_1 that occur between elements 30 and 31, 48 and 49, and 80 and 81 coincide with the completion of the inner d orbital and the addition of one p electron to the outer shell.

The reader is encouraged to verify these and other trends by a careful examination of the data in Table 2-4, Table 3-2, and Figure 3-6 and by a consideration of the competing effects of increasing nuclear charge and screening by inner electrons. It will become apparent that, whereas this superficial consideration provides reasonable explanations for most of the trends, a more sophisticated approach is required to explain some of the anomalies that are present.

An examination of the higher ionization potentials is also rewarding but will only be briefly referred to here. A general trend observed without exception in Table 3-2 is that $I_n > I_{n-1}$. This is to be expected, because the additional electrons are being removed from increasingly positive ions. We note that the largest increases occur at the electron configurations of the noble gases. Such large increases are to be noted, for instance, between I_3 and I_4 for boron and between I_2 and I_3 for magnesium.

Electron Affinities

The *electron affinity* of an atom is defined as the energy released when a neutral atom in the gaseous state captures an electron of zero kinetic energy to become a uninegative ion. Thus, the energy is that released by the reaction

$$X(g) + e^-(g) \rightarrow X^-(g) + \text{energy} \tag{3-2}$$

Comparison of Equations 3-1 and 3-2 indicates that the electron affinity of a unipositive ion is equal to the first ionization potential of the corresponding atom. Thus, all the values of I_1 listed in Table 3-2 may be considered to be the electron affinities of those unipositively charged species. (The electron affinity of a neutral atom would then correspond to the ionization potential of the uninegative ion.)

Unfortunately, very few electron affinities (*EA*) have been measured, owing to experimental difficulties. Table 3-3 lists those that are currently available. It is apparent that these numbers are much less than the typical values of I_1. This indicates that atoms are relatively stable and that only very small increases in stability occur upon the addition of an extra electron. Large amounts of energy are required to remove an electron from a neutral atom, on the other hand.

The negative values in Table 3-3 indicate that energy is required to add an electron. The five elements that are listed with negative values represent atoms with filled or half-filled orbitals, again demonstrating the special stability of such species.

The largest electron affinities are those of the halogens. This is to be expected, in that the addition of one electron yields the stable ns^2np^6 configuration of the noble gases. The electron affinities of the alkali metals Li and Na are appreciable—about 0.5 eV. On the other hand, the values of *EA* for the alkaline earth elements Be and Mg are somewhat negative. These observations again indicate the stability of the ns^2 configuration.

Electronegativities

The American physicist Robert S. Mulliken (b. 1896) proposed in 1934 that these two experimentally determined quantities, the ionization potential and the electron affinity, could be combined to yield a quantity proportional to the *electronegativity* of the element, a concept introduced two years earlier by Linus Pauling. Pauling had proposed that this quantity would be a rough measure of the ability of an atom *in a molecule* to attract electrons to itself from the other atoms. Mulliken reasoned that the arithmetic average of I_1 and *EA* would be a measure of this tendency to attract electrons. This would be the average of the energy required to remove an electron *from* the neutral atom, producing a positive ion, and the energy released upon capture of an additional electron *by* the neutral atom, producing a negative ion. The electronegativity is never zero or negative. Table 3-4 gives the data for the computation of the electronegativities of the first eleven elements.

Table 3-3. **Electron Affinities, X + e^- → X^- + energy.**

Atom	Orbital Electron Configuration	*EA*, Electron Volts	Orbital Electron Configuration of the Negative Ion
H	$1s$	0.747	He
F	(He)$2s^22p^5$	3.45	Ne
Cl	(Ne)$3s^23p^5$	3.61	Ar
Br	(Ar)$4s^23d^{10}4p^5$	3.36	Kr
I	(Kr)$5s^24d^{10}5p^5$	3.06	Xe
O	(He)$2s^22p^4$	1.47	(He)$2s^22p^5$
S	(Ne)$3s^23p^4$	2.07	(Ne)$3s^23p^5$
Se	(Ar)$4s^23d^{10}4p^4$	(1.7)	(Ar)$4s^23d^{10}4p^5$
Te	(Kr)$5s^24d^{10}5p^4$	(2.2)	(Kr)$5s^24d^{10}5p^5$
N	(He)$2s^22p^3$	(−0.1)	(He)$2s^22p^4$
P	(Ne)$3s^23p^3$	(0.7)	(Ne)$3s^23p^4$
As	(Ar)$4s^23d^{10}4p^3$	(0.6)	(Ar)$4s^23d^{10}4p^4$
C	(He)$2s^22p^2$	1.25	(He)$2s^22p^3$
Si	(Ne)$3s^23p^2$	(1.63)	(Ne)$3s^23p^3$
Ge	(Ar)$4s^23d^{10}4p^2$	(1.2)	(Ar)$4s^23d^{10}4p^3$
B	(He)$2s^22p$	(0.2)	(He)$2s^22p^2$
Al	(Ne)$3s^23p$	(0.6)	(Ne)$3s^23p^2$
Ga	(Ar)$4s^23d^{10}4p$	(0.18)	(Ar)$4s^23d^{10}4p^2$
In	(Kr)$5s^24d^{10}5p$	(0.2)	(Kr)$5s^24d^{10}5p^2$
Be	(He)$2s^2$	(−0.6)	(He)$2s^22p$
Mg	(Ne)$3s^2$	(−0.3)	(Ne)$3s^23p$
Li	(He)$2s$	(0.54)	(He)$2s^2$
Na	(Ne)$3s$	(0.74)	(Ne)$3s^2$
Zn	(Ar)$4s^23d^{10}$	(−0.9)	(Ar)$4s^23d^{10}4p$
Cd	(Kr)$5s^24d^{10}$	(−0.6)	(Kr)$5s^24d^{10}5p$

SOURCE: H. B. Gray and G. P. Haight, Jr., *Basic Principles of Chemistry,* W. A. Benjamin, New York, 1967.

Pauling's original proposal of the concept of electronegativity, made in 1932, was based on thermodynamic data. This approach has been more useful than Mulliken's because relatively few electron affinities have been measured. Pauling suggested that the difference between the bond energy D_{AB} of a heteronuclear diatomic molecule AB and the geometric mean of the bond energies of the homonuclear molecules AA and BB was proportional to the ability of each of the atoms to attract the shared electrons in the bond between A and B. If we let x_A and x_B represent the electronega-

Table 3-4. **Electronegativities.**

Atom	Ionization Potential, I_1	Electron Affinity, EA	Mulliken Electronegativity, $\frac{1}{2}(I_1 + EA)$	Pauling Electro-negativity
H	13.60	0.747	7.2	2.1
He	24.58	≈0	12.3	—
Li	5.39	(0.54)	3.0	1.0
Be	9.32	(−0.6)	4.4	1.5
B	8.30	(0.2)	4.3	2.0
C	11.26	1.25	6.3	2.5
N	14.54	(−0.1)	7.2	3.0
O	13.61	1.47	7.5	3.5
F	17.42	3.45	10.4	4.0
Ne	21.56	≈0	10.8	—
Na	5.14	(0.74)	2.9	0.9

tivities of elements A and B, respectively, then the difference in electronegativity is computed from the empirical equation

$$x_A - x_B = 0.18\,[D_{AB} - (D_{AA}D_{BB})^{1/2}]^{1/2} + \alpha$$

where the constants 0.18 and α are chosen arbitrarily. A partial set of Pauling's electronegativities is given in Figure 3-7.

Values of x will be used in subsequent chapters to make qualitative predictions of the percent ionic character of covalent bonds and to predict rough values for electric dipole moments. It will be very useful to remember the electronegativities of those elements having values of 3.0 or larger, the value of about 1.0 for the alkali metals, and the general trends displayed in Figure 3-7, namely, that electronegativities increase in going up a column and in going from left to right in a row.

3.3 TRENDS IN SEVERAL MEASURES OF ATOMIC SIZE

Atomic Volume

The only measure of the relative sizes of the elements that is poorly amenable to theoretical interpretation is the *atomic volume.* This quantity was defined by Meyer in 1869 as the atomic weight of the element divided

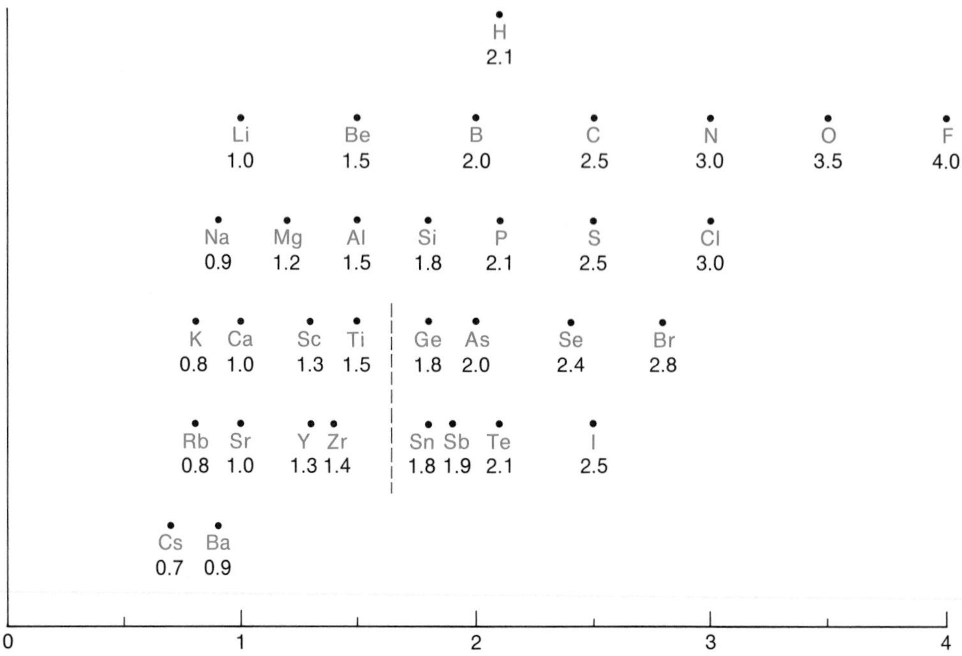

Figure 3-7. Electronegativities of the elements. The elements are arranged so as to correspond roughly to their positions in the periodic table. The dashed vertical line represents approximate values for the transition metals. (L. Pauling and P. Pauling, *Chemistry,* W. H. Freeman and Company. Copyright © 1975.)

by its density. The meaning of the atomic volume is readily apparent from a simple dimensional analysis:

$$\text{Atomic volume} = \frac{\text{g/mole}}{\text{g/cm}^3} = \text{cm}^3\text{/mole}$$

Thus, the atomic volume is the volume of 1 mole of the element. These values are subject to some variation because of the variation of the density with temperature and with crystal structure. For instance, the densities of graphite and diamond, the two allotropic forms* of carbon, are 2.25 and 3.51 g/cm³, respectively. Of the several possible densities of an element, the one generally chosen for a plot of atomic volume vs atomic number is that which fits most smoothly on the graph. A striking periodicity occurs

---

**Allotropes* are different crystalline or molecular forms of the same element. Their physical properties are often very different (as in graphite and diamond), and their chemical properties as well. Other examples of the phenomenon of *allotropy* are phosphorus, of which the best known forms are white phosphorus, red phosphorus, and black phosphorus, and oxygen, which exists as O_2 (ordinary oxygen) and O_3 (ozone).

in this graph, which is shown in Figure 3-8. It was the discovery of this periodicity, in fact, that led Meyer to propose his periodic table of the elements. One of the few interpretations that are obvious from the data is that the lone *ns* electron of the alkali metals makes these elements (except for lithium) much larger than their neighboring elements in the periodic table.

Atomic Radii

Of course, there is no such thing as an unequivocal atomic radius. The quantum mechanical description of the electrons in an atom indicates that there is no limit to the distance from the nucleus at which there is a finite probability of finding the electrons. We could compute the total electron density for all electrons in the atom and draw circles corresponding to the enclosing of 90 or 99% of the shading in figures such as Figure 2-27. These calculations are rather difficult, however, and would not be appropriate here.

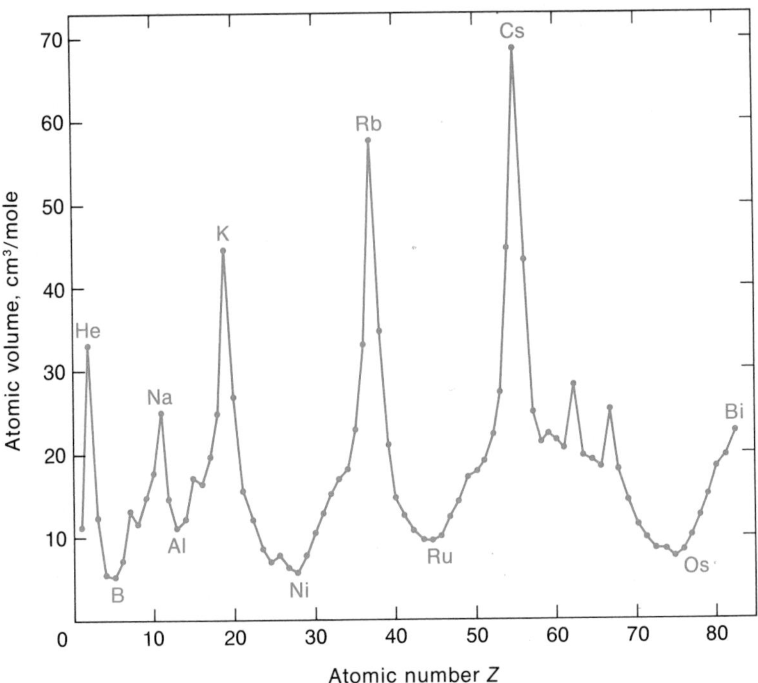

Figure 3-8. Variation of atomic volume with atomic number. The values for the gaseous elements are for the corresponding liquids. The anomalous value for helium is due to the very low density of liquid helium, one of the many strange properties of this substance. (B. H. Mahan, *College Chemistry*, Addison-Wesley, Reading, Mass., 1966.)

A simpler method, based on experimental data, is to define the radius of an atom as one-half the internuclear distance in homonuclear diatomic molecules, A_2. This procedure works well for those atoms that form diatomic molecules, but more complicated reference molecules are required for those atoms that do not.

The noble gases cause special problems. They form no stable diatomic molecules. Of the relatively few Kr, Xe, and Rn compounds that can be synthesized, some are quite unstable and the Rn compounds are highly radioactive. The values of the atomic radii of the noble gases given in Table 3-5 were computed by Sir John Lennard-Jones (1894–1954) as the distance of closest approach of identical atoms in the gaseous phase. These values are consistently "too large."

The relative sizes of the atoms of the second and third rows of the periodic table are depicted in Figure 3-9 to illustrate the trends in their values. It is apparent that the atomic sizes decrease when we traverse a row. For each of the rows shown, the electrons that are added go into orbitals of the same principal quantum number. Thus, the electron density within this shell increases with atomic number, but this does not increase the size of the shell. The observed decrease in atomic size must therefore be due to the increasing nuclear charge. The atoms in the lower row are slightly larger than those in the upper row. This indicates that the electrons entering successively larger shells are effectively screened from the increasing nuclear charge.

The accepted values (in angstroms) of the atomic radii of the first 39 elements are presented in Table 3-5. Periodicity is observed in the decreases in size in the elements of atomic numbers 3 through 9, 11 through 17, and 19 through 35. These sets of numbers correspond to the second, third, and fourth rows of the periodic table. (As mentioned above, the anomalous sizes of the noble gases were computed by a different method and should therefore be compared only with each other and not with those of the other elements.) The sizes of the transition elements in row four, atomic numbers 21 through 30, are nearly constant. The slight increase in size that follows the initial, gradual decrease is apparently due to the completion of the 3*d* orbital.

Ionic Radii

Table 3-5 also presents the ionic radii of the first 39 elements; they are given at the positions corresponding to the appropriate positive or negative oxidation states on the figure. These radii are computed by apportioning the measured interionic distances in crystals between the two ions in question. Again, if an absolute assignment is made for one ion, all other ionic radii

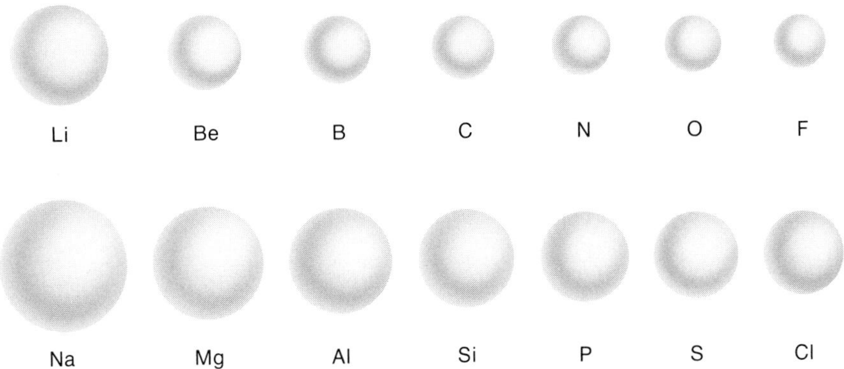

Figure 3-9. Variation of atomic radius in the second and third rows of the periodic table.

can be computed from appropriate pairings of the reference ion with other ions and then by extending the pairings of these ions to include still others. A simple method for obtaining the size of one ion, due to the German physicist Alfred Landé (b. 1888), is based on the structure of LiI shown in Figure 3-10. It is assumed that the *anion* (the negative ion) is so much larger than the *cation* (the positive ion) in this case that the geometry of the crystal is essentially fixed by the iodide ions' touching each other.

The nuclei of the ions can be very accurately located by means of x-ray diffraction. Thus, the length l, which is the side of the unit cell of the crystal (see Chapter 6, p. 293), can be determined. If we assume that the electron clouds of the iodide ions, represented by the larger circles, are actually touching, we can compute the radius of the iodide ion as $\sqrt{2}l/4 = (\sqrt{2} \times 6.05 \text{ Å})/4 = 2.14 \text{ Å}$.

Interesting and consistent trends are observed in Table 3-5. As we move across row two of the periodic table, the ions of maximum oxidation state that we encounter are Li^+, Be^{2+}, B^{3+}, C^{4+}, and N^{5+}, which are all isoelectronic with He. They all have two $1s$ electrons but they differ in having from three to seven protons in their nuclei. This large increase in positive charge density of the nucleus leads to a greatly increased attractive force and the radii are seen to decrease steadily.

Another trend is observed for those elements that have radii listed for more than one oxidation state. The best example is provided by the four different vanadium ions, atomic number 23. We observe a steady decrease in radius as the oxidation state increases from +2 to +5.

A trend is also clear for elements within columns of the periodic table. As would be expected, the size of the ions having the same oxidation state increases as we go down a column. The series of numbers 0.68, 0.97, 1.33, and 1.47 for Li^+, Na^+, K^+, and Rb^+ demonstrates this.

Table 3-5. **Correlation of Atomic Number with Atomic Radius and Oxidation State with Ionic Radius. The atomic radii are represented by the oxidation state 0. The known oxidation states of the elements are shown above the corresponding symbols; a dash denotes the existence of an oxidation state for which the ionic radius has not been determined. All radii are given in angstroms.**

Oxidation State															
+7															
+6															
+5							0.13								0.35
+4						0.16	—							0.42	
+3					0.23	—	0.16						0.51		0.44
+2				0.35		—	—					0.66			
+1	—		0.68			—	—				0.97				—
0	0.37	1.31	1.23	0.89	0.81	0.77	0.70	0.66	0.64	1.39	1.57	1.36	1.25	1.17	1.10
−1	—					—	—	—	1.33						
−2						—	—	1.40							
−3						—	—								—
−4						—								—	
Atomic number	1	2	3	4	5	6	7	8	9	10	11	12	13	14	15
Element	H	He	Li	Be	B	C	N	O	F	Ne	Na	Mg	Al	Si	P

Oxidation State															
+7		0.27								0.46					
+6	0.30								0.52						
+5		0.34						0.59							
+4	0.37						0.68	0.63		0.60					
+3		—				0.81	0.76	0.74	0.63	0.66	0.64	0.63			
+2					0.99		—	0.88	—	0.80	0.74	0.72	0.69	0.72	0.74
+1		—		1.33										0.96	
0	1.04	0.99	1.70	2.03	1.74	1.44	1.32	1.22	1.18	1.17	1.17	1.16	1.15	1.17	1.25
−1		1.81													
−2	1.84														
−3															
−4															

Atomic number	16	17	18	19	20	21	22	23	24	25	26	27	28	29	30
Element	S	Cl	Ar	K	Ca	Sc	Ti	V	Cr	Mn	Fe	Co	Ni	Cu	Zn
Oxidation State +7					—										
+6				0.42											
+5			0.46		0.47										
+4		0.53		0.50											
+3	0.62		0.58		—				0.92						
+2		0.73						1.12							
+1					—		1.47								
0	1.25	1.22	1.21	1.17	1.14	1.80	2.16	1.91	1.62						
−1					1.96										
−2				1.98											
−3			—												
−4		—													
Atomic number	31	32	33	34	35	36	37	38	39						
Element	Ga	Ge	As	Se	Br	Kr	Rb	Sr	Y						

SOURCE: Abstracted from M. J. Sienko, R. A. Plane, and R. E. Hester, *Inorganic Chemistry*, W. A. Benjamin, New York, 1965.

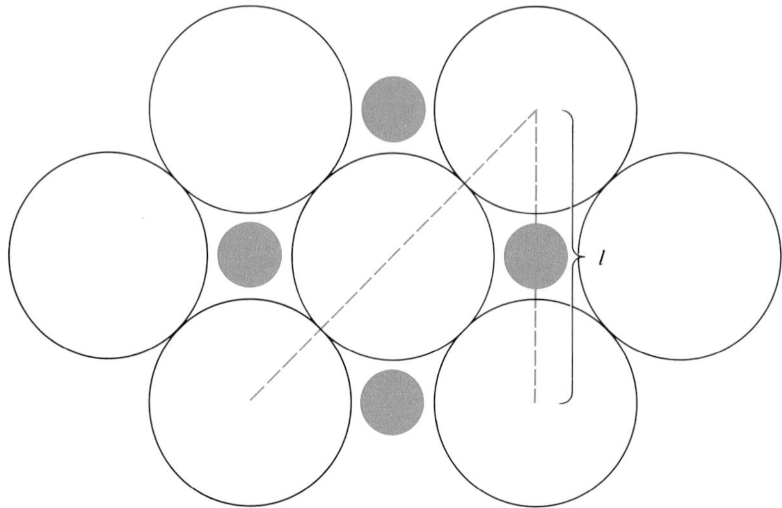

Figure 3-10. Assumed lattice structure of crystalline lithium iodide.

3.4 OTHER PHYSICAL PROPERTIES

Metallic Character

Metals are characterized by a number of distinctive properties. They are good conductors of heat and electricity. They are both malleable and ductile—that is, they can readily be hammered into sheets and drawn into wires. The few millionths-of-an-inch thickness of gold leaf and the nature of copper wire are familiar examples. Finally, metals are generally hard and shiny, and many of them display a striking luster.

As we will see in Chapter 6 in more detail, metals consist of a crystal structure with positive ions occupying the lattice sites. The intervening space is occupied by a "sea" of delocalized electrons. Only those elements with fairly low values of the lower ionization potentials can lose electrons in this manner. Thus, the metals are found on the left and middle of the periodic table. The nonmetals are on the right. The *metalloids,* those elements having quasimetallic properties, are located adjacent to the broken line in Figure 3-11, which represents the separation between metals and nonmetals. The elements generally considered to be metalloids are B, Si, Ge, As, Se, Sb, and Te. The locations of these classes of elements, together with the names of some of the particularly important groups of elements, are shown in Figure 3-11.

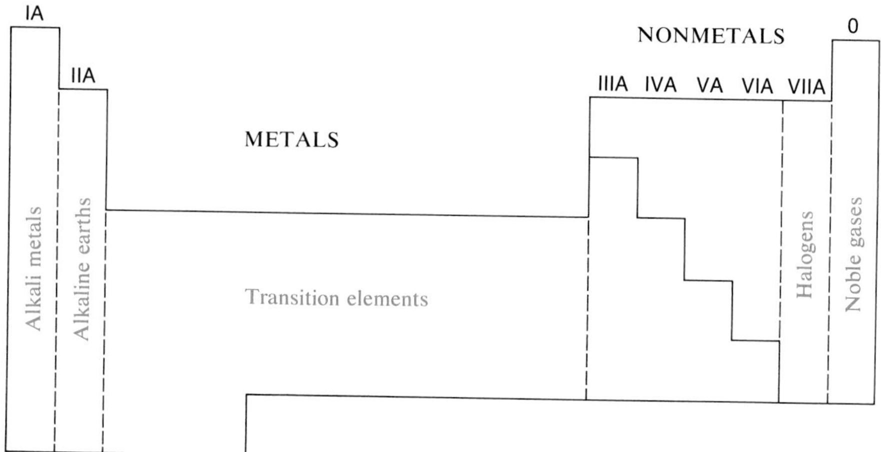

Figure 3-11. Variation of metallic character in the periodic table.

Color

The discussion in Chapter 2 may have led you to believe that all matter interacts with electromagnetic radiation by absorbing or emitting light only at discrete wavelengths. Actually, because of complex rotational and vibrational motions, many substances absorb radiation over a broad range of wavelengths. If an appreciable part of this range of the spectrum is in the visible, the substance is colored. A typical absorption spectrum of a colored solution is shown in Figure 3-12. This figure presents a plot of the absorbance A – which is the logarithm of the ratio of the intensity of light entering the solution to the intensity of light leaving the solution – versus the wavelength λ of the incident radiation, for a solution of potassium permanganate, $KMnO_4$. A strong absorption band is seen in the green-yellow portion of the spectrum. The transmitted light is violet and the permanganate ion, MnO_4^-, gives rise to intensely colored violet solutions.

A mildly entertaining diversion for a chemist on a dreary day is to assemble the best rainbow of colors he can find on his laboratory shelf. A rather attractive set of colors is obtained from solutions of the following compounds:

Compound	Color	Colored Species
$Co(NO_3)_2$	Red	$Co(H_2O)_n^{2+}$
$K_2Cr_2O_7$	Orange	$Cr_2O_7^{2-}$
K_2CrO_4	Yellow	CrO_4^{2-}
$Ni(NO_3)_2$	Green	$Ni(H_2O)_n^{2+}$
$Cu(NO_3)_2$	Blue	$Cu(H_2O)_n^{2+}$
$KMnO_4$	Violet	MnO_4^-

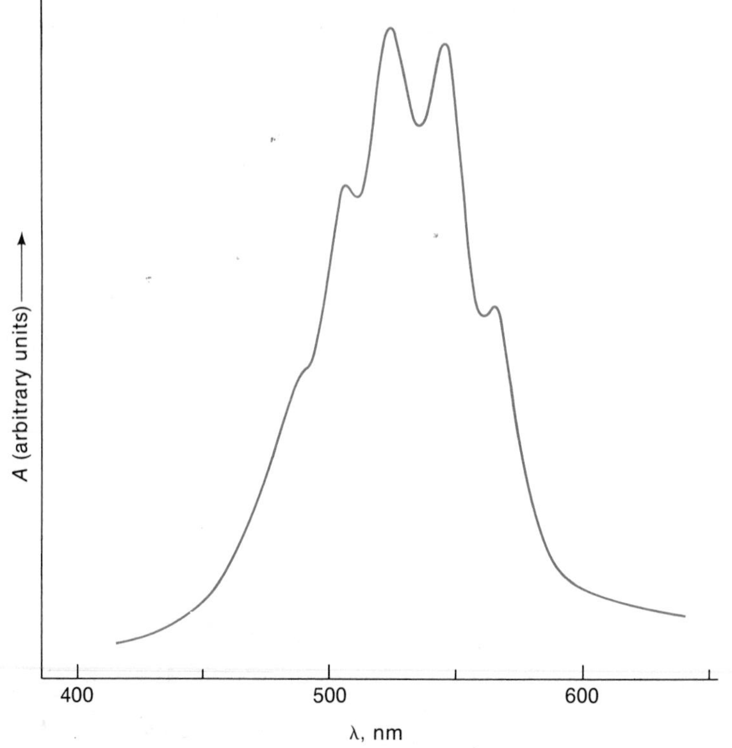

Figure 3-12. Absorption spectrum of potassium permanganate in the visible region. (J. Waser, *Quantitative Chemistry,* revised ed., W. A. Benjamin, New York, 1964.)

Can you spot the feature that all the elements responsible for the colors of the above solutions have in common? They are all transition elements. They happen to occupy the first row of the transition elements, where the 3*d* orbital is filling.

There are no firm rules regarding the periodicity of colors of the elements and their simple compounds. As we have seen in the above example, however, colored elements and colored oxide anions are generally to be found among the transition elements. Very few elements in the A groups of the periodic table display any color.

Density

It might be of interest to examine how the densities of the elements vary within the periodic table. This will give us a chance to examine the technique used by Professor R. T. Sanderson in his interesting text, *Chemical Periodicity.** Figure 3-13 is taken from this source. The solid circles

*Reinhold, New York, 1960.

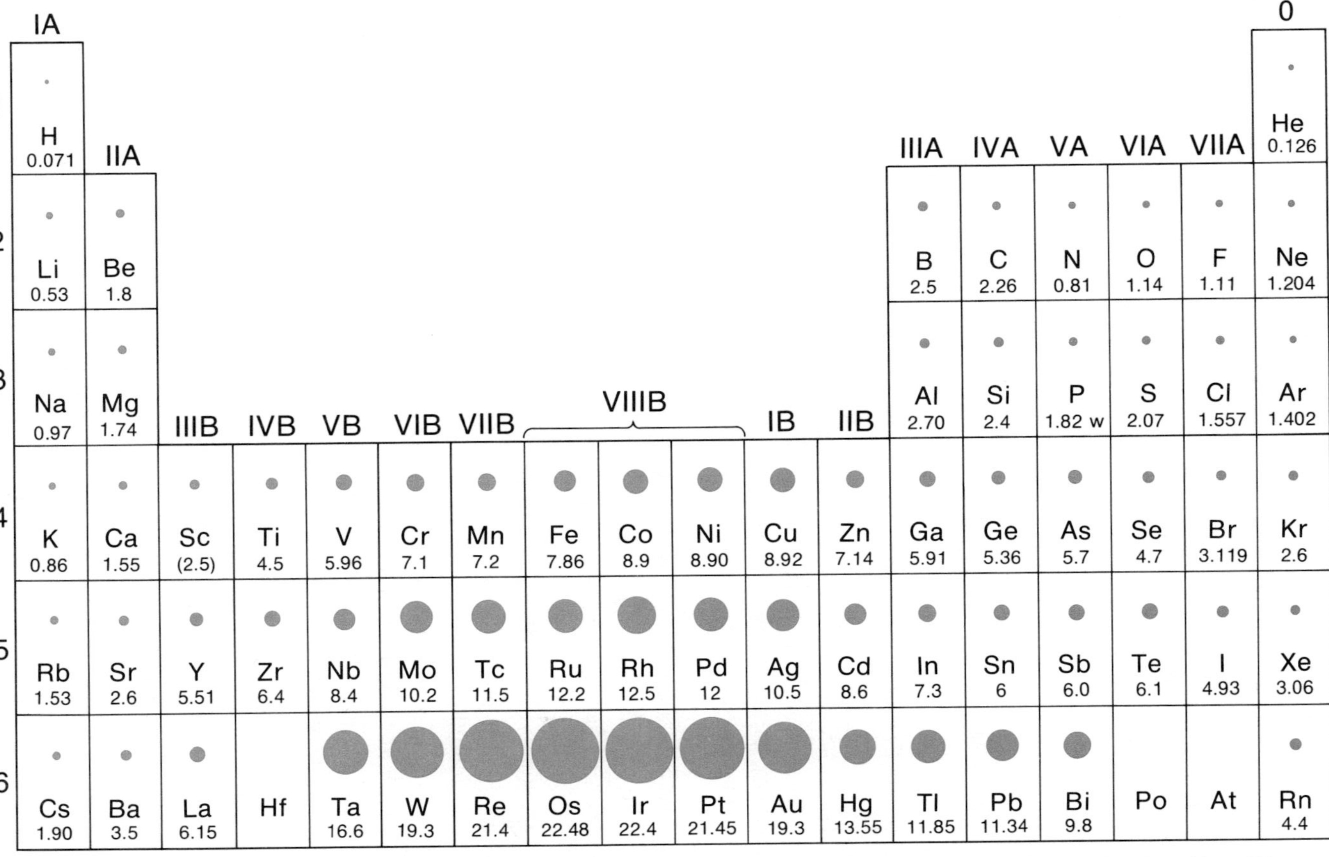

Figure 3-13. Variation of density in the periodic table. The diameters of the circles are not linearly proportional to the values of the densities, but they accurately illustrate the trends in those values. The "w" with the value for P denotes the white allotrope. (Adapted from R. T. Sanderson, *Chemical Periodicity*. © 1960 by Litton Educational Publishing, Inc. Reprinted by permission of Van Nostrand Reinhold Co.)

represent the relative values of the densities on a compressed, nonlinear scale. The actual values in grams per cubic centimeter are given below the circles. The densities of the gaseous elements are for the element in the liquid form at its boiling point.

Several features of Figure 3-13 deserve comment. The general notion that metals are denser than nonmetals requires reconsideration after comparing the densities of lithium and iodine. The idea that mercury and lead are the densest elements is seen to be incorrect. General trends that can be spotted are that density increases as we go down a column but peaks near the middle of a period.

3.5 CHEMICAL PROPERTIES

Oxidation States

Table 3-5 has already served to present the atomic and ionic radii of the first 39 elements. Its third, quite important function is to show the observed oxidation states of these elements. Some useful rules for the assignment of oxidation states are given below.

1. The oxidation state of all elements is 0.
2. The oxidation state of a monatomic ion equals the charge on the ion.
3. Oxygen has an oxidation state of -2 in all compounds except those designated as *peroxides,* in which its oxidation state is -1, and those designated as *superoxides,* in which its oxidation state is effectively $-\frac{1}{2}$.
4. Hydrogen has an oxidation state of $+1$ in all compounds except those designated as *metal hydrides,* in which its oxidation state is -1.
5. The sum of the oxidation states of the elements in a compound or ion equals the charge on that species.
6. The most electronegative element in a complex ion is assigned the negative oxidation state.

Example 3-1. Assign oxidation states to all atoms in the following species: Cl^-, Cl_2, Cl_2O_7, CrO_3, NaH (a metal hydride), MnO_4^-, CH_4.

Species	Oxidation State	Comments
Cl^-	-1	The oxidation state equals the charge on the ion.
Cl_2	0	Elements have oxidation states of 0.

Species	Oxidation State	Comments
Cl_2O_7	Cl: +7 O: −2	Oxygen has an oxidation state of −2 unless otherwise specified. In order for the chlorine heptoxide to have a charge of 0, each chlorine atom must be assigned a value of +7.
CrO_3	Cr: +6 O: −2	Comments analogous to those above apply.
NaH (metal hydride)	Na: +1 H: −1	Because this is a metal hydride, hydrogen is −1, which means that sodium has its usual value of +1. The electronegativity of H is 2.1 and that of Na is 1.0, which also supports this conclusion.
MnO_4^-	Mn: +7 O: −2	$4(-2) + (+7) = -1$. This satisfies rule number 5.
CH_4	H: +1 C: −4	The oxidation state of hydrogen is +1 unless otherwise specified. The sum of the oxidation states must equal 0.

Exercise 3-1. Assign oxidation states to all atoms in the following species: CO_2, Na_2SO_4, Al, P_2O_5, HCl, SF_6, C_6H_6, CH_3CH_2OH, CH_3Cl, H_2O_2 (a peroxide).

Now that we know how to make the assignments of oxidation states, it is appropriate to ask what they mean. In the case of an element or a monatomic ion, their meaning is quite clear: the oxidation state corresponds directly to the net electric charge on the atom. In the case of the permanganate ion, however, are we saying that each oxygen atom bears a charge of −2 and the central manganese atom has a charge of +7? The answer is definitely no. Quantum mechanical calculations can reveal the *actual* charge on each atom. The actual charges are only fractions of these two values. Then why bother with assigning numbers to atoms if these numbers do not have any physical significance? The value of these numbers will become more apparent as we study the chemical bond in the next chapter and especially in Chapter 10, when we consider *redox reactions,* that is, reactions in which electrons are transferred from one species to another. Oxidation states provide a very useful, albeit artificial, method of keeping track of electrons in such reactions.

Let us examine the properties of the so-called *representative elements,* those elements found in the A groups of the periodic table. Perusal of Table 3-5 reveals that the only oxidation states of groups IA, IIA, and IIIA are +1, +2, and +3, respectively. Beginning with group IVA, negative oxidation states are found. Except for the two smallest members of groups VIA and VIIA, oxygen and fluorine, the oxidation states of elements in groups IVA through VIIA range between a maximum negative value equal to the group number minus 8 and a maximum positive value equal to the group number.

The transition elements in the B groups exhibit a somewhat similar pattern with regard to the maximum positive oxidation states, namely, that these states correspond to the group number. Some striking exceptions are the common +2 oxidation state of Cu in group IB and the fact that, of the nine elements in group VIIIB, only Os and Ru ever attain the +8 oxidation state. Most of the transition elements have either +2 or +3 as one of their principal stable oxidation states. A notable difference between the representative elements and the transition elements is that the latter seldom exhibit negative oxidation states. The above comments about periodicity in oxidation states are summarized in Table 3-6.

Periodic Properties of Hydrogen Compounds

We will discuss the properties of the hydrogen compounds and the oxides in this section and the following section. This discussion is somewhat premature, in that we have not discussed molecules yet. However, we feel

Table 3-6. Trends in Oxidation States. (The 0 oxidation state available to all elements is not included.)

Representative elements (A groups)	I	II	III	IV	V	VI	VII	
Range of oxidation states	+1	+2	+3	+4	+5	+6	+7	
				⋮	⋮	⋮	⋮	
				−4	−3	−2	−1	
Transition elements (B groups)	**I**	**II**	**III**	**IV**	**V**	**VI**	**VII**	**VIII**
Range of most common oxidation states	+3	+2	+3	+4	+5	+6	+7	+8
	⋮	⋮		⋮	⋮	⋮	⋮	⋮
	+1	+1		+3	+2	+2	+2	+2

that it will be helpful to you to begin to see the names and formulas of some chemical compounds now,* and that these two sections will serve as a bridge between Chapter 2 on Atoms and Chapter 4 on The Chemical Bond. For now, we can treat a molecule simply as a well-defined group of atoms without paying too much attention to the detailed description of what makes the atoms stick together.

The hydrogen compounds of the representative elements display a very definite periodicity with atomic number, as shown in Figure 3-14. The number of hydrogen atoms combined with one atom of the element is seen to equal the group number or (8 − group number), whichever is smaller.

In addition to the variation in the number of hydrogen atoms found in each compound, the type of hydrogen compound varies considerably with position in the periodic table. Hydrides of the elements on the left side of the table have the properties of *salts:* they are solids consisting of a lattice of alternating metal cations, in this case, M^+ or M^{2+}, and nonmetallic anions, in this case, hydride ions, H^-. Solution of these compounds in water results in the evolution of hydrogen gas and the generation of a *basic* (or *alkaline*) solution, a solution containing an excess of hydroxide ions, OH^-:

$$MH + H_2O \rightarrow M^+ + OH^- + H_2\uparrow$$

The hydrogen compounds of the transition elements are more difficult to characterize, inasmuch as many of them are *interstitial compounds,* that is, the metal has hydrogen atoms dispersed through it.

The hydrogen compounds of the elements on the right are generally gaseous, which means they exist as discrete molecules. Some examples of these compounds are: CH_4, methane; NH_3, ammonia; H_2O, water (steam); HCl, hydrogen chloride; and H_2S, hydrogen sulfide. The compounds named are of very widespread occurrence, both inside and outside the chemical laboratory, and are hence of especial importance for us to remember and study further. Some of these compounds are quite soluble in water and react to give either an *acidic* solution (a solution containing an excess of hydrogen ions, H^+)† or a basic solution:

$$HCl + H_2O \rightarrow H_3O^+ + Cl^- \qquad \text{(Acidic)}$$

$$NH_3 + H_2O \rightarrow NH_4^+ + OH^- \qquad \text{(Basic)}$$

*See Appendix C for rules regarding the nomenclature of inorganic compounds.

†Actually, the species H^+ (a bare proton) does not exist as such in aqueous solution. Instead, it combines with a water molecule to form H_3O^+, the *oxonium ion* (a definite chemical species; when an *indefinite* degree of hydration of the proton occurs, the resulting species is called a *hydronium ion*). It is often more convenient, however, to use the simple symbol H^+.

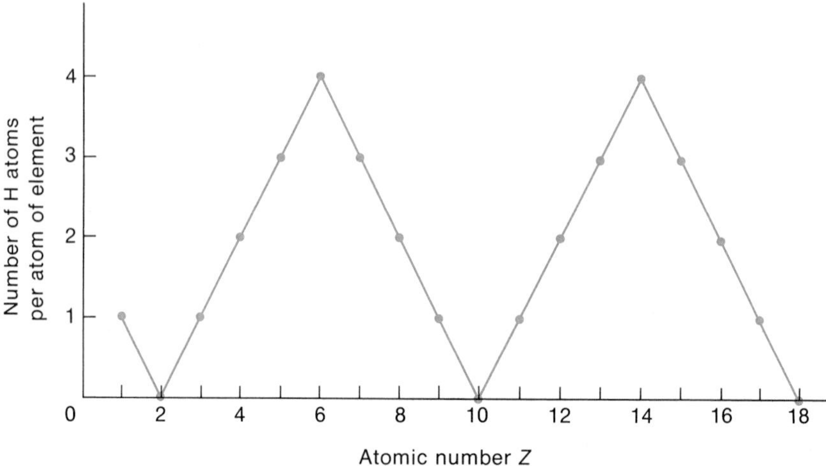

Figure 3-14. Variation of the composition of the hydrogen compounds of the elements in the first three rows of the periodic table.

Some of the hydrogen compounds, such as methane, are neither soluble in, nor reactive with, water. Figure 3-15 shows some of the trends in the properties of the hydrogen compounds (R stands for any element in a given group).

Periodic Properties of the Oxides

The compounds of oxygen are among the most widespread and important compounds on earth. Water, H_2O, covers 71% of the Earth's surface; carbon dioxide, CO_2, is the final product of respiratory processes and the combustion of fossil fuels, and is one of the reactants in the first step of photosynthesis. Most inorganic acids, bases, and other compounds contain oxygen; all organic acids, aldehydes, ketones, and ethers contain oxygen. And, as we will see at the end of this section, most of the compounds involved in the formation of photochemical smog are inorganic oxides. Figure 3-16 is analogous to Figure 3-15.

Our first use of the electronegativity concept, which we examined earlier in this chapter, is now at hand. The electronegativity of an element is a measure of the ability of an atom in a molecule to attract electrons from the other atoms. For now, we will accept without proof the idea that a chemical bond between two atoms consists of two electrons: one electron may originate with each atom, or both may originate with the same atom. Let us consider an alkaline earth oxide, MO, as an example. The electronegativity of oxygen is 3.5. Thus, there is a large electronegativity difference between

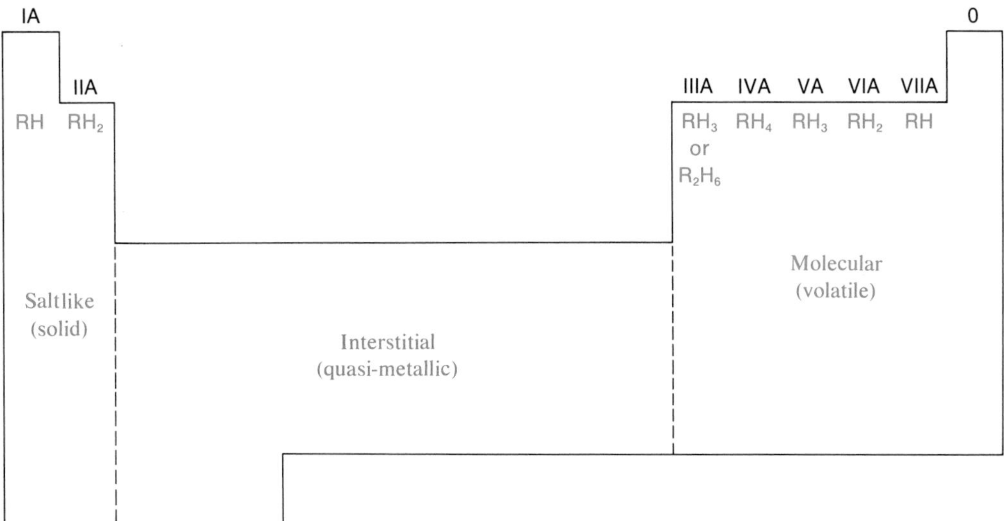

Figure 3-15. Variation of the properties of the hydrogen compounds of the elements in the periodic table.

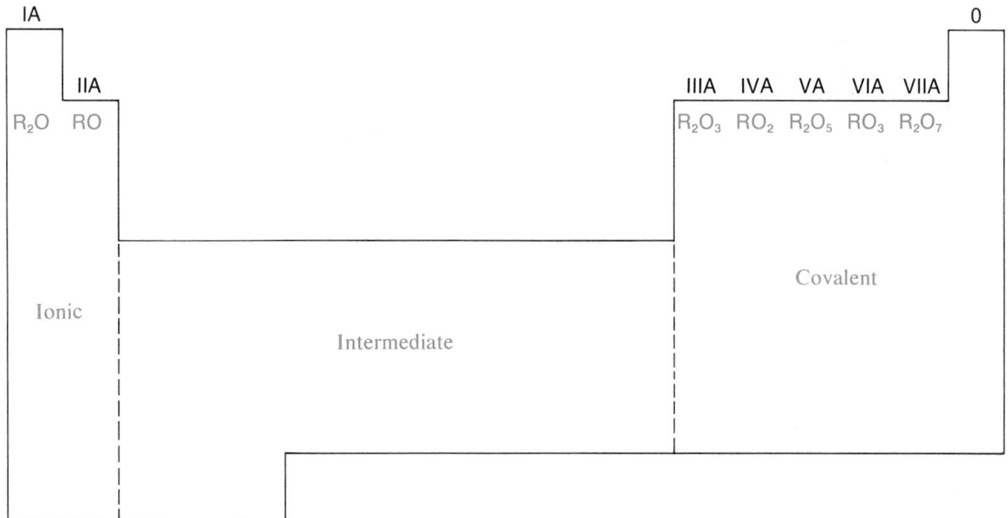

Figure 3-16. Variation of the properties of the oxides of the elements in the periodic table.

oxygen and the alkaline earths. The oxygen atom has such a great attraction for the two electrons available from the metal atom that the structure of the compound is essentially $M^{2+}O^{2-}$: the electrons are transferred completely from the metal atom to the oxygen atom. Such compounds are termed *ionic*. An unusual feature of the two groups on the left side of the periodic table is that both peroxides and superoxides are found there. These are compounds in which oxygen has oxidation states of -1 and $-\frac{1}{2}$, respectively. Some examples are: Na_2O_2, sodium peroxide; KO_2, potassium superoxide; and BaO_2, barium peroxide. A "peroxide blonde" achieves his or her desired hair tint by employing a dilute solution of hydrogen peroxide, H_2O_2.

The elements on the right side of the periodic table have electronegativities closer to that of oxygen. Hence, the electron pair between the element and the oxygen atom is shared nearly equally. Such compounds are termed *covalent* and are represented as, e.g.,

$$\text{M—O} \quad \text{or} \quad \text{M:O}$$

$$\text{O—M—O} \quad \text{or} \quad \text{O:M:O}$$

The electron pair can be represented by either a dash or a pair of dots.

A further property of the oxides that is of interest is the reaction of these compounds with water. The ionic oxides might be thought to dissolve in water to give

$$MO \xrightarrow{H_2O} M^{2+} + O^{2-}$$

However, the oxide ion is an extremely reactive species and instantly abstracts a proton from a water molecule:

$$O^{2-} + H_2O \rightarrow 2OH^-$$

Thus, the reaction of Na_2O with water is expected to be

$$Na_2O + H_2O \rightarrow 2Na^+ + 2OH^- \qquad \text{(Basic)}$$

Such compounds that produce hydroxide ions upon reaction with H_2O are said to be *basic*. An alternative classification of basic oxides is that their solubility, the amount of oxide that dissolves in a given amount of water, increases upon the addition of an acid but decreases upon the addition of a base.

An *acid* can be defined as a substance that releases H^+ upon reaction with water, or as a compound whose solubility increases upon the addition of base and decreases upon the addition of another acid. The covalent oxides on the right side of the periodic table react with water to give acids. An example is chlorine monoxide, Cl_2O, which gives hypochlorous acid:

$$Cl_2O + H_2O \rightarrow 2H\text{—}O\text{—}Cl$$

Because chlorine and oxygen differ in electronegativity by only 0.5, we do not expect the Cl—O bonds to ionize in water. Because the O and Cl are more electronegative than the H, however, the electrons in the entire molecule are withdrawn from the H. This causes a weakening of the H—O bond and the resultant loss of a proton:

$$H\text{—}O\text{—}Cl \rightarrow H^+ + OCl^- \qquad \text{(Acidic)}$$

There are two important periodic trends in the acid-base behavior of the oxides. The first is that, the higher the oxidation state of the central atom (for a given element), the more acidic the compound. The oxygen acids of chlorine—$HClO$, $HClO_2$, $HClO_3$, and $HClO_4$—provide a good example of this trend.* As the oxidation state of the Cl atom increases from +1 to +3 to +5 and finally to +7, the chlorine attracts the electrons of the molecule ever more firmly. This provides for a progressively easier release of the H^+. Thus, $HClO$ is a very weak acid—relatively few H^+ ions are released to the solution—whereas $HClO_4$, perchloric acid, is a very strong acid, being completely dissociated in water.

The other trend is that, within a given group of the periodic table, the oxides of the higher-atomic-weight elements are more basic. Thus, H_2SO_3 (sulfurous acid) is a weak acid, whereas H_2TeO_3 (tellurous acid) is *amphoteric* (possesses both acidic and basic properties); $Be(OH)_2$ (beryllium hydroxide) is amphoteric, whereas $Ba(OH)_2$ (barium hydroxide) is a fairly strong base. As the ionic radii increase, the bonding between the central atom and the oxygen becomes weaker, resulting in a greater tendency to lose OH^-. The distribution of several types of oxides with respect to acid-base properties is shown in Figure 3-17.

*The formulas $HOCl$, $HOClO$, $HOClO_2$, and $HOClO_3$ are more representative of the molecular structures of these species (note that the H atom is bonded to an O atom in all of them) but are more cumbersome to write. Chemists usually adopt a shorthand approach to the writing of formulas that enables large, complex molecules to be expressed clearly and succinctly with a minimum of symbols. For consistency with other compounds in a series, the correct order of the atoms in a triatomic molecule is sometimes juxtaposed (thus, $HClO$, etc.). It is essential, of course, to learn the principles of molecular structure so that shorthand formulas can be correctly interpreted.

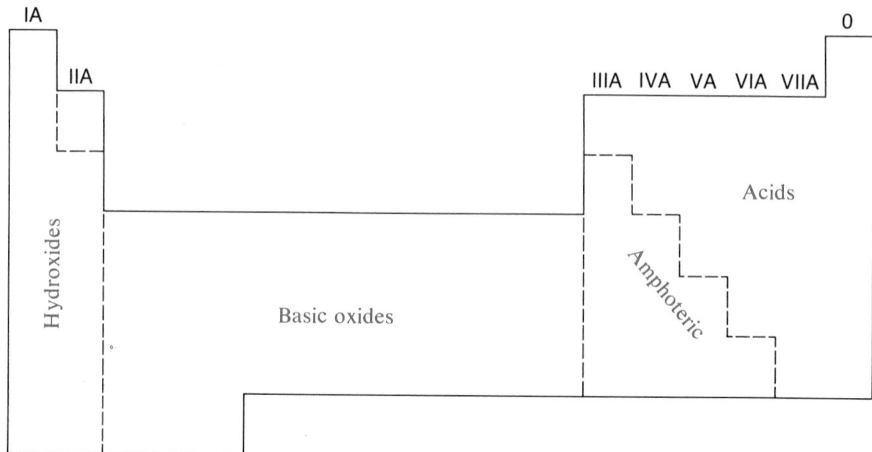

Figure 3-17. Variation of the acid-base properties of the oxides of the elements in the periodic table.

Table 3-7. Some Oxygen and Hydrogen Compounds of Nitrogen, Sulfur, and Carbon. The compounds shown in color are all of importance in the formation of photochemical smog.

Oxidation State	N		S		C	
+6			SO_3	Sulfur trioxide		
+5	N_2O_5	Dinitrogen pentoxide				
+4	NO_2	Nitrogen dioxide	SO_2	Sulfur dioxide	CO_2	Carbon dioxide
+3	N_2O_3	Dinitrogen trioxide				
+2	NO	Nitric oxide			CO	Carbon monoxide
+1	N_2O	Nitrous oxide				
0	N_2	Nitrogen	S	Sulfur	C	Carbon
−1	NH_2OH	Hydroxylamine				
−2	N_2H_4	Hydrazine	H_2S	Hydrogen sulfide	CH_3OH	Methanol
−3	NH_3	Ammonia				
−4					CH_4	Methane

Table 3-7 lists some oxygen and hydrogen compounds of the elements nitrogen, sulfur, and carbon, and most of the oxidation states available to them. In addition, it provides a good introduction to the topic considered in the Interlude at the end of this chapter, smog.

Bibliography

R. J. Puddephatt, **The Periodic Table of the Elements**, Oxford University Press, New York, 1972. This brief volume (94 pages) provides a nice overview of periodic properties. The first three chapters are an introduction to the electronic structures of atoms. The remainder of the book deals with the chemistry of the main-group elements, the transition elements, and the lanthanons and actinons.

Ronald L. Rich, **Periodic Correlations**, W. A. Benjamin, New York, 1965. This is the most advanced of the books listed here. It is part of a series on physical inorganic chemistry. However, the treatment is largely nonmathematical and well within the grasp of students of general chemistry. The topics of electron configuration, ionization energy, acidity and basicity, oxidation state, and color are particularly thorough and provide a nice extension of the present chapter.

Robert T. Sanderson, **Chemical Periodicity**, Reinhold, New York, 1960. This is the most thorough treatment of periodic properties among the three books mentioned here. The principles underlying the modern form of the periodic table are discussed. The major portion of the book is devoted to a discussion of the binary compounds of the elements found in the upper right corner of the periodic table. A valuable facet of Sanderson's treatment is the use of circles whose diameters, for a large number of elements or compounds arranged in a periodic table, are proportional to the magnitude of the physical or chemical property under study.

Problems

1. Each of the following electronic structures contains the configurations of the electrons having the two highest principal quantum numbers represented in that group of elements. Name the group and give the group number.
  a. s^2p^6, s^2
  b. $s^2p^6d^{10}$, s^2p^6
  c. $s^2p^6d^{10}f^{14}$, s^2p^5
  d. $s^2p^6d^5$, s^2

2. Which element in the following pairs has the higher ionization potential:
  a. N, O
  b. Mg, Si
  c. Na, K
  Give a brief explanation of the higher ionization potential in each case.

3. Arrange the following sets of atoms in order of increasing ionization potential. (Do not consult the table of ionization potentials.)
   a. Cs, Zn, Ga, K
   b. Cu, Se, As, Ge

4. Explain the fact that the second ionization potential of boron is greater than the second ionization potential of beryllium, whereas the first ionization potential of boron is less than the first ionization potential of beryllium.

5. Which atom has the lowest first ionization potential: Na, Cl, Rb, Xe? Why?

6. Calculate the ionization potential of the hydrogen atom, using the equations from Chapter 2. Compare the calculated value with the experimental value.

7. Calculate the fourth ionization potential of the beryllium atom, using the equations from Chapter 2. Compare the calculated value with the experimental value (see Equation 2-22).

8. Which atom has the largest electron affinity: B, O, C, N? Why?

9. Why is the electron affinity of P lower than those of both S and Si, which are on either side of it?

10. For the following diatomic molecules, state which atom has a partial negative charge:
    a. H—Cl
    b. Li—F
    c. N—N

11. In which of the following salts would you expect the spacing between adjacent nuclei to be the smallest, assuming that they have a common crystal structure: NaI, LiF, KCl, MgO, CaO?

12. The density of potassium is 0.85 g/cm^3. Calculate its atomic volume in cm^3/mole. Predict whether the atomic volumes of germanium and cesium are larger or smaller than that of K.

13. Arrange the following set of atoms in order of increasing atomic radius: Fr, Ca, K, Cs.

14. Arrange the following set of atoms in order of increasing atomic radius: Rb, O, Nb, Ga, Pd.

15. Which ion would be expected to have the larger radius, Li^+ or Be^{2+}? Why?

16. Assign oxidation states to all atoms in the following species: $Cr_2O_7^{2-}$, CaH_2, BaO_2 (a peroxide), SO_3^{2-}, Na_2O_2 (a peroxide), SrO, Ne.

17. In 1971 the top ten chemical products in the United States were H_2SO_4, NH_3, O_2, NaOH, Cl_2, CH_2CH_2, Na_2CO_3, HNO_3, NH_4NO_3, and H_3PO_4 [*Chemical and Engineering News* **50** (23), 12(1972)]. Assign oxidation states to each of the atoms in the above species. (*Hint*: NH_4NO_3 should be thought of as $NH_4^+NO_3^-$.)

18. Classify each of the following hydrogen compounds as salt hydride, interstitial hydrogen compound, or molecular hydrogen compound: CdH_2, PbH_4, BiH_3, H_2Te, RbH, HgH_2, Ga_2H_6, H_2Po.

19. For each group of compounds given below, state which compound is the strongest acid and which is the strongest base, in water.

a. MgO, $Si(OH)_4$, Na_2O, $HClO_4$, P_4O_{10}, $Al(OH)_3$

b. BaO, SrO, BeO, CaO

"Would you care to know, dear, that your 'hazy morn of an enchanted day in May' is composed of six-tenths parts per million sulfur dioxide, two parts per million carbon monoxide, four parts per million hydrocarbons, three parts . . ."

INTERLUDE

smog

Welcome, sulphur dioxide,
Hello, carbon monoxide,
The air, the air is ev'rywhere.
Breathe deep while you sleep, breathe deep.

Bless you alcohol blood stream,
Save me nicotine lung steam
Incense, incense is in the air.
Breathe deep while you sleep, breathe deep.

Cataclysmic ectoplasm,
Fallout atomic orgasm,
Vapor and fume, at the stone of my tomb,
Breathing like a sullen perfume,
Eating at the stone of my tomb.

Welcome, sulphur dioxide,
Hello, carbon monoxide,
The air, the air is ev'rywhere.
Breathe deep while you sleep, breathe deep,

(cough) deep, (cough) deep de deep, (cough).

"AIR," from *Hair*

Air pollution. Smog. The old cliché, "Everyone talks about it but no one does anything about it," is only half true. The rock'n'roll balladeers, the politicians, the housewives, the shoe salesmen, and even the kids are talking about it. Physicians are talking about it: they recently tried to talk 10,000 people into moving out of the Los Angeles basin because their very lives were endangered by Los Angeles' (in)famous smog.

And, indeed, people are really *doing* something about dirty air. Pittsburgh, Pennsylvania, is virtually unrecognizable after its antipollution crusade, which yielded clean skies and clean buildings. The federal government closed down 23 industries in Birmingham, Alabama, on November 18, 1971, to relieve an air pollution crisis. Electric companies' advertising has undergone a radical switch (even before the energy crisis) from urging us to buy more electric stoves and air conditioners to emphasizing ways in which homeowners can curb their use of electrical energy. Gasoline companies are literally "getting the lead out." Legislation has been introduced (but defeated) to ban the sale of cars using internal combustion engines in the state of California after 1975.

Now that you have encountered a few small molecules at the end of Chapter 3, but before you study the geometry and energy states of molecules in the next chapter, we would like to present some elementary facts of the chemistry of air pollution. As indicated by the compounds shown in color in Table 3-7, the molecules you are studying are of more than academic interest. You are undoubtedly inhaling a number of these molecules right now. The long-term effects of this on your health are somewhat unclear at this point, unless you are afflicted with heart disease or respiratory ailments. However, the dying vineyards, the millions of dying Ponderosa pines in national forests close to our urban centers, the reduced citrus crops, the regulations against children going out to recess when the smog reaches a given level, and the tears streaming down people's faces on a hot, smoggy afternoon in Los Angeles are quite clear.

If our concern is *dirty* air, we should begin with an understanding of what *clean* air is. Then we will learn what the pollutants in our environment are and where they come from. We will conclude with a discussion of the two distinct types of smog.

Let's begin with a look at your smog IQ. Answer the following five questions to see whether you even need to read the rest of this Interlude.

SMOG QUIZ

1. What is PAN?
   a. Perfluoroacetonitrile
   b. Peter's last name
   c. Peroxyacyl nitrate

2. What is the formula of the lead compound emitted in the exhaust of automobiles?
   a. $Pb(CH_2CH_3)_2$
   b. PbBrCl
   c. PbO_2

3. What is the effect on the temperature of the atmosphere of increased concentrations of
   a. CO_2: raise, lower
   b. Particles: raise, lower
4. Man produces a major fraction of the hydrocarbons (compounds containing only C and H) released into the atmosphere. True or false?
5. Name three of the American cities that made the Top Ten on the 1970 National Pollution Control Administration list for SO_2 levels.

The answers are given at the end of this Interlude.

The subject of discussion is air pollution. Before we say air is polluted, however, it would be well to find out what unpolluted air is like. The excellent book, *Cleaning Our Environment: The Chemical Basis for Action,* published by the American Chemical Society in 1969, gives us this information, which is shown in Table 1.

Although there is some fluctuation in these data in different "clean" locales, they clearly indicate several things about clean air. For example, 99% of clean air consists of the two small molecules nitrogen and oxygen, which are present in a ratio of about 4:1. Most of the remaining 1% consists of the noble gases and carbon dioxide. The trace compounds are mainly small hydrogen compounds and oxides of carbon, nitrogen, and sulfur. Apparently even clean air has small amounts of such dangerous compounds as sulfur dioxide, nitrogen dioxide, and carbon monoxide.

Before we go on to describe less-than-clean air, there are two terms we should understand. A *contaminant* is anything added to our environment that causes a local deviation from the mean geochemical composition given

Table 1. **Composition of Clean, Dry Air Near Sea Level.**

Compound	Formula	Volume %	Compound	Formula	Volume %
Nitrogen	N_2	78.09	Nitrous oxide	N_2O	0.000025
Oxygen	O_2	20.94	Hydrogen	H_2	0.00005
Argon	Ar	0.93	Methane	CH_4	0.00015
Carbon dioxide	CO_2	0.0318	Nitrogen dioxide	NO_2	0.0000001
Neon	Ne	0.0018	Ozone	O_3	0.000002
Helium	He	0.00052	Sulfur dioxide	SO_2	0.00000002
Krypton	Kr	0.0001	Carbon monoxide	CO	0.00001
Xenon	Xe	0.000008	Ammonia	NH_3	0.000001

SOURCE: *Cleaning Our Environment: The Chemical Basis for Action,* The American Chemical Society, Washington, D.C., 1969.

in Table 1. Thus, forest fires, volcanoes, and virtually all of man's technological activities produce contaminants. A *pollutant* is a contaminant that adversely affects something that man values. That happens to be an entirely anthropocentric definition but it is the one usually used. There are some problems with our definitions, however. For instance, water vapor is hardly a contaminant, but a dense fog created by a water cooling tower near a freeway can constitute a deadly pollutant.

Atmospheric pollution is very real, as any citizen of New York City, London, or Tokyo can attest. The Public Health Service has identified the five major pollutants, their sources, and their amounts for the year 1965. This information is shown in Table 2.

Several points that are apparent in these data deserve comment. Our cars are the worst polluters in America, especially in terms of CO, hydrocarbons, and nitrogen oxides. Industry and power plants account for most of the sulfur oxides and particulate matter that find their way into the air. Although CO_2 emissions exceed those of all these pollutants combined, CO_2 is not included in the table because it is generally regarded as a contaminant rather than a pollutant. Moreover, since the overwhelming majority of our transportation, energy-production, and temperature-control systems depend upon the combustion of carbonaceous fuels, CO_2 emissions cannot be controlled until we replace combustion with alternative power sources. Not everyone is so sanguine about the constant addition of CO_2 to our atmosphere, however. Mr. Edward Schuck of the Environmental Protection Agency predicts that, all else being equal, the increasing CO_2 level in the atmosphere will trap enough of the thermal energy that the Earth normally radiates back into space to start the polar ice caps melting significantly by the year 2000. Enough water is contained in this ice to raise the sea level 400 feet, which would have interesting consequences for most of the largest cities on earth.

The term *smog* is frequently used to describe air pollution. The use is somewhat indiscriminate, in that there are two very different kinds of smog.

Table 2. **National Air Pollutant Emissions, Millions of Tons Per Year, 1965.**

	CO	Sulfur Oxides	Hydrocarbons	Nitrogen Oxides	Particles	Total	%
Automobiles	66	1	12	6	1	86	60
Industry	2	9	4	2	6	23	17
Electric power plants	1	12	1	3	3	20	14
Space heating	2	3	1	1	1	8	6
Refuse disposal	1	1	1	1	1	5	3
	72	26	19	13	12	142	100

We will now consider the conditions under which they arise and discuss their principal constituents. We will include a few comments on how these constituents might be controlled.

LONDON SMOG

London smog has afflicted that city for many centuries—hence, the name. Of the various (and largely unsuccessful) attempts made to curb it, none was more drastic than the official edict in 1273 that prohibited the burning of coal in London—on pain of death! (The edict was eventually repealed by a more enlightened monarch.) After 700 years, the smog now carries its own death sentence, and not just in London. On October 26, 1948, the small town of Donora, Pennsylvania, became the victim of a well-documented and much-publicized catastrophe resulting from smog of this type. The town nestles in a deep valley and on that day a *temperature inversion* occurred. In the lower 5–10 miles of the atmosphere, the temperature normally falls steadily as the altitude increases, permitting the warm, less dense air near the ground to rise steadily (and carry off pollutants to other levels of the atmosphere or other regions of the Earth). As Figure 1b shows, an inversion is the meteorological condition in which a layer of warm air exists over the cooler, denser air below. This traps the cooler air, along with any pollutants that are being produced.

Donora had a large wire factory and zinc, sulfuric acid, and steel plants in operation on the Tuesday on which the inversion occurred. It lasted until Sunday, leaving 20 people dead, the lives of others undoubtedly shortened, and half the people of the town ill.

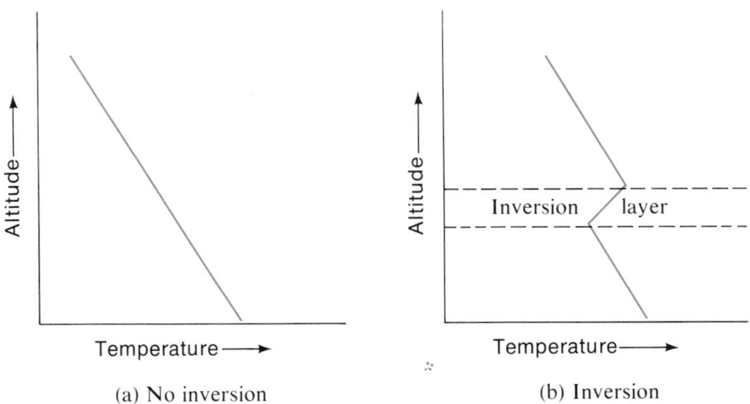

Figure 1. The idealized dependence of temperature on altitude above the Earth's surface in the troposphere: (a) under normal conditions and (b) when a temperature inversion occurs.

The same thing happened in London on December 5, 1952, causing some 4000 deaths. Similar disasters have occurred in Belgium and Mexico.

These air pollution crises are correctly termed smogs in that appreciable concentrations of both *sm*oke and f*og* must be present. They usually occur at night. The major constituents of London smog are particles, SO_2, and small amounts of CO and NO_x. (The subscript x in the nitrogen oxide compound indicates that a variety of oxides, especially NO and NO_2, are present.) This combination is both grimy and deadly. Deaths are due mainly to the conversion of SO_2 to sulfuric acid in the lungs. This is an especially serious situation for persons already suffering from asthma, bronchitis, emphysema, or any other respiratory ailment. Particles clog the tiny chambers (alveoli) of the lungs, where oxygen exchange should occur, and CO and NO_x reduce the oxygen-carrying capacity of the blood.

The major source of pollutants in London smog is the combustion of fossil fuels. Tables 3 and 4 give the composition of 1 ton of hard coal and the products of its combustion. The improbable result of obtaining 14 tons of products from the combustion of 1 ton of coal is due mainly to the 10 tons of nitrogen that pass through the furnace along with the oxygen. Except for the engineering problem of having to heat up all that nitrogen, most of it passes through unchanged.

Table 3. **Composition of 1 Ton of Hard Coal.**

Constituent	Quantity, lb
C	1840
H	60
S	60
Minerals	40
	2000

Table 4. **Products of Combustion of 1 Ton of Hard Coal.**

Reaction	Quantity of Products, lb
$C + O_2 \rightarrow CO_2$	6700
$4H + O_2 \rightarrow 2H_2O$	480
$S + O_2 \rightarrow SO_2$	120
$N_2 + O_2 \xrightarrow{1800\ K} 2NO$	28
Minerals $\rightarrow$ smoke + particles	20
$N_2 \rightarrow N_2$	20,000
2000 lb coal $\rightarrow$	27,348 lb products

Three of the products—CO_2, H_2O, and N_2—are either harmless or objects of little concern at present. The other three are the problem.

Sulfur Dioxide

Surprisingly, man is not the chief contributor of sulfur to the atmosphere. Nature puts 80% of the sulfur into the air through such natural processes as the decay of vegetable matter and the evaporation of sea spray. Much of this sulfur is released as H_2S, which is later converted to SO_2. Man's 20% contribution is due almost solely to the combustion of fossil fuels, and is usually confined to the areas near cities. This is basic to the problem.

There are several possible fates for this SO_2. One is the conversion to sulfuric acid:

$$SO_2 + \tfrac{1}{2}O_2 \xrightarrow{\text{Catalysts}} SO_3$$

$$SO_3 + H_2O \longrightarrow H_2SO_4$$

A *catalyst* is any substance that speeds up a chemical reaction without itself being appreciably altered by the reaction. The catalysts for the atmospheric oxidation of SO_2 to SO_3 are believed to be NO_2, hydrocarbons, and sunlight. The end product, sulfuric acid, is not good for living things—we have seen how it affects humans. Because it is an acid and is removed from the air by rainfall, lakes and streams are becoming more acidic in severely polluted regions of the world. Some streams in Sweden no longer support salmon runs because of their high acidity, caused by pollutants that drift north from the heavy manufacturing districts of central Europe.

Sulfur dioxide has the distinction of being the only major gaseous air pollutant that is not involved in the formation of photochemical smog. It "does its own thing."

One of the most promising ways to control SO_2 emissions is being investigated by Professor Leroy Bromley of Berkeley. His method is to spray seawater into the stacks of fuel-burning plants. The minerals in the water oxidize the SO_2 to SO_3, which dissolves to give H_2SO_4. The sulfuric acid can be recovered and sold to pay at least part of the cost of the additional equipment.

Particles

Particulate matter in the atmosphere (smoke, ash, grit, aerosols, etc.) is the most poorly understood part of air pollution. These particles are aggregates of both liquids and solids. They are very small, with diameters ranging from 0.001 μm (1 μm = 1 micrometer = 10^{-6} m) to 10 μm. To compound the difficulty, both their composition and size change erratically.

The surfaces of particles are the sites of many of the chemical reactions that produce air pollution. It is estimated that there are as many as 1 million particles per cubic centimeter on a smoggy day. Many of these particles can be removed from flue gases by placing electrostatic precipitators on industrial smoke stacks. However, a recent Caltech study has revealed that 25% of the particles are formed in the air. There would be a haze over many cities even if all flue gases were cleansed of particulate matter.

Nitrogen Oxides

At the end of the preceding chapter, eight compounds of nitrogen are listed. Two of them, plus N_2 itself, are shown in color, indicating that they are important in air pollution. The presence of nitrogen in pure air and the formation of NO from it were indicated earlier (Tables 1 and 4). Most NO in the atmosphere is manmade but, although it is rather toxic, it is not a serious problem in itself. The problem is that it is converted to NO_2, which is highly toxic and initiates the reactions leading to photochemical smog. The conversion reaction is

$$NO + \tfrac{1}{2}O_2 \rightarrow NO_2$$

This reaction has puzzled chemists for many years. At the concentrations of NO and at the temperatures found in automobile exhausts, the rate at which this reaction proceeds should be immeasurably slow. The fact that there *is* a lot of NO_2 in our air may be explainable on the basis of still other pollutants, which we will encounter in the next section.

As with the sulfur oxides, man is not solely responsible for the nitrogen oxides in the atmosphere. Global balances indicate that there may be biological sources as well. The problem (for people) again is that we put it into the air where we live.

One of the nitrogen oxides we have not mentioned is N_2O, nitrous oxide, also known as laughing gas. This compound is *not* manmade and is rather inert and innocuous. If we had our choice among the several nitrogen oxides, N_2O would surely be the preferable one.

PHOTOCHEMICAL SMOG

Photochemical smog is not a real smog, in the sense that, if large concentrations of either smoke or fog *are* present, this type of smog cannot form.

Table 5. **Composition of the Exhaust of an Internal Combustion Engine (Gasoline Fuel).**

Constituent	Volume %
N_2	71
H_2O	13
CO_2	10
CO	5
NO	0.0001–0.3
Unburned hydrocarbons	0.05–0.4

In contrast to London smog, it is found mainly in the daytime and in the summer. This suggests that direct sunlight and relatively high temperatures are required for its formation.

There is extensive documentary evidence that this form of smog has a devastating effect on Ponderosa pine trees, grape vines, and citrus trees. Direct medical effects on humans are harder to demonstrate. Nevertheless, in 1969, 60 faculty members of the medical school of UCLA signed a statement saying that air pollution had become such a severe problem in the Los Angeles area that all residents without compelling reasons for staying should leave the area in order to avoid chronic respiratory diseases.

The major constituents of photochemical smog are NO_2, O_3, hydrocarbons, and PAN (the correct answer was peroxyacyl nitrate). The major source of these pollutants is the internal combustion engine. Table 5 gives the typical composition of automobile exhaust gases.

We again find some nitric oxide, as with the burning of coal. Its production by an internal combustion engine can be regulated somewhat by adjusting the engine. If less air is introduced into the cylinder and the gas-air mixture is ignited at a lower temperature, less NO will result. Unfortunately, these two adjustments result in an increase in two other pollutants listed in Table 5, carbon monoxide and hydrocarbons.

Carbon Monoxide

Virtually all atmospheric CO is manmade. It arises from the incomplete combustion of carbon compounds:

$$C + \tfrac{1}{2}O_2 \rightarrow CO$$

Carbon monoxide is unpleasant for a variety of reasons. It is a systemic poison, competing with oxygen for coordination with the heme group of the hemoglobin molecule (see Chapter 11). Since it binds to the heme much more strongly than does oxygen, it eventually leads to suffocation. Furthermore, carbon monoxide may be involved in the production of PAN: in 1970 Dr. Donald Stedman, then of the Ford Motor Company, showed that CO may partially explain the mystery of the rapid oxidation of NO to NO_2. He worked out a series of reactions whose sum is

$$NO + CO + O_2 \rightarrow NO_2 + CO_2$$

It seems likely, however, that the compounds most responsible for speeding up the conversion of NO to NO_2, by a complex and not yet fully understood mechanism, are hydrocarbons.

Hydrocarbons

Gasoline consists of a mixture of hydrocarbons having formulas ranging from C_4H_{10} to $C_{14}H_{30}$. Certain amounts of these substances are not burned in the engine but are released unchanged in the exhaust. In addition, hydrocarbons are released in the processing and handling of petroleum and its products.

The principal natural sources of hydrocarbons are trees and vegetation. They release molecules known as *terpenes,* which are a special class of hydrocarbons having 40 carbon atoms or less and having several double bonds. One example is *limonene,* from lemon oil:

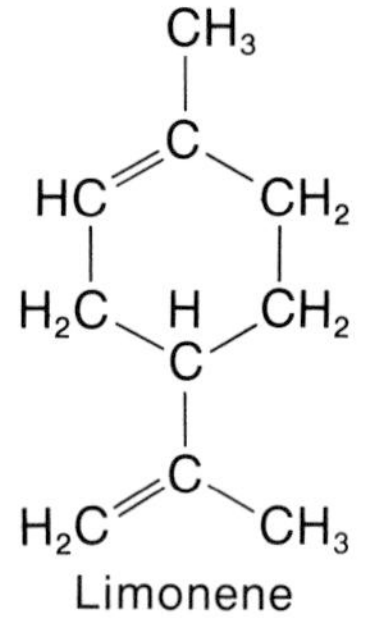

```
          CH3
          |
          C
       //   \
     HC       CH2
     |         |
    H2C   H   CH2
       \  |  /
          C
          |
          C
       //   \
    H2C       CH3
```

Limonene

PAN

These initials stand for *peroxyacyl nitrate,* which is not a single compound, but a class of compounds. The prefix *peroxy* denotes an —O—O— linkage and the last part of the name is familiar. The term *acyl* denotes an $R{-}C{\lesssim}^{O}$ group, e.g., the acetyl group, in which $R = CH_3$. PAN is the agent primarily responsible for the extensive damage to plants and trees observed in southern California. The details of the chemical reactions leading to its formation are still rather obscure, but the essential ingredients are known: hydrocarbons having double bonds, NO_2, lots of sunshine, and a giant "reactor." The latter is some combination of topographical and meteorological factors leading to a temperature inversion. A proposed sequence of reactions is given below. The dot following certain species denotes an

$$NO_2 \xrightarrow{h\nu} NO + O$$

$$O + O_2 \rightarrow \underset{\text{Ozone}}{O_3}$$

$$O_3 + \underset{\text{Propene}}{CH_3{-}CH{=}CH_2} \rightarrow \underset{\text{An ozonide}}{CH_3{-}CH(O{-}O{-}O)CH_2}$$

$$CH_3{-}CH(O{-}O{-}O)CH_2 \rightarrow \underset{\text{Formaldehyde}}{H_2C{=}O} + \underset{\text{Acetyl free radical}}{CH_3{-}\overset{O}{\overset{\|}{C}}\cdot} + \underset{\text{Hydroxyl free radical}}{HO\cdot}$$

$$CH_3{-}\overset{O}{\overset{\|}{C}}\cdot + O_2 \rightarrow \underset{\text{Peroxyacetyl free radical}}{CH_3{-}\overset{O}{\overset{\|}{C}}{-}O{-}O\cdot}$$

$$CH_3{-}\overset{O}{\overset{\|}{C}}{-}O{-}O\cdot + \cdot NO_2 \rightarrow \underset{\text{Peroxyacetyl nitrate (PAN)}}{CH_3{-}\overset{O}{\overset{\|}{C}}{-}O{-}O{-}NO_2}$$

unpaired electron. Such species are called *free radicals* and are highly reactive.

Air pollution is everybody's business. The solution to the problem lies with teams of chemists, meteorologists, sociologists, economists, engineers, governmental planners, and others working with an informed and concerned public, which must be willing to accept fundamental changes in its patterns of living.

ANSWERS TO THE SMOG QUIZ

1. c. Peroxyacyl nitrate.

2. b. PbBrCl. The halogens are added to gasoline to scavenge the lead.

3. a. CO_2: raise. The Earth normally radiates large amounts of thermal energy back into space. CO_2 molecules vibrate at a frequency that absorbs this energy, thereby trapping it and raising the temperature of the Earth's atmosphere.

b. Particles: lower. Increasing particulate matter blocks the passage of sunlight to the Earth, thus lowering the temperature.

Until about 1940 the temperature of the Earth rose steadily, perhaps owing to increasing concentrations of CO_2 in the atmosphere. Since then the temperature has slowly but steadily been dropping, perhaps reflecting an increasing amount of dirt in our air.

4. False. About 85% comes from natural sources, primarily forests, vegetation, and the bacterial decomposition of organic matter.

5. The Top Ten cities are all east of or on the Mississippi River. They are New York City, Huntington, W.Va., Philadelphia, Pittsburgh, Cleveland and St. Louis (tie), Washington, D.C., Detroit, and Providence.

"Looking down the ecological road ten years hence, let us keep in mind what our astronauts have let us verify for ourselves: We really *do* live on a relatively small spacecraft whirling through space. We always knew that our astronauts would perish in space if they exhausted their life-support systems, but we did not evidently heretofore fully comprehend that we are all in space and that if we exhaust the life-support systems of our planet spacecraft, we will all perish."

Samuel Yorty
Former Mayor of Los Angeles
April 3, 1970

4 the chemical bond

A molecule is a well-defined collection of atoms that are attracted to each other such that the whole collection may be thought of as a single unit. The attractive interaction between two atoms is called a *chemical bond.* Clearly the chemical bond must be something more specific than a simple attraction between atoms or we would have to refer to all the water in a glass as a single molecule, or to a table top as a single molecule. This specific property is really only one of degree, but generally if the attraction between two atoms is such that an energy of at least 10 kcal/mole* is required to move the two atoms an infinite distance apart, a chemical bond exists. The energy required to "break" the bond is called the *bond energy.* A special bond called the *hydrogen bond* (bond energy 3–6 kcal/mole) is discussed in this chapter even though single hydrogen bonds are often not strong enough to confine otherwise separate atoms to the same molecule. They represent a limiting case requiring special attention.

The idea of a chemical bond is an important and logical hypothesis that is overwhelmingly supported by experimental evidence. The assumption that a molecule of many atoms is a structure connected by bonds, each joining two atoms, is an oversimplification, although it is usually justified.

---

*One kcal (kilocalorie) is 1000 calories. One calorie is the energy required to increase the temperature of one gram of water from 14.5°C to 15.5°C.

For example, consider the water molecule, H_2O. The chemist designates the structure of the water molecule by

$$\mathrm{H}\diagup\overset{\mathrm{O}}{}\diagdown\mathrm{H}$$

where there is a single chemical bond between the oxygen atom and each of the hydrogen atoms. Yet there is no bond between the two hydrogen atoms. We must answer the following questions: What is a chemical bond? Why do chemical bonds form between atoms?

In principle, the Schrödinger equation and the postulates of quantum mechanics are all that is required to calculate all the properties of any molecule. In practice, however, the exact solution of the Schrödinger equation for complicated molecules has not been achieved, and experiments are required to determine the structure and behavior of molecules. Approximate methods of solution lead us to some important conclusions, however.

The major conclusions of the quantum mechanical analysis are generally in agreement with experimental observations:

1. Electrons constituting a bond are localized to the region of the bond and the two atoms it connects, except in special cases that will be discussed later. Shifting of electrons between bonds is not significant. The poor electrical conductivity of nonmetallic crystals such as diamond is an experimental demonstration of this fact. It is this localization of electrons that enables us to think of a chemical bond as being something real.
2. Electrons in bonds are described by *molecular orbitals*. These orbitals are one-electron wave functions that are approximated by weighted sums of atomic orbitals centered at the two atoms of interest. Such molecular orbitals (MO's) are extremely useful in developing a descriptive picture of the nature of molecules. It must be remembered that these sums, or linear (in the mathematical sense) combinations of atomic orbitals (LCAO's), are merely approximations, not exact representations.

4.1 THE COVALENT BOND

Let us first consider the simplest model for a covalent chemical bond. (All covalent bonds consist of two electrons acting as a pair.) The simplest stable molecule is the hydrogen molecule, for it contains only two protons and two electrons.

If we start with two hydrogen atoms an infinite distance apart, we can say that electron 1 is in a 1*s* state around proton *a* and electron 2 is in a 1*s* state around proton *b*. Our shorthand for this statement is

$$\psi_1 = 1s_a \qquad \psi_2 = 1s_b \tag{4-1}$$

where ψ_1 and ψ_2 represent the electron wave functions (atomic orbitals).* As the distance between the two atoms decreases, the electrostatic interactions between each electron and the nucleus of the other atom, as well as between the two electrons and between the two nuclei, become increasingly important. Eventually a chemical bond is formed.

An important feature of the interaction between the two hydrogen atoms concerns the identities of the electrons. In the discussion above we labeled the electrons, saying that one belonged to proton *a* and the other to proton *b*. However, these two electrons are indistinguishable once the bond has formed: there is no experiment that can be done to tell one from the other. We are therefore unable to say which is which and to which proton each electron "belongs."

We now use a linear combination of the 1*s* atomic orbitals to "construct" a molecular orbital for which each electron† is equally likely to be in the neighborhood of either proton. There are two, and only two, possible combinations that satisfy the condition that each electron be associated equally with each nucleus. Figure 4-1 shows plots of the molecular orbitals obtained by taking both the sum and the difference of the two 1*s* atomic orbitals. The figure also shows the electron probability distributions for these two molecular orbitals, and the corresponding electron density contour diagrams. These plots all apply to a single electron in the molecule, i.e., to the hydrogen molecule-ion H_2^+, but they would not look appreciably different for the neutral molecule H_2.

Figure 4-1a shows the addition of the two atomic orbitals when they are in phase (they are therefore both positive at the same time). As was stated in Chapter 2, such addition leads to constructive interference between the two wave functions in the region between the two nuclei and results in increased electron density in this region. The + signs and the one − sign in the figure refer to the signs of the wave functions, as in Figure 2-24. This orbital is called the *bonding molecular orbital.*

---

*Although we used capital Ψ to denote atomic orbitals in Chapter 2 (see Equation 2-19), we now use lower-case ψ and reserve the capital letter for molecular orbitals.

†Any atomic or molecular orbital (one-electron wave function) can be regarded as representing the energy state of *two* electrons. These electrons must be identical except for their spins, which must be antiparallel. Thus, the electrons have identical *n*, *l*, and *m* quantum numbers but different *s* quantum numbers. This is in accord with the Pauli exclusion principle, discussed in Chapter 2.

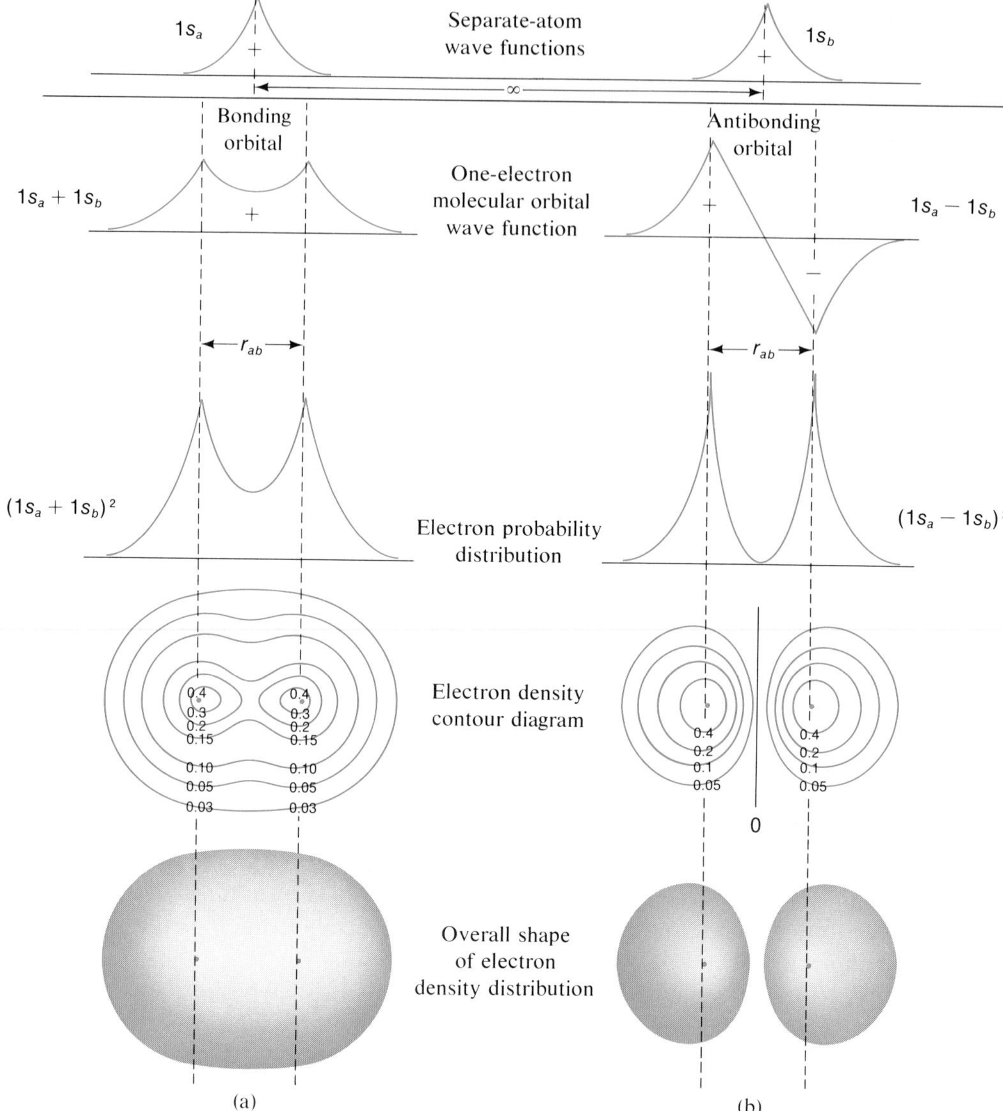

Figure 4-1. The atomic orbitals of separate hydrogen atoms and two trial one-electron molecular orbitals. The molecular diagrams apply to a *single* electron in the hydrogen molecule, i.e., they describe the species H_2^+. (a) Addition of the atomic wave functions gives constructive interference (considerable electron density) between the nuclei. (b) Subtraction of the atomic wave functions gives destructive interference (near zero electron density) between the nuclei.

Figure 4-1b shows the addition of the two atomic orbitals when they are out of phase (one is positive and one is negative). In this case there is destructive interference between the two wave functions in the region between the two nuclei, and the electron density is very low in this region (it is zero at the node). This orbital is called the *antibonding molecular orbital.*

The Schrödinger equation can now be used to calculate the energy E of the hydrogen molecule with its electrons described by the two molecular orbitals introduced in Figure 4-1. This energy is a function of the distance between the nuclei, r_{ab}. Figure 4-2 is a plot of E versus r_{ab} for the molecular orbitals $1s_a + 1s_b$ and $1s_a - 1s_b$. The arbitrary zero of energy is taken to be the energy of two hydrogen atoms infinitely far apart. The orbital $1s_a + 1s_b$ is called the bonding orbital because it predicts that a molecule in this state will have an energy less than that of the dissociated atoms, i.e., it will be a stable entity. The molecule has a *dissociation energy* E_0 and a bond length r_0. The orbital $1s_a - 1s_b$ is called the antibonding orbital because it predicts higher energies than that of the dissociated atoms for all values of r_{ab}. Therefore no stable molecule can exist in this state. All bonding molecular orbitals that are cylindrically symmetric about the center line joining the two bonded nuclei are called σ orbitals; all antibonding orbitals with the same cylindrical symmetry are called σ^* orbitals. One can see from Figures 4-1 and 4-2 that the bonding orbital $1s_a + 1s_b$, for which the system has a large negative potential energy, predicts a high electron density between the two protons. Here the electron(s) serves as the bonding agent. On the other hand, the antibonding orbital $1s_a - 1s_b$ predicts very little electron density between the two protons and the atoms therefore repel each other.

The calculations for each curve in Figure 4-2 were done with both elec-

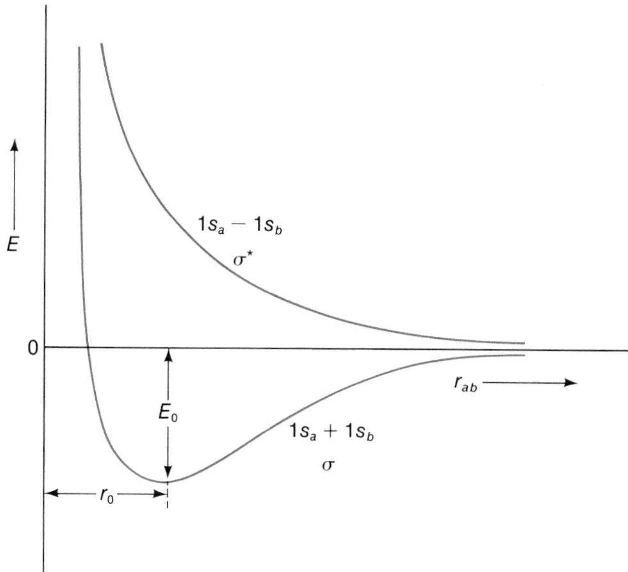

Figure 4-2. The energy of a neutral hydrogen molecule as a function of internuclear distance, with both electrons in either the bonding or the antibonding orbital.

trons in the same orbital. We know from the Pauli exclusion principle that no more than two electrons can "occupy" the same orbital. If one electron is in the bonding orbital and the other in the antibonding orbital, the net effect is a slight repulsion between the atoms, since antibonding orbitals are generally slightly more repulsive than bonding orbitals are attractive. The ground state of the hydrogen molecule is thus $(1\sigma)^2$, where the superscript denotes two electrons. There are excited states of the hydrogen molecule other than the $1\sigma^*$ state we have already discussed; some of these will be considered later.

An alternative way to view the two lowest energy levels in the hydrogen molecule is to plot the energies of the 1σ and $1\sigma^*$ states at the equilibrium distance r_0, as shown in Figure 4-3. (The atomic and molecular orbitals are shown schematically as circles, and arrows are used to denote electron occupancy.) The bonding orbital is more stable by the amount E_0 than the energy level of the separate atoms, whose atomic orbitals are shown on the left and right. The antibonding orbital has an energy somewhat greater than E_0 *above* the energy level of the separate atoms.

Let us now review our conclusions. The covalent bond is two electrons in a molecular orbital that predicts a high electron density in the region between the two nuclei that are bonded. Most covalent bonds are σ bonds, which have an electron probability density that is cylindrically symmetric about the center line between the two nuclei. The two electrons constituting the chemical bond occupy the valence orbitals of both atoms to a significant

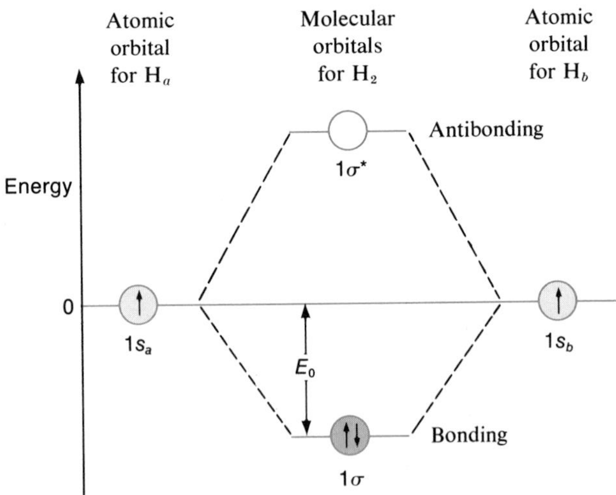

Figure 4-3. Comparison of the energies of atomic orbitals with those of bonding and antibonding molecular orbitals, for hydrogen.

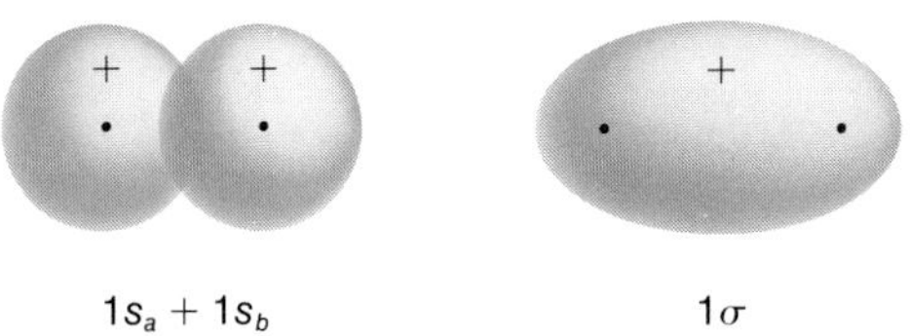

Figure 4-4. Conceptual representation of the 1σ molecular orbital in hydrogen.

extent; hence the word *covalent*. Molecular orbitals may be constructed by the method of linear combination of atomic orbitals, a mathematical technique that is very helpful in presenting a descriptive picture of the geometry and symmetry of the valence electron distributions in molecules. These molecular orbitals are rather imprecise in their ability to predict bond energies and other molecular properties quantitatively. In their simplest form they function as a graphic representation of the electron distribution in a bond. Figure 4-4 provides a conceptual picture of the overlap of two atomic orbitals and the resulting molecular orbital. The + signs signify that both wave functions are positive, and their addition is therefore constructive.

Even in polyatomic molecules, the electrons in σ bonds are localized largely to those bonds. It is thus possible to speak of the properties of covalent bonds, such as their dissociation energy E_0 and their length r_0, in more complicated molecules as well.

4.2 THE MOLECULAR ORBITALS OF HOMONUCLEAR DIATOMIC MOLECULES

We saw in the last section that a covalent bond results from the hypothetical overlap of two atomic orbitals. Although the addition of atomic orbitals does not provide a precise wave function, it is done because it is the most convenient method of describing bond formation. A similar technique is used for diatomic molecules containing many valence electrons.

Two $2s$ electron wave functions and two $2p$ electron wave functions can overlap as shown in Figure 4-5. The atomic s orbitals overlap to give molecular σ orbitals, which are cylindrically symmetric about the axis joining the nuclei, as we noted above for the $1s$ orbitals. Two possibilities exist, however, for the p orbitals. If two p orbitals that are perpendicular to the internuclear axis overlap side-by-side, as shown in the fifth and sixth rows, a bonding and an antibonding orbital that are *not* symmetric about the internuclear axis are generated. Rotation of either the 2π or the $2\pi^*$ orbital by 180° about this axis yields a wave function of identical shape but opposite

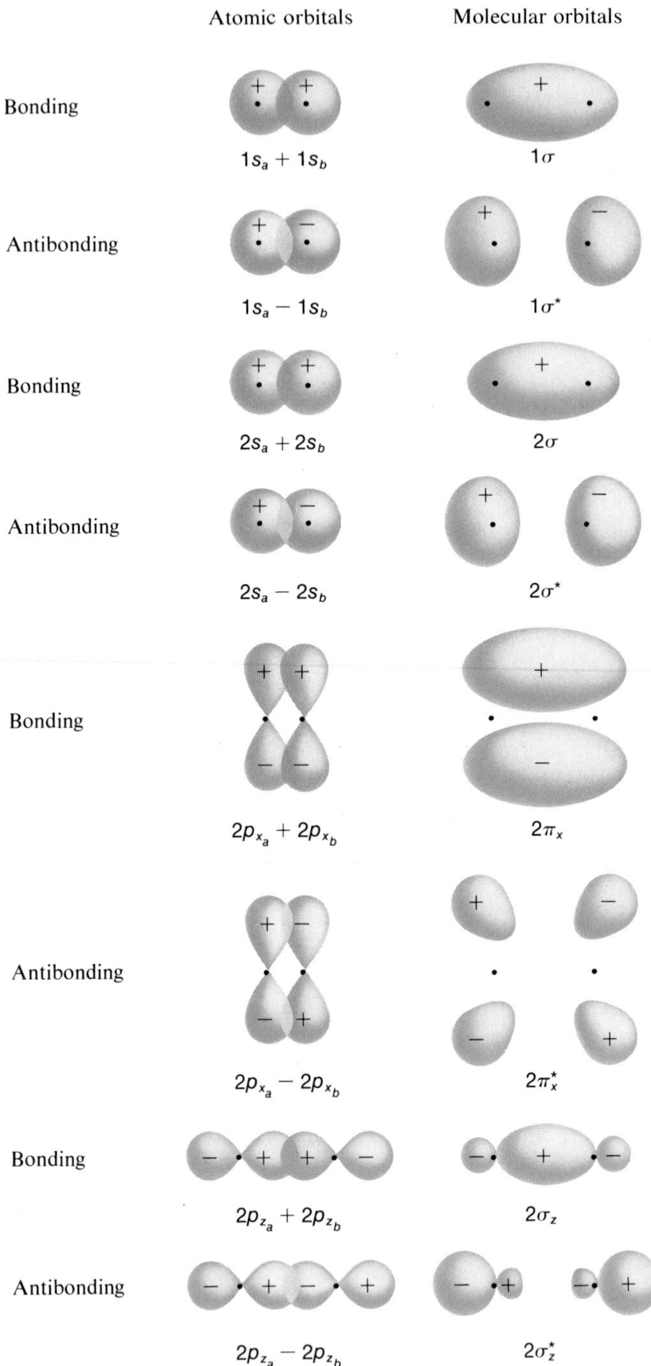

Figure 4-5. The interactions of some simple atomic orbitals and the resultant molecular orbitals.

sign. These molecular orbitals are called π orbitals to denote this property. If the p orbitals overlap in an end-to-end fashion, as shown in the last two rows of Figure 4-5, cylindrically symmetric bonding and antibonding orbitals called σ_z and $\sigma_z{}^*$ molecular orbitals are generated. The z is appended to the σ because the internuclear axis is generally selected as the z-axis in a Cartesian coordinate system.

It is observed experimentally that the energies of these molecular orbitals increase in the following order:

$$2\sigma < 2\sigma^* < 2\pi_x = 2\pi_y < 2\sigma_z < 2\pi_x{}^* = 2\pi_y{}^* < 2\sigma_z{}^*$$

Each of these states can account for two electrons of opposite spin.

Figure 4-6 is an extension of Figure 4-3 to show the relative energies of all the molecular orbitals resulting from combinations of the $1s$, $2s$, and $2p$ atomic orbitals. This diagram shows the electron occupancy of the atomic orbitals for oxygen atoms; the molecular orbitals display the electron configuration of O_2. The diagram is useful for predicting the electron configurations of all the second-row homonuclear diatomic molecules, as shown in Table 4-1. The individual molecular orbitals are invoked in increasing order of their energy to account for the valence electrons of the molecules. Note that both the Pauli exclusion principle and Hund's rule are obeyed in this scheme.

The quantity called *bond order* in Table 4-1 is equal to the number of electrons in bonding orbitals minus the number of electrons in antibonding orbitals, all divided by two. It therefore equals the number of covalent bonds in the molecule. For example, F_2 has a single bond, whereas N_2 has a triple bond. Note that the bond length decreases and the bond energy increases with increasing bond order. A molecule or atom containing one or more unpaired electrons is paramagnetic, which means that it is attracted by a magnetic field, with a force that increases in proportion to the field strength. Molecular orbital theory predicts that oxygen gas is paramagnetic (see Figure 4-6 and Table 4-1) and, indeed, it is. We will see later that a simple Lewis electron dot structure cannot be used to *predict* this fact, although it can easily depict it.

4.3 THE IONIC BOND

There is another type of bond that is very common in chemistry: it is the bond formed when two oppositely charged ions attract each other. This *ionic bond* is most common in the crystalline and liquid states of salts but is also important in the vapors of salts and in the three-dimensional structure

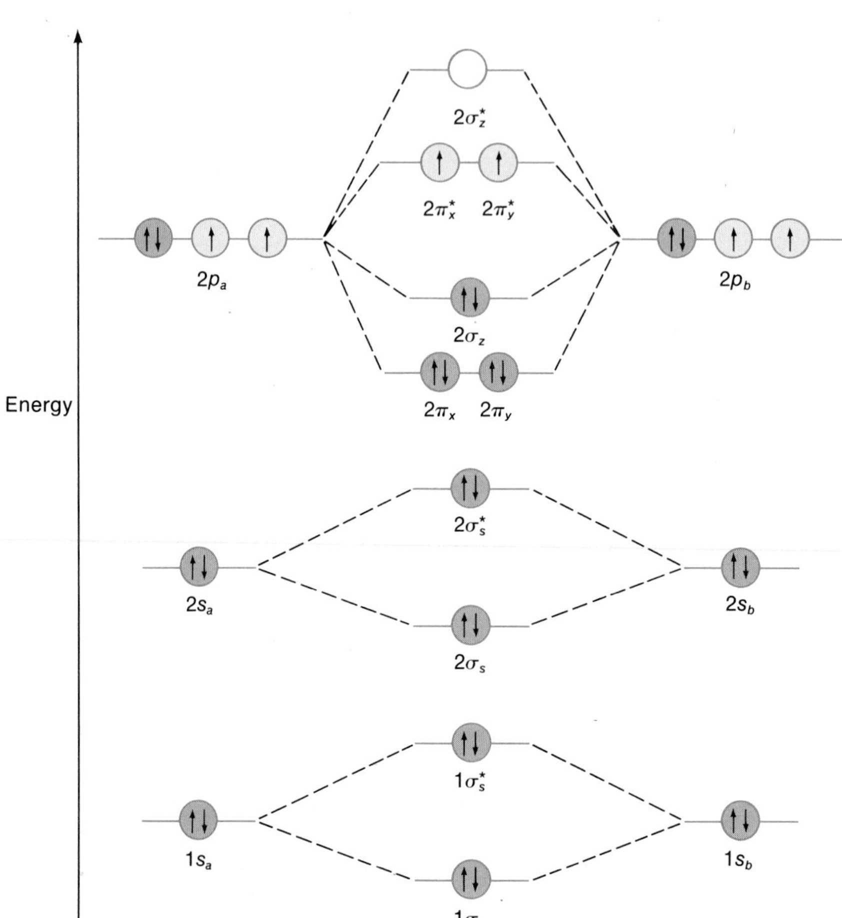

Figure 4-6. Relative energy levels of the atomic orbitals for separate oxygen atoms and the molecular orbitals for O_2.

of proteins. The attraction between a positive ion and a negative ion is quantitatively described by Coulomb's law when the ions are separated by relatively large distances. As the ions get so close to each other that their electron clouds overlap, however, repulsive forces between the electrons and between the nuclei become important and, eventually, dominant.

A plot of the potential energy of two oppositely charged ions as a function of internuclear distance r_{ab} is shown in Figure 4-7, in which the arbitrary zero of energy is taken to be the energy of the two ions infinitely far apart. The potential energy curve for an ionic bond has a minimum, just as with the

Table 4-1. Bonding in Gas-phase Diatomic Molecules.

		Unpaired electrons	Bond order	Bond energy, kcal/mole	Bond length, Å
Li_2	$(2\sigma)^2$	0	1	25	2.67
Be_2	$(2\sigma)^2(2\sigma^*)^2$ (Unstable)	0	0	—	—
B_2	$(2\sigma)^2(2\sigma^*)^2(2\pi_x)^1(2\pi_y)^1$	2	1	$\approx$67	1.59
C_2	$(2\sigma)^2(2\sigma^*)^2(2\pi_x)^2(2\pi_y)^2$	0	2	144	1.24
N_2	$(2\sigma)^2(2\sigma^*)^2(2\pi_x)^2(2\pi_y)^2(2\sigma_z)^2$	0	3	227	1.10
O_2	$(2\sigma)^2(2\sigma^*)^2(2\pi_x)^2(2\pi_y)^2(2\sigma_z)^2(2\pi_x^*)^1(2\pi_y^*)^1$	2	2	119	1.21
F_2	$(2\sigma)^2(2\sigma^*)^2(2\pi_x)^2(2\pi_y)^2(2\sigma_z)^2(2\pi_x^*)^2(2\pi_y^*)^2$	0	1	38	1.42
Ne_2	$(2\sigma)^2(2\sigma^*)^2(2\pi_x)^2(2\pi_y)^2(2\sigma_z)^2(2\pi_x^*)^2(2\pi_y^*)^2(2\sigma_z^*)^2$	0	0	Does not exist	

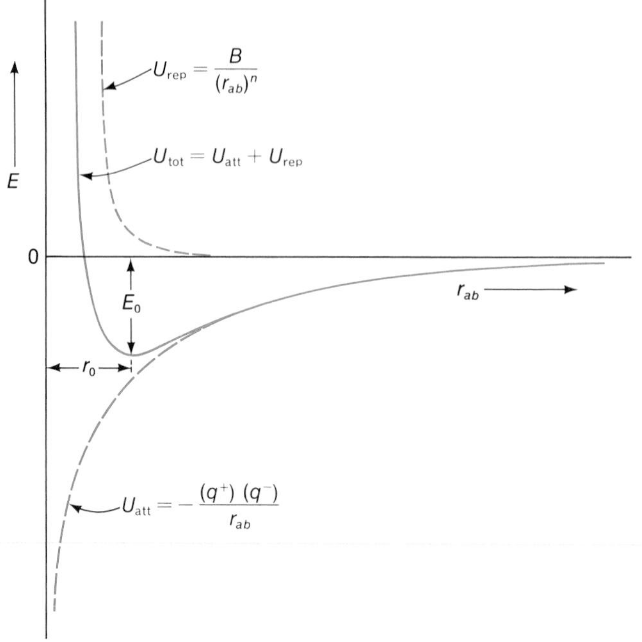

Figure 4-7. The potential energy of two oppositely charged ions as a function of internuclear distance. At large distances the attractive Coulomb potential dominates; at small distances the interionic repulsive potential dominates. The position of minimum potential energy corresponds to the length of the ionic bond, r_0.

covalent bond. For very small internuclear distances the two ions repel each other strongly, for large distances they attract each other weakly, and somewhere in between there is an equilibrium distance for which the attractive force is at a maximum. This distance, the ionic bond length r_0, corresponds to the minimum energy E_0 in the potential energy curve. The major feature that distinguishes the ionic bond from the covalent bond is that the ionic bond is not directional, being equally likely to form in all directions. This is not true of the covalent bond.

In Figure 4-7 the attractive potential energy U_{att} is the negative of the energy required to move the opposite charges infinitely far apart if they start at some finite separation distance r_{ab}. The ions are regarded as point charges (q^+ and q^-) for simplicity. There is a strong, short-range repulsive force between ions when their atomic orbitals begin to overlap, i.e., when the ions are squeezed too closely together. The repulsive potential energy U_{rep} is the energy released when the ions move from some very small separation distance—less than r_0, say—to infinity. Each of these potential energy functions exists independently of the other, but they cannot *act* independently. The total potential energy of the system at any given separation distance r_{ab} is their sum: $U_{tot} = U_{att} + U_{rep}$. The exponent n in the expression for U_{rep} is usually observed to be between 9 and 12, so U_{rep} rises very rapidly at small internuclear distances. The factor B is a constant.

4.4 ELECTRONEGATIVITY AND DIPOLE MOMENTS

We have discussed the covalent bond and the ionic bond. The covalent bond forms between two atoms of similar electronegativities, and the ionic bond forms between ions from atoms with very different electronegativities. Most chemical bonds have properties intermediate between these two extremes. In most bonds the electrons are attracted more strongly by one atom than the other. The electron density in the bond is therefore shifted toward one of the nuclei. When this occurs the bond is said to have *partial ionic character* (it exhibits polarity).

Lithium bromide, LiBr, is a good example of this phenomenon. From Figure 3-9 we see that the radius of the lithium atom is 1.23 Å and the radius of the bromine atom is 1.14 Å; the atoms are nearly identical in size. If the electrons of the bond in the free (gaseous) LiBr molecule were shared equally, the molecule would have the same "radius" at the Li nucleus as at the Br nucleus. The actual chemical bond between these two atoms has a large partial ionic character because of the electronegativity difference between them: $x_{Br} - x_{Li} = 1.8$ (see Figure 3-7). From Figure 3-9 we see that the radius of the lithium ion, Li^+, is 0.68 Å and the radius of the bromide ion, Br^-, is 1.96 Å. So, if LiBr were 100% ionic, the "radius" of the molecule at the Br nucleus would be almost three times as large as that at the Li nucleus.

The actual LiBr molecule is intermediate between these two extremes. Figure 4-8 shows the nearly equal atomic radii of Li and Br, the vastly different ionic radii, and the asymmetric distribution of electrons in the polar bond of the free lithium bromide molecule. The symbol δ (Greek delta) denotes effective partial charge at a nucleus, and the arrow points in the direction of the negative charge accumulation in the molecule.

Electronegativities have been established as a relative measure of the elements' abilities—in molecules—to attract electrons from the other elements. Table 4-2 shows the approximate relation between differences in electronegativity and the partial ionic character of the bond. From the electronegativities of Li and Br, the bond in the LiBr molecule is estimated to be 55% ionic.

A polar bond has a *bond dipole moment* associated with it because of the charge separation in the bond. If a positive charge $+q$ is separated from a

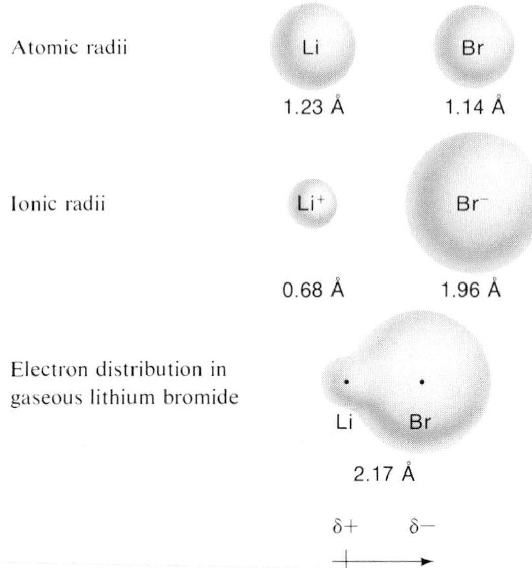

Figure 4-8. Distribution of the bonding electrons in LiBr.

negative charge $-q$ by a distance d, as shown in Figure 4-9, the *dipole moment* μ of this charge distribution is defined as qd. The dipole moment of a diatomic molecule is directed (by arbitrary convention) toward the negative charge, parallel to the bond. It is usually, but not always, large if the electronegativities of the bonded atoms differ greatly; it is zero if the two atoms are identical. Table 4-3 contains the dipole moments of three of the hydrogen

Table 4-2. **Relation Between Electronegativity Difference and Partial Ionic Character in a Chemical Bond.**

$x_A - x_B$	% Ionic character	$x_A - x_B$	% Ionic character
0	0	1.8	55
0.2	1	2.0	63
0.4	4	2.2	70
0.6	9	2.4	76
0.8	15	2.6	82
1.0	22	2.8	86
1.2	30	3.0	89
1.4	39	3.2	92
1.6	47		

SOURCE: L. Pauling, *General Chemistry*, 3rd ed., W. H. Freeman and Company. Copyright © 1970.

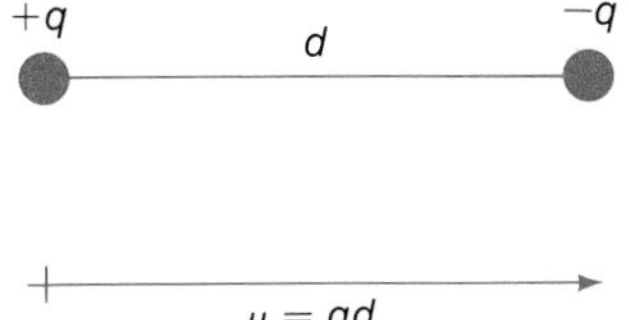

Figure 4-9. Representation of a dipole moment arising from charges $+q$ and $-q$ separated by distance d. The dipole moment $\mu = qd$ is represented by the arrow pointing from the positive end of the dipole to the negative end.

halide molecules. A comparison of these numbers with the corresponding electronegativity differences verifies the connection between the two. This simple relation is generally invalid, however, in molecules containing multiple bonds. The conventional unit of dipole moment is the debye (D), named after the Dutch physicist Pieter Debye (1884–1966); in terms of the electrostatic unit of charge, $1\ D = 10^{-18}$ esu cm. Bond dipole moments are *vectors,* i.e., quantities that have a specified direction as well as a magnitude. The dipole moment of a polyatomic molecule is the vector sum of all the bond dipole moments.

It is possible to calculate the partial ionic character of a bond from the bond dipole moment if it is assumed that the pure covalent bond has zero ionic character (not quite true) and zero dipole moment (true). For example, the bond length in HCl is 1.27 Å. If the bond were completely ionic, its dipole moment would be

$$\mu = ed = (4.80 \times 10^{-10})\,(1.27 \times 10^{-8}) = 6.10 \times 10^{-18} \text{ esu cm} = 6.10 \text{ D}$$

where e is the electron charge in electrostatic units. The observed dipole moment is 1.08 D, so the value of the partial ionic character is 1.08/6.10 = 18%.

Values of partial ionic character calculated by this method are generally in reasonable agreement with the values predicted from electronegativity differences (see Table 4-2). There are some cases for which the agreement is poor, however, because electronegativity and its relation to bond polarity are qualitatively useful but not quantitatively precise.

Table 4-3. Some Bond Dipole Moments.

Bond $\overset{+\!\longrightarrow}{\text{A—B}}$	Electronegativity difference $x_A - x_B$	Bond dipole moment, D
H—I	−0.4	0.44
H—Br	−0.7	0.82
H—Cl	−0.9	1.08

4.5 BOND ENERGIES AND BOND LENGTHS

The dissociation energies of a number of primarily covalent molecules are listed in Table 4-4. The general reaction considered is of the form

$$AB(g) \rightarrow A(g) + B(g)$$

The table contains some extremely large dissociation energies, such as those of N_2, O_2, NO, and CO. These molecules contain multiple bonds, the properties of which we will discuss later in this chapter. The single bonds represented in this table have energies between 11 and 104 kcal/mole, except for HF, which is a highly polar molecule. The large electronegativity difference between H and F gives the H—F bond a large value of partial ionic character, making it one of the strongest known single bonds in a diatomic molecule. Such large bond energies are a common property of very ionic bonds.

It is possible to measure the dissociation energies of bonds in polyatomic molecules as well. Systematic study of many compounds containing the same bond enables us to derive an average energy for this bond. Some of these *average bond energies* are shown in Table 4-5. The variation in a given bond energy from compound to compound is about $\pm 10\%$. Where there is bond multiplicity, the energy increases as the number of bonds between the atoms (bond order) increases. Note the difference between the dissociation energy of CO and the average bond energies for C—O and C=O in polyatomic molecules. These data suggest the existence of a triple bond in carbon monoxide.

Table 4-5 also contains the average bond lengths derived from experimental observations. For a given bond, the average value varies by about 4% from one compound to another. Bond lengths are generally small for high-energy bonds and large for low-energy bonds. This is strikingly exemplified by the carbon-carbon single, double, and triple bonds.

Table 4-4. **Dissociation Energies of Some Diatomic Molecules, kcal/mole.**

Li_2	25	H_2	104	N_2	227	LiH	57	HF	136
Na_2	18	F_2	38	O_2	119	NaH	48	HCl	103
K_2	13	Cl_2	58	NO	151	KH	44	HBr	87
Rb_2	12	Br_2	46	CO	257	RbH	40	HI	71
Cs_2	11	I_2	36			CsH	43		

Table 4-5. **Average Bond Energies and Bond Lengths for Bonds in Polyatomic Molecules.**

Bond	Bond energy, kcal/mole	Bond length, Å
C—C	83	1.54
C=C	146	1.33
C≡C	199	1.20
C—N	73	1.47
C=N	147	1.22
C≡N	213	1.16
C—O	85	1.40
C=O	170	1.18
N—O	53	1.36
N=O	145	1.21
C—H	98	1.09
C—F	110	1.41
C—Cl	80	1.76
O—H	111	0.96

4.6 LEWIS THEORY

We have already learned that the covalent chemical bond consists of two electrons that are shared by the valence shells of the bonded atoms. The bonding in molecules is usually such that the valence shells of all the atoms are filled. This is an especially valuable generalization for the elements in the first three rows of the periodic table. For example, the hydrogen atom has a complete valence shell when it has two electrons. The hydrogen atom therefore forms only one covalent bond to another atom. The electron dot formula for the hydrogen molecule, showing the completed valence shells of the two H atoms, is

$$H:H$$

Such formulas, of which this is the simplest, were invented by the great American chemist Gilbert N. Lewis (1875–1946).

The valence shells of the elements of the second row of the periodic table are complete with eight electrons, and these elements show a strong tendency to achieve this electron configuration through chemical reaction. This is closely related to the exceptional chemical nonreactivity of the noble gases, which is caused by the extreme stability of the octet configuration.

For Lewis structures the valence electrons are defined as the *s* and *p* electrons corresponding to the largest principal quantum numbers in the respective atoms. Therefore, Lewis structures are almost never written for compounds of the transition elements and inner transition elements, in which *d* electrons are important in the bonding. (In the actinon series, electrons of the 5*f* level also participate in bonding.) Lewis structures are especially useful for depicting the ground electronic states of the atoms of the second- and third-row elements. Table 4-6 shows these structures for the second-row elements.

Lewis's theory (1916), which is the basis of the modern electronic theory of valence, states that when atoms combine to form molecules, they tend to complete an octet of valence electrons by either transferring electrons or sharing electrons by forming electron pairs. Thus a typical ionic bond is formed by the electron-transfer reaction

$$\mathrm{Cs\cdot} + \cdot\ddot{\underset{..}{\mathrm{F}}}: \rightarrow \mathrm{Cs}^+ + :\ddot{\underset{..}{\mathrm{F}}}:^-$$

The typical covalent bond forms by the sharing of electrons such that both valence shells are filled, e.g.,

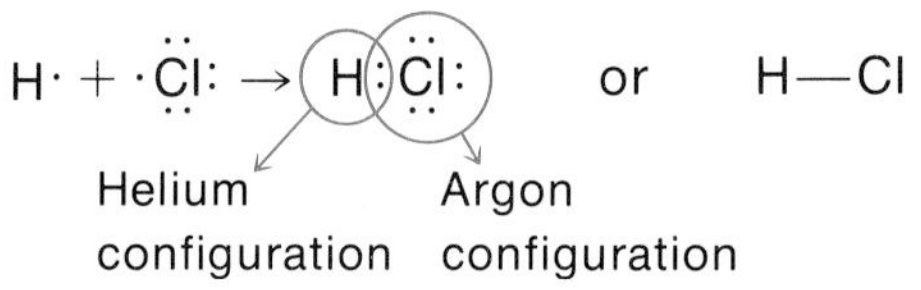

$$\mathrm{H\cdot} + \cdot\ddot{\underset{..}{\mathrm{Cl}}}: \rightarrow \mathrm{H}:\ddot{\underset{..}{\mathrm{Cl}}}: \quad \text{or} \quad \mathrm{H{-}Cl}$$

Helium configuration   Argon configuration

There are two classes of electrons in hydrogen chloride: bonding and nonbonding. The bonding electrons are those shown between the H and Cl atoms in the above formula. The electrons not participating in the chemical bond are the nonbonding electrons; they are also referred to as *n* electrons.

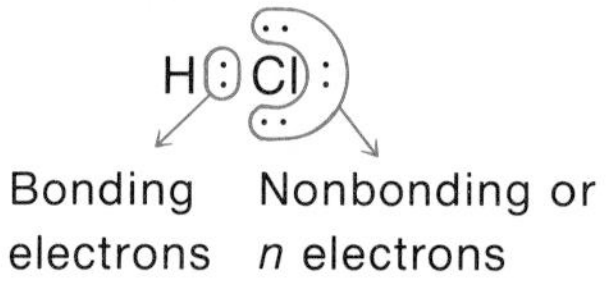

$$\mathrm{H}:\ddot{\underset{..}{\mathrm{Cl}}}:$$

Bonding electrons   Nonbonding or *n* electrons

Table 4-6. Lewis Structures for the Second-Row Elements.

Li	$(2s)^1$	$\cdot\dot{\underset{\cdot}{\mathrm{N}}}\cdot$	$(2s)^2(2p_x)^1(2p_y)^1(2p_z)^1$
$\ddot{\mathrm{Be}}$	$(2s)^2$	$\cdot\ddot{\underset{\cdot}{\mathrm{O}}}:$	$(2s)^2(2p_x)^2(2p_y)^1(2p_z)^1$
$\ddot{\mathrm{B}}\cdot$	$(2s)^2(2p_x)^1$	$\cdot\ddot{\underset{..}{\mathrm{F}}}:$	$(2s)^2(2p_x)^2(2p_y)^2(2p_z)^1$
$\ddot{\underset{\cdot}{\mathrm{C}}}\cdot$	$(2s)^2(2p_x)^1(2p_y)^1$	$:\ddot{\underset{..}{\mathrm{Ne}}}:$	$(2s)^2(2p_x)^2(2p_y)^2(2p_z)^2$

The rules for writing Lewis structures for simple molecules and ions are:

1. Write the name and chemical formula of the compound.
2. Decide whether the bonds are likely to be primarily ionic or covalent, on the basis of electronegativity differences or the positions of the elements in the periodic table. (Some compounds, such as Na_2SO_4, have both types of bonds.)
3. Draw a skeleton of the compound from previous knowledge or rational guesswork.
4. Count the total number of valence electrons.
5. Put these electrons into the skeleton, according to the following principles:
   a. the octet rule;
   b. all electrons must be paired, if possible;
   c. H, the alkali metal atoms, and the halogen atoms can form only one bond each;
   d. an O—O bond never forms except in O_2, O_3, and the relatively uncommon peroxides, superoxides, and ozonides.

Lewis structures are useful for representing the electron configurations in both reactions and molecular structures.

Reactions

Many simple chemical reactions result in the formation of electron pairs and the closed electron configurations of the noble gases. In the examples that follow, the covalent bond is represented as a straight line, as an alternative to the electron dot diagrams.

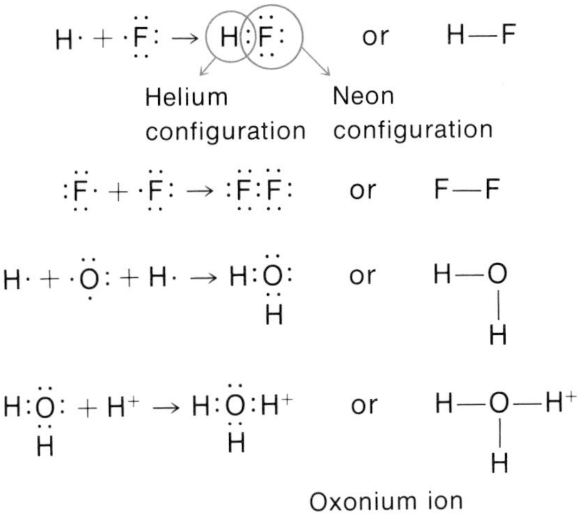

H· + ·F: → H:F: or H—F

Helium configuration (H) Neon configuration (F)

:F· + ·F: → :F:F: or F—F

H· + ·O: + H· → H:O: (with H below O) or H—O—H (second H below O)

H:O: (H below) + H^+ → H:O:H^+ (H below) or H—O—H^+ (H below O)

Oxonium ion

The bond formed in the last of these examples is called a *coordinate covalent bond* because both electrons forming the bond are donated by the same atom. The same kind of bond is made in the formation of the ammonium ion:

$$\mathrm{H{:}\underset{\cdot\cdot}{\overset{\mathrm{H}}{\overset{\cdot\cdot}{N}}}{:}} + \mathrm{H^+} \rightarrow \mathrm{H{:}\underset{\mathrm{H}}{\overset{\mathrm{H}}{\overset{\cdot\cdot}{\underset{\cdot\cdot}{N}}}}{:}H^+} \quad \text{or} \quad \mathrm{H{-}\underset{\mathrm{H}}{\underset{|}{\overset{\mathrm{H}}{\overset{|}{N}}}}{-}H^+}$$

Ammonium ion

The classification of this bond as distinct from any other type of covalent bond is rather artificial, because once the bond is formed, the origin of the electrons cannot be determined.

Structures

Most of the examples of Lewis structures shown below are written such that the origin of the valence electrons is clear. Once the molecule has formed, such a distinction cannot actually be made, but it is a convenient bookkeeping device.*

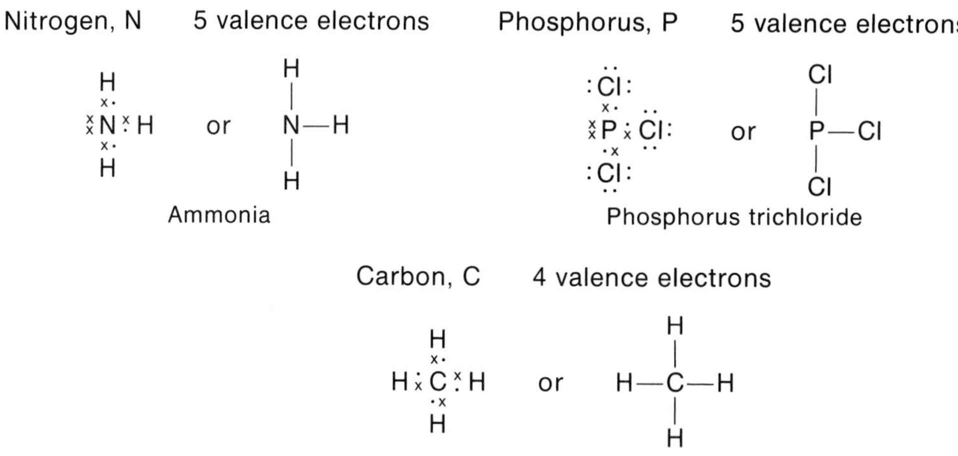

Note that, although the electron configuration of the carbon atom is $\underset{\cdot}{\ddot{\mathrm{C}}}\cdot$, carbon always forms four covalent bonds. You will understand this after learning about hybridization.

*See Appendix C for rules regarding the nomenclature of inorganic compounds.

Sulfur, S   6 valence electrons

```
xx  ..
xS x Cl :      or    S—Cl
 x.   ..             |
: Cl :               Cl
  ..
```

Sulfur dichloride

Ethanol, or ethyl alcohol, C_2H_5OH

$(2 \times 4) + 6 + 6 = 20$ valence electrons

```
                              H   H
  H H                         |   |
  .. ..  ..
H:C:C:O:H        or       H—C—C—OH
  .. ..  ..
  H H                         |   |
                              H   H
```

Ethane, C_2H_6 $(2 \times 4) + 6 = 14$ valence electrons

```
                         H   H
  H H                    |   |
  .. ..
H:C:C:H        or    H—C—C—H
  .. ..
  H H                    |   |
                         H   H
```

Ethylene, or ethene, C_2H_4 $(2 \times 4) + 4 = 12$ valence electrons

```
  .  .
H:C:C:H   ?
  .. ..
  H H
```

There are two problems with the last structure shown above. First, there are two unpaired electrons. However, ethylene is not paramagnetic and therefore has no unpaired electrons. Second, each carbon has only a heptet of electrons around it instead of an octet. We must move the two unpaired electrons between the carbons to form a second covalent bond:

```
H.      .H              H         H
  :C::C:         or      \       /
H'      'H                C==C
                         /       \
                        H         H
```

This double bond is our first example of a multiple bond. Another example is acetylene, which contains a triple bond:

Acetylene, or ethyne, C_2H_2 $(2 \times 4) + 2 = 10$ valence electrons

H:C:::C:H   or   H—C≡C—H

We have already seen in Table 4-4 that the carbon-carbon bond length decreases as the number of bonds increases and that the energy of the bond increases as the number of bonds increases. This is true of all sets of multiple bonds. For example, the oxygen-oxygen distance in hydrogen peroxide, H—O—O—H, is 1.48 Å, whereas in the oxygen molecule, O=O, which contains a double bond, it is 1.21 Å.

The ozone molecule, O_3, provides an example of another property that is common in compounds containing multiple bonds. Our rules for writing Lewis structures suggest the following:

Ozone, O_3 $3 \times 6 = 18$ valence electrons

O O O or O=O—O

This structure provides an octet to all three oxygen atoms and predicts no unpaired electrons, which is consistent with experimental measurements on ozone. A problem remains because such a structure makes the bonds non-equivalent and predicts bond lengths as shown below:

1.21 Å (O=O)
1.49 Å (O—O)

The actual O—O bond lengths, however, are observed to be identical and equal to 1.28 Å. We are forced to conclude that the true representation of the molecule would be intermediate between two extremes, the Lewis structures of which are written as

{ O=O—O   O—O=O }

These structures are called *resonance structures*. This is an unfortunate terminology but it has become well-established. We will see in the section on hybridization that the actual structure of ozone is not the result of resonance between two forms; it is better represented as

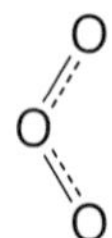

which implies that the two O—O bonds are equivalent and are "1½ bonds."

The octet rule is especially useful in organic chemistry, where it is almost universally obeyed. There are, however, many inorganic compounds for which this generalization does not apply. Any element in a molecule can be found with less than a closed valence shell. This is especially true of compounds containing very polar bonds, and boron compounds. Two examples are shown below.

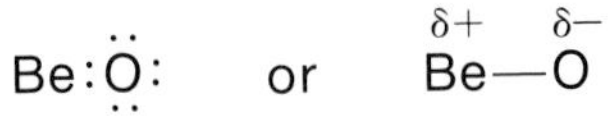

Be:O: or $\overset{\delta+}{\text{Be}}—\overset{\delta-}{\text{O}}$

Beryllium oxide

BCl_3 (Lewis dot structure) or B bonded to three Cl

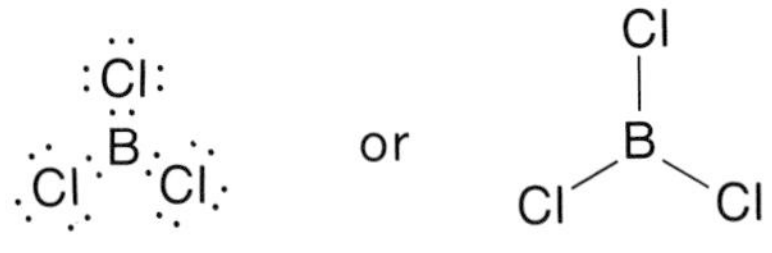

Boron trichloride

The elements of the third row and beyond in the periodic table are frequently observed to form molecules with more than eight electrons in their valence shells. Such molecules are formed by use of the 3*d* atomic orbitals and the *d* and *f* orbitals of the higher periods. Some examples of such expanded valence shells are shown below.

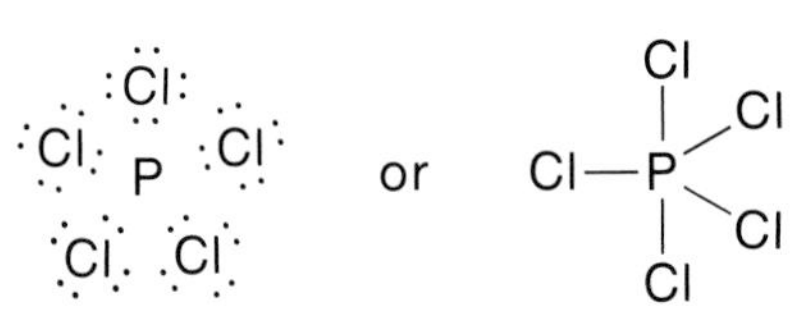

PCl_5 (Lewis dot structure) or P bonded to five Cl

Phosphorus pentachloride

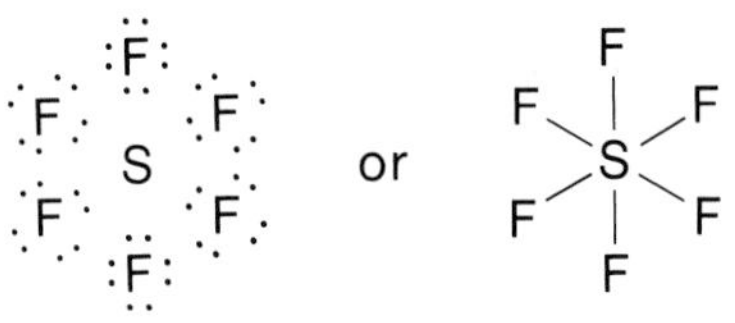

SF_6 (Lewis dot structure) or S bonded to six F

Sulfur hexafluoride

⁻:O:P::O: with :O:⁻ above and below P, or P bonded to three O⁻ and =O

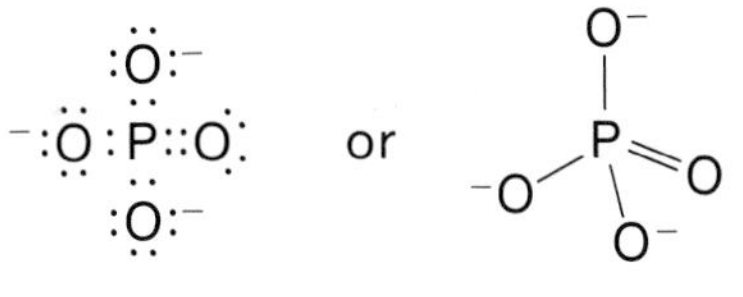

Phosphate ion

Exercise 4-1. Draw the other resonance structures of the phosphate ion. How many are there?

Formal Charges and Resonance Structures

The concept of *formal charge* is introduced to estimate the electric charge imbalance in the neighborhood of an atom in a molecule. The formal charge on a given atom is calculated by subtracting the number of nonbonding valence electrons and half the number of bonding electrons from the number of valence electrons in the free neutral atom. (Note that the number of valence electrons is equal to the group number of the element in the periodic table.) Thus all nonbonding electrons in the given atom and half the electrons in bonds to other atoms are considered to be "property" of that atom.

$$\begin{aligned}\text{Formal charge} = {} & (\text{number of valence electrons}) \\ & - (\text{number of nonbonding electrons} \\ & + \tfrac{1}{2}\ \text{number of bonding electrons})\end{aligned}$$

We have so far seen only one compound with hypothetical Lewis structures having formal charge, namely, ozone:

$$\left\{ \overset{\oplus}{:}\text{O}(=\ddot{\text{O}}:)(-\overset{\ominus}{\ddot{\text{O}}}:) \qquad \overset{\oplus}{:}\text{O}(-\overset{\ominus}{\ddot{\text{O}}}:)(=\ddot{\text{O}}:) \right\}$$

There are three general principles about formal charge to remember when deciding which of the possible structures are likely to be important:

1. The lowest possible formal charge is the most probable.
2. The most electronegative atoms are the most likely to have negative formal charge because they have the greatest affinity for additional electrons.
3. There is a tendency toward structures in which like atoms do not have opposite formal charge, if such structures are possible.

There are numerous other examples of molecules for which resonance structures with formal charge are important, e.g.,

Carbon dioxide, CO_2

$(2 \times 6) + 4 = 16$ valence electrons   Not a peroxide

$$\left\{ \ddot{\text{O}}{=}\text{C}{=}\ddot{\text{O}} \qquad :\overset{\ominus}{\ddot{\text{O}}}{-}\text{C}{\equiv}\overset{\oplus}{\text{O}}: \qquad :\overset{\oplus}{\text{O}}{\equiv}\text{C}{-}\overset{\ominus}{\ddot{\text{O}}}: \right\}$$

All three of these structures are acceptable Lewis structures. The last two, having opposite formal charge on like atoms, are less probable than the first, although they do have some significance in describing the properties of the molecule. The first structure is the most probable, of course, because it has no formal charge at all.

Nitrous oxide, N_2O $(2 \times 5) + 6 = 16$ valence electrons

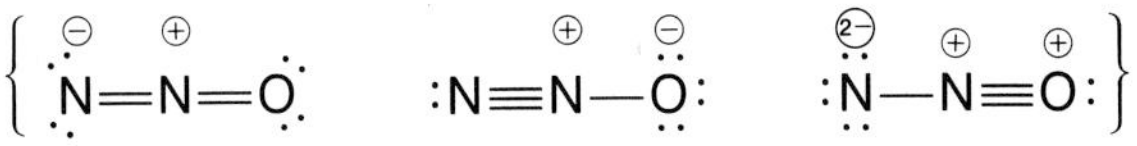

$$\left\{ \overset{\ominus}{\ddot{N}}=\overset{\oplus}{N}=\ddot{O} \qquad :N\equiv\overset{\oplus}{N}-\overset{\ominus}{\ddot{O}}: \qquad :\overset{2\ominus}{\ddot{N}}-\overset{\oplus}{N}\equiv\overset{\oplus}{O}: \right\}$$

Again all three structures are acceptable Lewis structures, with the center one being the most favored according to the three rules given above, and the right-hand one being the least favored.

When expanded valence shells are possible and the octet rule breaks down, the choice between probable structures becomes much more difficult. For example, the following four sets of structures for sulfur trioxide all appear in different chemistry textbooks:

Sulfur trioxide, SO_3 24 valence electrons Not a peroxide

1. 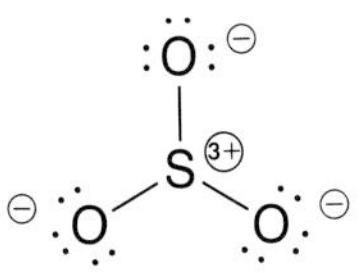 S (3+) singly bonded to three O (each ⊖)

2. { S (2+): one S=O, two S—O⊖ } (three resonance structures)

3. { S (+): two S=O, one S—O⊖ } (three resonance structures)

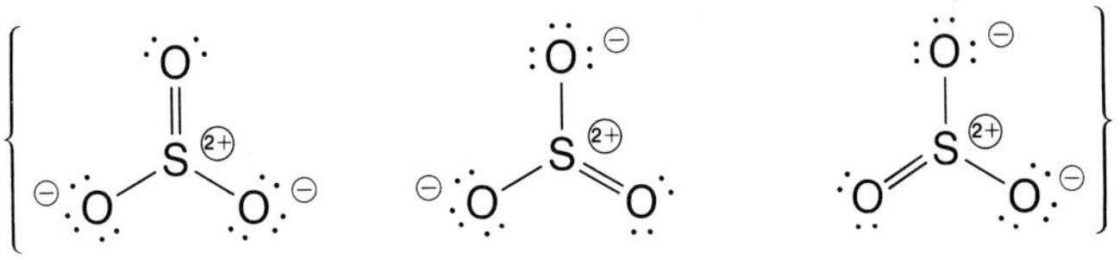

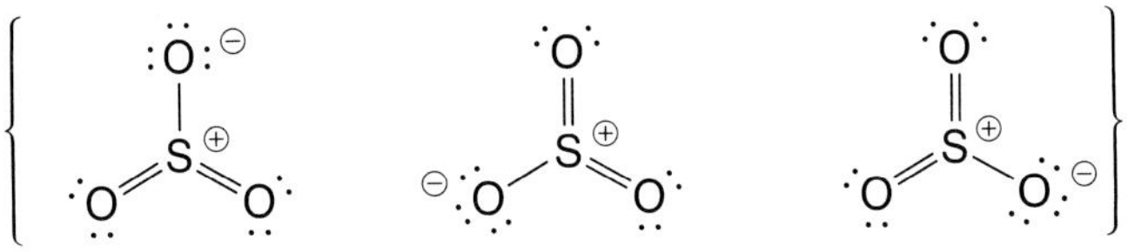

4. 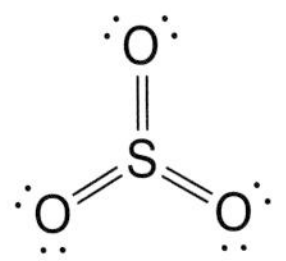 S with three S=O (No formal charge)

Without sufficient data on the bond lengths, a choice between these structures cannot be made, because we have no rules regarding the relative merits of expanding the valence shell beyond eight electrons (or shrinking it to less than eight), or the existence of formal charge under such circumstances. The first structure is unlikely because it does not have an octet around the sulfur atom and because of the high formal charge on this atom. Information on the bond lengths suggests that structure 4 is the most probable.

The most important example of resonance in organic chemistry is benzene, the structure of which is a planar hexagon with all carbon-carbon bonds identical:

Benzene, C_6H_6 $\quad (6 \times 4) + 6 = 30$ valence electrons

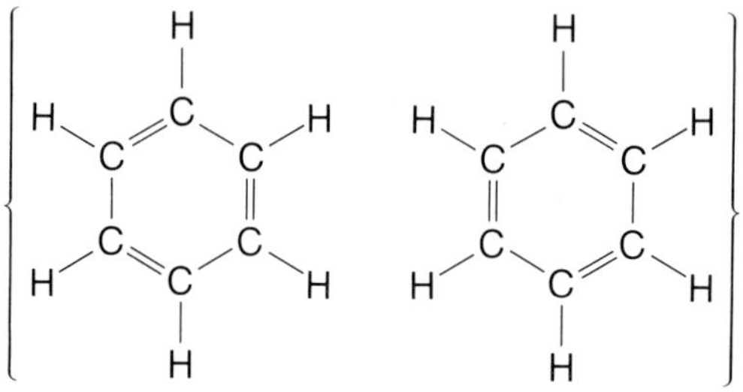

The carbon-carbon bond length in benzene is 1.39 Å; we see from Table 4-4 that this is intermediate between the average C—C single bond and the C=C double bond. All positions on the ring are chemically and physically equivalent, so the bonds must be "1 ½ bonds," as in ozone. Benzene is often represented as

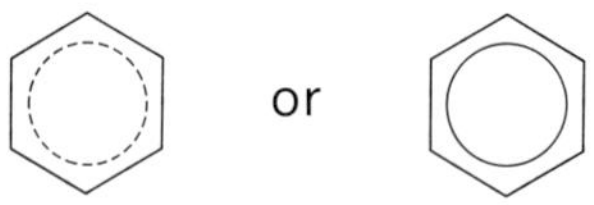

or

Benzene

The carboxylate ion, found in organic acids and soaps, is

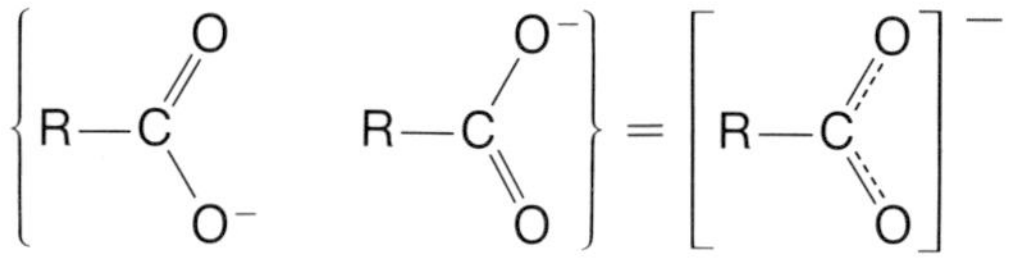

where R denotes any organic group (such as CH_3 or C_2H_5) covalently bound to the carbon atom. The carbonate ion, found in limestone, is

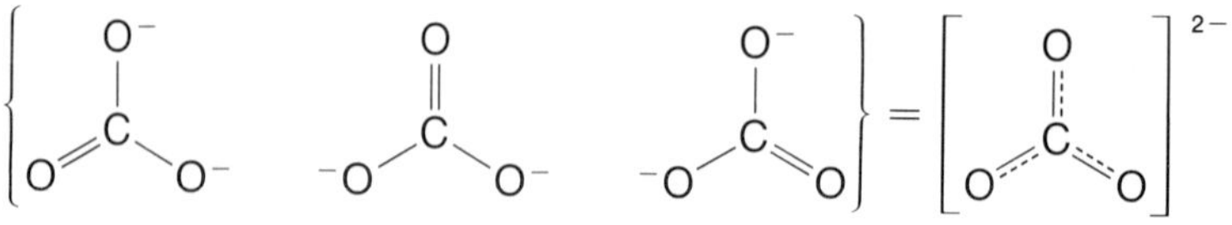

in which the carbon-oxygen bonds are "1 ⅓ bonds."

We see that Lewis structures are useful for predicting covalence, or combining capacity. They also provide information regarding bond order and therefore make predictions about bond lengths and energies possible.

Lewis also introduced a theory that is especially useful in dealing with nonaqueous solutions, which are common in organic chemistry. A *Lewis acid* is defined as an electron-pair acceptor in a chemical reaction and a *Lewis base* is defined as an electron-pair donor. This is analogous to aqueous acid-base reactions such as

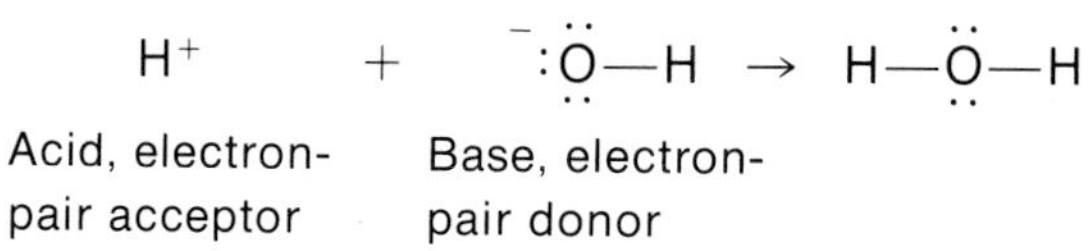

$$H^+ \quad + \quad {}^{-}\!:\ddot{\underset{..}{O}}-H \rightarrow H-\ddot{\underset{..}{O}}-H$$

Acid, electron-pair acceptor	Base, electron-pair donor

An example of a Lewis acid-base reaction is the formation of the tetrafluoroborate ion:

$$BF_3 + :\ddot{\underset{..}{F}}:^- \rightarrow [BF_4]^-$$

Boron trifluoride, a Lewis acid	Fluoride ion, a Lewis base	Tetrafluoroborate ion

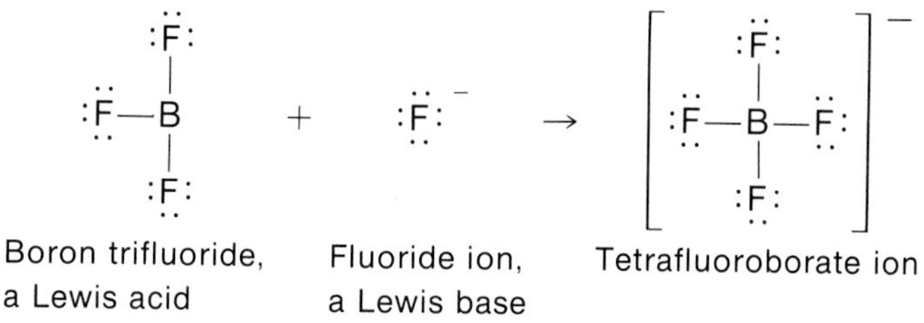

4.7 ATOMIC ORBITAL THEORY

Bond Formation by Overlap of Atomic Orbitals

Covalent bonds can be represented rather well by the overlap of atomic orbitals. In this simplified description of bond formation, a chemical bond forms when two atomic orbitals overlap and two electrons occupy (i.e., spend *most* of their time in) the region of the overlap. A more accurate description, of course, is that the two atomic orbitals "coalesce" into a molecular orbital encompassing both nuclei. The difference in the two approaches is evident from the two diagrams for the formation of the hydrogen molecule shown in Figure 4-10.

The dots in the molecular orbital description show the positions of the nuclei. In the atomic orbital description, they represent the valence electrons. It must be appreciated that these dots, implying discrete positions for the electrons, are purely schematic and serve mainly for bookkeeping purposes. The usefulness of this naive approach will be seen shortly.

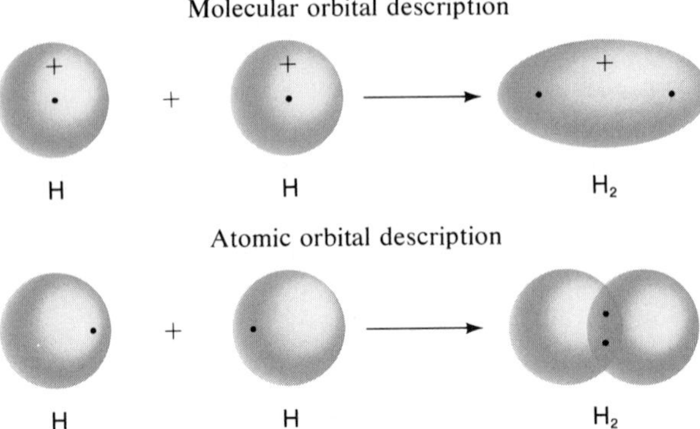

Figure 4-10. Formation of the hydrogen molecule. The dots do not have the same meaning in both diagrams (see text at bottom of page 159).

It is instructive to see representations of the simplest hydrogen compounds of the second-row elements. We will proceed from right to left in the second row, beginning with the element fluorine. Hydrogen fluoride, HF, exists as a stable diatomic molecule. The Lewis and atomic orbital representations of its formation are shown in Figure 4-11.

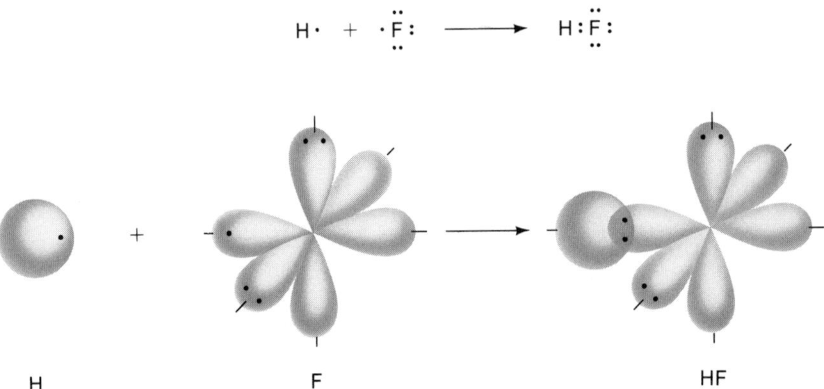

Figure 4-11. Formation of the hydrogen fluoride molecule.

For clarity in these diagrams, we show the *p* orbitals of the second-row elements much thinner than they actually are and eliminate the inner, filled 1*s* and 2*s* orbitals. The 2*s* electrons are included in the Lewis representations, of course, and account for the apparent discrepancies in the number of electrons in the two representations. Either representation obviously predicts the only possible geometry for a diatomic molecule: linear.

The simplest hydrogen compound of the next element, moving to the left, is the familiar water molecule, H_2O. Our two representations of it are shown in Figure 4-12. The H—O—H bond angle predicted from this diagram is 90°. The actual angle is 104.5°. A difference of 14.5° between theory and experiment is alarmingly large and indicates that modifications to the theory are required.

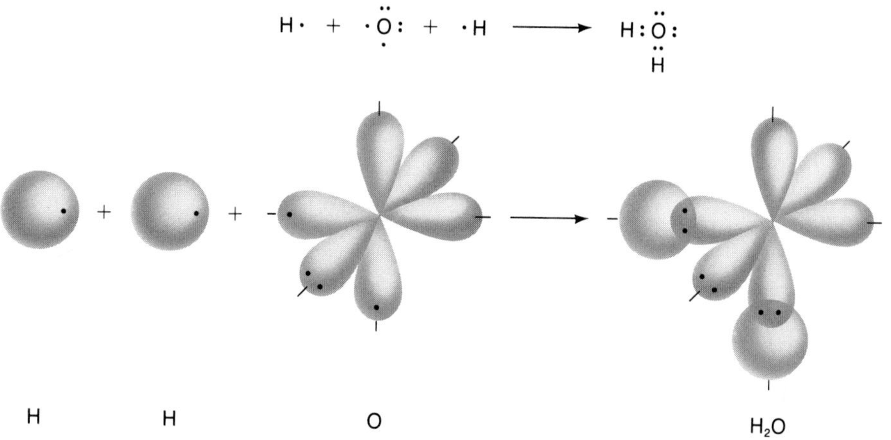

Figure 4-12. Formation of the water molecule.

The difficulties are even more severe with ammonia, NH_3, which is shown in Figure 4-13. All the H—N—H angles are expected to be 90°. They are actually 107.3°.

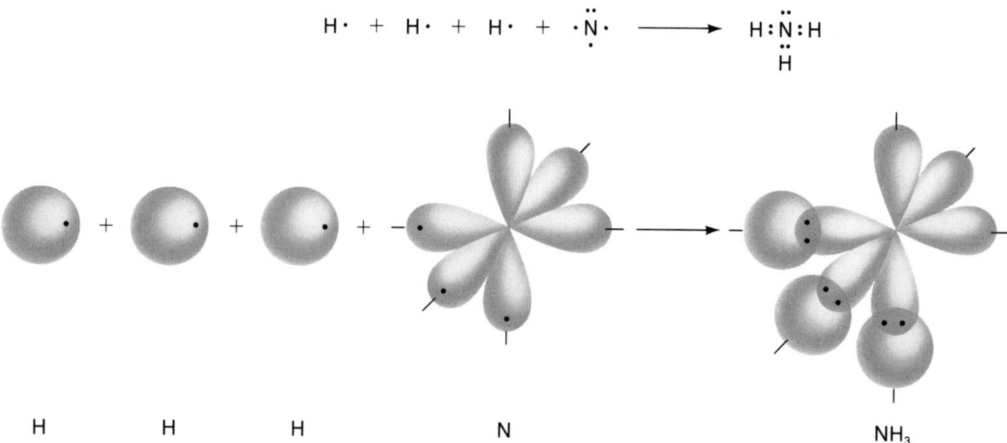

Figure 4-13. Formation of the ammonia molecule.

The H_2O and NH_3 cases can be rationalized by asserting that the electron clouds of the hydrogen atoms repel each other, causing a spreading of the *p* orbitals. Distortions of 15° and 17° are still difficult to swallow, however. Our simple scheme grinds to a complete halt with the next hydrogen compound, methane, CH_4. The ground state of the carbon atom is $(1s)^2(2s)^2(2p_x)^1(2p_y)^1$, which seems to provide only two electrons that are available for bond formation. The problem is illustrated in Figure 4-14.

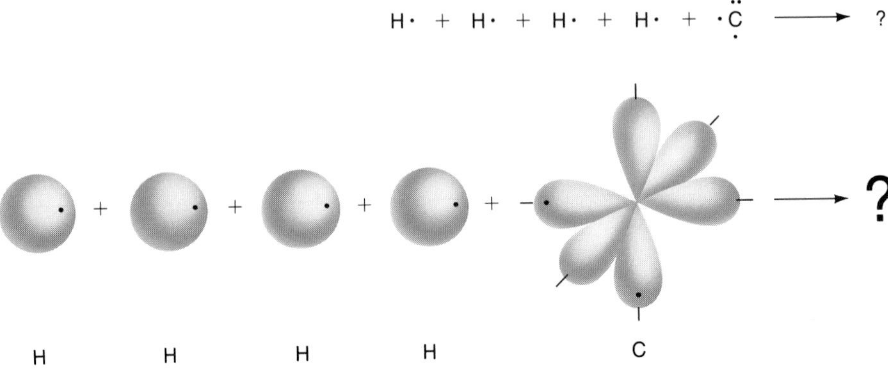

Figure 4-14. Failure of the simple Lewis and atomic orbital descriptions to account for the formation of the methane molecule.

We can form CH_2, but neither representation leads to the formation of CH_4. Yet methane and literally hundreds of thousands of other compounds with carbon bonded to four other atoms do exist. Not only does carbon routinely form four bonds, these bonds are known to be identical in length in CX_4 compounds such as methane. The geometry is the same in all such carbon compounds, namely, *tetrahedral.* Three representations of tetrahedral geometry are shown in Figure 4-15, along with the bond properties of methane. It is essential that you obtain a firm grasp of the three-dimensionality of this form.

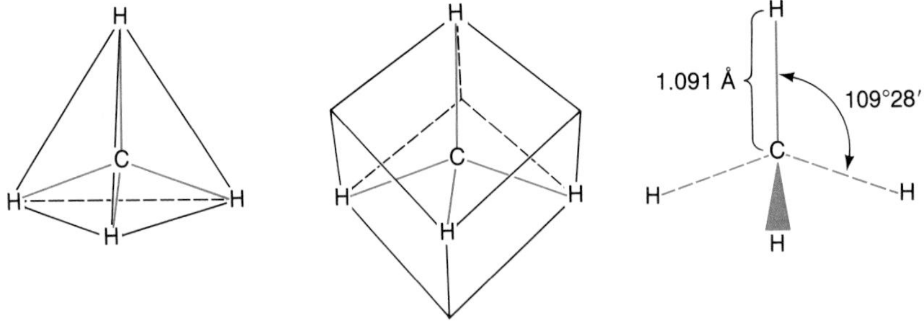

Figure 4-15. Three representations of the tetrahedral structure of methane.

The validity of the tetrahedral structure can be proved in many ways, but perhaps the simplest is to consider derivatives of methane in which one or more of the hydrogens are replaced by chlorine atoms. Such compounds can be made by the following substitution reactions:

$$CH_4 + Cl_2 \rightarrow \underset{\text{Chloromethane}}{CH_3Cl} + HCl$$

$$CH_3Cl + Cl_2 \rightarrow \underset{\text{Dichloromethane}}{CH_2Cl_2} + HCl$$

$$CH_2Cl_2 + Cl_2 \rightarrow \underset{\text{Chloroform}}{CHCl_3} + HCl$$

$$CHCl_3 + Cl_2 \rightarrow \underset{\text{Carbon tetrachloride}}{CCl_4} + HCl$$

Chemists have been able to isolate only one kind of chloromethane. This suggests that all four positions about the carbon atom in methane are equivalent. We might, however, imagine these four equivalent positions to be at the corners of a square instead of the apexes of a tetrahedron. But we find that there is also only one kind of dichloromethane. This proves that the square model cannot be correct and that the tetrahedral model, shown in Figure 4-16, must be correct. Naturally the perfect tetrahedral symmetry of CH_4 and CCl_4 is not present in the "mixed" compounds because of the different bond lengths and the slight distortion arising from repulsive interactions between the substituent atoms. The basic geometry of four-substituent carbon compounds, however, is always tetrahedral.

Figure 4-16. Tetrahedral models of CH_3Cl and CH_2Cl_2.

Exercise 4-2. Draw all possible structures of the molecule CH_2Cl_2 for square and tetrahedral configurations to verify the above statements regarding these configurations. Some pipe-cleaner models may be quite helpful.

Thus the experimental facts are that carbon always forms four bonds and that, in CX_4 compounds, all the bonds are identical and all the bond angles are the tetrahedral angle, 109°28′. Because carbon has only three $2p$ orbitals, however, some way must be found to incorporate the $2s$ orbital in the bonding scheme. The concept introduced (by Pauling) to account for the simultaneous use of the $2s$ and the three $2p$ orbitals is called *hybridization.* It is helpful to consider this concept from the viewpoints of both the energies and the shapes of orbitals.

Hybridization of Atomic Orbitals

Figure 4-17 is an artificial but useful representation of how we might view the energetics involved in the formation of four identical bonds. In order to have four unpaired electrons, one of the s electrons is "promoted" into a p orbital, creating an excited state of the C atom, s^1p^3. The energy required for this process is 96 kcal/mole. The blending of these four singly occupied orbitals into four identical orbitals, which are called *sp^3 hybrid orbitals,* would theoretically require an additional 55 kcal/mole of energy. However, this "hybridization energy" is not real because a free carbon atom in a hybridized electronic state cannot exist. Hybridization occurs only during bond formation, in the course of which the system is stabilized by the release of much more energy than would be required for the hypothetical hybridization of the free atom. This stabilization arises from the large volume of overlap that is possible between sp^3 orbitals and other atomic orbitals, owing to the high degree of directionality of the hybrid orbitals.

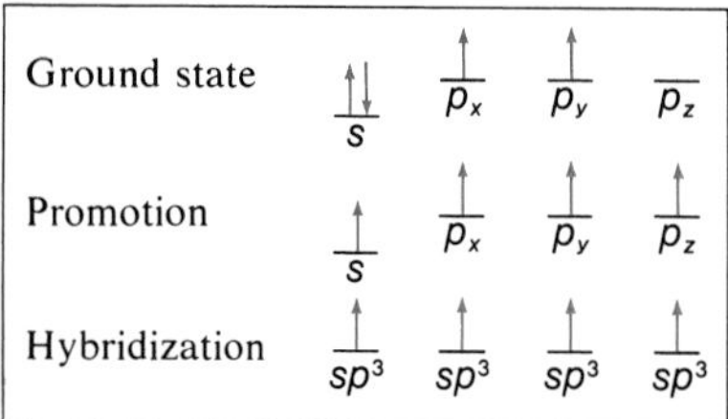

Figure 4-17. Representation of the hypothetical process by which four equivalent sp^3 hybrid orbitals of a carbon atom are formed from one s orbital and three p orbitals. Each hybrid orbital has one electron.

The shape of the sp^3 hybrid orbital is illustrated in Figure 4-18. In analogy with the particle-in-a-box problem in Chapter 2, the larger volume for the electron pair to occupy (spend most of its time in) results in a lowering of the energy of the system.

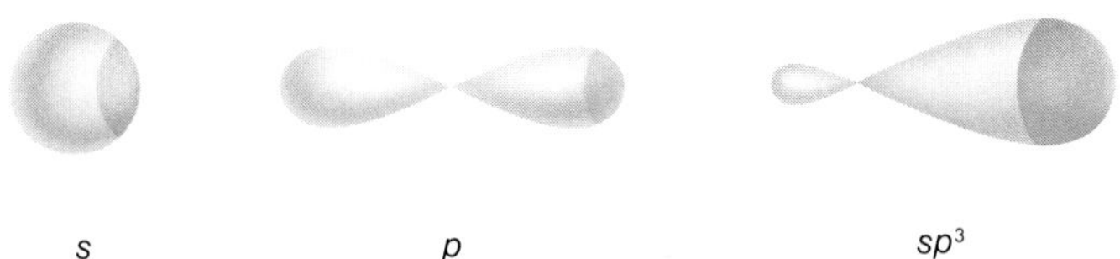

Figure 4-18. The shapes of the s and p atomic orbitals and the sp^3 hybrid atomic orbital. Note the high degree of directionality (preponderance of electron density in one direction) of the hybrid orbital. The heavily shaded areas represent the overlap regions associated with bond formation.

The hybrid atomic orbitals are formulated mathematically by taking a linear combination of the four simple atomic orbitals, in a manner similar to that which we used when formulating molecular orbitals. Four equivalent sp^3 hybrid orbitals in a tetrahedral configuration can be formed as shown in Figure 4-19. (Bear in mind that s, p_x, p_y, and p_z are merely shorthand symbols for the various atomic one-electron wave functions that we call atomic orbitals.)

The sp^3 hybrid orbitals are very useful for describing carbon-containing molecules, and some others as well. We will use them frequently in this book to represent the tetrahedral configurations in carbon compounds. In methane the bonds between carbon and hydrogen form by the overlap of the carbon sp^3 hybrid orbitals and the hydrogen s orbitals, as shown in Figure 4-20. In order to keep the drawings simple, the small lobes of the hybrid orbitals are not shown in this and most of the later figures. They are always present, however.

Hybrid Orbitals in the Description of Molecules

We now return to the troublesome cases of ammonia and water. The valence shell of nitrogen is complete with eight electrons and the molecule contains one pair of nonbonding electrons. If we put these eight electrons into four sp^3 hybrid orbitals, they would form bond angles of 109.5°, with the nonbonding electron pair occupying one orbital. This tetrahedral configuration is shown in Figure 4-21.

In fact, the bond angles in ammonia have been measured as 107.3°, which is very close to our prediction. A reasonable explanation for the small decrease in angle from the perfectly symmetric tetrahedral configuration is that the nonbonding pair constitutes a negative cloud that is not neutralized by a bonded positive nucleus. This portion of the molecule is relatively negative and repels the bonding electrons, forcing the H—N—H bond angle to a smaller value. Ammonia has a dipole moment of 1.47 D in the

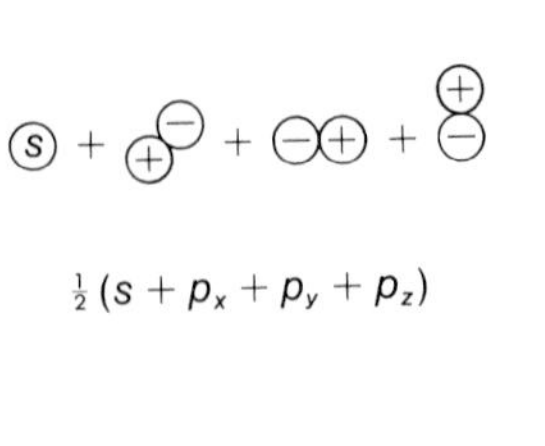

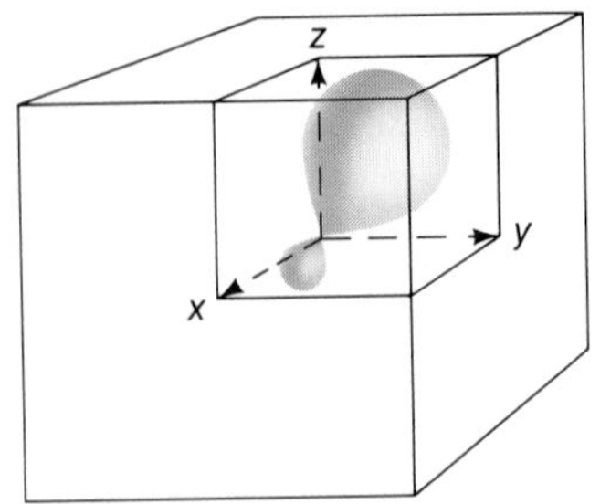

$\frac{1}{2}(s - p_x - p_y + p_z)$

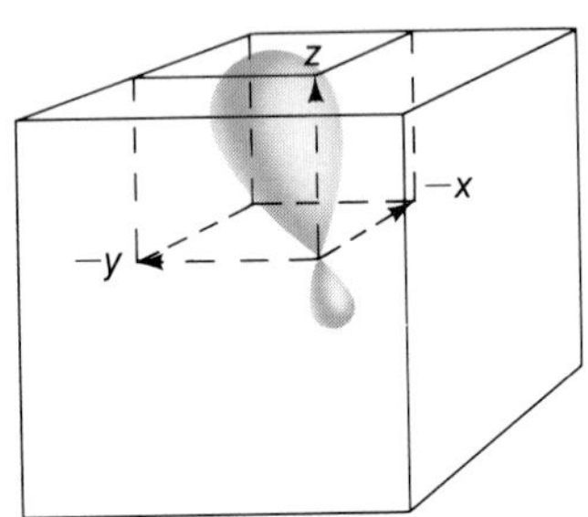

$\frac{1}{2}(s + p_x - p_y - p_z)$

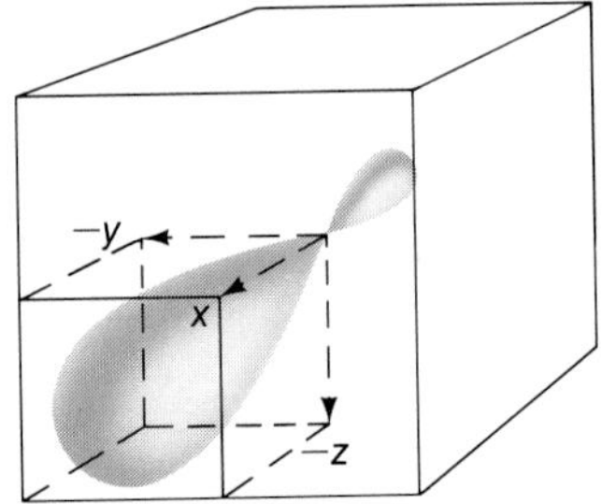

$\frac{1}{2}(s - p_x + p_y - p_z)$

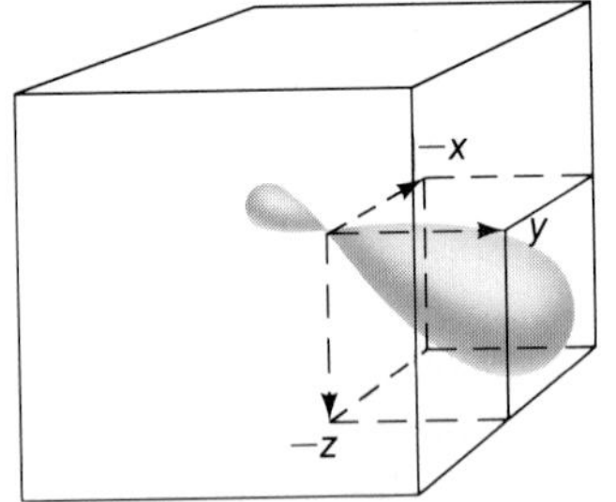

Figure 4-19. Formation of sp^3 hybrid orbitals. The s and three p atomic orbitals are combined mathematically in four different phase arrangements to give the four sp^3 hybrid atomic orbitals. The hybrid orbitals are directed at the four corners of a tetrahedron.

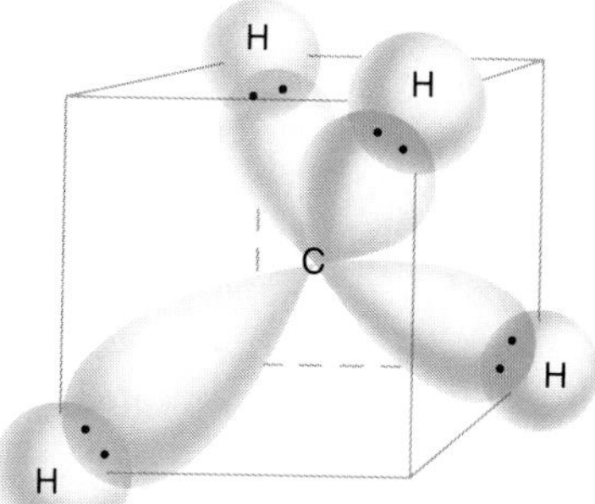

Figure 4-20. The structure of methane.

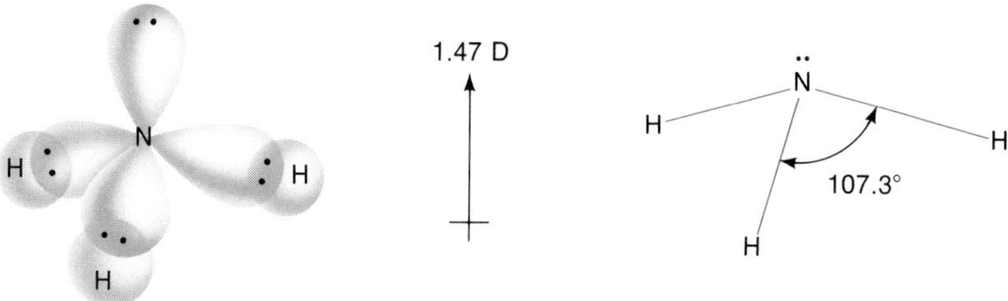

Figure 4-21. The structure and dipole moment of ammonia.

direction of the nonbonding orbital, suggesting that the basis for this explanation is correct.

Water may be treated in a similar manner, but in this case the molecule contains two pairs of nonbonding electrons. The orbital picture (Figure 4-22) again shows nearly tetrahedral symmetry, but the bond angle is actually 104.5°, suggesting even stronger repulsion by the nonbonding electrons than in ammonia. The dipole moment of water is 1.86 D.

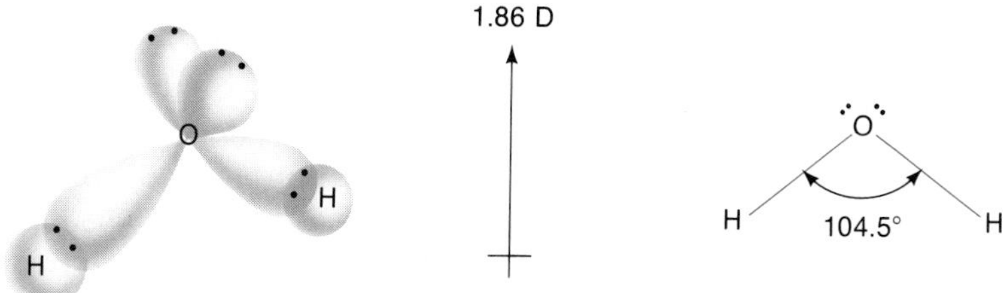

Figure 4-22. The structure and dipole moment of water.

The concept of sp^3 hybridization has satisfactorily explained the structures of H_2O, NH_3, and CH_4. As we continue to move to the left in the second row of the periodic table, new forms of hybridization occur. If we apply the same ideas of promotion and hybridization as worked above for methane to borane, BH_3 (an unstable compound that has been discovered but not isolated), perhaps we will be able to predict the geometry of the borane molecule. The ground state of boron is $(1s)^2(2s)^2(2p_x)^1$. In this state it can bond to only one hydrogen atom. According to the promotion-hybridization scheme shown below, however, three equivalent bonds can be formed, using a new kind of hybrid orbital called an sp^2 orbital. The sp^2 orbitals lie in a plane at 120° to each other. We can therefore predict that BH_3 should have a trigonal planar structure, as shown in Figure 4-23, and experimental evidence indicates that this is so. Although the unoccupied p_z orbital in Figure 4-23 plays no role in the bonding, you will soon see that it is important in molecules having multiple bonds.

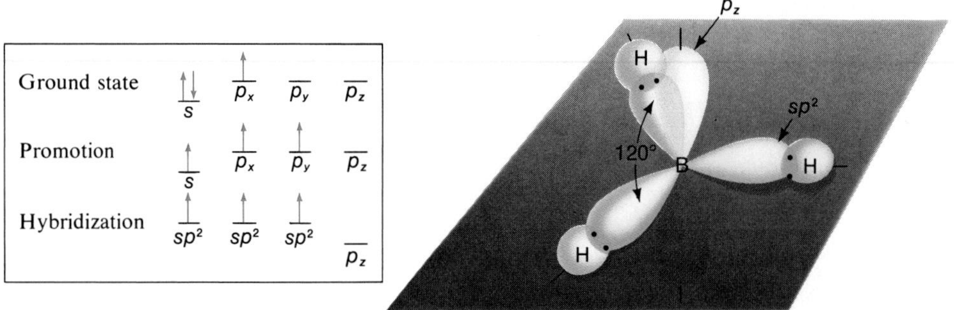

Figure 4-23. Promotion and sp^2 hybridization in boron, and the structure of borane.

Beryllium hydride, BeH_2, is known to be linear. Because the two valence electrons of Be are paired in the 2*s* orbital, promotion and hybridization are again required to form two identical bonds. The formation of these *sp* hybrid orbitals by linear combinations of the normal atomic orbitals is particularly easy to visualize, as shown in Figure 4-24.

Just as the sp^3 and sp^2 orbitals are called *tetrahedral* and *trigonal* orbitals, respectively, the *sp* orbitals are called *digonal* orbitals, for obvious reasons. The diagram for BeH_2 is shown in Figure 4-25.

Lithium hydride, LiH, is highly ionic. To the extent that a covalent Li—H bond does exist, however, it is a σ bond formed by the overlap of the Li 2*s* orbital and the H 1*s* orbital. Clearly, no hybridization is required for this bond to exist.

We have seen *sp*, sp^2, and sp^3 hybrid orbitals, which form two, three, and four identical bonds, respectively. Now we will look at molecules in which

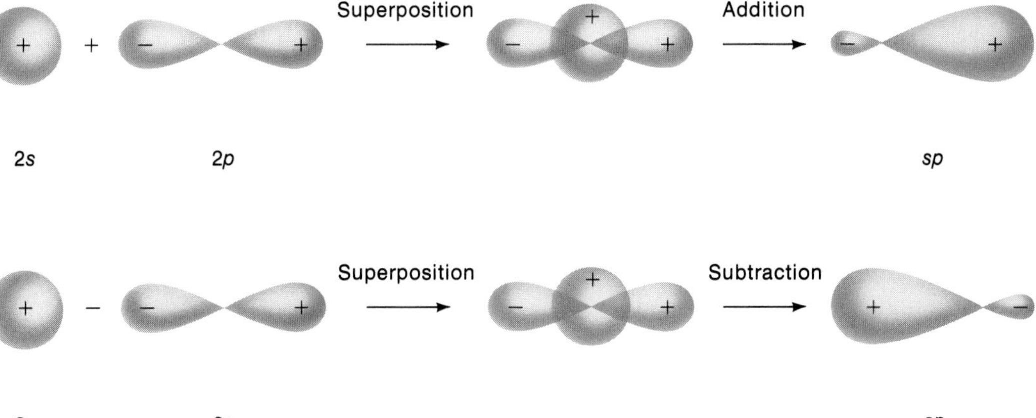

Figure 4-24. Formation of *sp* hybrid orbitals from pure *s* and *p* orbitals.

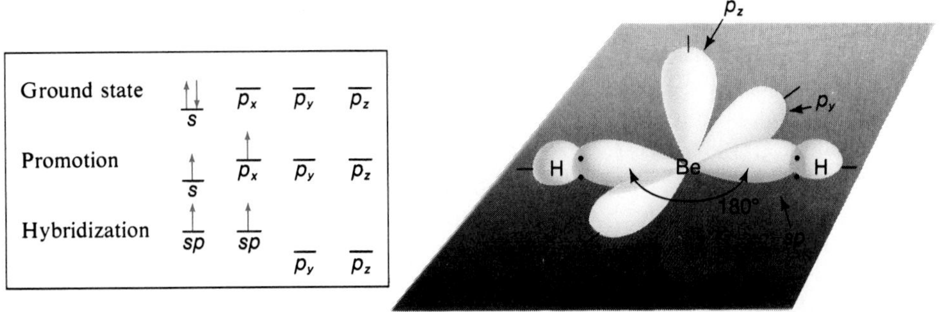

Figure 4-25. Promotion and *sp* hybridization in beryllium, and the structure of beryllium hydride.

the central atom has more than eight electrons, i.e., molecules whose structures cannot be represented by Lewis diagrams because more than an octet is required for the central atom. To further demonstrate the usefulness of hybridization, we will choose compounds of the noble gases as examples. Lewis diagrams are impotent to describe such compounds because one must *begin* with the noble gas structure $:\ddot{\underset{..}{X}}:$.

From the time of their discovery (the late nineteenth century) until 1962, the group 0 elements were usually called the *inert gases*. This reflected the almost universal belief that it was impossible for He, Ne, Ar, Kr, Xe, and

Rn to form chemical compounds. But in 1962 the English chemist Neil Bartlett (b. 1932), working in Canada, observed that the ionization potentials of the Xe atom and the O_2 molecule were almost identical (slightly over 12 eV). Since he had just discovered $O_2^+PtF_6^-$, an ionic compound in which the highly electronegative PtF_6 molecule performs the remarkable feat of pulling an electron off an O_2 molecule, he tried to synthesize the analogous compound $Xe^+PtF_6^-$—and succeeded! This was a milestone discovery in the history of chemistry because it overthrew one of the most sacred principles of that science, namely, that the noble gases never react. Stimulated by Bartlett's work, other chemists abandoned their tradition-bound thinking regarding the noble gases and began synthesizing more new compounds. Within a few years a variety of fluorides, oxides, and oxyfluorides of the heavy noble gases krypton, xenon, and radon were discovered, confirming Pauling's brilliant prediction made in 1933 (see Chapter 3). We will describe possible bonding schemes for several of these compounds, using the concept of hybridization.

Xenon difluoride, XeF_2, is a linear molecule. The valence electrons of Xe are $5s^25p^6$. The techniques used above can be employed here to generate five sp^3d orbitals, yet another type in our growing list of hybrid orbitals. The sp^3d hybrid orbitals are directed toward the apexes of a trigonal bipyramid. These are the only hybrid orbitals that are nonequivalent: the two orbitals directed above and below the plane containing the three orbitals at 120° angles are slightly longer and have only one electron each. Because XeF_2 is known to be linear, we show these two orbitals (Figure 4-26) overlapping the p orbitals of the F atoms to form the chemical bonds.

One of the real bonuses of this representation is that it leads to the prediction of both the composition and the geometry of several oxygen derivatives of this molecule. XeF_2 has three unshared electron pairs, which can form coordinate covalent bonds with either one, two, or three oxygen atoms. Thus we can predict that XeF_2O, XeF_2O_2, and XeF_2O_3 should exist and have geometries corresponding to that shown in Figure 4-26. All three of these compounds do, in fact, exist but their structures are not yet known with certainty.

The existence of xenon tetrafluoride, XeF_4, can be explained by promoting two $5p$ electrons into $5d$ orbitals, giving four unpaired electrons. Hybridization then yields the common sp^3d^2 hybrid orbitals. These six orbitals are identical in shape and are directed toward the apexes of a regular octahedron. If we invoke the criterion, employed earlier, that lone electron pairs occupy a larger region of space than bonded pairs, we should place them in orbitals pointing toward opposite apexes of the octahedron to allow them to be as far apart as possible. The four remaining electrons are in four orbitals lying in a plane, correctly predicting the square planar

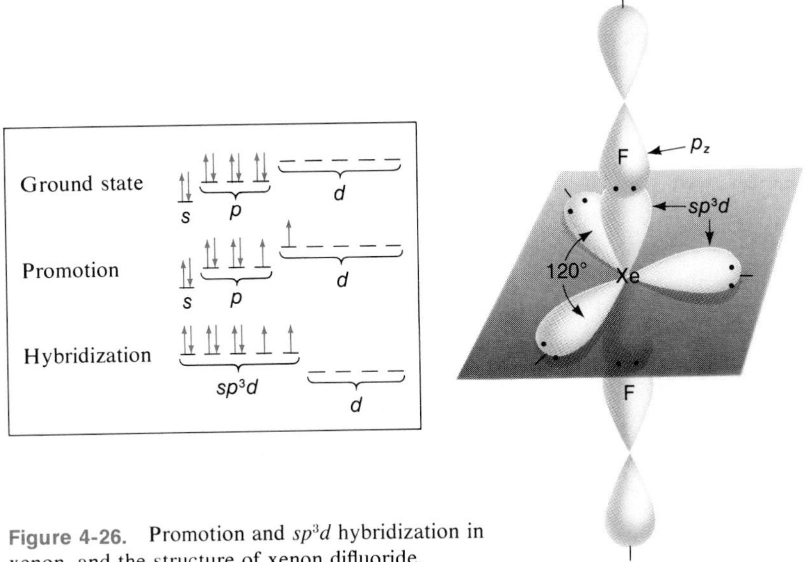

Figure 4-26. Promotion and sp^3d hybridization in xenon, and the structure of xenon difluoride.

structure of XeF_4, which is shown in Figure 4-27. As before, we can predict the existence of XeF_4O as a square pyramid and XeF_4O_2 as an octahedron. The former does exist and is a square pyramid; the latter is not yet known. Figure 4-28 summarizes the properties of the types of pure and hybrid orbitals discussed so far.

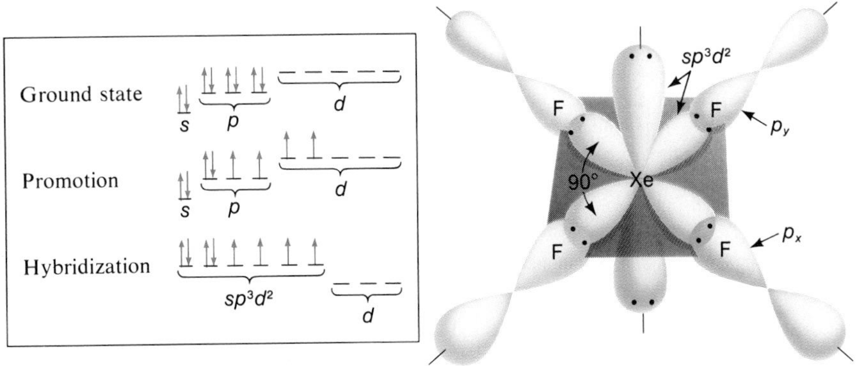

Figure 4-27. Promotion and sp^3d^2 hybridization in xenon, and the structure of xenon tetrafluoride.

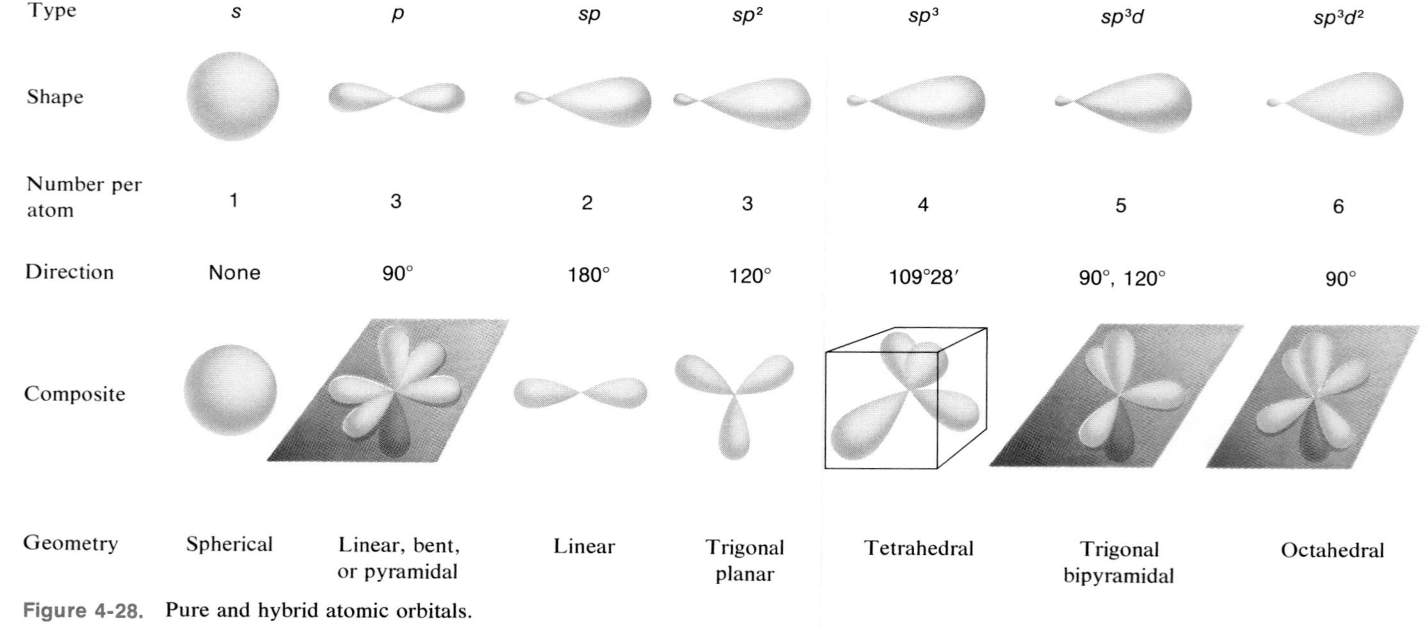

Type
s
p
sp
sp²
sp³
sp³d
sp³d²
Shape
Number per atom
1
3
2
3
4
5
6
Direction
None
90°
180°
120°
109°28′
90°, 120°
90°
Composite
Geometry
Spherical
Linear, bent, or pyramidal
Linear
Trigonal planar
Tetrahedral
Trigonal bipyramidal
Octahedral

Figure 4-28. Pure and hybrid atomic orbitals.

The last challenge for our description of chemical bonds by means of atomic orbitals is multiple bonds. We previously saw the Lewis structure of ethylene, C_2H_4, as

```
H.     .H
  :C::C:
H'     'H
```

Because each carbon atom is bonded to three other atoms and because the bond angles are close to 120°, we might guess that the basic structure (we will refer to it as the skeleton) of the molecule is determined by a set of sp^2 bonds (see Figure 4-28 for all the possibilities)—and it is. The fourth bond, between the carbon atoms, is formed by the side-by-side overlap of p orbitals, as shown in Figure 4-5. As long as the signs of the wave functions coincide, orbitals can overlap side-by-side to give π bonds. The symbol π is used, as with molecular orbitals, to denote orbitals that are unsymmetric with respect to rotation about the z-axis (arbitrarily chosen as the internuclear axis). The end-to-end overlap found in all bonds described thus far *is* symmetric with respect to rotation about the axis, and these bonds are therefore σ bonds. Several representations of π bonds are shown in Figure 4-29. The last two representations are the more accurate ones because they show the overlap of atomic orbitals to form molecular orbitals. We will generally use the second representation for convenience in drawing our diagrams, however.

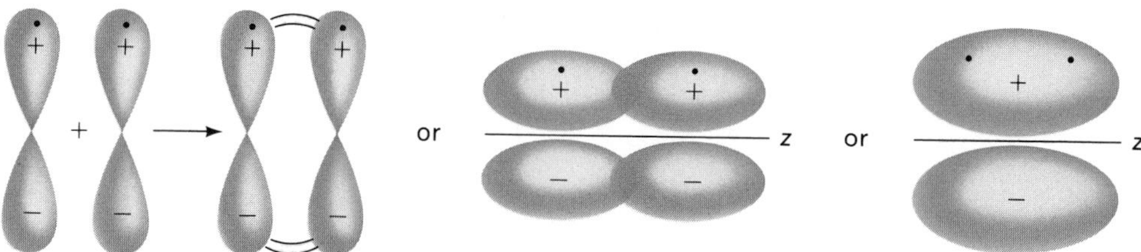

Figure 4-29. Side-by-side overlap of p atomic orbitals, giving π molecular orbitals.

We can now summarize the above discussion of energy considerations and the structure of the ethylene molecule in the diagram shown in Figure 4-30. The observed bond angles for α (H—C—H) and β (H—C═C) are 117° and 121.5°, respectively, in satisfactory agreement with the predicted values of 120°.

The structure of acetylene, C_2H_2, is derived in an identical way. The triple bond is generated by one σ bond and two π bonds, all of which are orthogonal (at right angles to each other). The linear molecule H—C≡C—H is expected from the known properties of sp orbitals, and is represented by the diagram shown in Figure 4-31.

Ground state	↑↓ s	↑ p_x	↑ p_y	p_z
Promotion	↑ s	↑ p_x	↑ p_y	↑ p_z
Hybridization	↑ sp^2	↑ sp^2	↑ sp^2	↑ p_x

Figure 4-30. Promotion and sp^2 hybridization in carbon, and the structure of ethylene.

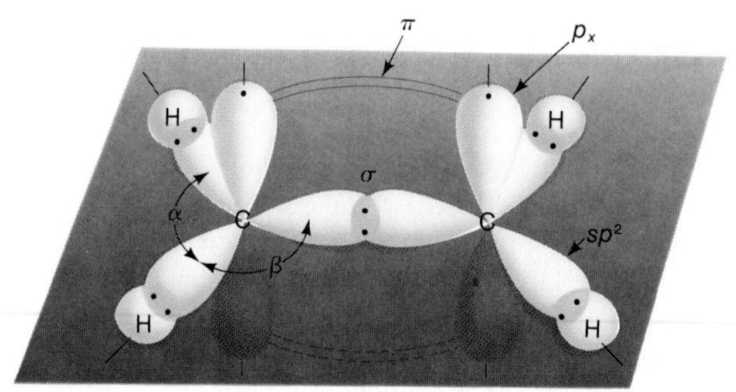

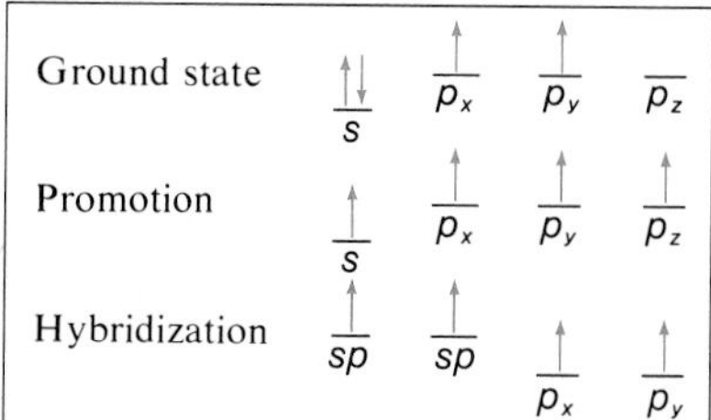

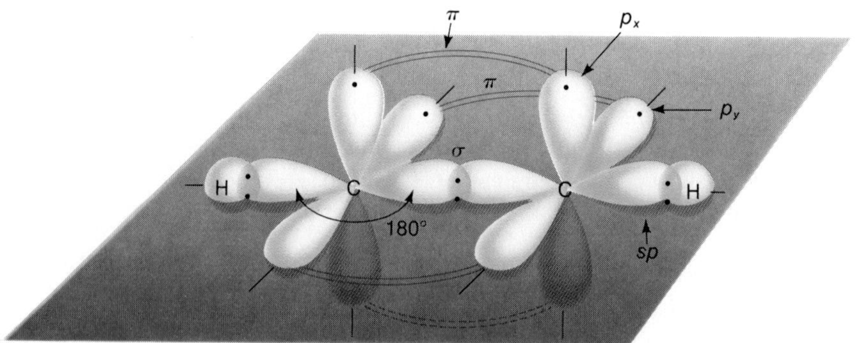

Figure 4-31. Promotion and sp hybridization in carbon, and the structure of acetylene.

Benzene, C_6H_6, provides an interesting test for this type of representation. Each carbon is bonded to two other carbons and one hydrogen, at 120° angles. This indicates that the skeleton of the molecule is determined by σ bonds between sp^2 orbitals, as shown in Figure 4-32.

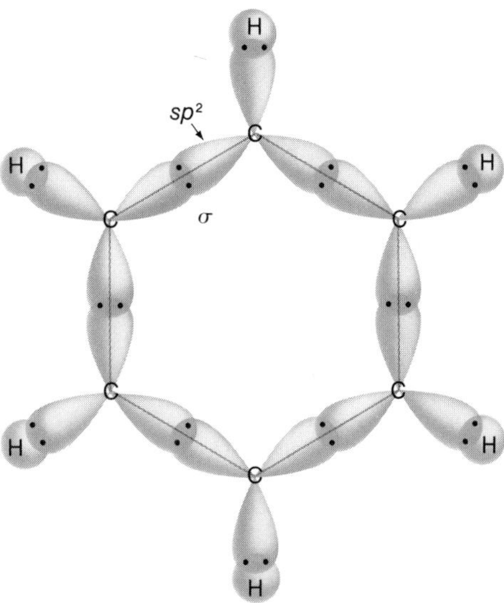

Figure 4-32. The σ bonds in benzene.

These σ bonds provide the correct hexagonal framework (often called a *ring*, for convenience). As in the case of ethylene, the fourth bond, between the carbon atoms, is a π bond formed by the half-occupied p orbitals, of which there is one on each C atom. The only difficulty rests in deciding where to draw the three sets of curved lines that indicate localized π bonds. As we found with the Lewis diagrams, two such structures are equally probable and indistinguishable, and we show them as resonance structures in Figure 4-33.

This representation of the benzene molecule is inferior to a molecular orbital representation because it does not show the overall symmetry of the distribution of π electrons in the molecule. The six p orbitals can be combined in different ways to give six molecular orbitals of different shapes and energies. The most stable molecular orbital, shown in Figure 4-34, has only one node (surface of zero electron density), which is in the plane of the ring. Orbitals of higher energy have an increasing number of nodes. (This is the same behavior noted with the particle in a box; see Figure 2-16.)

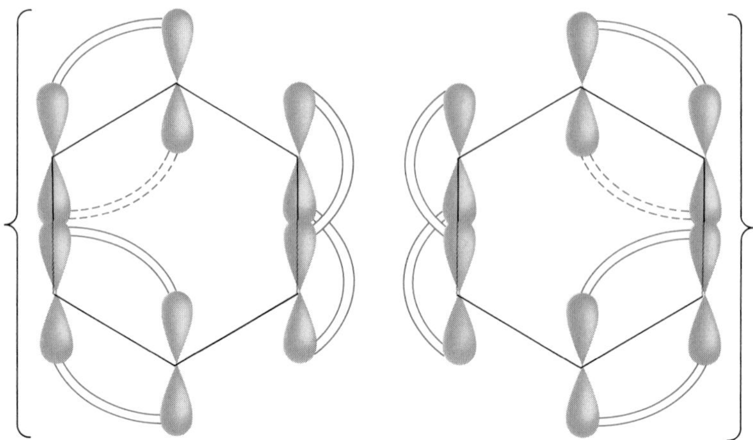

Figure 4-33. The π bonds in the two resonance structures of benzene.

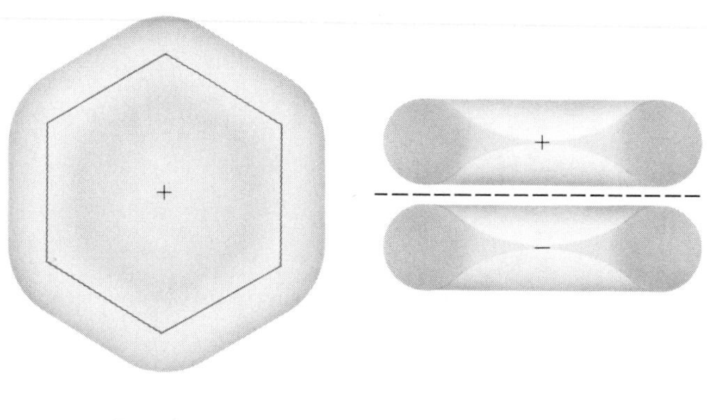

Figure 4-34. The most stable molecular orbital of benzene. The side view is a cross section through the molecule.

4.8 METALS AND THE METALLIC BOND

Metals provide an extreme example of electron delocalization. The metallic elements all appear on the left and middle of the periodic table. They are characterized by two features that dominate their chemical and physical properties and distinguish them from the nonmetals. First, they all have low ionization potentials, low electron affinities, and, therefore, low electronegativities (see Tables 3-2, 3-3, and 3-4), indicating that they have only a weak attraction for electrons. They form weak covalent bonds, as indicated by the low dissociation energies of their homonuclear diatomic molecules. (Table 4-4 shows that the dissociation energies of the diatomic molecules of the alkali metals in the gaseous phase are between 11 and 25 kcal/mole.)

Second, all metals have more valence orbitals than they have valence electrons, i.e., they all have some empty valence orbitals. When their crystals are formed, the valence electrons do not form covalent bonds between the metal atoms; instead, they occupy a large number of energetically similar states. These are molecular orbitals (or *crystal orbitals*) as large as the piece of metal itself, so the valence electrons are highly delocalized with respect to the atoms in which they originated. Figure 4-35 is a schematic representation of the electronic structure of a metal. The positive ions represent the metal nuclei plus their inner-shell electrons, and the valence electrons constitute a "sea" occupying the entire piece of metal.

We might suspect that these free valence electrons would have allowed energy levels (states) similar to those shown in Figure 2-16 for a particle in a box. This is partially correct, although in this case we have a three-dimensional "box," which means that the spacing between levels will not increase with principal quantum number in the same way that it did before. The regular array of positive ions also affects the shape and energy of the wave functions, but to a first approximation our particle-in-a-box model is all right. The most important feature of the solution to the particle-in-a-box problem is that the spacing between energy levels is proportional to 1/(length of box)^2. The length of this metallic "box" is enormous compared to atomic dimensions, so the energy spacings are extremely small. A mole of any metal contains on the order of 10^{23} or more valence electrons, and these closely spaced energy levels are filled in pairs to satisfy the Pauli exclusion prin-

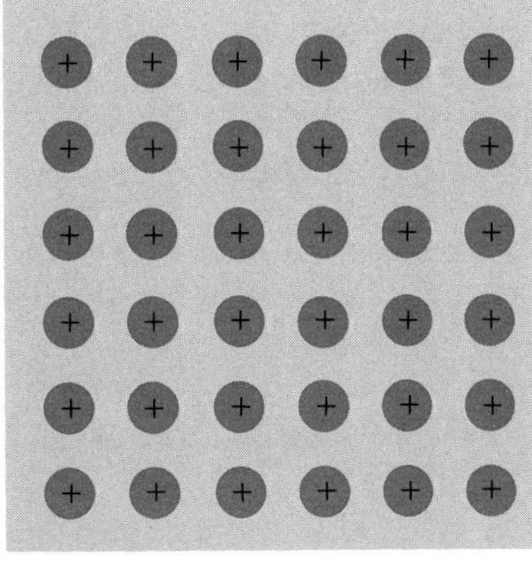

Figure 4-35. Schematic representation of the electronic structure of a metal. The ionized metal atoms at their lattice sites are surrounded by a sea of delocalized valence electrons.

ciple. This results in a "band" of filled states, as shown in Figure 4-36 for lithium. Note that the band itself is only partly filled, a key feature of the electronic structure of metals.

From this model of a metal, we can rationalize several of its properties. Conduction of electrons comes about by exciting them into the unoccupied states in the band of Fig. 4-36. The spacings are so small that even an extremely small electric potential difference can easily cause this excitation. The malleability and ductility of metals are easily explained by the absence of covalent bonds, which would tend to hold atoms firmly in their places. The shininess or luster of metallic surfaces results from the interaction of light with the sea of free electrons.

Bands of different energies result from the combination of atomic energy levels of different energies. Thus, even though beryllium contains two 2*s* valence electrons and its 2*s* band is therefore filled, it is still a metal because the bottom of the 2*p* band overlaps the top of the 2*s* band, providing a continuum of excited energy levels for population by electrons. The highest *occupied* energy band is thus only partly filled, as is necessary for metallic properties to exist. Figure 4-37 shows the band structure for beryllium.

Insulators differ from conductors in two respects: their higher-energy band (the *conduction band*) does not overlap the lower-energy band (the *valence band*)—so there is an energy gap between them—and the lower-energy band (which is the highest *occupied* band) is completely filled. Conduction is therefore impossible. The details of the origins of orbitals in such nonmetallic crystals are complicated by the presence of covalent bonds with various hybrid-orbital geometries, but bands of very closely spaced energy states are also present in such crystals. For example, diamond is a crystalline insulator composed entirely of tetrahedrally bonded carbon atoms. The energy-band structure for diamond is shown in Figure 4-38.

The energy gap in the diamond structure is 120 kcal/mole, which is large enough to make it a true insulator. On the other hand, the analogous energy-

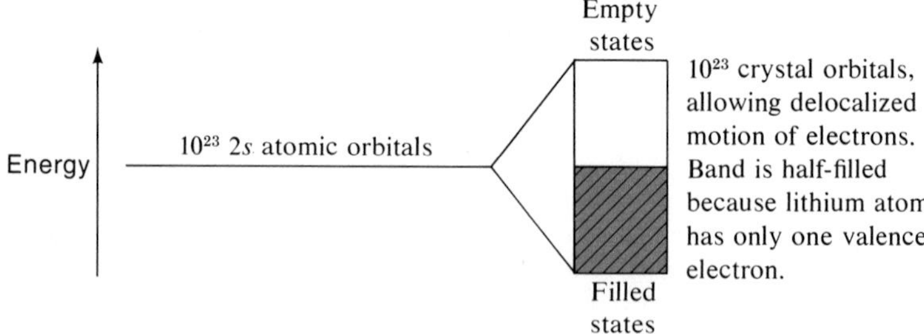

Figure 4-36. The atomic and crystal orbitals of a mole of lithium atoms.

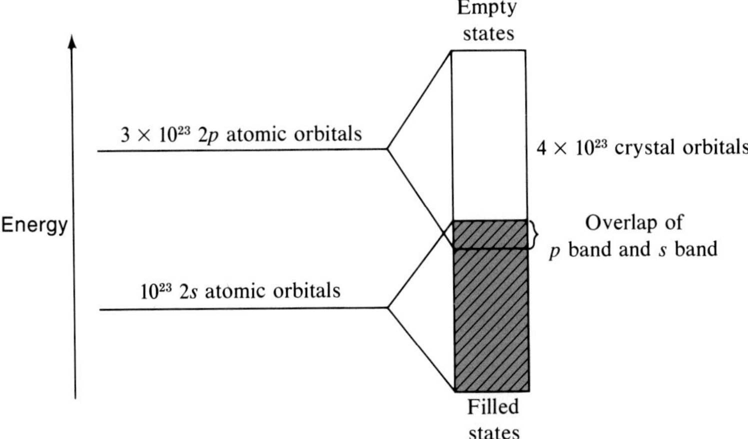

Figure 4-37. The atomic and crystal orbitals of a mole of beryllium atoms.

band structures for Si and Ge have gaps of 25 and 14 kcal/mole, respectively. These crystals are called *semiconductors* because their electrical conductivity is intermediate between those of true conductors and true insulators. Thermal energy or weak electric fields are sufficient to cause some of the electrons in semiconductors to "jump" the energy gap and enter the conduction band.

For use in solid-state electronics, crystals of semiconductors such as Si and Ge are generally doped with traces of impurity atoms such as B or P. The boron causes a deficiency of electrons in the valence band and the phosphorus causes an excess of electrons, which must populate the conduc-

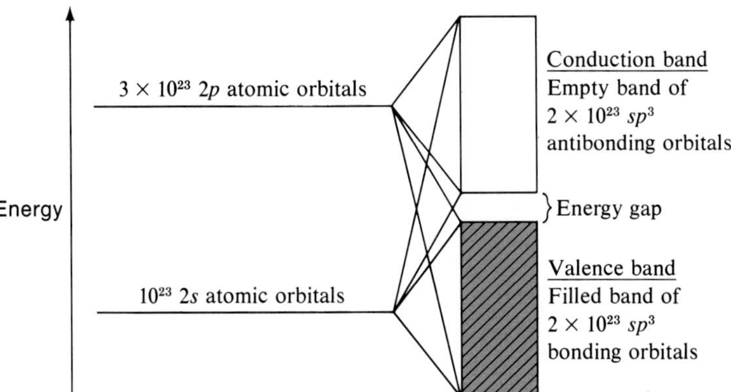

Figure 4-38. The atomic and crystal orbitals of a mole of carbon atoms in the diamond lattice.

tion band. Such *doped crystals* are now weak conductors, the one conducting *holes* (or positions deficient in electron density) and the other conducting electrons. They are used in the production of transistors and many other electronic components. (See the Interlude following Chapter 6.)

4.9 VALENCE SHELL ELECTRON PAIR REPULSION

Orbital hybridization suggests the generalization that the valence electrons in molecules distribute themselves around the atoms as symmetrically as possible. We may view the electrons in the valence shell as distributing themselves in pairs on a spherical surface, with the pairs as far apart as possible because of charge repulsion. This theory was advanced in the early 1960s by R. J. Gillespie (b. 1924) and is called the *valence shell electron pair repulsion* theory (VSEPR). There is abundant evidence that, to a first approximation, the theory is correct. We first ignore all transition element compounds that do *not* have either empty or completely filled *d* orbitals (d^0 or d^{10}). We will discuss such compounds in the next section, but for now we recognize that they contain such large numbers of valence electrons in distributions of uncertain shape that we choose to treat them in a special way. Those transition metal compounds and ions that *do* have empty or completely filled *d* orbitals are regarded as spherically symmetric, provided that any pairs of nonbonding electrons are included, as directed orbitals, in the structural representation. In such species the *d* orbitals therefore have little effect on the geometry.

Figure 4-39 shows the various symmetric geometries in which the valence electron pairs are arranged as far from each other as possible. These structures are tabulated according to the number of valence electron pairs around the central atom. The last column in Figure 4-39 contains the hybrid orbital types having the same symmetry. Although the use of VSEPR theory does not require knowledge of orbital hybridization, the two concepts are so similar that their relationship is presented here.

In the formulas describing these structures, A is the central atom, X is any atom covalently bonded to the central atom, and E is a pair of nonbonding electrons. The lobes in the structures of Figure 4-39 represent the orbitals occupied by pairs of nonbonding electrons.

Table 4-7 contains many examples of each of these structures. The relation between Figure 4-39 and Table 4-7 is illustrated by the example of $SnCl_2$. Tin, Sn, has the electron configuration $1s^2 2s^2 2p^6 3s^2 3p^6 4s^2 3d^{10} 4p^6$-$5s^2 4d^{10} 5p^2$. Because Sn contains only completely filled *d* orbitals, we can apply the VSEPR theory. The Sn atom contributes four valence electrons and each Cl atom contributes one valence electron, for a total of six valence electrons, or three pairs of valence electrons. The $SnCl_2$ molecule there-

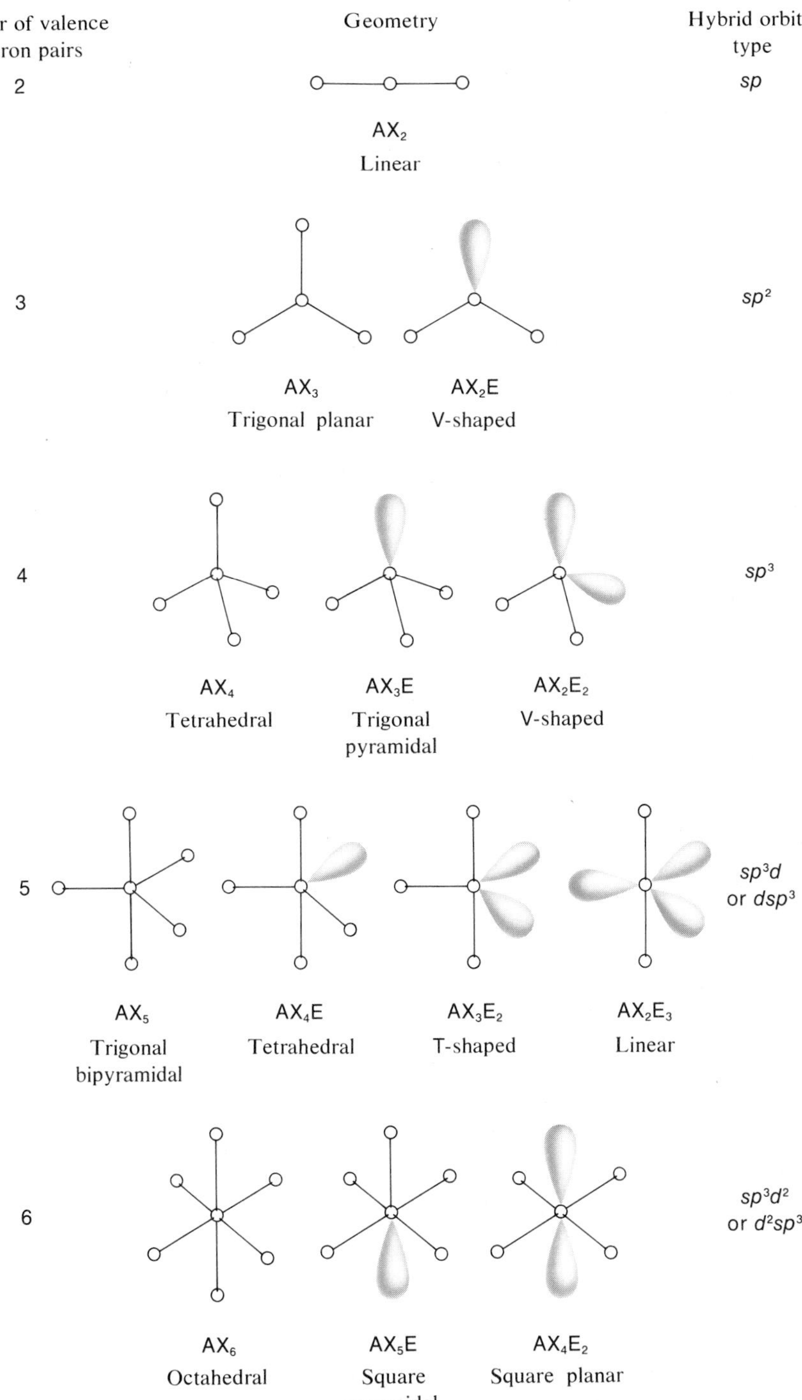

Figure 4-39. Prediction of molecular geometry from valence shell electron pair repulsion theory. [R. J. Gillespie, *Journal of Chemical Education* **40**, 295 (1963).]

fore contains two covalent bonds and one pair of nonbonding valence electrons, which places it in the AX_2E category.

Table 4-7. Examples of Molecules Classified According to Valence Shell Electron Pair Repulsion Theory.

TWO valence electron pairs	
AX_2:	$HgCl_2$, $HgBr_2$, HgI_2, $CdCl_2$, $CdBr_2$, CdI_2, ZnI_2, $Zn(CH_3)_2$, $Cd(CH_3)_2$, $Hg(CH_3)_2$, $Ag(CN)_2$, $Au(CN)_2$
THREE valence electron pairs	
AX_3:	BF_3, BCl_3, BI_3, $B(CH_3)_2F$, GaI_3, $In(CH_3)_3$
AX_2E:	$SnCl_2$, $SnBr_2$, SnI_2, $PbCl_2$, $PbBr_2$, PbI_2
FOUR valence electron pairs	
AX_4:	CX_4, NX_4^+, BX_4^-, BeX_4^{2-}, SiX_4, GeX_4, SnX_4, PbX_4, AsX_4^+, ZnX_4^{2-}, HgX_4^{2-}
AX_3E:	NX_3, OH_3^+, PX_3, AsX_3, SbX_3
AX_2E_2:	OX_2, SX_2, SeX_2, TeX_2
FIVE valence electron pairs	
AX_5:	PCl_5, PF_5, PF_3Cl_2, $SbCl_5$, $Sb(CH_3)_3Cl_2$
AX_4E:	SF_4, SeF_4, $Se(CH_3)_2Cl_2$, $TeCl_4$, $Te(CH_3)_2Cl_2$
AX_3E_2:	ClF_3, BrF_3
AX_2E_3:	ICl_2^-, I_3^-, XeF_2
SIX valence electron pairs	
AX_6:	SF_6, SeF_6, TeF_6, PCl_6^-, PF_6^-, $Sb(OH)_6^-$, SbF_6^-, AsF_6^-, $Sn(OH)_6^{2-}$, $Sn(Cl)_6^{2-}$, $Pb(Cl)_6^{2-}$, AlF_6^{3-}
AX_5E:	BrF_5, IF_5
AX_4E_2:	ICl_4^-, BrF_4^-, XeF_4

4.10 BONDING AND STRUCTURE OF TRANSITION METAL COMPOUNDS

The transition metals have a very diverse chemistry, partly because of the many oxidation states in which these elements are found. Some examples of this are the common cations, oxides, and oxyanions of chromium, manganese, iron, and cobalt, which are listed in Table 4-8.

In addition to having variable oxidation number, the transition metals and their ions are able to use their unfilled orbitals to form *coordination compounds* with atoms, molecules, or ions called *ligands*. The number of ligands bonded to the central (metal) atom is called the *coordination number*

Table 4-8. **Typical Oxidation States of Some Transition Metals.**

Oxidation number				
7		Mn_2O_7, MnO_4^-		
6	CrO_3, CrO_4^{2-}, $Cr_2O_7^{2-}$	MnO_4^{2-}		
5				
4	CrO_2	MnO_2		
3	Cr^{3+}, Cr_2O_3	Mn^{3+}, Mn_2O_3	Fe^{3+}, Fe_2O_3	Co^{3+}, Co_2O_3
2	Cr^{2+}, CrO	Mn^{2+}, MnO	Fe^{2+}, FeO	Co^{2+}, CoO
1				
0	Cr	Mn	Fe	Co

of that atom; it is usually 2 to 9. One example of such a species is a *complex ion* between nickel and ammonia:

$$[Ni(NH_3)_6]^{2+}$$

The oxidation number of the nickel atom in this species is +2; we therefore call the compound of which this is the cation a Ni(II) compound. The ligand is NH_3 (a neutral molecule), and the coordination number of the nickel atom is 6. Generally the ligand contains a pair of nonbonding electrons that form a coordinate covalent bond with an empty atomic orbital of the central atom. The ligand thus acts as an electron-pair donor in such compounds and is therefore a Lewis acid. The electron-pair acceptor is the metal, which is a Lewis base. Typical ligands are H_2O, OH^-, NH_3, Cl^-, and CN^-. Complex ions have some of the most intense colors found in inorganic chemistry.

The names and formulas of such coordination compounds, or complexes, are determined by the following set of rules:

1. Cations appear first in chemical names and formulas, and anions appear second.
2. Ligand names precede that of the central atom in the name of a complex. The order is reversed in the formula.
3. Ligand names end in *o* if negative and are unchanged if neutral or positive. (The two exceptions to this rule are water and ammonia, which are designated *aquo* and *ammine,* respectively.)
4. Greek prefixes specify the number of ligands.
5. Roman numerals specify the oxidation state of the central atom.

6. If a complex ion is negative, its name ends in *ate*.
7. If several ligands are present, they are named in the following order: negative, neutral, positive.

Some examples of such complexes are given below. Note that the formulas of complex ions are always enclosed in brackets.

$[Co(NH_3)_6]^{3+}$	Hexaamminecobalt(III) ion
$[Fe(CN)_6]^{4-}$	Hexacyanoferrate(II) ion
$[Fe(CN)_6]^{3-}$	Hexacyanoferrate(III) ion
$[Co(NO_2)_2(NH_3)_4]^+$	Dinitrotetraamminecobalt(III) ion

The bonding in these complex ions may be represented in several ways. There are three basic schools of thought, all of which have certain advantages. *Crystal field theory* considers the interaction between central atom and ligand to be totally ionic, which must be wrong since neutral ligands are known, as well as complexes in which the central atom has oxidation state 0. It is successful, however, in explaining the colors of such compounds, which will be discussed briefly in Chapter 5, and many other properties as well. *Valence bond theory,* which we will describe below, successfully explains the geometries and is partly successful in predicting the number of unpaired electrons in the complex. *Molecular orbital theory,* which is beyond the scope of this book, can rationalize most of the known properties of these compounds, although the arguments are rather complex and sophisticated.

In octahedral complexes the six ligands donate electron pairs to form coordinate covalent bonds to the central atom. Valence bond theory explains the structure of these complexes by the formation of sp^3d^2 hybrids derived from the atomic orbitals of the central atom.

For example, consider the hexafluorocobaltate(III) ion, $[CoF_6]^{3-}$. The electron configuration of the central Co^{3+} ion is $1s^22s^22p^63s^23p^64s^23d^4$. The orbitals that can participate in bonding are the $3d$, $4s$, $4p$, and $4d$ orbitals. The electron configuration of the complex ion is shown below.

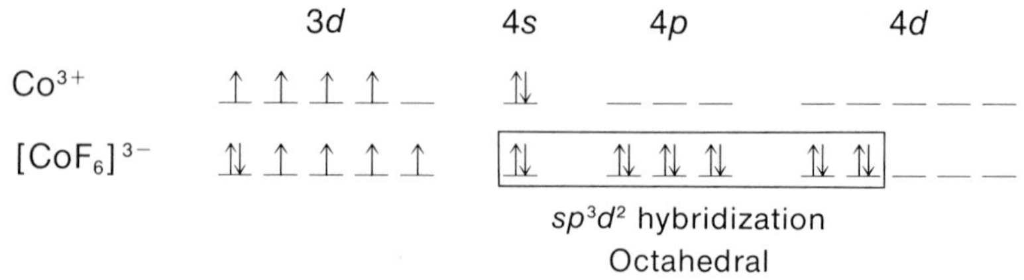

	$3d$	$4s$	$4p$	$4d$
Co^{3+}	↑ ↑ ↑ ↑ _	⇅	_ _ _	_ _ _ _ _
$[CoF_6]^{3-}$	⇅ ↑ ↑ ↑ ↑	⇅	⇅ ⇅ ⇅	⇅ ⇅ _ _ _

sp^3d^2 hybridization
Octahedral

The six ligands contribute twelve valence electrons to the complex. The sp^3d^2 hybrids formed in this complex lead to six equivalent bonds directed along the *x*-, *y*-, and *z*-axes in the positive and negative directions. These hybrids are formed from the $4d_{x^2-y^2}$ and $4d_{z^2}$ orbitals (see Chapter 2) together with the 4*s*, $4p_x$, $4p_y$, and $4p_z$ orbitals. The d_{xy}, d_{xz}, and d_{yz} orbitals are directed along the diagonals between the *x*-, *y*-, and *z*-axes and are not used for bonding. This description of the hexafluorocobaltate(III) ion predicts four unpaired electrons, which is consistent with experimental observation. The complex is called a *high-spin* complex because of these unpaired electrons in its structure. Hund's rule requires the four spins to have the same direction.

Another cobalt complex, the hexaamminecobalt(III) ion, $[Co(NH_3)_6]^{3+}$, is found to have no unpaired electrons. Its electron configuration is of the *low-spin* form, in which the inner 3*d* orbitals are used to form d^2sp^3 hybrids:*

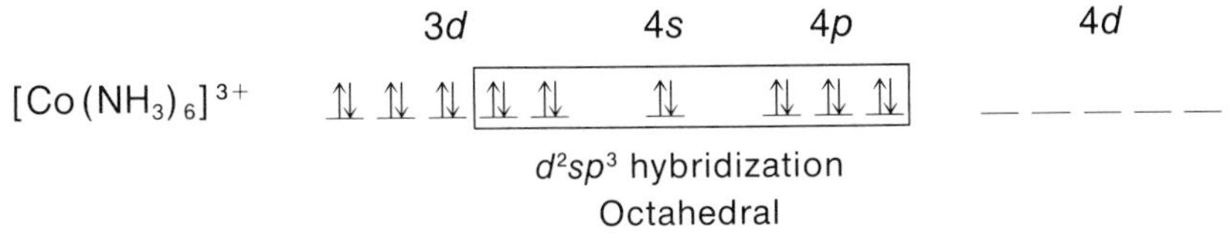

	3*d*		4*s*	4*p*	4*d*
$[Co(NH_3)_6]^{3+}$	⇅ ⇅ ⇅	[⇅ ⇅	⇅	⇅ ⇅ ⇅]	— — — — —

d^2sp^3 hybridization
Octahedral

The orbitals used for these hybrids are the $3d_{x^2-y^2}$ and $3d_{z^2}$ orbitals together with the 4*s* and 4*p* orbitals. The absence of unpaired electrons is consistent with experimental observation.

The valence bond description does not enable us to predict with certainty whether a complex is of the high-spin or low-spin form. The high-spin form is usually found when the bonds between ligand and central metal atom have a high percentage of ionic character. In this case the number of unpaired electrons in the complex is the same as that in the metal ion itself.

Tetra-ligand complexes are found in two geometries, tetrahedral and square planar. Tetrachlorozincate(II) ion, e.g., is tetrahedral:

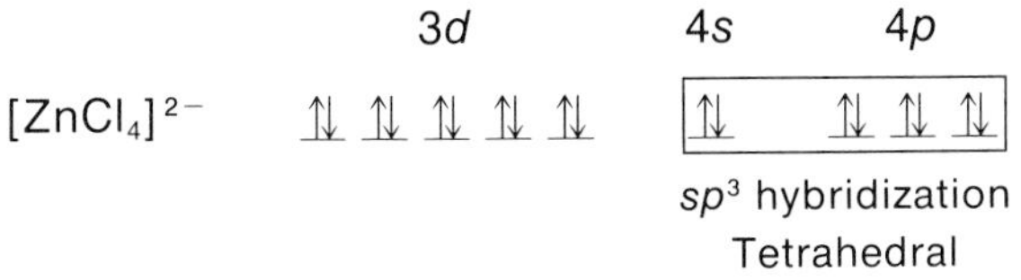

	3*d*	4*s*	4*p*
$[ZnCl_4]^{2-}$	⇅ ⇅ ⇅ ⇅ ⇅	[⇅	⇅ ⇅ ⇅]

sp^3 hybridization
Tetrahedral

*Note that the inner hybrid is designated d^2sp^3 whereas the outer hybrid, which uses the outer 4*d* orbitals, is designated sp^3d^2. In the energy level diagrams, the orbitals used in complex formation are enclosed in a box.

Tetrachloroplatinate(II) ion, on the other hand, is square planar:

	5d	6s	6p
$[PtCl_4]^{2-}$	⇅ ⇅ ⇅ ⇅ [⇅	⇅	⇅ ⇅] ⇅

*dsp*² hybridization
Square planar

Such heavy metal complexes have been extensively studied because of their varied colors and the great variety of structural features and chemical reactivity that they exhibit. For example, geometric isomers occur in the coordination compound dichlorodiammineplatinum(II), $[PtCl_2(NH_3)_2]$, which is shown in Figure 4-40. *Isomers* are compounds with identical chemical formulas but different structures. *Geometric isomers* differ in the positions of two different chemical groups in the molecule, in this case the NH_3 and Cl^- groups. If two groups of the same kind are on the same side of the molecule, the molecule is said to be in the *cis* configuration and is called the *cis* isomer of that compound. If the two groups are on opposite sides of the molecule (i.e., on opposite ends of a diagonal through the molecule, in this case), the molecule is called the *trans* isomer of that compound. Because *cis* and *trans* isomers have distinguishably different structures, they are, in fact, different compounds with different physical and chemical properties. These differences are sometimes difficult to detect and of little consequence, but they are often pronounced and very important.

There is a class of reagents of considerable chemical and biochemical importance that have more than one complexing group in the same ligand molecule. These are called *chelating agents* (from the Greek *chēlē,* claw).

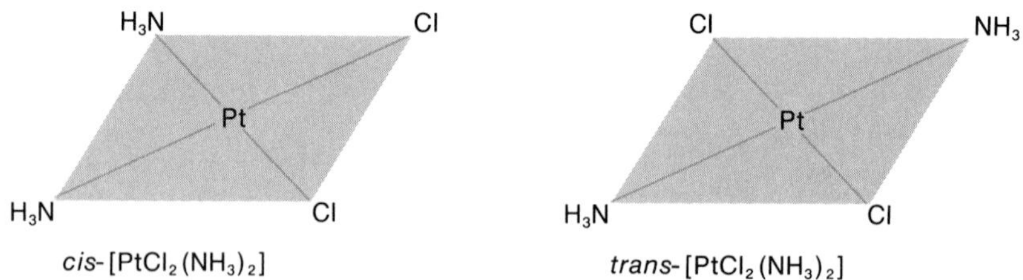

Figure 4-40. The geometrical isomers of dichlorodiammineplatinum(II).

Two examples are ethylenediamine, $H_2NCH_2CH_2NH_2$, and the oxalate ion, $C_2O_4^{2-}$, the structures of which are

Ethylenediamine Oxalate ion

Such compounds are called *bidentate* ligands because they have two coordinating groups per molecule. An example of a chelated structure is the tris(ethylenediamine)cobalt(III) ion, $[Co\ en_3]^{3+}$, which has two optical isomers, shown in Figure 4-41. *Optical isomers* are mirror-image compounds that lack a plane of symmetry and are therefore not superimposable upon one another even though the points of attachment of the groups in the molecules are at identical relative positions. The properties of such isomers will be discussed further in Chapters 5 and 11.

A more spectacular example of a chelating agent is the ethylenediaminetetraacetate ion (EDTA), which has six coordinating groups in the same molecule and a geometry that permits coordination in an octahedral configuration. This compound is widely used for forming complexes with heavy metal ions in aqueous solutions. Its structure is shown on the following page, as is that of the ethylenediaminetetraacetatoferrate(III) ion (Figure 4-42).

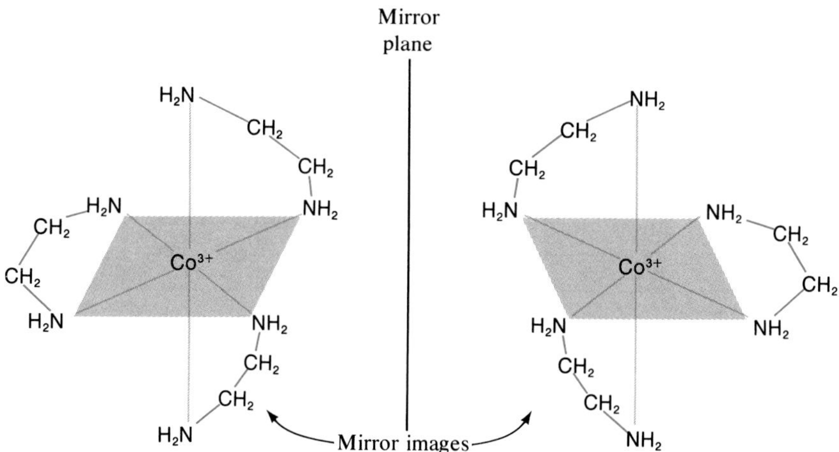

Figure 4-41. The optical isomers of tris(ethylenediamine)cobalt(III).

Ethylenediaminetetraacetate ion

There are many examples of the importance of coordination chemistry in living cells. Many proteins, such as the hemoglobins (the oxygen carriers in the blood of mammals), the myoglobins (the oxygen-storing molecules in mammals), the cytochromes (the chemical energy transducers in oxidative phosphorylation), and the catalases (the peroxide-destroying enzymes), contain heavy-metal complexes with a planar compound called a *porphyrin ring*. Near its center this planar ring has four strategically

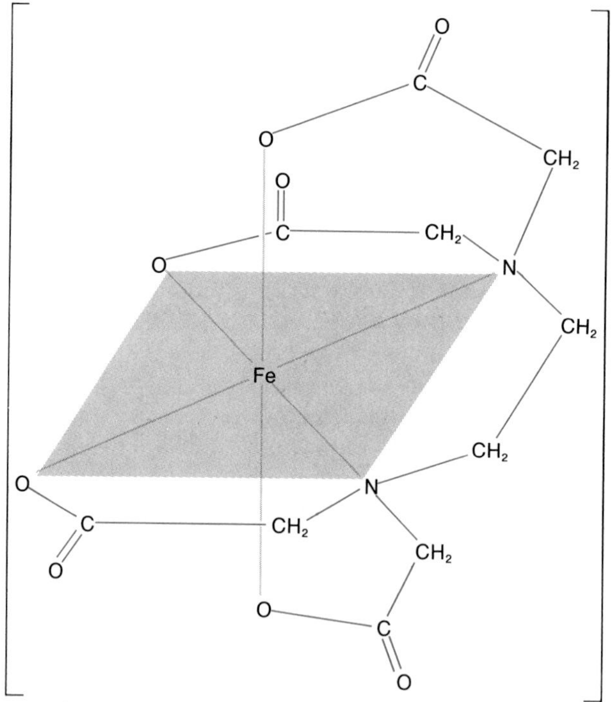

Figure 4-42. The six-fold coordination of an iron(III) ion by the ethylenediaminetetraacetate ion.

placed nitrogen atoms that can coordinate with metal ions. We will see the actual structure of the porphyrin ring in Chapter 5, but here we will focus only on the coordination of the tetradentate porphyrin with Fe^{2+} to form heme, the O_2-binding part of hemoglobin. The reaction is

N: HN N(H) N: + Fe^{2+} → N—Fe—N (N, N) + $2H^+$

Porphyrin ring   Heme

The octahedrally coordinated iron atom in heme has two remaining coordination sites, above and below the plane of the ring. One is complexed to a histidine molecule (an amino acid attached to another part of the protein structure) via a pair of nonbonding electrons on a nitrogen atom in the histidine. The other binds an O_2 molecule, which is delivered by the arterial bloodstream to a cell somewhere in the body. In the process of cellular respiration, the O_2 is given off by the heme and replaced by a CO_2 molecule at the same coordination site. The CO_2 is then transported by the venous bloodstream to the lungs, from which it is exhaled as the cycle begins anew.

Bibliography

Fred Basolo and Ronald C. Johnson, **Coordination Chemistry**, W. A. Benjamin, New York, 1964. Basolo and Johnson give a brief but substantive presentation of the chemistry of coordination compounds. The first chapters present the historical development of the subject, a discussion of several theories of the coordinate covalent bond, a discussion of the stereochemistry of coordination compounds, and a description of the preparation and reactions of these compounds.

Audrey L. Companion, **Chemical Bonding**, McGraw-Hill, New York, 1964. This brief volume contains a wide-ranging discussion of many types of chemical bonds and most of the present theories of chemical bonding. The molecular orbital and valence bond theories are treated about equally. Ionic, metallic, and transition metal bonding are presented in the second half of the book.

Harry B. Gray, **Chemical Bonds**, W. A. Benjamin, Menlo Park, Calif., 1973. Much of this book is derived from the well-known text by Dickerson, Gray, and Haight.

Professor Gray has used many tables and diagrams to provide a thorough explanation of molecular orbitals. Some mention is made of the valence bond and VSEPR theories but the bulk of the discussion centers on molecular orbital theory. There is some discussion of transition metal complexes, including the complete molecular structure of cytochrome c.

Linus Pauling, **The Chemical Bond**, Cornell University Press, Ithaca, N.Y., 1967. Linus Pauling is the father of the chemical bond. His 1935 classic, *The Nature of the Chemical Bond,* now in its third edition, is the basic work from which all subsequent descriptions of chemical bonding have been derived. *The Chemical Bond* is an abridged version of it. The text reflects Pauling's views on the different methods of describing the chemical bond. The hybridization of atomic orbitals, Lewis diagrams, and resonance are heavily emphasized. Molecular orbital theory is discussed only briefly.

Linus Pauling and Roger Hayward, **The Architecture of Molecules**, W. H. Freeman, San Francisco, 1964. This is undoubtedly one of the most unusual and beautiful chemistry books ever published. It represents a notable rapprochement between science and art. Linus Pauling, the world-renowned chemist, and Roger Hayward, a gifted artist, have combined their talents to show the three-dimensional structures of a wide variety of molecules. They have been careful to make sure that the reader fully understands the properties of a number of geometric forms that are observed in molecules. The result is a chemical and aesthetic delight. The drawings of diamond and Prussian blue are accurate and unforgettable representations of these crystal structures.

Problems

1. Agree or disagree with the following statement: The sharing of an electron by two atoms is the only necessary condition for forming a stable bond. Support your answer.

2. Using molecular orbitals, predict the bond orders for CN^+ and CN^-.

3. Using molecular orbitals, predict which of the following molecules and molecular ions exhibit paramagnetism: CO, B_2, NO^-, NO^+, C_2^+. Why?

4. In Table 4-1, Be_2 and Ne_2 are listed as unstable and nonexistent molecules, respectively. Explain why.

5. Explain why, in removing an electron from O_2, we obtain a shorter bond length in O_2^+ (1.12 Å vs 1.21 Å), whereas in removing an electron from N_2 we obtain a longer bond length in N_2^+ (1.12 Å vs 1.09 Å).

6. Using Pauling's electronegativity scale, arrange the following compounds in order of increasing ionic forces between atoms: CsF, SiF_4, PI_3, PH_3, NaI.

7. Write Lewis structures for the following atoms and ions: Mg, S, K^+, Cl^-, C, Ar.

8. Write Lewis structures for the following diatomic molecules and ions: HCl, Li_2, MgO, Br_2, CN^-.

9. Write Lewis structures for the following molecules and predict which ones are paramagnetic: CO, O_2, NO, H_2CO.

10. Write Lewis structures for the following polyatomic molecules and ions, assuming that all atoms are bonded to a central atom: AlF_3, $BeCl_2$, BF_3, NO_3^-, $SOCl_2$, ClO_4^-. Which of these species do not obey the octet rule?

11. Determine the electronic structure of NO by using molecular orbitals. Is the molecule paramagnetic? Does this agree with the prediction from the Lewis structure?

12. Write the resonance structures for the sulfate ion, SO_4^{2-}, carbon dioxide, CO_2, and carbon monoxide, CO.

13. Select the important electron dot structures for the thiocyanate ion, SCN^-. State what is wrong or unlikely in the other structures.

a. :S̤̈—C≡N:⁻ c. :S̤̈=C=N:⁻

b. :S≡C—N̤̈:⁻ d. ⁚S=C=N⁚⁻

14. The nitrate ion, NO_3^-, contains a central nitrogen atom surrounded by three oxygen atoms. The nitrogen-oxygen bonds are equivalent and at 120° angles. Give the resonance structures for this ion.

15. Decide which structure is more important in each of the following pairs:

a. H—N≡N—N̤̈: and H—N̈=N=N⁚

b. ⁚N=N=N⁚⁻ and :N̤̈—N≡N:⁻

Explain your answers.

16. Using molecular orbitals, predict which of the following molecules has the greatest bond energy: CO, CN, NO.

17. Which of the following molecules have a dipole moment: CCl_4, $BeCl_2$, NH_3, I_2.

18. What are the percent ionic characters of HI and BrF? (The bond lengths of HI and BrF are 1.61 and 1.76 Å, respectively, and the observed dipole moments are 0.38 and 1.29 D, respectively.)

19. The CO_2 molecule has a dipole moment of zero, whereas SO_2 has a dipole moment of 1.61 D. Account for this in terms of their structures.

20. Draw the three possible isomers of $C_2H_2F_2$. If a compound of this molecular formula is prepared and found to have a dipole moment of zero, which structure does it have?

21. Label the following as Lewis acids or Lewis bases: $AlCl_3$, H^+, H_2O, BF_3, PCl_3.

22. What type of hybrid atomic orbitals are associated with the following geometries:
    a. Tetrahedral
    b. Trigonal bipyramidal
    c. Linear
    d. Octahedral
    e. Trigonal planar

    Which hybrid orbital has the greatest percentage of *s* character? As the *s* character of a hybrid orbital decreases, what happens to the bond angle?

23. Predict the geometries and types of hybridization in the following molecules and ions: $CaCl_2^-$, F_2O, SO_4^{2-}, SiF_6^{2-}.

24. Aqueous solutions of formaldehyde, H_2CO, are used for preserving anatomical specimens. Draw an orbital diagram of formaldehyde and state the type of hybridization, the number of sigma bonds, and the number of pi bonds.

25. Discuss the bonding and shape of the hydrogen peroxide molecule, H_2O_2.

26. Assign names to the following complex ions: $[PtCl(NH_3)_5]Cl_3$, $K[PtCl_5(NH_3)]$, $[CrCl_2(H_2O)_4]Cl \cdot 2H_2O$.

27. Write the formulas for the following coordination compounds: potassium hexacyanoferrate(II), potassium trichloro(ethylene)platinate(II), trichlorotriammineplatinum(IV) bromide.

28. When the city of East Providence, R.I., cleaned its water lines, rust and copper salts were found to discolor the water. To facilitate cleaning the resultant stains in sinks and bathtubs, oxalic acid was recommended. Explain this.

29. The sap of the plant *Rhus toxicodendron* (poison oak) contains 3-pentadecylcatechol,

    OH

    OH

    $(CH_2)_{14}CH_3$

    which is a strong skin irritant. The treatment is to apply an ointment containing a metal ion such as Zn^{2+}. How does this work?

INTERLUDE

Perfume
PASSION TROPICALE
EXTASE
NUIT D'AMOUR
REASONABLE EXPECTATIONS
Richter

the chemistry of smell

The shapes of molecules are crucial in determining many of the characteristics of our world. Two thousand years ago the Roman poet-philosopher Lucretius, in his great didactic epic, *De Rerum Natura* (On the Nature of Things), advanced a theory to explain the sense of smell, among others. The theory included a description of minute nasal pores of different sizes and shapes. These pores selectively admitted the minute particles, or molecules, that were the source of the odor. The odor perceived was determined by which pores the particles entered. The modern theory of smell is essentially the same as this uncannily accurate guess, which was unsupported by any evidence whatsoever!

Table 1 contains the names of what are thought to be the seven primary odors, and examples of each. Five of these odors have been related to the basic shape of the molecules creating the odor. The olfactory area in the nose is believed to contain receptor sites of definite geometry that sense molecules of a certain basic shape but cannot sense others. Figure 1 shows the location of the olfactory area and Figure 2 shows the proposed shapes of the various receptor sites. The two remaining primary odors, pungent and putrid, appear to be sensed not on the basis of molecular shape, but rather molecular charge.

Table 1. **The Seven Primary Odors.**

Primary Odor	Chemical Example	Familiar Substance
Camphoraceous	Camphor	Moth repellent
Musky	Pentadecanolactone	Angelica root oil
Floral	Phenylethyl-methyl-ethyl carbinol	Roses
Pepperminty	Menthone	Mint candy
Ethereal	Ethylene dichloride	Dry-cleaning fluid
Pungent	Formic acid	Vinegar
Putrid	Butyl mercaptan	Bad eggs

SOURCE: J. E. Amoore, J. W. Johnston, Jr., and M. Rubin, "The Stereochemical Theory of Odor."

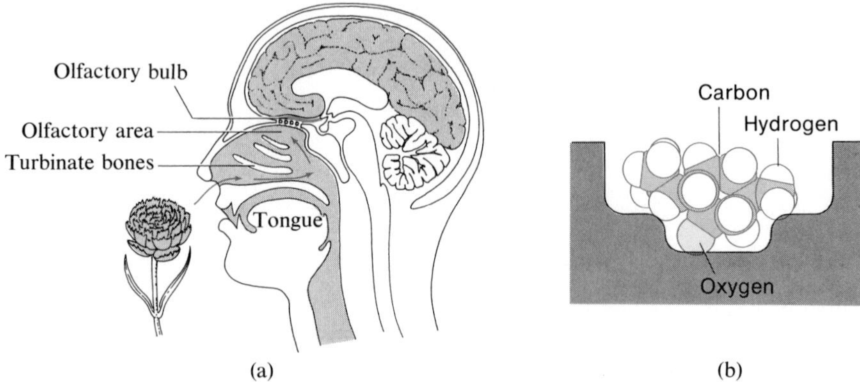

Figure 1. The sense of smell is illustrated in these drawings. (a) Air carrying odorous molecules passes by the three turbinate bones to the olfactory area, in which are embedded the endings of large numbers of olfactory nerves. According to the stereochemical theory, different olfactory nerve cells are stimulated by different molecules on the basis of the size and shape or the charge of the molecule; these properties determine which of various pits and slots on the olfactory endings the molecule will fit. (b) A molecule of *l*-menthone is shown fitted into the pepperminty site. (Adapted from J. E. Amoore, J. W. Johnston, Jr., and M. Rubin, "The Stereochemical Theory of Odor." Copyright © 1964 by Scientific American, Inc. All rights reserved.)

There are many examples of molecules that are very dissimilar in their chemical properties but that have similar odors, apparently because they have roughly the same shape. Figure 3 provides some examples of this. Furthermore, molecules with complex odors generally fit into more than one of the proposed receptor sites, a phenomenon that is illustrated in Figures 4, 5, and 6.

We will see later in this book that molecular shape is important for the specificity of almost all of life's chemical processes. For example, it under-

Figure 2. Proposed olfactory receptor sites are shown for the seven primary odors, together with molecules representative of each odor. The shapes of the first five sites are shown in perspective and, with the molecules silhouetted in them, from above and the side; known dimensions are given in angstroms. The molecules are (left to right) hexachloroethane, xylene musk, α-amylpyridine, *l*-menthol, and diethyl ether. Pungent (formic acid) and putrid (hydrogen sulfide) molecules are sensed on the basis of charge, not shape. (J. E. Amoore, J. W. Johnston, Jr., and M. Rubin, "The Stereochemical Theory of Odor." Copyright © 1964 by Scientific American, Inc. All rights reserved.)

Camphoraceous

Musky

Floral

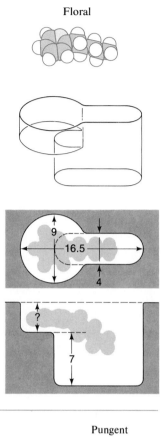

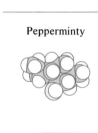

7.5

9

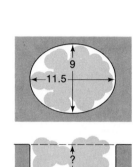

9

11.5

9

16.5

4

4

?

?

7

Pepperminty

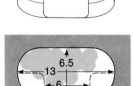

Ethereal

Pungent

δ+

−

6.5

13

6

5

?

Putrid

δ−

+

4

2

4

lies the extreme specificity of enzymes, the catalysts of most biochemical reactions. Stereochemistry, the study of the effects of molecular shape on chemical properties, explains the chemical mechanism for the replication of genetic information. It is important that you be able to visualize the three-dimensional shapes of molecules.

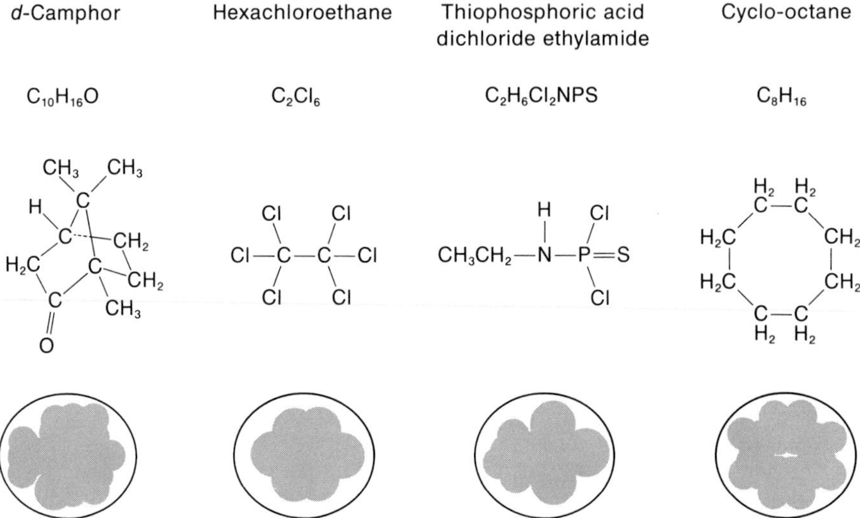

Figure 3. These unrelated molecules with camphorlike odors show no resemblance in their empirical formulas and little in their structural formulas. Yet, because their overall sizes and shapes are similar, they all fit the bowl-shaped receptor for camphoraceous molecules. (Adapted from J. E. Amoore, J. W. Johnston, Jr., and M. Rubin, "The Stereochemical Theory of Odor.")

H COOCH$_3$ / C / $H_3COOC—CH_2—CH_2$ $CH_2—CH_2—COOCH_3$

H_3C COOCH$_3$ / C / $H_3COOC—CH_2—CH_2$ $CH_2—CH_2—COOCH_3$

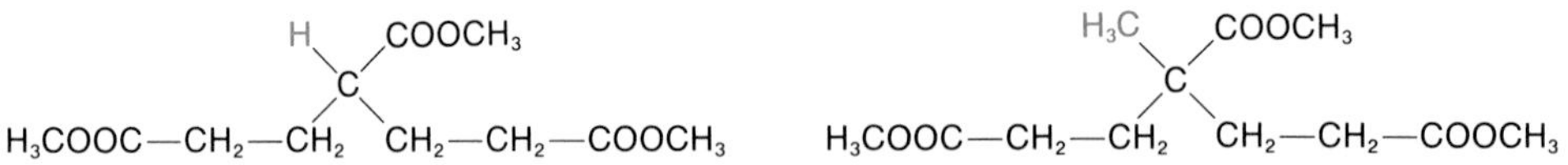

Figure 4. Changing a molecule slightly caused a change in its odor. The molecule at left smelled fruity because it fitted into three sites. When it was modified (right) by the substitution of a methyl group for a hydrogen atom, it smelled somewhat ethereal. Presumably the methyl group made it fit two of the original sites less well but allowed it still to fit the ethereal site. (J. E. Amoore, J. W. Johnston, Jr., and M. Rubin, "The Stereochemical Theory of Odor.")

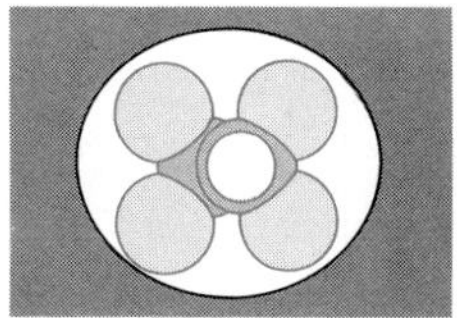

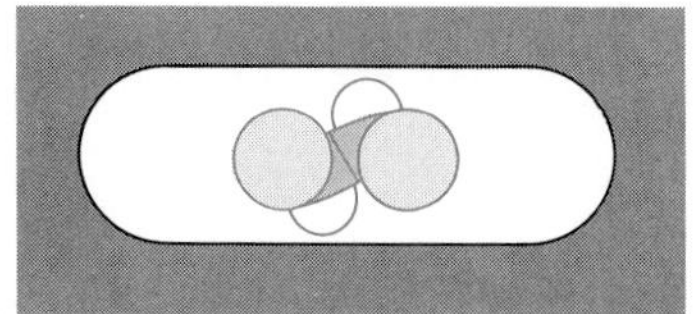

Figure 5. A single chemical has more than one primary odor if its molecule can fit more than one site. Acetylene tetrabromide, for example, is described as smelling both camphoraceous and ethereal. It turns out that its molecule can fit either site, depending on how it lies. (J. E. Amoore, J. W. Johnston, Jr., and M. Rubin, "The Stereochemical Theory of Odor.")

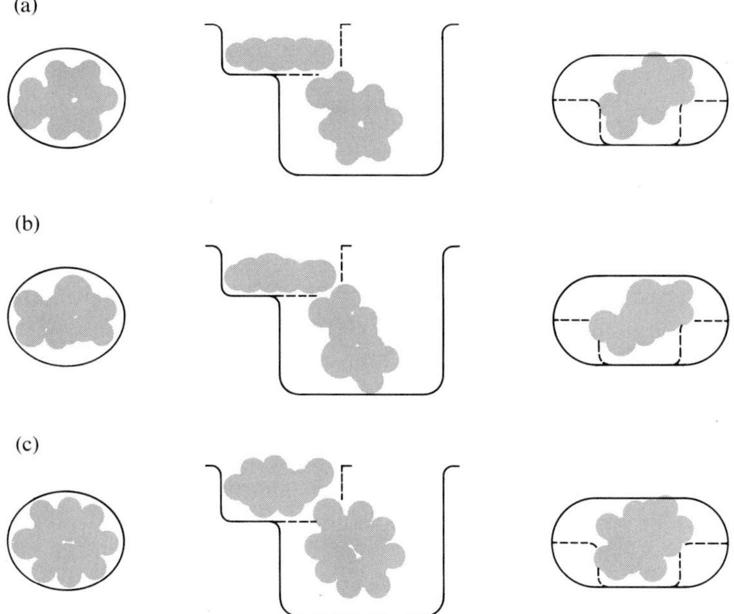

Figure 6. Complex odors consist of several primaries. Three molecules with an almond odor are illustrated: (a) benzaldehyde, (b) α-nitrothiophene, and (c) cyclo-octanone. Each of them fits (left to right) camphoraceous, floral (with two molecules) and pepperminty sites. (J. E. Amoore, J. W. Johnston, Jr., and M. Rubin, "The Stereochemical Theory of Odor.")

5 the properties of molecules

The properties of all matter are related to the molecular structure of its component parts. In principle, if we know the structure of a molecule, all other properties, such as its melting point, its boiling point, its viscosity as a liquid and as a gas, its hardness and malleability as a solid, its solubility in other substances, and its color, should be predictable. Such predictions are far from easy and the successes of the theoretical chemists are still rather modest. There are, however, many generalizations that enable the chemist to use his intuition to make valid qualitative predictions. Since your everyday interactions with materials make you familiar with a tremendous variety of phenomena, you may already have asked yourself most of the questions posed in this chapter—and in some cases you may have found a satisfactory answer. Such questions include the following:

1. Why are hydrogen, oxygen, nitrogen, fluorine, and chlorine gases, whereas bromine is a liquid and iodine a solid? Why are cellulose and polyethylene solids?
2. Why are ethyl alcohol, ethylene glycol (antifreeze), and sucrose (table sugar) soluble in water, whereas gasoline, ether, vegetable oil, and human skin are not?
3. Why are leaves green and blood red? What structural features are necessary for a molecule to absorb light? What happens to molecules when they absorb light? What is the chemistry of vision?

This chapter deals with all of these topics. Some of the tools for determining chemical structure are discussed, for they are important in themselves and provide illustrative examples of what scientists do.

5.1 PROPERTIES OF PURE NONPOLAR COMPOUNDS

The melting and boiling temperatures of a pure compound are measures of the attraction between the molecules in the crystalline and liquid states, respectively, of the compound. The higher the melting point, e.g., the stronger the attractive forces between the molecules in the solid. Several interesting trends in these properties emerge when a tabulation of related compounds is made. Table 5-1 shows the effect of molecular weight in three families of nonpolar substances.

Table 5-1. **The Effect of Molecular Weight on Melting Points and Boiling Points of Nonpolar Substances.**

Substance	Formula	Atomic or Molecular Weight	Melting Point, °C	Boiling Point, °C
Noble gases				
Helium	He	4	−272 at 26 atm	−269
Neon	Ne	20	−249	−246
Argon	Ar	40	−189	−186
Krypton	Kr	84	−157	−152
Xenon	Xe	131	−112	−107
Radon	Rn	222	−71	−62
Tetrahalomethanes or carbon tetrahalides				
Methane	CH_4	16	−182	−164
Tetrafluoro-methane	CF_4	88	−150	−129
Tetrachloro-methane	CCl_4	154	−23	77
Tetrabromo-methane	CBr_4	332	90	190
Tetraiodo-methane	CI_4	520	171 (decomposes)	
Linear hydrocarbons				
Methane	CH_4	16	−182	−164
Ethane	CH_3CH_3	30	−183	−89
Propane	$CH_3CH_2CH_3$	44	−190	−42
Butane	$CH_3CH_2CH_2CH_3$	58	−138	−1
Pentane	$CH_3CH_2CH_2CH_2CH_3$	72	−130	36
Hexane	$CH_3(CH_2)_4CH_3$	86	−95	69
Heptane	$CH_3(CH_2)_5CH_3$	100	−91	98

Table 5-1. **(continued)**

Substance	Formula	Atomic or Molecular Weight	Melting Point, °C	Boiling Point, °C
Octane	$CH_3(CH_2)_6CH_3$	114	−57	126
Nonane	$CH_3(CH_2)_7CH_3$	128	−51	151
Decane	$CH_3(CH_2)_8CH_3$	142	−30	174
Undecane	$CH_3(CH_2)_9CH_3$	156	−26	196
Dodecane	$CH_3(CH_2)_{10}CH_3$	170	−10	216
Tridecane	$CH_3(CH_2)_{11}CH_3$	184	−6	235
Tetradecane	$CH_3(CH_2)_{12}CH_3$	198	6	254
Pentadecane	$CH_3(CH_2)_{13}CH_3$	212	10	271
Hexadecane	$CH_3(CH_2)_{14}CH_3$	226	18	287
Heptadecane	$CH_3(CH_2)_{15}CH_3$	240	22	302
Octadecane	$CH_3(CH_2)_{16}CH_3$	255	28	316
Nonadecane	$CH_3(CH_2)_{17}CH_3$	269	32	330
Eicosane	$CH_3(CH_2)_{18}CH_3$	283	37	343
Hexacontane	$CH_3(CH_2)_{58}CH_3$	844	104	250 at 10^{-5} Torr
Polyethylene	$CH_3(CH_2)_nCH_3$ ($n \cong 800$)	$\approx$11,000	140 (Softening temperature)	Decomposes

In each of these three families, or *homologous series,* of compounds, we see that, as the molecular weight increases, the melting and boiling points increase. This is generally true of nonpolar compounds. Some features of their structures other than molecular weight must be considered, however. If we compare the melting temperature of radon (−71°C), which has a molecular weight of 222, with that of carbon tetrachloride (−23°C), which has a molecular weight of 154, we see that the substance with the lower molecular weight has the higher melting temperature. Similarly, undecane has a molecular weight of 156, almost identical to that of carbon tetrachloride, yet its boiling temperature is 196°C, or 119° higher than that of carbon tetrachloride. Thus the correlation works well in a homologous series, but composition and structure remain important factors. We must explain why this is so.

Exercise 5-1. Using the data in Table 5-1, show that the order of strength of intermolecular forces for a given molecular weight is: noble gases < carbon tetrahalides < hydrocarbons.

Before developing an explanation for the trends evident in Table 5-1, a brief digression on the uses of hydrocarbons may be of interest. Linear hydrocarbons are the major constituents of petroleum. Table 5-2 shows how some of the various components, or fractions, are used in our society as energy sources and for other purposes. Polyethylene, a synthetic plastic familiar to everyone, is made from ethylene by a process called *polymerization* (the repetitive chemical linking of a large number of identical small molecules to form a giant molecule, or *polymer*). We will discuss polymerization in Chapter 11.

5.2 LONDON FORCES

We see from Table 5-1 that nonpolar molecules and even atoms such as neon must attract each other, since otherwise they would not liquefy at low temperatures. Neon atoms are spherically symmetric, so the origin of this attraction is perhaps not obvious. The forces responsible for it were first explained in 1930 by the German physicist Fritz London (1900–1954), in whose honor they are called *London forces.*

Figure 5-1 is a representation of two neon atoms very near each other. Although the time-averaged distributions of electrons in both atoms are spherically symmetric (Figure 5-1a), statistical fluctuations occur (Figure 5-1b) that can result in an instantaneous dipole moment in either atom. Such an instantaneous dipole moment can induce a similar dipole moment in the other atom (or any neighboring atom). Both atoms are then slightly polarized in the same direction, and they therefore attract each other electrostatically.

Table 5-2. **Some Petroleum Fractions.**

Approximate Boiling-Point Range, °C	Name of Fraction	Common Uses
0–30 (gas)	Natural gas	Domestic and industrial heating; power generation
40–150 (liquid)	Gasoline	Automobiles
175–325 (liquid)	Kerosene and fuel oils	Domestic and industrial heating; power generation; diesel and jet engines
325 and up (viscous liquid)	Lubricating oils	Lubrication
Residue (solid or semisolid)	Asphalt	Paving; roofing

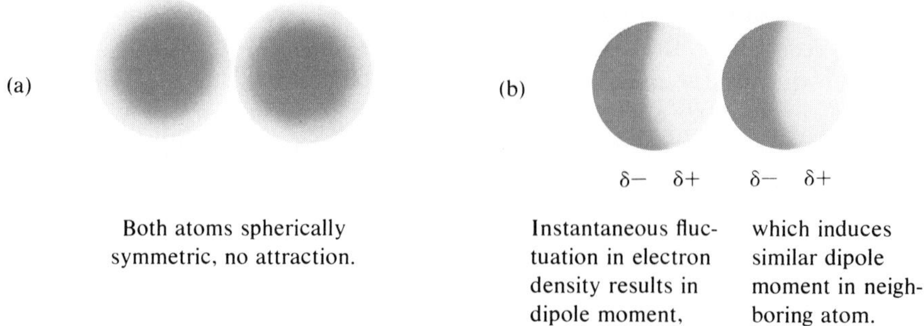
(a)
(b)
δ− δ+
δ− δ+
Both atoms spherically symmetric, no attraction.
Instantaneous fluctuation in electron density results in dipole moment,
which induces similar dipole moment in neighboring atom.

Figure 5-1. The origin of London forces. (a) Two spherically symmetric neon atoms. (b) Two instantaneously asymmetric neon atoms.

London forces are always attractive. The *polarizability* α of an atom or molecule is defined as the ratio of the induced dipole moment μ_{ind} in the atom or molecule to the intensity of the electric field **E** that induces it. (The electric field intensity at a point in space is the force exerted by the field on a unit charge at that point.)

$$\text{Polarizability} = \alpha = \frac{\mu_{\text{ind}}}{\mathbf{E}} \tag{5-1}$$

The ease with which a fluctuation in charge density can occur in a molecule is proportional to its polarizability. The ease with which the resulting instantaneous dipole moment can induce a dipole moment in a neighboring molecule is proportional to the polarizability of the neighboring molecule. For identical molecules we therefore expect the energy of interaction through London forces to be proportional to α^2. The equation for the potential energy of interaction due to London forces is

$$U_{\text{London}} = -\,(\text{const})\,\frac{\alpha^2}{r^6} \tag{5-2}$$

where r is the distance between the centers of the two molecules. The polarizability α has the dimensions of a volume and is usually expressed in cubic angstroms. The strong dependence on distance, r^{-6}, means that London forces are significant only at very small distances, i.e., on the order of atomic diameters.

It is mainly the distortion of the valence-shell electron distribution that contributes to the polarization, since the valence electrons are farthest from the nucleus and therefore the most susceptible to the effects of external fields. This means that atoms of high atomic number are more polarizable than those of lower atomic number. They also have a larger number of electrons, again making them easier to polarize. Since the strength of

Table 5-3. **Polarizabilities and Boiling Points of Some Nonpolar Substances.**

Substance	α, Å^3	T_{bp}, °C	Substance	α, Å^3	T_{bp}, °C
He	0.2	−269	C_2H_6	4.5	−89
Ne	0.39	−246	C_3H_8	6.3	−42
Ar	1.63	−186	C_4H_{10}	8.1	−1
Kr	2.46	−152	C_5H_{12}	10.0	36
Xe	4.00	−107	C_6H_{14}	11.8	69
			C_7H_{16}	13.6	98
CH_4	2.61	−164	C_8H_{18}	15.4	126
CF_4	—	−128	$C_{10}H_{22}$	19.1	174
CCl_4	10.5	77	$C_{12}H_{26}$	22.8	216
CBr_4	—	190			

London forces increases with polarizability, this explains the increase in melting and boiling points with atomic number in the noble gases. It also explains the increase in melting and boiling points in the homologous series CH_4, CF_4, CCl_4, CBr_4, and CI_4, since the polarizability of these molecules increases as the molecular weight increases. We see in Table 5-3 that $\alpha_{CCl_4} = 10.5$ Å^3 (mol wt of $CCl_4 = 154$), whereas $\alpha_{Xe} = 4.00$ Å^3 (mol wt of Xe = 131). Although they have similar molecular weights, carbon tetrachloride is about 2.6 times more polarizable than xenon. This accounts for the greater intermolecular attraction in the carbon tetrahalides than in the noble gases. In general, valence electrons in chemical bonds, or molecular orbitals, are more polarizable than electrons in atomic orbitals. This is because the bonding electrons are less localized (the molecular orbitals are larger in volume than the atomic orbitals).

We have so far restricted ourselves to atoms and essentially spherical molecules. However, the energy of interaction of two molecules due to London forces also depends on the surface area of interaction between the molecules. In the linear hydrocarbons interaction can occur along the entire lengths of adjacent carbon chains, so attractive forces are exerted over the entire length of the molecule. This accounts for the increase in melting and boiling points of the hydrocarbons with molecular weight, as well as the greater attraction between hydrocarbon molecules than between either carbon tetrahalides or noble gases of comparable molecular weight.

5.3 DIPOLE MOMENTS AND MOLECULAR INTERACTIONS

In the last chapter we saw that bonds between unlike atoms have dipole moments. A molecule containing such bonds also has a dipole moment,

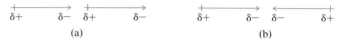

Figure 5-2. Dipole-dipole interactions. (a) Attraction. (b) Repulsion.

which is the vector sum of all the bond moments. (With the appropriate molecular composition and symmetry, this vector sum can, of course, be zero.) Molecules with nonzero dipole moments are called polar molecules; their properties are in many ways dominated by the direction and strength of their polarity. Polar molecules interact with one another through dipole-dipole interactions, which are either attractive or repulsive, depending on the orientations of the two dipoles (see Figure 5-2).

The potential energy of interaction of two dipoles with dipole moments μ_1 and μ_2 is

$$U_{\text{d-d}} = -\frac{\mu_1\mu_2}{r^3} f(\theta_1, \theta_2, \psi) \tag{5-3}$$

where r is the distance between the two dipoles and $f(\theta_1, \theta_2, \psi)$ is a function of the angles determining the orientation of the dipoles with respect to each other. The maximum value of $f(\theta_1, \theta_2, \psi)$ is $+2$, when the dipoles are arranged as in Figure 5-2a, for which the attraction is greatest (minimum potential energy). The minimum value is -2, when the dipoles are arranged as in Figure 5-2b, for which the repulsion is greatest (maximum potential energy). The dipole-dipole interaction has an orientational effect on polar molecules that is important in crystal packing and in the three-dimensional structure of proteins.

A molecule with a permanent dipole moment is able to induce a dipole moment in a neighboring molecule, which always results in an attractive force between them. If molecule 1 has dipole moment μ_1 and molecule 2 has polarizability α_2, the potential energy of this attraction is

$$U_{\text{ind}} = -\frac{(\mu_1^2)\alpha_2}{r^6} \tag{5-4}$$

There are thus three major types of electrostatic interactions between uncharged molecules: dipole-dipole, dipole-induced dipole, and London forces. The three together are called *van der Waals interactions*, after the Dutch physicist Johannes van der Waals (1837–1923). They determine the magnitude of the constant a in the van der Waals equation of state, which you will see in Chapter 6. The average sum is always attractive and is the source of the forces bringing about the condensation of gases and the freezing of liquids. The energy range for van der Waals interactions is about 0.1 to 1.0 kcal/mole, far less than the energies of chemical bonds.

5.4 DIPOLE MOMENTS, MOLECULAR STRUCTURE, AND CHEMICAL REACTIVITY

The data in Table 4-3 indicate that some chemical bonds are very polar and therefore have large dipole moments associated with them. This is especially true of certain multiple bonds, such as $>C{=}O$, which has a bond dipole moment of 2.5 D, and $-C{\equiv}N$, which has a bond dipole moment of 3.3 D. Compounds that contain a $>C{=}O$ bond are called *carbonyl compounds* because $>C{=}O$ is the *carbonyl group*. There are two major types of carbonyl compounds, called *aldehydes* and *ketones*:

$$\underset{\text{Aldehyde}}{R-\overset{\overset{\displaystyle O}{\|}}{C}-H} \qquad \underset{\text{Ketone}}{R_1-\overset{\overset{\displaystyle O}{\|}}{C}-R_2}$$

Here the R's can be any of innumerable different organic groups. For example, they might be *alkyl groups*, which are hydrocarbon compounds from which a hydrogen atom has been removed (methyl group, CH_3; ethyl group, C_2H_5; etc.).

The chemistry of the carbonyl compounds is dominated by the fact that there is a relative negative charge on the oxygen atom and a relative positive charge on the carbon atom. The carbonyl carbon, therefore, has a strong tendency to react with nonbonding electron pairs on other molecules. Two examples of such reactions are shown below.

$$\underset{\text{Acetaldehyde}}{CH_3-\overset{\delta+}{C}(=\overset{\delta-}{O})-H} + \underset{\text{Ethyl alcohol}}{H-\ddot{O}-C_2H_5} \longrightarrow \underset{\text{Acetaldehyde hemiacetal}}{CH_3-\underset{\displaystyle H}{\overset{\displaystyle OH}{C}}-OC_2H_5}$$

$$\underset{\text{Acetaldehyde}}{CH_3-\overset{\delta+}{C}(=\overset{\delta-}{O})-H} + \underset{\text{Sodium bisulfite}}{:\underset{\displaystyle O}{\overset{\displaystyle OH}{S}}-O^-Na^+} \longrightarrow \underset{\text{Acetaldehyde sodium bisulfite}}{CH_3-\underset{\displaystyle H}{\overset{\displaystyle OH}{C}}-\underset{\displaystyle O}{\overset{\displaystyle O}{S}}-O^-Na^+}$$

The class of compounds known as *ethers* has the general formula $R_1—O—R_2$ (bent molecules). Because the C—O bond has a rather low dipole moment (0.8 D), ethers are weakly polar molecules. *Alcohols* have the general formula R—OH and *carboxylic acids* have the general formula $R—C\begin{smallmatrix}{\nearrow}O\\{\searrow}OH\end{smallmatrix}$. Both these classes of compounds are intermediate in polarity between ketones and ethers. Table 5-4 contains the boiling points and dipole moments of several organic compounds with nearly identical molecular weights. From this table we conclude that intermolecular interactions are weakest in the hydrocarbons and generally increase in the order: ethers, aldehydes, ketones, alcohols, and carboxylic acids. This order is consistent with the increasing dipole moments of the compounds through the ketones, but the alcohols and carboxylic acids appear to have stronger interactions than are predictable from dipole moments alone.

Table 5-4. Boiling Points and Dipole Moments of Several Organic Compounds of Nearly Identical Molecular Weight.

Compound	Formula	Molecular Weight	T_{bp}, °C	μ, D	Class of Compound
2-Methylpropane (isobutane)	$CH_3CH(CH_3)CH_3$	58	−10	0.1	Hydrocarbon
Methylethyl ether	$CH_3OCH_2CH_3$	60	8	1.2	Ether
Propanal	$CH_3CH_2C(=O)H$	58	48	2.5	Aldehyde
Acetone	$CH_3C(=O)CH_3$	58	57	2.9	Ketone
2-Propanol (isopropanol)	$CH_3CH(OH)CH_3$	60	82	1.7	Alcohol
Acetic acid	$CH_3C(=O)OH$	60	118	1.7	Carboxylic acid

5.5 HYDROGEN BONDS

Two things are required for *hydrogen bonding*: (1) The hydrogen atom in question must be covalently bonded to a highly electronegative atom. Such hydrogens can serve as electron-pair acceptors. (2) There must be a nonbonding electron pair on another highly electronegative atom, which serves as the donor. The most common hydrogen bonds arise from hydrogen atoms covalently bonded to either oxygen or nitrogen atoms, although other elements, notably fluorine and chlorine, also participate in hydrogen bonding. The requisite pair of nonbonding electrons can be donated by an atom of the same element or a different element. These electrons form the weak hydrogen bond to the hydrogen atom. An example of a hydrogen bond between two methyl alcohol (methanol) molecules is shown below.

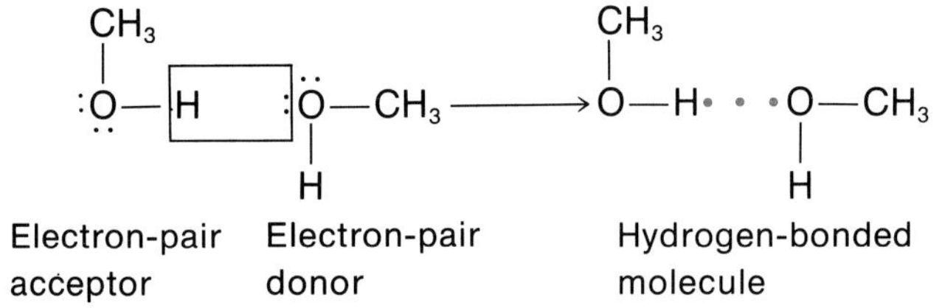

$$\underset{\text{Electron-pair acceptor}}{CH_3\text{—}\ddot{O}\text{—}H} \quad \underset{\text{Electron-pair donor}}{:\ddot{O}(H)\text{—}CH_3} \longrightarrow \underset{\text{Hydrogen-bonded molecule}}{CH_3\text{—}O\text{—}H\cdots O(H)\text{—}CH_3}$$

The most common hydrogen bonds (often called simply H bonds) in biological systems are shown in Figure 5-3. All these H bonds have energies between 3 and 7 kcal/mole. The hydrogen bond is therefore about one-tenth as strong as a typical covalent bond. This fact is of crucial importance to the chemistry of life because, as we will see in Chapter 11, it allows for weak but highly specific intermolecular interactions that can be altered without affecting the covalently bonded structures of the molecules in question.

Table 5-5 contains a list of several molecules, with the number of acceptor and donor sites given for each molecule. From this table we can see why aldehydes and ketones do not hydrogen bond to themselves: they have no acceptor sites. Hydrogens bonded to carbon do not serve as electron-pair

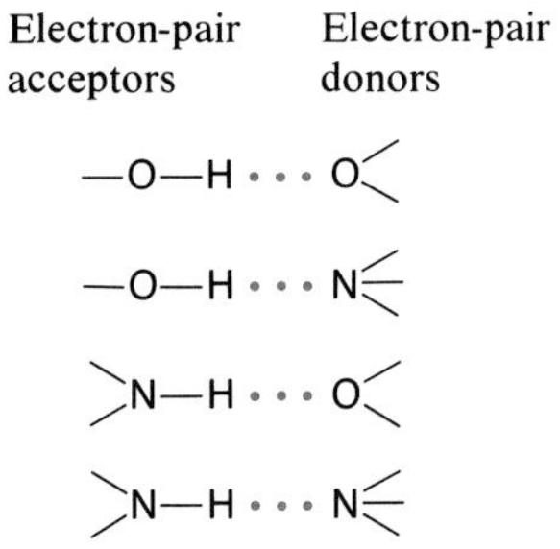

Electron-pair acceptors	Electron-pair donors
—O—H	··· O<
—O—H	··· N≤
>N—H	··· O<
>N—H	··· N≤

Figure 5-3. The biologically important hydrogen bonds.

Table 5-5. **Hydrogen Bonding Sites in Various Molecules.**

Compound	Formula	Number of Acceptor Sites	Number of Donor Sites
Water	$H_2\ddot{O}:$ (H–O–H, two lone pairs on O)	2	2
Ammonia	$:NH_3$ (N bonded to three H, one lone pair)	3	1
Hydrogen fluoride	$H—\ddot{F}:$ (three lone pairs on F)	1	3
Ethanol	$CH_3CH_2—\ddot{O}:—H$	1	2
Acetone	$CH_3C(=\ddot{O}:)CH_3$	0	2
Acetic acid	$CH_3C(=\ddot{O}:)—\ddot{O}—H$	1	4
Acetaldehyde	$:\ddot{O}=C(H)CH_3$	0	2

acceptors because carbon is not sufficiently electronegative to permit H bonding to occur. On the other hand, simple alcohols (such as ethanol) and carboxylic acids (such as acetic acid) can hydrogen bond to themselves and therefore have abnormally high melting and boiling points.

Hydrogen bonding is common in liquids and has also been observed in some gases. The existence of the hydrogen bond in crystals has been verified by the techniques of x-ray diffraction and—more successfully—neutron diffraction, which is much more sensitive to the presence of hydrogen atoms.

The Anomalous Properties of Hydrogen-Bonded Substances

It is clear from Table 5-5 that ammonia, water, and hydrogen fluoride are all capable of hydrogen bonding to themselves. The consequences of

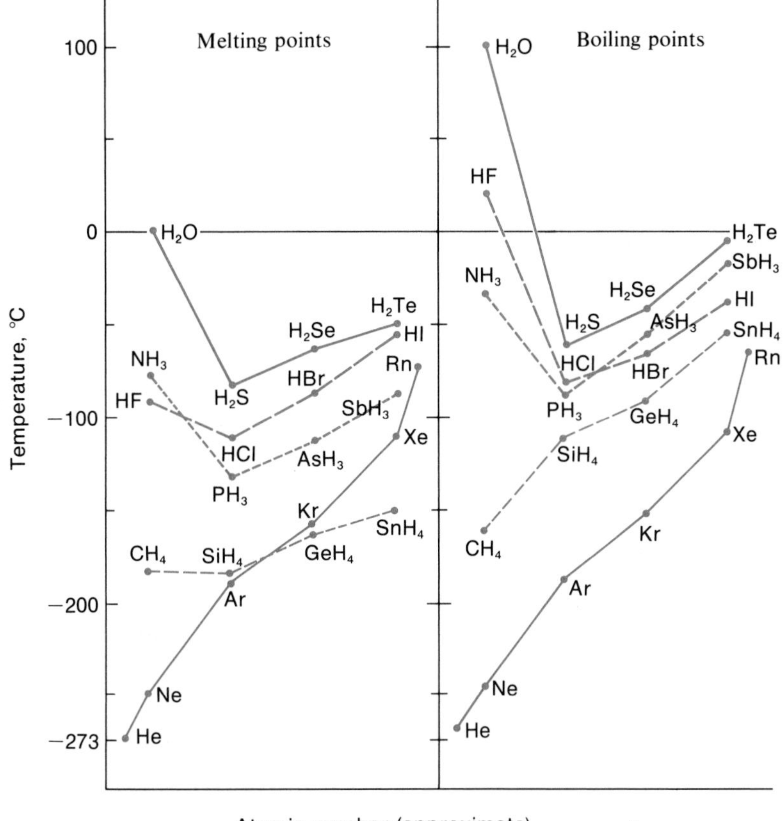

Figure 5-4. Melting and boiling points of the hydrogen compounds of a number of nonmetallic elements. Note the abnormally high values for hydrogen fluoride, water, and ammonia, due to hydrogen bonding in these compounds. (L. Pauling and P. Pauling, *Chemistry,* W. H. Freeman and Company. Copyright © 1975.)

this fact are varied and of great importance, especially for water. For example, the melting and boiling points of these compounds are abnormally high relative to other hydrogen compounds in the same groups. This is clearly evident in Figure 5-4, which shows that the general trend is for melting and boiling points to increase regularly with increasing molecular weight in a group series.

The H bonding in NH_3, H_2O, and HF is obviously very significant in the condensed phases, and can even carry over into the gaseous phase, especially in HF. In fact, gaseous samples of hydrogen fluoride have been shown to contain polymers of H—F such as

$$\text{H—F}\cdots\text{H—F} \qquad H_2F_2$$

$$\text{H—F}\cdots\text{H—F}\cdots\text{H—F} \qquad H_3F_3$$

and others such as H_4F_4, H_5F_5, and H_6F_6 (all shown in Figure 5-5). The latter has special stability because it is apparently a ring compound:

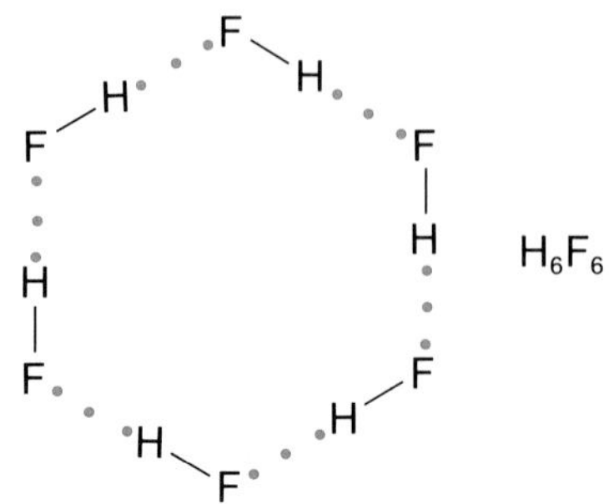

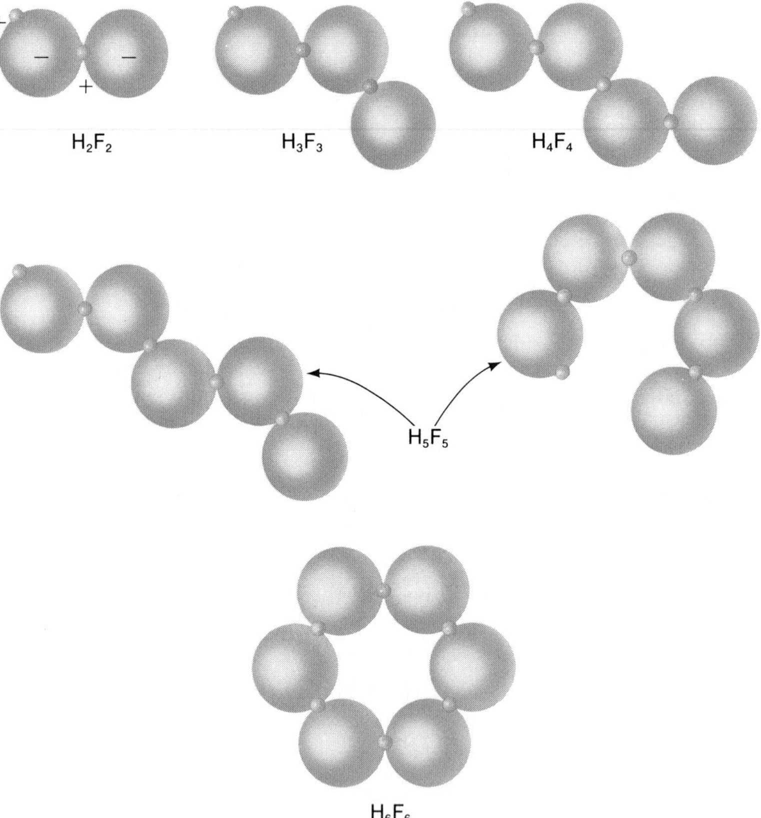

Figure 5-5. Several polymers of hydrogen fluoride. The fluorine atoms are held together by hydrogen bonds. The small balls represent the hydrogen atoms between the large, electronegative fluorine atoms. (L. Pauling and P. Pauling, *Chemistry,* W. H. Freeman and Company. Copyright © 1975.)

Rings of this sort are especially stable because the hydrogens can shift their positions along the FHF axes (by a shift in the bonding), providing an extra stabilization somewhat analogous to resonance stabilization.

Other molecules are also found to be H bonded in the gaseous phase. When the molecular weight of gaseous acetic acid is measured, it is found to be twice the expected value of 60. This means that gaseous acetic acid exists as a hydrogen-bonded dimer,

$$\mathrm{CH_3C}\begin{matrix} \diagup\!\!\diagup \mathrm{O}\cdots\mathrm{H{-}O} \diagdown \\ \diagdown \mathrm{O{-}H}\cdots\mathrm{O} \diagup\!\!\diagup \end{matrix}\mathrm{CCH_3}$$

indicating that the H-bond binding energy of 3–7 kcal/mole is sufficient to hold some molecules together at room temperature.

Hydrogen bonding determines the crystal structure of ice, in which each donor and each acceptor is utilized and there is essentially tetrahedral symmetry about each oxygen atom (see Figure 5-6). The fact that the structure of ice is so open causes ice to have a lower density than water. Therefore ice floats in water—a most unusual property, since almost all solids are denser than their corresponding liquids. Hydrogen bonds are also very important in the structure of water itself. Almost all molecules of such low molecular weight are gases, yet water, the most common liquid in our environment, is a liquid because of the hydrogen bond. Note also that hydrogen atoms can shift readily from one oxygen atom to another in both water and ice. One consequence of this is a high electrical conductivity of water if hydrogen ions are present in it.

5.6 THE SOLUBILITY OF SUBSTANCES IN LIQUIDS

A great many salts are soluble in water, whereas they are insoluble in most other common solvents. The high solubility of salts in water is the result of two closely related properties of water. First, water has an extremely high dielectric constant (80), so the force of attraction between oppositely charged ions in water, given by Coulomb's law (Equation 2-2), is only 1/80 of the force that would exist between them in a vacuum (or in air, which also has a dielectric constant of 1). Thus, a great many salts dissociate into ions upon contact with water, i.e., they dissolve. Second, water has a large dipole moment (1.86 D), which interacts strongly with the ions, especially the cations, to form *hydrated ions*. This process is accompanied by the release of thermal energy. Typical hydrated ions are $Be(OH_2)_4^{2+}$, $Mg(OH_2)_6^{2+}$, $Al(OH_2)_6^{3+}$, and $Cu(OH_2)_6^{2+}$, in which the relatively negative end of the water molecule (the oxygen) interacts with the positive central ion.

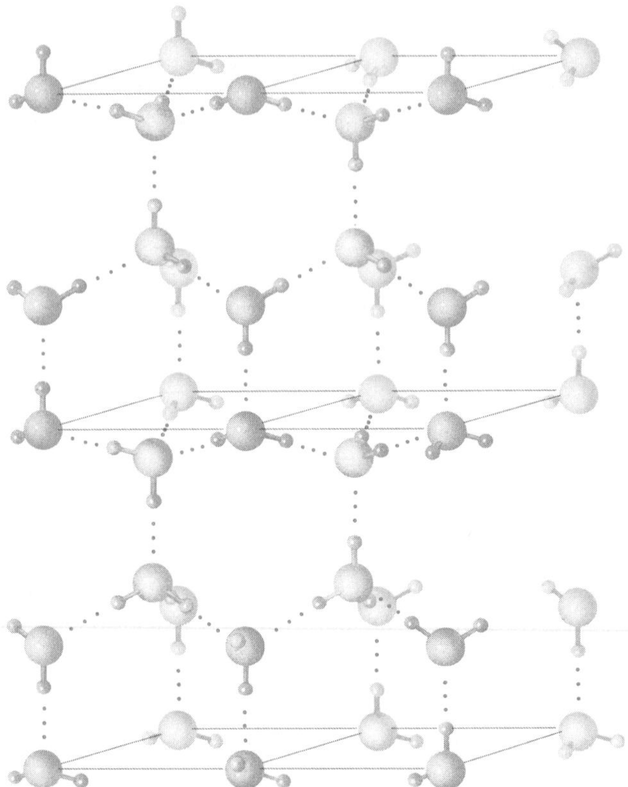

Figure 5-6. The ice lattice. The small and large balls represent hydrogen and oxygen atoms, respectively. Covalent bonds are represented by thick lines, and the hydrogen bonds are represented by three dots. (Adapted from L. Pauling and P. Pauling, *Chemistry*, W. H. Freeman and Company. Copyright © 1975.)

A few other liquids are ionizing solvents like water, resulting in salt solutions that conduct electricity because of the mobile ions in the solution. Some such liquids are hydrogen peroxide, hydrogen cyanide, liquid ammonia, and liquid hydrogen fluoride. They all have high dielectric constants and large dipole moments, and are therefore called *polar solvents*.

A common and usually valid generalization is that polar substances are miscible, or soluble, in one another, and nonpolar substances, such as benzene and gasoline, are soluble in one another. Polar and nonpolar substances do *not* usually dissolve in one another, however. Thus, gasoline does not dissolve in water. There are, of course, many intermediate cases. For example, ethanol (CH_3CH_2OH) is somewhat polar because of the OH group. It dissolves readily in both water and gasoline.

Hydrogen bonding is also very important for solubility. For example, although aldehydes and ketones cannot hydrogen bond to themselves because of their lack of acceptor sites, they dissolve well in water because they have donor sites for hydrogen bonds to water. Hydrogen bonding is, in fact, the factor primarily responsible for the solubility of organic compounds in water.

Soaps and Detergents

Soaps and detergents provide a particularly nice example of the principles of solubility. Soaps are made by reacting natural fatty acids (alkyl carboxylic acids) with sodium hydroxide in a process called *saponification.* A typical soap molecule is sodium laurate:

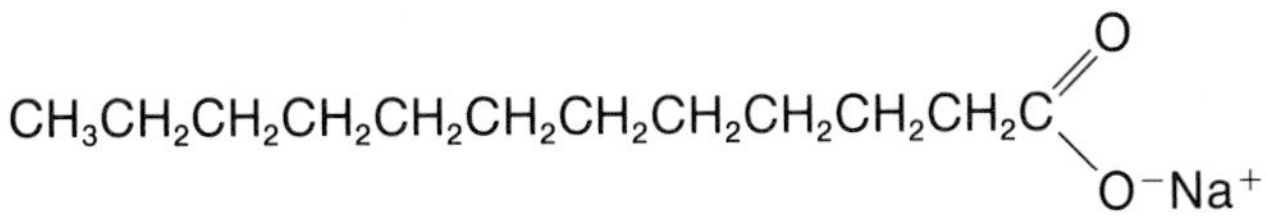

$$CH_3CH_2CH_2CH_2CH_2CH_2CH_2CH_2CH_2CH_2CH_2C\begin{matrix} \nearrow O \\ \searrow O^-Na^+ \end{matrix}$$

or, in chemical shorthand, $CH_3(CH_2)_{10}COO^-Na^+$. The long carbon chain in this molecule is nonpolar and does not dissolve well in water. We call this end of the molecule *hydrophobic* ("water-hating"). The polar end of the molecule, $—COO^-Na^+$, is a salt and dissolves well in water. It is *hydrophilic* ("water-loving") and therefore gives the soap its solubility. Since the hydrophobic "tails" would rather interact with themselves than with the water (which would mean the disruption of hydrogen bonds in the water), special soap structures, called *micelles,* are formed in solution. These structures are aggregates of soap molecules with all the hydrophobic tails interacting with each other and all the hydrophilic heads interacting with the water. A soap micelle is shown in Figure 5-7, together with the structure of a soap bubble and the mechanism for dissolving oil (a hydrophobic substance) in the soap solution. We will see in Chapter 11 that hydrophobic interactions are very important in protein folding and other biological structures.

Detergent is the name used for synthetic substances that are not soaps but have similar properties. In the most common type, the polar carboxylate end of the soap molecule is replaced by a sulfate group. A typical detergent molecule is sodium dodecylsulfate:

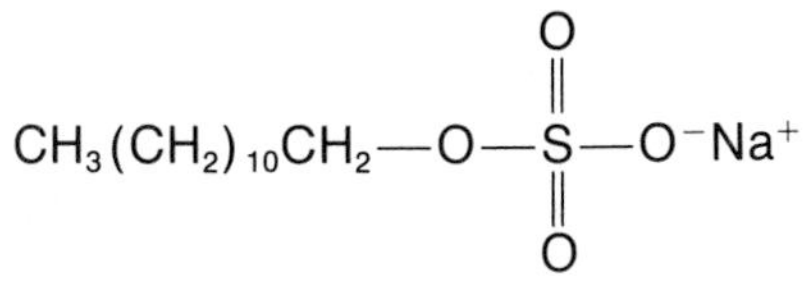

$$CH_3(CH_2)_{10}CH_2—O—\overset{\overset{\displaystyle O}{\|}}{\underset{\underset{\displaystyle O}{\|}}{S}}—O^-Na^+$$

Whereas soaps tend to precipitate in hard water (i.e., water containing bicarbonates of such divalent metals as Mg or Ca), leaving a ring of Mg soap

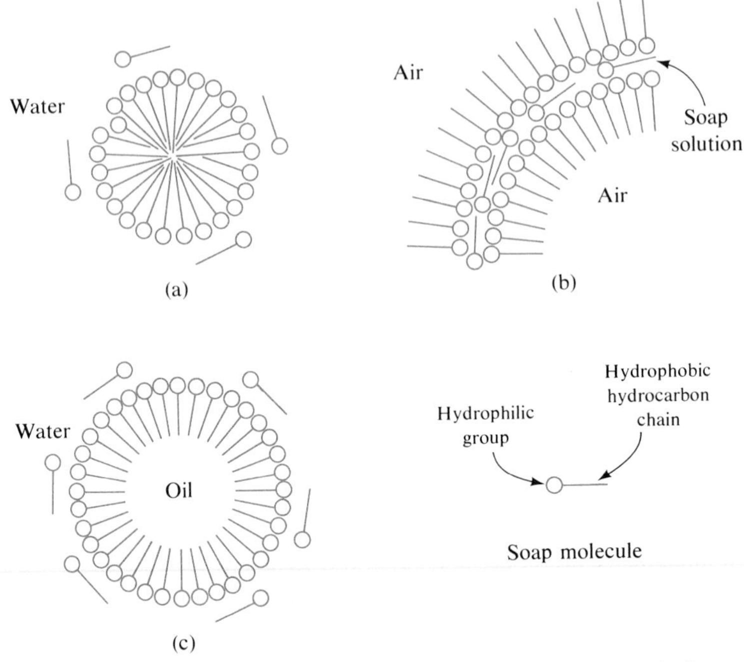

Figure 5-7. Micelle formation in soaps and detergents. (a) Soap micelle. (b) Section of soap bubble. (c) Oil droplet in soap solution.

or Ca soap around a bathtub, most detergents do not, and are therefore superior for use in hard water. Most detergents are not biodegradable, however, so in recent years their environmental impact has become the source of increasing concern. The detergent industry is responding by seeking new chemical formulations to circumvent this problem.

5.7 OPTICAL ACTIVITY AND OPTICAL ISOMERISM

A *polarizer,* such as Polaroid sunglasses, transmits only light waves whose electric field vectors are oriented in a specific direction, absorbing the waves whose electric vectors deviate significantly from this direction. Light emerging from a polarizer is called *plane-polarized* light. Figure 5-8 depicts this process.

When polarized light (or any light) passes a molecule, it interacts with the electrons of that molecule, because light is just an oscillating electromagnetic field. When this interaction results in a rotation of the plane of polarization of the light, the process is called *optical rotation.* Molecules

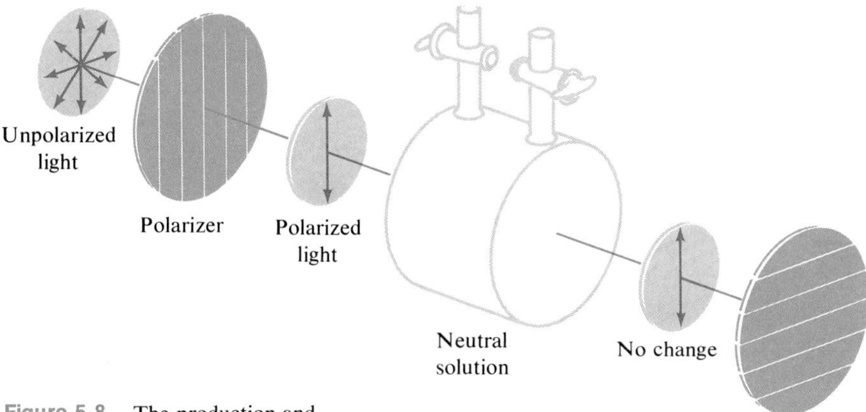

Figure 5-8. The production and nature of plane-polarized light.

that have the ability to rotate the plane of polarization of light are called *optically active* molecules. Figure 5-9 shows that the amount of rotation can be measured by placing a second polarizer in the light beam after the sample and finding the angle to which it must be rotated to transmit the maximum or minimum amount of light. This second polarizer is called the *analyzer* and the entire instrument is called a *polarimeter.*

Only molecules having a very special structural characteristic rotate the plane of polarized light, namely, molecules that are not superimposable on their mirror images. (Rotation occurs, however, only if the molecule and its mirror-image molecule are present in unequal amounts.) An example of nonsuperimposable mirror images is your right and left hands. A "right-handed" molecule rotates the plane of polarized light in the direction opposite to that characteristic of a "left-handed" molecule.

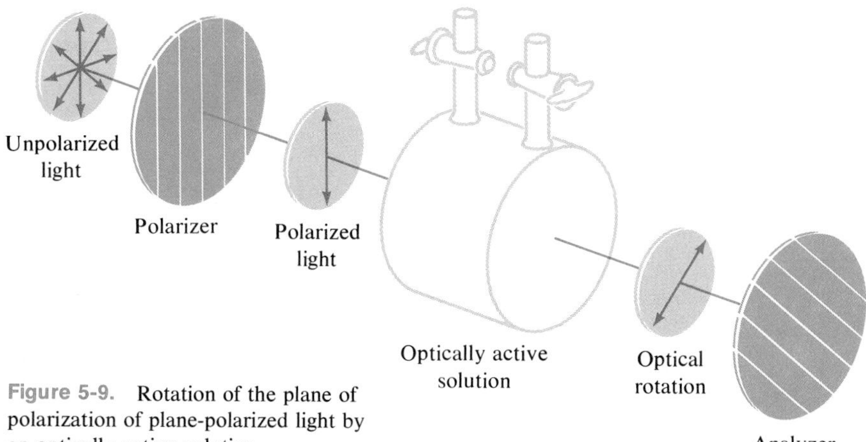

Figure 5-9. Rotation of the plane of polarization of plane-polarized light by an optically active solution.

Most molecules have superimposable mirror images. For example, consider chlorobromomethane, CH_2ClBr, which is shown in Figure 5-10. If we rotate the molecule on the right by 120° about the carbon-bromine bond, we see that it is identical to the molecule on the left, as shown in Figure 5-11. The CH_2ClBr molecule is therefore optically inactive.

In fact, all molecules of the forms CX_4, CX_3Y, CX_2Y_2, and CX_2YZ (X, Y, and Z representing different atoms or groups of atoms single-bonded to the carbon) have mirror images identical to themselves, so they are all optically inactive. Only if all four atoms or groups of atoms are different, as in CWXYZ, are the mirror images different. This leads to the rule, for simple carbon compounds, that *to be optically active a molecule must contain a carbon atom that is attached to four different groups.* This carbon atom is called an *asymmetric center.* Thus, 2-propanol is not optically active, but 2-butanol is:

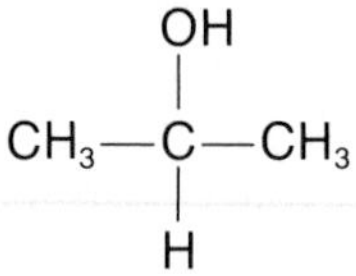

OH
|
CH_3—C—CH_3
|
H

2-Propanol

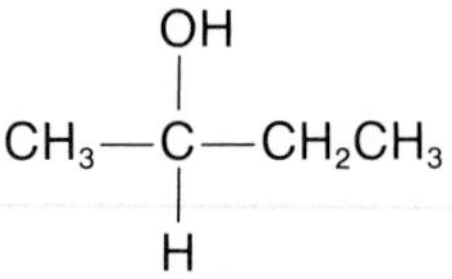

OH
|
CH_3—C—CH_2CH_3
|
H

2-Butanol

Ball-and-stick models are very useful for visualizing these structural features.

Two three-dimensional formulas for the 2-butanol molecule are shown in Figure 5-12 (asterisks are used to label the asymmetric centers). These are different molecules: they are not superimposable by any set of rotations. The two structures are called *optical isomers,* or *enantiomorphs.* A compound that rotates the plane of polarized light to the right as one looks down the beam toward the light source is given the symbol (+) and is said to be *dextrorotatory.* A compound that rotates the plane to the left is given the symbol (−) and is said to be *levorotatory.* The two 2-butanols represent these two classes of optically active compounds.

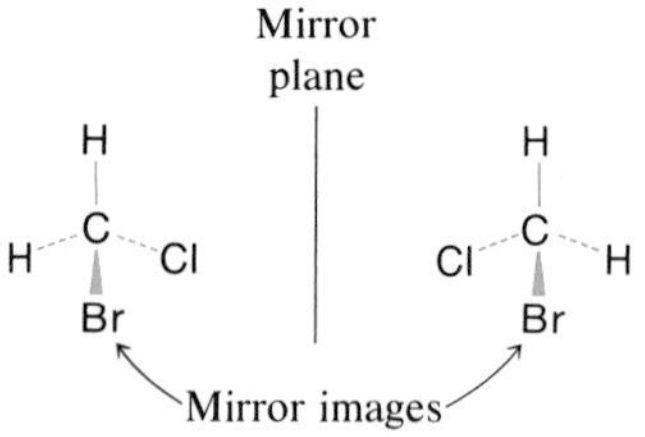

Mirror plane
H
H C Cl
Br
H
Cl C H
Br
Mirror images

Figure 5-10. The structures of chlorobromomethane and its mirror image.

H
Cl C H
120°
Br
→
H
H C Cl
Br

Figure 5-11. Equivalence of the mirror-image forms of chlorobromomethane.

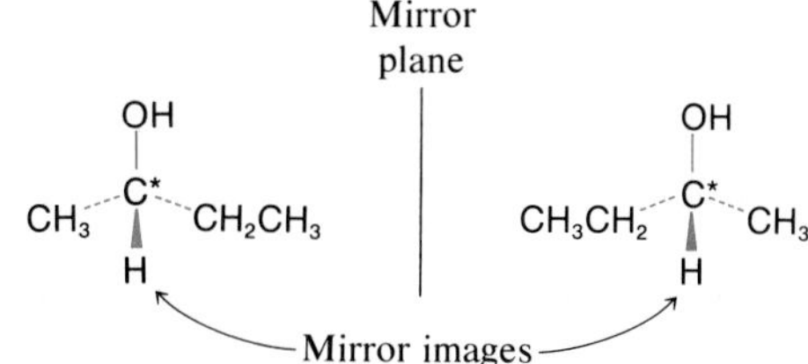

Figure 5-12. The structures of 2-butanol and its mirror image. This compound is optically active.

Synthesis of a molecule with an asymmetric center from reagents that are not themselves optically active produces a mixture containing equal amounts of the (+) and (−) molecules, called a *racemic mixture.* A racemic mixture does not rotate the plane of polarized light because there are equal numbers of molecules rotating the plane to the left and to the right. We will see in Chapter 11 that, generally speaking, living cells synthesize and use only one of the two possible optical isomers. The reason for this intrinsic biological asymmetry is unknown and is one the great mysteries of nature.

Typical examples of this phenomenon are the *amino acids,* which are the building blocks of proteins. Amino acids have the general chemical formula

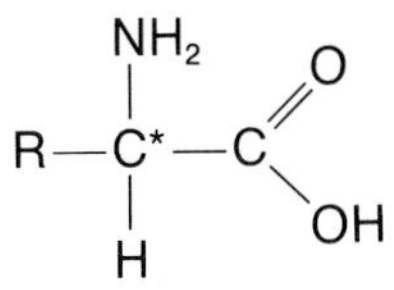

$$R-\overset{\overset{\displaystyle NH_2}{|}}{\underset{\underset{\displaystyle H}{|}}{C^*}}-C\begin{matrix}\nearrow\!\!\!\!\nearrow O\\ \searrow OH\end{matrix}$$

where R can be one of many different organic groups. Most of the amino acids have an asymmetric center and all such optically active amino acids exist in proteins in what is called the L-configuration. The L-configuration is that stereochemical arrangement of groups around the asymmetric center that is shown in Figure 5-13. The structure shown on the right is called a *Fischer projection formula* after the great German organic chemist Emil Fischer (1852–1919). It is important to understand that the geometry about the central carbon atom is the same as that specified in the two structures on the left.

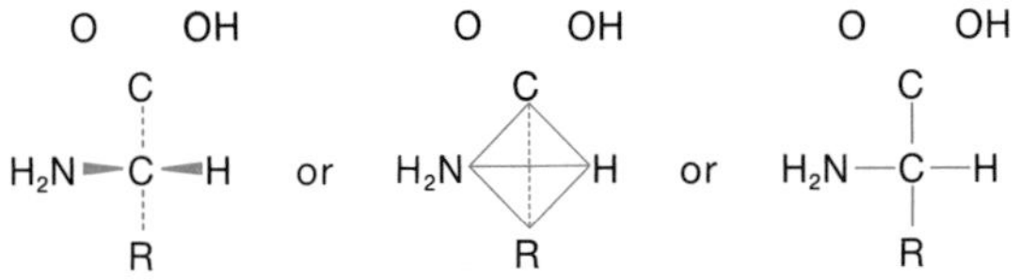

O OH / C / H_2N—C—H / R or O OH / C / H_2N ◇ H / R or O OH / C / H_2N—C—H / R

Figure 5-13. The absolute configuration of an L-amino acid.

The direction of rotation of the plane of polarized light is a function of the way light interacts with the groups attached to the asymmetric center. Therefore some L-amino acids are dextrorotatory (+) and some are levorotatory (−), even though they have the same configuration. It is important to remember that the molecules of life have a handedness and that the enzymes of the cell can distinguish between the right- and left-handed molecules.

In Chapter 4 we learned a little about coordination compounds. These provide another example of optical isomerism but the rules of their existence are not as simple as in carbon compounds with asymmetric centers. In general, you should draw the mirror images and test for superimposability in each case. One example of optical isomerism, the tris(ethylenediamine)-cobalt(III) ion, is shown on page 187.

Another example is dibromochloroaquodiamminecobalt(III), $[CoBr_2Cl(H_2O)(NH_3)_2]$. There are many possible structures with this name. One of them, with its mirror image, is shown in Figure 5-14. These two compounds are not superimposable, so they are enantiomorphs. To help you visualize such structures, you can make a simple model by inserting a pencil through the center of a cardboard square. Pieces of colored tape can be attached to the four corners of the cardboard and the ends of the pencil to represent the different ligands bound at the corners of an octahedron.

Exercise 5-2. Draw some other geometric isomers (isomers with different relative arrangements of the ligands) of $[CoBr_2Cl(H_2O)(NH_3)_2]$ and test them for the existence of optical isomerism.

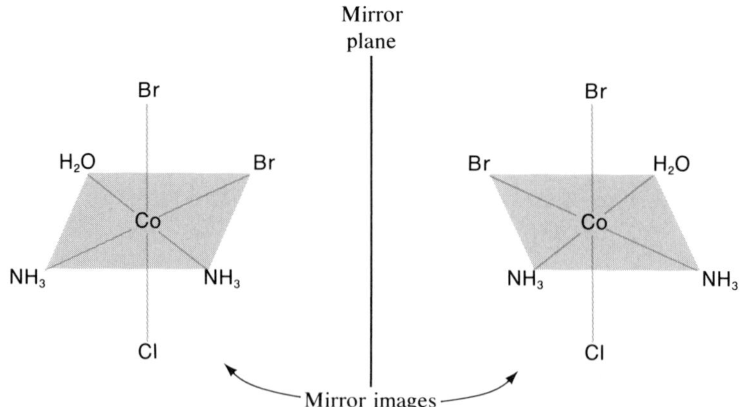

Figure 5-14. Two stereoisomers of dibromochloroaquodiamminecobalt(III).

5.8 INFRARED SPECTROSCOPY

Molecules are groups of atoms connected by chemical bonds. All the atoms in all molecules vibrate with respect to each other, all the time. This is a fundamental fact of nature, for which there is no a priori explanation. In considering the vibrations of atoms in a molecule, it is convenient to consider each bond as a spring connecting two balls. If the balls move apart, the spring is stretched, causing a restoring force that counteracts the divergence of the balls. Similarly, if the balls move toward each other, the restoring force of the spring is outward. A system consisting of two balls and a spring can be caused to vibrate at a frequency characteristic of that system by the application of an external oscillating force of the same characteristic frequency. This frequency is called the *natural frequency of vibration*. If the external force oscillates at the natural frequency of the system, the system absorbs mechanical energy with maximum efficiency.

The natural frequency of vibration of such a system is given by the relation

$$\nu = \frac{1}{2\pi}\sqrt{\frac{k}{\mu}} \tag{5-5}$$

where k is the *force constant* of the spring and μ is the *reduced mass* of the balls. The force constant is the proportionality constant between the restoring force F and the displacement x of the spring from its equilibrium position. This relation, $F = -kx$, is known as *Hooke's law*, after the English physicist Robert Hooke (1635–1703). The reduced mass of two balls of individual masses m_1 and m_2 is $m_1m_2/(m_1 + m_2)$. The stronger the spring, the greater are both its force constant and its natural frequency of vibration. As m_1 and m_2 increase, the frequency of vibration decreases.

Analogously, two chemically bonded atoms have a natural frequency at which the system absorbs energy most efficiently. If the bond is polar to any degree, i.e., if the atoms have partial electric charges relative to each other, the oscillating electric field of a light wave will alternately "push" and "pull" on the charged atoms. If the frequency of the light is the same as the natural frequency of the bond, a photon can be absorbed. When this happens, electromagnetic energy is converted into kinetic and potential energy of the vibrating molecule. (Note that the criterion of bond polarity, i.e., the presence of a permanent electric dipole moment, excludes the homonuclear diatomic molecules, such as H_2, N_2, O_2, and F_2, from consideration here.)

The strengths of chemical bonds vary widely. Therefore the effective force constants of different bonds are different, and the bonds absorb electromagnetic radiation of different frequencies. The radiation frequencies corresponding to the natural frequencies of vibration of chemical bonds are in the *infrared* region of the electromagnetic spectrum, i.e., that region that is adjacent to the red end of the visible spectrum. Infrared frequencies are lower than visible frequencies, so the corresponding wavelengths are longer than visible wavelengths.

Each type of bond has a characteristic frequency at which it absorbs energy. A molecule containing many bonds absorbs energy at a number of frequencies, and a plot of absorbance versus wave number (see Chapter 2) over the range of frequencies in question is called the *absorption spectrum* of the compound. Since the number of possible combinations of absorption frequencies (which often interact with each other in complex ways) is virtually infinite, no two compounds have the same infrared spectrum. The spectrum of any compound may thus be thought of as a "fingerprint" of the molecule, and unknown compounds can often be identified by their spectra.

A *spectrophotometer* is an instrument for measuring absorbance as a function of frequency. Figure 5-15 is a schematic representation of a typical infrared spectrophotometer, which compares the intensities of two beams of infrared radiation. One beam is passed through a cell containing a sample of the compound, generally either in its pure gaseous or liquid state or dissolved in some solvent. The other beam, called the *reference beam,* is

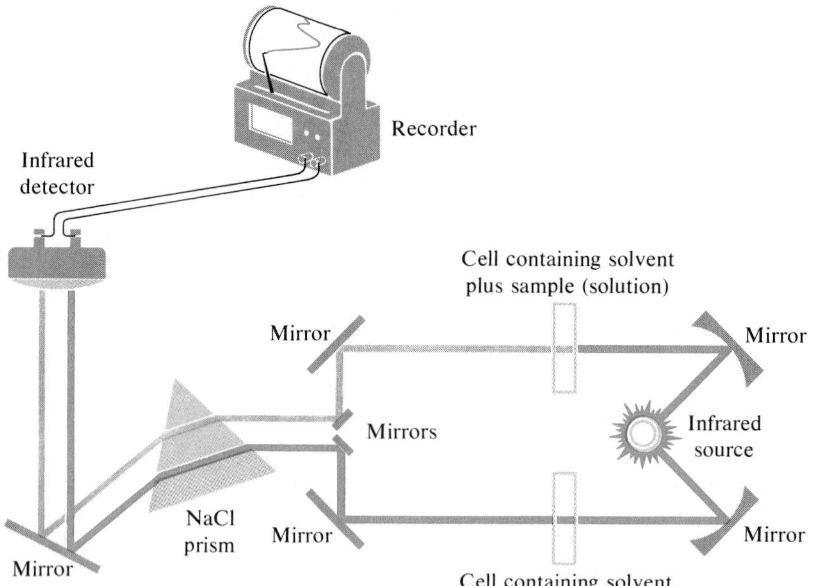

Figure 5-15. Diagram of an infrared spectrophotometer.

passed either through air (which, because it consists almost entirely of N_2 and O_2, has no infrared spectrum) or an identical quantity of the solvent. If the sample absorbs radiation at some frequency, the intensity I of the radiation of that frequency passing through the sample is less than the intensity I_0 passing through the reference substance. The spectrophotometer then determines the ratio as the *percent transmission* (equals $100 \times I/I_0$) over a range of frequencies and plots it automatically.

Table 5-6 is a compilation of the characteristic absorption frequencies

Table 5-6. Characteristic Absorption Frequencies of Some Organic Functional Groups.

Bond		Wave Number of Absorption, cm^{-1}
$-\overset{\vert}{\underset{\vert}{C}}-H$	(stretch)	2700–3100
$-\overset{\vert}{\underset{\vert}{C}}-H$	(bend)	1350–1450
—O—H	(Alcohols)	3500–3700
>C=O	(Aldehydes, ketones, acids)	1660–1870
$-\overset{\vert}{\underset{\vert}{C}}-\overset{\vert}{\underset{\vert}{C}}-$	(Alkanes)	600–1500
>C=C<	(Alkenes)	1600–1680
—C≡C—	(Alkynes)	2200–2260
$-\overset{\vert}{\underset{\vert}{C}}-O-$	(Ethers)	1000–1300
$-\overset{\vert}{\underset{\vert}{C}}-F$		1000–1400
$-\overset{\vert}{\underset{\vert}{C}}-Cl$		600–800
$-\overset{\vert}{\underset{\vert}{C}}-Br$		500–600

of several types of bonds associated with various organic groups. Note that for each type of bond a range of wave numbers is given because any one bond in a molecule is generally affected by the surrounding bonds. One might, e.g., expect small differences in the C—H stretching frequencies in ethane, CH_3—CH_3, and ethylene, CH_2=CH_2, because it is actually the entire molecules that are vibrating. The presence of the carbon-carbon double bond instead of the single bond therefore affects the frequency of radiation absorbed by the C—H bond. From Table 5-6 we see that, as the mass of the halogen atom increases, the wave-number range of absorption of the carbon-halogen bond decreases. This is exactly what one expects from the decreasing force constant (decreasing bond energy) and the increasing reduced mass.

In addition to the stretching vibrations considered thus far, molecules also undergo other types of vibrations, e.g., bending vibrations. The diagrams below illustrate these two types of motion.

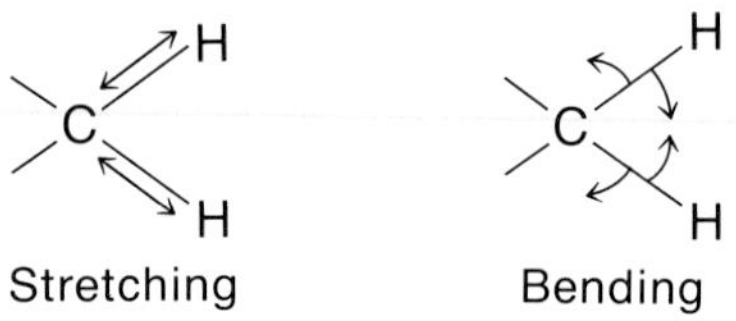

Stretching

Bending

Bending modes of vibration also absorb infrared radiation, further complicating the spectra of molecules.

Figure 5-16 shows the infrared absorption spectra of four compounds. Compare the absorption peaks, or *bands,* in these spectra with the data in Table 5-6, assigning organic functional groups to the various bands. We have already assigned the bands for methanol in Figure 5-16a.

The quantum mechanical model of molecular vibrations is not very different, in principle, from the spring model suggested above. The potential energy curve for a diatomic molecule is shown in Figure 5-17 together with the vibrational energy levels, which designate the only vibrational energy states allowed to the molecule according to the laws of quantum mechanics. The vibrational quantum number v specifies which vibrational energy level the molecule is in. Note that at the bottom of the well the levels are spaced nearly equally. The difference in energy between any two adjacent levels is $\Delta E = h\nu$, where ν is related to the force constant of the bond by Equation 5-5. The energy levels are close together if the force constant is small and far apart if it is large.

Figure 5-16. Low-resolution infrared spectra of four compounds. The four principal absorption bands in methanol are identified. (C. J. Pouchert, *The Aldrich Library of Infrared Spectra,* Aldrich Chemical Company, Inc., Milwaukee, Wisc.)

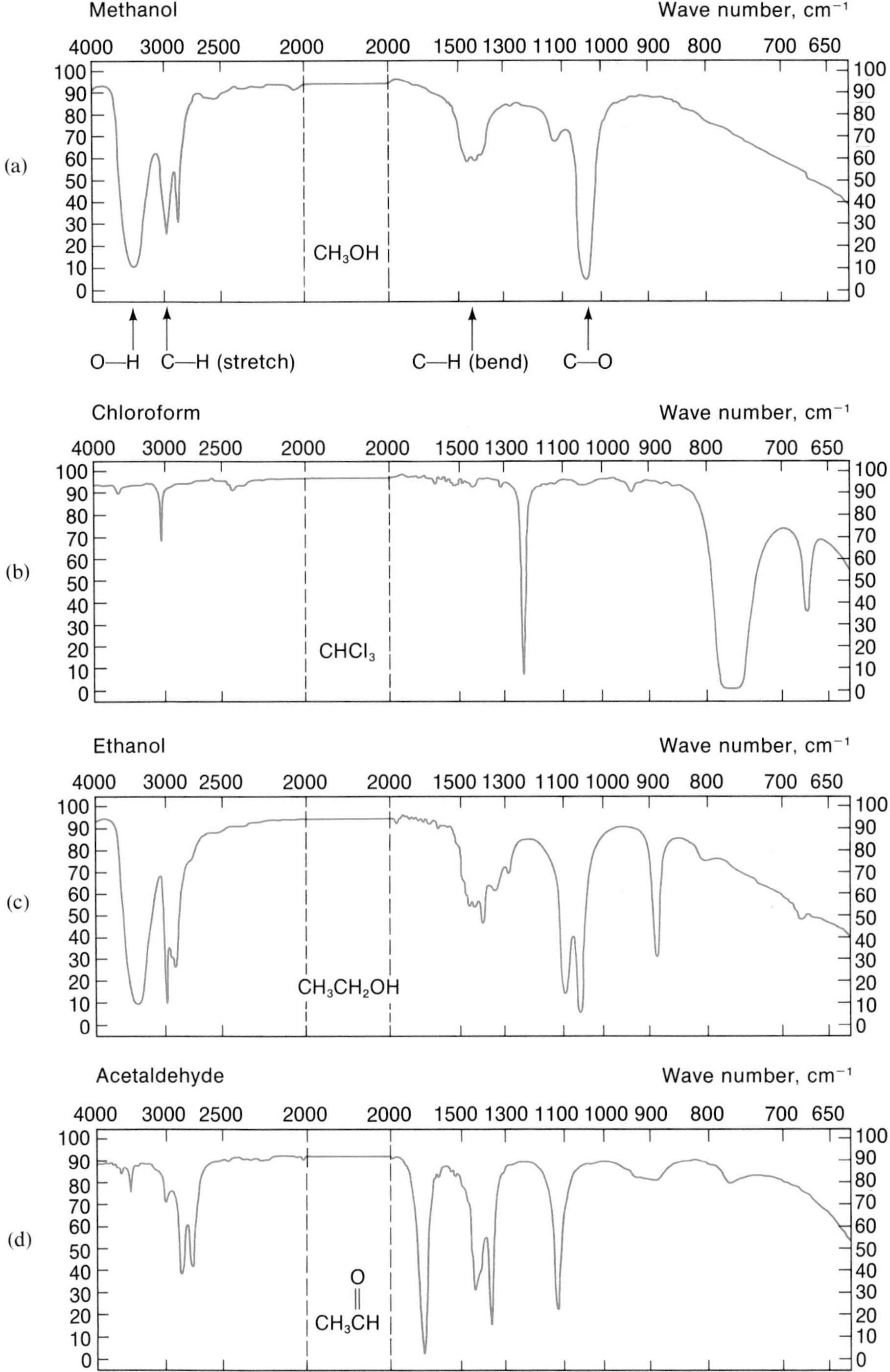

(a)
Methanol
Wave number, cm⁻¹
4000 3000 2500 2000 2000 1500 1300 1100 1000 900 800 700 650
100 90 80 70 60 50 40 30 20 10 0
CH₃OH
O—H
C—H (stretch)
C—H (bend)
C—O
(b)
Chloroform
Wave number, cm⁻¹
4000 3000 2500 2000 2000 1500 1300 1100 1000 900 800 700 650
100 90 80 70 60 50 40 30 20 10 0
CHCl₃
(c)
Ethanol
Wave number, cm⁻¹
4000 3000 2500 2000 2000 1500 1300 1100 1000 900 800 700 650
100 90 80 70 60 50 40 30 20 10 0
CH₃CH₂OH
(d)
Acetaldehyde
Wave number, cm⁻¹
4000 3000 2500 2000 2000 1500 1300 1100 1000 900 800 700 650
100 90 80 70 60 50 40 30 20 10 0
O
‖
CH₃CH

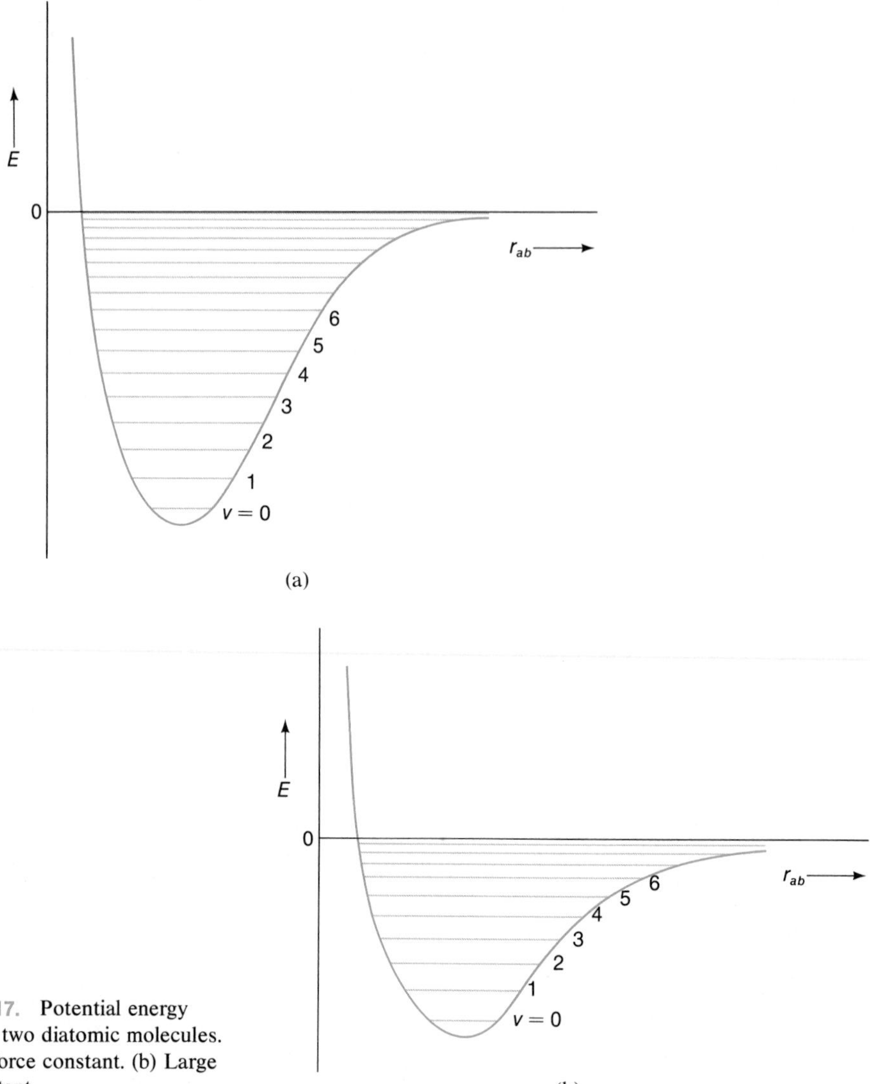

Figure 5-17. Potential energy curves for two diatomic molecules. (a) Small force constant. (b) Large force constant.

5.9 ABSORPTION OF VISIBLE LIGHT

Light in the visible region of the electromagnetic spectrum, although more energetic than infrared radiation, is still not energetic enough to break most covalent bonds. Molecules that absorb visible light do so by being excited to higher electronic energy levels. Only those molecules for which the spacing between the ground electronic state and an excited state is equal to the energy of a photon of visible light can absorb visible light. In the world of organic molecules, the only molecules that can absorb visible light are those with delocalized electrons, i.e., those having *conjugated* structures, as described in Chapter 4. Such molecules contain alternating carbon-carbon double bonds and carbon-carbon single bonds.

A typical example of such a molecule is the red pigment in tomatoes, lycopene. The carbon skeleton of lycopene is shown below. The hydrogens in the compound are omitted for convenience; for complete accuracy they would have to be added so as to provide each carbon with four bonds.

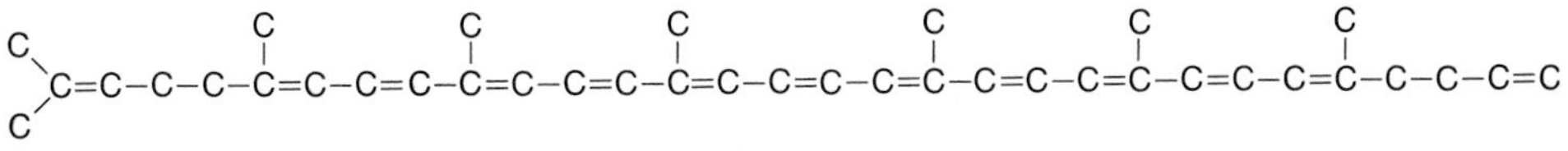

```
C               C           C           C           C           C           C
 \              |           |           |           |           |           |
  C=C−C−C−C=C−C=C−C=C−C=C−C=C−C=C−C=C−C=C−C=C−C=C−C=C−C−C−C=C
 /
C
```

Lycopene

The 11 conjugated double bonds in lycopene are all *trans* double bonds, which means that the carbon chain propagates on opposite sides of the double bond, as shown in Figure 5-18.

Lycopene is red because it transmits red light and absorbs most of the remaining light in the visible spectrum. The electrons that absorb light in lycopene are in the molecular orbitals of this conjugated system. The energy levels of these delocalized electrons can be predicted by solving the wave equation for the electron in a one-dimensional box. Lycopene has 11 conjugated double bonds and therefore has 22 delocalized electrons. These fill the lowest 11 energy states of the particle-in-a-box model. Absorption of light occurs when an electron is excited from the 11th to the 12th energy level, as shown in Figure 5-19.

The color of a compound such as lycopene is determined not just by the number of conjugated double bonds in the molecule. A cultivated variety of yellow tomato called the "tangerine tomato" contains a geometric isomer of lycopene called prolycopene. This compound contains alternating *cis* and *trans* double bonds in the conjugated system, although its chemical formula is, of course, identical to that of lycopene. The structure of the conjugated portion of prolycopene is shown in Figure 5-20. The color of this compound is different from that of lycopene because the delocalized electrons must travel a different path in this molecule, which changes the

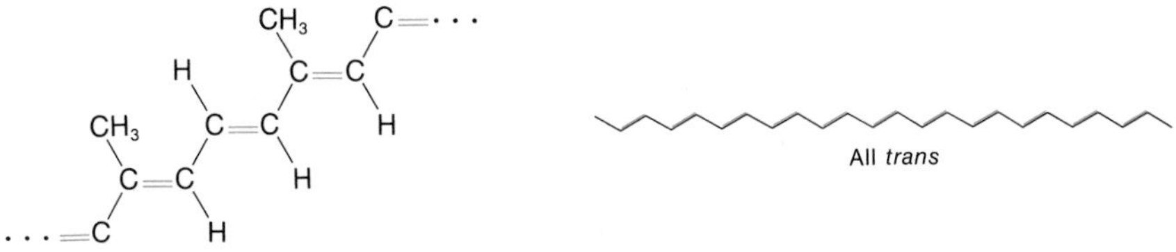

```
                    CH3     C=···
                      \    /
            H          C=C
             \        /   \
  CH3         C=C          H
    \        /   \
     C=C          H
    /   \
···=C    H
```

All *trans*

Figure 5-18. The all-*trans* conjugated structure of lycopene. A portion of the molecule is shown, together with a schematic of the complete backbone.

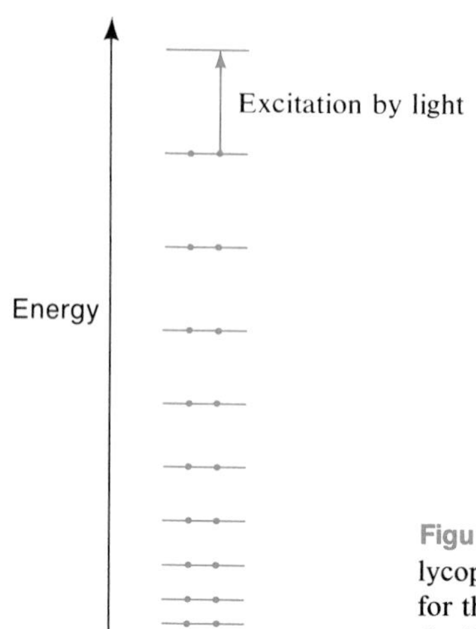

Figure 5-19. The particle-in-a-box energy levels for lycopene. The electron populations of the first eleven levels for the ground state of the molecule are shown by dots on the lines.

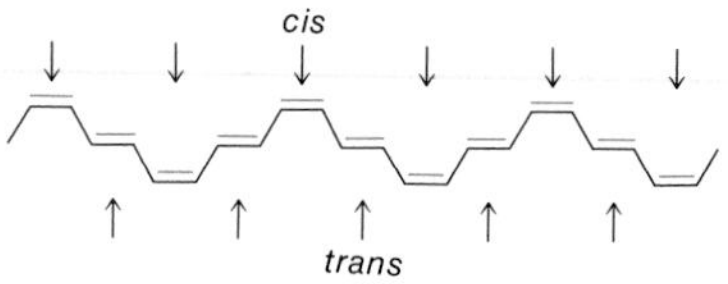

Figure 5-20. The conjugated structure of prolycopene. Only the backbone is shown.

spacings between the energy levels. Lycopene and prolycopene are wonderful examples of how significant changes in nature can result from rather subtle structural changes at the molecular level.

Two of the most familiar things in nature that absorb light are blood and green plants. The former derives its color from heme, the latter from chlorophyll. Both compounds are porphyrins (see Figure 5-21 for the structure of the porphyrin ring); they differ only in the substituents on the ring, and the central metal ion coordinated to the nitrogens. But this is enough to make one red and the other green. The absorption of light by the porphyrin-Fe(II) in hemoglobin is more complicated than that in a conjugated π-electron system because the d electrons in the Fe^{2+} ion are important in the optical transitions that determine the color of the compound. The numerous optical transitions associated with the d orbitals of the metal ions in complex ions are the reason for the wealth of color in these compounds. Ligand field theory explains these transitions as a transfer of an electron from a d orbital between the ligand-metal bonds (a d_{xy}, d_{yz}, or d_{xz} orbital for octahedral complexes) to a d orbital directed at a ligand (a $d_{x^2-y^2}$ or d_{z^2} orbital for octahedral complexes). An analysis of the importance of the metal ion relative to the π electrons in the porphyrin ring in the absorption of light is beyond the scope of this book.

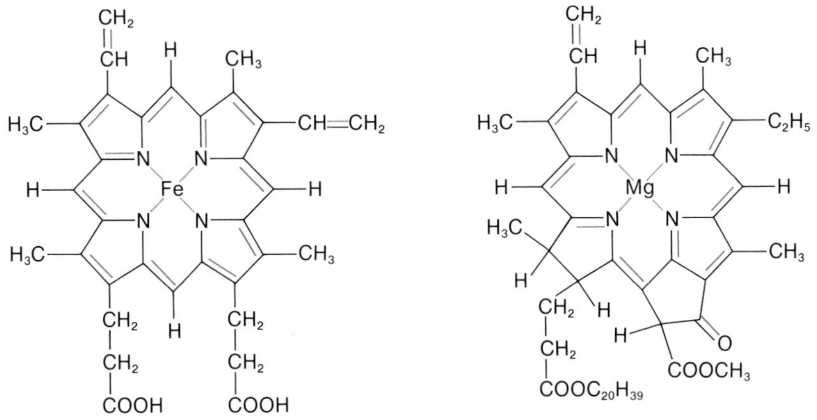

Heme (red) Chlorophyll a (green)

Figure 5-21. The porphyrin rings of heme and chlorophyll a. Note the slight differences between the two structures: they differ only in the central atom and the nature of the side chains.

Bibliography

Gordon M. Barrow, **The Structure of Molecules**, W. A. Benjamin, New York, 1963. Barrow is a spectroscopist who has written an excellent text entitled *Introduction to Molecular Spectroscopy.* The book cited here is essentially an elementary version of the more advanced text. It is relatively nonmathematical and very clearly written. The emphasis is on rotational and vibrational spectra, with some discussion of electronic spectra.

Wallace S. Brey, Jr., **Physical Methods for Determining Molecular Geometry**, Reinhold, New York, 1965. This brief volume is one in the series, *Selected Topics in Modern Chemistry*, published by Reinhold. It describes, in an interesting and concise manner, how molecular properties are determined. The discussions of the electrical properties of molecules and molecular vibrations are particularly pertinent to the material in the present chapter. The section on diffraction methods will be especially useful as a supplement to Chapter 6.

Problems

1. The following four carboxylic acids belong to a homologous series:

Propanoic acid	CH_3CH_2COOH
Dodecanoic acid	$CH_3(CH_2)_{10}COOH$
Pentanoic acid	$CH_3CH_2CH_2CH_2COOH$
Octanoic acid	$CH_3(CH_2)_6COOH$

   a. List them in order of decreasing boiling point.
   b. List them in order of decreasing solubility in water. Give the reasons for the solubility order that you have chosen.

2. The following four alcohols belong to a homologous series:

1-Hexanol	$CH_3(CH_2)_4CH_2OH$
Ethanol	CH_3CH_2OH
1-Decanol	$CH_3(CH_2)_8CH_2OH$
1-Butanol	$CH_3(CH_2)_2CH_2OH$

a. List them in order of increasing boiling point.
b. List them in order of decreasing solubility in water. Give the reasons for the solubility order that you have chosen.

3. The magnitude of the dipole moment of α-amino acids can be estimated from the distance between the positively and negatively charged groups. This distance is approximately 3 Å. The charge of an electron is 4.8×10^{-10} esu. Calculate the approximate dipole moment of an α-amino acid in Debye units.

4. Consider the following compounds: methanol, CH_3OH; sulfuric acid, H_2SO_4; acetamide, CH_3CONH_2; benzoic acid, C_6H_5—COOH; benzenesulfonic acid, C_6H_5—SO_3H; glycine, NH_2CH_2COOH; propane, $CH_3CH_2CH_3$; ethyl chloride, CH_3CH_2Cl.
a. For each molecule give the number of electron-pair donor sites and electron-pair acceptor sites for hydrogen bonding.
b. Which of the molecules form hydrogen bonds with other molecules in the same family?
c. Which of the molecules form hydrogen bonds with water?

5. Diethyl ether has a much lower boiling point (34.5°C) than 1-butyl alcohol (117°C), which has the same molecular weight. How do you account for this?

6. Which of the following pairs of compounds form hydrogen bonds to each other:
a. Chloroform, $CHCl_3$, and trimethylamine, $(CH_3)_3N$.
b. Ethanol, CH_3CH_2OH, and pentane, $CH_3(CH_2)_3CH_3$.
c. Polyethylene glycol, $(CH_2—CH_2—O)_n$, and methanol, CH_3OH.
Draw a detailed diagram of each hydrogen-bonded system that you have identified.

7. The structure of a common constituent of household detergents, sodium dodecylsulfate, is

$$CH_3CH_2CH_2CH_2CH_2CH_2CH_2CH_2CH_2CH_2CH_2CH_2—O—\overset{\overset{\displaystyle O}{\|}}{\underset{\underset{\displaystyle O}{\|}}{S}}—O^-Na^+$$

Detergents make dirt and grease water-soluble. Relate the above chemical structure to its ability to serve this function.

8. Which of the following compounds have enantiomers: chloroform, $CHCl_3$;

2-butanol, $CH_3CHOHCH_2CH_3$; 2-chloropentane, $CH_3CHClCH_2CH_2CH_3$; 3-chloropentane, $CH_3CH_2CHClCH_2CH_3$; tris(ethylenediamine)cobalt(III), $[Co\ en_3]^{3+}$?

9. Draw the enantiomers of the following molecules:

a. BrCHICl

b. Cl—C(H)(I)—SO_3H (H above C, I below C)

c. $C_2H_5CHClCH_3$

10. Assign an absolute configuration (D or L) to each of the molecules shown below, and draw its mirror image, i.e., the other stereoisomer.

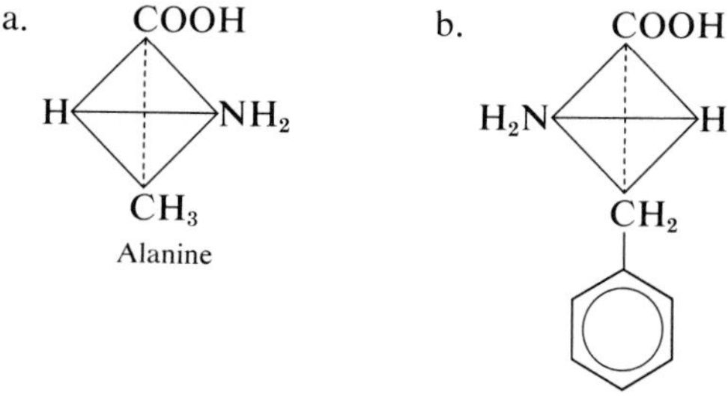

Alanine

Phenylalanine

11. For each of the following octahedral cobalt complexes, sketch the isomers and state the number of optical isomers and geometric isomers.

a. $[Co(NH_3)_6]^{3+}$

b. $[Co(C_2O_4)_3]^{3-}$ (oxalate is a bidentate ligand)

c. $[CoCl_2(NH_3)_4]^+$

d. $[CoCl_3(NH_3)_3]$

12. Draw the possible isomers of the following molecules:

a. $[CoCl_2(NH_3)_2]$, which has a tetrahedral configuration;

b. $[PtCl_2(NH_3)_2]$, which has a square planar configuration.

13. The force constants for a single and double bond are approximately 5.0×10^5 and 1.0×10^6 dyn/cm, respectively. Calculate the natural vibration frequencies (in s^{-1}) for the following bonds:

a. Carbon-carbon double bond

b. Carbon-oxygen double bond

c. Carbon-hydrogen single bond

(For the purposes of this problem, assume that the reduced mass μ of the two bonded atoms can be used, despite the presence of many other atoms in the molecules.) Compare your results with the data in Table 5-6 by converting the wave-number ranges to frequency ranges.

14. Two atoms, A and B, are connected by a double bond. If all other factors are held constant, would the natural vibration frequency of the A=B stretching vibration increase, decrease, or remain the same when

a. the atomic masses of A and B are tripled;

b. the force constant of the bond is reduced by one half;

c. the wave number of the radiation entering the sample is $10\ \text{cm}^{-1}$ less than that corresponding to the natural absorption frequency of the bond.

15. Which of the following compounds have conjugated electrons (delocalized electrons)? Which do you think might absorb visible light?

a. Cyclopentene,

b. Octatetraene, $CH_2{=}CH{-}CH{=}CH{-}CH{=}CH{-}CH{=}CH_2$

c. Octane, $CH_3CH_2CH_2CH_2CH_2CH_2CH_2CH_3$

d. Cyclopentane,

e. Vitamin A,

INTERLUDE

S.GROSS

the molecular basis of night vision

The retina of the human eye contains two types of structural elements that are sensitive to light: the rods and the cones. The rods are used for night vision and the cones for day, or color, vision. In bright light the photosensitive pigment in the rods is bleached, making the rods unable to function as visual receptors. When you enter a dark room, your eyes require two or three minutes to adapt to the low light levels at which the cones are ineffective and the rods are effective.

The light-absorbing molecule in the rods is vitamin A aldehyde, or *retinal.* The structure of the carbon skeleton of retinal is

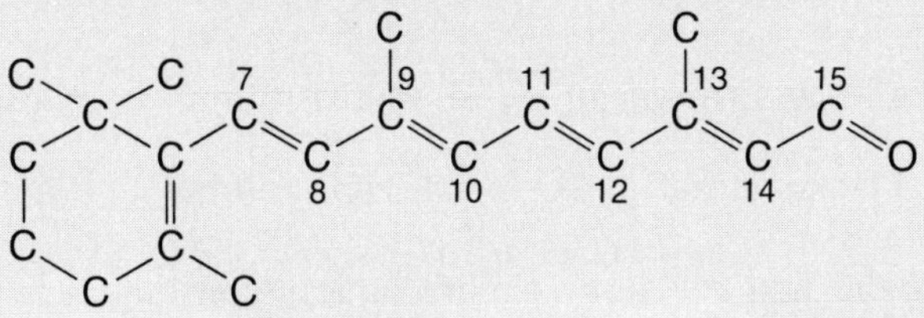

It absorbs visible light because it has a fairly long series of conjugated double bonds. The eye's acclimatization to darkness is due to the action of an enzyme called *retinal isomerase*, which converts the all-*trans* retinal shown above to the somewhat less stable 11-*cis*-retinal,

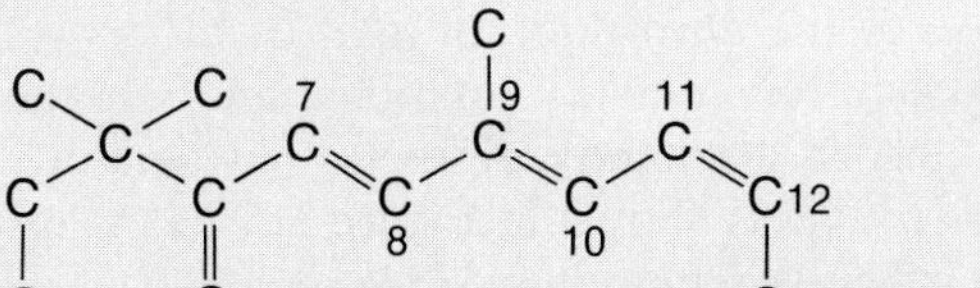

(so named because the double bond starting at position 11 in the molecule is in the *cis* configuration). This isomer, which also absorbs light, is responsible for night vision. As it is generated, it binds to a protein molecule called opsin to form *rhodopsin*, a complex that is extremely sensitive to light. Since visible light (780–380 nm) has energies of 37–75 kcal/einstein,* it can disrupt the *cis* double bond in 11-*cis*-retinal enough to allow rotation of the end of the molecule about this bond. When this happens the retinal reverts to the more stable all-*trans* configuration and simultaneously disengages itself from the opsin molecule. By a mechanism not yet fully understood, this photochemical reaction affects the optic nerve endings in the retina and causes an electric impulse to be sent to the brain. The brain's integration and spatial interpretation of all such signals produces the sensation of vision. This system is so sensitive that, under optimum conditions, the human eye can sometimes detect as few as five photons at a time.

In bright light almost all the 11-*cis*-retinal is converted to the *trans* form. The enzymes in the rods cannot convert it back to the *cis* form as fast as the light converts it to the *trans* form, however, and only about 1% of the retinal is in the *cis* form at any instant. Night vision is therefore effectively bleached out, and the less sensitive but more versatile cones take over the visual function. This bleaching can be thought of as a competition between two reactions:

$$\text{all-}\mathit{trans}\text{ retinal} \xrightarrow{\text{Enzyme}} \text{11-}\mathit{cis}\text{-retinal} \qquad \text{(Slow)}$$

$$\text{11-}\mathit{cis}\text{-retinal} \xrightarrow{h\nu} \text{all-}\mathit{trans}\text{ retinal} \qquad \text{(Fast)}$$

The rate of the first reaction is more or less constant, and two to three minutes are required for the reaction to go to completion. The rate of the second depends on the intensity of the light, but the response of 11-*cis*-retinal to an absorbed photon is instantaneous. Of course, if there is a deficiency of retinal in the eyes, owing to a dietary vitamin A deficiency, the rods cannot function properly; the result is a condition called *night blindness*.

One of the characteristics of the absorption of light by molecules that is distinct from those of the absorption of light by atoms is the broadness of the bands. This is due to the many rotational and vibrational states of the molecule, which absorb at slightly different wavelengths. The absorption spectrum of 11-*cis*-retinal is shown in Figure 1. The major peak is centered at 387 nm, which is barely within the visible region. The compound absorbs longer-wavelength violet and blue light so strongly, however, that its color is yellow. It is interesting to see the change in the spectrum (and hence the color) of retinal when it is bound to opsin. Figure 2 shows that the maximum absorption in rhodopsin is at 498 nm, which is in the middle of the green region of the visible spectrum. The color of this complex is therefore

*One einstein is Avogadro's number of photons, or a mole of photons.

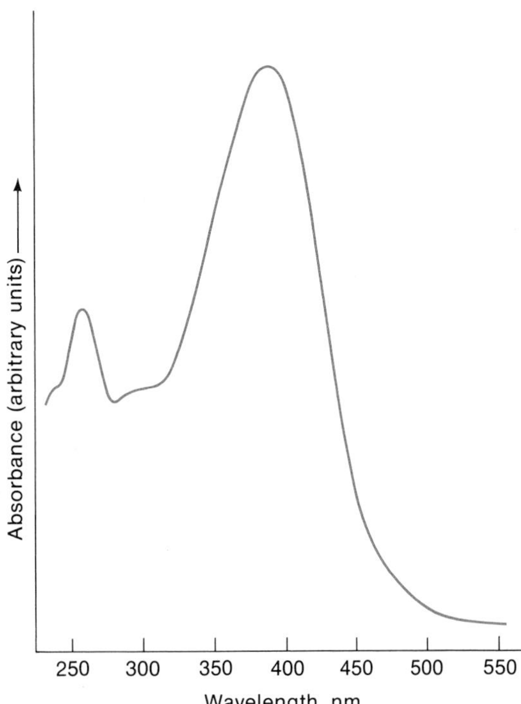

Figure 1. The absorption spectrum of 11-*cis*-retinal. (Courtesy of Prof. Wayne Hubbell, Department of Chemistry, University of California, Berkeley.)

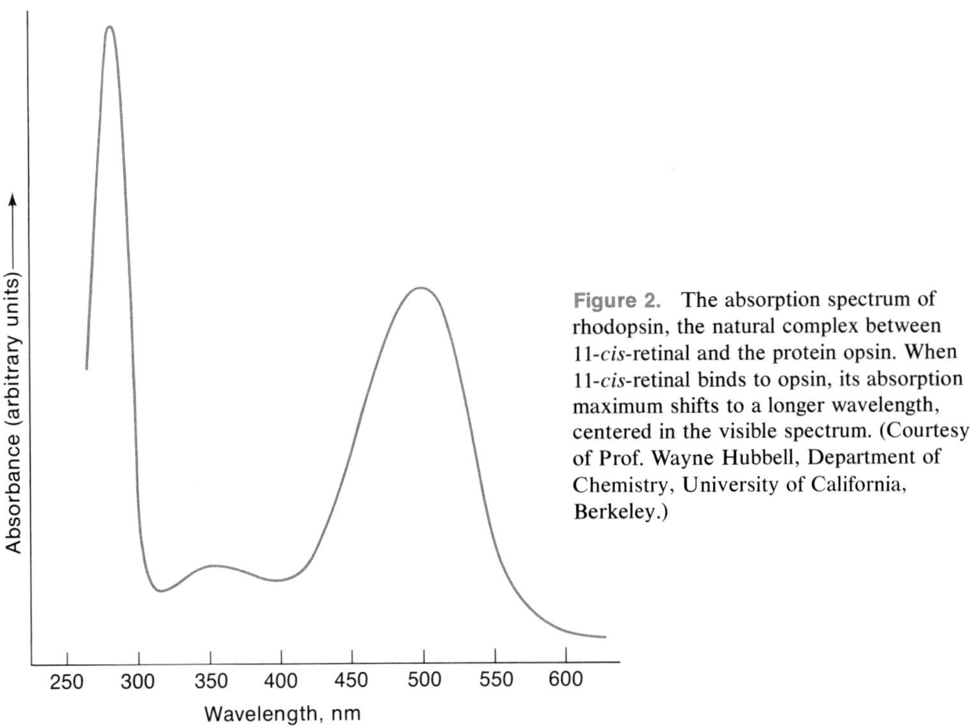

Figure 2. The absorption spectrum of rhodopsin, the natural complex between 11-*cis*-retinal and the protein opsin. When 11-*cis*-retinal binds to opsin, its absorption maximum shifts to a longer wavelength, centered in the visible spectrum. (Courtesy of Prof. Wayne Hubbell, Department of Chemistry, University of California, Berkeley.)

magenta, the result of a mixture of light from the red and violet ends of the spectrum. (Rhodopsin is often called *visual purple*, although it is not really purple.)

Since rhodopsin absorbs green light so strongly, it would not be surprising if the dark-adapted human eye were far more sensitive to green light than to any other kind. Indeed, this has been verified experimentally. The curve of spectral sensitivity of the eye in night (rod) vision matches the absorption spectrum of rhodopsin almost perfectly. Thus, although the dark-adapted eye is totally colorblind (all objects appear as different shades of gray), an individual being tested with different wavelengths of visible but very weak radiation will consistently pick those near 500 nm (green) as being the brightest in intensity.

6 the states of matter

An extraterrestrial visitor speculating about Spaceship Earth might wonder how many different forms of matter he would encounter here. He might be surprised to find that there are basically only three. There is no a priori reason that this should be so. Granted that there exist a liquid (liquid helium) that crawls up out of beakers, and solids (e.g., lead and glacier ice) that flow—very slowly, to be sure. But our everyday experience indicates that there are only three states in which matter ordinarily exists on Earth: solid, liquid, and gaseous. (Liquids and gases are often referred to jointly as *fluids*.)

Let us again consider the substance H_2O. Residents of Missoula, Montana, view water in January rather differently than do surfers in Hawaii in June, and these forms of water are quite different from that which powers Old Faithful in Yellowstone National Park. Fortunately (or unfortunately, as your viewpoint dictates), scientists are rather more systematic in their investigations than to traipse around the Earth looking at matter under a variety of conditions. Apart from the romance, all the fundamental aspects of each of the three forms of matter, as well as their interconversions, can be rigorously examined in the chemists' or physicists' laboratories.

The conditions under which the several states of matter exist can be examined in a cylinder in which the pressure is varied by moving a piston and the temperature is varied by bringing the cylinder into contact with a variable-temperature bath. Consider such a cylinder containing 18.02 g

H_2O, 1 mole. Let us first examine what happens when we lower the temperature from, say, 225°C at a constant pressure of 1 atm (this quantity, approximately equal to normal atmospheric pressure at sea level, will be defined below). We find that all the water is initially in the gaseous phase and occupies a volume of 40,900 ml. As the temperature is decreased, we observe that the piston must be moved inward to maintain the 1-atm pressure. This behavior continues with all the water remaining in the gaseous phase until the temperature reaches 100°C.

We observe liquid water for the first time as the temperature is lowered even slightly below 100°C. When the conversion is complete, the volume of the water is only 18.81 ml. As the temperature is lowered further, only a very slight increase in pressure on the piston is required to maintain the pressure at 1 atm. If very careful measurements of the volume of the water are made, it is found that the volume decreases steadily until the temperature 4°C is reached, at which the volume is 18.02 ml. A further slight *increase* in volume is noted as the temperature is lowered to the freezing point of water, 0°C.

As the third form of water—ice—is produced at this point, the volume is seen to increase sharply again, to the value 19.65 ml. Further lowering of the temperature produces only very small further decreases in volume.

Experience tells us what would happen if we were to place each of these forms of H_2O into another container, perhaps a 5-ℓ* box: the ice would retain its shape, the water would conform to the shape of the bottom of the box, and the water vapor would totally fill the box. These observations are important clues that will be helpful in our subsequent discussion of the magnitudes of the forces operative in each phase.

Another way to examine transitions between the several states of matter is to hold the temperature fixed with a constant-temperature bath and vary the pressure. In order again to observe all three phases, we must be careful in our selection of the temperature. If we choose a temperature of 0.001°C and a pressure of 1 Torr (1 Torr is 1/760 atm and will be defined more precisely below), our water sample will again be gaseous. As the pressure is raised, however, the next phase formed is ice, at about 4 Torr, followed by the appearance of liquid water at nearly atmospheric pressure. A continuation of such measurements under a wide variety of conditions would yield the data shown in Figure 6-1. Carefully performed measurements demonstrate that the states of water can be represented by the diagram shown in Figure 6-2, which is a refinement and extension of Figure 6-1.

---

*To avoid confusion with the numeral 1, we use the script ℓ rather than l or l. to designate the liter as a unit of volume. Abbreviations of derived units, such as ml and μl, are printed with the roman l because they do not cause confusion. The liter was redefined in 1964 to be exactly equal to one cubic decimeter (1 dm^3); 1 ℓ is therefore exactly equal to 1000 cm^3 = 1000 ml.

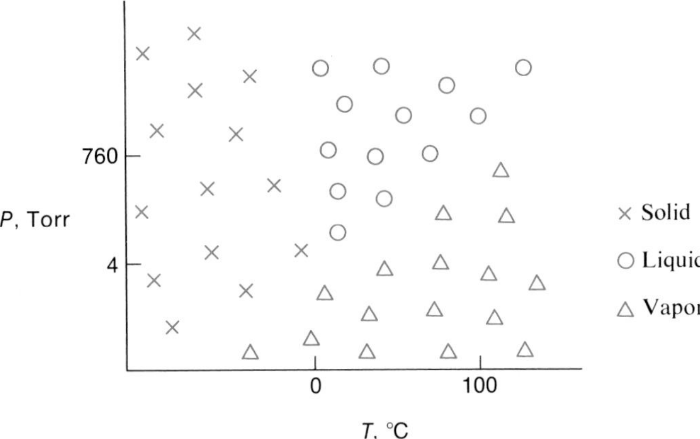

Figure 6-1. States of matter in H_2O, as determined by the pressure and temperature.

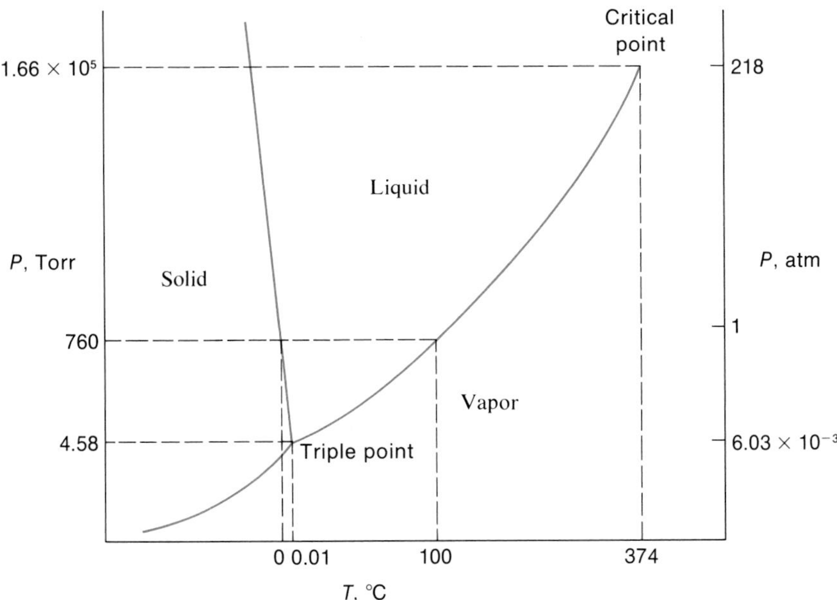

Figure 6-2. Phase diagram for H_2O at relatively low pressures (not drawn to scale).

Figure 6-2 is called a *phase diagram.* A *phase* is a portion of matter that is homogeneous throughout and separated from other parts of the system by physically distinct boundaries. Few phase diagrams are as simple as Figure 6-2. The complete diagram for H_2O is far more complicated, because eight different forms of ice have been found to exist at low temperatures and high pressures. A complete diagram is given in Figure 6-3. Because the ordinate must extend to such high pressures, the vapor region cannot be displayed.

Several features of these diagrams merit comment. At one point in the low-pressure diagram and at many points in the higher-pressure diagram, there appear points where three lines intersect. It is only for those precise values of the temperature and pressure that the three phases indicated in the three adjacent areas are able to coexist in equilibrium. The slightest deviation in either the temperature or the pressure causes one of the phases

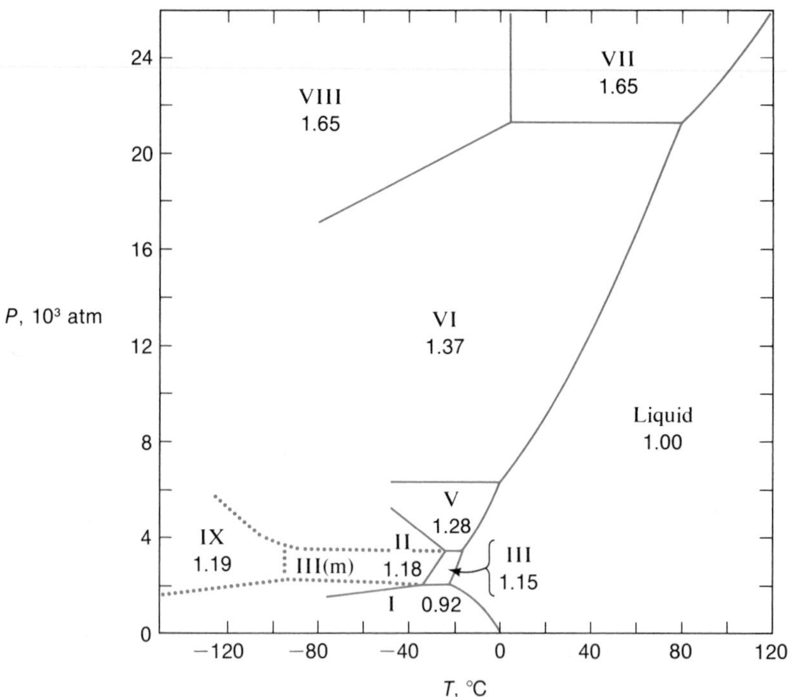

Figure 6-3. Phase diagram for H_2O at high pressures. The diagram shows that eight different forms of ice have been discovered (there is no ice IV, and ice III(m) is a metastable form of ice III). The numerals inside the diagram are densities in g/cm³. (L. Pauling, *General Chemistry,* 3rd ed., W. H. Freeman and Company. Copyright © 1970.)

to be transformed into one of the other two. Thus ice, water, and water vapor can coexist indefinitely only at 4.58 Torr and 0.01°C. This temperature is called the *triple point* of water.

Another point of interest is the *critical point*, at which the equilibrium line between the liquid and vapor phases ends. Above this temperature there is no discontinuous transition between the two phases, and only one fluid phase can be observed.* For water, the values of the critical temperature and critical pressure are 374°C and 218 atm.

Finally, the lines themselves are of great importance. They are useful in considering such concepts as phase transformations (sublimation, melting, freezing, evaporation, condensation), melting and boiling points, the variation of vapor pressure with temperature, and even the reason those people in Missoula are able to ice-skate.

There are many questions we want to ask about the behavior of matter represented by each region of a phase diagram. What forces exist between molecules in each phase? What theories and equations are available to describe these phases? What experimental techniques have been useful in measuring the several properties of each phase? What models do we use to describe these phases?

6.1 GASES

It is reasonable to begin with gases, not because we do more chemistry in the gaseous phase but because our understanding of it is the most nearly complete. We currently have available two simple and satisfactory descriptions of this state, an experimental one and a theoretical one. Their combination yields an interesting relation in our study of gases. The experimental law is one of the most basic laws of chemistry; the theoretical result is of great intrinsic interest.

Experimental Description of Ideal Gases

You are probably already familiar with the three basic gas laws. Thus we will present them only briefly. Their description will be followed by their combination to yield the ideal gas equation. Before we begin, however, we should examine the units of pressure that are commonly used.

*Although the terms *vapor* and *gas* are often used interchangeably, we can now draw a distinction between them; a *vapor* is a gas that is below its critical temperature, i.e., one that can be liquefied by the application of sufficient pressure. Broadly, the vapor phase is the gaseous state of a substance that is a liquid or solid under ordinary conditions.

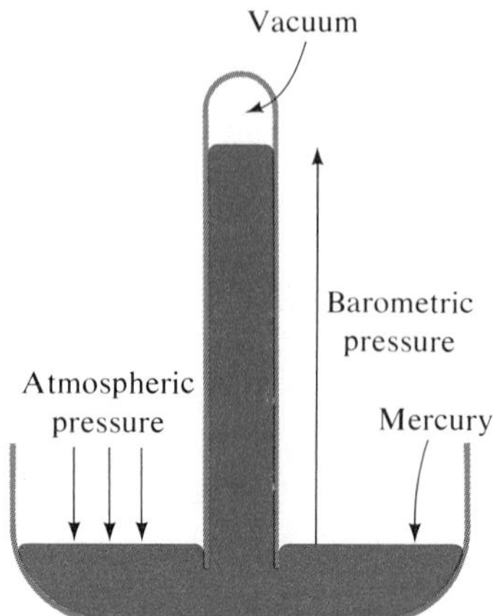

Figure 6-4. A Torricellian barometer. This device can be used to measure atmospheric pressure.

The most familiar unit of pressure is the *atmosphere,* atm. One atmosphere is defined as that pressure that will push a column of mercury in a mercury barometer exactly 760 mm above its reservoir when the mercury is at 0°C and the acceleration due to gravity is 980.665 cm/s^2. Note that there is no restriction on the diameter or shape of the tube. This pressure closely approximates the pressure observed under ordinary conditions at sea level.

A simple barometer can be easily constructed from a long glass tube sealed at one end and a dish of mercury. The tube is filled to the brim with mercury, a thumb placed across the top, the tube inverted, the open end placed under the surface of the mercury in the dish, and the thumb released.* A simple Torricellian barometer [named after its inventor, the Italian physicist Evangelista Torricelli (1608–1647)] is sketched in Figure 6-4.

It may seem strange that a distance is used as a measure of pressure. Note that the distance (usually given in millimeters) is always accompanied by the phrase ". . . of mercury." This is because the common unit of pressure, a mmHg, is really a measure of the mass of mercury supported by the pressure of the gas. Consider a column one square centimeter in cross section. The mass of mercury (at 0°C) contained in 760 mm of this tubing is the product of its volume and its density:

$$m = V\rho = (76 \text{ cm})\,(1 \text{ cm}^2)\,(13.5951 \text{ g/cm}^3) = 1033.23 \text{ g} \qquad (6\text{-}1)$$

**Warning*: Mercury, especially its vapor, is highly toxic, and adequate care must be taken to avoid spillage and ensure proper ventilation.

The force that this mass exerts on the mercury pool is the product of the mass and the acceleration due to gravity:

$$F = mg = (1033.23\ \text{g})\ (980.665\ \text{cm/s}^2) = 1.01325 \times 10^6\ \text{dyn} \qquad (6\text{-}2)$$

The pressure exerted is the force per unit area. Because we have chosen a column of 1-cm^2 cross section, the pressure is

$$P = \frac{F}{A} = 1.01325 \times 10^6\ \text{dyn/cm}^2 = 1.01325 \times 10^5\ \text{N/m}^2 \qquad (6\text{-}3)$$

This pressure is defined as exactly 1 atm, or 760 mmHg, or 760 Torr.* You should repeat this calculation for columns of different shape and cross section to verify that, when the pressure is 1 atm, the mercury level is always 76.0 cm (760 mm) above the reservoir. Of course, actual atmospheric pressures near sea level fluctuate about the value 760 Torr, depending on prevailing meteorological conditions, and the mean atmospheric pressure decreases with increasing altitude. Other pressure units, seldom used in scientific work, are pounds per square inch and inches of water. Why do you suppose water barometers are not very common?

Two other factors that affect the measurement of pressure should be carefully considered because they arise in many practical situations. The first is *hydrostatic pressure*; this is the pressure caused by the difference in elevation between the surfaces of connected columns of liquid. As with the barometer example, the *only* parameters of interest in the measurement are the density of the liquid and the difference in elevation. A mercury manometer is a simple example of this phenomenon. Two others are gas burettes and canal locks.

The second factor that must be remembered in many problems dealing with gases is that they are frequently collected over liquids, such as water. As we will see in the section on liquids in this chapter, most liquids have appreciable vapor pressures. This means that molecules pass from the liquid to the gaseous phase—as long as any liquid remains—until a definite pressure, characteristic only of the liquid and the temperature, is reached. Thus, the volume above the liquid now contains *two* gases, but we may require a value for the pressure of only one of them. Of course, we can only measure the *total* pressure. The difficulty is resolved by the application of *Dalton's*

*The unit "torr" is preferable to the rather cumbersome "mmHg," to which it is identical. The word torr is capitalized whenever it designates a numerical value of the pressure, as in 20 Torr.

law—provided that the gases can be considered to be ideal. (An *ideal gas* is one that exactly obeys the three basic gas laws: those of Boyle, Charles and Gay-Lussac, and Avogadro. Under normal conditions of temperature and pressure, most gases can be considered to be ideal, to a good approximation.) This important law states that the total pressure of a mixture of ideal gases is equal to the sum of the pressures that each gas would exert if it were present alone. The latter pressure is termed the *partial pressure* of that gas and is generally denoted by a subscript. Dalton's law, which was discovered by the English chemist John Dalton (1776–1844), can be expressed as

$$P = P_1 + P_2 + P_3 + \cdots + P_n = \sum_{i=1}^{n} P_i \tag{6-4}$$

where P_i is the partial pressure of gas i and the Greek capital sigma stands for the summation of all these pressures through the nth gas that is present.

Our problem in measuring the pressure of a gas collected over a liquid is thus solved: the total pressure is measured and the vapor pressure of the liquid at that temperature, obtainable from a variety of handbooks, is subtracted from it.

Boyle's Law

Robert Boyle (1627–1691) was an English physicist, chemist, and theologian who performed a classic set of simple experiments in 1662. They were classic in that they were of outstanding importance and lasting significance, and simple perhaps only because we are viewing them through the perspective of three centuries. Boyle added mercury to the open arm of the J-tube shown in Figure 6-5, trapping a small amount of air in the sealed arm. The difference h in height between the mercury levels in the open and sealed arms is a measure of the difference between the pressure of the gas and atmospheric pressure, or, as Boyle termed it, "the spring of the air." The measure of the pressure of the gas sample is $h + h_0$, where h_0 is the ambient barometric pressure in units of "height of mercury." Atmospheric pressure is exerted on the mercury in the open arm of the J-tube. If the tube is of uniform bore, the distance from the mercury meniscus to the sealed end of the tube is a measure of the volume of the gas. The pressure was varied by adding increasing amounts of mercury to the open arm of the tube and observing the corresponding decrease in the volume of the trapped gas sample. The new pressure was indicated by the larger difference in height h'. Some of the original data for P and V obtained by Boyle in this manner are also given in Figure 6-5.

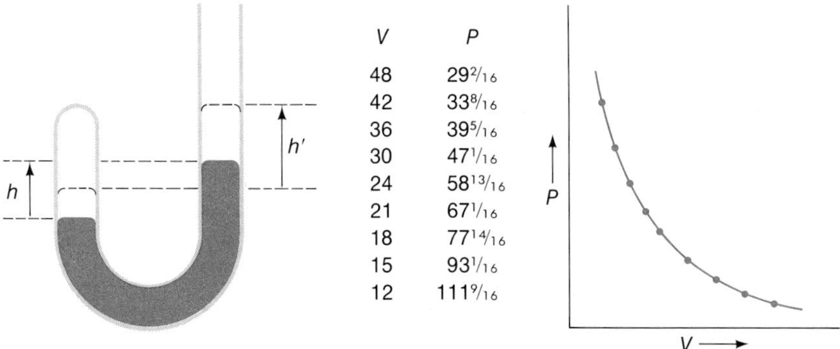

h
h'
V P
48 29²/₁₆
42 33⁸/₁₆
36 39⁵/₁₆
30 47¹/₁₆
24 58¹³/₁₆
21 67¹/₁₆
18 77¹⁴/₁₆
15 93¹/₁₆
12 111⁹/₁₆
P
V

Figure 6-5. A simple manometer. The liquid generally employed is mercury. The space above the left-hand column is filled with the gas under study. Also shown are some of Boyle's original data and a plot of these data. The units of pressure are inches of mercury and the units of volume are graduated marks on a tube of uniform bore.

The other two variables—the amount of gas and the temperature—must also be considered. The mass of the gas within the tube is constant, provided that we do not liquefy some of it by going to very high pressures or low temperatures (see the phase diagrams in the introduction to this chapter). The temperature was not carefully regulated in Boyle's experiments but we may assume that it was within a few degrees of 15°C.

Boyle's data are plotted on the right-hand side of Figure 6-5. The curved shape of the plot suggests a hyperbola, which, indeed, it is. This now suggests what the algebraic relation between P and V is, namely, $PV = \text{constant}$, or, more accurately,

$$PV = A(m,T) \tag{6-5}$$

which should be read, "The product of the pressure and the volume of a gas is equal to a constant whose value depends upon the mass m and the temperature T of the gas." This is *Boyle's law*.

There are a number of other ways to treat these data. It would be instructive for you to try several such plots, using the data given in Figure 6-5. A plot of P vs $1/V$ is a straight line that passes through the origin with positive slope $A(m,T)$. This demonstrates that, as V approaches infinity, P goes to zero. Similarly, a plot of V vs $1/P$ suggests that, as $P \to \infty$, $V \to 0$. (The explanation of the paradox inherent in the latter case is that the gas becomes nonideal and liquefies long before V could go to zero, so the equation no longer holds.)

Example 6-1. A 17.1-g sample of an ideal gas occupies 7.60 ℓ at 0°C and 1.00 atm pressure. Calculate the volume of the gas at 0°C and a pressure of 930 Torr.

The temperature and mass of the gas are fixed. Therefore the pressure and volume values are determined by Boyle's law. We can write

$$P_1V_1 = P_2V_2$$

or

$$V_2 = V_1 \frac{P_1}{P_2} = (7.60\ \ell)\ \frac{(1\ \text{atm})\ (760\ \text{Torr/atm})}{930\ \text{Torr}} = 6.21\ \ell$$

Charles and Gay-Lussac's Law

More than a century after Boyle's work, the French scientists Jacques Charles (1746–1823) and, later, Joseph Gay-Lussac (1778–1850) independently studied the variation in the volume of a gas when the pressure is kept constant but the temperature is varied. (Experience tells us that warming a child's balloon at a birthday party can lead to tear-streaked faces.) Before Charles and Gay-Lussac could describe the relation between volume and temperature quantitatively, however, they had to define a temperature scale. To do this they used a technique developed as early as 1654 by the Grand Duke Ferdinand II of Tuscany, and refined in the meantime by Gabriel Fahrenheit (1686–1736) in Holland and Anders Celsius (1701–1744) in Sweden.

An evacuated capillary tube was partially filled with liquid and the end was sealed. A mark was placed at the position of the liquid column after this tube (now called a thermometer) was immersed in a mixture of ice and water at one atmosphere pressure. Another mark was placed at the point reached by the liquid column after the lower reservoir was placed in a mixture of water and steam at one atmosphere. Following Celsius, the two positions were marked as 0° and 100°. Ninety-nine evenly spaced marks were placed in between and each division described as *one degree Celsius,* °C. (The term *centigrade* is obsolete and should not be used.)

This technique, and variations of it, are still used today. The liquid most commonly used in scientific thermometers is mercury. The red liquid contained in most household thermometers is alcohol with a dye dissolved in it for better visibility. It should be noted that, if two extremely precise thermometers, one filled with mercury and one with alcohol, are placed in the same thermal bath, very slight differences are observed. This is because the procedure described above for the calibration of these thermometers presupposes the perfect linear expansion of the liquids with increasing temperature. Neither liquid expands exactly linearly, however, and the deviations from linearity do not occur in identical ways for the two liquids.

This can lead to differences of more than 0.1°C in certain regions of the temperature scale. Thus the type of thermometer employed must be stated in very precise work.

The most accurate thermometers of all, however, are gas thermometers, and this brings us back to the work of Charles and Gay-Lussac. They found that, at any fixed, low pressure, the volume of a given amount of gas expands exactly linearly with the temperature:

$$V = V_0(1 + \alpha t) \tag{6-6}$$

where V is the volume at temperature t, V_0 is the volume at 0°, and α is a proportionality constant. For the Celsius temperature scale, α is found to have the value 0.003661 per °C for all gases at low pressures. This equation can be inverted to express the idea that, all else being constant, $t = t(V)$. This should be read, "The temperature t is a function of the volume V."

$$t = \frac{V - V_0}{V_0 \alpha} \tag{6-7}$$

Equation 6-7 has been used as a means of defining various temperature scales. For any given scale, all gases at low pressures obey this relation exactly, using the same proportionality constant α. However, this equation is a little awkward and we would like to cast it into a more convenient form, such as the compact form of Boyle's law. To do this we begin by rewriting Equation 6-6 as

$$V = V_0 \frac{1/\alpha + t}{1/\alpha} \tag{6-8}$$

Consider a given sample of gas with a particular value of V_0. If we measure its volumes V_1 and V_2 at two temperatures t_1 and t_2 other than 0°C, we can write

$$\begin{aligned} V_1 &= V_0 \frac{1/\alpha + t_1}{1/\alpha} \\ V_2 &= V_0 \frac{1/\alpha + t_2}{1/\alpha} \end{aligned} \tag{6-9}$$

Division of the first equation by the second permits us to cancel the term $V_0(1/\alpha)^{-1}$:

$$\frac{V_1}{V_2} = \frac{1/\alpha + t_1}{1/\alpha + t_2} \tag{6-10}$$

The form of this relation suggests that we define a new temperature scale, in which the volume is directly proportional to this new temperature. Thus if we set $T = 1/\alpha + t$, which is actually

$$T = 273.15 + t \tag{6-11}$$

we can then write

$$\frac{V_1}{V_2} = \frac{T_1}{T_2} \tag{6-12}$$

from which, by a simple rearrangement of terms, it is clear that the volume is directly proportional to the temperature. This concept can be expressed in more general terms as

$$V = T \cdot B(m,P) \tag{6-13}$$

where B is a proportionality constant whose value depends on the mass m and the pressure P of the gas. This is *Charles and Gay-Lussac's law,* first published (by the latter) in 1802.* Note the similarity (and the difference) in form between Equations 6-5 and 6-13. Can you state the Boyle's-law analogue of Equation 6-12?

The new temperature scale defined by T is called the *Kelvin,* or *absolute, temperature scale* and its units are now called *kelvins* (symbol K), not "degrees Kelvin" (°K). It is named in honor of its inventor, the Scottish physicist Lord Kelvin (1824–1907). By international agreement, the term kelvin now denotes the numerical value of a temperature on the absolute scale (as in 300 kelvins, or 300 K) and also the magnitude of the temperature interval formerly known as "a degree Kelvin" or "a degree Celsius" (the interval is the same in both scales). It will always be clear from the context whether a specific temperature or a temperature interval is meant. For example, the freezing point of water is 273.15 K and the boiling point is 373.15 K; the difference between them is 100 K.

Consider the implications of the statement that, for a fixed amount of an ideal gas of mass m held at a constant pressure P, the volume varies directly with the absolute temperature. Experience indicates that certain quantities can never be negative, even if they can be zero. Such quantities are, e.g., the population of a country, the score of a football team, and (we

*Actually, this law was a rediscovery of the work published in 1699—and ignored for a century thereafter—of the French physicist Guillaume Amontons (1663–1705). Such instances of premature discovery, undeserved oblivion, subsequent rediscovery, and misdirected credit are, unfortunately, all too common in the history of science, right up to the present.

hope) the score on an examination. The variable, volume, must also be included in this category. Of course, the volume of a given mass of gas cannot even go to zero if the temperature is decreased indefinitely, because the gas eventually liquefies and the law no longer holds. Nevertheless, an extrapolation to (but no further than) this hypothetical zero of volume forces a rather interesting and unusual conclusion, namely, that there is a limit to the lower value that the absolute temperature can have, that value being 0 K. Later we will discuss some of the implications of this finding and some of the remarkable properties that matter displays as it is cooled this far. Suffice it to say for now that, not only has 0 K never been achieved, but theoretical reasons have been advanced as to why it never can be. (Scientists *have* come to within one-millionth of a kelvin of absolute zero, however.)

No such limit on the values of t°C is implied by Equation 6-6. Temperature on the Celsius scale can be expressed as negative numbers and this relation places no restriction on how large such negative numbers can become. However, Equation 6-11 and the above observation about 0 K do impose such a restriction. We find that, when $T = 0$ K, $t = -273.15$°C.

The relation between these temperature scales, their lower limits, and gas data that demonstrate this law are displayed in Figure 6-6. Clearly, such measurements cannot be made below the temperature at which the gas liquefies. This problem demands rather long extrapolations, leading to appreciable errors in the value of the absolute zero of temperature on the Celsius scale. Other methods are required for great accuracy. Nevertheless, simple gas experiments demonstrate that the V vs T line does indeed cross the abscissa in the vicinity of −270°C.

The slope of the line in Figure 6-6 is the constant $B(m,P)$. Its value depends only on the amount of gas present and the pressure. This means

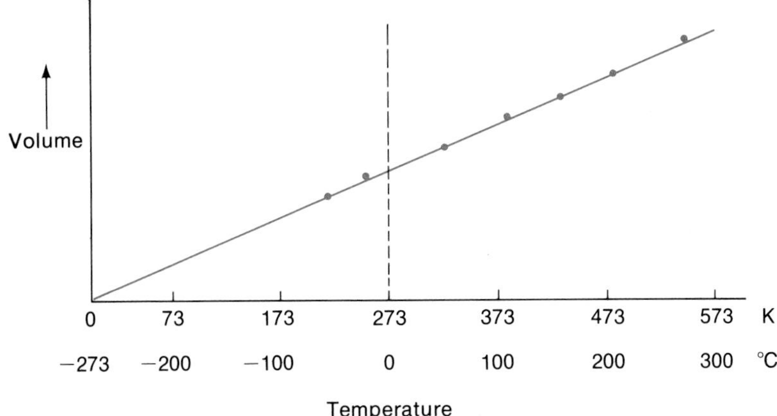

Figure 6-6. Plot of the volume of an ideal gas versus temperature.

that a series of measurements of the variation of the volume with temperature on several gas samples would yield a family of straight lines of different slope, which intersect the dashed line at different points but which all extrapolate to −273°C. The several values of V at the 0°C intercepts of these lines are the values of V_0 in Equation 6-6. It is obvious from both Equation 6-13 and Figure 6-6 that no upper limit is placed on either V or T.

Avogadro's Law

The third and final important empirical law leading to the ideal gas law was formulated in 1811 by Avogadro. Gay-Lussac had been studying gas-phase chemical reactions for some time. He had demonstrated that the ratio of the volumes of gaseous species at constant temperature and pressure that would either react with each other or be formed in the course of a chemical reaction was always the ratio of small whole numbers. This experimental observation plus Avogadro's bold suggestion that gaseous elements could be polyatomic led Avogadro to propose that "Equal volumes of gases at the same temperature and pressure contain equal numbers of molecules." This statement is now known as *Avogadro's law*. It can be stated as

$$V = \frac{m}{M} \cdot C(P,T) \tag{6-14}$$

where m and M are the mass of the gas and its molecular weight, respectively, and $C(P,T)$ is a constant. The quotient m/M is the number of moles n.

As we have learned, or remembered, from the previous two sections, the volume of a gas depends on both the temperature and the pressure. Therefore, if the measured volumes of any gases are to be meaningful, it is necessary to define a set of standard conditions under which the measurements are to be made, or to which the actual conditions of the measurements can be compared. Scientists have selected this *standard temperature and pressure,* or STP, as 0°C and 1 atm pressure. A corresponding, derived quantity is the *standard molar volume,* which is the volume occupied by one mole of any ideal gas at STP. Its value is 22.414 ℓ/mole.

Avogadro has an even more important quantity named for him: *Avogadro's number,* which you should recall from Chapter 2. This is the number of atoms or molecules contained in that 22.414 ℓ, or, more specifically, the number of particles contained in exactly 12 g of the pure isotope ^{12}C. By a variety of procedures to be discussed later, Avogadro's number has been found to be 6.022×10^{23} particles/mole.

The three gas laws are summarized as follows:

Boyle's law:	$PV = A(m,T)$
Charles and Gay-Lussac's law:	$V = T \cdot B(m,P)$
Avogadro's law:	$V = \frac{m}{M} \cdot C(P,T)$

They may be combined into a single law called the *ideal gas law.* Since Charles and Gay-Lussac's law states that volume is proportional to absolute temperature at constant pressure and mass of gas, and Boyle's law states that PV is constant at constant temperature and mass of gas, PV must be proportional to T at constant mass of gas:

$$PV = T \cdot D(m) \tag{6-15}$$

Finally, Avogadro's law requires that $D(m)$ be proportional to m/M, or the number of moles of gas n. Therefore

$$PV = nRT \tag{6-16}$$

where R is a constant.

Equation 6-16 is the *ideal gas equation.* It applies with exactness only to ideal gases, but is approximately correct in almost all applications except gas experiments at very high pressures. It will be used repeatedly in the discussion of thermodynamics to follow. This equation and other, more sophisticated equations that specify the behavior of a given amount of gas in terms of the variables P, V, and T are called *equations of state.*

The constant R is called the *gas constant* (a fundamental physical constant), and has the units of energy per degree per mole. The "degree" to be used is the kelvin. The currently accepted values of R are listed in Table 6-1 for several different energy units. Note the useful fact that energy is equivalent to pressure times volume. Thus, pressure can be thought of as energy per unit volume.

We will learn from the kinetic theory of gases, which follows, that the ideal gas law applies only to noninteracting atoms or molecules. For a mixture of gases, it applies to each gas individually. Thus we can write

$$P_iV = n_iRT \tag{6-17}$$

where P_i is the partial pressure of gas i, which has already been defined, and n_i is the number of moles of gas i in the volume V. Summing Equation

Table 6-1. Values of the Gas Constant R in Various Units.

Pressure	Volume	Temperature	Gas Constant R
atm	ℓ	K	0.08206 ℓ atm/K mole
atm	ml	K	82.06 ml atm/K mole
Torr	ℓ	K	62.36 ℓ Torr/K mole
—	—	K	1.987 cal/K mole
N/m²	m³	K	8.314 J/K mole
dyn/cm²	cm³	K	8.314×10^7 erg/K mole

6-17 over all components i gives back the ideal gas equation because $\Sigma P_i = P$, the total pressure, and $\Sigma n_i = n$, the total number of moles of gas in the sample.

Example 6-2. A sample of oxygen gas is bubbled through water and collected until the total pressure in a 675-ml volume is 720 Torr, at 25.0°C. The vapor pressure of water at 25.0°C is 24 Torr. Calculate the number of moles of O_2 in the sample, assuming ideal behavior.

The ideal gas equation is applied separately to each gas present in the gas mixture. Thus we can write

$$P_{O_2}V = n_{O_2}RT$$

$$n_{O_2} = \frac{P_{O_2}V}{RT}$$

The pressure of the O_2 is given by Dalton's law:

$$P_{total} = P_{O_2} + P_{H_2O}$$

$$P_{O_2} = P_{total} - P_{H_2O} = 720 - 24 = 696 \text{ Torr}$$

and the temperature is 273 + 25 = 298 K. We therefore obtain

$$n_{O_2} = \frac{(696 \cancel{\text{Torr}})(0.675 \cancel{\ell})}{(62.4 \cancel{\ell}\ \cancel{\text{Torr}}/\cancel{\text{K}}\ \text{mole})(298 \cancel{\text{K}})}$$

$$= \frac{6.96 \times 6.75 \times 10}{6.24 \times 2.98 \times 10^3} \text{ mole}$$

$$= 2.53 \times 10^{-2} \text{ mole}$$

Dividing Equation 6-17 by the ideal gas equation shows us that the partial pressure of a gas in a mixture is related in a simple way to the mole fraction of that gas in the sample:

$$\frac{P_iV}{PV} = \frac{n_iRT}{nRT}$$

$$\frac{P_i}{P} = \frac{n_i}{n} \tag{6-18}$$

The *mole fraction* of gas *i*, defined as

$$X_i = \frac{n_i}{n} \tag{6-19}$$

is just the fraction of total molecules that are *i* molecules. It follows that the partial pressure of gas *i* is the mole fraction of that gas times the total pressure:

$$P_i = X_iP \tag{6-20}$$

Example 6-3. A carbon monoxide concentration 2000 ppm (parts per million) is generally lethal if breathed for about one hour or more. A gaseous mixture consists of 0.40 mole O_2, 0.90 mole N_2, 0.20 mole CO_2, 0.10 mole H_2O, and 0.010 mole CO. Calculate the mole fraction of CO to see whether this mixture would be lethal. Also calculate the partial pressure of this gas if the total pressure is 1.1 atm.

$$n = n_{O_2} + n_{N_2} + n_{CO_2} + n_{H_2O} + n_{CO}$$

$$n = 0.40 + 0.90 + 0.20 + 0.10 + 0.01 = 1.61 \text{ moles}$$

$$X_{CO} = \frac{n_{CO}}{n} = \frac{0.010}{1.61} = 0.0062$$

The lethal concentration of CO mentioned above is 2000 ppm, which corresponds to the mole fraction

$$X_{CO} = \frac{2000}{1{,}000{,}000} = \frac{2.0 \times 10^3}{1.0 \times 10^6} = 2.0 \times 10^{-3} = 0.0020$$

The mole fraction of CO in the mixture exceeds this value and it would be lethal to breathe it for very long. The partial pressure is given by

$$P_{CO} = X_{CO}P = 0.0062\,(1.1 \text{ atm}) = 0.0068 \text{ atm}$$

The Kinetic Theory of Ideal Gases

The kinetic theory attempts to explain the pressure-volume relations in gases, as well as virtually all other aspects of the behavior of gases, strictly in terms of the microscopic properties of atoms and molecules, i.e., the properties exhibited by these particles acting as separate entities rather than ill-defined aggregates of bulk matter. The spectacular success of the kinetic theory in explaining and predicting the behavior of gases is one of the triumphs of physical science. It reached its full flower in the middle of the nineteenth century, and more than anything else, it served to establish unequivocally the correctness of the atomic hypothesis of the structure of matter. Because of the physical simplicity and mathematical exactness of this theory, the gaseous state is understood better than any other.

It is of great importance for us to be aware of the procedure we will follow and to realize its limitations, because this procedure is characteristic of most theoretical derivations in the natural sciences. A physical model of the system under investigation is hypothesized and the logical consequences of the postulates employed are derived by means of whatever known physical, chemical, or mathematical relations prove useful. Generally the physical-chemical derivation has few, if any, flaws in it. Frequently the mathematical derivation is totally rigorous, with the possible exception of a few minor and generally nearly exact approximations. The major difficulty usually lies in the model itself. It is difficult to find simple conceptual models of nature that approach physical reality closely enough that subsequent calculations based on them lead to reasonably correct descriptions.

The model employed in this derivation is described by the following four assumptions:

1. Gases consist of molecules (or atoms) that are small relative to the distances traveled between collisions.
2. Gas molecules exert no forces upon each other except during collisions.
3. Gas molecules are spherical and in constant, rapid motion. Collisions of the molecules with the walls and with each other are totally elastic (i.e., the molecules are hard spheres).
4. It is the collisions of the molecules with the walls of the container that give rise to the pressure of a gas.

So we do indeed have a simple model. It amounts to a large number of tiny "billiard balls" rapidly bouncing around inside a container and bumping into each other and the walls. Too simple? There are two ways to judge. Examination of the postulates shows them to be not unreasonable. The first is a direct consequence of atomic theory and the second is reasonable in terms of the extremely low density of gases. (The volume in which a gas molecule is free to move under ordinary conditions is about a thousand times larger than the volume of the molecule itself.) There is no a priori reason for the remaining two postulates. Their merit must be judged pri-

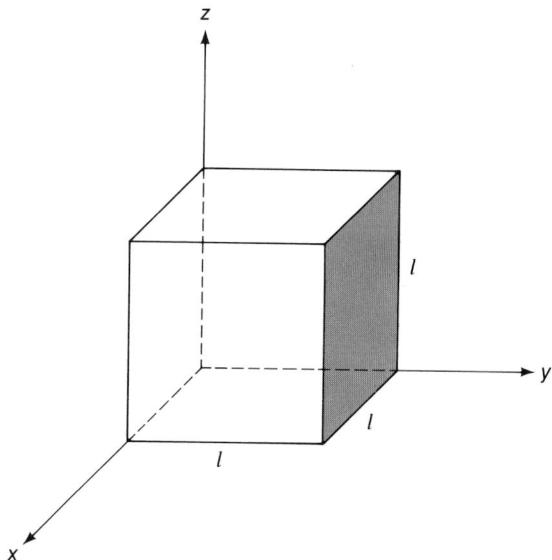

Figure 6-7. A coordinate system and hypothetical cube employed in the kinetic theory of gases.

marily by the second criterion we will apply to this theory, namely, does it yield a description of nature that is consistent with reality?

The derivation presented here is a very simple one. Considerably more sophisticated derivations that utilize more elegant geometrical and mathematical models are available. But we will enjoy the simplicity of this method and not be too disconcerted by a small measure of inelegance, recognizing that the present derivation yields the same results as do more complicated ones.

Consider a cube of side l filled with those lively, minuscule billiard balls (see Figure 6-7). Let us focus on just one of the identical molecules, of mass m: it is moving with velocity $\mathbf{v}$. This is a vector, i.e., a quantity having both magnitude and direction.* We will assume that the molecule moves in a straight line, without collisions with other molecules, until it hits a wall. At any particular instant this motion can be represented by the vector $\mathbf{v}$ and its three components, as shown in Figure 6-8.

For the present we need only consider motion in one direction; we choose the positive y-direction. The molecule will strike the shaded face of the

*In order to emphasize the important property of *directionality* in several physical units of measure, the symbols for these units (all vectors) will be printed in boldface throughout this section. This convention does not apply, however, to the symbols for vector components along the coordinate axes.

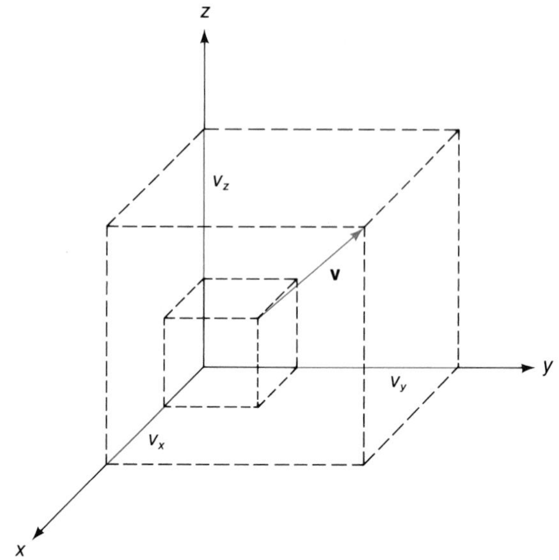

Figure 6-8. The velocity vector **v** and its three components.

cube in Figure 6-7 with the normal (perpendicular) component of its velocity, which is equal to v_y. Because our model requires the collision to be elastic (most of us have watched a billiard ball strike the side of a table), the molecule rebounds with velocity $-v_y$.

This change in velocity must now be related to a pressure. You already know that pressure is equal to force per unit area, and you may have learned that force is given by the time rate of change of momentum, $d\mathbf{p}/dt$. The latter statement is readily derived from Newton's second law of motion (which, of course, has no derivation):

$$\mathbf{F} = m\mathbf{a} \tag{6-21}$$

where **F** is the force required to impart an acceleration **a** to an object of mass m. The acceleration is the time rate of change of velocity, $\mathbf{a} = d\mathbf{v}/dt$, so we now have

$$\mathbf{F} = m\mathbf{a} = m\frac{d\mathbf{v}}{dt} = \frac{d(m\mathbf{v})}{dt} = \frac{d\mathbf{p}}{dt} = \frac{\Delta\mathbf{p}}{\Delta t} \tag{6-22}$$

We already know the change in momentum per collision. The molecule strikes the wall with momentum mv_y and rebounds from it with momentum

$-mv_y$. Thus the change in momentum is $2m|v_y|$, where the vertical bars denote the absolute value of the quantity between them.

Now to obtain $\Delta\mathbf{p}/\Delta t$ from $\Delta\mathbf{p}$/collision, we see by dimensional analysis that we must multiply the latter quantity by collisions/Δt, or the number of collisions in unit time. This is readily found from the y-component of the velocity and the dimensions of the box. After a collision the molecule must travel a distance $2l$ before it strikes the shaded surface again. Since it is traveling at $\pm v_y$ cm/s, it will cover this distance $|v_y|/2l$ times per second. The force that the molecule exerts on the one wall is thus given by

$$F_y = \frac{\Delta p_y}{\Delta t} = \frac{\Delta p_y}{\text{collision}} \frac{\text{collisions}}{\Delta t} = 2m|v_y| \frac{|v_y|}{2l} = \frac{m(v_y)^2}{l} \tag{6-23}$$

Recognizing that we have more than one particle striking the wall, we multiply this force exerted by one molecule by the number of molecules N contained in the box:

$$F_{y,\text{total}} = \frac{Nm(v_y)^2}{l} \tag{6-24}$$

But now we have overlooked an important point: not all these molecules have the same velocity v_y when they hit the wall. There is a statistical distribution of velocities, with some molecules having larger and some smaller y-components of their velocity. An average value of v_y is clearly needed. For reasons that we need not go into here, the best value is the average of the square of all the velocity components $v_{y,i}$, which is the velocity in the y-direction for molecule i. We denote this average value by $\overline{(v_y)^2}$ and define it as

$$\overline{(v_y)^2} = \frac{(v_{y,1})^2 + (v_{y,2})^2 + (v_{y,3})^2 + \cdots + (v_{y,N})^2}{N}$$

$$= \frac{\sum_{i=1}^{N} (v_{y,i})^2}{N} \tag{6-25}$$

We can now write that the average force exerted by the total number N of the molecules is given by

$$F_{y,\text{total}} = \frac{Nm\overline{(v_y)^2}}{l} \tag{6-26}$$

To obtain the average pressure, we need only divide this force by the area on which it is exerted:

$$P_y = \frac{F_{y,\text{total}}}{l^2} = \frac{Nm\overline{(v_y)^2}}{l^3} \tag{6-27}$$

Because l^3 is just the volume V of the cube, we can substitute and rearrange to yield

$$P_y V = Nm\overline{(v_y)^2} \tag{6-28}$$

The next step is to recognize that there can be no preferential direction in the cube and that we should be using the average value of the square of the total velocity **v** of each molecule. By a generalization of Pythagoras's theorem to three dimensions and by averaging over all the molecular velocities, it can be shown that the mean square value of the velocity vector (see Figure 6-8) is given by

$$\overline{\mathbf{v}^2} = \overline{(v_x)^2} + \overline{(v_y)^2} + \overline{(v_z)^2} \tag{6-29}$$

Since we assume that the molecular motions are statistically equivalent in all three dimensions, we have

$$\overline{(v_x)^2} = \overline{(v_y)^2} = \overline{(v_z)^2} \tag{6-30}$$

Therefore

$$\overline{\mathbf{v}^2} = 3\overline{(v_y)^2} \tag{6-31}$$

Substituting Equation 6-31 into Equation 6-28 now gives*

$$PV = \tfrac{1}{3}Nm\overline{v^2} \tag{6-32}$$

The right-hand side of Equation 6-32 may look vaguely familiar. It should (remember that pressure times volume equals energy), because the expression for the translational kinetic energy *KE* of a moving particle is

$$KE = \tfrac{1}{2}mv^2 \tag{6-33}$$

Substitution into Equation 6-32 thus yields

$$PV = \tfrac{2}{3}N(\overline{KE}) \tag{6-34}$$

---

*We abandon the boldface vector notation here because we are now dealing with *bulk* values of the pressure and velocity, i.e., values for which no particular direction is specified.

which states that the product of the pressure and the volume of a gas described by our model is directly proportional to the number of molecules in the gas and to the average kinetic energy of those molecules. If you think about it, this makes good sense and is, in fact, true for ideal gases.

You may have noticed that this entire derivation was carried out without ever mentioning collisions of the molecules with each other. We assumed initially that there were no such collisions, but of course this is not true—all the molecules constantly collide with each other and hence move in a completely random manner. The statistical effect of these collisions, however, is to leave the *average* values of the x-, y-, and z-components of the velocity of each molecule unchanged, so the assumption was not even necessary! (Without it, however, the mathematics would have been much more complicated.) Furthermore, it can be shown that the assumption of perfectly elastic collisions with the walls of the cube was also unnecessary (as was the assumption of spherical molecules), although again very helpful. And finally, there is nothing special about a cube except its simplicity; the result obtained is equally valid for an ideal gas in a container of any arbitrary shape.

Thus, the kinetic theory has provided us with a result of great generality and usefulness, as will become apparent in the next section. Since this is as far as our simple theoretical investigation can go without additional input, we must now turn again to the experimental description of gases in order to gain further insight into the behavior of ideal gases.

Combination of Experimental and Theoretical Descriptions of Ideal Gases

The two distinct approaches to a description of gases have provided the relations

$$PV = \tfrac{2}{3}N(\overline{KE}) \tag{6-35}$$

and

$$PV = nRT \tag{6-36}$$

These two equations clearly yield

$$\tfrac{2}{3}N(\overline{KE}) = nRT \tag{6-37}$$

Thus the average kinetic energy of a gas molecule—any gas molecule—is

$$\overline{KE} = \frac{3}{2}\frac{nRT}{N} \tag{6-38}$$

The quantity N is the number of gas molecules in the sample. This is given by the number of moles n of the gas times Avogadro's number

$$\overline{KE} = \frac{3}{2}\frac{nRT}{nN_A} = \frac{3}{2}\frac{R}{N_A}T = \tfrac{3}{2}kT \tag{6-39}$$

where $k = R/N_A$, the gas constant on a per molecule basis, is the *Boltzmann constant,* named after the Austrian theoretical physicist Ludwig Boltzmann (1844–1906). It is a fundamental physical constant of great importance and widespread occurrence in molecular theories. In terms of ergs, its value is 1.381×10^{-16} erg/K.

Equation 6-39 is a fascinating result. It states that the average kinetic energy of an ideal gas molecule is determined solely by the absolute temperature. Neither the mass, shape, nor chemical composition of the molecule has any effect on the kinetic energy as long as the molecule behaves ideally. (Apparently the heavier molecules move more slowly at any given temperature.) This result is important in statistical considerations of molecular properties and in predicting the heat capacities of compounds as a function of temperature. We will deal with those topics in the next chapter.

It can now be seen that Equation 6-34 represents a theoretical derivation of the laws of Boyle, Charles and Gay-Lussac, and Avogadro. On the right-hand side of the equation are the quantities N and KE. The former is the sole measure of the mass m of the gas sample (i.e., $m \propto N$), and from Equation 6-39 we now know that the kinetic energy of the gas depends on the temperature alone ($KE \propto T$). Therefore, if Equation 6-34 is cast successively into the forms of Equations 6-5, 6-13, and 6-14 and the appropriate substitutions are made mentally, it will be seen that the latter three equations are, indeed, all subsumed by the former.

Another result is directly available from the comparison of the ideal gas equation with the expression for Boyle's, etc., law derived from the kinetic theory of gases: it is the *root-mean-square* velocity of a gas. We will first derive this quantity and then consider some other ways of expressing the velocity of a gas. The reasons for this examination are that gas velocity is the dominant factor in such transport processes as diffusion and effusion and that it is of fundamental importance in determining the rates of gas-phase reactions.

We have obtained the result that the average kinetic energy $\overline{KE}$ of a molecule of an ideal gas is equal to $\frac{3}{2}kT$. Thus we have

$$\overline{KE} = \tfrac{1}{2}m\overline{v^2} = \tfrac{3}{2}kT \tag{6-40}$$

It follows directly that

$$\overline{v^2} = \frac{3kT}{m} \tag{6-41}$$

Multiplying the right-hand side by N_A/N_A (in order to convert to molar quantities) and taking the square roots of both sides gives us the root-mean-square value of the molecular velocity:

$$v_{rms} = (\overline{v^2})^{1/2} = \sqrt{\frac{3RT}{M}} \tag{6-42}$$

where $M = mN_A$ is the molecular weight of the gas. For gases at room temperature, v_{rms} is thus given approximately by

$$v_{rms} \cong \frac{3 \times 10^5}{M^{1/2}} \text{ cm/s} \tag{6-43}$$

Therefore gases with molecular weights in the range 10 to 1000 have molecular velocities of the order of 10^5–10^4 cm/s. That's pretty fast! Had you realized that the atmospheric oxygen and nitrogen molecules bumping into your skin and even into the delicate cornea of your eye are going some 1000 mph? (Fortunately, molecules are rather small.) Now, do you know what the approximate speed of sound in air is? Do you see a connection here?

The root-mean-square velocity is only one of several kinds of average value. Another is simply the average velocity, the sum of the velocities of a group of molecules divided by the number of molecules. The fact that these averages are not identical is seen in the following example.

Example 6-4. Calculate the root-mean-square velocity and the average velocity of three molecules moving at 3×10^4, 4×10^4, and 5×10^4 cm/s.

$$v_{av} = \bar{v} = \frac{(3 + 4 + 5) \times 10^4}{3} = 4.0 \times 10^4 \text{ cm/s}$$

$$v_{rms} = (\overline{v^2})^{1/2} = \left(\frac{3^2 + 4^2 + 5^2}{3}\right)^{1/2} \times 10^4 = 4.1 \times 10^4 \text{ cm/s}$$

Thus we see that the two numbers are different and that the rms value is slightly larger. This is always true, because the net contribution to the rms average from squaring the number(s) larger than the common average is always greater than the contribution from squaring the smaller number(s). These considerations are of very general usefulness whenever we need to consider an average value of a set of numbers.

So far we have said nothing about the actual distribution of velocities, which has been determined both experimentally and theoretically. A typical example is shown in Figure 6-9. It represents the theoretical equation known as the *Maxwell distribution law,* after the Scottish mathematical physicist James Clerk Maxwell (1831–1879). The *most probable* velocity v_{mp}, corresponding to the peak of the curve, is that characterizing the largest fraction of the molecules. Because the area under the curve is larger on the right side of the peak than on the left side, the average velocity $\bar{v}$ is larger than v_{mp}. For the reason given above, v_{rms} is even larger than $\bar{v}$.

We will need to refer to Figure 6-9 a number of times in subsequent discussions. Because velocity is one measure of molecular energy, this figure also provides an illustration of the molecular kinetic energy distributions of gases. Such distributions are important because, in order for chemical reactions to occur, existing chemical bonds must generally be broken, and this requires a source of energy. Among the several ways this energy is supplied is the transfer of kinetic energy during molecular collisions.

The final topics in our discussion of the molecular properties of ideal gases are the two closely related transport phenomena, diffusion and effusion. *Diffusion* is the movement of gas molecules from one region of space to another region, in which the concentration of the gas is initially lower. The analysis of diffusion is based on the existence of intermolecular collisions, which occur in all gases. If two different gases are separated by a porous barrier and allowed to diffuse through each other, their respective rates of diffusion can be calculated on the basis of known molecular properties. *Effusion* is somewhat more straightforward in that it is the passage of gas

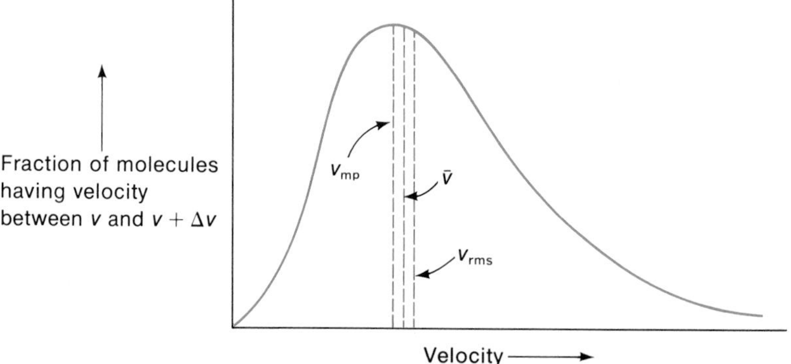

Figure 6-9. The distribution of molecular velocities in a gas. The scale of the ordinate is the probability of finding a molecule having a velocity within an arbitrarily narrow range of values. The values of the most probable velocity v_{mp}, the average velocity $\bar{v}$, and the root-mean-square velocity v_{rms} are indicated.

molecules through a very small orifice. Scientists who work with high-vacuum lines are well aware of effusion because pinholes are both hard to find and, surprisingly, hard to seal.

We will forego a rigorous discussion of these two phenomena. We state without proof, but with a little intuition, that the rates of both diffusion and effusion are directly proportional to the molecular velocity. Given this premise and Equation 6-42 for the root-mean-square velocity, we can deduce the following relations:

$$\frac{\text{Rate of diffusion of gas 1}}{\text{Rate of diffusion of gas 2}} = \frac{\text{Rate of effusion of gas 1}}{\text{Rate of effusion of gas 2}} = \sqrt{\frac{M_2}{M_1}} \qquad (6\text{-}44)$$

Exercise 6-1. A professor with a somewhat warped sense of humor releases some laughing gas of molecular weight 44.02 in the front row of his classroom and a lachrymator of molecular weight 140.6 in the last row at the same moment. If the classroom is twenty-five rows deep, in which row will the students begin laughing and crying simultaneously?

The reason that there is so little hydrogen and helium in the Earth's atmosphere is that these extremely light, high-velocity gases (remember Equation 6-43) diffuse rapidly upward through the atmosphere and gradually "leak" off into space. The only force holding the atmosphere to the Earth is gravity, but when an atom or molecule in the outer reaches of the atmosphere happens to exceed the Earth's escape velocity of 11.2 km/s (the same for any object, be it atom or spaceship), it is gone forever. Over the eons since the Earth was formed, the lightest gases have thus escaped while the heavier, slower gases remain. If the Earth were much smaller, its escape velocity would be smaller also, and the Maxwell distribution of velocities of oxygen and nitrogen would allow these gases to leak into space at an appreciable rate. The consequences are not hard to imagine.

The Behavior of Real Gases

The entire discussion of gases to this point has centered on the behavior of ideal gases. In reality, many gaseous systems at low temperatures and at pressures below atmospheric do conform to the ideal gas equation to within a few percent. However, we should examine the behavior of gases at higher pressures to see whether this equation is still useful. If not, we may need to look for other, more sophisticated models of how gases behave and subsequently to derive new equations of state.

Gases at High Temperature and High Pressure

One way to examine the behavior of gases at high temperature and high pressure is to plot the experimental data for PV as a function of P, for a given value of T. Ideal gas behavior would correspond to the dashed line in Figure 6-10. The solid curves are obtained from the data for one mole each of three real gases at 0°C.

Similar, rather peculiar curves are observed if PV is plotted against P for one gas as a function of the temperature. As the temperature increases, the curves tend more and more to the horizontal line representing ideality. Clearly, we are forced to recognize that our model for the behavior of gases is too simple and that the ideal gas equation is adequate to deal with real gases only at low pressure and high temperature.

Gases at Relatively Low Temperature and High Pressure

A P vs V plot is useful to see how real gases behave at relatively low temperature and high pressure. Figure 6-11 presents the data for CO_2 at a num-

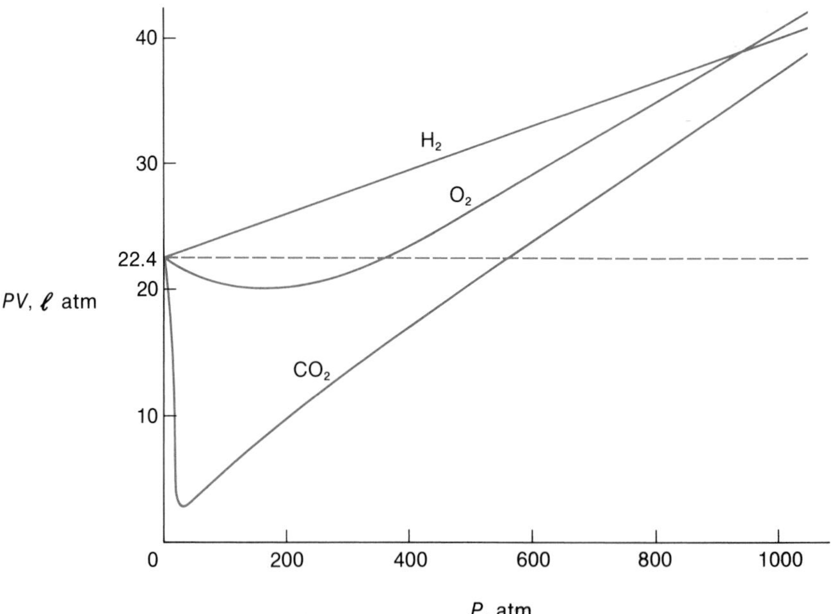

Figure 6-10. The behavior of several real gases at 0°C and high pressure. The value of PV for one mole of an ideal gas at 0°C is 22.4 ℓ atm and does not depend upon the pressure.

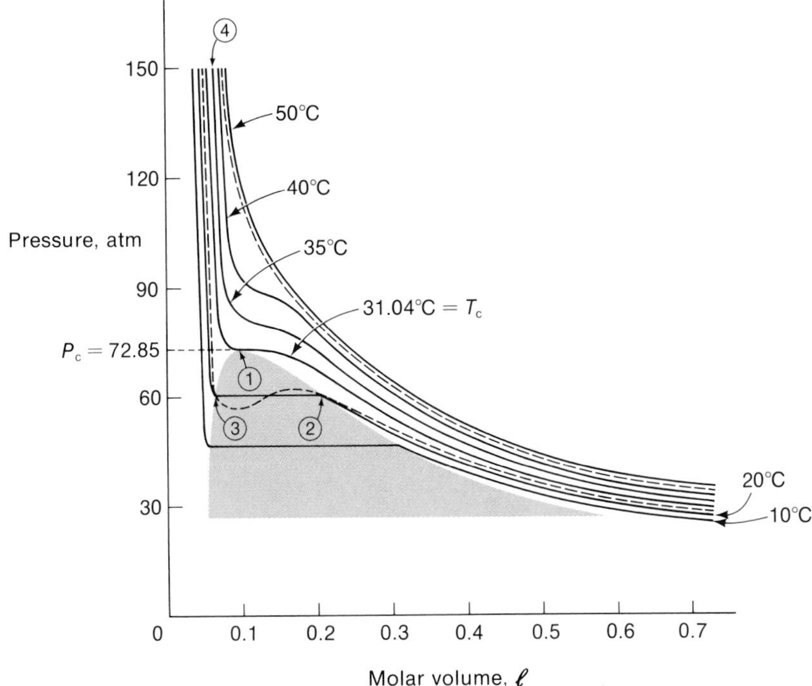

Figure 6-11. The behavior of carbon dioxide at low temperature and high pressure (T_c is the critical temperature). The colored area represents the vapor-liquid region. The dashed lines were calculated from the van der Waals equation (explained on pages 268–270).

ber of temperatures. Each curve is called an *isotherm,* meaning "at the same temperature." The isotherm for 50°C nearly follows ideal gas behavior, which would be represented by a perfect hyperbola. The isotherms for 40°C and 35°C have a peculiar and unpredicted hump in them. Point 1 on the 31.04°C isotherm is the critical point. All the curves for temperatures lower than the critical temperature have hypothetical horizontal regions within the colored area; this area represents the vapor-liquid region.

The several curves are best understood if we traverse them one at a time. Only the gaseous phase is observed if we raise or lower the pressure along the isotherms above 31.04°C. However, if we begin at a very low pressure on the 20°C curve and gradually raise the pressure, it is found that vapor begins to condense at point 2. Condensation continues along the line 2–3. This pressure is the vapor pressure of liquid CO_2 at this temperature. At point 3 all the gas has been converted to liquid and the curve now rises very steeply because, unlike gases, liquids are almost completely incompressible.

A final comment on these curves deals with discontinuities. Each time one of the curves intersects the colored area, a discontinuous transition (phase change) from liquid to gas or gas to liquid occurs. In the procedure described above for the traverse of the 20°C curve, we would expect to see some liquid droplets become visible at point 2 and the volume of the liquid phase become larger and larger until, at point 3, no vapor was present. At all times when the two phases were present, a distinct interface would be visible between them. A different situation exists if a different route is chosen between points 2 and 3. If the gas were heated at a constant molar volume of 0.6 ℓ to a temperature of, say, 35°C, which is above the critical temperature, and the pressure were then increased while the temperature was held constant, the system could be brought to point 4 with no phase changes occurring. Then, lowering of the temperature at this new constant molar volume would result in the transformation of all the gas into liquid at point 3, without there being any instant when a distinct boundary could be seen between gas and liquid. This interesting phenomenon can be observed by sealing the appropriate amount of a liquid into a tube of the right volume, followed by gradual heating and cooling.

Equations of State for Real Gases and Molecular Interpretations

It would be helpful, in dealing with real gases, if we had an equation of state to describe them. Such an equation was derived almost exactly one century ago by van der Waals, who began by re-examining the model used for ideal gases. This model assumes that the gas molecules themselves have negligible volume and that there are no attractive forces between these molecules. At high pressures, however, all gases have large positive deviations of the product PV from the ideal gas value, as shown in Figure 6-10. This is because the first assumption above is wrong: the contribution of the molecular volume to the total volume at pressures of the order of 1000 atm *is* very large. (At lower pressures most heavy gases initially have negative deviations.) Also, as we have seen from Figure 6-11, gases do condense to the liquid state under moderate pressures. These two factors indicate that the second assumption is wrong as well: there *are* appreciable attractive forces between molecules. Let us start with the ideal gas equation and rewrite it, taking these two factors into account.

We begin with the volume correction. If one mole of gas occupies volume V, then the volume in which the molecules are actually free to move, V_{actual}, is V minus the volume of the molecules themselves. Thus if the volume excluded by the molecules in one mole of gas is b, we should use

$$V_{actual} = (V - b) \tag{6-45}$$

in the ideal gas equation. For n moles of gas, we can write

$$P(V - nb) = nRT \quad (6\text{-}46)$$

Values of the van der Waals constant b are usually obtained empirically by finding values that best represent the behavior of real gases according to Equation 6-46 and its refinement, Equation 6-47 (see below). Values of b for some common gases are given in Table 6-2.

The second correction to the ideal gas equation must be an additive term for the pressure. The measured pressure is too low because of the attractive forces between the molecules. We can arrive at an expression containing several of the state variables of the gas by considering what factors determine the forces the molecules can exert on each other. The predominant factor must surely be how close the molecules are to each other. This is determined by the density, which can be expressed by n/V, the number of moles per unit volume. The attraction of molecule A_1 for molecule A_2 is proportional to n/V and the attraction of A_2 for A_1 is similarly proportional to n/V. The net attraction must therefore be proportional to n^2/V^2:

$$(A_1) \xleftrightarrow{\text{Force} \propto n^2/V^2} (A_2)$$

Because the proportionality constant a is generally obtained empirically, the nature of the force, whether it be London, dipole-dipole, dipole-induced dipole, or some other kind, is "built in" and therefore immaterial. (Such intermolecular forces are often lumped together under the general name *van der Waals forces*.) Thus the additive pressure term is an^2/V^2. A few values of the van der Waals constant a are given in Table 6-2. Note the correlation between the magnitudes of these values and the strengths of the intermolecular forces to be expected in each compound.

The ideal gas equation has now become the nonideal gas equation widely known as the *van der Waals equation*,

$$(P + an^2/V^2)(V - nb) = nRT \quad (6\text{-}47)$$

Table 6-2. Van der Waals Constants for Some Common Gases.

Gas	a, ℓ^2 atm/mole^2	b, cm^3/mole
H_2	0.244	26.6
N_2	1.39	39.1
O_2	1.36	31.8
CO_2	3.59	42.7
H_2O	5.46	30.5

Table 6-3. Volume of One Mole of CO_2 at 320 K as a Function of Pressure.

Pressure, atm	Volume, ℓ		
	Observed	Van der Waals Equation	Ideal Gas Equation
1	26.2	26.2	26.3
10	2.52	2.53	2.63
40	0.54	0.55	0.66
100	0.098	0.10	0.26

The dashed curves in Figure 6-11 were calculated for the temperatures 20°C and 50°C by means of this equation, using the constants for CO_2 given in Table 6-2. Although all the calculated curves display the same kind of maximum and minimum in the phase transition region, the agreement of this simple equation with the experimental data for most gases is astonishing. The data shown in Table 6-3 indicate that the van der Waals equation is a great improvement over the ideal gas equation, especially at high pressures.

Plasmas

At the beginning of this chapter, we said that our everyday experience indicates that there are only three ways in which matter ordinarily occurs on Earth. "Ordinarily"? And what about the rest of the universe? We tend to forget that our planet is but a tiny, rather frigid speck whirling through the unimaginable immensity of space. The truth is that the three familiar states of matter that we have discussed are quite uncommon in the universe taken as a whole. Almost all the matter in the universe exists in the form of *plasma*, the fourth state of matter.

A plasma is any gas in which an appreciable number of the atoms or molecules are ionized and in which free electrons are the predominant negatively charged species. This ionization leads to very distinctive physical and chemical properties, the most important of which is high electrical conductivity. (All un-ionized gases, by contrast, are very poor conductors.) The ionization is generally brought about by high temperatures—anywhere from a few thousand kelvins, as in most flames, to a few billion, as in the cores of the hottest stars. With continuously increasing temperature, *all* matter is first vaporized (gas), then partially ionized (weak plasma, in which some of the atoms have lost at least one electron), and eventually fully ionized (strong plasma, in which every atom has lost at least one electron). At some stage during this process, depending on the nature of the starting material, all molecules are destroyed by the heat, leaving only atoms.

All flames are weak plasmas. (If you have lit a match recently, you have seen a plasma.) Strong plasmas occur on Earth in lightning bolts, nuclear fireballs, and some research laboratories. Not all plasmas are hot, however. In fact, you may well be reading these words by the light of a very weak, cool plasma—the kind found in all fluorescent lamps—after it is converted from ultraviolet to visible by the inorganic phosphors coating the inside walls of the tubes. And if you happen to have an AM radio on, the broadcast signals may have reached it by being bounced off the Earth's built-in plasma, the ionosphere, in which large numbers of positive ions and free electrons are produced by the action of ultraviolet radiation from the sun. Can you think of any other examples of plasmas, hot or cool, in your surroundings? There are quite a few!

Because there are large numbers of discrete charges in plasmas, a flowing plasma constitutes an electric current and is thus strongly susceptible to magnetic fields. This accounts for most of the unusual properties of plasmas and is the basis for the study of *magnetohydrodynamics* (MHD), a field of great importance for nuclear energy. In fact, the single most promising source of electric power, controlled thermonuclear fusion, is the ultimate goal of plasma physicists: it is literally a harnessing of the energy of the stars. Other research areas in which plasmas are very important are gaseous electronics, rocket propulsion, laser studies, and upper-atmospheric phenomena. In chemistry, plasmas are studied for their intrinsic interest and are used as media for the synthesis of numerous compounds, both in research laboratories and in the chemical process industries. A detailed discussion of this remarkable state of matter, however, is beyond the scope of this book.

6.2 LIQUIDS

Qualitative Description

Liquids are unique. They are rare, they are very important, and they are not well understood.

Liquids are relatively rare at STP: there are only two liquids among the 105 elements. Of the few hundred thousand known inorganic compounds, only a few hundred are liquids under ordinary conditions, and of these, water is the only one we meet in our everyday experience. There are a small number of liquid inorganic carbon compounds, such as CCl_4 and CS_2, and hydrogen compounds, such as H_2SO_4 and H_2O_2, that are common in the chemical laboratory. Most of the remaining inorganic liquids are polyhalogen or oxyhalogen compounds of metalloids and nonmetals of medium atomic weight, e.g., Si, P, and S. The situation is somewhat different among the organic compounds: there are many liquids among the several million organic compounds, but their number is still far fewer than the number of solids. The hydrocarbons in Table 5-1 illustrate the point under consideration here, in that the first four members of the series are gases, the next eleven are liquids, and all higher members are solids.

Liquids, especially water, are of great importance. The life processes that occur on Earth are totally dependent upon water. All cellular activity directed toward the biosynthesis of the vital proteins and nucleic acids occurs in an aqueous environment. In recent years all of us have become acutely aware of the need to limit the pollution of the Earth's fresh and salt water resources to levels that are ecologically tolerable, technologically feasible, and economically realistic. (To eliminate pollution *completely*, as desirable as that would be, is, unfortunately, not a realistic goal in any highly industrialized society.) Such limitation is necessary because all plant and animal life depends on water for its sustenance, and desirable because of man's aesthetic and recreational values.

Liquids play a major role in the science of chemistry. The great majority of chemical reactions are conducted in aqueous solutions or organic solvents such as acetone, benzene, and alcohol. This is partly because of the fundamental properties of liquid solutions and partly because liquids are a medium in which precisely controlled quantities of reactants can be easily brought together. These quantities are frequently so small as to be extremely difficult to measure in any way other than as a volume of a very dilute solution. The liquid itself often plays a crucial role in the reaction. For example, it may be a reactant or a product, or it may alter the reaction conditions by solvating the reactant molecules, or it may change the electrostatic interactions between the reactant molecules through a changed dielectric constant of the medium. All the reactions for the synthesis of the biopolymers described in Chapter 11 include the removal of a water molecule from the reactants (an H from one and an OH from the other, forming H_2O).

Having said all this regarding the rarity and importance of liquids, it is disappointing to note that the liquid state is the least well-understood of the three states of matter. To date there is no simple equation of state for liquids, as there is for gases, nor is there as precise and rigorous a theory for the structure of liquids as there is for that of solids. We will next consider some of the observable properties of liquids and then present a simple picture of what liquids probably look like.

Observable Properties of Liquids

Although no simple theoretical expression to describe liquids is at hand yet, there are a great many experimental observations regarding liquids. We should become familiar with the most important of these because any theory that attempts to describe liquids must not make predictions that contradict any of these observations.

Vapor Pressure

Any liquid allowed to stand for a sufficiently long time in a sealed container at temperature T will evaporate until a fixed pressure in the gaseous phase is reached (providing that enough liquid was initially present that some of it remains as a liquid after this constant pressure is achieved). This pressure is called the *vapor pressure* of the liquid. Its numerical value depends only on the nature of the liquid and the temperature of the system.

Figure 6-12 shows the vapor-pressure curve for water. This line is the same as the right-hand line in Figure 6-2. Only for values of P and T corresponding to points on this line can the liquid and vapor phases exist in contact with each other *at equilibrium* (a condition achieved when there

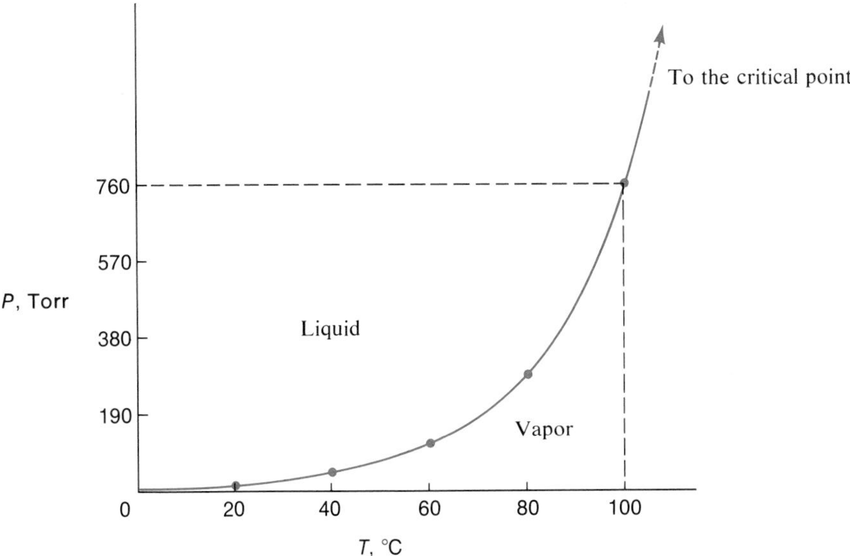

Figure 6-12. The vapor pressure of water as a function of temperature.

are no observable changes with time). It is true that liquid and vapor can also coexist under many other conditions, e.g., puddles in the street and water vapor in the air after a hot shower. But they are not in equilibrium with each other; in the first instance the puddle will evaporate and in the second the water vapor will condense, as these systems move toward equilibrium.

Note in Figure 6-12 that, if the pressure were held constant at, say, 200 Torr (by means of a hypothetical, nonequilibrium amount of water vapor) and the temperature were slowly increased from 0°C, the liquid phase would continue to be observed until the temperature 66°C was reached. At this point more heat would be added without any corresponding change in temperature as all the liquid was converted to vapor. The amount of heat required to effect this phase change at a given pressure is called the *heat of vaporization* and is characteristic of the substance. Finally, the vapor could be heated further at this pressure (assuming what of the volume?).

Another, perhaps more interesting, route to follow in the diagram would be to go up the equilibrium line itself, in a closed system. The vapor pressure of the liquid (i.e., the actual pressure in the system) would increase rather slowly at first and then climb very rapidly. It would increase indefinitely as the temperature was increased, with the distinction between liquid and vapor disappearing at the critical point. At any other point along the line, liquid and vapor are in equilibrium, but their relative amounts depend on the thermal energy of the system. The line can be thought to have an imaginary "thickness" represented by the heat of vaporization of the liquid. To pass through the line from the "left" side (water plus a small amount of water vapor) to the "right" side (all water vapor) at any point, the system must be open and the heat of vaporization of water must be added to it (the temperature does not change during this process, nor does the pressure).

If we heated a quantity of water in a pressure cooker, the pressure would increase along the equilibrium line until the safety valve opened (typically at about 1.5 atm). The further addition of heat at that pressure and temperature (about 112°C) would cause the water to vaporize rapidly, i.e., boil. A liquid in an open container boils when its vapor pressure reaches the same value as the pressure of the surrounding atmosphere. At this point the vapor pressure within the liquid is great enough to push the liquid away, forming bubbles of vapor that rise and escape from the surface. Because the pressure of the Earth's atmosphere varies from about 760 Torr at sea level to about 240 Torr at the tops of the highest mountains, boiling points of liquids vary considerably [much to the consternation of the mountaineer—see the *Journal of Chemical Education* **45**, 556 (1968) for the problems of boiling an egg at the top of Pikes Peak]. For this reason a standard value has been chosen for the pressure at which the boiling points of liquids are recorded.

This pressure is 760 Torr and the corresponding temperature is called the *normal boiling point* of the liquid. The point at which the water curve reaches 760 Torr is defined as exactly 100°C (373.15 K). The heat of vaporization of water at this temperature—540 cal/g—is unusually large because of the strong hydrogen bonding in water.

Freezing Curves

The transition between the liquid state and the remaining state of matter, the solid state, is generally studied by means of freezing curves. If a substance such as naphthalene is melted completely and then allowed to cool slowly with thorough stirring, a plot of temperature versus time such as that shown in Figure 6-13 results. Initially the temperature falls rapidly to 80.55°C, as heat is given off. At this point solid begins to appear and no further drop in temperature is noted until all the liquid has solidified, even though heat continues to be given off. The temperature drop thereafter is less rapid, owing to the slower transfer of heat from the solid to its surroundings than was true for the liquid.

The most important point about the diagram is that the portion of the curve in which solid and liquid are in equilibrium is horizontal, because the conversion of liquid to solid occurs at constant temperature. This fact is an important clue for our predictions of a model for liquids and, later in this chapter, a model for solids.

The reverse of the process described above is a classical method for aiding in the identification of solid organic compounds. The compound, or often a purified derivative of it, is slowly heated, either in a simple capillary tube

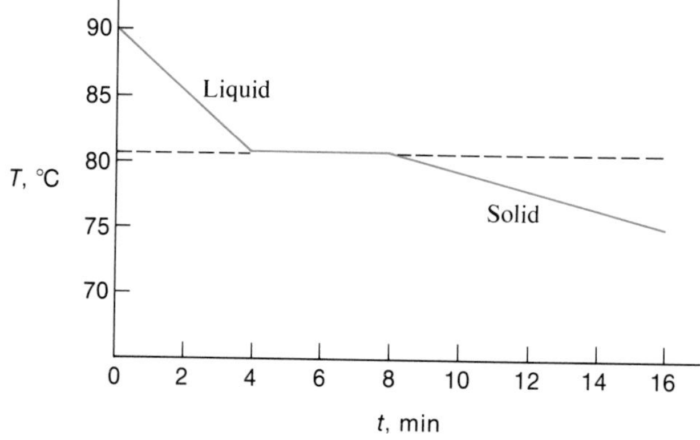

Figure 6-13. The freezing curve of naphthalene.

immersed in an oil bath, or on a more sophisticated apparatus called a melting-point stage, until the first crystals begin to melt. The melting point thus obtained is characteristic of the compound being observed and is therefore useful for identification. The heat that must be added to the substance to melt it completely (at constant temperature) is called the *heat of fusion* and is analogous to the heat of vaporization of a liquid.

Density

Another characteristic property of matter is *density,* which is the mass per unit volume, m/V, of the substance. It can be crudely determined with any device (such as a graduated cylinder) that can measure the volume of the substance, and a balance to determine the mass contained in that volume. Very accurate values can be obtained with a *pycnometer*, a small glass vessel of definite, known volume that can be precisely filled in a thermostated bath and then weighed on an analytical balance. One of the more sophisticated of the many other techniques for measuring density utilizes the resonant frequency of a small glass U-tube filled with the liquid.

A closely related concept is *specific gravity,* which is the ratio of the mass of a given amount of a substance to the mass of an identical volume of water at the same temperature or at some standard temperature (usually 4°C or 20°C). Because it is a *ratio* of two masses, the units cancel and specific gravity itself is a dimensionless quantity. When densities are expressed in grams per cubic centimeter (or grams per milliliter, which is identical but more meaningful with respect to fluids than to solids), the numerical values are essentially equal to the specific gravities because the density of water at ordinary temperatures is 1.000 g/ml. Thus, the density of benzene is 0.879 g/ml and its specific gravity is 0.879. (If greater accuracy is desired, the temperatures must be specified and slight differences between density and specific gravity may occur because of the unequal expansions of various substances with increasing temperature.)

The density of an ideal gas is easily calculable as follows:

$$PV = nRT \tag{6-48}$$

$$\frac{n}{V} = \frac{P}{RT} \tag{6-49}$$

$$\frac{m/M}{V} = \frac{P}{RT} \tag{6-50}$$

$$\text{Density} = \rho = \frac{m}{V} = \frac{PM}{RT} \tag{6-51}$$

For oxygen, $M = 32.0$ g/mole, so at STP we have

$$\rho = \frac{(1\ \cancel{\text{atm}})\ (32.0\ \text{g}/\cancel{\text{mole}})}{(82.1\ \text{ml}\ \cancel{\text{atm}}/\cancel{\text{K}}\ \cancel{\text{mole}})\ (273\ \cancel{\text{K}})} = 0.00143\ \text{g/ml} \qquad (6\text{-}52)$$

In contrast, the densities of most nonmetallic liquids lie between 0.7 and 3 g/ml. Mercury, a metal that happens to be liquid at room temperature, is extremely dense. Table 6-4 lists the physical properties of some common liquids.

The variation of density with temperature provides a valuable insight into the structure of matter. Let us again look at the important molecule, water. Figure 6-14 shows the density of water as a function of temperature for each of the three states of matter at 1 atm pressure. It is clear that the structure of liquid water more closely resembles the structure of ice than that of water vapor.

The fact that ice has a lower density than water has far-reaching implications. (It causes considerable grief to maritime industries in northern latitudes but great delight for ice skaters.) There would be no fish, and very little other aquatic life, in the Earth's polar and cool temperate zones if

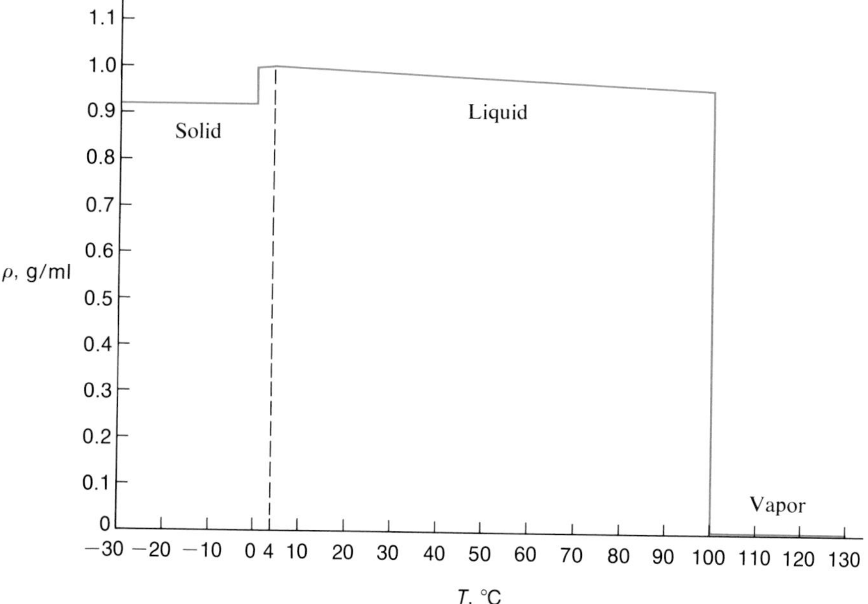

Figure 6-14. The density of water as a function of temperature. The very slight rise in liquid density from 0 to 4°C is exaggerated in this diagram for easier visualization. $P = 1$ atm.

Table 6-4. **Physical Properties of Some Common Liquids.**

Substance	Formula	Vapor Pressure at 20°C, Torr	Melting Point, °C	Boiling Point, °C	Density at 20°C, g/ml
Water	H_2O	17.535	0.0	100.0	0.99823
Ethyl alcohol	CH_3CH_2OH	44	−117.3	78.5	0.7893
Chloroform	$CHCl_3$	151	−63.5	61.7	1.4832
Acetone	CH_3COCH_3	170	−95.35	56.2	0.7899
Mercury	Hg	0.001201	−38.87	356.58	13.5939
Carbon disulfide	CS_2	286	−110.8	46.3	1.2632

the winter ice continuously forming on the surfaces of lakes and streams sank to the bottom, there to build up steadily until the entire body of water froze solid. The same thing would happen to the far northern and southern oceans, and the net result would have such profound effects on the global ecology, topography, and climate that life on Earth would be very different from what it is. On such slender threads does our existence hang.

Figure 6-14 is *not* typical. The densities of almost all solids are greater than those of the corresponding liquids. Thus we see again that water is a most unusual substance, and we will re-examine this anomalous behavior in the section on solids.

Brownian Motion

A fundamental observation on the behavior of liquids was made by a Scottish botanist in 1827. Robert Brown (1773–1858) was studying pollen grains on the surface of water, using a microscope, when he observed that they were continually jostling about in a completely random manner and, on occasion, even changing shape. After carefully ruling out fluid movement, capillary forces, evaporation, and forces between the solid particles, he initially concluded that the motion was somehow due to the particles themselves, because of the life within them. When he repeated his experiment using a variety of finely ground, inanimate particles, however, he observed the same phenomenon, and could not explain it. Only after the development of the kinetic theory of gases half a century later did the idea first arise that this *Brownian motion* of any fine particle in any liquid (or gas) could be due to thermal agitation of the molecules in the liquid. Until this idea was eventually proved correct, numerous scientists, who did not believe in atoms or molecules, refused to accept it.

Brownian motion was the first—and it remains the most striking—demonstration that molecules in liquids undergo ceaseless translational motion, just as they do in gases. It has been observed on dust particles in tiny liquid inclusions that have been trapped inside quartz crystals for countless millennia. Even more important, however, is that Brownian motion was the first bit of evidence for the atomic theory of matter that was a direct observation rather than a deduction from a theory. As such, it was a milestone discovery in the history of science, which Brown himself never realized.

The complete theoretical explanation of Brownian motion was given by Einstein in 1905 (the same remarkable year in which he published both the quantum theory of the photoelectric effect, for which he received the Nobel Prize, and the special theory of relativity!). His theory, based on the idea of random fluctuations in the velocities of molecules in a fluid, predicted a specific relation between the sizes of molecules and the Brownian motion* they should produce. It also predicted that the mean kinetic energy of *any* particle suspended in a fluid, regardless of its size, is given by $\frac{3}{2}kT$. In this respect, therefore, pollen grains or dust particles undergoing Brownian motion behave like gigantic gas molecules. In 1908 the French physicist Jean Perrin (1870–1942) performed an ingenious series of experiments by which he was able to corroborate Einstein's theory and calculate the size of water molecules. Perrin also calculated a value for Avogadro's number on the basis of Brownian motion. His first result (later improved upon by himself and others) was about 7×10^{23}, in good agreement with the then accepted value of about 6×10^{23}. This destroyed any remaining doubts regarding the validity of the atomic theory.

Summary of the Physical Properties of Liquids

Table 6-4 lists some of the properties of one element and five common organic and inorganic compounds that exist as liquids at room temperature. A perusal of this table, in conjunction with some thought about what these orders of magnitude mean, may well serve as a useful way to review the discussion of the preceding sections before we seek a model for the structure of liquids.

Some questions you might ponder are: Why does ethyl alcohol have a higher vapor pressure than water even though it has a higher molecular weight? What is the effect of molecular weight on the freezing and boiling points of liquids? Can you explain the exceptions? In view of the discussion of interatomic forces in Chapter 5, can you explain the anomalous behavior of mercury?

---

*Initially, Einstein did not refer to it as such, however, because at the time he formulated his theory, he had not even heard of Brownian motion—he had "invented" it by his own genius!

A Simple Model of Liquid Structure

As we have previously seen, the distinction between the gaseous and liquid states vanishes above the critical point. We might therefore suppose that a model for the liquid state can be derived by beginning with our model for real gases. When the molecules are compressed by a sufficiently large external pressure and the attractive forces between molecules become large enough, properties identifiable as those of a liquid are observed. Thus, the condensation of a gas to a liquid corresponds to the simple process pictured in Figure 6-15.

Our model for liquid structure contains the following assumptions (it may be of interest to you to compare these assumptions with those given on page 256 for an ideal gas):

1. Liquids consist of molecules (or atoms) that can be large relative to the distances traveled between collisions.
2. Molecules in liquids exert attractive forces upon each other, but interpenetration of electron clouds does not occur.
3. Molecules in liquids are hard, spherical, and in constant, rapid motion. Collisions between molecules are totally elastic. Molecules possess a range of velocities determined by the Maxwell-Boltzmann velocity distribution for gases.
4. Molecules in liquids slip past each other with an ease that depends upon the relative magnitudes of the potential energy of attraction, which binds

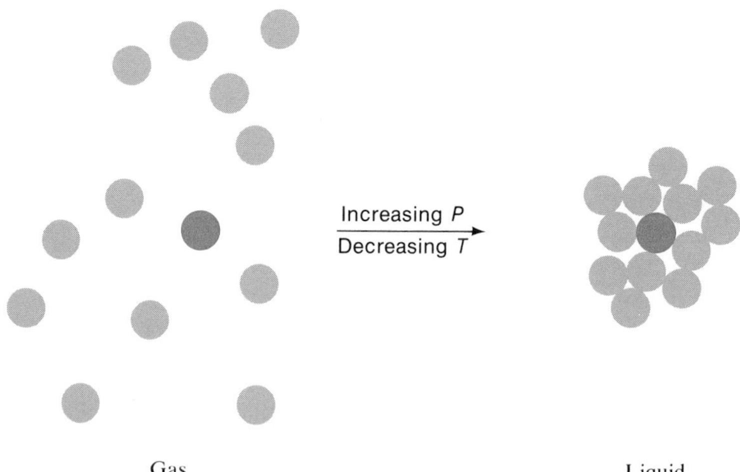

Figure 6-15. Schematic representation of the gas-to-liquid phase transition. A reference molecule is shown darker than the rest.

the molecules together, and the kinetic energy of the molecules, which tends to separate them.

Let us now consider the several properties of liquids discussed earlier and see how well (or whether) this model fits. According to assumption 3 of the model, the vapor pressure of the liquid is expected to increase as the temperature is raised. As T increases, more and more molecules acquire sufficient energy to escape the liquid because the root-mean-square velocity of the molecules increases and the distribution shown in Figure 6-9 peaks at higher velocities. Many of the molecules in the high-velocity tail of this distribution are able to break free of the surface of the liquid and enter the gaseous phase, thereby causing an increase in the vapor pressure.

The next property of liquids we discussed was their freezing at a definite temperature. This can be understood from a comparison of Figures 6-15 and 6-30 for a liquid and a solid, respectively, and from assumptions 2, 3, and 4 of our model.* The crystal structure shown in Figure 6-30a is highly regular, with six coplanar molecules touching any one molecule. A small increment of thermal energy at the melting point is sufficient to introduce disorder into a small region of this crystal structure. This disorder might correspond to the cluster of molecules shown in Figure 6-15, in which the coordination number about one molecule has been reduced to five. It can be shown graphically that this minor irregularity results in long-range disorder throughout the crystal; namely, when the coordination number of one molecule is reduced to five, it is then impossible to draw other, neighboring molecules with a coordination number of six. The one irregularity is "contagious." This model helps to explain the sharpness of both melting and freezing points.

The density of a liquid must be much higher than that of a gas because of the packing of the molecules assumed in our model. Increasing the pressure on a gas results in a corresponding increase in its density. Increasing the pressure on a liquid does not significantly increase its density, however, because the molecules are already packed nearly as closely as possible. Liquids are therefore considered to be incompressible, for all practical purposes, and it is this property that provides the basis for hydraulic engineering.

An interpretation of Brownian motion follows directly from our third assumption. Even while the liquid appears to be quiescent, a great deal of thermal agitation is occurring. The molecules in a liquid undergo continuous oscillations that, as predicted by our model, are analogous to those of gas molecules. Because these oscillations are completely random, there are statistical fluctuations that occasionally cause a cluster of molecules to

*Attractive and repulsive intermolecular forces were discussed in Chapter 5.

move in essentially the same direction at once. If such a cluster is large enough, it can perceptibly jostle even a much more massive pollen grain or some other fine particle suspended in the liquid. This perpetual jostling, or Brownian motion, is thus a visible manifestation of molecular dynamics.

It appears that our simple model of liquid structure is not inconsistent with some of the observed physical properties of liquids. However, it is too simple to be of significant value to the scientist because of other, more complex aspects of the behavior of liquids that it does not adequately explain. Given some basic information, such as the size, geometry, and mass of the molecule and the nature of the intermolecular forces, an adequate theory should be able to predict values of most of the physical properties of the liquid in question, as well as a variety of thermodynamic properties to be discussed in the next chapter. To date, no theory has such a capability.

6.3 SOLIDS

If we should momentarily discount the oxygen and nitrogen of the atmosphere and the water of the oceans, lakes, and rivers, about all that would remain in, on, and above the mantle of the Earth is solids. The earth, asphalt, and concrete we walk on, the furniture in our homes (not counting waterbeds!), most of the food we eat—all are solids. (Even the atmosphere contains vast amounts of solids as ultrafine particles, the overwhelming majority of them coming from natural sources.) In fact, solids have given their name to the decade of the sixties—the *solid-state decade*. The discovery of the properties of extremely pure silicon and germanium with trace amounts of other elements (*dopants*) added has produced the solid-state diode and transistor, which have revolutionized modern technology, from radios to computers, from manufacturing to medicine. Without them, manned space flight would be inconceivable.

We will first examine several of the techniques that have been useful in the study of solids. Then we will consider some of the general properties of the crystalline state. We will conclude with a look at the four main types of solids and a discussion of closest packing of atoms or molecules in these structures.

Experimental Methods

X-Ray Diffraction

Of the four techniques that we will discuss for the study of the structure of solids, x-ray diffraction has been by far the most informative. Therefore

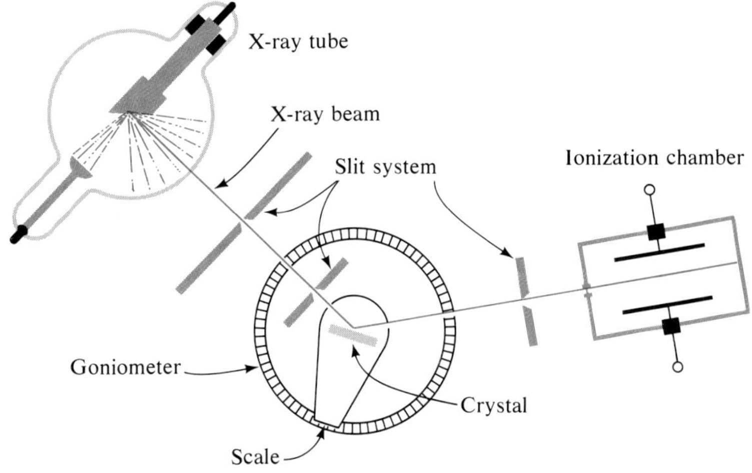

Figure 6-16. Diagram of a Bragg x-ray spectrometer.

we will devote the greatest portion of our time to this method. Some of the groundwork for our present discussion was laid in Chapter 2, where we discussed the nature of electromagnetic radiation and several types of diffraction experiments using slits. It might be helpful for you to review this material before continuing.

An *x-ray spectrometer* is shown schematically in Figure 6-16. This is an instrument for analyzing the detailed structure of a crystal by means of the patterns of x rays diffracted by the regular arrays of atoms within the crystal. The diffraction occurs only when the crystal is oriented at certain characteristic angles with respect to the incident x-ray beam. Measurement of these angles is accomplished by mounting the crystal on a high-precision, rotating platform called a *goniometer*. As the crystal is rotated, signals are received by the ionization chamber (a device for detecting x rays) whenever diffraction of the x rays occurs. This powerful experimental technique, conceived in 1912 by the German physicist Max von Laue (1879–1960) and pioneered by the English physicists Sir William Lawrence Bragg (1890–1971) and his father, Sir William Henry Bragg (1862–1942), has been of tremendous importance* in determining the exact atomic and molecular structures of a vast array of pure crystalline compounds (as well as metals and alloys), ranging from the relatively simple structure of NaCl all the way to those of gigantic protein molecules such as hemoglobin and lysozyme. (X-Ray diffraction is also used in the study of liquid structures, albeit with far less success.)

*The value of x-ray diffraction was promptly recognized with Nobel Prizes for von Laue in 1914 and the Braggs in 1915. The younger Bragg, who had taken the initiative in the work, was twenty-five at the time!

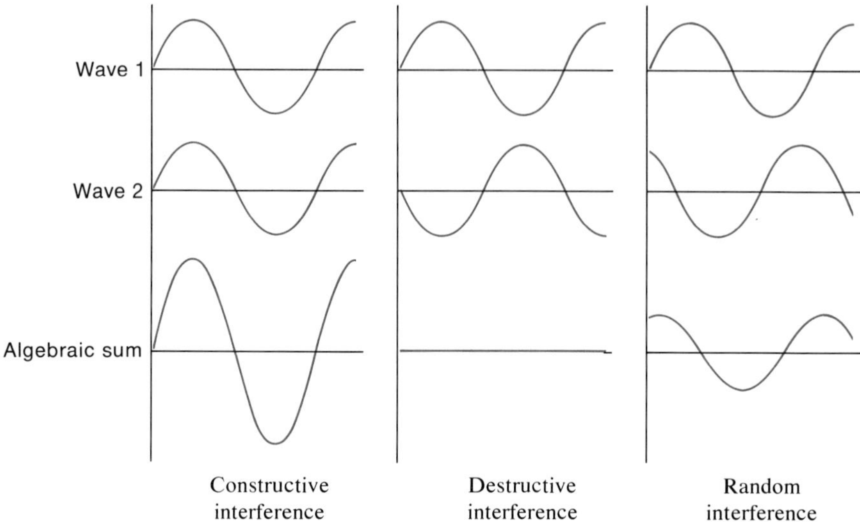

Figure 6-17. Interference phenomena between two waves.

In order to understand something of the origin of this phenomenon and to see how valuable structural data (bond lengths and bond angles) are obtained from it, we begin with a look at some of the properties of electromagnetic waves. Figure 6-17 illustrates what happens when two parallel waves are added in several ways. If the two waves propagate together in such a way that their maxima and minima always appear at the same relative positions at the same instant, the waves are said to be *in phase* and their algebraic sum is a new wave of identical wavelength but twice the amplitude. If one wave is shifted π radians (180°) relative to the other, the waves exactly cancel each other—they are *out of phase*—and total interference occurs.

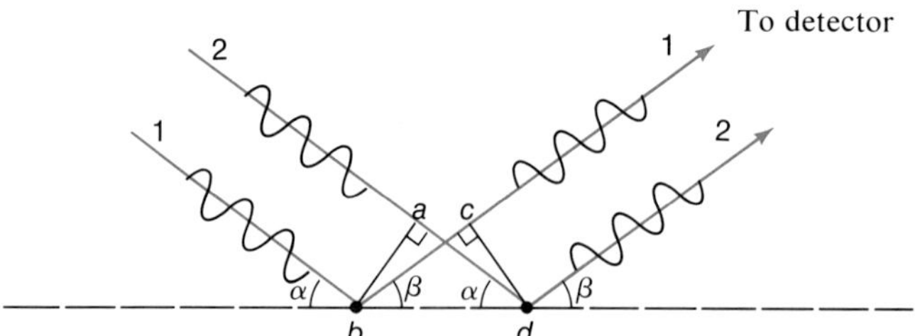

Figure 6-18. Diffraction of two waves from two atoms. In this drawing $\alpha = \beta$.

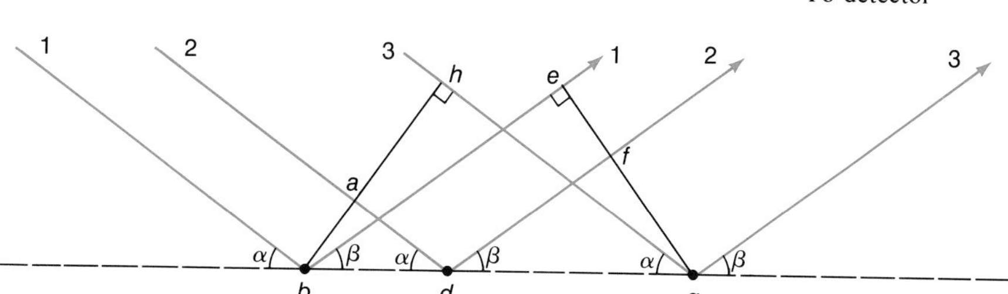

Figure 6-19. Diffraction of three waves from three atoms.

Of course there are infinitely many variations between these two extremes, as suggested by the last set of waves in Figure 6-17.

What happens to a set of such parallel, in-phase waves when they hit a crystal? Let us begin by looking at the simplest situation: two waves hitting two atoms (Figure 6-18). The two waves are initially in phase. Their arrival at the detector depends upon their also being in phase after they have both been reflected.* If they are, they will again interfere constructively and a signal will be detected. If not, they will interfere destructively to some extent and a weaker signal will be detected, if any. Of course, it does not matter whether the waves have the *same* phase as they did when they struck b and d or whether one has advanced 2π radians ahead of the other; constructive interference will still result.

The way the two waves and their diffracted trajectories have been drawn (with $\alpha = \beta$), the waves will be in phase at the detector because $bc = ad$. With any arbitrary relation between α and β, it can also be shown that the waves will be in phase if

$$bd(\cos \alpha - \cos \beta) = m\lambda \qquad m = 0, 1, 2, 3, \ldots \tag{6-53}$$

where bd is the interatomic spacing and λ is the wavelength of the radiation. The diffraction observed in Figure 6-18 corresponds to $m = 0$.

Let us now add one more atom (Figure 6-19) and again examine the diffraction. Our third atom is at g, which is farther from d than d is from b. The three waves are in phase at the line bah. To reach the line efg, ray 1 then travels the distance be, ray 2 travels $ad + df$, and ray 3 travels hg. It may be possible to find values of α and corresponding values of β such that the three waves are in phase at efg. However, we have greatly restricted our opportunities to observe constructive interference, not because we have added a third atom, but because we have added a third atom at a distance

*The spots that constitute an x-ray diffraction pattern are generally called *reflections*.

dg that does not equal bd. If $dg = bd$, an analysis similar to that given above would show that Equation 6-53 was still valid. In fact, it is valid for an infinite number of diffracting centers—provided that they are equally spaced.

We have extended our crystal in one dimension. Now let us extend it in a second dimension. Because we already know that maximum reinforcement will result only from equally spaced scattering centers, our array in Figure 6-20 is a regular one.

The main reason for examining such a two-dimensional array of atoms is to show how two-dimensional structural information can be derived from it by means of x-ray diffraction. (It can, of course, be generalized to three dimensions.) Waves will be reflected by each atom in the top row, as in Figure 6-18. But reflection also occurs at a number of rows underneath. Figure 6-20 shows two rays impinging on atoms on a vertical line. As before, the requirement for constructive interference is that the difference between the distances traveled by the two rays must be an integral number of wavelengths $n\lambda$, where $n = 1, 2, 3, \ldots$. This difference is the distance abc, which can be calculated as follows:

$$abc = 2ab = 2d \sin \beta \tag{6-54}$$

where d is the spacing between the rows of atoms (or layered arrays of rows, in a three-dimensional structure) and β is equal to the angle of incidence θ because β and θ are angles with mutually perpendicular sides. Therefore

$$abc = 2d \sin \theta \tag{6-55}$$

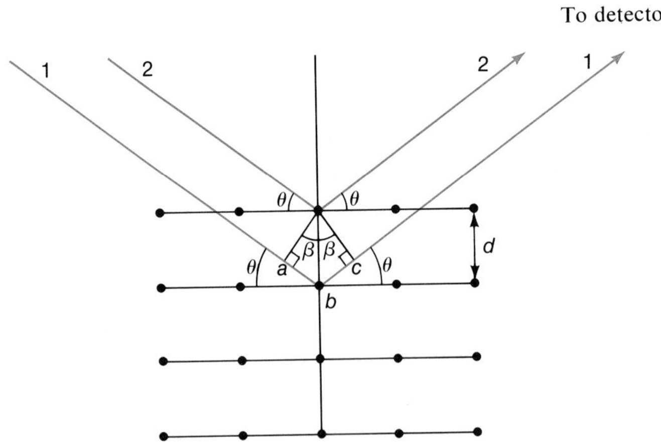

Figure 6-20. Diffraction of two waves from a two-dimensional array of atoms.

and since the distance *abc* must equal $n\lambda$ for constructive interference to occur, we arrive at the famous *Bragg equation*:

$$n\lambda = 2d \sin \theta \tag{6-56}$$

This equation can be used in either of two ways. It originally served for calculating the wavelengths of x rays, using crystals whose interatomic spacings were known. Now that the wavelengths of the x rays from a number of sources are very accurately known, it is used to determine interatomic spacings with correspondingly high accuracy.

One feature of the Bragg equation deserves our special attention. For a given value of $n\lambda$, $\sin \theta$ must increase as d decreases. Because $\sin \theta$ can never exceed 1, there is therefore a minimum to the observable spacing between rows of scattering centers in crystals. Clearly, this minimum must occur when $n = abc/\lambda = 1$, since that condition represents the smallest of the spacings for which diffraction can occur. We thus have $\lambda = 2d$ and the minimum spacing is given by

$$d_{\text{min}} = \frac{\lambda}{2} \tag{6-57}$$

Therefore if dimensions on an atomic scale are to be examined by diffraction methods, the wavelength of the radiation employed must itself be of atomic dimensions, i.e., of the order of an angstrom or so. In the electromagnetic spectrum, only x rays satisfy this criterion. The wavelength of the most commonly used x rays in diffraction analysis is 1.540562 Å.

Following our discussion of the unit cell, we will find that an interesting use of the Bragg equation is to provide an accurate value of Avogadro's number.

The analysis of crystals more complex than that of our model is vastly more complicated than a simple application of the Bragg equation, and requires the use of large, high-speed computers. With gigantic protein molecules, such as myoglobin, cytochrome c, and aldolase, as many as 10,000 reflections must be measured and interpreted before even a 3- or 2-Å resolution is achieved. To date, rather complete x-ray structures at a resolution of 2–3 Å have been obtained for six proteins, and work is well underway on perhaps a dozen others.

Figure 6-21 is the experimental diffraction pattern of myoglobin, the first protein whose three-dimensional structure was determined in virtually complete detail. Such photographs provide a wealth of information for protein and enzyme chemists (all enzymes are proteins), such as how the amino acids are distributed in the protein structure and where other molecules can be attached to the enzyme.

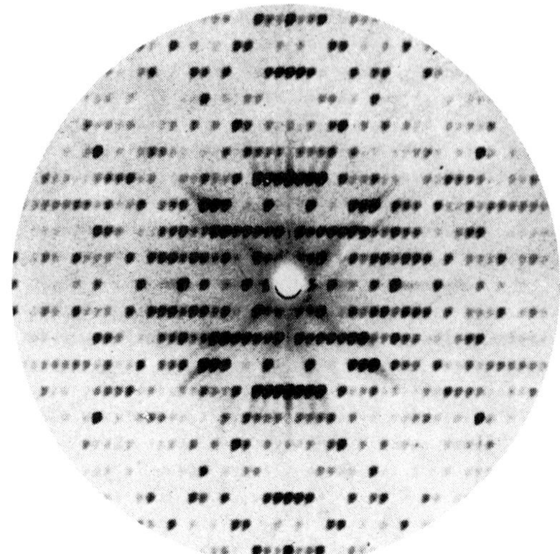

Figure 6-21. X-Ray diffraction pattern of the protein myoglobin. [F. H. C. Crick and J. C. Kendrew, *Advances in Protein Chemistry* **12**, 133 (1957).]

Other Experimental Techniques for the Study of Solids

We have dwelt at length on x-ray diffraction because it has provided us with the most detailed knowledge of the structures of solids. We now briefly mention a few other techniques.

The magnetic properties of many metals—especially the transition metals—and of their complex ions are of interest because of the information they give regarding electron configurations and molecular and crystal structures. If a vial containing the substance is placed in a magnetic field, it can be determined whether the substance is attracted by the field or repelled by it (see Figure 2-34 and the discussion of electron spin in Chapter 2). The attraction may be due to *paramagnetism* or *ferromagnetism* (e.g., Fe, Co, and Ni) and the repulsion is due to *diamagnetism*. Substances that are attracted by the field must have at least one unpaired electron. All substances are diamagnetic, but this property is conspicuous only in those substances that do not simultaneously exhibit some other type of magnetic property.

A variety of microscopic techniques is now available. At some time or other, almost everyone has peered through a light microscope at squirming bugs in a drop of pond water. Even a few hundred-fold magnification reveals many striking new insights into the microscopic world around and within us. Light microscopes can now be obtained in a variety of types and with magnification factors up to 2500×. Features of the gross structures in unicellular organisms and unique surface irregularities of crystals can be easily seen with light microscopes. However, there is a limit to the resolving power of optical microscopy and that limit is approximately half the wavelength of the low end of the visible spectrum, or about 2000 Å. Because atomic dimensions are several thousand times smaller than this, we cannot hope to resolve the fine details of molecules by using radiation that the eye can see.

One obvious solution would be to use radiation that we cannot see (except by its effect on photographic film or some other kind of detector) but that has the proper wavelength. What about x rays? A splendid idea, except that no one has found a way to construct a good lens to focus x rays. Some lenses have been made, but their resolution is no better than about 20 Å.

The most potent kind of microscope at present is the *electron microscope*. In Chapter 2 we stated that electrons have the characteristics of waves as well as particles. If this is true, they should display wave-like properties, such as diffraction and interference, in addition to their material properties, such as mass and charge. Figure 2-13b shows that this *is* true.

In an electron microscope an electromagnetic lens system creates shaped magnetic fields that direct and focus a high-energy beam of electrons, much as glass lenses direct and focus photons. The energies of the electrons used in most commercial electron microscopes are between 20,000 and 100,000 electron volts. Such electrons have very short wavelengths (see Example 2-4), which means that electron microscopes are, in principle, capable of resolving individual atoms. In practice the best commercial instruments achieve a resolution of ≈3 Å and a maximum magnification of ≈800,000×. The resolution is limited not by the wavelength of the electrons, but by the limits of accuracy within which the iron pole pieces in the electromagnetic lenses can be machined.

The electron microscope is similar to the light microscope in several ways: there is an electron source that is analogous to the lamp, there are lenses that project an image of the specimen under study, and there is a fluorescent screen for transforming the electron image into light that is visible to the human eye. Figure 6-22 is a schematic representation of a commercial electron microscope.

In order to avoid interaction of the electrons with molecules not in the specimen itself, the entire electron beam and the specimen must be in a very high vacuum. Because of this the electron microscope is used almost exclusively for studying solid objects. It is used, e.g., by metallurgists for studying crystal imperfections, by physicists for analyzing the structures of composite materials, by biologists for determining cell structures, and by chemists for learning about subunit structures and symmetries in

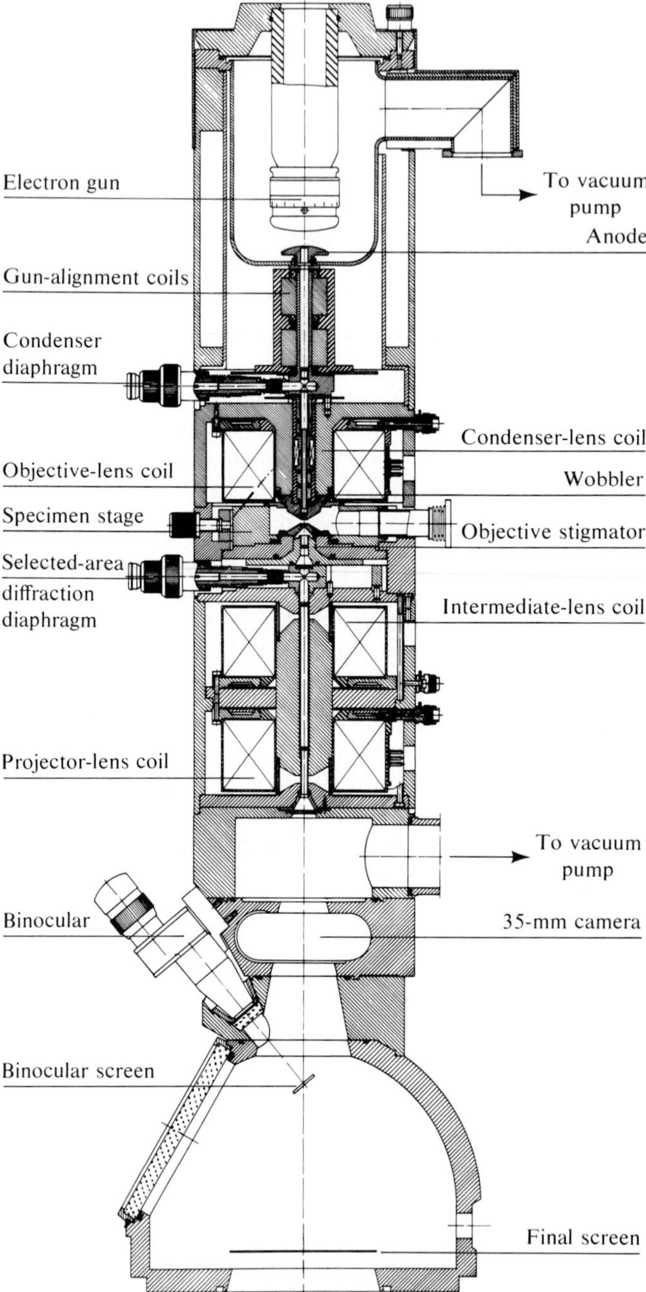

Figure 6-22. Diagram of the Philips EM 201 electron microscope. (Courtesy of Philips Electronic Instruments.)

proteins and nucleic acids. The electron microscope has provided vital information regarding the structure of cell membranes and the genetic information carrier, deoxyribonucleic acid (DNA).

Because the penetrability of solids by electron beams is small, the specimens must be extremely thin, of the order of 200–1000 Å. Figure 6-23 is a spectacular electron micrograph of a thin section of the nucleus of a human cell that has been infected with an adenovirus. The regular arrays of the virus particles clearly demonstrate the phenomenon of crystallization and suggest that virus microcrystals exist in living cells. The electron microscope thus provides the biologist with information regarding the site and mechanism of virus production and also about the sizes and shapes of the viruses. For example, adenoviruses are 600–900 Å in diameter and have icosahedral symmetry.

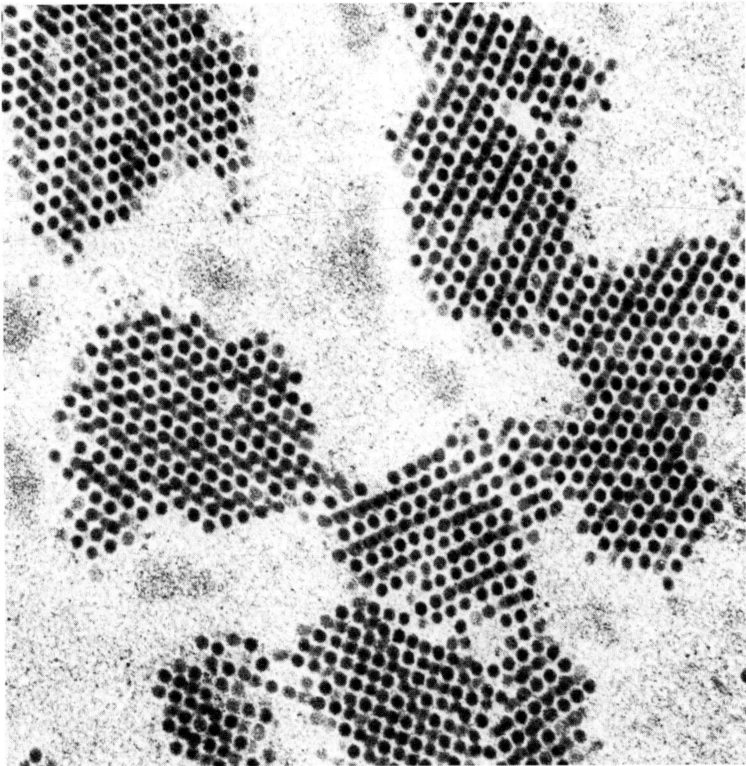

Figure 6-23. Electron micrograph of a thin section of the nucleus of a human KB cell, showing crystalline masses of human adenovirus # 5. Magnification 33,000×. (Courtesy of Prof. Daniel De Kegel, Vrije Universiteit Brussel and Institut Pasteur du Brabant, Brussels.)

Finally, it might interest you to learn a little more about these viruses. There are at least 31 different human adenoviruses, all of which consist of a core of double-stranded DNA (11–13% of the virus) and a coat, or *capsid*, of protein molecules (the rest of the virus). They were originally found in cultures of human adenoids (hence, their name), and are also often found in the tonsils. They are present during many acute viral respiratory infections, although they do not necessarily cause these infections. Some of the human adenoviruses induce tumor formation when injected into experimental animals. Historically they are the first such tumor-inducing viruses isolated from the human. They are being studied intensively at present because of their possible relation to cancer.

Crystal Lattices and Unit Cells

The x-ray diffraction technique described above reveals the spacings between atoms in the crystal structure and the overall geometry of the structure. What geometries have been found? What geometries are possible? How do you describe such arrays?

A *crystal* is a three-dimensional array of atoms, ions, or molecules that exhibits periodically repetitive structural features, i.e., it exhibits *symmetry*. A *crystal structure* is the complete array of atoms in the crystal; by definition, it has a certain amount of symmetry. Some crystals have much more symmetry than others. From the mathematical point of view, however (and therefore for the purposes of x-ray diffraction analysis), the symmetry of crystals is best considered not in terms of the crystal structure itself, but rather in terms of the *crystal lattice*. This is an imaginary array of points

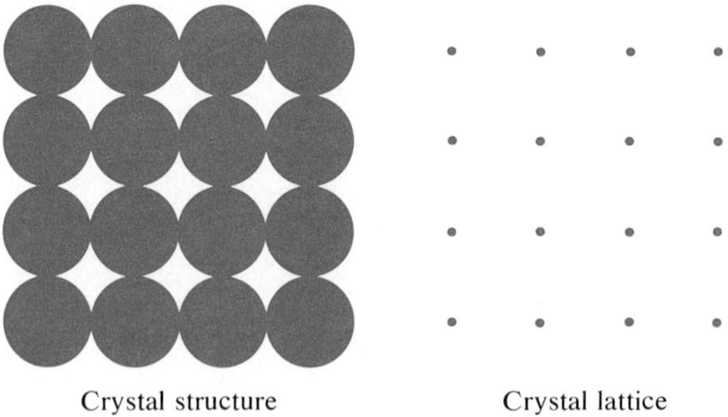

Figure 6-24. Representations of a hypothetical crystal structure and crystal lattice.

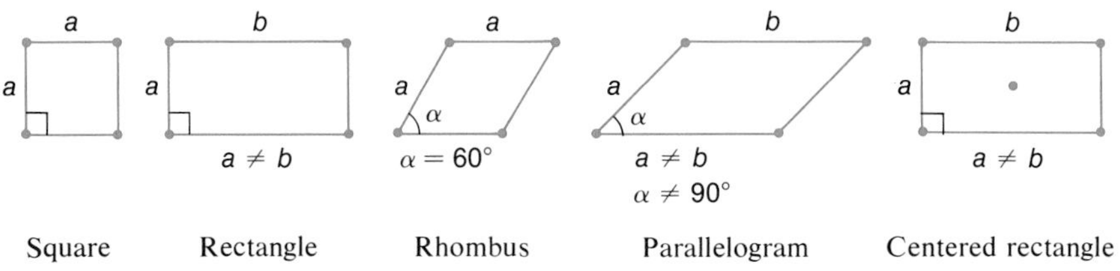

Figure 6-25. The five two-dimensional unit cells.

in space, each point corresponding to one of the corners of each of the unit cells in the crystal.

The *unit cell* is the basic conceptual building block of the crystal. It is the smallest structural entity in the crystal that, when translated parallel to itself in three dimensions (i.e., moved along rectilinear coordinates) without rotation, fills all the available space and reproduces the entire crystal. Unit cells—and therefore crystal lattices, which are defined on the basis of unit cells—are extremely useful because they simplify and systematize the enormous variety and complexity of crystal structures. There are 230 different crystal structures possible, according to the mathematics of symmetry, but each can be classified within one of the only fourteen possible types of crystal lattice. These are often called *Bravais lattices* after the French physicist Auguste Bravais (1811–1863), who discovered them, oddly enough, during the course of an investigation in phyllotaxy (the arrangement of leaves on a stem). The difference between a crystal structure and a crystal lattice is shown schematically in Figure 6-24.*

You are already familiar with a number of two-dimensional lattices. Some examples are the patterns of tiles in many old bathroom floors, the surface of a honeycomb, and the trees in an apple orchard. All two-dimensional lattices can be generated by rectilinear translations of only five unit cells, which are shown in Figure 6-25.

Each of the fourteen space lattices (Bravais lattices) has six faces, and each corresponds to a crystallographic unit cell. Of these, however, only six are *primitive,* meaning that they have lattice points only at their eight corners. These primitive unit cells define the six *crystal systems*: triclinic, monoclinic, orthorhombic, tetragonal, hexagonal, and cubic. All fourteen lattices are shown in Figure 6-26.

---

*In real crystals the positions of the atoms are called *lattice sites.* Although these sites often correspond to the imaginary lattice points (as in Figure 6-24 and the examples to be presented below), this is not always true.

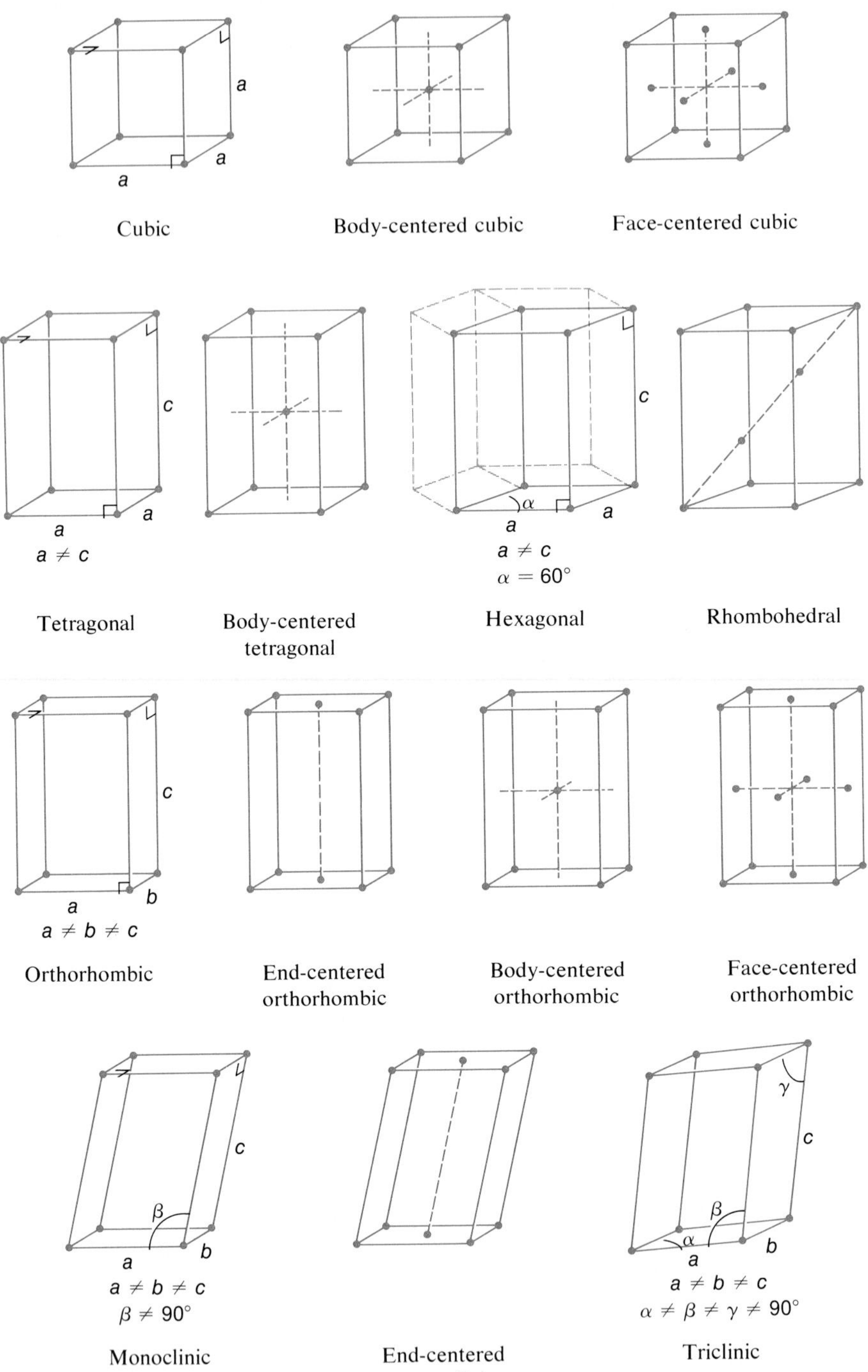

Figure 6-26. The fourteen three-dimensional unit cells, or Bravais lattices. The lines, both solid and dashed, have no significance except to make the drawings intelligible; they are irrelevant to the crystal structure and symmetry.

The eight nonprimitive unit cells have the same shapes as the primitive cells, but each has one or more additional lattice points arranged as follows: if there is a lattice point in the center of the cell, the cell is *body-centered*; if there are lattice points in the centers of only one pair of opposite faces, the cell is *end-centered*; if there are lattice points in the centers of all the faces, the cell is *face-centered*; and if there are two lattice points one-third and two-thirds of the way along one diagonal of a hexagonal cell, the cell is *rhombohedral*. The three types of cubic crystals are shown again in Figure 6-27, together with the corresponding space-filling models. It is important to remember that all fourteen unit cells are required to generate all possible crystal lattices, not just the six primitive unit cells that define the crystal systems.

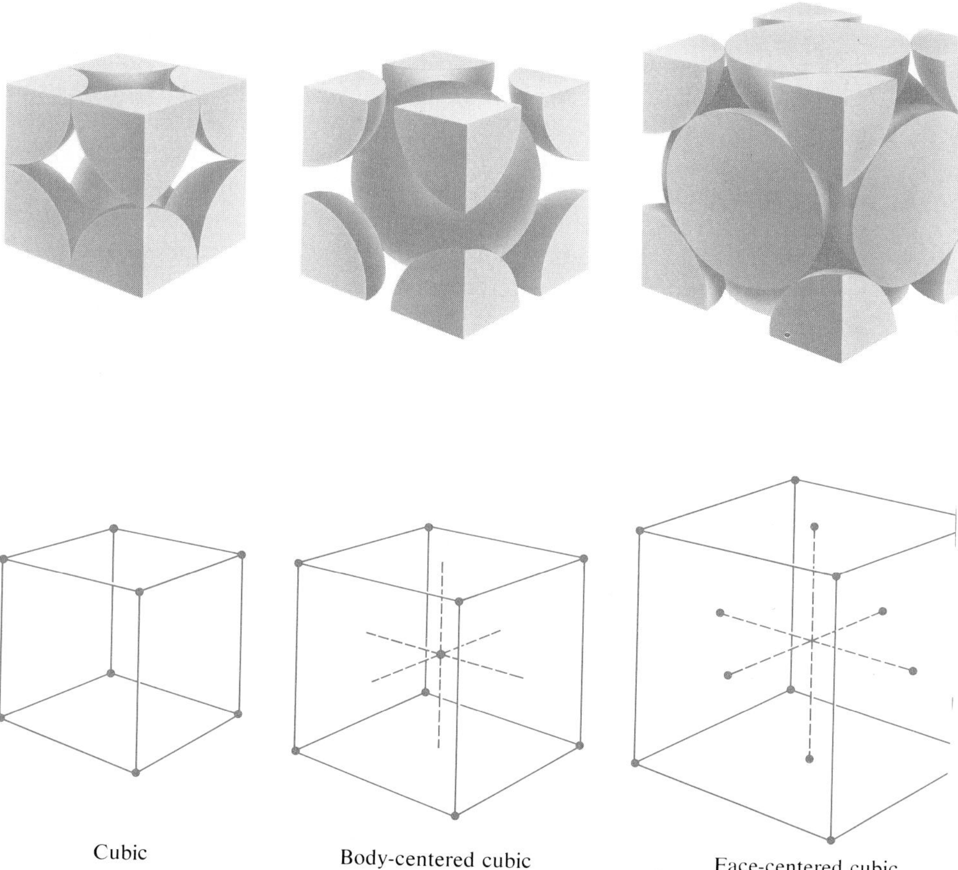

Figure 6-27. The three cubic unit cells and the corresponding space-filling atomic models; the latter are cut in the planes of the former.

To summarize, there are six crystal systems (primitive unit cells), fourteen Bravais lattices (six primitive unit cells plus eight nonprimitive unit cells), and 230 crystal structures. These are theoretical numbers, and all the experimental evidence indicates that they are correct.

Types of Solids

It is useful to classify solids in the same way that it is useful to consider the periodic properties of the elements. Solids can be classified in four main groups, depending upon the type of particle that occupies the lattice sites in the crystal structure. Table 6-5 presents this classification. The nature of each of the forces existing in the types of bonding listed has been described in the preceding two chapters.

The molecular solids differ sharply from the other three types. As discussed earlier, van der Waals and hydrogen bonds are an order of magnitude weaker than covalent or ionic bonds. Thus the forces holding these solids together are rather weak. In fact, several of the substances listed as molecular solids in Table 6-5 are gases at room temperature.

The ions occupying the lattice sites in ionic crystals must be of two types, cations and anions. Alternating lattice sites in the NaCl crystal, e.g., are occupied by Na^+ ions and Cl^- ions. Figure 6-28 shows the crystal lattice and the crystal structure of the unit cell of NaCl. It can be envisioned as two interpenetrating, face-centered cubic lattices of Na^+ and Cl^- ions. We are already familiar with some of the properties of such crystals from our use of rock salt, which is used to melt ice on streets and for making home-

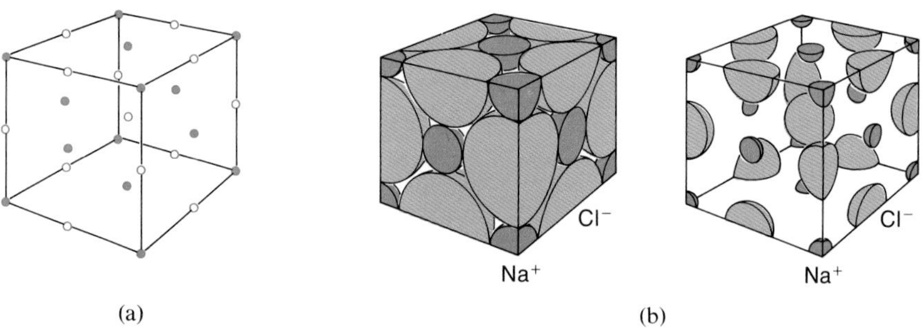

Figure 6-28. The crystal lattice and crystal structure of the unit cell of NaCl. (a) The crystal lattice, consisting of two interpenetrating face-centered cubic lattices. (b) Two views of the crystal structure, showing Na^+ ions at each corner and each face center, and Cl^- ions on each edge and one in the center. The unit cell content is $4(\text{NaCl}) = [8(\frac{1}{8}) + 6(\frac{1}{2})](\text{Na}^+) + [12(\frac{1}{4}) + 1](\text{Cl}^-)$. [(b) from J. A. Campbell, *Chemical Systems: Energetics, Dynamics, Structure,* W. H. Freeman and Company. Copyright © 1970.]

Table 6-5. Four Main Types of Solids.

	Molecular	Ionic	Metallic	Covalent
Particles occupying lattice sites	Molecules or noble-gas atoms	Ions	Positive ions	Atoms
Types of bonding	Van der Waals Dipole-dipole Dipole-induced dipole London dispersion Hydrogen bonding	Coulomb force	Resonating shared electrons	Shared electrons
Typical properties	Low melting points	High melting points	Moderately low to high melting points	Very high melting points
	Soft	Hard and brittle	Hard or soft	Very hard
	Poor electrical conductors	Poor electrical conductors	Good electrical conductors	Poor electrical conductors
Examples	H_2O, CO_2, H_2, HCl, XeF_6, UF_6, Ar, any organic compound	NaCl, CaF_2, KNO_3, ZnS, Cu_2O, TiO_2, Al_2O_3	Na, Al, Fe, Ag, W, Hg, U	B, Si, C(diamond), SiC(carborundum), SiS_2

made ice cream. Salt is hard, yet it cleaves easily. This is because of the alternating planes of Na^+ and Cl^- ions, which permit a clean fracture from a sharp blow.

In Chapter 4 it was pointed out that metals consist of metal cations occupying the lattice sites and a sea of delocalized electrons surrounding them. These electrons are the "glue" that holds the positive ions together, via Coulomb forces. Because of their high degree of delocalization, they are easily transported through the lattice, making this type of solid the only good electrical conductor. Copper is not the best conductor (silver is) but it represents the best compromise between conductivity and cost.

Apart from the ionic and metallic minerals, there are a limited number of covalent solids. Diamond and graphite are two classic examples. Diamond is extremely hard because each atom is bonded to four others by strong, sp^3 hybrid bonds to form the tetrahedral network shown in Figure 6-29a. The high degree of directionality of these bonds makes them very difficult to deform. Graphite also consists of pure carbon but it is soft and slippery. This is because it consists of giant layers of atoms, each layer being held together by three σ bonds per atom, arising from sp^2 hybridization, and one delocalized π bond per atom. Thus each C atom is bonded to three other C atoms by bonds having an effective bond order of $1\frac{1}{3}$. The layers shown in Figure 6-29b are loosely stacked at a separation (3.40 Å) too large to allow covalent bonds to exist between them. They are held together only by weak van der Waals forces and hence can readily slip past each other (this is why graphite is a good lubricant). Two other covalent solids that are notably atypical in their physical properties are phosphorus and sulfur. Both are soft and have low melting points.

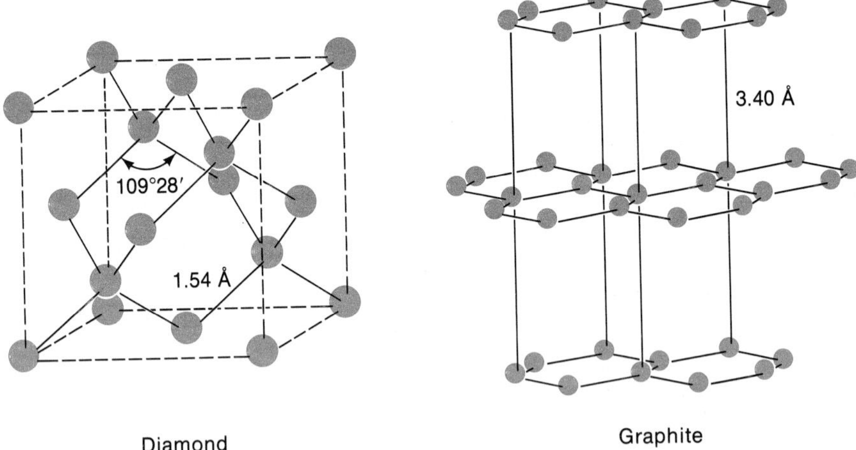

Figure 6-29. The crystal structures of the two allotropes of carbon.

Closest Packing in Crystal Structures

You can learn a surprising amount about crystal structure (and have a very pleasant time in the process) by getting a number of balls and sitting down and playing with them. Styrofoam balls are the easiest to handle, but any balls will do—except snowballs. If you don't happen to have a few dozen balls handy, you can refer to the diagrams below.

Atoms, ions, and molecules can be represented as spheres of definite radius. Thus we could represent the Cu^{2+} ion as a sphere of one-inch radius. Obviously no other Cu^{2+} ion can occupy the same space as the first Cu^{2+} ion. We could label the first ball somehow and begin packing other balls around it. We find that exactly six will fit in the same plane in contact with the first ball (Figure 6-30a). This is called *closest packing*. If we now place a second layer on top of the first so that the balls fit as snugly as possible, each ball in the second layer nestles in a depression of the first (Figure 6-30b). The placement of the third layer offers two choices, however. If each ball in the third layer is placed directly over a ball in the first layer, the structure can be described as ABABAB . . . , where the letters refer to successive layers. Such a structure is called *hexagonal closest packing*, or *hcp* (Figure 6-30c) and is based on the hexagonal unit cell. If the third layer is rotated 60° so that each ball is directly over one of the still unoccupied depressions in the first layer, the structure is ABCABCABC . . . and is called *cubic closest packing*, or *ccp* (Figure 6-30d). It is based on the face-centered-cubic unit cell and is generally referred to as *fcc* rather than *ccp*.

Both of these structures are closest packed, which means that the maximum density (for spheres) has been achieved. Surprisingly, this leaves 26% of the crystal structure empty. The holes responsible for this much empty space provide the explanation for a variety of crystal structures described below. Note also that the *coordination number*—the number of balls touching any one ball—is 12 in both cases. Thus the density of the balls, packed in either way, is the same.

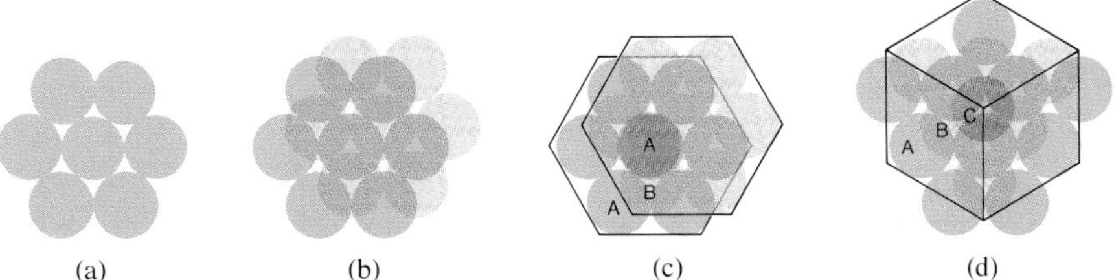

Figure 6-30. The closest packing of spheres. (a) Closest packing in one plane. (b) Most stable superposition of two closest-packed layers. (c) Hexagonal closest packing. (d) Cubic closest packing.

Many molecular and metallic solids fit one or both of these two models. (The multiplicity, if any, is due either to allotropy or to temperature ranges.) Figure 6-31 presents the crystal structures of most of the metals. The abbreviation *bcc* stands for *body-centered cubic,* a non-closest-packed structure in which 32% of the space is empty. All the alkali metals crystallize as this lattice. As shown in Figure 6-27b, there are only eight nearest neighbors in this packing system. Hydrogen and oxygen crystallize as *hcp* lattices and methane and hydrogen chloride crystallize as *fcc* lattices. There is no simple way to use the properties of the molecules themselves to predict which type of lattice such substances will crystallize as.

It is a pleasant surprise to find that we can not only describe crystal structures by the concept of closest packing, but also predict where the cations

Li *bcc* *hcp*	Be *hcp*												
Na *bcc*	Mg *hcp*											Al *fcc*	
K *bcc*	Ca *fcc* *hcp* *bcc*	Sc *hcp* *fcc*	Ti *hcp* *bcc*	V *bcc*	Cr *bcc* *fcc* *A12* (cubic)	Mn *A12* (cubic) and others	Fe *bcc* *fcc*	Co *hcp* *fcc*	Ni *fcc*	Cu *fcc*	Zn *hcp*	Ga *A11* (ortho- rhombic)	
Rb *bcc*	Sr *fcc* *hcp* *bcc*	Y *hcp*	Zr *hcp* *bcc*	Nb *bcc*	Mo *bcc* *fcc*	Tc *hcp*	Ru *hcp*	Rh *fcc*	Pd *fcc*	Ag *fcc*	Cd *hcp*	In *A6* (tetra- gonal)	Sn *A4* (diamond) *A5* (tetra- gonal)
Cs *bcc*	Ba *bcc*	La *hcp* *fcc* *bcc*	Hf *hcp*	Ta *bcc*	W *bcc*	Re *hcp*	Os *hcp*	Ir *fcc*	Pt *fcc*	Au *fcc*	Hg *A10* (rhombo- hedral)	Tl *hcp* *bcc*	Pb *fcc*

Ce *fcc* *hcp* *bcc*	Pr *hcp* *fcc*	Nd *hcp* *bcc*	Pm ?	Sm Rhombo- hedral	Eu *bcc*	Gd *hcp*	Tb *hcp*	Dy *hcp*	Ho *hcp*	Er *hcp*	Tm *hcp*	Yb *fcc*	Lu *hcp*

Figure 6-31. The crystal structures of most metals, as displayed in the periodic table. The abbreviations used are: *fcc,* face-centered cubic; *bcc,* body-centered cubic; *hcp,* hexagonal closest packing. (The *A* designations for some of the elements refer to more complex crystal structures.)

will be found in the lattices. From the values of the ionic radii given in Table 3-5, we find that the anions are typically much larger than the cations. We can envision the structure of the salt M^+X^- to consist of a closest-packed arrangement of the X^- ions. The M^+ ions will fit in the holes that constitute the empty 26% of the structure. The question now is: What kind of holes are there in such a structure and how many of them are there? The question is best answered by looking at a set of closest-packed balls. Alternatively we can look at a small portion of the crystal structure itself. It turns out that there are three kinds of holes: trigonal, tetrahedral, and octahedral. The crucial factor determining which holes the cations will occupy is the relative sizes of the cation and the anion. The following example illustrates this.

Example 6-5. Calculate the ratio of the size of the largest sphere (cation) that will fit in an octahedral hole to the size of the spheres (anions) that define this hole.

The hole has the geometry shown below. The octahedral figure shows the relative positions of the spheres. The diagram next to it shows the four coplanar spheres (of radius r_2) just touching each other and enclosing the largest sphere (of radius r_1) that will fit in the middle.

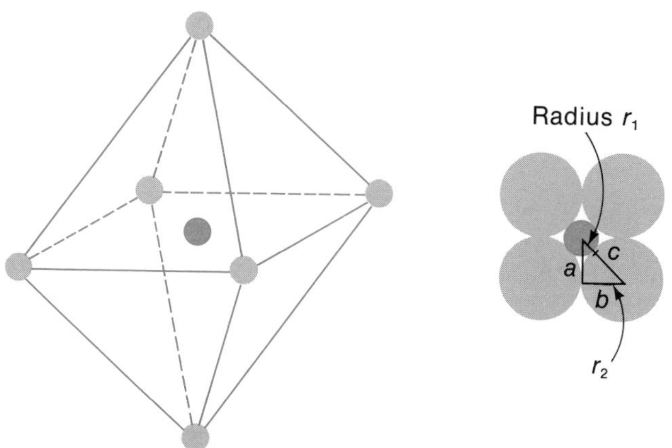

The triangle outlined is a right triangle $(a = b = r_2)$. Therefore

$$a^2 + b^2 = c^2$$

or

$$r_2^2 + r_2^2 = (r_1 + r_2)^2$$

Then

$$\sqrt{2}r_2 = r_1 + r_2$$

$$r_1 = (\sqrt{2} - 1)r_2 = 0.414r_2$$

$$\frac{r_1}{r_2} = 0.414$$

The maximum ratio of the cationic to anionic radii that will permit a snug fit is thus 0.414.

As a result of this example, we might expect that any smaller radius ratio would permit the cations to occupy octahedral holes. It turns out to be just the opposite: any radius ratio *larger* than 0.414 but smaller than the next-larger hole's radius ratio permits the cations to be in octahedral holes. Apparently the repulsion of the negatively charged anions when they are in contact is so great that the system is more stable if it expands slightly, with the anions forming a not quite closest-packed structure.

Gray and Haight have generated a nice figure that summarizes these ideas (Figure 6-32). The radius ratios are computed as above. You should find it amusing and challenging to verify the numbers of holes per anion and the radius ratios listed there. (Note the additional types of holes—linear and cubic—for non-closest-packed spheres.)

But what is all this talk about holes good for? Two examples may help answer that question.

Example 6-6. Predict the crystal structure of NaCl, given that the ionic radii are

$$r_{Na^+} = 0.97\ \text{Å} \qquad r_{Cl^-} = 1.81\ \text{Å}$$

We first calculate the radius ratio: $r_+/r_- = 0.54$. This falls between the ratios for octahedral and cubic holes, so we predict that the Na^+ ions occupy octahedral holes. One Na^+ is required for every Cl^- for the stoichiometry to come out right. Because there is one octahedral hole per Cl^-, every hole is occupied. The radius ratios have correctly predicted the structure given in Figure 6-28.

Example 6-7. Predict the crystal structure of wurtzite, ZnS, given that

$$r_{Zn^{2+}} = 0.74\ \text{Å} \qquad r_{S^{2-}} = 1.84\ \text{Å}$$

The radius ratio $r_+/r_- = 0.40$. This value happens to be right on the borderline. If the zinc ions occupied the octahedral holes, the sulfide ions would very nearly

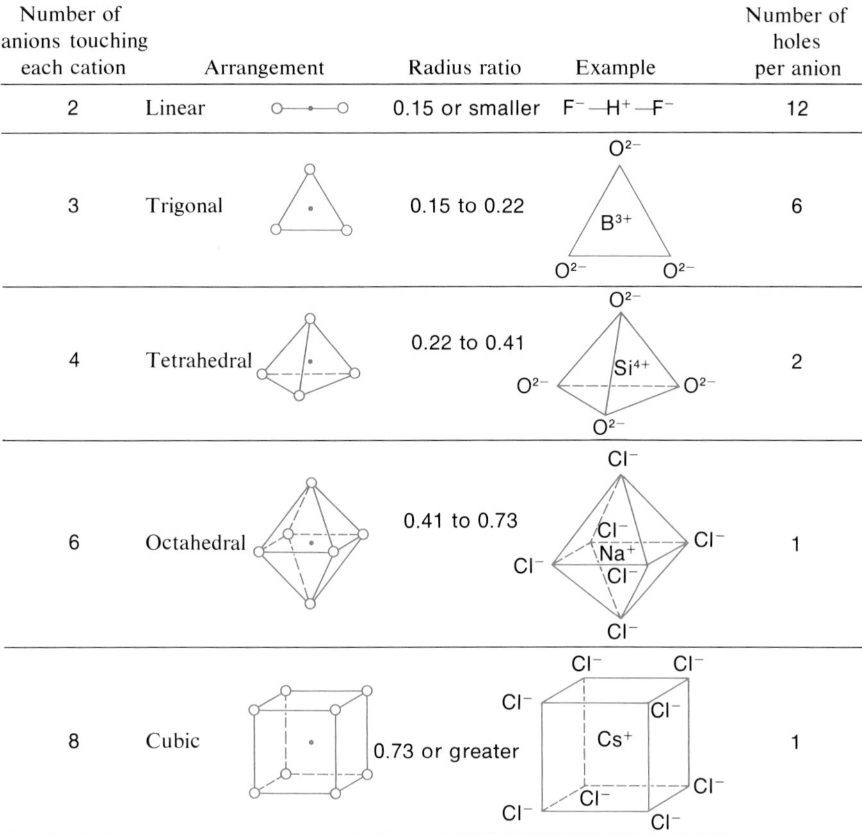

Number of anions touching each cation	Arrangement	Radius ratio	Example	Number of holes per anion
2	Linear	0.15 or smaller	F^-—H^+—F^-	12
3	Trigonal	0.15 to 0.22	B^{3+}, O^{2-}	6
4	Tetrahedral	0.22 to 0.41	Si^{4+}, O^{2-}	2
6	Octahedral	0.41 to 0.73	Na^+, Cl^-	1
8	Cubic	0.73 or greater	Cs^+, Cl^-	1
12	Does not occur, since spheres would have to be identical, thus not allowing positive and negative ions to exist in the same crystal.			

Figure 6-32. The dependence of crystal structure upon radius ratio. (Adapted from H. B. Gray and G. P. Haight, Jr., *Basic Principles of Chemistry,* W. A. Benjamin, New York, 1967.)

touch. Thus we predict that the zinc ions will be in tetrahedral holes instead. The lattice will be rather spread apart. But there are two tetrahedral holes per sulfide ion, whereas the stoichiometry demands a 1 : 1 ratio. Thus we predict that only half the tetrahedral holes will be occupied by Zn^{2+} ions. X-ray studies confirm this and also show that the S^{2-} ions are in an *hcp* configuration, something we still cannot predict.

Avogadro's Number from Crystal Density

One of the most accurate ways to calculate Avogadro's number is by the accurate measurement of crystal densities and unit cell dimensions. In this method the dimensions of the unit cell and the number of molecules per unit cell are used to calculate the volume per molecule. The density and the molecular weight yield the volume per mole. The ratio of these two volumes yields the number of molecules per mole, which is Avogadro's number N_A. The following example illustrates the calculation.

Example 6-8. The unit cell of KBr is the same as that of NaCl, given in Figure 6-28. The dimensions of the KBr unit cell are 6.60 Å per side. The density of the crystal is 2.75 g/cm³ and the molecular weight of KBr is 119 g/mole. Calculate Avogadro's number.

The volume of the unit cell is $(6.60 \times 10^{-8}\ \text{cm})^3 = 287.5 \times 10^{-24}\ \text{cm}^3$. Therefore the volume per molecule is $(287.5 \times 10^{-24}\ \text{cm}^3)/4 = 71.9 \times 10^{-24}\ \text{cm}^3/\text{molecule}$ because there are four KBr molecules per unit cell. The volume of 1 mole of KBr molecules is $(119\ \text{g/mole})/(2.75\ \text{g/cm}^3) = 43.3\ \text{cm}^3/\text{mole}$. Therefore

$$N_A = \frac{43.3\ \text{cm}^3/\text{mole}}{71.9 \times 10^{-24}\ \text{cm}^3/\text{molecule}} = 6.02 \times 10^{23}\ \text{molecules/mole}$$

Bibliography

Terrell L. Hill, **Lectures on Matter and Equilibrium,** W. A. Benjamin, New York, 1966. This book was written to accompany an honors-level general chemistry course. Thus the treatment is somewhat more advanced than that of the present chapter. However, much of the material on ideal gases, real gases, liquids, and solids is written in a very understandable and elementary manner. The discussion of intermolecular forces represents a nice extension of the material in Chapter 5. The discussions of thermodynamics and chemical equilibrium will be useful supplementary reading for Chapters 7 and 9, respectively.

Walter J. Moore, **Seven Solid States,** W. A. Benjamin, New York, 1967. This excellent book gives a solid introduction to the chemistry of the solid state. Although some of the material may be somewhat difficult for you, you will find most of the qualitative discussions and the diagrams both interesting and understandable. The book provides detailed descriptions of seven crystalline solids: salt, gold, silicon, steel, nickel oxide, ruby, and anthracene. These common but very different crystals (two elements, three compounds, and two solid solutions) were chosen to include the four main types of solids and cover a very broad spectrum of solid-state properties.

Problems

1. A quantity of gas weighing 9.2 g at 1 atm and 25°C occupies a volume of 4.0 ℓ. What is the volume of this gas at STP?

2. Calculate the density of CO_2 at STP.

3. The density of O_2 at STP is 1.43 g/ℓ. What is its density at 325 K and 710 Torr?

4. A gaseous compound with the formula $C_xH_yCl_z$ is mixed with the amount of oxygen required for complete combustion. The burning of 8 volumes of the original compound produces 16 volumes of CO_2, 16 volumes of H_2O vapor, and 8 volumes of Cl_2 at the same temperature and pressure. What is the molecular formula of the compound?

5. A 560-ml bulb is filled with the vapor of a volatile compound at 100°C and 740 Torr. The mass of the vapor is found to be 2.53 g. Analysis of the compound shows it to contain 84.4% carbon and 15.6% hydrogen by weight. Calculate the molecular weight and molecular formula of the compound.

6. A mixture of 33% by weight of helium and 67% by weight of oxygen is often used as an artificial atmosphere for divers and others working under pressure, since helium is less soluble in the blood than nitrogen and does not produce the "bends." If a diver's tank contains this mixture at a gauge pressure of 740 Torr (i.e., 740 Torr above atmospheric pressure), what is the partial pressure of each gas in the mixture?

7. While doing some remodeling, a homeowner decides to install a central air-conditioning unit. Calculate the volume of 1 mole of air at the air conditioner if the air temperature there is 5°C and atmospheric pressure that day is 718 Torr. Calculate the volume of this mole of air when it emerges from the duct in one of the new rooms if its temperature there is 10°C.

8. A sample of gas weighing 0.0302 g is introduced into the space at the top of a barometer and the mercury drops 118 mm below its original level. If the volume of the gas is measured as 73.2 ml and the temperature of the tube is 30.0°C, what is the molecular weight of the gas?

9. A stopcock between a 3.00-ℓ bulb containing O_2 at 100 Torr and 300 K and a 2.00-ℓ bulb containing N_2 at 400 Torr and 300 K is opened. After equilibration of the pressure and an increase in temperature to 400 K, what is the final pressure in torr?

10. A gaseous mixture of 4.69 g CCl_4 and 1.34 g C_2H_4 at 427°C is contained in a 30.0-ml metal bomb.
    a. What is the number of moles of each compound in the bomb?
    b. What is the partial pressure in atmospheres of each of the gases, assuming they are ideal?
    c. What is the total pressure in atmospheres of gas inside the bomb?

11. A gaseous mixture in a bulb has the following composition: 3.0 parts Ar, 2.0 parts O_2, 4.0 parts CO, and 6.0 parts SO_2. How many grams of oxygen are there in 300 ℓ of this mixture at STP?

12. To a gaseous mixture of ethane (C_2H_6) and propane (C_3H_8) at a total pressure of 78 Torr, enough oxygen is added for complete combustion. The products of the combustion are CO_2 and H_2O, and the CO_2 is collected in a container of the same size as the one that contained the original mixture. If the CO_2 pressure is 210 Torr, what fraction of the original mixture was ethane?

13. During World War II a process was invented in which gaseous uranium hexafluoride (UF_6, mol wt = 352) was allowed to diffuse through thousands of porous barriers to separate the uranium isotopes. At what rate would deuterated ^{14}C-methyl fluoride ($^{14}CD_3F$, mol wt = 39) diffuse compared to UF_6?

14. The time required for a given volume of oxygen to effuse through an orifice is 45.0 s. If an equal volume of an unknown gas requires 67.2 s to pass through the same orifice, what is that gas's molecular weight?

15. Calculate the total kinetic energy in calories of 1 mole of an ideal gas at 50°C.

16. Calculate the root-mean-square velocity v_{rms} of the nitrogen molecule N_2 in a sample of air that is at 600 K and 25 atm. What is the molecule's average kinetic energy at this temperature?

17. Cu has a face-centered cubic structure and Na has a body-centered cubic structure. Draw a diagram of each of these structures and give the coordination number and the number of atoms per unit cell. Describe the relation between the two types of closest-packed structures.

18. The density of liquid argon is 1.40 g/cm³ at −186°C, whereas the density of solid argon is 1.65 g/cm³ at −233°C. The argon atom can be assumed to be a sphere of radius 1.70×10^{-8} cm.
    a. Calculate the percentage of the total volume occupied by 1 mole of gaseous argon at STP that is empty space.
    b. Calculate the percentage of liquid argon that is empty space at −186°C.
    c. Calculate the percentage of solid argon that is empty space at −233°C.
    d. Compare the calculated empty space in the argon crystal structure with that predicted for a closest-packed crystal. Comment on any apparent discrepancy.

19. Ag has a face-centered cubic structure with a density of 10.5 g/cm³.
    a. Calculate the mass of the unit cell of Ag metal.
    b. Calculate the length of an edge of the unit cell of Ag metal.

20. Au has a face-centered cubic structure with a unit-cell length of 4.07 Å and a density of 19.3 g/cm³. Calculate Avogadro's number from these data.

21. Consider the solid substances Na(*c*), Cl_2(*c*), and NaCl(*c*). For each of these three substances, state (a) the type of solid, (b) the species occupying the lattice sites, (c) the type of bonding force, and (d) the type of unit cell predicted.

22. Consider the three crystalline substances $MgBr_2$, KBr, and CsI. The radii of the anions and cations are as follows:

Ion:	Mg^{2+}	Br^-	K^+	Cs^+	I^-
Ionic radius (Å):	0.66	1.96	1.33	1.69	2.16

a. Predict the type of hole that the cations will occupy in each structure.
b. Calculate the fraction of holes of that type that are occupied in the closest-packed lattices of anions.

"Amazing! It would take four thousand mathematicians four thousand years to make a mistake like that!"

INTERLUDE

those incredible transistors

The 1960s have been called the "solid-state decade." During those ten years we saw (and heard) the boom in transistor radios and TV, the exponential growth of computer capabilities, the excitement of space travel, and incredible achievements in microminiaturization. All of these developments, and many more, were made possible by tiny crystals of very slightly contaminated silicon and germanium. Now that we have studied the properties of the solid state and have been introduced to the concept of the metallic bond (in Chapter 4), we are in a position to understand some of the basic chemistry and physics of those tiny metalloid chips.

We have seen the energy-band structures of the electrons in lithium, beryllium, and diamond in Figures 4-36, 4-37, and 4-38, respectively. Each band consists of a huge number of extremely closely spaced electron energy states (crystal orbitals). Lithium and beryllium have more valence orbitals than they have valence electrons, and are therefore metals (conductors). Diamond, however, is a nonmetal (insulator). In nonmetals, and in metalloids (semiconductors), there are two bands of crystal orbitals separated by a *band gap* of varying "width," i.e., of varying energy per mole. The lower-energy band, called the *valence band,* is completely populated by electrons, with two electrons in each of the many closely spaced levels. Electrical conductivity in this band is impossible because the electrons cannot move to other energy levels within the band. The higher-energy band, called the *conduction band,* is completely empty of electrons unless the crystal is subjected to heat or radiation of sufficient energy, or to an applied electric field of sufficient strength, to make some of the electrons in the valence band "jump the gap" and become conduction electrons.

Nonmetals are insulators because their band gaps are very wide and therefore difficult to bridge by means of heat, radiation, or electric fields. For example, diamond, a good insulator, has a band gap of 120 kcal/mole (5.2 eV). Most metalloids are semiconductors because their band gaps are relatively narrow and therefore easy to bridge. For example, silicon's band

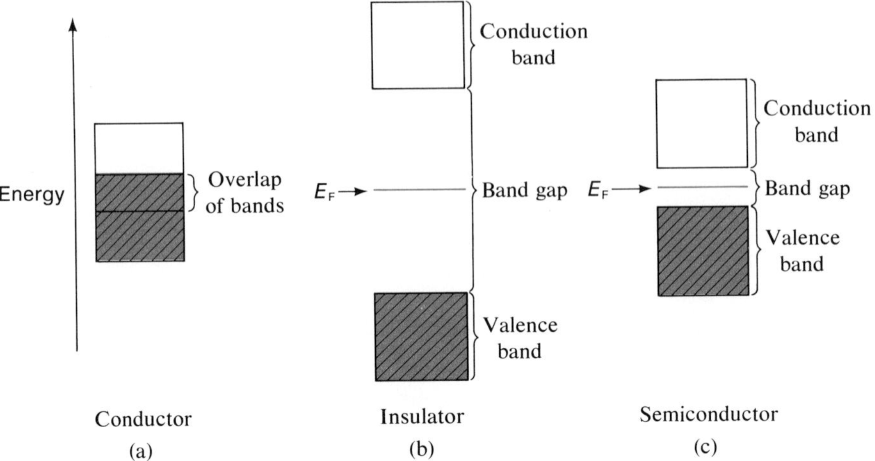

Figure 1. The energy-band structures of (a) a metallic conductor, (b) a nonmetallic insulator, and (c) a metalloid semiconductor. The colored lines represent the Fermi energy.

gap is only 25 kcal/mole (1.1 eV), which means that thermally excited electrons can jump this gap. At 1000 K about 1 in every 600 electrons in pure Si becomes a conduction election. The energy-band structures of the three kinds of crystals are shown in Figure 1.

One particular energy in the energy-band structure of an insulator or semiconductor has special significance. This is the *Fermi energy, E_F*, named for the great Italian-American physicist Enrico Fermi (1901–1954). It is the energy of that crystal orbital for which the probability of electron occupancy at ordinary temperatures (below ≈1000 K, say) is exactly 0.5. At ordinary temperatures the probability of occupancy of most of the orbitals in the valence band is asymptotically 1, and of most of the orbitals in the conduction band, asymptotically 0. Most of the orbitals for which the probability of occupancy is significantly different from 1 or 0 (i.e., almost all values in between) have energies within the band gap and are therefore only hypothetical. The Fermi energy, at the center of the band gap, corresponds to one of these hypothetical orbitals. We will see below why this is important.

An analogy for the behavior of electrons in an insulator or semiconductor can be made by envisioning a theater with a peculiar seating arrangement and a dictatorial policy. The front rows of seats represent the valence band and are completely filled with patrons. The back rows represent the conduction band and are completely empty. In between is a large section with no seats at all. Suppose we "quantize" the motion of the patrons by forbidding them to do anything but move from the seat they are in to any empty seat, *in one hop*. In an average crowd of theater-goers, there will obviously be no motion at all in the front section because all those seats are full. If, however, we have an unusually athletic patron who can summon the strength to take a flying leap into the back section, two things will happen: (1) Our athlete will have great freedom of motion and will hop around rapidly among all the empty seats at his disposal. (2) The remaining, less vigorous patrons will take turns hopping into the one empty seat at *their* disposal. Since this constantly changing empty seat constitutes a "hole" in the front section, the net effect is that *the hole hops around in the front section.* The hole hops much more slowly than the athlete, however. This is because there are always so many eager but clumsy patrons trying to hop into the one empty seat that they often collide with each other and bounce back to their own seats, resulting in a momentary stalemate.

The athlete is, of course, an energetic electron in the conduction band. The hole, being the absence of an electron in a sea of electrons, is, in effect, a unit positive charge in the valence band. What do physicists call this mobile positive charge in the valence band of a semiconductor? They call it a hole.

The band gap in a semiconductor can be narrowed considerably by doping the metalloid crystal with traces (e.g., a few parts per billion) of an impurity having the right properties. The techniques of fabricating ultrahigh-purity silicon and germanium crystals and doping them with precisely the right amounts of certain atoms were vital to the development of solid-state electronics.

The crystal structure of silicon is the same as that of diamond (Figure 6-29). Although it does considerable violence to the actual, tetrahedral geometry, we can represent the silicon structure in two dimensions (see Figure 2), with four sp^3 bonds per atom. If a P atom is introduced into the structure as shown in Figure 2a, it takes the place of one of the Si atoms. However, phosphorus has five valence electrons, which is one more than silicon has. This extra electron cannot be accommodated in the four-covalent-bond system and is therefore highly mobile. Furthermore, because all the energy levels of the valence band are filled, the electron goes into a

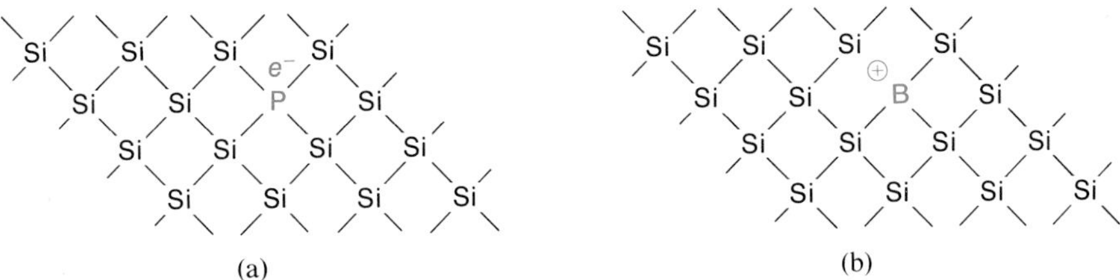

Figure 2. Doped silicon crystals. (a) A P atom introduces one extra electron. (b) A B atom introduces a deficiency of one electron (a hole), which is equivalent to a unit positive charge.

higher energy level. This level lies only 0.012 eV (0.28 kcal/mole) below the conduction band, so the extra electron is very easy to excite into the conduction band.

With many P atoms there are many electrons close to the conduction band. The resulting crystal is called an *n-type semiconductor* because the predominant species available for carrying charge are *n*egative electrons. The impurity atoms are called *donors* and the new energy levels resulting from their electrons are called *donor levels* (see Figure 3a). The ease with which the electrons are excited from the donor levels into the conduction band greatly enhances the conductivity of the doped crystal relative to that of the pure crystal. Note that the Fermi energy, the energy corresponding to that hypothetical level that has a 50% chance of being occupied by an electron, has now moved up to a position just beneath the conduction band because the total number of electrons in the crystal has increased.

The silicon can also be doped with an atom such as boron (see Figure 2b), which has only three valence electrons, or one *less* than silicon has. This means that one of the sp^3 bonds in the silicon structure does not form. The impurity atoms are now called *acceptors*. The result of their introduction into the crystal is a new set of energy levels called *acceptor levels*, which lie just above the valence band (see Figure 3b). For boron in silicon, the acceptor levels are 0.010 eV (0.23 kcal/mole) above the valence band. Electrons are easily excited into the acceptor levels, thereby creating holes in the valence band. The resulting crystal is called a *p-type semiconductor* because the predominant species available for carrying charge are *p*ositive holes. These mobile holes, like their electron counterparts in *n*-type semiconductors, greatly enhance the conductivity of the crystal. Again we find that the Fermi energy has shifted, in this instance down to a position just above the valence band because the total number of electrons in the crystal has decreased.

We now have the basis for understanding two of the most important solid-state electronic components in use today, the *diode* and the *transistor*.

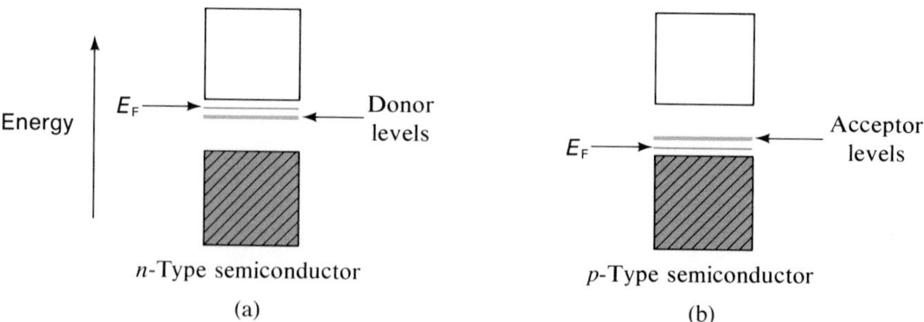

Figure 3. The energy-band structures of (a) an *n*-type semiconductor and (b) a *p*-type semiconductor. (W. J. Moore, *Seven Solid States*, W. A. Benjamin, New York, 1967.)

The diode is a *rectifier,* i.e., a device that has a much higher resistance to current in one direction than in the other. Such a device converts an ac current to a dc current by allowing electrons to flow in one direction but not the other. This function is essential in many kinds of circuits and is useful in protecting sensitive meters from ac currents. A diode can be made by bringing a *p*-type semiconductor into contact with an *n*-type semiconductor, forming a *p–n junction*. This is shown schematically in Figure 4a (note that the figure shows the energy bands, not the actual crystals, but that the vertical black lines denote the junction). The *n*-type has an excess of negative electrons and the *p*-type has an excess of positive holes. When the two crystals are brought together, we might expect electrons to flow to the left and holes to flow to the right in the diagram in Figure 4a, in order to set up a potential difference sufficient to equalize the Fermi energies. This is exactly what happens (see Figure 4b). Whenever a junction is made between two crystals with different Fermi energies, electrons flow "downhill" and holes flow "uphill" until the "hill" disappears. In the process, the valence and conduction bands become distorted in the region of the junction and further charge transfer across the junction is blocked by the potential difference that is set up.

Now let us consider what happens when we connect a battery across the junction and apply a positive potential to the *p* side. The setup is shown in Figure 5a, in which, for illustrative convenience, we "attach" the battery leads to the conduction bands of the two semiconductors. The system is now said to be under *forward bias*. The external potential overcomes the potential barrier established by the initial flow of electrons and holes and causes a current to flow. Electrons from the almost empty conduction band of the *n* section flow "down the Fermi hill" into the completely empty conduction band of the *p* section. At the same time, holes from the valence

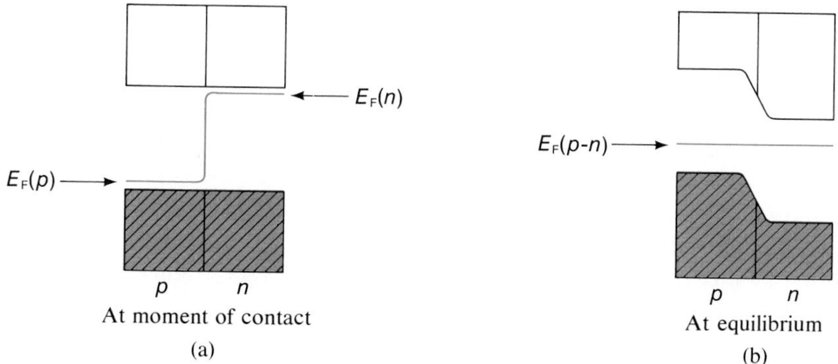

Figure 4. The energy-band structure of a *p-n* junction diode (a) at the moment of contact of the two crystals and (b) at electrical equilibrium, after equalization of the Fermi energies. At equilibrium there is a small potential difference across the diode. (W. J. Moore, *Seven Solid States,* W. A. Benjamin, New York, 1967.)

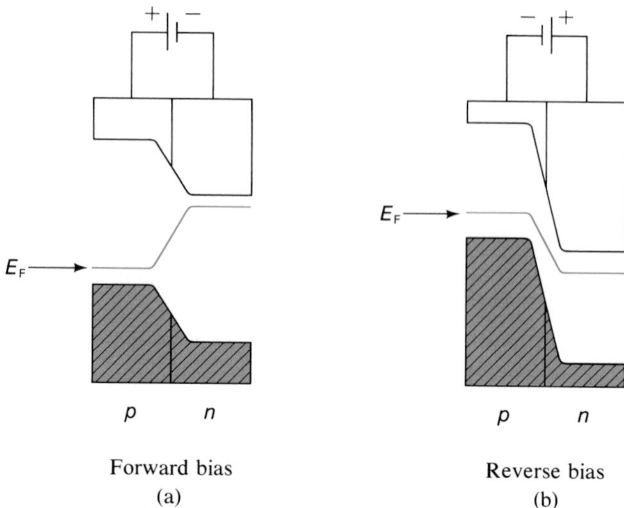

Figure 5. The energy-band structure of a *p-n* junction diode with (a) a forward-bias voltage across it and (b) a reverse-bias voltage across it. (W. J. Moore, *Seven Solid States,* W. A. Benjamin, New York, 1967.)

band of the *p* section flow "up the Fermi hill" into the valence band of the *n* section, where there were no holes. (In reality electrons from the valence band of the *n* section flow into the valence band of the *p* section and "fill" the holes; the two descriptions are equivalent. Thus there is a net flow of electrons from the *n* section to the *p* section. As the potential across the junction is increased, the current increases rapidly. This characteristic is shown in Figure 6.

If the battery is connected in the opposite sense, i.e., with the positive potential applied to the *n* side, the system is under *reverse bias.* This is shown in Figure 5b. Now the shape of the Fermi energy curve is such that electrons are expected to flow "downhill" from the *p* section to the *n* section. Indeed, the tendency of the external potential is to cause the exact reverse of the flows described in the preceding paragraph. But there are no electrons to flow from the empty conduction band of the *p* section into the *n* section, and there are no holes to flow from the filled valence band of the *n* section into the *p* section. There is therefore no current, except for a negligibly small one resulting from thermal effects (see Figure 6).

Whereas the diode is a rectifier, the transistor is a solid-state *amplifier,* a more versatile and important device by far. This device, invented in 1948, is a tiny fraction of the size and weight of the triode vacuum tube, which it has replaced in countless electronic applications. It is also almost unbreakable, more reliable, and generates far less heat. One kind of transistor is

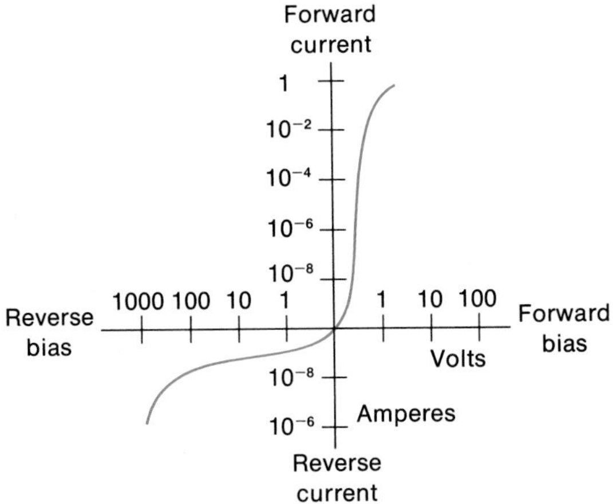

Figure 6. The current-voltage characteristic curve for a *p-n* junction diode. (W. J. Moore, *Seven Solid States,* W. A. Benjamin, New York, 1967.)

made by sandwiching a very thin sheet of *n*-type semiconductor between two pieces of *p*-type semiconductor, as shown in Figure 7. If this *p–n–p* junction transistor is unbiased, the energy-band structure is that shown in Figure 8a. This is just what we would expect from a simple extension of Figure 4b.

If the transistor is part of the circuit shown in Figure 7, in which a forward bias is applied to the left side and a reverse bias to the right side, the energy-band structure may look like that shown in Figure 8b. Again this is just what

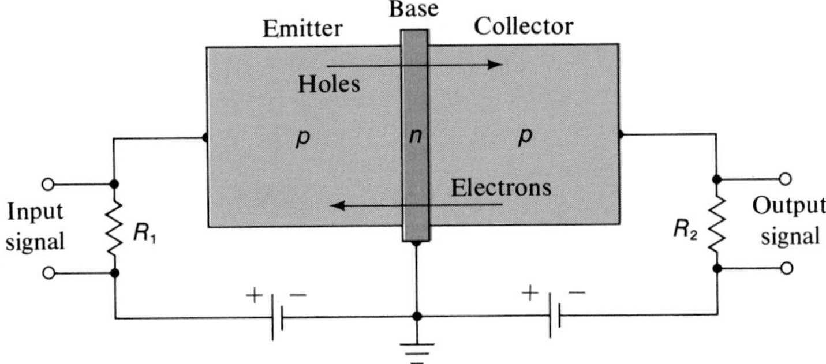

Figure 7. A *p-n-p* junction transistor in a typical circuit. The three segments of the transistor are given the same names as the three operational components of a triode vacuum tube because of the similarity of their functions. (W. J. Moore, *Seven Solid States,* W. A. Benjamin, New York, 1967.)

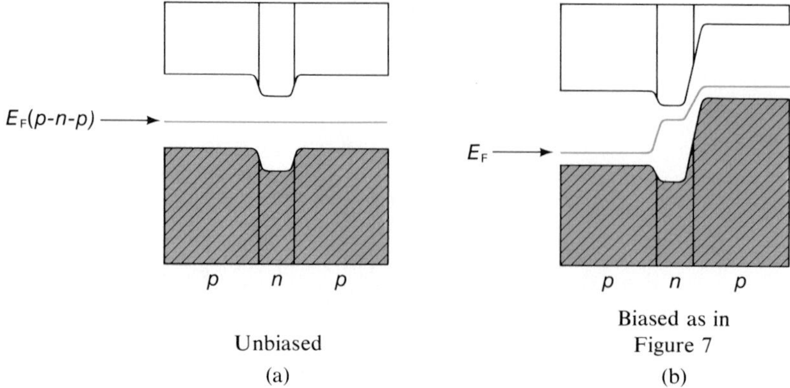

Figure 8. The energy-band structure of a *p-n-p* junction transistor that is (a) unbiased and (b) biased as in the circuit shown in Figure 7. (W. J. Moore, *Seven Solid States*, W. A. Benjamin, New York, 1967.)

we would expect from a simple extension of Figure 5. Remembering the directions of flow of electrons and holes on the Fermi hill, we expect a net flow through the transistor of electrons from right to left and holes from left to right. But we also remember why there is almost no current in a *p-n* (or *n-p*) junction diode under reverse bias. Why then does the right side of the transistor conduct current?

The answer is that the emitter section (*p*) is more heavily doped than the base section (*n*), so the predominant charge carriers in the transistor are positive holes. Thus the forward bias on the left side causes a large flux of holes to be emitted into the base, which is exceedingly thin—so thin that most of the holes easily pass through it into the collector section (*p*) before having a chance to combine with electrons. Note that, in the reverse-biased *p-n* junction diode, there *were* no holes to flow from the *n* section into the *p* section. Here, however, there *are* holes. When the holes reach the collector, they easily flow across it under the influence of the reverse bias.

By careful selection of the dopant concentrations and bias voltages, current can be made to flow from the low-resistance region R_1 to the high-resistance region R_2. The current I is a constant but, to a good approximation, the input voltage E_1 is amplified to the output voltage E_2 by the factor R_2/R_1 because

$$E_1 = IR_1 \qquad E_2 = IR_2$$

$$\frac{E_2}{E_1} = \frac{R_2}{R_1}$$

The power is also amplified by the same factor because electric power is the product of voltage and current:

$$P_1 = E_1 I \qquad P_2 = E_2 I$$

$$\frac{P_2}{P_1} = \frac{E_2}{E_1} = \frac{R_2}{R_1}$$

With a different circuit, current amplification can also be achieved. The essence of the transistor effect is that the current leaving the emitter is controlled much more strongly by the voltage of the base than of the collector, so a change in the input signal results in a corresponding but much greater change in the output signal.

The success of the transistor depends more than anything else on the extreme purity (except for the desired dopants) and near perfection of the crystals used, in order that the crucial condition for its operation can be met. This condition is that free negative electrons and positive holes can coexist in close proximity in a crystal for times long enough to allow the conduction of current across the crystal.

A transistor of the opposite kind, i.e., an *n-p-n* junction transistor, can be made just as easily. Here the predominant charge carriers are electrons rather than holes, but the principles are the same. There is great variety in the composition of transistors and the kinds of circuits in which they are used. The result is a device of such enormous utility and versatility that it can easily be regarded as one of the most important inventions in the history of man. The credit goes to three physicists at the Bell Telephone Laboratories: William Shockley (b. 1910), John Bardeen (b. 1908), and Walter Brattain (b. 1902), who shared the Nobel Prize for their work. (Bardeen has since won *another* Nobel Prize, thereby joining the select company of Marie Curie and Linus Pauling.)

The vacuum tube initiated the modern age of electronics; the transistor has revolutionized it. Surely further major breakthroughs lie ahead of us. What do you expect to see and what would you like to see in the future?

7 chemical thermodynamics

Thermodynamics is one of the three cornerstones upon which modern physical chemistry is built, the other two being quantum mechanics and statistical mechanics. Thermodynamics is the oldest of the three disciplines, dating to the middle and end of the nineteenth century. Literally, thermodynamics means the dynamics of thermal changes. It is that branch of physical science that deals with the relations between heat and other forms of energy and that is used to describe and predict the changes, both physical and chemical, that matter undergoes. It is always concerned with relatively large collections of atoms or molecules and is therefore related only to the macroscopic properties of matter. Many of the variables of thermodynamics are familiar to us, for they include pressure (P), volume (V), temperature (T), heat (q), work (w), and energy (E). An important part of thermodynamics is definitions, which must be understood clearly. For example, a *system* to a thermodynamicist is a part of the universe upon which he wishes to focus special attention, while the *surroundings* are everything else, that is, the rest of the universe. Since the chemist is most interested in the transfer of heat, work, and energy between a system and its surroundings, precise definitions are necessary to avoid ambiguity.

7.1 SYSTEMS

There are three types of systems of importance to us. An *isolated system* is one that has no interaction with its surroundings. There is therefore no transfer of either matter or energy between the isolated system and its surroundings. The contents of a sealed, ideally functioning (i.e., perfectly insulating) thermos bottle represent an isolated system, and the universe as a whole is believed to be an isolated system.

A *closed system* is one that cannot transfer matter to or from its surroundings but *is* capable of transferring energy in the form of heat, work, and radiation to and from its surroundings. The contents of any sealed, real container represent a closed system because heat can pass through the walls of the container. The gas in a balloon is a closed system because it can perform work on its surroundings (the balloon and the rest of the universe) by expanding or contracting as a result of heat transfer through the balloon. Finally, any sealed, transparent container (e.g., a stoppered glass flask) allows the transfer of radiant energy to and from the closed system within.

An *open system* is one that can transfer both energy *and* matter to and from its surroundings. Some examples of open systems are a beaker of water on the laboratory bench, the Earth, and the human body. Most systems in real life are open. Most laboratory systems of chemical interest are closed or may be regarded as closed for all practical purposes. Many are deliberately held open, however, and some are very nearly isolated, although none are perfectly isolated.

Thermodynamic Properties

An *intensive property* of a system is a property that is independent of the size of the system. Some examples of intensive properties are pressure (P), temperature (T), density (ρ), and concentration (c). If the overall temperature of a glass of water (our system) is 20°C, then any drop of water in that glass also has a temperature of 20°C. Similarly, if the concentration of salt, NaCl, in the glass of water is 0.1 mole/ℓ, then any drop of water from the glass also has a salt concentration of 0.1 mole/ℓ.

An *extensive property* of a system is a property that does depend on the size, or mass, of the system. Some examples of extensive properties are volume (V), number of moles (n), internal energy (E), enthalpy (H), entropy (S), and Gibbs free energy (G). Some of these properties are unfamiliar to you now, but they will be defined and illustrated in this chapter. Consider the glass of water again. If all the intensive properties of our system remain constant, then doubling the mass of the water doubles the volume, the number of moles, the internal energy of the system, and every other extensive property. Since extensive properties depend on the amount of matter being considered, they are usually based on a specific quantity of the substance. In Chapter 4, for example, bond energies were tabulated in units of kcal/mole. Thus, it requires 104 kcal to dissociate 1 mole of H_2 into 2 moles of H atoms, 208 kcal to dissociate 2 moles of H_2 into 4 moles of H atoms, etc.

Thermodynamic State Functions

A *thermodynamic state function* is a property, either intensive or extensive, that depends only upon the instantaneous state of the system. It follows that the history of the system is irrelevant to the state function. Some examples of thermodynamic state functions are temperature, pressure, internal energy, and entropy.

As an analogy, consider the position of a city on the Earth's surface. The latitude and longitude of any point in the city are precisely determinable; they are "state functions" describing its position. If you travel from Fisherman's Wharf in San Francisco to the Empire State Building in New York, the changes in your latitude and longitude are exact numbers that depend only upon your origin and destination. The number of miles traveled is *not* a state function, however, because there are infinitely many paths that you could take. For example, you could fly directly to New York, but you could also travel via the North Pole or via Tokyo, Moscow, and London. Your actual travel is thus a *process* that is determined not just by the starting and ending points, but also by every point in between, i.e., by your route. When you arrive, the difference between your old and new coordinates does not reveal *how* you got there, but only that you *did.*

The *internal energy* E of a system is the sum of the kinetic energies and potential energies of all the atoms in the system. All changes in the internal energy of the system then represent changes in the energies of the atoms of which the system is composed. This internal energy (or total energy) depends only upon the instantaneous state of the system, not upon its history. Thus, internal energy is a state function. For *all* processes the *change* in a state function equals the value of the function in the final state minus the value of the function in the initial state. The nature of the path from initial state to final state is irrelevant.

7.2 CONSERVATION OF ENERGY—THE FIRST LAW OF THERMODYNAMICS

There are many ways of transferring energy between a closed system and its surroundings. Some examples are the transfer of heat and light, and the performance of electrical work and mechanical work. In this chapter we will restrict ourselves to the first and last of these processes, but in Chapter 10 the thermodynamics of systems that do electrical work will be discussed.

The *heat absorbed by a closed system* from its surroundings is given the symbol q. Therefore q is positive when heat is absorbed by the system. Such a process is called *endothermic*. If heat is *given off* by the system, i.e., if heat is absorbed by the surroundings from the system, then q is negative.

Such a process is called *exothermic*. If no heat is transferred during a process, $q = 0$. The term *adiabatic* is used to describe such processes. The *work done by a closed system* on its surroundings is given the symbol w. It is positive when the system does work on the surroundings and negative when the surroundings do work on the system.

The law of the conservation of energy can now be written as

$$\Delta E = q - w \tag{7-1}$$

This is the *first law of thermodynamics*, which states that the change in the internal energy of a closed system equals the heat absorbed by the system from the surroundings minus the work done by the system on the surroundings. "Energy in" minus "energy out" equals the net change in energy.

The heat absorbed by the system, q, and the work done by the system, w, are *not* state functions, but processes analogous to following a variable path on the surface of the Earth. To verify this for yourself, consider two beakers, each holding exactly 1000 g of water at exactly 20.0°C (see Figure 7-1). Neglecting such factors as evaporation and accidental contamination,

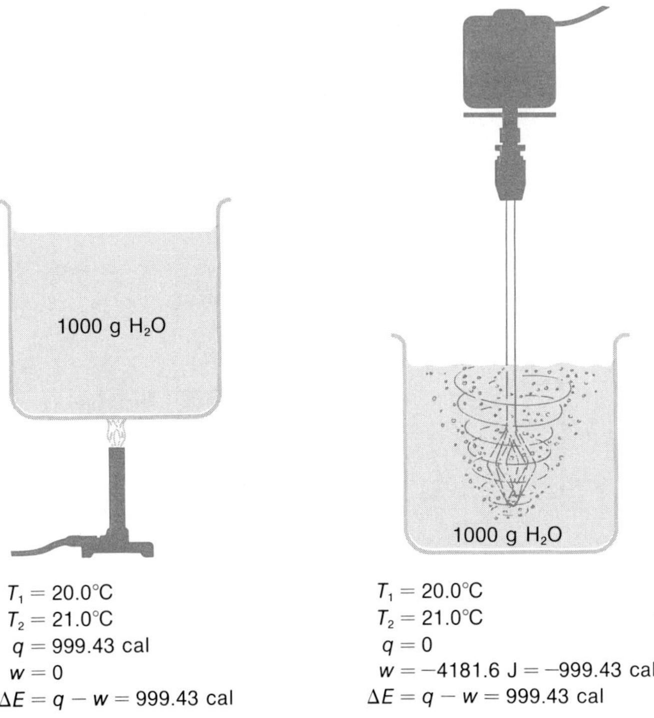

Figure 7-1. The equivalence of heat and work.

each sample of water may be regarded as a closed system. The water in the first beaker is heated exactly 1.0°C to 21.0°C with a Bunsen burner, which is part of the surroundings. The amount of heat q that must be absorbed by the system in order to bring about this change in temperature is 999.43 cal.

The water in the second beaker is stirred by an electric stirrer, which is also part of the surroundings. The stirrer is run until the temperature of the water rises exactly 1.0°C to 21.0°C. The amount of mechanical work done by the surroundings on the system ($-w$) is 4181.6 J, which, upon converting units, gives 999.43 cal.

The initial and final states in both experiments are identical. Because E is a state function, it is therefore no surprise that ΔE is the same for both experiments. We see, however, that the paths are entirely different and that there are infinitely many other paths that could be taken to get from state 1 to state 2. Thus, the heat absorbed and work done by the system determine the path, but they are not state functions.

Pressure-Volume Work

Consider the system consisting of a sample of gas trapped in a sealed cylinder by a piston, as shown in Figure 7-2. The external force exerted by the surroundings on the system is given the symbol F_{ext}. The work done by the system on the surroundings is the external force exerted on the piston times the distance through which the piston moves, or

$$w = F_{ext}(x_2 - x_1)$$

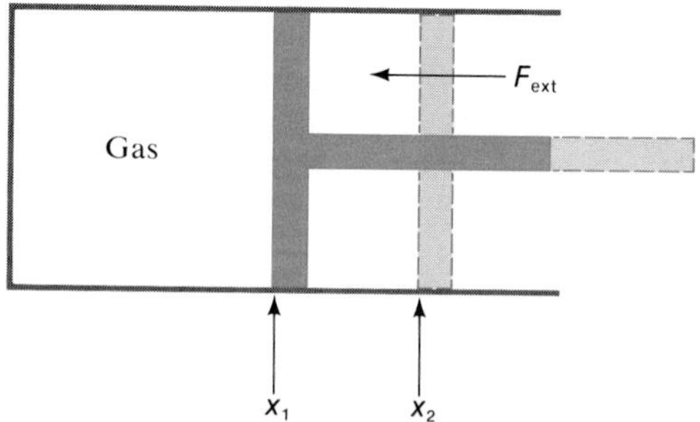

Figure 7-2. Pressure-volume work. The system consists of the gas confined by the piston. The process described in the text corresponds to the movement of the piston outward from x_1 to x_2.

In terms of differential calculus,

$$dw = F_{ext}dx$$

If the cross-sectional area of the piston is A and if the distance x changes by an infinitesimal amount dx, then $A\ dx = dV$, the differential change in the volume of the system.

We can now integrate the equation for dw to obtain the work done by the system on the surroundings:

$$w = \int_{x_1}^{x_2} F_{ext}dx = \int_{x_1}^{x_2} \frac{F_{ext}}{A}\ A\ dx = \int_{V_1}^{V_2} P_{ext}dV \qquad (7\text{-}2)$$

Thus, the mechanical work of a thermodynamic system is usually expressed as *pressure-volume work*, or *PV* work. Gas expansions that occur along more than one coordinate do not require that the changes in volume along the *x*-, *y*-, and *z*-coordinates be specified independently.

Pressure-volume work is path-dependent. For example, consider the expansion of a gas at constant external pressure and the subsequent cooling of the gas at constant volume (see Figure 7-3a). The work done by the system on the surroundings in going from point A to point B in the diagram is represented by the colored area. Now suppose that a different path is taken between points A and B: the gas is first cooled at constant volume and then expanded at constant external pressure (Figure 7-3b). The work done on the surroundings is again represented by the colored area.

In the second example far less work is done on the surroundings than in the first. Yet, since the starting points and ending points are the same in each case, $E_B - E_A = \Delta E = q - w$ must be the same. Therefore the differences in q along the two paths must just compensate for the differences in w. Energy is conserved and the first law of thermodynamics is obeyed.

Reversible Processes

A system at *equilibrium* is one in which there is no net change in any measurable property with time. A *reversible process* is one that is so close to equilibrium throughout its entire course that its direction can be reversed by an infinitesimal change in the state of the system. A reversible work process is one in which the internal pressure (the pressure of the system) differs from the external pressure (the pressure of the surroundings) by only an infinitesimal amount at any instant. For all practical purposes, they are equal:

$$P_{int} = P_{ext}$$

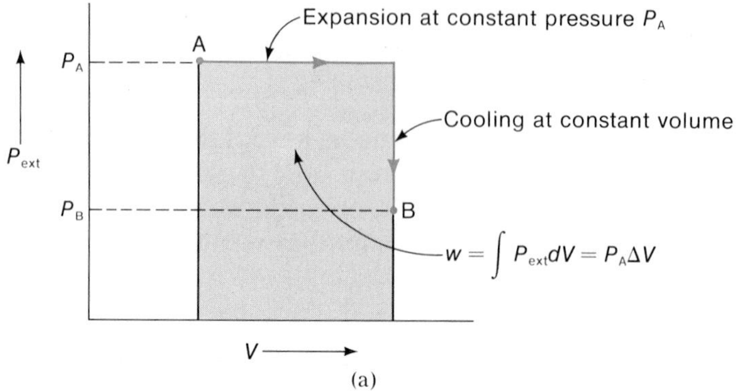

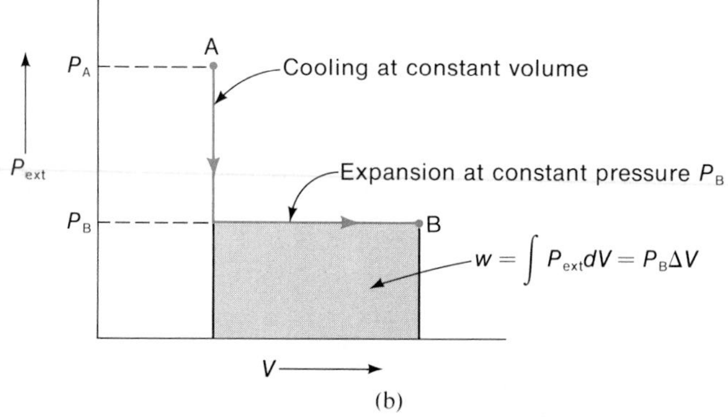

Figure 7-3. Path dependence of pressure-volume work. (a) Expansion at constant pressure followed by cooling at constant volume. (b) Cooling at constant volume followed by expansion at constant pressure.

Thus, if the system is expanding reversibly, P_{int} is only infinitesimally greater than P_{ext}. The reverse process (the reversible compression of the system) occurs if P_{ext} is infinitesimally greater than P_{int}.

Because an infinite number of steps would be required to carry out a reversible process, such a process would occur infinitely slowly. Clearly, reversible processes are not real. They are idealized processes that are very useful to define and analyze, however, because of their mathematical simplicity and the physical insights they provide. All processes that actually occur (real processes) are *irreversible*, i.e., they occur in a finite time and deviate from equilibrium by a finite amount. However, some real processes—such as the expansion of a gas—if allowed to occur sufficiently slowly, can be held so *close* to equilibrium at all points between the initial and final states that they are, for all *practical* purposes, reversible.

For a reversible work process, such as the reversible expansion of a gas,

$$\Delta E = q - w = q - \int_{V_1}^{V_2} P\, dV \tag{7-3}$$

where P is the internal pressure of the system.

Heat Absorbed During a Process

Chemists are often interested in the amount of heat that is absorbed or liberated by a system during a chemical reaction or a phase transformation. It is therefore useful to define a state function that, under the appropriate conditions, changes in direct proportion to the amount of heat transferred during the process. For example, we have seen (Equation 7-3) that the change in internal energy for a reversible process in which PV work is the only form of work done is $q - \int P\, dV$. This shows that, if the volume of the system does not change during the process,

$$\Delta E = q_V$$

where the subscript V denotes a constant-volume process. The change in the internal energy equals the heat absorbed by the system. Thus, for chemical reactions performed in sealed containers, the heat of the reaction is equal to the change in internal energy, ΔE, that accompanies the reaction.

Most chemical reactions are not performed at constant volume, however. They are usually performed in open containers at constant pressure (atmospheric pressure) and the volume is allowed to change freely. A new state function, the change in which is equal to the heat transferred at constant pressure, can be defined by recognizing that the sum of two state functions is also a state function. Thus, the *enthalpy* H, which is often simply called *heat*, is defined by the following equation:

$$H = E + PV \tag{7-4}$$

By the methods of calculus, it can be shown that

$$\Delta H = q + \int_{P_1}^{P_2} V\, dP \tag{7-5}$$

Compare this equation with Equation 7-3. Equation 7-5 shows that if the pressure is constant

$$\Delta H = q_P$$

where the subscript P denotes a constant-pressure process. For all constant-pressure processes, the change in the enthalpy of the system, ΔH, equals the heat absorbed by the system from the surroundings.

Example 7-1. This photograph* of the Matterhorn was taken near Zermatt, Switzerland. The beautiful cloud apparently hanging from the peak is present when the winds are from the northwest. Use the first law of thermodynamics to explain the presence of this cloud.

Several assumptions based upon actual experience at the mountain and upon the laws of thermodynamics are required. We assume that a moisture-laden wind is blowing across the mountain. If we consider a given mass of warm, humid air at the base of the windward side of the mountain, we can envision its being blown up one side of the peak and down the other:

The thermodynamic assumption is that the air mass moves adiabatically across the peak, i.e., it exchanges no heat with its surroundings. This means that $q = 0$ for this process, which is reasonable because heat transfer in gases is very slow. The change in internal energy of the air mass is thus

$$\Delta E = q - w = -w$$

As the air mass moves up the mountain, the pressure of the surrounding atmosphere decreases. The air mass therefore expands and pushes back its surroundings, thereby doing PV work. The sign of w is positive because the system (the air mass) does work on the surroundings (the atmosphere). Therefore, ΔE must

*Courtesy of Prof. Ignacio Tinoco, Department of Chemistry, University of California, Berkeley.

be negative:

$$\Delta E = -w < 0$$

We saw in Chapter 6 that the average kinetic energy of an ideal gas is directly proportional to the absolute temperature. If we equate kinetic energy with internal energy,

$$\Delta E = \tfrac{3}{2}R\Delta T$$

(for 1 mole of gas). If ΔE is negative, ΔT is also negative, which means that the temperature falls as the air mass rises and does work. Therefore, as the air mass rises up the mountain, both its pressure and temperature decrease and the water vapor it contains undergoes the change in state represented in the following phase diagram for H_2O:

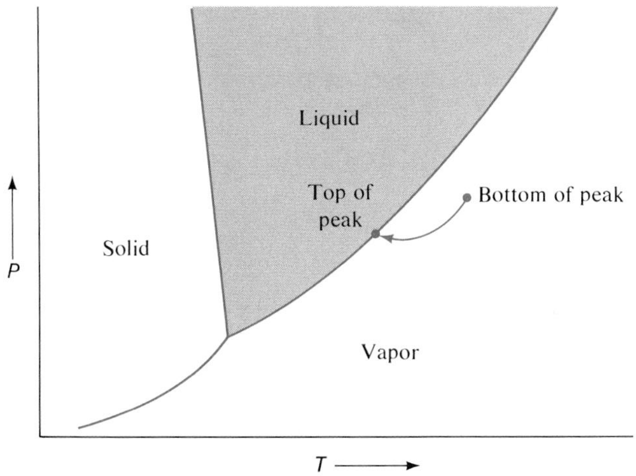

When the liquid-vapor equilibrium curve is reached, the liquid phase is in equilibrium with the vapor phase and fog, or a cloud, appears. On the opposite side of the mountain, the air mass warms up as it descends because the atmosphere is now doing work on it. The state of the air mass at any instant is now described by a point on the arrow moving *away* from the liquid-vapor equilibrium curve. The liquid droplets vaporize, giving clear, moist air on the leeward side of the mountain.

We conclude that the Matterhorn cloud is not a static entity and is not held to the mountain by some special affinity of clouds for the rocks of the Matterhorn. The cloud exists dynamically by the continuous, steady-state process of being formed on one side and disappearing on the other. The first law of thermodynamics provides the basis for understanding this beautiful phenomenon.

7.3 THERMOCHEMISTRY—HEATS OF REACTION

The existence of state functions simplifies thermodynamic calculations enormously. An example of their usefulness is shown in the following three reactions.

1. The burning of graphite (coal) to carbon dioxide can be studied and the amount of heat liberated can be easily measured:

$$\text{C(graphite)} + \text{O}_2(g) \rightarrow \text{CO}_2(g) \qquad \Delta H_1 = -94.05 \text{ kcal/mole}$$

2. The burning of graphite to carbon monoxide is harder to control—invariably, some CO_2 is produced. The heat liberated in the following reaction is harder to measure:

$$\text{C(graphite)} + \tfrac{1}{2}\text{O}_2(g) \rightarrow \text{CO}(g) \qquad \Delta H_2 = ?$$

3. On the other hand, carbon monoxide can be easily purified and burned, and the heat liberated can be measured:

$$\text{CO}(g) + \tfrac{1}{2}\text{O}_2(g) \rightarrow \text{CO}_2(g) \qquad \Delta H_3 = -67.64 \text{ kcal/mole}$$

We notice that Equation 2 plus Equation 3 equals Equation 1, a fact depicted in Figure 7-4.

We have already noted that the change in a thermodynamic state function in going from any state of a system to any other state is independent of the path, so such changes must be additive over any number of individual steps that constitute the overall process. This extremely valuable generalization holds for all state functions. When applied to enthalpy changes, it is called *Hess's law of heat summation*; by discovering it, the Swiss-Russian chemist Germain Hess (1802–1850) founded the field of *thermochemistry*. In the

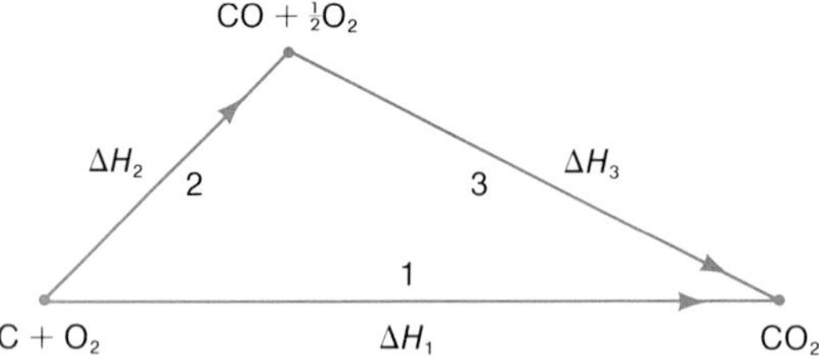

Figure 7-4. Diagram illustrating Hess's law. The two different paths from $C + O_2$ to CO_2 illustrate the equivalence of different paths in deriving changes in state functions. Note that the reaction $C + O_2 \rightarrow CO + \frac{1}{2}O_2$ is equivalent to the reaction $C + \frac{1}{2}O_2 \rightarrow CO$.

example given above, the large amount of heat liberated in Reaction 1 is thus the sum of the lesser amounts of heat liberated in Reactions 2 and 3:

$$\Delta H_1 = \Delta H_2 + \Delta H_3$$

$$\Delta H_2 = \Delta H_1 - \Delta H_3 = -94.05 - (-67.64) = -26.41 \text{ kcal/mole}$$

This example demonstrates that, knowing ΔH_1 and ΔH_3, we can calculate ΔH_2 without doing the experiment, and, by Hess's law, the answer must be exactly correct. We can, of course, do similar calculations with any of the other state functions.

Standard Enthalpies of Formation, ΔH_f°

The zero of internal energy or enthalpy in thermodynamics is arbitrary, just as the zeros of latitude and longitude are arbitrary. In order to simplify the consideration of changes in state functions, chemists have chosen the *standard state* of any pure element or compound to be the most stable physical form of the substance at 1 atm and 298 K. These conditions were chosen for their obvious convenience. The following rules now apply:

1. By convention, *each element in its standard state is assigned an enthalpy of zero.*
2. The enthalpy change that accompanies the formation of one mole of a compound in its standard state *from the elements* in their standard states is called the *standard enthalpy of formation* of that compound, or, colloquially, the standard heat of formation.

Most standard enthalpies of formation are negative. For example, methane can be formed by the reaction of graphite with hydrogen gas at standard conditions, with the liberation of 17.889 kcal of heat per mole of CH_4 formed (an exothermic reaction, for which ΔH is negative):

$$\text{C(graphite)} + 2\text{H}_2(g) \rightarrow \text{CH}_4(g) \qquad \Delta H = -17.889 \text{ kcal/mole}$$

The standard enthalpy of formation of methane is therefore −17.889 kcal/mole. In Appendix E of this book, we see the entry

$$\text{CH}_4 \qquad \Delta H_f^\circ = -17.889 \text{ kcal/mole}$$

(The superscript degree sign means "standard state" and the subscript f means "formation.")

Similarly, graphite reacts with oxygen at standard conditions to produce carbon dioxide, with the release of 94.052 kcal per mole of CO_2 produced:

$$\text{C(graphite)} + O_2(g) \rightarrow CO_2(g) \qquad \Delta H = -94.052 \text{ kcal/mole}$$

$$CO_2 \qquad \Delta H_f^\circ = -94.052 \text{ kcal/mole}$$

Note that the tables in Appendix E contain two entries for water, one for $H_2O(l)$ and one for $H_2O(g)$. Although H_2O is a liquid at standard conditions, many reactions occur in which it is produced as a gas.

$$H_2O(l) \qquad \Delta H_f^\circ = -68.317 \text{ kcal/mole}$$

$$H_2O(g) \qquad \Delta H_f^\circ = -57.798 \text{ kcal/mole}$$

The difference between these numbers $[-57.8 - (-68.3)]$ is just the heat required to evaporate one mole of water at standard conditions; it is called the *standard heat of vaporization*:

$$H_2O(l) \rightarrow H_2O(g) \qquad \Delta H_{vap}^\circ = 10.5 \text{ kcal/mole}$$

The usefulness of standard enthalpies of formation is illustrated by the following example.

Example 7-2. A public-utility engineer wants to know how much heat is liberated by burning natural gas (in this example, methane) at standard conditions. The reaction that occurs in your stove is

$$CH_4(g) + O_2(g) \rightarrow CO_2(g) + H_2O(g)$$

First he balances the equation:

$$CH_4(g) + 2O_2(g) \rightarrow CO_2(g) + 2H_2O(g)$$

Now he "constructs" this reaction by adding three other reactions that are either reactions forming compounds from the elements or the reverse of such reactions:

a. $CH_4(g) \rightarrow \text{C(graphite)} + 2H_2(g)$

$$\Delta H_a^\circ = -\Delta H_f^\circ[CH_4(g)] = +17.889 \text{ kcal/mole}$$

b. $\text{C(graphite)} + O_2(g) \rightarrow CO_2(g)$

$$\Delta H_b^\circ = \Delta H_f^\circ[CO_2(g)] = -94.052 \text{ kcal/mole}$$

c. $2H_2(g) + O_2(g) \rightarrow 2H_2O(g)$

$$\Delta H_c^\circ = 2\Delta H_f^\circ[H_2O(g)] = -115.596 \text{ kcal/mole}$$

Sum: $CH_4(g) + 2O_2(g) \rightarrow CO_2(g) + 2H_2O(g)$

$$\Delta H_{a+b+c}^\circ = -191.759 \text{ kcal/mole}$$

Note that this is a very exothermic reaction.

Standard enthalpies of formation can be used in a similar way to calculate the enthalpy change associated with any chemical reaction, provided that the value of ΔH_f° is known for each reactant and product. The following equation is all that is needed:

$$\Delta H^\circ_{\text{reaction}} = \sum n_p \Delta H_f^\circ \text{ (Products)} - \sum n_r \Delta H_f^\circ \text{ (Reactants)} \qquad (7\text{-}6)$$

where the coefficients n_p and n_r denote the stoichiometric coefficients of the products and reactants, respectively, in the balanced chemical reaction. This equation, which can be easily derived from Hess's law, states that the change in enthalpy associated with any chemical reaction equals the sum of the standard enthalpies of formation of the products, each one being multiplied by the appropriate stoichiometric coefficient, minus the sum of the standard enthalpies of formation of the reactants, each one being similarly multiplied. For example, consider the generalized chemical reaction

$$\text{aA} + \text{bB} \rightarrow \text{cC} + \text{dD}$$

where A, B, C, and D are chemical species and a, b, c, and d are their stoichiometric coefficients. Equation 7-6 tells us that

$$\Delta H^\circ_{\text{reaction}} = \text{c}\Delta H_f^\circ(\text{C}) + \text{d}\Delta H_f^\circ(\text{D}) - \text{a}\Delta H_f^\circ(\text{A}) - \text{b}\Delta H_f^\circ(\text{B})$$

Substitution of the numerical values of ΔH_f° in this equation then gives the desired enthalpy change for the reaction.

Bond Energies

In Chapter 4 we presented a table of bond energies (Table 4-5). It is important to recognize that each of these energies is the *average* energy of the bond in many different compounds. For example, the dissociation of hydrogen atoms from methane and ethane requires different amounts of energy:

$$CH_4 \rightarrow CH_3 + H \qquad \Delta H^\circ = 102 \text{ kcal/mole}$$

Methane → Methyl radical

$$\text{H}-\underset{\text{H}}{\overset{\text{H}}{\text{C}}}-\underset{\text{H}}{\overset{\text{H}}{\text{C}}}-\text{H} \rightarrow \text{H}-\underset{\text{H}}{\overset{\text{H}}{\text{C}}}-\underset{\text{H}}{\overset{\text{H}}{\text{C}}} + \text{H} \qquad \Delta H^\circ = 96 \text{ kcal/mole}$$

Ethane — Ethyl radical

If a survey of a large number of organic molecules is made, it is found that the bond energies for carbon-hydrogen single bonds are generally between 96 and 102 kcal/mole. Note, in Table 4-5, that the average C—H bond energy is 98 kcal/mole.

Bond energies can be used to calculate *approximate* enthalpy changes for chemical reactions. As an example, the hydrogenation of ethylene is treated below.

$$\text{H}_2\text{C}{=}\text{CH}_2 + \text{H}-\text{H} \rightarrow \text{H}-\underset{\text{H}}{\overset{\text{H}}{\text{C}}}-\underset{\text{H}}{\overset{\text{H}}{\text{C}}}-\text{H}$$

Ethylene — Ethane

Calculating the approximate ΔH of this reaction requires that the bonds broken and the bonds formed in the reaction be counted:

Bonds broken: 1 C=C, 1 H—H

Bonds formed: 1 C—C, 2 C—H

For any two species A and B, the symbol for the A—B bond energy is $E_{\text{A-B}}$. Since energy is required to break a bond and is released when a bond is formed,

$$\begin{aligned}\text{Bond energy in} &= E_{\text{C=C}} + E_{\text{H-H}} \\ &= 147 + 104 = 251 \text{ kcal/mole} \\ \text{Bond energy out} &= E_{\text{C-C}} + 2E_{\text{C-H}} \\ &= 83 + 2(98) = 279 \text{ kcal/mole}\end{aligned}$$

$$\Delta H = (\text{bond energy in}) - (\text{bond energy out}) \tag{7-7}$$

$$\Delta H = 251 - 279 = -28 \text{ kcal/mole}$$

The *exact* ΔH° measured for this reaction, or obtained from tables of standard enthalpies of formation, is −32.7 kcal/mole. This is a good example of the kind of accuracy that can be expected from enthalpy calculations based on bond energies.

Note that Table 4-5 does not contain a bond energy for the H—H bond. This number can be obtained from the table of dissociation energies of diatomic molecules (Table 4-4). It is also found to be equal to twice the standard enthalpy of formation of the hydrogen atom:

$$\Delta H_f^\circ(\text{H atom}) = 52.089 \text{ kcal/mole}$$

$$E_{\text{H-H}} = 2(52.1) = 104 \text{ kcal/mole}$$

Both of these sources provide exact numbers for $E_{\text{H-H}}$.

Another problem arises if a solid is involved in the chemical reaction, since bonding in solids is complicated. The reaction that forms methane from the elements in their standard states is

$$\text{C(graphite)} + 2\text{H}_2(g) \rightarrow \text{CH}_4(g)$$

What are the bonds broken? First, graphite must be dissociated into a gas of atomic carbon:

a. $\text{C(graphite)} \rightarrow \text{C}(g) \qquad \Delta H_f^\circ = 171.698 \text{ kcal/mole}$

The enthalpy of this reaction is the standard enthalpy of formation of gaseous carbon atoms. The rest of the reaction is

b. $\text{C}(g) + 2\text{H}_2(g) \rightarrow \text{CH}_4(g)$

$$\text{Bond energy in} = 2E_{\text{H-H}} = 208 \text{ kcal/mole}$$

$$\text{Bond energy out} = 4E_{\text{C-H}} = 4(98) = 392 \text{ kcal/mole}$$

For Reaction b, $\Delta H = 208 - 392 = -184$ kcal/mole. The total $\Delta H = \Delta H_a + \Delta H_b$:

$$\Delta H = 172 - 184 = -12 \text{ kcal/mole}$$

The exact ΔH° for this reaction is just the standard enthalpy of formation of methane:

$$\Delta H_f^\circ(\text{CH}_4) = -17.889 \text{ kcal/mole}$$

This again demonstrates the kind of accuracy to be expected from enthalpy calculations based on bond energies.

7.4 HEAT CAPACITIES

The *molar heat capacity* C of a pure substance is defined as the amount of heat required to increase the temperature of 1 mole of the substance 1°C. The mathematical way of saying this is that the amount of heat absorbed by 1 mole of the substance during a change in temperature equals the heat capacity times the change in temperature:

$$q = C\Delta T \qquad (7\text{-}8)$$

There are two kinds of heat capacity, one measured for constant-volume processes, the other for constant-pressure processes:

$$\text{Constant volume:} \quad C_V = \frac{dq_V}{dT} = \frac{dE}{dT} \qquad (7\text{-}9)$$

$$\text{Constant pressure:} \quad C_P = \frac{dq_P}{dT} = \frac{dH}{dT} \qquad (7\text{-}10)$$

When a gas is heated at constant pressure, it expands and does work on its surroundings. When it is heated at constant volume, it does no work. Because of this difference, more heat is required to increase the temperature of a gas by any given amount at constant pressure than at constant volume. For all gases, therefore, C_P is always greater than C_V.

Exercise 7-1. Given the relation between H and E and the definitions of the two heat capacities, derive the following equation for the relation between the two molar heat capacities of an ideal gas:

$$C_P = C_V + R \qquad \text{(Ideal gas)} \qquad (7\text{-}11)$$

From Equation 7-9 we see that for a constant-volume process the change in internal energy associated with a change in temperature is

$$\Delta E = \int_{T_1}^{T_2} nC_V dT \qquad \text{(Constant } V\text{)} \qquad (7\text{-}12)$$

where n is the number of moles in the system. Since heat capacities are usually weak functions of temperature, C_V may be considered to be constant for small temperature changes, so

$$\Delta E = nC_V(T_2 - T_1) \qquad \text{(Constant } V\text{, small } \Delta T\text{)} \qquad (7\text{-}13)$$

Similar equations can be derived for the enthalpy change at constant pressure:

$$\Delta H = \int_{T_1}^{T_2} nC_P dT \qquad \text{(Constant } P\text{)} \qquad (7\text{-}14)$$

$$\Delta H = nC_P(T_2 - T_1) \qquad \text{(Constant } P\text{, small } \Delta T\text{)} \qquad (7\text{-}15)$$

Example 7-3. Calculate the heat required to raise the temperature of 50 gallons of water from 18°C to 65°C in your hot-water heater. The value of C_P for water is 18 cal/K mole.

There are 3.8 ℓ/gal and 1000 g H_2O/ℓ, so

$$\text{Grams } H_2O \text{ in 50 gal} = (50 \text{ gal})(3.8 \ \ell/\text{gal})(10^3 \text{ g}/\ell) = 1.9 \times 10^5 \text{ g}$$

$$n = \text{moles } H_2O \text{ in 50 gal} = \frac{1.9 \times 10^5 \text{ g}}{18 \text{ g/mole}} = 1.1 \times 10^4 \text{ moles}$$

$$q_P = nC_P\Delta T = (1.1 \times 10^4 \text{ moles})(18 \text{ cal/K mole})[(65 - 18) \text{ K}]$$

$$= 9.3 \times 10^6 \text{ cal} = 9.3 \times 10^3 \text{ kcal}$$

Example 7-4. Skiing has become such a popular sport that many ski-lift operators find it profitable to use snow-making machines to extend the length of the season. Use the first law of thermodynamics to show why compressed air must be used in the snow-making process.

The process entailed in making snow is

$$H_2O(l) \rightarrow H_2O(c) \qquad \Delta H° = -1436 \text{ cal/mole}$$

Imagine that 1 mole of H_2O (18 g) is sprayed into 1 mole of air (22.4 ℓ at STP; normal atmospheric pressures are near 1 atm and an evening temperature of 0°C is plausible). The molar heat capacity C_P of air is 7.0 cal/K mole. We can now calculate the temperature rise to be expected for the air if this mole of H_2O were to freeze:

$$C_P = \frac{\Delta H}{\Delta T}$$

$$\Delta T = \frac{\Delta H}{C_P} = \frac{1436 \text{ cal/mole}}{7.0 \text{ cal/K mole}} = 205 \text{ K above } 0°\text{C} = 205°\text{C!}$$

(Note that the enthalpy change for the air is equal in magnitude but opposite in sign to the enthalpy change for the water, since the heat is transferred from the water to the air during the freezing process.) The net result of our labors would be to produce 1 mole of very hot air that would, of course, quickly melt any snow crystals formed in the first place.

Some clever person with either good intuition or good thermodynamics realized that two hoses were needed, one to spray the water and another simultaneously to spray compressed air. The compressed air expands very rapidly. This process is essentially adiabatic because the system (the air-water mixture leaving the combined nozzles) has no time in which to exchange any heat with its surroundings (the atmosphere). From here on the thought process is identical to that used in our consideration of the Matterhorn cloud:

$$\Delta E = q - w = -w$$

Work is done by the system on the surroundings, giving $w > 0$. This means that $\Delta E < 0$. Because E is directly proportional to T, the temperature of the system falls. The cooling of the expanding air is sufficient to freeze the water vapor in the spray.

The average kinetic energy of one molecule of an ideal gas is equal to $\frac{3}{2}kT$. For one mole, therefore, the relation is

$$KE = \tfrac{3}{2}RT$$

For atomic gases such as He, Ne, Ar, Kr, and Xe, the only kinds of energy possible are kinetic energy due to the motion of the atoms, sometimes called translational energy, and electronic energy resulting from the excitation of electrons to excited states. Since the spacing between electronic energy levels is quite large, most atoms and molecules are always in their ground electronic state at room temperature. If kinetic energy is the only kind of energy that a sample of an atomic gas has, then the molar internal energy is

$$E = \tfrac{3}{2}RT$$

and the molar heat capacity at constant volume is

$$C_V = \frac{dE}{dT} = \tfrac{3}{2}R$$

Table 7-1 confirms that monatomic gases have only translational, or kinetic, energy at 25°C. The heat capacities for diatomic molecules are interesting because they start at approximately $2.5R$ and increase as the bond in the diatomic molecule becomes weaker. We would expect a diatomic molecule to have rotational and vibrational energy as well as translational energy. The molecule has two rotational degrees of freedom (two angles are required to specify the direction of the bond in space). It is observed

Table 7-1. Molar Heat Capacities at Constant Volume of Various Gases at 298 K.

Gas	C_V
Atoms	
He	1.50R
Ne	1.50R
Diatomic Molecules	
H_2	2.47R
N_2	2.50R
O_2	2.53R
F_2	2.78R
Cl_2	3.08R
Br_2	3.33R
I_2	3.43R

to have a rotational energy equal to RT, with $\frac{1}{2}RT$ of rotational energy per degree of freedom, just as there is $\frac{1}{2}RT$ of translational energy for each translational degree of freedom. Nonlinear molecules, e.g., H_2O or CH_4, have three rotational degrees of freedom and are observed to have $\frac{3}{2}RT$ of rotational energy in the gaseous phase, the additional $\frac{1}{2}RT$ being due to the additional degree of freedom.

The remaining energy is vibrational energy. For a given temperature, this energy is greater, the weaker the bond and the heavier the molecule. This is because the natural frequency of vibration of the molecule decreases with decreasing force constant and increasing reduced mass (see Equation 5-5). The lower the natural frequency, the less energy is required to excite the vibration.

For the diatomic molecules,

$$C_V = \underset{\text{Translation}}{3(\tfrac{1}{2}R)} + \underset{\text{Rotation}}{2(\tfrac{1}{2}R)} + \underset{\text{Vibration}}{\text{vib}} \tag{7-16}$$

Note that H_2, N_2, and O_2 have essentially no vibrational excitation at 25°C, indicating that the energy spacing between vibrational states in these molecules is too large for anything but the ground vibrational state to be populated at this temperature. The value of C_V for these molecules is therefore approximately 2.5R. For F_2, Cl_2, Br_2, and I_2, there is an appreciable contribution from vibrational excitation. In I_2, for example, 1.50/3.43 = 43.7% of the energy absorbed by the molecule increases the translational energy, 1.00/3.43 = 29.2% increases the rotational energy, and the remaining 27.1% increases the vibrational energy.

7.5 SYSTEMS OF MANY PARTICLES—STATISTICAL MECHANICS

In previous chapters we have given considerable attention to the properties of single particles, either atoms or molecules. The question we address here is, "What can we say about the behavior of large numbers of molecules if we know the properties of the individual molecules?" Our objective is to develop a method of predicting which processes will occur in nature, i.e., which processes occur spontaneously. We are not concerned with the *rate* at which they occur, just with whether they will or will not occur.

One of the most important things we know is that molecules minimize their potential energy whenever possible, just as a ball does in rolling down a hill. The ball starts at a high potential energy. If there were no friction, all the potential energy would become kinetic energy, leaving the ball's total energy unchanged. When it reached the bottom of the hill, therefore, it would continue to roll—for example, up the next hill. And so on. If there *is* friction—as there always is—the ball eventually comes to rest when it reaches the lowest accessible point and its energy has been transferred to the surface of the hill.

Two hydrogen atoms collide and form a bond, thereby reducing their total potential energy, like the ball at the bottom of the hill. Alone, however, they cannot stay bound together because their kinetic energies are great enough to break the bond. This corresponds to the ball's rolling up a second hill. The only way they *can* stay bound together is if a third particle participates in the collision and removes some of the excess kinetic energy. Thus, the potential energy of the system of particles tends toward a minimum, provided that some energy-dissipating mechanism for achieving this minimum is available. The presence of a third body to remove the excess kinetic energy in the reaction of two hydrogen atoms is analogous to the friction in the example of the rolling ball.

On the basis of this argument, we expect that, when a chemical reaction or any other process occurs spontaneously in an isolated system of many particles (one that can exchange neither matter nor energy with its surroundings), the system will reach a lower potential energy and the particles in the system will have greater kinetic energy. Therefore, the temperature of the system should rise. When a spontaneous change occurs in a closed system, we expect heat to be transferred to the surroundings. (If the rate of heat transfer is great enough to allow the temperature of the system to remain effectively constant, the system is said to be *isothermal*.) In either case—isolated system or closed system—the change is exothermic.

This argument, which we have constructed by analogy with the behavior of single particles (or balls), is not always valid. There are spontaneous changes known that do *not* give off heat, but absorb heat: they are endothermic. An example is the dissolution of table salt in water. There are also many spontaneous processes in nature that result in essentially *no* change in the potential energy of the substances involved. Some common examples are the dispersal of sewage water in the rivers and, eventually, the oceans, the dispersal of smoke from a cigarette or smokestack, and the diffusion of ink molecules as they spread from a drop of ink in a glass of water. There must, therefore, be something else to be considered before we, as chemists, can predict which processes will occur in nature and which will not.

Let us construct a simple model for the dispersal processes mentioned above. Consider a box that is divided in two by an imaginary plane. If we start with four balls—A, B, C, and D—in the left half of the box and allow them to be jostled about at random, we find that a random succession of many configurations can be observed, including all balls on the left, all balls on the right, three on the left and one on the right, and so on. All possible configurations for this system are listed in Table 7-2.

There are $2 \times 2 \times 2 \times 2 = 16$ different configurations for this system because there are two ways to pick a position for ball A, two ways for ball B, and so forth. Since we expect each of these configurations to be equally probable, the probability of finding one particular configuration at any

Table 7-2. All Possible Configurations of a System of Four Balls in a Bisected Box.

Ball A	B	C	D	Number on Left	Number on Right
L	L	L	L	4	0
L	L	L	R	3	1
L	L	R	L	3	1
L	R	L	L	3	1
R	L	L	L	3	1
L	L	R	R	2	2
L	R	L	R	2	2
L	R	R	L	2	2
R	R	L	L	2	2
R	L	R	L	2	2
R	L	L	R	2	2
R	R	R	L	1	3
R	R	L	R	1	3
R	L	R	R	1	3
L	R	R	R	1	3
R	R	R	R	0	4

instant is 1/16. We notice that there are six configurations with two balls on the left and two on the right, whereas there is only one configuration with all four balls on the left. We therefore conclude that finding two on the left and two on the right is six times more probable than finding four on the left and none on the right. If we plot the frequency of occurrence against the number of balls on the left, we obtain the frequency distribution curve shown in Figure 7-5. (The analogous curve for the number of balls on the right would, of course, be identical.)

If the number of configurations for a given number of balls on one side is multiplied by the probability of each configuration (1/16), we obtain the probability of each of the five distributions. Note that, when we speak of a *configuration* we are still distinguishing between balls A, B, C, and D, but when we use the term *distribution* the balls are considered to be indistinguishable. The distribution refers merely to the number of balls on the left and on the right. Thus, there are 16 possible configurations of the system of four balls, but only five distributions of the balls, one corresponding to each of the numbers 0, 1, 2, 3, and 4 for the number of balls on a given side. The probability distribution curve is shown in Figure 7-6.

As the number of balls in the box is increased, the probability distribution curve assumes more and more the shape of a curve known as the *normal distribution*, or *Gaussian distribution.* This curve is of great importance in science because it describes a wide variety of natural phenomena that are based on statistical effects. The probability distribution curve for a

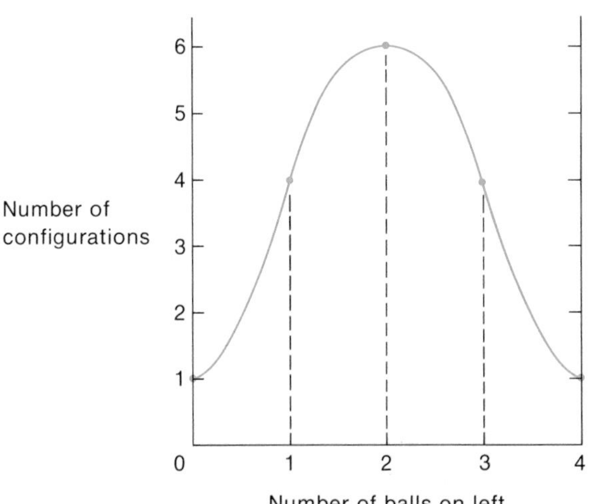

Figure 7-5. The frequency distribution of four balls in the two halves of a box.

system of 20 balls in the box is shown in Figure 7-7. The probability of finding any particular ball on the left is 1/2, so the probability of finding all 20 balls on the left is $(1/2)^{20}$, which equals 1/1,048,576, a very small number. Note that the curve of Figure 7-7 appears sharper than that of Figure 7-6.

Now we consider the same problem for the box containing Avogadro's number (6×10^{23}) of gas molecules. The probability of all the molecules' spontaneously gathering in the left half of the box is $2^{-6\times10^{23}}$, which is an infinitesimal number. The probability is one in about $10^{10^{23}}$, which is a one with a mole of zeros after it. This is a number incomparably larger than the presumed age of the universe in microseconds or any other number with physical meaning (number of H atoms in the observable universe?). So, although in principle there is no reason why such a rare event cannot occur, it is safe to say that it will never be observed. Our observations are restricted to the *most probable* states of systems when we deal with such large numbers of particles. Clearly, "most probable" is here synonymous with "most random," a fact that underlies the statistical basis of the behavior of all systems of many particles. Indeed, the concept of the most probable (most random) distribution of states in bulk matter is at the heart of *statistical mechanics*, that branch of physics that deals with the prediction of the macroscopic properties of matter from a knowledge of the properties of atoms and molecules and the forces between them.

Statistical mechanics is an inherently mathematical theory that provides a bridge between mechanics (classical or quantum) and thermodynamics.

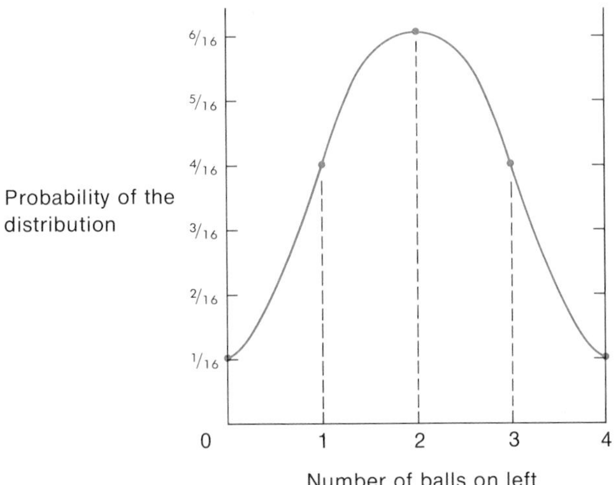

Figure 7-6. The probability distribution of four balls in the two halves of a box.

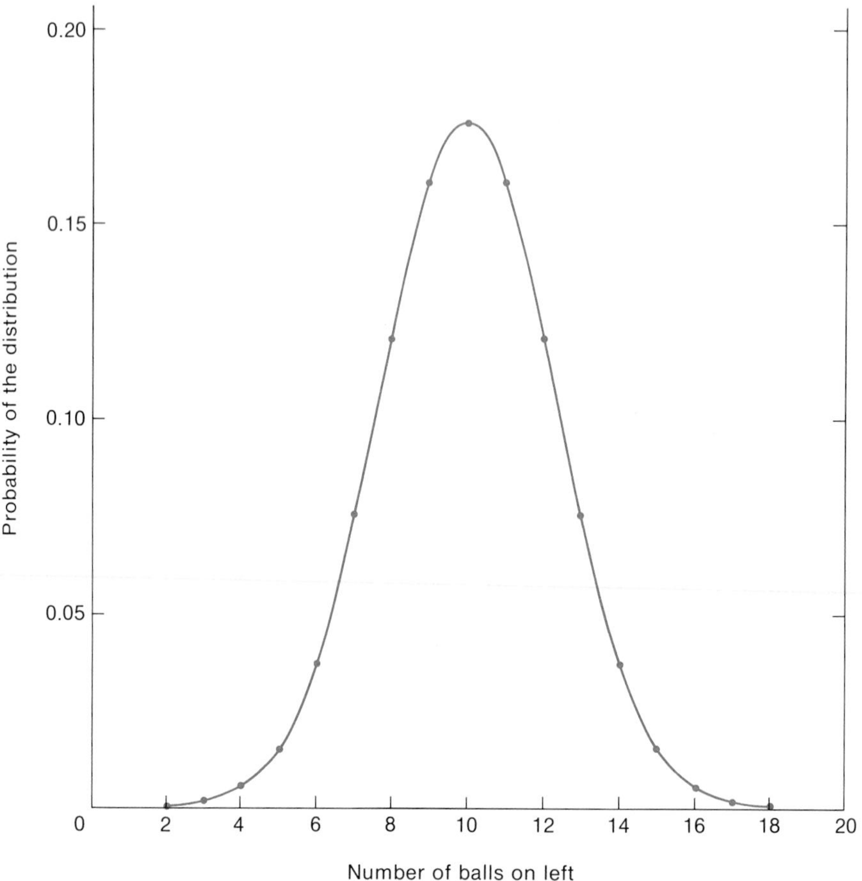

Figure 7-7. The probability distribution of 20 balls in the two halves of a box. The curve represents a Gaussian distribution.

Classical mechanics deals with such particulate concepts as mass, position, velocity, momentum, force, and kinetic and potential energy, whereas thermodynamics deals with such bulk concepts as internal energy, enthalpy, entropy, free energy, and heat capacity. The attempt to explain thermodynamics in terms of the behavior of atoms and molecules led to the development of statistical mechanics, largely by Boltzmann and the great American theoretical physicist J. Willard Gibbs (1839–1903), toward the end of the nineteenth century—at a time when the very existence of atoms and molecules was still disputed by many scientists! (With the subsequent advent of quantum mechanics and the profound conceptual changes it rep-

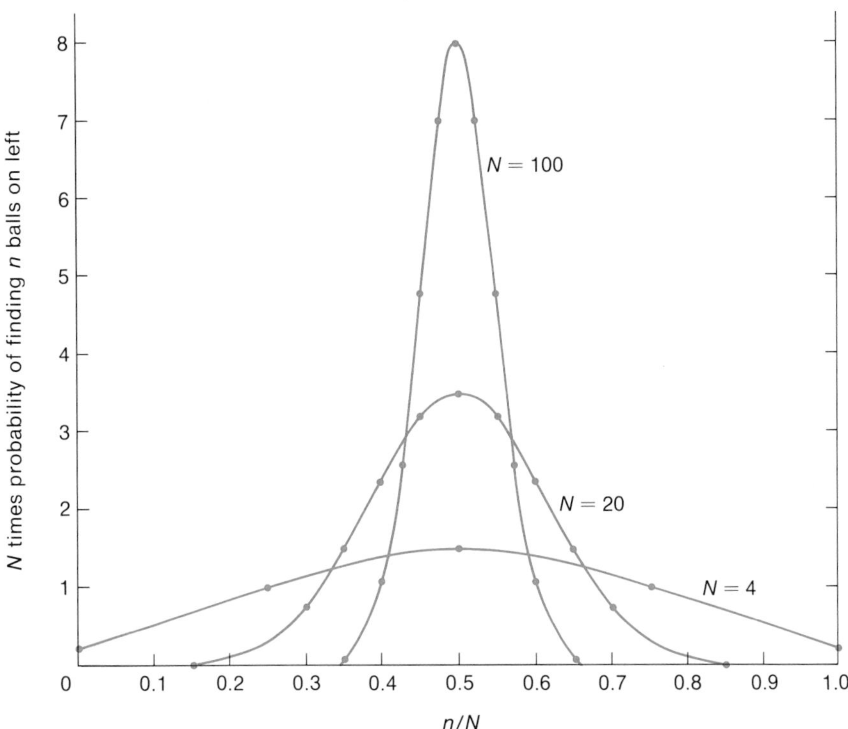

Figure 7-8. Probability distributions as a function of the number of particles.

resented, statistical mechanics changed accordingly, but the basic principles remained just as valid as before.) It was eventually realized that all of thermodynamics can be rigorously derived from the basic principles of statistical mechanics. The importance of this fact becomes clear when we realize that, whereas mechanics tells us only what is *possible* in the behavior of bulk matter in nature, statistical mechanics tells us what is *probable* and, therefore, what *is*.

Let us now return to our hypothetical box. Figure 7-8 shows three probability distribution curves for $N = 4$, 20, and 100 balls in the box, respectively. The symbol n represents the number of balls in the left half of the box. As N increases the curve becomes much sharper, indicating that fractional deviations from the *average* number of balls on the left become smaller as the total number of balls becomes larger. We conclude that for all the molecules in a mole of gas to collect spontaneously in one half of the box is energetically possible but statistically improbable in the extreme. Only if we could wait and watch *forever* could we be certain of seeing it happen. The system is best described as being in its most random (most probable) state. The gas molecules are therefore uniformly distributed in the box at equilibrium.

Spontaneous Processes with No Change in Potential Energy

Let us now consider the same bisected box that we discussed in the previous section, with the additional feature of a movable partition between the two equal halves of the box. We start with all the particles in the left half, and remove the partition. At the instant of removal of the partition, the particles are in one of the very large number of possible configurations of the system, but it is the least probable of all configurations. In time the moving particles enter the right half of the box, and at equilibrium they are once again uniformly distributed in the box.

The process described above is a model representation of the expansion of a gas. Many such spontaneous processes occur in nature, *always* in the direction of increased randomness. We must conclude that, in systems of many particles, spontaneous processes occur that are not predictable from energy considerations alone. There is a second factor, which scientists call *entropy*, that is related to the randomness of the system. Before discussing entropy, however, we will consider another useful example that is particularly appropriate to chemistry.

If one mole of hydrogen gas is mixed with one mole of deuterium gas, what would you expect to happen? Since the deuterium atom is just a hydrogen isotope with an atomic mass of 2 instead of 1 (the nucleus contains one neutron in addition to the proton), the chemical bonds between two hydrogen atoms, two deuterium atoms, and one hydrogen atom and one deuterium atom are nearly identical; for the sake of this example, we will assume that they *are* identical. If the two gas samples are mixed at a high enough temperature, the following reaction is observed:

$$H_2 + D_2 \rightarrow 2HD$$

Within the limitations of the above assumption, how much HD will be formed? There are four configurations that are equally probable:

$$H_2 \qquad D_2 \qquad HD \qquad DH$$

but, since HD and DH are indistinguishable, the final (equilibrium) relative concentrations of the three forms of hydrogen should be

$$H_2 : D_2 : HD$$
$$1 : 1 : 2$$

These ratios have been verified experimentally.

The hydrogen example is completely analogous to that of two particles in a box, one half of which we call H and the other half D. It illustrates the importance of the concept of randomness in explaining chemical reactions. We have seen that such maximum randomness results in a chemical reaction that, at equilibrium, entails the coexistence of both reactants and products.

The flow of heat from a high-temperature region of a system to a low-temperature region is explained by a similar randomization process, but in this case it is the kinetic, rotational, and vibrational energies of the molecules that become uniformly distributed throughout the system.

7.6 THE RELATION BETWEEN ENTROPY AND RANDOMNESS

Entropy is a thermodynamic state function. This means that if a system is clearly defined its entropy can be calculated as a definite number, and a change in entropy depends only on the initial and final states of the system. It is therefore possible to discuss the entropy of a system in the same way that we discuss the energy or enthalpy of a system. The methods used for calculating entropies will be discussed in the next section. Our purpose here is to explain why a concept like entropy is necessary when discussing spontaneous processes in nature.

Entropy is related to the number of distinguishable microscopic states (the number of molecular arrangements) that a system can have and still be in a well-defined macroscopic state (one defined by a set of state functions). Thus, the number of microscopic states we considered for four particles in a bisected box was 16, since there were 16 possible configurations of the system. In our discussion, of course, we ignored the multitude of differences in microscopic states caused by the positional variability of the particles. But there is a way to count all these microscopic states, and it is found that the relation between the entropy of a mole of particles, S, and this number of microscopic states, W, is

$$S = R \ln W \tag{7-17}$$

7.7 THE SECOND LAW OF THERMODYNAMICS—ENTROPY CALCULATIONS

The *second law of thermodynamics* defines the state function, entropy. It also provides a method for calculating changes in entropy in terms of thermodynamic variables instead of the statistical variables introduced in the last section. According to the second law, there is a state function, entropy, that is a function of the degree of randomness, or disorder, of a system. In an irreversible process the entropy of the universe increases. In a reversible process the entropy of the universe remains constant. At no time does the entropy of the universe decrease.

The entropy change in going from state 1 to state 2 of an isolated or closed system can be calculated from the equation

$$\Delta S = \int_1^2 \frac{dq_{rev}}{T} \qquad (7\text{-}18)$$

where the integration must be performed over a reversible path, even though the actual process is irreversible. For isothermal (constant-temperature) processes the calculation of entropy changes is especially easy, since Equation 7-18 becomes

$$\Delta S = \frac{q_{rev}}{T} \qquad \text{(Constant } T\text{)} \qquad (7\text{-}19)$$

An example of such an isothermal process is the evaporation of 1 mole of water at its normal boiling point, 373 K ($P = 1$ atm). Since this is also a constant-pressure process,

$$q_{rev} = \Delta H_{vap,373} = 9.77 \text{ kcal/mole}$$

$$\Delta S_{vap} = \frac{9770 \text{ cal/mole}}{373 \text{ K}} = 26.2 \text{ cal/K mole}$$

We see that vaporization increases the entropy of the system (the water), as we would expect, since molecules in the gaseous phase are more randomly distributed than those in the liquid phase.

Another example of an isothermal process is the reversible isothermal expansion of an ideal gas. The energy of an ideal gas depends only on the temperature. Thus, in an isothermal process,

$$\Delta E = q - w = 0$$

For a reversible process, therefore,

$$q = w = \int P_{ext} dV = \int P\, dV$$

For an ideal gas, $P = nRT/V$, so

$$q_{rev} = w_{rev} = \int_{V_1}^{V_2} \frac{nRT}{V} dV = nRT \ln \frac{V_2}{V_1}$$

and

$$\Delta S = \frac{q_{rev}}{T} = nR \ln \frac{V_2}{V_1} \qquad \text{(Constant } T\text{)} \tag{7-20}$$

Note that, as the volume of an ideal gas increases at constant temperature, its entropy increases. If the volume of the system decreases owing to the application of an external pressure, the entropy of the *system* decreases. (Is this a violation of the second law?)

7.8 THE THIRD LAW OF THERMODYNAMICS

Entropy is a state function related to the degree of randomness of a system. There is a very logical zero point for entropy, namely, the point at which there is *no* randomness in the system. The absence of randomness implies that all atoms are ordered in a perfect crystal structure, that all electrons are in their ground states, and that all molecules are in their ground vibrational and rotational states. Such a condition exists only at the absolute zero of temperature, 0 K. At this temperature only one microscopic state of the system is possible ($W = 1$), so the entropy $S = R \ln W$ must equal zero.

The *third law of thermodynamics* states that the entropy of a perfect crystal of any pure element or compound is zero at the absolute zero of temperature:

$$S_0^\circ = 0$$

With this definition of a standard state for entropy, we see that, at constant pressure, for one mole of the element or compound,

$$S = \int_0^T \frac{C_P dT}{T} \tag{7-21}$$

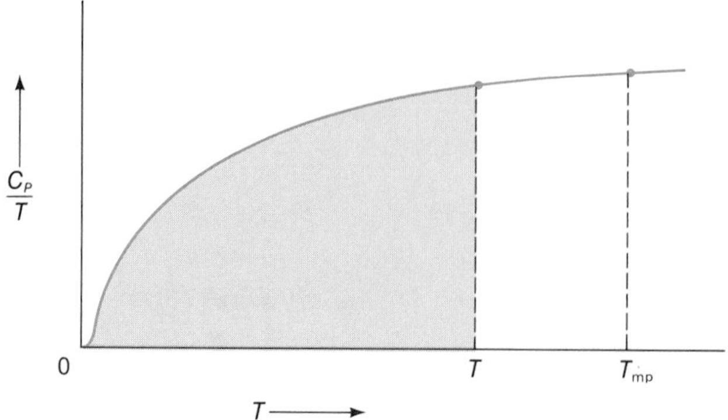

Figure 7-9. A plot of C_P/T versus T. Such plots are useful in the determination of the entropy of a substance.

since $dq_{rev} = dH = C_P dT$ at constant pressure. This equation is valid only over temperature ranges in which no phase changes occur. When they do occur, Equation 7-19 must be used to calculate the additional increases in entropy. Equation 7-21 provides a method for determining absolute entropies experimentally. The integral is just the area under the curve shown in Figure 7-9.

7.9 FREE ENERGY, INDICATOR FOR THE DIRECTION OF ALL SPONTANEOUS PROCESSES IN NATURE AT CONSTANT TEMPERATURE AND PRESSURE

We have already concluded that all spontaneous processes in an isolated system, i.e., one that does not exchange matter or energy with its surroundings, are accompanied by an increase in entropy. Since the universe is believed to be an isolated system, we conclude that the entropy of the universe is constantly increasing.

Most of the processes in which chemists and biologists are interested, however, do not occur in isolated systems, but in closed (or essentially closed) systems that exist at conditions of approximately constant temperature and pressure. Such systems exchange energy with their surroundings, so entropy alone cannot be used to predict their spontaneity. In the course of his monumental work in thermodynamics and statistical mechanics, Gibbs defined a new state function, now called the *Gibbs free energy*, that *always* decreases when spontaneous processes occur in a closed system at constant temperature and pressure. You may recall that for single particles the processes that occur spontaneously, e.g., a ball rolling down a hill, always result in a decrease in the potential energy of the particle. In a many-particle system, the Gibbs free energy is analogous to the potential energy in a single-particle system. Thus, chemical reactions that occur spontaneously, e.g., the burning of a candle, are always associated with a decrease in the Gibbs free energy of the system (reactants plus products). The analogy between a ball rolling down a potential-energy hill and a chemical reaction "rolling" down a Gibbs-free-energy hill is very useful for chemists and biologists.

The Gibbs free energy is of the utmost importance because it enables us to predict which reactions will occur and which will not. The process of life itself is a "downhill" free-energy process that demonstrates clearly the importance of food for animal life and the importance of sunlight for plant life and, ultimately, all life.

Like enthalpy, the Gibbs free energy* G is defined in terms of other state functions, so it is also a state function. It is defined by the equation

$$G = H - TS \tag{7-22}$$

The equation for the change in the Gibbs free energy during any process that occurs in a closed system at constant temperature and pressure is

$$\Delta G = \Delta H - T\Delta S \tag{7-23}$$

where the enthalpy change ΔH is the heat absorbed by the system from the surroundings during such a process, T is the absolute temperature of the system, and ΔS is the change in entropy. A process that is exothermic and that therefore transfers heat to the surroundings has a $\Delta H < 0$. A process that also entails an increase in the entropy of the system has a $\Delta S > 0$. For such a process the change in Gibbs free energy is $\Delta G < 0$ and the process is therefore "downhill," or spontaneous.

You may recall that we have already mentioned an endothermic reaction that is spontaneous. When salt is dissolved in water, the solution must absorb heat from the surroundings in order to remain at constant temperature. Here $\Delta H > 0$ but the process is still spontaneous because $T\Delta S > \Delta H$, so $\Delta G < 0$. Similarly, it is possible for a process occurring at constant temperature and pressure to be associated with a decrease in the entropy of the system and still be spontaneous, if the enthalpy decreases by more than enough to offset the decrease in entropy. In other words, if the absolute value of the enthalpy change is greater than $T\Delta S$ in such a process, then $\Delta G < 0$ and the process is spontaneous.

*In the older literature the now obsolete symbol F is used for the Gibbs free energy.

Standard Free Energies of Formation

The free-energy change associated with a chemical reaction is crucial information because it indicates whether or not the reaction is spontaneous. This is equivalent to knowing if the reaction will go in the "forward" direction or the "reverse" direction.

The standard state for free energies of formation, ΔG_f°, is the same as that defined for standard enthalpies of formation, ΔH_f°: the most stable physical form of the substance at 1 atm and 298 K. The free energies of the elements at standard conditions are assigned a value of zero. The *standard free energy of formation* of a compound is then taken to be the free-energy change that accompanies the formation of 1 mole of that compound from the elements in their standard states. For example, H_2O in its standard state is formed by the reaction

$$H_2(g) + \tfrac{1}{2}O_2(g) \rightarrow H_2O(l) \qquad \Delta G = -56.690 \text{ kcal/mole}$$

(Here the phase change from $H_2O(g)$ to $H_2O(l)$ is included in the reaction as written, and hence in the value of ΔG as well.) Thus, the standard free energy of formation of water is

$$H_2O(l) \qquad \Delta G_f^\circ = -56.690 \text{ kcal/mole}$$

Note that the above reaction is associated with a large negative free-energy change and is thus a "very spontaneous" reaction. This should be no surprise, for the reaction between premixed H_2 and O_2, once initiated, is explosive.

Standard free energies of formation can be used to calculate the free-energy changes of more complex reactions by the same thermochemical methods used to calculate the enthalpy changes. In analogy with Equation 7-6, we have

$$\Delta G^\circ_{\text{reaction}} = \sum n_p \Delta G_f^\circ \text{ (Products)} - \sum n_r \Delta G_f^\circ \text{ (Reactants)} \qquad (7\text{-}24)$$

For the generalized chemical reaction

$$aA + bB \rightarrow cC + dD$$

we therefore obtain the equation

$$\Delta G^\circ_{\text{reaction}} = c\Delta G_f^\circ(C) + d\Delta G_f^\circ(D) - a\Delta G_f^\circ(A) - b\Delta G_f^\circ(B)$$

Example 7-5. Calculate the change in the Gibbs free energy accompanying the burning of natural gas under standard conditions.

$$CH_4(g) + 2O_2(g) \rightarrow CO_2(g) + 2H_2O(l)$$

Making use of Equation 7-24 and the numerical values of ΔG_f°, we obtain

$$\begin{array}{lll} CO_2(g) & \Delta G_f^\circ & = -94.260 \text{ kcal/mole} \\ 2H_2O(l) & 2\Delta G_f^\circ & = -113.380 \text{ kcal/mole} \\ \hline \multicolumn{2}{l}{\Sigma n_p \Delta G_f^\circ \text{ (Products)}} & = -207.640 \text{ kcal/mole} \end{array}$$

$$\begin{array}{lll} CH_4(g) & \Delta G_f^\circ & = -12.140 \text{ kcal/mole} \\ 2O_2(g) & 2\Delta G_f^\circ & = 0 \text{ kcal/mole} \\ \hline \multicolumn{2}{l}{\Sigma n_r \Delta G_f^\circ \text{ (Reactants)}} & = -12.140 \text{ kcal/mole} \end{array}$$

$$\Delta G^\circ_{\text{reaction}} = -207.640 + 12.140 = -195.500 \text{ kcal/mole}$$

Standard Free-Energy Changes and Equilibrium

In a mixture of ideal gases, the free energy of a mole of any one of the components at constant temperature is a simple function of its partial pressure in the mixture. If we let $\overline{G}$ represent the free energy *per mole* at any arbitrary pressure P, the change in free energy at constant temperature associated with a change in pressure from the standard-state pressure P° to any other pressure P is given by the following equation:

$$\overline{G} - \overline{G}^\circ = RT \ln \frac{P}{P^\circ} \tag{7-25}$$

Since $P^\circ = 1$ atm, we obtain

$$\overline{G} = \overline{G}^\circ + RT \ln P \tag{7-26}$$

Thus, at constant temperature, the free energy of an ideal gas depends upon the partial pressure of that gas.

Now, for the same generalized chemical reaction that we have used twice before, the change in the Gibbs free energy is given by

$$\Delta G = \sum n_p \overline{G} \text{ (Products)} - \sum n_r \overline{G} \text{ (Reactants)} \tag{7-27}$$

$$\Delta G = c\overline{G}_C + d\overline{G}_D - a\overline{G}_A - b\overline{G}_B$$

For a reaction involving only ideal gases,

$$\Delta G = c\overline{G}^\circ_C + d\overline{G}^\circ_D - a\overline{G}^\circ_A - b\overline{G}^\circ_B + cRT \ln P_C + dRT \ln P_D - aRT \ln P_A - bRT \ln P_B \tag{7-28}$$

But the first four terms on the right side of this equation give the *standard* free-energy change for the reaction:

$$\Delta G^\circ = c\overline{G}^\circ_C + d\overline{G}^\circ_D - a\overline{G}^\circ_A - b\overline{G}^\circ_B$$

and therefore, collecting terms,

$$\Delta G = \Delta G^\circ + RT \ln \frac{(P_C)^c(P_D)^d}{(P_A)^a(P_B)^b} \tag{7-29}$$

At equilibrium there is no net change in the system (the forward and reverse reactions effectively cancel each other), so there can be no change in the free energy:

$$\Delta G = 0 \qquad \text{(Equilibrium)} \tag{7-30}$$

Therefore, *at equilibrium*,

$$\Delta G^\circ = -RT \ln \left[\frac{(P_C)^c(P_D)^d}{(P_A)^a(P_B)^b}\right]_{\text{equilibrium}} \tag{7-31}$$

$$\Delta G^\circ = -RT \ln K \tag{7-32}$$

$$K = e^{-\Delta G^\circ/RT} = 10^{-\Delta G^\circ/2.3RT} \tag{7-33}$$

Equation 7-32 demonstrates the existence of an *equilibrium constant K* for an ideal-gas reaction at constant temperature. We will see in Chapter 9 that equilibrium constants exist for *all* chemical reactions and that the relation (Equation 7-33) between the equilibrium constant and the standard free-energy change for a reaction is general for all reactions.

7.10 COLLIGATIVE PROPERTIES

Colligative properties are properties of a solution that depend only upon the concentration of solute present in the solution, not upon any detailed molecular properties of the solute. (The solute is the minor component(s) of a solution and the solvent is the major component—e.g., sugar and water, respectively.) There are only four commonly studied colligative properties; we will discuss each of them in turn.

Vapor-Pressure Lowering

If a nonvolatile solute is added to a solvent, the vapor pressure of the solvent above the solution decreases. *Raoult's law*, named for the French

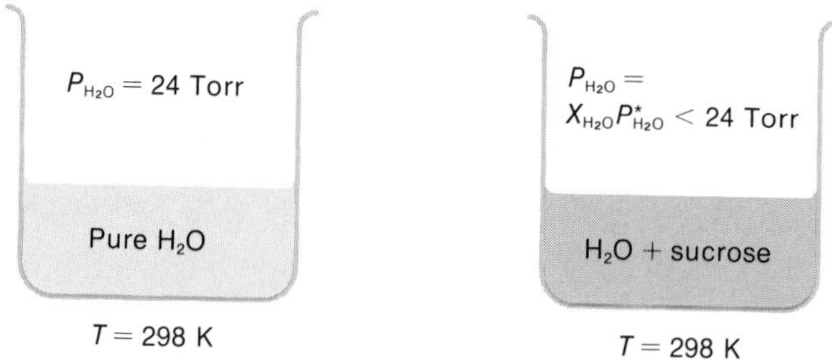

Figure 7-10. Schematic representation of the effect of adding a nonvolatile solute to a solvent upon the vapor pressure of the solvent above the solution. Raoult's law applies to the solution if it is very dilute.

chemist François Raoult (1830–1901), states that the partial pressure of a solvent in equilibrium with a solution at a given temperature is proportional to the mole fraction of the solvent in the solution:

$$P_A = X_A P_A^* \tag{7-34}$$

where A denotes the solvent, P_A is the partial pressure of the solvent, and P_A^* is the vapor pressure of the *pure* solvent at the same temperature. The dimensionless quantity X_A is the *mole fraction* of solvent in the solution:

$$X_A = \frac{n_A}{n_A + n_B} \tag{7-35}$$

where n_A and n_B are the numbers of moles of solvent and solute, respectively. The effect of a solute on the partial pressure of the solvent above a solution is shown in Figure 7-10.

Example 7-6. Two beakers, each containing 500 g of water, are placed in a closed container at 20°C. If beaker 1 contains pure water and beaker 2 contains a solution of 50.0 g of LiCl, what is the water-vapor pressure above each of the beakers? What will happen to this closed system in time?

The vapor pressure of pure water in beaker 1 at 20°C is $P^*_{H_2O} = 17.5$ Torr. The mole fraction of water in beaker 2 may be calculated as though Li^+ and Cl^- were identical, because colligative properties depend only on the numbers of solute particles present, not their identities.

$$\text{Moles Li}^+ = \text{moles Cl}^- = \frac{50.0 \text{ g}}{6.94 \text{ g/mole Li}^+ + 35.5 \text{ g/mole Cl}^-} = 1.18 \text{ moles}$$

$$\text{Moles H}_2\text{O} = \frac{500 \text{ g}}{18.0 \text{ g/mole H}_2\text{O}} = 27.8 \text{ moles}$$

The mole fraction of H_2O in the LiCl solution is therefore

$$X_{H_2O} = \frac{27.8}{27.8 + 2(1.18)} = \frac{27.8}{30.2} = 0.921$$

Thus, the water-vapor pressure above beaker 2 can be obtained from Raoult's law as

$$P_{H_2O} = X_{H_2O}P^*_{H_2O} = 0.921\ (17.5\ \text{Torr}) = 16.1\ \text{Torr}$$

The vapor pressure of water in equilibrium with the LiCl solution is less than that of pure water at the same temperature, so water will evaporate from beaker 1, which contains pure water, and condense in beaker 2, which contains the LiCl solution. This process will continue until all the water in beaker 1 is gone. The LiCl solution will be diluted but its vapor pressure will never be as high as that of pure water at the same temperature.

Raoult's law applies strictly only to very dilute solutions, which are considered "ideal" in the same sense that gases at high temperature and low pressure are considered ideal. In analogy with deviations from the ideal gas law, deviations from Raoult's law can be interpreted in terms of the behavior of *real* solutions. The greatest practical value of Raoult's law, however, is that it provides a method for calculating the molecular weights of dissolved substances. Note that, because colligative properties depend only on the numbers of dissolved particles, the vapor-pressure lowering in a 0.01 M solution of LiCl (an ionic solid) is twice as great as that in a 0.01 M solution of sucrose (a molecular solid). What would you expect for a 0.01 M solution of $Fe(NO_3)_3$?

Boiling-Point Elevation

The addition of a nonvolatile solute to a solvent increases the boiling point of the solvent. This is a colligative property related only to the *molality* of the solution (moles solute/1000 g solvent). It can be easily explained by considering the phase diagram for water, shown in Figure 7-11.

Addition of a solute reduces the vapor pressure of the solvent above the solution in accord with Raoult's law. The dashed curve in Figure 7-11 represents the liquid-vapor equilibrium curve in the presence of a certain amount of solute. The reduction in vapor pressure above the solution means that, for the solution to boil at an external pressure of 1 atm, the temperature must be raised by an amount ΔT_b above the normal boiling point T_b in order to raise the vapor pressure of the solvent to 1 atm. Since the reduction in vapor pressure depends only on the mole fraction of solvent, the

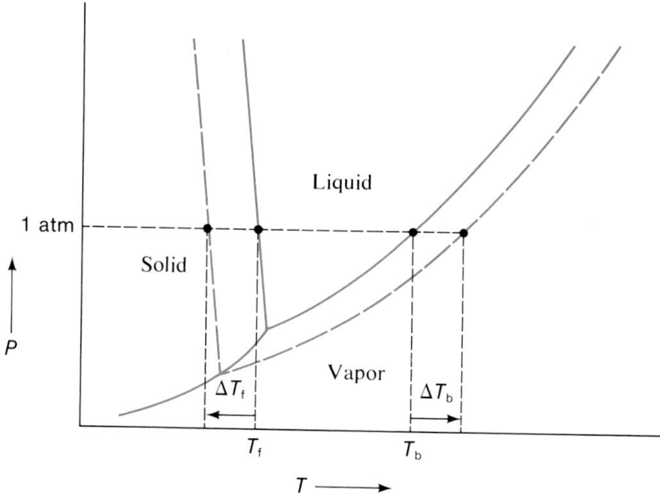

Figure 7-11. The origin of the freezing-point depression and boiling-point elevation in an aqueous solution.

boiling-point elevation depends, to a first approximation, only on the molality *m* of the solution:

$$\Delta T_b = K_b m \tag{7-36}$$

where the proportionality factor K_b is the *molal boiling-point constant* of the solvent, in units of kelvins per molal, [i.e., K/(moles solute/1000 g solvent)]. Table 7-3 provides molal boiling-point constants for a number of solvents.

Table 7-3. **Molal Freezing-Point and Boiling-Point Constants.**

Solvent	K_f, K/molal	K_b, K/molal
Acetic acid	3.90	2.93
Benzene	5.12	2.53
Ethanol	1.83	1.22
Naphthalene	6.8	—
Water	1.86	0.512

Freezing-Point Depression

The addition of a nonvolatile solute to a solvent also decreases the freezing point of the solvent. The use of rock salt to melt ice and antifreeze to prevent the water in your car's radiator from freezing are practical applications of this fact. The explanation is again found in Figure 7-11, which shows that the reduction in vapor pressure above the solution due to the addition of solute requires a reduction in temperature by an amount ΔT_f below the normal freezing point T_f for freezing to occur. In analogy with the boiling-point constant, there is a *molal freezing-point constant* K_f that relates the freezing-point depression of the solvent to the molality of the solution:

$$\Delta T_f = -K_f m \tag{7-37}$$

Some molal freezing-point constants are shown in Table 7-3.

Example 7-7. A fountain near the chemistry department at the University of California, Berkeley, has a pool of water 1.00 ft deep, 15.0 ft wide, and 25.0 ft long. A recent winter was very severe (by Berkeley standards) and night temperatures fell to 22.0°F. How much salt, NaCl, would you have added to the fountain to prevent it from freezing? (The value of K_f for H_2O is 1.86 K/molal.)

To prevent freezing ΔT_f must be at least 22°F − 32°F = −10°F.

$$\Delta T_f = (-10.0°\text{F}) \frac{1\ \text{K}}{1.80°\text{F}} = -5.56\ \text{K}$$

$$\Delta T_f = -K_f m$$

We can now solve for the required molality:

$$m = -\frac{-5.56\ \cancel{\text{K}}}{1.86\ \cancel{\text{K}}/\text{molal}} = 2.99\ \text{molal in Na}^+ \text{ and Cl}^- \text{ ions}$$

$$m_{\text{NaCl}} = \frac{2.99\ \cancel{\text{moles ions}}/\text{kg H}_2\text{O}}{2\ \cancel{\text{moles ions}}/\text{mole NaCl}} = 1.50\ \text{molal}$$

The fountain contains

$$(1.00 \times 15.0 \times 25.0\ \cancel{\text{ft}^3})\ [(12.0\ \cancel{\text{in.}}/\cancel{\text{ft}})\ (2.54\ \text{cm}/\cancel{\text{in.}})]^3 = 1.06 \times 10^7\ \text{cm}^3\ \text{H}_2\text{O}$$

Using a density of 1.00 g/cm³ for H_2O, the mass of H_2O in the fountain is

$$(1.00\ \cancel{\text{g}}/\cancel{\text{cm}^3})\ (1.06 \times 10^7\ \cancel{\text{cm}^3}) \frac{1\ \text{kg}}{10^3\ \cancel{\text{g}}} = 1.06 \times 10^4\ \text{kg}$$

The number of moles of NaCl to be added to the fountain is

$$n_{NaCl} = (1.50\ \text{moles}/\cancel{\text{kg H}_2\text{O}})\ (1.06 \times 10^4\ \cancel{\text{kg H}_2\text{O}}) = 1.59 \times 10^4\ \text{moles}$$

which is equal to

$$(1.59 \times 10^4\ \cancel{\text{moles}})\ (58.5\ \cancel{\text{g}}/\cancel{\text{mole}})\ \frac{1\ \text{kg}}{10^3\ \cancel{\text{g}}} = 930\ \text{kg}$$

or, in pounds,

$$(930\ \cancel{\text{kg}})\ (2.20\ \text{lb}/\cancel{\text{kg}}) = 2050\ \text{lb}$$

Osmotic Pressure

Figure 7-12 shows a simple experimental setup for demonstrating the effect of the *osmotic pressure* of a sucrose solution. A semipermeable membrane is one that is permeable to solvent—in this case, water—while it is impermeable to solute—in this case, sucrose. Because of the sucrose in the solution above the membrane, the free energy of water in this solution

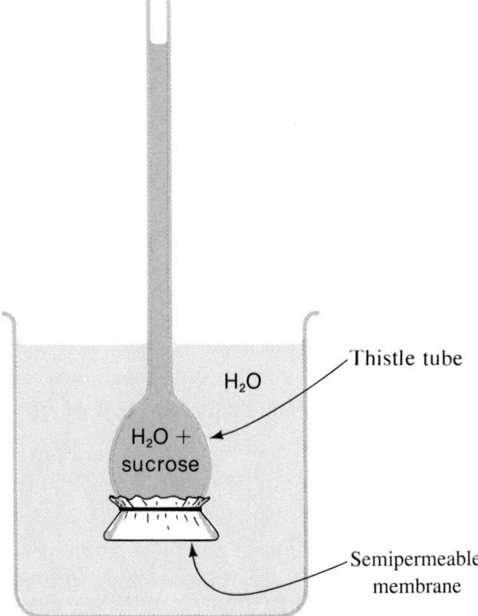

Figure 7-12. Simple apparatus for demonstrating the effect of osmotic pressure.

is less than that of the pure water in the beaker, so water passes through the membrane into the solution, building up hydrostatic pressure until equilibrium is achieved. At equilibrium the molar free energy of the water in the solution under pressure is equal to the molar free energy of the pure water on the other side of the membrane. If the membrane were permeable to sucrose, the sucrose would pass through the membrane into the beaker until its concentration was equal on both sides of the membrane. The osmotic pressure across a membrane with equal concentrations of solute on both sides is, of course, zero. It is therefore imperative that the membrane be semipermeable.

The osmotic pressure Π can be calculated from the equation

$$\Pi V = n_s RT \tag{7-38}$$

where V is the volume of the solution (in Figure 7-12, the sucrose solution) and n_s is the number of moles of solute in that volume. Note the similarity between this equation and the ideal gas law.

Example 7-8. Osmotic pressure is the only colligative property suitable for the study of large biological molecules. George Scatchard has used this technique [*Journal of the American Chemical Society* **68**, 2320 (1946)] to determine the molecular weight of bovine serum albumin, a protein found in the serum of cattle. Some of his data are

Protein concentration	= 9.63 g protein/kg H_2O
Osmotic pressure	= 2.65 Torr
Temperature	= 25°C

Calculate the molecular weight M of this protein.

Equation 7-38 can be recast in terms of M:

$$\Pi = \frac{n_s}{V} RT = \frac{m_s}{V} \frac{RT}{M}$$

where m_s is the mass of solute of molecular weight M in volume V. Therefore

$$M = \frac{cRT}{\Pi}$$

where $c = m_s/V$ is the concentration of solute in grams per liter of solution. We can substitute the concentration in grams per kilogram of solvent for grams per liter of solution because the solution is dilute and 1 kg H_2O = 1 ℓ H_2O. Solving for M, we obtain

$$M = \frac{(9.63\ \text{g}/\cancel{\ell})\ (0.0821\ \cancel{\ell}\ \cancel{\text{atm}}/\cancel{\text{K}}\ \text{mole})\ (298\ \cancel{\text{K}})}{(2.65\ \cancel{\text{Torr}})\ \dfrac{1\ \cancel{\text{atm}}}{760\ \cancel{\text{Torr}}}}$$

$$= 67{,}600\ \text{g/mole}$$

This value agrees reasonably well with the accepted molecular weight of bovine serum albumin of 69,000.

Bibliography

Henry A. Bent, **The Second Law,** Oxford University Press, New York, 1965. Henry Bent is a noted thermodynamicist. This book is essentially a junior-level text on thermodynamics and statistical mechanics. However, we have included it here because the first few chapters are not too difficult and because of the delightfully fresh and exciting style of the author. The classic cartoon by Steinberg, reproduced on the frontispiece, is reason enough for you to go to the library and find the book.

Hugo F. Franzen and Bernard C. Gerstein, **Rudimentary Chemical Thermodynamics,** D. C. Heath, Lexington, Mass., 1971. This text is used by general chemistry students and by students preparing for their preliminary examinations in graduate school. This reflects the fact that the book presents not only the basic concepts of thermodynamics but also some of their more advanced applications, such as activity coefficients, electromotive force, and phase diagrams—and all of it in a clear, understandable way. You may be interested to read about the Carnot cycle, a concept of great historical importance from which much of thermodynamics, including the concept of entropy, was deduced.

Bruce H. Mahan, **Elementary Chemical Thermodynamics,** W. A. Benjamin, New York, 1963. This short text provides an excellent introduction to chemical thermodynamics. The detailed treatment of the first two laws is followed by a discussion of the applications of thermodynamic principles to some problems of interest to chemists.

Leonard K. Nash, **Elements of Chemical Thermodynamics,** Addison-Wesley, Reading, Mass., 1962. The level of this text is similar to that of Mahan's. A broad but brief introduction to the principles of chemical thermodynamics is presented, followed by an interesting short epilogue on "Science and the Social Order."

Jürg Waser, **Basic Chemical Thermodynamics,** W. A. Benjamin, New York, 1966. Professor Waser's text is intended for use in an honors course in general chemistry. Thus the level of mathematical sophistication is somewhat higher than that of the present chapter. However, the clarity of the writing and the rigor of the arguments make this one of the best introductions to chemical thermodynamics available.

Problems

1. The daily food requirement for an average man is about 2500 nutritional Calories. If all this energy were available to do useful work, what would be the average power output in watts of this average man? (1 nutritional Calorie = 1 kilocalorie; 1 W = 1 J/s.)

2. If a gas is held in a bomb calorimeter (constant volume), how much does its internal energy change when 300 cal of heat is added to it? Describe how this energy is absorbed by He and by O_2.

3. State whether each of the following assertions is true or false. If it is false, state how it can be made true.
   a. The equation $\Delta E = q - w$ is applicable to any microscopic process, provided that no electric work is done by the system on its surroundings.
   b. A reversible process can proceed in either direction. However, a reversal in direction requires a large amount of energy.
   c. When a system undergoes an isothermal change in state, the enthalpy change depends on the process in question.
   d. A typical example of a reversible process is the expansion of a gas into a vacuum.

4. Six moles of an ideal gas are expanded reversibly at 27°C from 40 to 90 ℓ. Find w in calories and liter-atmospheres.

5. Four moles of an ideal gas are expanded at 25°C such that the volume is increased by a factor of 6 and work against a constant external pressure of 4.0 atm is performed. The initial internal pressure is 24 atm. Calculate w, q, ΔE, and ΔH for this process.

6. Two moles of an ideal gas at 27°C and 5 atm are expanded adiabatically against a constant external pressure of 2.0 atm to a final internal pressure of 2.0 atm. Calculate the final temperature of the gas. Assume $C_V = 3.0$ cal/K mole.

7. Two of the oxides of nitrogen are major contributors to the production of photochemical smog. Calculate $\Delta H°$ for the oxidation of nitric oxide, NO, to nitrogen dioxide, NO_2, from the following data:

$$\tfrac{1}{2}N_2(g) + \tfrac{1}{2}O_2(g) \rightleftharpoons NO(g) \qquad \Delta H° = 21.60 \text{ kcal/mole}$$

$$\tfrac{1}{2}N_2(g) + O_2(g) \rightleftharpoons NO_2(g) \qquad \Delta H° = 8.09 \text{ kcal/mole}$$

8. Calculate $\Delta H°$ for each of the following reactions:
   a. $CaCO_3(c) \rightarrow CaO(c) + CO_2(g)$
   b. $2SO_2(g) + O_2(g) \rightarrow 2SO_3(g)$
   c. $NH_3(g) + HCl(g) \rightarrow NH_4Cl(c)$
   d. $H_2(g) + CO_2(g) \rightarrow H_2O(l) + CO(g)$

9. Assuming ideal gas behavior, find ΔH and ΔE at 25°C for the following reaction at 1 atm and 15 atm:

$$NH_3(g) + HCl(g) \rightarrow NH_4Cl(c)$$

10. The geometric isomers of 2-butene are

$$\underset{\textit{cis}\text{-2-Butene}}{\begin{array}{ccc} CH_3 & & CH_3 \\ & C{=}C & \\ H & & H \end{array}} \quad \text{and} \quad \underset{\textit{trans}\text{-2-Butene}}{\begin{array}{ccc} CH_3 & & H \\ & C{=}C & \\ H & & CH_3 \end{array}}$$

Calculate the enthalpies of formation of these isomers from average bond energies (see Table 4-5) and compare them with the tabulated values of −1.36 kcal/mole and −2.40 kcal/mole for the *cis* and *trans* isomers, respectively. Explain why the enthalpy of formation of the *trans* isomer has a larger negative value.

11. Using average bond energies, calculate the heats of formation of gaseous dichloromethane, CH_2Cl_2, and 1-butene, $CH_2{=}CH{-}CH_2{-}CH_3$.

12. Oxyacetylene torches are used by metal workers to cut through iron and steel. Using average bond energies, calculate $\Delta H°$ for the reaction

$$H{-}C{\equiv}C{-}H(g) + \tfrac{5}{2}O_2(g) \rightarrow 2CO_2(g) + H_2O(g)$$

Looking at the standard heats of formation in Appendix D, what can you say about the value for acetylene relative to those for other carbon compounds? How is your observation related to the use of acetylene in torches? (Neglect the resonance stabilization of CO_2 for this problem.)

13. Estimate the maximum temperature difference in 5 g of water after it drops through the 167 ft of Niagara Falls. The value of C_P for H_2O is 18 cal/K mole.

14. A lead bullet weighing 12 g is heated to 200°C and then dropped into a beaker containing 200 g of water at 20°C. What is the final temperature at equilibrium? The heat capacities of H_2O and Pb are 18 cal/K mole and 6.3 cal/K mole, respectively.

15. After joining the Navy, an ex-chemistry student decides to apply his knowledge to a crap game, which is played with two dice. He knows that if he throws a 2, 3, or 12 he loses his bet unless he has a point. If he throws a 7 while trying to make a second point, he is out of the game. The student decides to calculate the probability of each of the above throws. What are his results? Which of the above throws has the highest entropy?

16. Four molecules (A, B, C, and D) can share two photons of energy among themselves. Assume that two photons can be captured by the same molecule.

a. Write all the possible distributions of the photons among the molecules.
b. Calculate the probabilities of finding molecule B with 2 photons, 1 photon, and 0 photon.
c. What is the average energy in photons of molecule B?

17. The entropy of a mixture of ideal gases is equal to the sum of the entropies that each gas would have if it alone occupied the same volume at the same temperature. Calculate the entropy change ΔS when 6.0 g of O_2 gas is mixed with 2.0 g of H_2 gas at the same temperature and pressure.

18. Calculate the entropy change ΔS when 2.0 g of ice in a glass of Coca-Cola melts at 0°C:

$$H_2O\,(c,\ 0°C) \rightarrow H_2O\,(l,\ 0°C) \qquad \Delta H_{\text{fusion}} = 1.4 \text{ kcal/mole}$$

19. Calculate the standard free-energy change $\Delta G°$ for the reaction

$$N_2(g) + 3H_2(g) \rightarrow 2NH_3(g)$$

Is the reaction spontaneous at 25°C? Calculate $\Delta H°$ and $\Delta S°$. Which of these factors predominates in determining the spontaneity of the reaction?

20. Calculate $\Delta G°$ and K at 25°C for the reaction

$$2HI(g) + Cl_2(g) \rightarrow 2HCl(g) + I_2(g)$$

Is the reaction spontaneous as written?

21. Consider the reaction

$$H_2(g) + Cl_2(g) \rightarrow 2HCl(g) \qquad \Delta G° = -45.54 \text{ kcal/mole}$$

What is the direction of the reaction at 25°C if the initial partial pressures of H_2, Cl_2, and HCl are 0.050 atm, 0.10 atm, and 0.60 atm, respectively?

22. An important biochemical reaction is the oxidation of glucose, $C_6H_{12}O_6$. Calculate $\Delta H°$, $\Delta G°$, and $\Delta S°$ for this reaction:

$$C_6H_{12}O_6\,(c) + 6O_2\,(g) \rightarrow 6CO_2\,(g) + 6H_2O\,(l)$$

(For glucose, $\Delta H_f° = -301$ kcal/mole, $\Delta G_f° = -218$ kcal/mole, and $S° = 64$ cal/K mole.) Is the equilibrium constant for this reaction greater than or less than 1? Is the reaction driven primarily by entropy or enthalpy?

23. Pure water has a vapor pressure of 24 Torr at 25°C. If 20 g of sucrose, $C_{12}H_{22}O_{11}$, is dissolved in 70 g of water, what is the vapor pressure of the solvent above the solution?

24. Ethylene glycol, CH_2OHCH_2OH, is used as antifreeze in automobiles. If a skier going to the Sierra wants to protect his car to temperatures as low as

14.0°F (−10.0°C), what concentration of ethylene glycol must he have in his radiator? Express your answer in g/1000 g H_2O.

25. If 14.0 g of an unknown substance dissolved in 150 g of benzene raises its boiling point by 1.20°C, what is the approximate molecular weight of the unknown substance? (See Table 7-3.)

26. The freezing point of pure *d*-camphor is 180°C and the molal freezing-point constant is $K_f = 40$ K/molal. When 0.80 g of a compound with empirical formula CH is dissolved in 25 g of *d*-camphor, the solution freezes at 152°C. What is the molecular formula of the compound?

27. The fluid in certain protoplasts has an osmotic pressure of 5.0 atm at 30°C. What is the concentration of nondiffusible particles inside these cells?

"The power's off. What do we do now?"

INTERLUDE

energy sources for society

The first law of thermodynamics states that the energy of the universe is a constant. It is the second law that tells us that spontaneous processes are always associated with an increase in the entropy of the universe and that the universe is "running downhill." Since the only form of energy that is useful to us is that which will promote spontaneous changes, in our constant pressure and temperature environment, the only useful energy sources are those that have free energies greater than their products. This limits us to three primary sources of energy: the sun, the interior of the Earth, and the tides, which are due to the gravitational forces exerted by the moon and the sun. There are many examples of secondary energy sources, e.g., the fossil fuels (coal, oil, and natural gas), which are derived from solar energy through photosynthesis. Hydroelectric energy, wind energy, and marine thermal energy are also secondary sources that are derived from the sun's energy.

Geophysical energy is energy derived from radioactivity in the interior of the Earth and from nuclear reactions in substances taken from the Earth's crust and oceans. Nuclear fission reactors are a reality, but thermonuclear fusion reactors are still several decades in the future. Geothermal energy is the heat generated by natural radioactivity deep within the Earth. At present there is little hope of tapping this huge source of energy except through the small number of volcanic areas and steam vents where it comes to the surface. Table 1 shows the major energy sources used in the United States in 1968.

Table 1. The Major Energy Sources Used in the United States in 1968.

Crude petroleum	40.7%
Natural gas	32.1%
Coal	21.9%
Natural-gas liquids	3.8%
Hydroelectric energy	1.3%
Nuclear energy	0.2%
Total (17.8×10^{12} kWh)	100.0%

SOURCE: N. H. Brooks, "Energy and the Environment," in *Science, Scientists, and Society,* ed. by W. Beranek, Jr., Bogden & Quigley, Tarrytown, N.Y., 1972.

It is now generally accepted that energy sources such as tidal energy, geothermal energy, and wind energy will not contribute significantly to the needs of technological societies in the foreseeable future, if ever. Approximately 12% of the hydroelectric energy capacity of the United States has been developed. Thus it is easy to conclude that hydroelectric energy will not be a major factor in our long-range energy consumption. Of the 17.8 trillion kilowatt-hours* consumed in the United States in 1968, only 1.33 trillion (7.5%) were in the form of electricity. Hydroelectric energy represented only 16.7% of the electric energy generated, or 1.25% of the total energy.

Before discussing the likely energy sources of the future, let us examine the magnitude of our energy needs. Primitive man required roughly 2000 kilocalories (2000 nutritional Calories) per day for his existence, and so do you. This is equivalent to

$$(2.0 \times 10^3 \text{ kcal/person day})(365 \text{ days/yr})(1.2 \times 10^{-3} \text{ kWh/kcal} = 880 \text{ kWh/person yr}$$

Early in man's history he learned to control fire, his first energy source. As civilization developed, he domesticated animals and they became a valuable source of mechanical as well as nutritional energy. Water and wind were used as sources of energy. In modern times, numerous kinds of engines—steam, internal combustion, diesel, turbine, jet, rocket, and nuclear—have provided enormous amounts of useful work from nature's fuels.

---

*One kilowatt-hour (kWh) is the total energy developed by a power of one kilowatt acting for one hour.

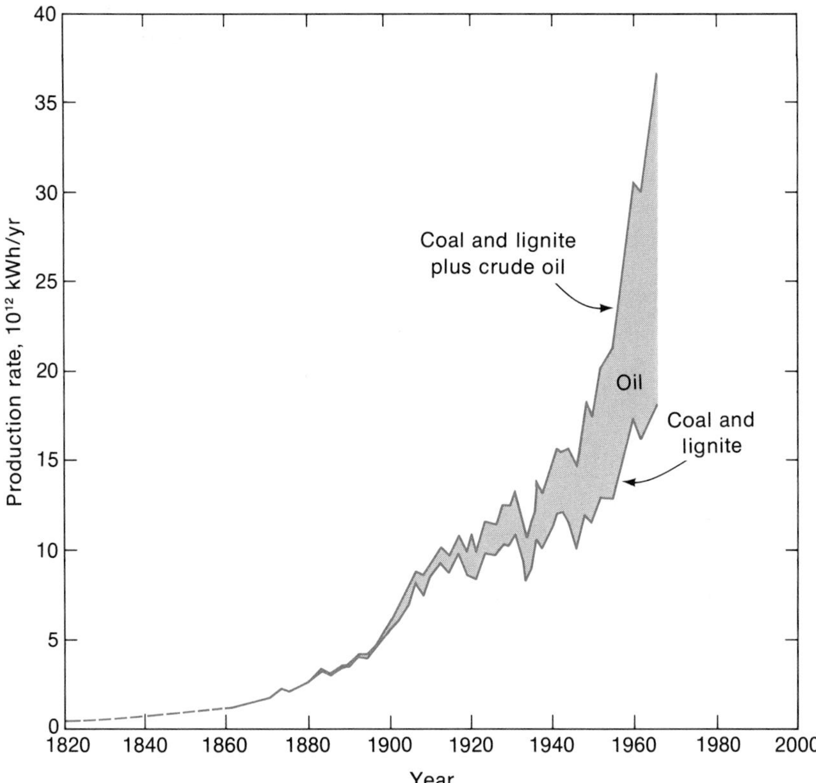

Figure 1. World production of thermal energy from coal and lignite plus crude oil. (Committee on Resources and Man, National Academy of Sciences–National Research Council, *Resources and Man,* W. H. Freeman and Company. Copyright © 1969 by the National Academy of Sciences.)

Figure 1 shows the increase in the world production of thermal energy from coal, lignite, and oil in the last century. In 1968 about 37 trillion kWh were produced from coal and crude oil. The United States, with about 5% of the world's population of 4 billion people, accounted for about 30% of this total. The per capita energy consumption in the United States in 1968 was

$$\frac{17.8 \times 10^{12} \text{ kWh/yr}}{2.02 \times 10^{8} \text{ persons}} = 88{,}000 \text{ kWh/person yr}$$

or about 100 times the amount required for nutrition alone. With increasing technological sophistication, this factor continues to grow.

Coal is primarily graphite; the chemical reaction describing its combustion is

$$\text{C(graphite)} + \text{O}_2(g) \rightarrow \text{CO}_2(g) \qquad \Delta H^\circ = -94 \text{ kcal/mole}$$

Therefore coal should provide a thermal energy of (94 kcal/mole)/(12 g/mole) = 7.8 kcal/g. In fact, because of impurities such as sulfur and water, a typical grade of coal provides 7.0 kcal/g.

$$(7.0 \text{ kcal/g})(1.2 \times 10^{-3} \text{ kWh/kcal}) = 8.4 \times 10^{-3} \text{ kWh/g}$$

The amount of coal equivalent to the per capita energy consumption in the United States in 1968 is thus

$$\frac{8.8 \times 10^4 \text{ kWh/person yr}}{8.4 \times 10^{-3} \text{ kWh/g}} = 1.0 \times 10^7 \text{ g/person yr}$$

$$= 10{,}000 \text{ kg/person yr} = 10 \text{ tons/person yr}$$

Note that these are metric tons, which happen to be roughly equivalent to English long tons.*

The world energy reserves are shown in Table 2 in terms of their life expectancy estimated on the basis of two extreme assumptions, chosen so as to bracket a reasonable range of values. First, it is assumed that the world population will remain constant at its 1968 level of 3.5 billion and that the energy-consumption rate of this population will remain constant at the estimated 1968 value of A ($A = 50 \times 10^{12}$ kWh/yr). Second, it is assumed that the world population will eventually reach 7 billion and that this population will consume energy at a per capita rate of 1.17×10^5 kWh/yr (about 33% higher than the present U.S. rate), giving a total world energy-consumption rate of

$$(1.17 \times 10^5 \text{ kWh/person yr})(7 \times 10^9 \text{ persons}) = 820 \times 10^{12} \text{ kWh/yr}$$

or $16.4A$. (A commonly projected world energy-consumption rate for the year 2000 is $6A$.) Based on the rate $16.4A$, it would take only 8 years to consume all of the world's known fossil-fuel reserves (coal, oil, and natural gas) and only about 160 years to consume all the potential reserves. Even at the present rate A, the world's known fossil-fuel reserves would be consumed in about 130 years.

It is clear that our present sources of fossil-fuel energy are inadequate

*1 metric ton = 1000 kg = 2205 lb = 0.984 English long ton; 1 English long ton = 2240 lb = 1.016 metric ton; 1 English short ton = 2000 lb = 0.907 metric ton.

Table 2. **World Energy Reserves.**

		Life Expectancy of Known Reserves (Years)		Life Expectancy of Potential Reserves (Years)		Life Expectancy of Total Reserves (Years)	
		At Rate A	At Rate $16.4A$	At Rate A	At Rate $16.4A$	At Rate A	At Rate $16.4A$
Finite energy sources	Fossil fuels (coal, oil, natural gas)	130	8	2700	160	2800	170
	More accessible fission fuels (U at $5–30 per pound of U_3O_8 burned at 1.5% efficiency)	66	4	66	4	130	8
	Less accessible fission fuels (U at $30–500 per pound of U_3O_8 burned at 1.5% efficiency)	43,000	2600	130,000	7900	170,000	11,000
"Infinite" natural energy sources	Hydroelectric, tidal, geothermal, and wind energy	Insufficient		Insufficient		Insufficient	
	Solar radiation	10 billion	10 billion	—	—	10 billion	10 billion
	Fusion fuels (deuterium from ocean)	45 billion	2.7 billion	—	—	45 billion	2.7 billion
"Infinite" artificial energy sources (elements transmuted from other elements by neutron bombardment)	Fission fuels (^{239}Pu from ^{238}U; ^{233}U from ^{232}Th)	8.8 million	540,000	21 million	1.3 million	30 million	1.8 million
	Fusion fuels (tritium from Li) On land In ocean	48,000 120 million	2900 7.3 million	Unknown	Unknown	48,000+ 120 million	2900+ 7.3 million

SOURCE: Adapted from W. C. Gough and B. J. Eastlund, "The Prospects of Fusion Power."

to sustain an indefinite increase in the world population at the present average standard of living. The only known or presently anticipated feasible alternatives for large-scale energy production are nuclear fission reactors, thermonuclear fusion reactors, and solar-energy converters. We will consider each of these in turn, but first let us continue our examination of Table 2. Current fission-converter reactors use only 1–2% of the uranium's potential energy content, since the component of the ore that is "burned" as fuel is primarily high-grade, or easily fissionable, ^{235}U. The world fission-fuel reserves were derived by multiplying the U.S. reserves by the ratio of the world land area to the U.S. land area ($\approx 16.2:1$). The figure for known world lithium reserves is based on a study carried out in 1970 by James Norton of the U.S. Geological Survey. The potential lithium reserves are unknown because there has been no exploration program comparable to that undertaken for, say, uranium. Lithium, however, is 5–15 times more abundant in the Earth's crust than uranium. Finally, the life expectancy of the Earth—and hence that of potentially useful solar radiation—is predicted to be at most 10 billion years.

NUCLEAR FISSION

Nuclear binding energies were discussed briefly in Chapter 2, where it was shown that the energies per atom associated with nuclear binding are about 100 million times greater than those associated with chemical binding. The average binding energy per nucleon (neutron or proton) is the total binding energy E_b divided by the mass number A. Figure 2 shows the dependence of the binding energy per nucleon on mass number, with the units of the related mass defect per nucleon shown on a separate scale on the right. The curve shows that the ratio E_b/A increases with increasing mass number until A equals approximately 55, and then decreases with increasing mass number.

Certain unstable nuclei of very high mass number, e.g., uranium-235, split spontaneously into two smaller nuclei with mass numbers still greater than 55, plus one or more neutrons. When this happens, a great deal of energy is released in the form of heat. Such a *fission reaction* also occurs when an unstable nucleus is bombarded with neutrons. For example, when a neutron strikes a ^{235}U nucleus, the result is often a nuclear fission yielding barium, krypton, and three more neutrons:

$$^{235}_{92}U + ^{1}_{0}n \rightarrow ^{139}_{56}Ba + ^{94}_{36}Kr + 3^{1}_{0}n$$

The three neutrons can be captured by other ^{235}U nuclei, and a *chain reaction* occurs, with the total number of neutrons increasing very rapidly.

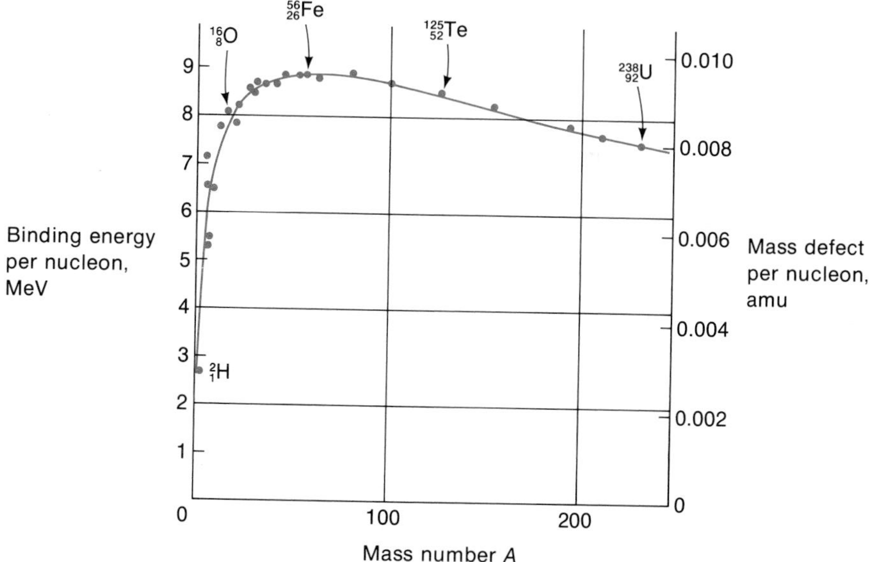

Figure 2. The binding energy per nucleon as a function of mass number. The units of the related mass defect per nucleon are shown on the right.

This reaction is utilized in the atomic bomb, which requires a *critical mass* of fairly pure ^{235}U (or ^{239}Pu). When this mass is exceeded, the rate of capture of neutrons within the mass, relative to the rate of their loss to the surroundings, is so great that the chain reaction builds up to an explosive rate within microseconds. In a nuclear fission reactor, on the other hand, the rate of neutron capture is carefully controlled so that it is just great enough to sustain the reaction at a steady rate. The rate is controlled by inserting cadmium rods, which absorb neutrons, to variable depths within the assembly of fuel rods.

Almost all nuclear fission reactors use the isotope uranium-235 as fuel. The amount of energy released by the fission of one such nucleus is about 200 million electron volts, or 200 MeV. In more familiar units, this becomes 200 MeV/atom $= 7.7 \times 10^{-12}$ cal/atom $= 4.6 \times 10^{9}$ kcal/mole. Compare this with the heat of combustion of coal! This means that one gram of ^{235}U is equivalent to about 2.7 metric tons of coal. That 10 metric tons of coal per person per year is thus equivalent to nearly 4 grams of ^{235}U per person per year. Nuclear energy clearly reduces the need to handle enormous masses of fuel. Nevertheless, the readily accessible U_3O_8 will last only four years at a use rate of 16.4A, according to Table 2, because uranium is a rather rare element. This problem is compounded by the fact that ^{235}U, the fissionable isotope, constitutes only 0.72% of natural uranium. A schematic of the fission reaction is shown in Figure 3.

FISSION POWER REACTION

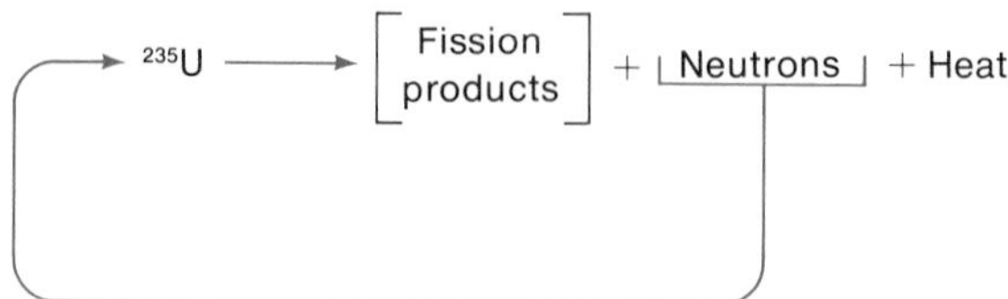

Figure 3. Diagram of the nuclear power reaction from the fission of uranium-235. (M. K. Hubbert, *Energy Resources*, a report to the Committee on Natural Resources, National Academy of Sciences–National Research Council, Publ. 1000-D, Washington, D.C., 1962.)

The dwindling supply of ^{235}U can be augmented by the use of *breeder reactors*, which produce more nuclear fuel than they consume! In a breeder reactor large neutron fluxes from fissioning nuclei (starting with ^{235}U) bombard the relatively abundant but nonfissionable isotopes ^{238}U and ^{232}Th. These nuclei capture one neutron each and then undergo spontaneous radioactive decay to the fissionable isotopes ^{239}Pu and ^{233}U, respectively. The breeder reactions are

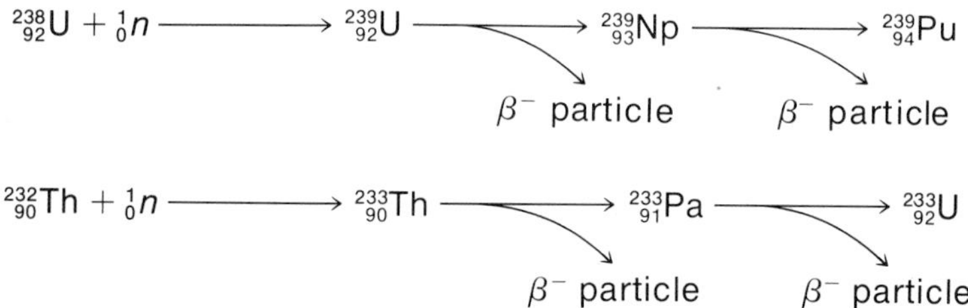

The final plutonium and uranium products are nuclear fuels. If the reactor design is optimized, more than one fissionable nucleus is produced for each nucleus that undergoes fission. This is possible because each fission reaction releases not just one, but usually two or three, neutrons. Thus the reactor "breeds" its own fuel—without violating the second law of thermodynamics. A schematic of the breeder reaction is shown in Figure 4.

The first fission reactor to generate commercial electricity began operating in the Soviet Union in 1954. Now there are several hundred commercial reactors, in many countries. The first prototype commercial breeder reactor began operating in Scotland in 1959, but progress since then has been relatively slow, especially in the United States. It now appears likely that significant commercial energy production by breeder reactors will not occur until the late 1970s in Europe and the Soviet Union, and not until sometime in the 1980s in the United States. When this does happen, however, the

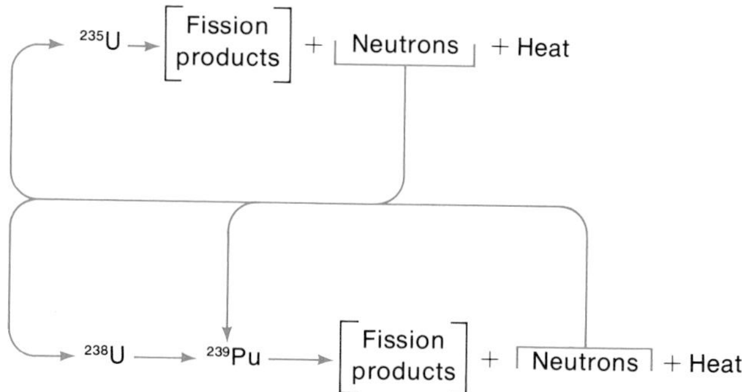

Figure 4. Diagram of the breeder reaction for converting uranium-238 to plutonium-239. (M. K. Hubbert, *Energy Resources,* a report to the Committee on Natural Resources, National Academy of Sciences–National Research Council, Publ. 1000-D, Washington, D.C., 1962.)

potential accessibility of fission fuels will be extended for hundreds of thousands, perhaps millions, of years (see Table 2).

The potential hazards associated with nuclear power plants have received a great deal of publicity in recent years and are the subject of ongoing study, not to mention dispute. We will not attempt to present the pros and cons of the various arguments that have been put forth, but will merely list the most obvious problems: (1) accidental release of radioactivity in any of several possible ways, including *non*-nuclear explosion; (2) accidents occurring during the transportation of radioactive fuels or wastes; (3) disposal of radioactive wastes; (4) thermal pollution due to the huge cooling requirements of the power plants; (5) sabotage. It is the growing opinion of experts in the field that sabotage (theft, hijacking, and the possibility of the nuclear blackmail of an entire population) represents by far the most serious and potentially intractable problem. The problem of an accidental nuclear explosion is no problem at all, because that is a physical impossibility in a nuclear power plant. The nuclear fuel is never, under any circumstances, sufficiently concentrated in fissionable nuclei to be able to undergo a nuclear explosion.

THERMONUCLEAR FUSION

Figure 5 shows that the *fusion,* or combining, of nuclei of very light atoms also releases huge amounts of energy, the energy of the sun and stars. The initiation of fusion reactions requires extremely high temperatures (as much

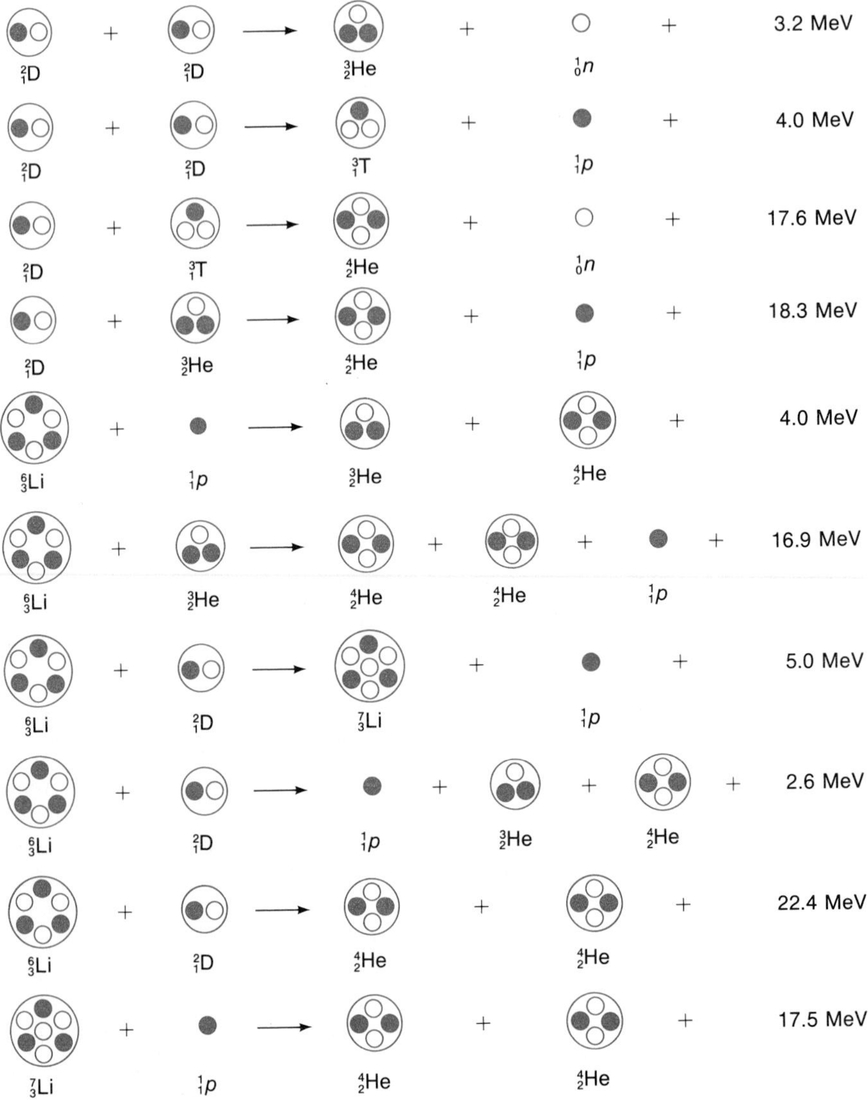

Figure 5. The fusion reactions regarded as potentially useful for generating electricity. (D = deuterium, the stable heavy isotope of hydrogen; T = tritium, the radioactive heavy isotope.) The two possible D-D reactions occur with equal probability. The D-T fuel cycle is considered particularly attractive because this mixture has the lowest ignition temperature known (about 40 million kelvins). The energy released by each reaction is given at the right in mega-electron volts (MeV). (W. C. Gough and B. J. Eastlund, "The Prospects of Fusion Power."

as 100 million kelvins), as well as particle densities and times of containment of the reacting nuclei such that their product equals 10^{14} particle-second/cm^3 or greater. The high temperatures are required so that the collision of two positively charged nuclei can occur with sufficient force to bring about fusion. At temperatures as high as these, all matter exists as a fully ionized plasma consisting of positive ions or bare nuclei, plus free electrons. No material can possibly contain such a plasma, but it is possible to contain it with strong magnetic fields. One such "magnetic bottle" is a torus with the plasma on its circular axis. The plasma can be heated by passing a very large electric current through it, as well as by other methods.

An alternative to magnetic confinement is laser-induced fusion, in which a small pellet of fuel (solid deuterium or tritium) is irradiated with a very short pulse of light from an extremely high-power laser. This approach is newer and its potential is uncertain, though promising. Because the plasmas formed are much more concentrated than those in the magnetic bottles, the times required for fusion to occur are much shorter.

Fusion research is being actively supported because fusion reactors will have certain advantages that would make them the ultimate terrestrial energy source. First, the deuterium fuel is available in virtually unlimited amounts in the oceans and can be extracted with relative ease. The environmental problems associated with its extraction are almost nil. Second, the fusion reactions produce no radioactive products except small amounts of tritium, which has a half-life of only 12.3 years, so disposal of radioactive wastes is a minor problem. Third, any malfunction in the reactor will tend to quench the fusion reactions instantly rather than allow them to get out of control and cause a potentially disastrous overheating, as is possible (though highly improbable) in a fission reactor. Fourth, fusion technology has none of the political and military liabilities associated with the proliferation of nuclear fission fuels, which can be converted for use in nuclear weapons.

For these reasons, and others, thermonuclear fusion is by far the most attractive of all the realistic possibilities for large-scale, economical generation of electricity in the future. The future, in terms of the fusion energy potentially available to us, will last a long time—at least a few billion years.

SOLAR ENERGY

The proponents of solar energy are not necessarily against nuclear energy. They just want to exploit the largest (and cheapest) fusion reactor known to us: the sun. The sun produces prodigious amounts of energy by nuclear fusion of hydrogen atoms. In 24 hours, 64 million tons of hydrogen are converted to 60 million tons of helium. The remaining 4 million tons of matter are converted to energy, the amount being given by the Einstein equation.

Only a tiny portion of this energy reaches the Earth as electromagnetic radiation, but it still amounts to 1.72×10^{17} J/s, or 1.72×10^{17} W. The magnitude of this solar power can be appreciated by considering the power input from other sources. Conduction of heat from the interior of the Earth amounts to only 32×10^{12} W, the tides yield 3×10^{12} W, and volcanoes and hot springs yield 0.3×10^{12} W. Thus, the sun contributes virtually 100% of the energy (or power) influx to the Earth's surface.

What becomes of all this energy? About 47% of it is absorbed by the Earth, converted to heat, and eventually reradiated to space as infrared radiation; about 30% is lost immediately by reflection; and about 23% evaporates surface water and is later released to the atmosphere when the water vapor precipitates. A very small amount produces wind, waves, and currents. Only about 0.02% is used by plants for photosynthesis, and of this amount, a negligibly small portion is eventually converted to energy in the form of fossil fuels.

You may be wondering why we have been able to harness only a tiny portion of this free, inexhaustible, and pollution-free energy. Two obvious characteristics of solar radiation suggest the answer: it is extremely "dilute," being spread out over half the Earth's surface at any given time, and it is intermittent even during the day, owing to variable weather conditions. Nevertheless, people have been harvesting solar energy on a local, small scale for some time. A solar still was built in Chile in 1872 and produced 6000 gallons of fresh water daily from salt water. A 50-hp solar-energy pump was constructed in 1913 near Cairo and was used for many years to pump water out of the Nile River for irrigation. Solar stoves have been used for some time in the Far East. Between mid-morning and mid-afternoon, such stoves can cook a complete meal of lentils and rice in 20 minutes.

The sun can also provide some of the energy needs of individual homes. Solar water heaters are available commercially and are widely used in Australia and Japan. Several dozen homes have been built in America in the last two decades with a significant portion of the space-heating and cooling requirements being provided by solar energy. The heart of most such systems is the flat-plate collector, a rooftop structure that operates on the *greenhouse effect*. Inside a greenhouse on a sunny day, the temperature is higher than outside because the glass transmits the visible and some of the ultraviolet radiation from the sun, but absorbs the infrared reradiated by the objects inside. The glass then reradiates some of this infrared back into the greenhouse, thereby heating it. In the flat-plate collector mentioned above, one or two large, plate-glass covers are used to trap the sun's energy, which heats a fluid—generally water—flowing through a system of metal tubes mounted just beneath the glass. This heated fluid is then pumped to a heat-storage unit, e.g., a large, insulated water tank. In mid-America, an average home could receive three-fourths of its annual heating requirements from a 500–800-ft^2 solar collector and a 1000-gallon storage tank. Figure 6 illustrates how such a system would operate.

As with many of the presently less developed but promising energy sources, one of the main problems with solar energy is the economic factor. The capital investment in a system such as that shown in Figure 6 is quite large, and it may not be possible to make solar energy economically competitive with conventional sources unless it can be harnessed on a very large scale. This, in turn, raises severe technological problems, the chief one being the efficient collection of sunlight over huge areas.

Various imaginative proposals have been put forth in the past few years on how to solve this problem. Perhaps the most ambitious proposal is that of Aden and Marjorie Meinel, two astronomers from Arizona. They envision a vast array of solar farms covering a total of 5000 square miles spread across the 130,000 square miles of desert along the Colorado River

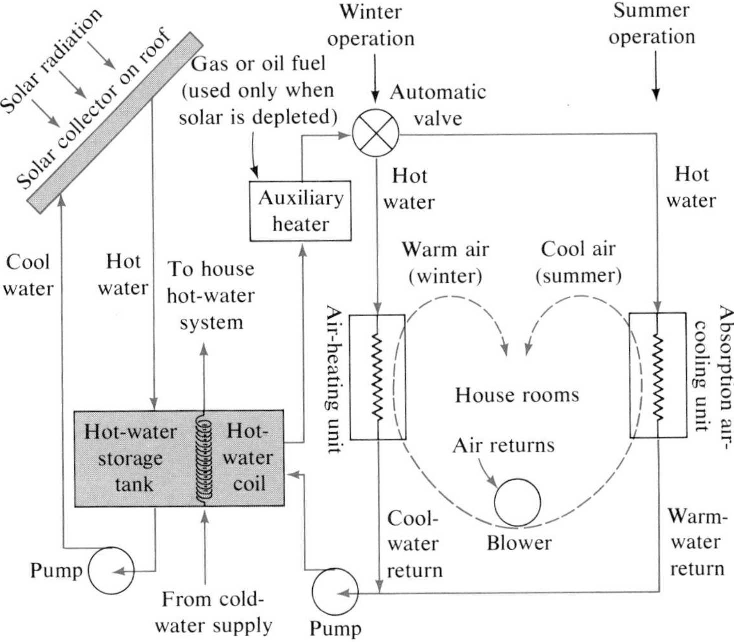

Figure 6. Residential heating and cooling with solar energy. (P. Donovan et al., *An Assessment of Solar Energy as a National Energy Resource,* prepared for the NSF/NASA Solar Energy Panel under the Research Applied to National Needs Program of the National Science Foundation, published by the University of Maryland, College Park, Md., 1972.)

in southern California and Arizona. The most astonishing aspect of their proposal is that they predict that such a system could provide *all* of America's power needs for the year 2000! It is estimated that this quantity will be about 1 million megawatts, or 1 trillion watts. This vast amount of power could be provided by 1000 solar farms, each having a 1000-megawatt capacity and each, unfortunately, costing about $1 billion to build (about $1 per watt).

In 1974 Congress passed the Solar Heating and Cooling Demonstration Act. This law authorizes $60 million for demonstrations of the use of solar energy in heating and cooling systems over a five-year period. Although this is not a large sum of money for research and development of this enormously complex and difficult problem, it is a significant step in the right direction. It is to be hoped that improvements in the small-scale applications of solar energy and the development of pilot programs for the large-scale utilization of solar energy will result.

What kinds of changes in our energy production and utilization are likely within your lifetime? Nuclear power plants will certainly become ever more important as primary sources of electricity. We must hope that fission reactors will eventually give way to fusion reactors—perhaps early in the twenty-first century, if all goes well. In addition, it does not appear unreasonable to expect that solar energy will make significant contributions to the requirements of many individual homes located in favorable climates by the end of the century. As petroleum products become scarcer and more expensive, the internal combustion engine may have to yield to batteries or synthetic fuels such as hydrogen. Electric cars require a great deal of battery storage capacity, which requires a large amount of nonferrous metals. This will require a large capital investment, although it will help to relieve the air pollution problem. An alternative would be to use nuclear or solar power to electrolyze water and use the hydrogen as fuel. Hydrogen is a dangerously explosive gas, but it is likely that ways of using it safely will be devised.

It may be that the solution to the energy crisis will be to severely restrict population growth and/or energy consumption so that we never reach the 16.4*A* level (or worse). In view of the present population growth rate and our voracious appetite for energy, however, this prospect does not seem very likely on a purely voluntary basis. But eventually, *something* will have to give, because indefinite growth and consumption is a biological impossibility. The solutions to the complex problems of population and energy will require that people like you be well-informed on the technological, environmental, economic, and political factors involved, so that you can make the *right* decisions.

8 chemical kinetics

The study of chemical thermodynamics in the preceding chapter has provided us with a powerful tool for predicting what will happen when chemicals are mixed. Given the temperature, the concentration of each species, and the requisite thermodynamic data, we can now predict accurately whether a given reaction will "go" to any appreciable extent. Two major questions remain in our examination of chemical reactions: How fast is the reaction? What is the equilibrium concentration of each species? We will consider the first question in this chapter and the second in the chapter to follow.

The importance of chemical kinetics (the study of the rates of chemical reactions) is readily seen in a variety of situations. For example, the manager of a plant that produces polymers is in a precarious situation if he embarks on the production of a new synthetic polymer just because his resident thermodynamicist has determined that the reactions will go. To be safe, he must also be assured that the reactions will go at a realistic rate, i.e., one that is commercially acceptable. It is difficult to produce ammonia industrially from nitrogen and hydrogen, even though $\Delta G_f^\circ(NH_3) = -3.97$ kcal/mole. Only when the optimum conditions of about 400 atm and 500°C and the best mixed-metal catalyst* were found could this compound

---

*A *catalyst* is a substance that increases the rate of a chemical reaction without itself being consumed in the process. Even tiny amounts of a catalyst can sometimes increase a reaction rate by many orders of magnitude.

be made rapidly enough to satisfy the world's gigantic appetite for fertilizers. This famous example of astute chemical engineering is called the *Haber process* after the German chemist Fritz Haber (1868–1934).

Living systems carry out a staggering array of metabolic reactions. Typically, these reactions occur in dilute aqueous solutions at atmospheric pressure and at temperatures below 40°C. When biochemists began to understand the reactions occurring in living cells (*in vivo*), they attempted to reproduce these reactions in glass laboratory apparatus (*in vitro*). The reactions took weeks and months. Most cells, however, cannot wait nearly that long to digest one meal, utilize the nutrients produced, and be ready for the next. Only when the role of the amazing biochemical catalysts called *enzymes* was discovered could this discrepancy be partially resolved. We will discuss a few of the elementary aspects of enzyme catalysis in this chapter. Much is known about this vital process, yet it is just a beginning.

The principal value of kinetics to the chemist is that it is the most powerful tool available for determining reaction mechanisms, the actual, step-by-step processes by which chemical reactions occur. Such mechanisms are often surprisingly complex, even for seemingly simple reactions. Their unraveling requires the utmost in experimental ingenuity and theoretical insight, and provides a great deal of useful information that can be applied to other systems.

Before we look at the factors that affect the rate of a reaction, we must find out what this rate actually means. Quite simply, the rate of a reaction is the speed with which reactant molecules are converted to product molecules, at a given set of conditions. The usual units are concentration per unit time, c/t. If the concentration changes with time as shown in Figure 8-1, the rate of reaction is a constant equal to the slope of the straight line:

$$\text{Rate} = \frac{\Delta c}{\Delta t} = k \tag{8-1}$$

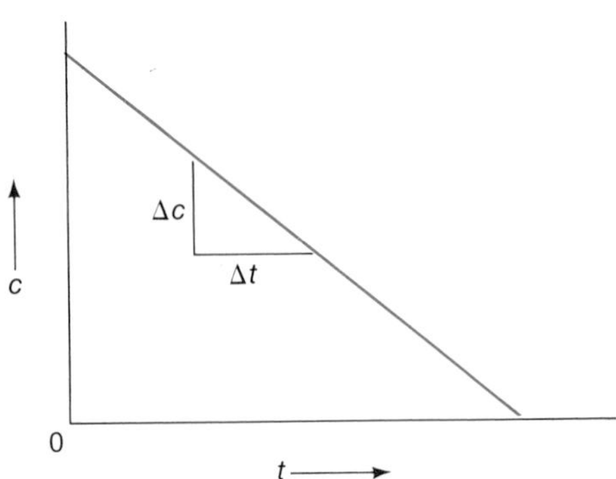

Figure 8-1. Straight-line plot of concentration versus time.

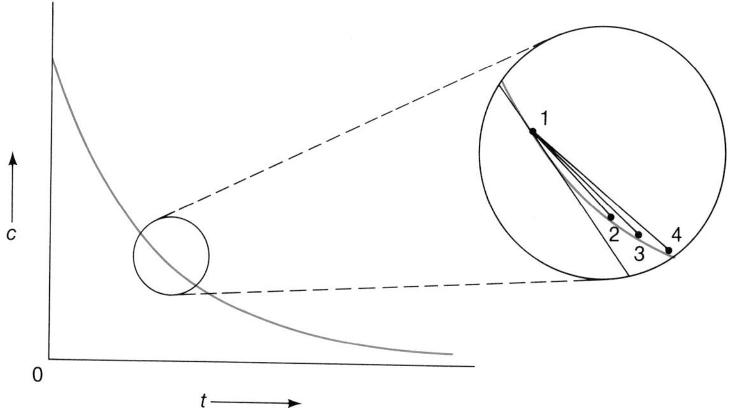

Figure 8-2. Curved-line plot of concentration versus time. The steepest line in the detail is the tangent to the curve at point 1.

However, if the concentration changes with time as shown in Figure 8-2, then the rate of reaction changes also. This is typical of many reactions.

How can such a variable rate be determined? Again, a slope is required, but its selection is not as simple as in the first case. The problem is apparent if we examine the magnified portion of Figure 8-2, in which chords are drawn through points 1 and 2, 1 and 3, and 1 and 4. The slopes of these chords become progressively smaller. Clearly, an unambiguous definition of the rate at point 1 is needed. The only way to resolve the difficulty is to use differential notation to define the rate in terms of infinitesimal increments in the concentration and the time:

$$\text{Rate} = \lim_{\Delta t \to 0} \frac{\Delta c}{\Delta t} = \frac{dc}{dt}$$

This expression gives the slope of the steepest line in Figure 8-2. It is the tangent to the curve at point 1.

Consider the following hypothetical reaction, which occurs exactly as shown:

$$\text{A} + 2\text{B} \rightarrow \text{P} \tag{8-2}$$

One molecule of A and two molecules of B collide simultaneously and coalesce to give one product molecule P. The plots of concentration versus time for the three species are shown in Figure 8-3. The problem apparent in this figure is that there are three different rates that could be used to describe

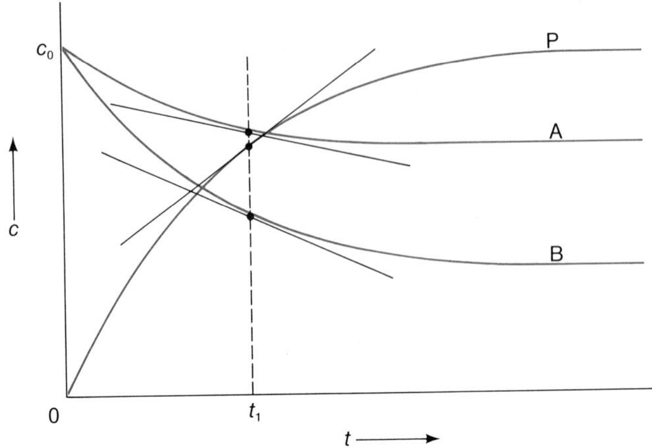

Figure 8-3. Variation of the concentrations of the species A, B, and P with time, for the reaction A + 2B → P.

the reaction at any instant. The tangents to the curves at a given time t_1 all have different slopes. The slopes of curves A and B are both negative, indicating a decrease in the concentrations of reactants with time. For the same initial concentration c_0 of A and B, the slope of curve B is twice that of curve A. The slope of curve P is positive, indicating an increase in the concentration of product with time. A standard convention is to set the rate of reaction equal to the rate of change in the concentration of the component(s) with stoichiometric coefficient equal to one. Minus signs are used for the reactants because their concentrations decrease with time; thus, all rates are positive. In the present example,

$$\text{Rate} = \frac{-d[\text{A}]}{dt} = -\frac{1}{2}\frac{d[\text{B}]}{dt} = \frac{d[\text{P}]}{dt}$$

where [A] signifies the concentration of A (usually in moles per liter), and similarly for B and P.

The product P can decompose, of course, yielding the reactants again:

$$\text{P} \rightarrow \text{A} + 2\text{B} \tag{8-3}$$

If P is present in the system, the rate measured is a *net rate*, given by

$$\text{Rate} = \text{rate}_{\text{forward}} - \text{rate}_{\text{reverse}}$$

This complicates kinetic analyses, and conditions are generally imposed that tend to minimize the effect of the reverse reaction on the rate. For example, measurements are generally made early in the course of the reaction, before significant amounts of product are formed.

We are now ready to ask what the factors are that determine the rate at which a reaction goes. For any chemical system there are three factors: the concentrations of the reactants, the temperature, and the presence of catalysts. We will discuss these three factors in turn in this chapter.

8.1 UNITS OF CONCENTRATION

We must now be more specific about units of concentration, because the numerical value of the rate of a reaction depends on the units we choose. You should become thoroughly familiar with the various units of concentration; they will be employed extensively in the remaining chapters of this book.

When dealing with gases and gas reactions, the most commonly used "concentration" unit is the partial pressure of the gas in atmospheres. The partial pressure can be used thus because it is directly proportional to the concentration in moles per liter of the gas (see Equation 6-20). For constant-total-pressure processes, mole fraction can, of course, also be used to express concentration, even though it is dimensionless.

When dealing with liquid solutions, the most commonly used concentration units are moles per liter of solution, which define the *molarity* of the component in question. (The symbol M denotes molarity.) For example, your blood contains approximately 0.15 moles of NaCl per liter. This means that the molarity of NaCl in blood is 0.15, i.e., blood is 0.15 M (0.15 molar) in NaCl. Because NaCl is completely dissociated in water (and therefore in blood, which is an aqueous solution-suspension), we can also say that blood is 0.15 M in Na^+ and 0.15 M in Cl^-.

We have already introduced mole fraction and molality as useful concentration units in dealing with solutions. These and other concentration units are summarized in Table 8-1.

Table 8-1. **Units of Concentration of a Species A.**

Concentration	Definition	Symbol	Units
Molarity	Moles of A/ℓ solution	M_A or [A]	mole/ℓ
Molality	Moles of A/kg solvent	m_A	mole/kg
Mole fraction	Moles of A/total moles	X_A	Dimensionless
Weight fraction	Wt of A/total wt	w_A	Dimensionless
Partial pressure	Pressure of A	P_A	atm

8.2 THE EFFECT OF CONCENTRATION ON RATE

The general nature of the effect of concentration on reaction rates can be easily observed in the classical "clock reaction." This is a reaction between sulfurous acid, H_2SO_3 (which has never been isolated and is formed only *in situ* in solution), and the iodate ion, IO_3^-, in the presence of starch. If 50 ml of 0.01 *M* $NaHSO_3$ (sodium bisulfite), with some added H_2SO_4 (sulfuric acid) and starch, are mixed with 50 ml of 0.005 *M* KIO_3 (potassium iodate), the colorless solution abruptly turns deep blue after about 18 s. The first reaction that occurs is the formation of I^-:

$$IO_3^- + 3H_2SO_3 \rightarrow I^- + 3SO_4^{2-} + 6H^+ \tag{8-4}$$

This is followed by two further reactions, in which I_2 is formed and immediately converted back to I^-:

$$IO_3^- + 5I^- + 6H^+ \rightarrow 3I_2 + 3H_2O \tag{8-5}$$

$$3I_2 + 3H_2SO_3 + 3H_2O \rightarrow 6I^- + 3SO_4^{2-} + 12H^+ \tag{8-6}$$

The sulfurous acid is present in limiting amounts and is eventually consumed entirely. Reaction 8-6 stops at that point and excess I_2 accumulates from Reaction 8-5. This immediately turns the solution almost ink black because of the reaction of molecular iodine (in the presence of the iodide ion) with the amylose molecules of the starch to form a helical complex that can be represented as

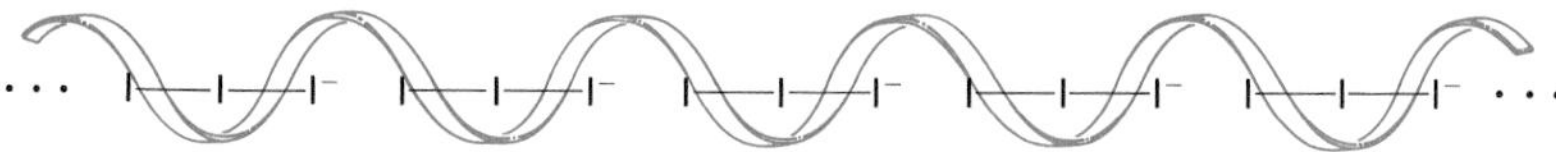

Division of the original molarity of the H_2SO_3 by the number of seconds elapsed gives an average value for the rate of the reaction. The effect of the concentrations of the reactants on this rate can be seen by diluting each of the starting solutions by a factor of two. The time required now is very close to 108 s, i.e., the rate has been reduced by a factor of twelve. It is apparent that the rate of the reaction depends strongly on the concentrations of the reactants.

A more clear-cut example is provided by the much-studied reaction between hydrogen and iodine:

$$H_2(g) + I_2(g) \rightarrow 2HI(g) \tag{8-7}$$

This reaction follows the rate expression

$$\text{Rate} = \frac{-dP_{H_2}}{dt} = kP_{H_2}P_{I_2} \tag{8-8}$$

where k is called the *rate constant* of the reaction. If the volume of the reaction vessel is halved, the pressure of each gas doubles. The rate should therefore increase by a factor of four—and it does. Similarly, doubling the number of moles of each reactant at constant volume and temperature quadruples the rate.

The reaction of hydrogen with bromine might be expected to follow a similar rate expression because of the similarity of the halogens. Surprisingly, the rate expression of the reaction

$$H_2(g) + Br_2(g) \rightarrow 2HBr(g) \tag{8-9}$$

is

$$\text{Rate} = \frac{-dP_{H_2}}{dt} = \frac{kP_{H_2}(P_{Br_2})^{1/2}}{1 + k'P_{HBr}/P_{Br_2}} \tag{8-10}$$

This result illustrates the important fact that nothing conclusive can be determined about the form of the rate expression from an examination of the overall stoichiometry of the reaction alone. Apparently, the H_2-Br_2 reaction does not proceed by the same series of steps as does the H_2-I_2 reaction.

You should develop some facility in examining the effect of changing the reactant concentrations on the rate of a reaction. The following example is a bit more complicated than the two previous ones.

Example 8-1. The reaction

$$5Br^- + BrO_3^- + 6H^+ \rightarrow 3Br_2(aq) + 3H_2O$$

proceeds according to the rate expression

$$\text{Rate} = \frac{-d[BrO_3^-]}{dt} = k[Br^-][BrO_3^-][H^+]^2$$

where the brackets always mean "the concentration of." What is the effect on this rate of: (1) increasing $[BrO_3^-]$ by a factor of two, (2) decreasing $[H^+]$ by a factor of ten, and (3) diluting a solution of the reactants at constant $[H^+]$ by a factor of two?

1. The rate is directly proportional to $[BrO_3^-]$, so it must double.
2. The rate varies as the square of $[H^+]$, so it must decrease by a factor of 100.

3. There is no effect on the rate from the $[H^+]$ term because it does not change. The rate decreases by a factor of four because $[Br^-]$ and $[BrO_3^-]$ both decrease by a factor of two, and the rate is directly proportional to each.

The net equation for a chemical reaction shows the overall stoichiometry of the reaction. None of the three reactions discussed above proceeds as shown by the net equation, however. Because the rate of the reaction between hydrogen and iodine depends linearly on the concentration of each, it seems reasonable to assume that the reaction proceeds via the mechanism of an H_2 molecule's colliding with an I_2 molecule and immediately forming two HI molecules. As we will see later in this chapter, however, this has been shown not to be true. Rarely, in fact, does a reaction actually proceed as shown by the net equation that describes it. It stretches our credulity a little too far to imagine the aqueous bromide-bromate reaction's occurring by the simultaneous collision of 12 particles! Three-body collisions are rare and four-body collisions are rarer still; to conceive of twelve bodies arriving at the same place at the same time, not to mention with the correct energies and orientations for a reaction to occur, is simply impossible.

Most chemical reactions proceed via a series of two or more simple reactions in which two (or sometimes three) particles collide, react, and separate as products. These individual steps are called *elementary processes.* The entire series of elementary processes that yields the net reaction is called the *mechanism of the reaction.*

A *general rate law* that describes the kinetics of every elementary process has been found empirically. It states that the rate of an elementary process is proportional to the concentration of each *reactant* raised to its stoichiometric coefficient. The general rate law thus provides the rate equation for the generalized reaction

$$aA + bB \rightarrow \text{Products} \tag{8-11}$$

$$\text{Rate} = -\frac{1}{a}\frac{d[A]}{dt} = -\frac{1}{b}\frac{d[B]}{dt} = k[A]^a[B]^b \tag{8-12}$$

Note that the rate equation gives the rate of the forward reaction only; for that reaction the rate depends strictly on the concentrations of the *reactants.* As products are formed and the reverse reaction begins to occur at an appreciable rate, the net rate is affected by the concentrations of the products as well. But regardless of what the products and their concentrations might be, the rate of the reaction always depends solely on the concentrations of the reactants *for the direction under consideration.*

The superscripts a and b denote the *order* of the reaction with respect to A and B, respectively. For example, if $a = 1$, the reaction is said to be *first-order* in A. Note that the order of the reaction with respect to a given species corresponds to the stoichiometric coefficient of that species in the elementary process only, and may or may not correspond to the stoichiometric coefficient in the net equation.

The order of a reaction with respect to a particular reactant can generally be determined as follows. The concentrations of all other reactants are held constant and the rate is measured as a function of the concentration of the reactant in question. For instance, if the concentration of B in Reaction 8-11 is held constant and the rate measured as a function of [A], the value of the exponent a can be determined. Thus, if the rate quadruples when [A] is doubled, we conclude that $a = 2$.

A more rigorous way to determine the exponent a is to plot the rate data. If we take logarithms of Equation 8-12, we can write

$$\begin{aligned} \log \text{rate} &= \log k + \log[\text{A}]^a + \log[\text{B}]^b \\ &= a \log[\text{A}] + \log c \end{aligned} \tag{8-13}$$

where $c = k[\text{B}]^b$ is a constant as long as [B] does not vary. Equation 8-13 shows that a log-log plot of rate versus [A] should give a straight line of slope a. This is a good way to analyze data, especially when there is a fair amount of scatter in the data.

The sum of the exponents in the general rate law ($a + b$ in the example above) is called the *order of the reaction*. This is the quantity we will refer to subsequently when we discuss zero-, first-, and second-order reactions, for which $a + b + \cdots$ equals 0, 1, and 2, respectively.

Another useful concept in describing kinetic behavior is the *molecularity* of elementary processes. (This concept applies *only* to such processes, not to overall reactions that consist of a sequence of them.) The molecularity of an elementary process is equal to the number of reactant particles (atoms, molecules, or ions). Thus an elementary process in which two particles collide and react is called a *bimolecular reaction*. *Termolecular* reactions are those in which three particles collide and react. Note carefully that the molecularity and the order of an elementary process are identical. However, it is often *not* true that the molecularity of an elementary process is identical to the order of the overall reaction.

The proportionality constant k in a rate expression is called the rate constant, as we mentioned earlier. Its value obviously depends on the chemical reaction under consideration. For a given reaction, however, the value of k depends only on the temperature, as does that of the equilibrium constant K. We will soon see that these two quantities, k and K, are directly related; we will also see exactly how they depend on T.

It should become evident from the amount of attention we will devote to a description of rate constants that they are very important in kinetic studies. They are often used to describe the rate of a reaction, almost without reference to the concentration terms.

8.3 THE MECHANISM OF A REACTION

The effect of concentration on the rate of a reaction and the order of a reaction are the principal tools employed in the study of reaction mechanisms. As mentioned earlier, reaction mechanisms are one of the primary goals of kinetic investigations.

The kinetics of the oxidation of Br^- by hydrogen peroxide has been thoroughly studied. The net reaction is

$$H_2O_2 + 2Br^- + 2H^+ \rightarrow 2H_2O + Br_2 \qquad (8\text{-}14)$$

If the net reaction were also the elementary process, the rate expression would be

$$\text{Rate}_{\text{proposed}} = \frac{-d[H_2O_2]}{dt} = k_1[H_2O_2][Br^-]^2[H^+]^2$$

The mechanism implied by such an expression would require a five-body collision: the reaction would be fifth-order. No such reactions have been found, however. So we turn to the experimentally determined rate expression,

$$\text{Rate}_{\text{actual}} = \frac{-d[H_2O_2]}{dt} = k_2[H_2O_2][Br^-][H^+]$$

which indicates that the reaction does not proceed as shown by Reaction 8-14. Rather, it proceeds via a series of elementary processes, of which the slowest is the reaction of an H_2O_2 molecule with a Br^- ion and an H^+ ion:

$$H_2O_2 + Br^- + H^+ \xrightarrow{k_2} H_2O + HBrO \qquad \text{(Slow)} \qquad (8\text{-}15)$$

In order to obtain the net reaction, we must now add one or more reactions to Reaction 8-15. This means that one more each of H^+ and Br^-, as well as the intermediate HBrO, must be consumed, and a Br_2 and another H_2O produced. These requirements can all be satisfied in one plausible equation (which must represent a relatively fast reaction, since it is known that the overall rate is determined by Reaction 8-15):

$$H^+ + Br^- + HBrO \xrightarrow{k_3} H_2O + Br_2 \qquad \text{(Fast)} \qquad (8\text{-}16)$$

It is always advisable to add up the entire series of reactions to be certain that the net equation is obtained:

$$H_2O_2 + Br^- + H^+ \xrightarrow{k_2} H_2O + \cancel{HBrO} \qquad \text{(Slow)}$$

$$H^+ + Br^- + \cancel{HBrO} \xrightarrow{k_3} H_2O + Br_2 \qquad \text{(Fast)}$$

$$H_2O_2 + 2Br^- + 2H^+ \xrightarrow{k_2} 2H_2O + Br_2 \qquad \text{(Slow)}$$

A symbolic representation of the postulated mechanism of this reaction has been proposed by Jürg Waser (b. 1916), and is shown in Figure 8-4. The reactants for the first reaction (which are the same as those for the net reaction) are introduced into the upper bulb. The relatively low rate at which

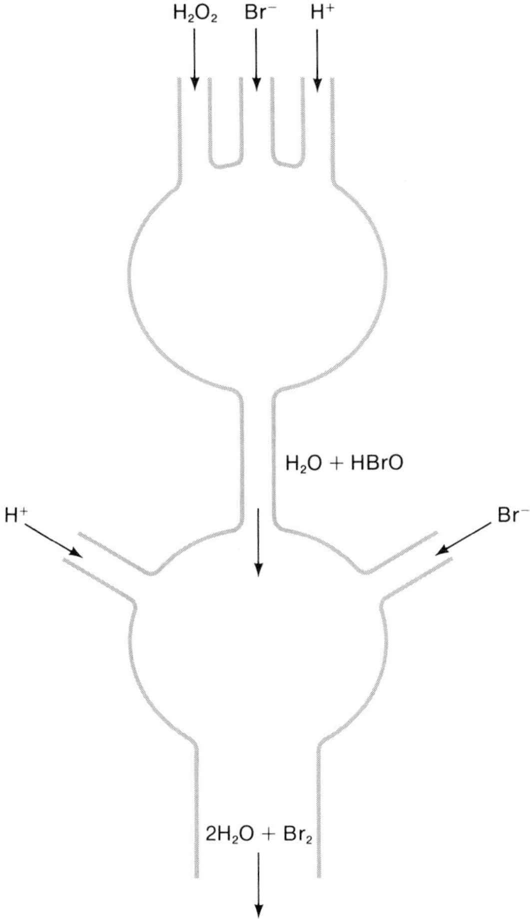

Figure 8-4. Mechanical analogue of the mechanism of the net reaction $H_2O_2 + 2Br^- + 2H^+ \rightarrow 2H_2O + Br_2$. (J. Waser, *General Chemistry*, unpublished.)

the products of this reaction become available for further reactions is represented symbolically by the narrow diameter of the exit tube of the bulb. The much greater rate of the second reaction is represented by the wider diameter of the exit tube of the lower bulb. Clearly, the overall reaction can go no faster than the first reaction, which acts like a kinetic bottleneck. Similar conceptually useful mechanical analogues can be constructed (at least in our minds) for all reaction mechanisms.

When the slowest elementary process in a reaction mechanism has been determined, its rate expression can be written. If this limiting rate is much smaller than the rates of all the other elementary processes in the mechanism (as is usually true), then it is, in effect, the observed overall rate of the reaction. The elementary process in question is then called the *rate-determining step*. The concept of a rate-determining step is easy to understand via a simple analogy: The rate at which a telegram is transmitted from the sender to the recipient is, for all practical purposes, the rate at which the delivery boy rides his bicycle.

One further example will help to illustrate how reaction mechanisms are deduced from kinetic information. An extremely important reaction occurs in the ozonosphere (an atmospheric region that is essentially the same as the stratosphere, about 10–50 km from the Earth's surface), namely, the photochemical conversion of ozone to oxygen:*

$$2O_3(g) \xrightarrow{h\nu} 3O_2(g) \tag{8-17}$$

This reaction is vitally important because the radiation absorbed is solar ultraviolet radiation in the wavelength range 2200–3000 Å. If a great deal of this radiation reached the Earth's surface, it would seriously damage most plant and animal life through genetic mutation or outright destruction. (Ultraviolet radiation of even shorter wavelengths is absorbed in the 40–50-km altitude range by oxygen molecules; the resulting photochemical reaction is what forms the ozone in the first place.)

As before, we look at the net reaction and write the rate expression that it implies:

$$\text{Rate}_{\text{proposed}} = -\frac{1}{2}\frac{dP_{O_3}}{dt} = k(P_{O_3})^2 \tag{8-18}$$

Oddly enough, the actual rate is

*It is ironic that this reaction protects life on Earth, whereas a different photochemical reaction involving ozone, but also hydrocarbons and oxides of nitrogen, imperils life. The latter is the photochemical smog reaction discussed in the Interlude following Chapter 3.

$$\text{Rate}_{\text{actual}} = \frac{k(P_{O_3})^2}{P_{O_2}} \tag{8-19}$$

Not only does this rate expression show that the reaction is of order −1 in O_2, but the O_2 is not even a reactant for the direction we are considering!

The difficulty is rather nicely resolved by the following mechanism, which is presently accepted as correct:

$$O_3(g) \underset{k_{-1}}{\overset{k_1}{\rightleftharpoons}} O_2(g) + \cancel{O(g)} \qquad \text{(Fast equilibrium)} \tag{8-20}$$

$$\cancel{O(g)} + O_3(g) \xrightarrow{k_2} 2O_2(g) \qquad \text{(Slow)} \tag{8-21}$$

$$2O_3(g) \xrightarrow{k} 3O_2(g) \qquad \text{(Slow)}$$

Note that positive subscripts on the rate constants are generally employed for forward reactions (i.e., those proceeding from left to right) and negative subscripts for the reverse reactions. "Fast equilibrium" means that both the forward and reverse reactions are fast.

As usual, we equate the overall rate of the reaction (Equation 8-19) with the rate of the slowest step:

$$\text{Rate} = -\frac{1}{2}\frac{dP_{O_3}}{dt} = k_2 P_{O_3} P_O \tag{8-22}$$

where $k_2 \neq k$. Note that we retain the factor 1/2 in the rate expression because every time Reaction 8-21 occurs, a second molecule of O_3 is consumed in Reaction 8-20. (Thus, the rate of decrease of O_3 due to Reaction 8-21 represents only half the total rate of decrease of O_3.) This rate expression is unsatisfactory on two counts: It contains the partial pressure of an intermediate species, O, that does not appear in the net reaction, and it is not the same as the actual rate expression.

Both problems can be resolved when we realize that Reaction 8-20 is an *equilibrium* reaction. This means that we can write an equilibrium-constant expression for it:

$$K = \frac{P_{O_2} P_O}{P_{O_3}} \tag{8-23}$$

which, upon rearrangement, gives

$$P_O = \frac{K P_{O_3}}{P_{O_2}} \tag{8-24}$$

Substitution of this relation into Equation 8-22 yields the desired expression:

$$\text{Rate} = k_2 P_{O_3} P_O = k_2 P_{O_3} \frac{K P_{O_3}}{P_{O_2}} = \frac{k_2 K (P_{O_3})^2}{P_{O_2}} = \frac{k (P_{O_3})^2}{P_{O_2}} \tag{8-25}$$

This rate expression has the proper pressure dependences for O_3 and O_2. Note, however, that the *apparent* rate constant k is just that: it is actually the product of a rate constant and an equilibrium constant.

It may seem strange that what began as a photochemical reaction has ended with a rate expression that does not contain a term for the intensity of the radiation. The radiation stimulates Reaction 8-20 in both directions so that the equilibrium concentrations are reached sooner, but their relative concentrations, as expressed by the equilibrium constant, are unaffected.

Exercise 8-1. Determine the units of k_2 and K and show that their product is equal to the units of k.

We have now seen how the experimental rate expressions have been instrumental in determining the reaction mechanisms for two very different reactions. It must be remembered that, although it is a *necessary* condition that the mechanism be consistent with the rate expression, this is not a *sufficient* condition to prove that the proposed mechanism is the *correct* one. Several alternative mechanisms can generally be proposed to account for a given rate expression. Sometimes energy considerations or comparisons with similar reactions of known mechanism are sufficient to indicate the correct mechanism. The only unambiguous test is to identify experimentally one or more of the transitory reaction intermediates while the reaction is in progress.

8.4 EXPERIMENTAL METHODS

A wondrous variety of experimental methods has been developed for the study of chemical kinetics. The problems encountered, particularly those posed by very fast reactions in solution and in the gaseous and plasma states, tax the chemist's ingenuity to the utmost, and have led to the invention of many remarkably sophisticated instruments.

The experimental parameters that must be determined are temperature, time, and concentration. A wide variety of temperature-measuring devices (thermometers, thermocouples, thermistors, etc.) is available for determining the temperature accurately, and a similar variety of constant-temperature baths is available for precise control of the temperature. The time can be measured with almost any precision desired, ranging from a few tenths of a second with a stopwatch down to about one picosecond (10^{-12} s) with very sophisticated spectroscopic and electronic instruments. The biggest problem is the concentration. The methods used to measure concentration as a function of time can be grouped in three broad categories. We will discuss each of these very briefly and list a few of the important methods in each category.

Quenching Methods

Quenching methods are based on the withdrawal of a sample at time *t* and the immediate stopping of the reaction by some means. The concentration of the substance of interest can then be determined at leisure. These methods are not used as extensively as those discussed in the following two sections, because they are more difficult. Some of the methods of quenching are:

1. Lowering the temperature. If the reaction is sufficiently temperature-dependent, plunging the sample into an ice or liquid-nitrogen bath completely arrests ("freezes") the reaction.
2. Removing the radiation source. Some reactions, such as the $H_2 + Cl_2$ reaction and the O_3 decomposition, depend on radiation for their progress. The reaction can be stopped by cutting off the source of photons of the necessary wavelength(s).
3. Adding a quenching agent. In some cases a specific compound can be found that, when added to the sample, immediately stops the reaction without altering the concentration of the substance of interest (aside from the dilution effect from adding the compound).

Continuous-Monitoring Methods

Continuous monitoring is more widely used than quenching because it is often more accurate, continuous changes can be recorded, and automatic instruments can do almost all the work. Some property of a reactant or product is selected, the reaction mixture is placed in the instrument at a controlled temperature, and the property is monitored as it decreases or increases with time. Some of the continuous-monitoring methods are:

1. Absorption of radiation. Commonly, one of the reactants or products absorbs radiation at a particular wavelength at which the other species

do not absorb. Some examples are the biologically important molecule diphosphopyridine nucleotide, DPN, which absorbs at 340 nm, and the tri-iodide ion, I_3^-, which absorbs at both ends of the visible spectrum. This method is probably the most widely used in kinetic studies.

2. Optical rotation. Provided that one species in the reaction mixture is optically active (rotates the plane of polarized light to the left or right, as discussed in Chapter 5), the measurement of the optical rotation of the solution is a convenient measure of the progress of the reaction.
3. Gas pressure. This method is useful in studying gas-phase reactions, provided that the reaction entails a change in the number of moles.
4. Dilatometry. A number of reactions that occur in solution result in a change in the volume of the solution. The most important example is the change occurring upon the ionization of acids and bases. A dilatometer is a simple, λ-shaped device that permits the two solutions to be mixed at the junction of the lower arms; the volume change is measured in the upper arm, which is a graduated capillary.
5. Conductivity. The conductivity of a solution is a measure of the number of charged particles present per unit volume. Provided that this number changes with time, this is a convenient way to study the reaction.

Fast-Reaction Methods

Fast-reaction methods embody many of the aspects of continuous monitoring but, because the time scale is so compressed, special techniques have had to be developed to conduct these studies. If the reaction is essentially complete in 100 ns, one simply cannot pipet 1 ml of the reaction mixture and quench it. The two major types of fast-reaction methods are discussed below.

Flow Methods

Figure 8-5 shows an apparatus suitable for measuring reactions that go to completion in a time of the order of a millisecond. The application of pressure to both syringe plungers simultaneously results in the solutions' mixing in the reaction chamber and flowing down the tube through the spectrophotometer, which is adjusted to a characteristic absorption wavelength of the species of interest. For a given amount of pressure on the plungers, the flow rate is a constant and the spectrophotometer analyzes the solution at a constant time after mixing has occurred. A series of times and concentrations can thus be recorded by varying the flow rate.

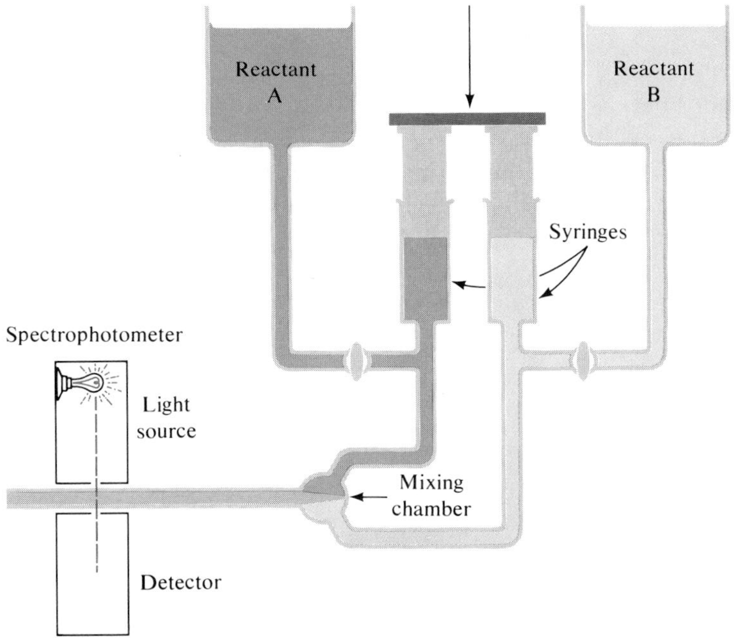

Figure 8-5. Stopped-flow apparatus for measuring the kinetics of very fast reactions. (E. L. King, *How Chemical Reactions Occur,* W. A. Benjamin, New York, 1964.)

Even faster reactions can be studied by the "stopped-flow" technique. Approximately the same apparatus is employed, except that the mixing chamber (now the reaction chamber) is sealed on the downstream side. A short stroke of the plungers injects the reactants into the chamber and the spectrophotometer monitors the reaction *in situ.* Much shorter reaction times, down to the nanosecond range, can be measured by this technique. It is presently being used extensively for the study of very fast, enzyme-catalyzed reactions.

Relaxation Methods

Some very fast reactions can be followed as the system "relaxes" following a drastic, sudden change in temperature, pressure, or some other property of the system. A temperature-jump apparatus utilizes a charged capacitor to send a tremendous burst of electricity through a solution in perhaps a few microseconds, resulting in a 5 or 10°C rise in temperature. In a very different technique, a shock tube is used to alter the temperature and pressure of a reactive system very rapidly. It consists of a long tube with a driving gas (usually an inert gas) at some convenient pressure in one end, and the gas or gas mixture of interest at a much lower pressure in the other end, the two ends being separated by a diaphragm. Rupture of the diaphragm sends the driving gas rushing down the tube at supersonic speed, creating a shock wave in which the temperature can be many thousands—even tens of thousands—of kelvins. In both the temperature-jump and shock-tube experiments, some spectroscopic method is generally employed to follow the reaction as the system relaxes to an equilibrium state from the highly nonequilibrium state created by the very rapid rise in temperature.

8.5 COMPARISON OF EXPERIMENTAL DATA WITH THEORETICAL RATE EQUATIONS

A problem has surreptitiously worked its way into our study of kinetics. On the one hand, we have stressed the importance of rate equations as a tool for the investigation of reaction mechanisms, and have also stressed that these equations are almost always determined by experiment. On the other hand, we have just concluded a brief discussion of some of the methods employed to examine kinetic phenomena experimentally. All these methods, however, merely yield the concentration as a function of time, not a rate equation. The dichotomy is apparent when we compare the following equations for a first-order reaction:

$$\text{General rate law: } -\frac{dc}{dt} = kc \tag{8-26}$$

$$\text{Experimental data: } c = c(t) \tag{8-27}$$

The two equations must be compared—the question is how it can best be done. To go from Equation 8-27 to Equation 8-26 would require one of two approaches. The data could be plotted and the slopes, dc/dt, could be obtained manually. This is both cumbersome and potentially inaccurate. Alternatively, the exact analytical dependence of c on t could be determined and that expression differentiated. This is difficult to do in practice.

The most accurate and efficient way to compare the rate equations with the experimental data is to integrate the rate equations. By now you should have some practice in handling simple integrations, and we will restrict ourselves to two simple cases that you are already familiar with. A bonus from the mathematical excursion below is that we will discover a powerful tool for determining the order of a reaction with respect to a particular reactant.

Zero-Order Reactions

Zero-order reactions are described by the following rate law:

$$\text{Rate} = -\frac{dc}{dt} = k \tag{8-28}$$

Such equations are important in heterogeneous (more than one phase) reactions, such as the decomposition of gaseous HI on a gold surface. Our main interest in this chapter is in homogeneous (one phase) reactions, however. Although zero-order reactions never occur in homogeneous solutions, many reactions are zero-order with respect to a particular reactant—say, A. If, when the concentrations of all the other reactants are held constant as [A] is varied, the rate is found to be constant, the reaction is zero-order in A. (Thus, [A] is not part of the rate equation, since $[A]^0 = 1$.) As mentioned earlier, this is a common method for determining the order of a reaction with respect to a particular reactant.

Integration* of Equation 8-28 yields

$$c = -kt + c_0 \tag{8-29}$$

where c_0 is the concentration at time $t = 0$. In each of the three cases that we will study, we will obtain such an equation for $c = c(t)$. In the first two cases, we will plot this equation to see how the concentration changes with time, as is done for the zero-order reaction in Figure 8-6. Here the concentration decreases linearly with time, eventually going to zero. The slope of the line is the negative of the rate constant, $-k$, and the intercept is the initial concentration c_0.

An important concept in kinetics is indicated in Figure 8-6. The time $\tau_{1/2}$, which is the time required for the concentration to fall to one-half its initial value, is called the *half-life* of the reaction. We will see that the measurement of $\tau_{1/2}$ for several values of c_0 is very useful for determining the order of a reaction, or the order of a reaction with respect to a given reactant.

In order to use $\tau_{1/2}$ in this way, we need to find the relation between $\tau_{1/2}$ and c_0. If we substitute $c = c_0/2$ and $t = \tau_{1/2}$ into Equation 8-29, we immediately have the relation we seek:

$$c_0/2 = -k\tau_{1/2} + c_0 \tag{8-30}$$

$$\tau_{1/2} = \frac{c_0}{2k} \tag{8-31}$$

---

*A brief discussion of how this integration is done is given in Appendix E for those of you who have taken, or are taking, a course in calculus.

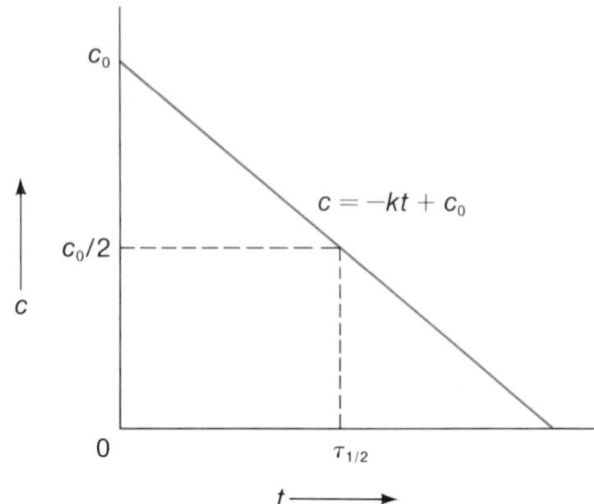

Figure 8-6. Plot of concentration versus time for zero-order kinetics. The relation between the concentration and the half-life of the reaction is shown.

The half-life is directly proportional to the initial concentration. Thus, a very good test for a zero-order reaction is to measure $\tau_{1/2}$ for several different initial concentrations. If there is a linear relation between the two quantities, the reaction is zero-order.

Exercise 8-2. Verify Equation 8-31 graphically by constructing several plots of c vs t for several values of c_0 but the same straight slope of $-k$, and verify that $\tau_{1/2}$ varies in the same way that c_0 does.

First-Order Reactions

First-order reactions are very common. They follow the rate law

$$\text{Rate} = -\frac{dc}{dt} = kc \tag{8-32}$$

Integration of this equation yields

$$-\ln\frac{c}{c_0} = kt \tag{8-33}$$

We should examine two different forms of Equation 8-33 because together they will help us understand first-order kinetics better than either one separately. Equation 8-33 can be easily rearranged to give the following two equations, which are plotted in Figures 8-7 and 8-8, respectively:

$$c = c_0 e^{-kt} \tag{8-34}$$

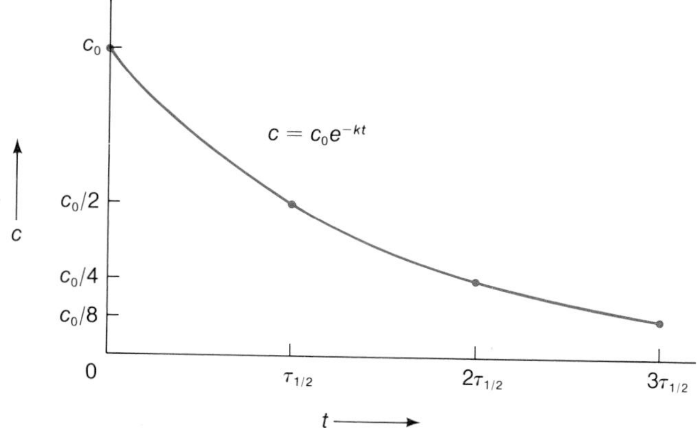

Figure 8-7. Plot of concentration versus time for first-order kinetics. The fact that the half-life of the reaction is independent of initial concentration can be determined by inspection.

$$\log c = \frac{-k}{2.303}t + \log c_0 \tag{8-35}$$

Figure 8-7 shows the exponential decay of concentration with time in a first-order reaction. Figure 8-8 shows that a plot of log c vs t yields a straight line of slope $-k/2.303$ and intercept log c_0. Figure 8-7 also shows the behavior of $\tau_{1/2}$ for such a reaction. We see that the times required for the concentration to fall from c_0 to $c_0/2$, from $c_0/2$ to $c_0/4$, and from $c_0/4$ to $c_0/8$ are

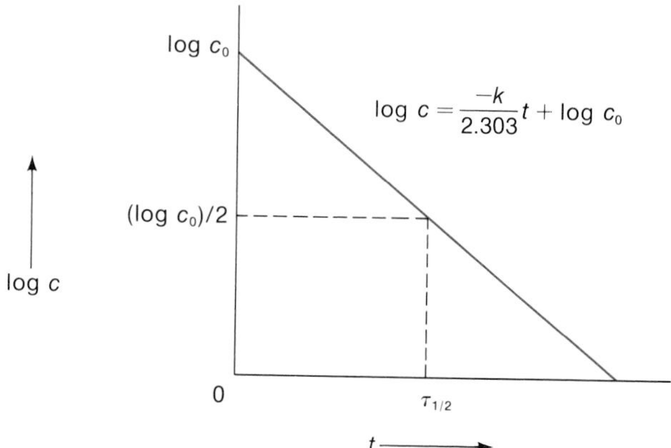

Figure 8-8. Logarithmic concentration plot for a first-order reaction.

all the same and are equal to $\tau_{1/2}$. This demonstrates that $\tau_{1/2}$ is independent of c_0 for a first-order reaction. The actual expression is

$$\tau_{1/2} = \frac{0.693}{k} \tag{8-36}$$

Exercise 8-3. Verify Equation 8-36 by making the appropriate substitutions into Equation 8-35.

It is worthwhile for us to compare these two types of kinetic behavior—zero-order and first-order—for the decomposition of dinitrogen pentoxide:

$$2N_2O_5 \xrightarrow{CCl_4} 4NO_2 + O_2(g) \tag{8-37}$$

(The symbol CCl_4 above the arrow denotes the solvent in which the reaction occurs.) The oxygen liberated is not soluble in the carbon tetrachloride, and a continuous monitoring of the pressure above the solution, as described earlier, provides a convenient means of following the course of the reaction.

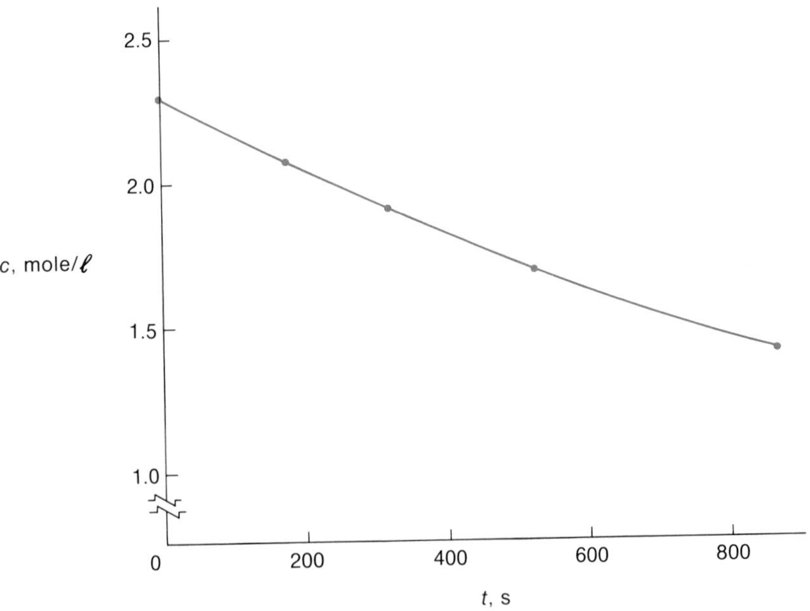

Figure 8-9. Plot of concentration versus time for the decomposition of N_2O_5 at 318 K.

We are interested in determining the order of the reaction with respect to N_2O_5.

As illustrated above, it is useful to make two plots, one of c vs t and one of log c vs t. This is done in Figures 8-9 and 8-10, respectively. If the former plot is a straight line, the reaction is zero-order; if the latter plot is a straight line, the reaction is first-order. Of course, if neither plot is straight, the reaction follows some other kinetics.

The data are good enough to show that the reaction is first-order. The c-vs-t plot is very slightly curved and the log c-vs-t plot is linear. The obvious visual point, however, is that the curves are really not very dissimilar, and only a small amount of scatter in the data would be needed to make impossible the distinction we have just made.

However, the order of the reaction can be established easily and rigorously by means of the half-life concept. Extrapolation of the curve in Figure 8-9 to the value $c = 2.3/2 = 1.2$ mole/ℓ gives a half-life of 1100 s. If the reaction were studied until $c = 1.2/2 = 0.60$ mole/ℓ, another 1100 s would be required, because this is a first-order reaction and $\tau_{1/2}$ is independent of c_0. If the reaction were zero-order and the half-life for $c_0 = 2.3$ mole/ℓ were 1100 s, the half-life for $c_0 = 1.2$ mole/ℓ would be $1100/2 = 550$ s. Thus, an

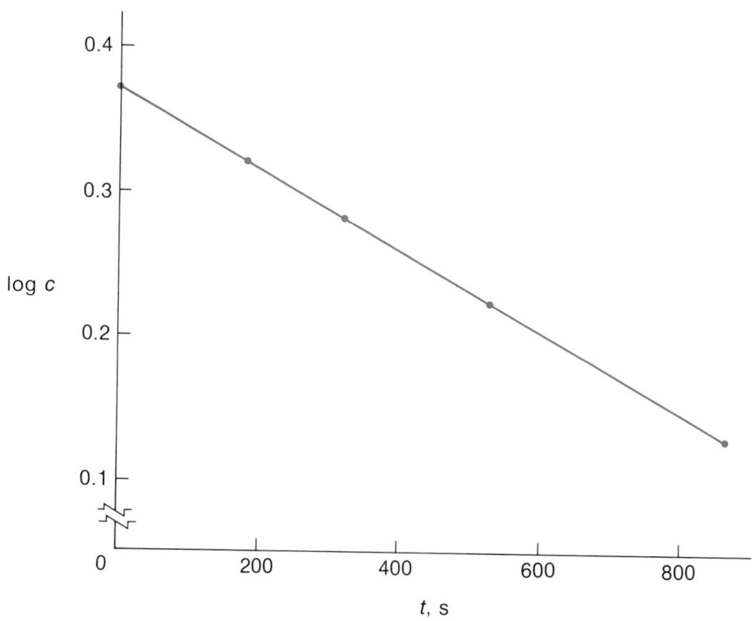

Figure 8-10. Logarithmic concentration plot for the decomposition of N_2O_5 at 318 K.

Table 8-2. **Half-Lives of Some Radioactive Isotopes.**

Isotope	Decay Process	Half-Life
^{32}P	$^{32}P \longrightarrow {}^{32}S + \beta^-$	14.3 days
^{60}Co	$^{60}Co \rightarrow {}^{60}Ni + \beta^-$	5.26 years
^{3}H	$^{3}H \longrightarrow {}^{3}He + \beta^-$	12.26 years
^{90}Sr	$^{90}Sr \rightarrow {}^{90}Y + \beta^-$	28.1 years
^{14}C	$^{14}C \longrightarrow {}^{14}N + \beta^-$	5730 years
^{238}U	$^{238}U \rightarrow {}^{234}Th + \alpha$	4.51×10^9 years

experimental uncertainty of even 25% would still allow the two cases to be distinguished clearly. In certain cases of enzymatic and heterogeneous catalysis, the half-life method is the only reasonable way to determine the order of the reaction, because of the experimental uncertainties.

An important class of first-order reactions is characterized predominantly by the half-life concept, namely, radioactive decay reactions. The rate of spontaneous emission of helium nuclei (α particles), electrons (β^-), or positrons (β^+) is governed solely by the amount of the substance present. Thus, these processes obey Equation 8-32 and are characterized by half-lives given by Equation 8-36. The half-life of a radioactive isotope is especially important because it indicates the length of time that the radiation will be emitted at appreciable intensities. This knowledge is necessary for the constructive use of such substances in radioisotope tracer experiments or for their destructive use in nuclear weapons. Table 8-2 lists the half-lives of a number of radioisotopes and their modes of decay.

Second-Order Reactions

Second-order reactions follow the rate law

$$\text{Rate} = -\frac{dc}{dt} = kc^2 \tag{8-38}$$

Integration of this equation yields

$$\frac{1}{c} - \frac{1}{c_0} = kt \tag{8-39}$$

Thus, for a second-order reaction, a plot of $1/c$ vs t is a straight line of slope k and intercept $1/c_0$. The half-life expression for a second-order reaction is

$$\tau_{1/2} = \frac{1}{c_0 k} \tag{8-40}$$

We thus have the gratifying result that the dependence of the half-life on the initial concentration of reactant is different in the three classes of reactions. In zero-order reactions, $\tau_{1/2}$ is directly proportional to c_0; in first-order reactions, $\tau_{1/2}$ is independent of c_0; and in second-order reactions, $\tau_{1/2}$ is inversely proportional to c_0. Even with rather erratic data, the order of the reaction can be determined readily by measuring $\tau_{1/2}$ for several values of c_0. This method is clearly superior to making plots of c, $\log c$, and $1/c$ vs t and trying to decide which one is linear.

8.6 RELATION OF KINETICS TO EQUILIBRIUM

Your first reaction to the heading of this section may be that it is paradoxical. However, we recall from our discussions of equilibrium phenomena that equilibrium is not a static condition but a dynamic process with chemical reactions occurring in both the forward and reverse directions simultaneously. The special feature of *equilibrium* is that these reactions proceed at the *same rate*. Also, Figure 8-3 suggests that kinetic phenomena are simply processes occurring as the system approaches equilibrium. We will now show how the two parameters that quantitatively describe each case—the rate constant k and the equilibrium constant K—are related.

If a chemical reaction occurs by the following, single elementary process,

$$\mathrm{A} + 2\mathrm{B} \underset{k_r}{\overset{k_f}{\rightleftharpoons}} \mathrm{C} + 3\mathrm{D} \tag{8-41}$$

then the rate of the reaction is given by

$$\text{Rate} = \frac{d[\mathrm{C}]}{dt} = k_f[\mathrm{A}][\mathrm{B}]^2 - k_r[\mathrm{C}][\mathrm{D}]^3 \tag{8-42}$$

If this same system is at equilibrium, the concentrations of all species, including C, must not change with time. We can set Equation 8-42 equal to zero for this case:

$$k_f[\mathrm{A}][\mathrm{B}]^2 - k_r[\mathrm{C}][\mathrm{D}]^3 = 0 \tag{8-43}$$

or

$$k_f[A][B]^2 = k_r[C][D]^3 \tag{8-44}$$

The *principle of microscopic reversibility* states that all elementary processes occur in both directions, the rate of the forward reaction being equal to the rate of the reverse reaction in any system at equilibrium. This applies to each step in a series of elementary processes (a reaction mechanism).

We can now rearrange Equation 8-44 to yield the following two equations:

$$\frac{k_f}{k_r} = \frac{[C][D]^3}{[A][B]^2} \qquad \frac{k_r}{k_f} = \frac{[A][B]^2}{[C][D]^3} \tag{8-45}$$

Both sets of fractions are constant, of course, for this chemical system at a given temperature. The equation on the left should look familiar because it is the same as the equilibrium constant K for this reaction. We can thus relate K directly to the rate constants of the reaction:

$$K = \frac{k_f}{k_r} = \frac{[C][D]^3}{[A][B]^2} \tag{8-46}$$

The same method for establishing the relation between the equilibrium constant and the rate constants can be applied to cases in which the reaction consists of more than one elementary process. This is demonstrated in the following example.

Example 8-2. Show that an expression identical to that for the equilibrium constant results from the application of the principle of microscopic reversibility to the reaction

$$A + B \rightarrow 2C \tag{8-47}$$

which is known to occur via the following mechanism:

$$A \underset{k_{-1}}{\overset{k_1}{\rightleftharpoons}} C + D \tag{8-48}$$

$$B + D \underset{k_{-2}}{\overset{k_2}{\rightleftharpoons}} C \tag{8-49}$$

$$\overline{A + B \longrightarrow 2C}$$

The principle of microscopic reversibility states that at equilibrium the rates of the forward and reverse reactions are the same for each of these steps:

$$k_1[A] = k_{-1}[C][D] \tag{8-50}$$

$$k_2[B][D] = k_{-2}[C] \tag{8-51}$$

Multiplication of the two equations yields

$$k_1k_2[\mathrm{A}][\mathrm{B}][\varnothing] = k_{-1}k_{-2}[\mathrm{C}]^2[\varnothing] \tag{8-52}$$

or

$$K = \frac{k_1k_2}{k_{-1}k_{-2}} = \frac{k_f}{k_r} = \frac{[\mathrm{C}]^2}{[\mathrm{A}][\mathrm{B}]} \tag{8-53}$$

Another approach to the relation between the equilibrium constant and the rate constants is demonstrated in the following example.

Example 8-3. The reaction

$$3\mathrm{HNO_2} \underset{k_r}{\overset{k_f}{\rightleftharpoons}} \mathrm{H^+} + \mathrm{NO_3^-} + 2\mathrm{NO}\,(g) + \mathrm{H_2O} \tag{8-54}$$

very probably does not proceed in either direction by the process shown, because of the large number of species involved. In fact, a satisfactory mechanism for this reaction has not yet been proposed. However, the experimental rate expression is available:

$$\frac{d[\mathrm{NO_3^-}]}{dt} = k_f\frac{[\mathrm{HNO_2}]^4}{(P_{\mathrm{NO}})^2} - k_r[\mathrm{H^+}][\mathrm{NO_3^-}][\mathrm{HNO_2}] \tag{8-55}$$

This proves that a different mechanism is operative. At equilibrium there can be no net production of the nitrate ion. We therefore set the rate equal to zero and rearrange the terms to obtain

$$K = \frac{k_f}{k_r} = \frac{[\mathrm{H^+}][\mathrm{NO_3^-}][\mathrm{HNO_2}]}{[\mathrm{HNO_2}]^4/(P_{\mathrm{NO}})^2} = \frac{[\mathrm{H^+}][\mathrm{NO_3^-}](P_{\mathrm{NO}})^2}{[\mathrm{HNO_2}]^3} \tag{8-56}$$

Again, K is equal to the ratio of the rate constants for the forward and reverse reactions.

The derivations of Equations 8-46, 8-53, and 8-56 actually constitute a demonstration of the existence of an equilibrium constant whose value for a given reaction ultimately depends only on the absolute temperature. These equations show that K depends only on the several values of k, and we know that the latter depend only on T. Thus, we now have three approaches to the existence of an equilibrium constant. It can be demonstrated experimentally, it can be derived rigorously from the laws of thermodynamics, and it can be derived from the established principles of chemical kinetics. The thermodynamic approach is the most powerful and rigorous, but it helps our understanding of the meaning of K if we are familiar with the other approaches as well.

We now turn our attention to the statement just made about the dependence of k on T. We seek the form of this dependence and ways in which we can use the variation of k with T in order to understand more about the reactions we are studying.

8.7 THE EFFECT OF TEMPERATURE ON RATE

We will begin by taking the same approach as that used at the end of the last section, namely, we will examine a thermodynamic relation in order to predict a kinetic one.

Two of the important results of our study of thermodynamics in Chapter 7 are expressions for ΔG°, the change in the Gibbs free energy of a system undergoing a reaction in which the concentrations of all species remain at 1 atm or 1 mole/ℓ:

$$\Delta G^\circ = -RT \ln K \tag{8-57}$$

and

$$\Delta G^\circ = \Delta H^\circ - T\Delta S^\circ \tag{8-58}$$

These expressions can therefore be equated and rearranged to give the dependence of K on T:

$$RT \ln K = -\Delta H^\circ + T\Delta S^\circ \tag{8-59}$$

and

$$\ln K = -\frac{\Delta H^\circ}{RT} + \frac{\Delta S^\circ}{R} \tag{8-60}$$

or

$$K = e^{\Delta S^\circ/R} e^{-\Delta H^\circ/RT} \tag{8-61}$$

If ΔS° is independent of T, we can define a constant $C = e^{\Delta S^\circ/R}$ and write

$$K = Ce^{-\Delta H^\circ/RT} \tag{8-62}$$

The brilliant Swedish chemist Svante Arrhenius (1859–1927) was aware of this relation and, in 1889, made a daring proposal based on it. He reasoned that, if the equilibrium constant K varied with T in this way and if K was directly related to the rate constants k, as we have just shown, then perhaps k would depend on T in the same way. His famous proposal, now called the *Arrhenius equation*, was that

$$k = Ae^{-E_a/RT} \tag{8-63}$$

where A is a constant called the *pre-exponential factor* and E_a is called the *activation energy* of the reaction. We will examine the form of the pre-exponential factor in more detail later in this chapter.

Transition-state theory, which is one of the major theoretical approaches to chemical kinetics, focuses primarily on the activation energy. By analogy with Equation 8-62, E_a can be thought of as an enthalpy. It is the energy that must be made available to the reactants in order for them to form an unstable complex called the *activated complex* (i.e., the transition state). This complex has an energy higher than the combined energies of the reactants by the amount E_a. It can dissociate to give either the products of the reaction or the original reactants. In the latter case, of course, there is no net reaction. The meaning of E_a is best seen in Figure 8-11.

The ordinate in Figure 8-11 is the enthalpy of the reacting system. The abscissa, however, is something new: *reaction coordinate* does not generally have a precisely defined meaning. For a very simple reaction, such as H + D—D → H—D + D, the reaction coordinate can be specified as the distance between the two deuterium atoms. As the reaction proceeds, this distance clearly increases. In general, however, we will use this term simply

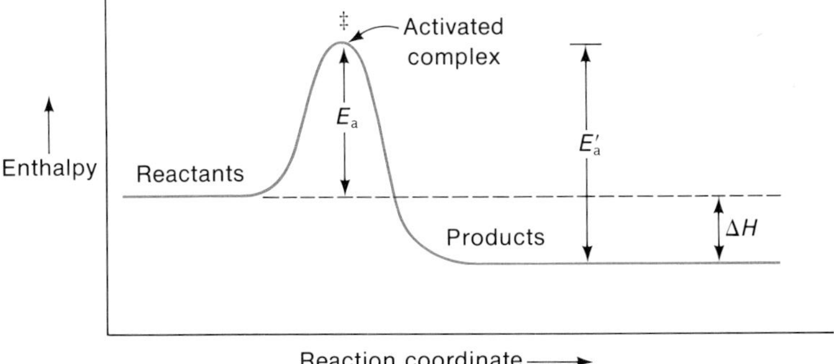

Figure 8-11. Hypothetical plot of enthalpy as a function of reaction coordinate, the extent of completion of the reaction. The reaction is exothermic.

in the qualitative sense of its being an indication of how far the reaction has proceeded toward completion.

Figure 8-11 shows that an energy E_a must be added to the reactants for them to be able to form the activated complex, which is customarily denoted by the double-dagger symbol ‡. (Thus, an activated complex A is represented as $A^\ddagger$, and the same superscript is used on the symbols for all of its properties.) This complex can then revert to reactants or proceed to products, in the latter case with the release of energy E'_a. The overall change in the thermodynamic enthalpy is then given by $\Delta H = E_a - E'_a$. In the present example there is a decrease in the enthalpy, so the reaction is exothermic. Unless there is also a large decrease in the entropy, the reaction will proceed in this direction spontaneously, since $\Delta G = \Delta H - T\Delta S$ is negative.

The reverse reaction requires an activation energy E'_a. Since $E'_a > E_a$ in the present example, the reactants in the reverse reaction must surmount a higher energy barrier. The reaction is therefore slower because proportionally fewer molecules are in the high-energy "tail" of the energy distribution curve (see Figure 6-9) and possess energy $E \geqslant E'_a$.

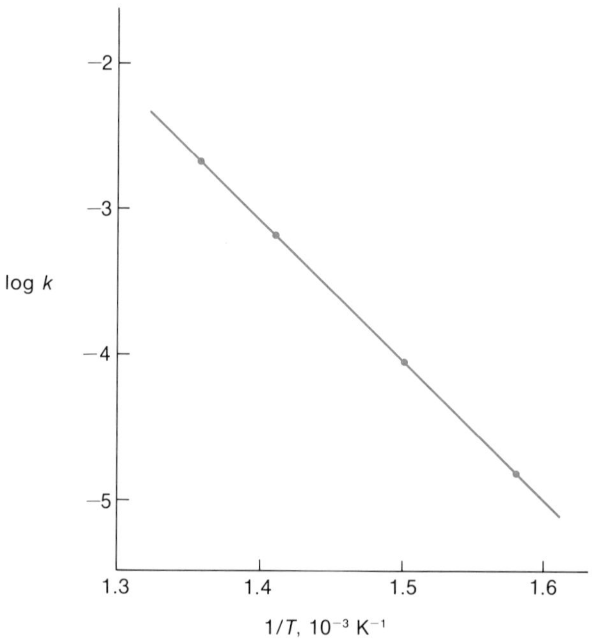

Figure 8-12. Arrhenius plot of the logarithm of k, the rate constant, versus $1/T$, the reciprocal of the absolute temperature, for the reaction $2HI(g) \rightarrow H_2(g) + I_2(g)$. [J. H. Sullivan, *Journal of Chemical Physics* **30**, 1292 (1959).]

The activation energy of a reaction is of great interest because it tells us a good deal about the energy requirements of the reaction. How can it be measured? Equation 8-63 can be rearranged to give a convenient expression for determining E_a. Taking natural logarithms,

$$\ln k = \ln A - E_a/RT \tag{8-64}$$

Conversion to base 10 gives

$$\log k = \log A - E_a/2.303RT \tag{8-65}$$

We see that a set of measurements of the rate of a reaction to yield k at several temperatures is all that is required. A plot of log k vs $1/T$, which is called an *Arrhenius plot,* should yield a straight line of slope $-E_a/2.303R$ and intercept log A. Figure 8-12 shows such a plot for the second-order reaction

$$2HI(g) \xrightarrow{k} H_2(g) + I_2(g) \tag{8-66}$$

The values of k were computed by assuming the reaction mechanism to consist of this one reaction. The plot is, indeed, linear, and the slope yields the value 43.8 kcal/mole for E_a. Values of A and E_a for some typical reactions are given in Table 8-3.

If we remember that the rate of a reaction is directly proportional to the rate constant, we can predict from Equation 8-65 that, as the temperature increases, so does the rate. As T increases, $1/T$ decreases, and Figure 8-12 shows that k increases. Thus, provided that E_a is positive, as is almost universally true, reactions go faster, the higher the temperature.

The paper you are reading from provides one example of this phenomenon. The reaction for the oxidation of cellulose, the principal ingredient of paper, is

Table 8-3. Values of A and E_a for Several Second-Order Reactions.

Reaction	A, cm³/mole s	E_a, kcal/mole
$2HI \rightarrow H_2 + I_2$	6×10^{13}	43.8
$CH_3I + HI \rightarrow CH_4 + I_2$	1.6×10^{15}	33.4
$2NOCl \rightarrow 2NO + Cl_2$	9×10^{12}	24.0
$NO + O_3 \rightarrow NO_2 + O_2$	8×10^{11}	2.5
$2C_2H_4 \rightarrow C_4H_8$	7.1×10^{10}	37.7
$2C_3H_6 \rightarrow C_6H_{12}$	1.6×10^{10}	38.0

$$(C_6H_{10}O_5)_n + 6nO_2 \rightarrow 6nCO_2 + 5nH_2O \qquad \Delta G^\circ \ll 0 \qquad (8\text{-}67)$$

The room is full of O_2, so thermodynamics tells us that the reaction should proceed spontaneously—and, in fact, it does. But we know that, although paper does "age" (oxidize slowly), it does not spontaneously burst into flame at room temperature. We also know that paper does burn at elevated temperatures. As the temperature goes up, so does the rate.

Another example is the clock reaction, which we discussed earlier. At room temperature the reaction is complete in about 18 s. At 0°C more than 100 s are required, and at 40°C the reaction is complete in about 5 s.

Why? What is the basis of the effect of temperature on reaction rate? The best explanation lies in Figure 8-13, which shows the Maxwell distributions of the velocities of nitrogen molecules at two temperatures (refer also to Figure 6-9). The type of gaseous molecular energy that is most likely to provide the activation energy required for a reaction is translational, or kinetic, energy. As we know,

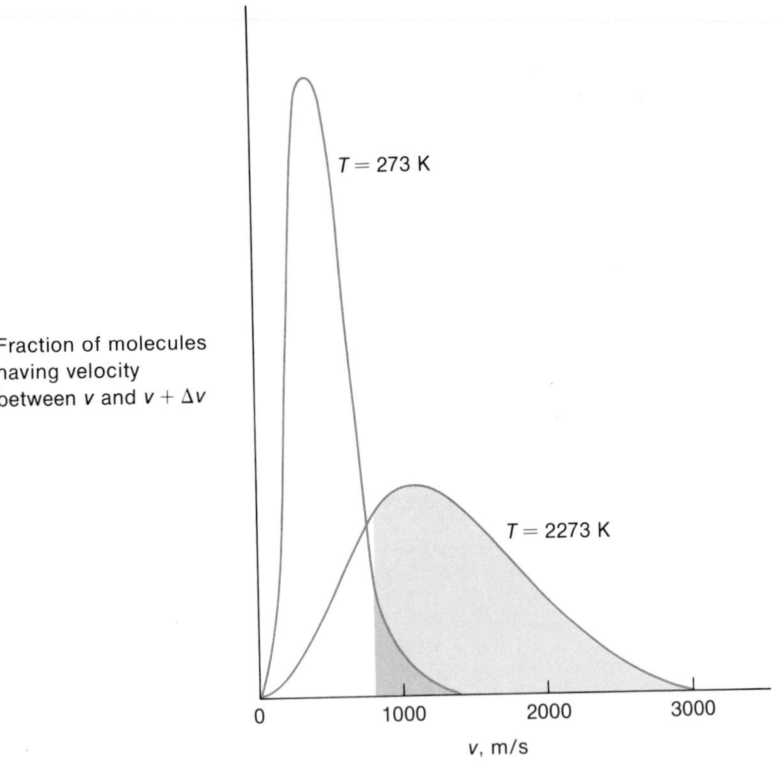

Figure 8-13. Maxwell velocity distributions of N_2 at two temperatures.

$$KE = \tfrac{1}{2}mv^2 \tag{8-68}$$

Thus, we can specify a minimum value of v necessary to provide the required value of E_a. Of course, molecules with larger values of v are also effective in causing the reaction to occur, the excess energy being transferred to the products as translational, electronic, vibrational, or rotational energy. For a given reaction involving nitrogen molecules to occur, a minimum value of perhaps 800 m/s might be required. The colored region in Figure 8-13 shows the relative numbers of molecules, at the two temperatures, that have this minimum velocity—hence, energy. It is apparent that at higher temperatures more molecules have energy equal to or greater than E_a. The same principles described here for gases are valid for liquids and solids also, although the energetics (and the mathematics) are much more complex.

Quantitative Relation between Reaction Rate and Temperature

You may be familiar with the chemical proverb, "The rate of a chemical reaction doubles for every 10°C rise in temperature." We are now in a position to investigate this approximation to see *why* it is true when it *is* true, and why it frequently fails.

We can write Equation 8-64 for two different temperatures as follows:

$$\ln k_1 = \ln A - E_a/RT_1 \tag{8-69}$$

$$\ln k_2 = \ln A - E_a/RT_2 \tag{8-70}$$

Subtracting the first of these equations from the second gives

$$\ln \frac{k_2}{k_1} = -\frac{E_a}{RT_2} + \frac{E_a}{RT_1} = \frac{E_a}{R}\left(\frac{T_2 - T_1}{T_1T_2}\right) \tag{8-71}$$

$$E_a = \frac{RT_1T_2}{T_2 - T_1}\ln\frac{k_2}{k_1} \tag{8-72}$$

We see that, for given values of T_1 and T_2, the logarithm of the ratio of the two rate constants is proportional to E_a.

Figure 8-14 shows the relation between E_a and the ratio k_2/k_1 for the 10-K temperature difference between $T_1 = 300$ K and $T_2 = 310$ K. We see from the plot that, if $E_a = 0$, $k_1 = k_2$: there is no effect of temperature on reaction rate. When $k_2/k_1 = 2$, the rate has doubled with the 10-K increase in

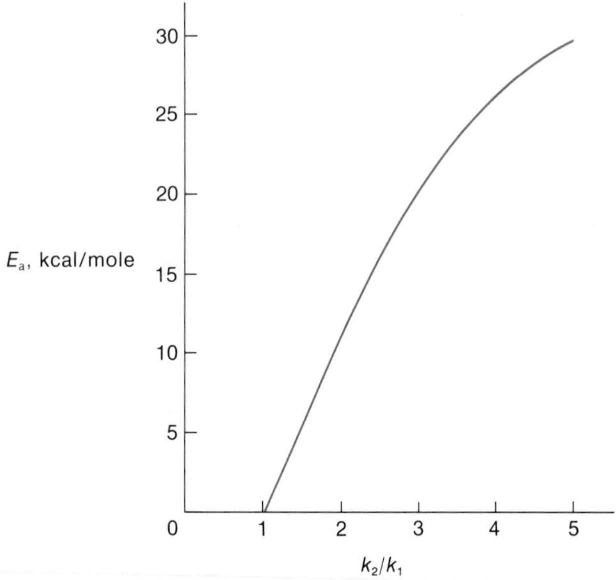

Figure 8-14. Plot of activation energy E_a versus k_2/k_1, the ratio of rate constants at temperatures $T_2 = 310$ K and $T_1 = 300$ K.

temperature. We see that the requirement for a rate's doubling at 300 K with an increase of 10 K is that $E_a = 12.8$ kcal/mole, a fairly low value (see Table 8-3). The figure illustrates that with a 10-K temperature increase the rate of a reaction may remain virtually constant or it may increase by a factor of up to 5 or more, depending upon the activation energy for that particular reaction.

Other Temperature Dependencies of Reaction Rates

Many chemical reactions obey the Arrhenius equation approximately, but there are many others that do not. Figure 8-15 displays two such types of kinetic behavior in addition to the Arrhenius behavior, which is the most typical (Diagram a). Diagram b represents the kinetic behavior for an explosion. The rate increases steadily until the ignition temperature is reached and then increases vertically as the explosion occurs. Diagram c is typical of enzyme-catalyzed reactions. As we will see later on, enzymes are large, comparatively fragile protein molecules that can undergo subtle changes as their environment is changed. The rate of the enzyme-catalyzed reaction increases up to a point but then diminishes owing to rearrangements in the protein structure, which lead to a loss of catalytic activity.

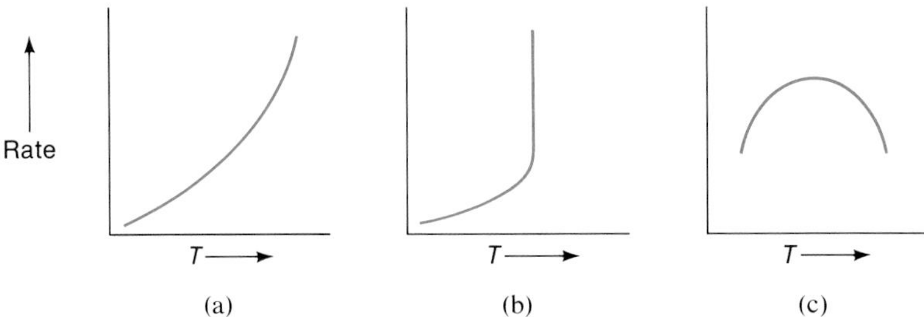

Figure 8-15. Plots of rate versus temperature for three different classes of reaction. (a) Behavior of a system obeying the Arrhenius equation. (b) An explosion. (c) Enzymatic catalysis.

The Classical $H_2 + I_2$ Reaction

We call the $H_2 + I_2$ reaction the "classical bimolecular reaction" because it has been used as a model for a great deal of theoretical work. For many years scientists believed that this reaction occurred according to the one-step mechanism

$$H_2(g) + I_2(g) \rightarrow 2HI(g)$$

This mechanism agreed with the experimental rate expression (Equation 8-8) and, despite a slight curvature in the log k-vs-$1/T$ plot, there appeared to be no evidence to suggest another mechanism.

Indeed, the reaction appeared to be so simple that it was used as a model for the calculation of the pre-exponential factor A. Among the quantities needed to calculate A is the factor p, the *steric factor,* which is the probability that reaction will occur when two molecules collide with a specific mutual orientation. Thus we would expect the collision in Figure 8-16a to be very effective and have a large value of p, whereas we would expect the collision in Figure 8-16b to be quite ineffective. Suitable statistical averaging of all such reaction geometries, combined with the appropriate energy data, yielded nearly correct values for A. The tenuous nature of these calculations became apparent only in 1967, when John Sullivan (b. 1919) of the Los Alamos Scientific Laboratory showed* that the reaction does not even proceed by this mechanism! Before we discuss Sullivan's work, let us backtrack briefly.

The original rate data of Max Bodenstein (1871–1940), obtained in the late nineteenth century, showed an unusual temperature dependence. This

Journal of Chemical Physics* **46, 73 (1967).

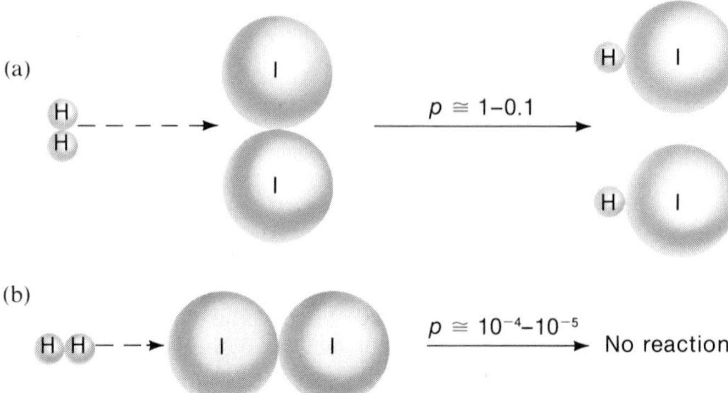

Figure 8-16. Two possible reaction geometries for the $H_2 + I_2$ reaction. (a) Broadside. (b) Head-on.

problem was resolved in 1955 by Sidney Benson (b. 1918) and R. Srinivasan (b. 1929), of the University of Southern California, who proposed* that the I_2 molecule dissociated to two I atoms as a part of the reaction mechanism. The problem that *this* raised, however, and that remained unresolved for another decade, was which of the following reactions was the correct elementary process:

$$H_2(g) + I_2(g) \rightarrow 2HI(g) \tag{8-73}$$

or

$$H_2(g) + 2I(g) \rightarrow 2HI(g) \tag{8-74}$$

Although the two reactions seem dissimilar enough, they are kinetically indistinguishable. The rate expression for Reaction 8-73 is

$$\frac{1}{2}\frac{dP_{HI}}{dt} = kP_{H_2}P_{I_2} \tag{8-75}$$

and, for Reaction 8-74,

$$\frac{1}{2}\frac{dP_{HI}}{dt} = k'P_{H_2}(P_I)^2 \tag{8-76}$$

But, because the I atoms are in thermal equilibrium with molecular I_2, we have

*_Journal of Chemical Physics_ **23**, 200 (1955).

$$I_2 \rightleftharpoons 2I \tag{8-77}$$

for which

$$K = \frac{(P_I)^2}{P_{I_2}} \tag{8-78}$$

Substitution into Equation 8-76 gives

$$\frac{1}{2}\frac{dP_{HI}}{dt} = k'KP_{H_2}P_{I_2} = k''P_{H_2}P_{I_2} \tag{8-79}$$

which is identical in form to Equation 8-75. Thus, there is no way to distinguish Reaction 8-73 from Reaction 8-74 on the basis of traditional kinetic methods for determining the order of a reaction.

Henry Eyring (b. 1901), at the University of Utah, and Nikolai Semenov (b. 1896), at Moscow State University, both proposed in 1965 that the termolecular mechanism (Equation 8-74) was correct. Two years later Sullivan proved it, by doing something fairly simple. (Many breakthroughs result from novel thinking that, in retrospect, seems rather obvious.) Sullivan measured the rate of this reaction under conditions at which only the termolecular mechanism was possible, then measured it under conditions at which either mechanism was possible, and compared the results. The kinetic behavior was identical in both cases. Therefore, he concluded, the termolecular mechanism occurs under all conditions. The technique he used was to measure reaction rates as a function of temperature; let us see how he did it.

First he studied the reaction in the temperature range 417–520 K, where the bimolecular reaction—the reaction resulting from the collision of H_2 and I_2—does not occur because so few molecules possess the kinetic energy necessary for the activation step. Instead, he initiated the reaction by shining a high-intensity lamp on the reaction mixture at several temperatures. The 5780-Å light from this lamp has sufficient energy to photochemically dissociate I_2 molecules to I atoms. He measured k as a function of T for the ensuing reaction, which, under these conditions, can only be termolecular.

Next he studied the reaction in the temperature range 633–738 K, where the bimolecular reaction *can* occur—if it occurs at all. Again he measured k as a function of T, this time *assuming* the termolecular mechanism. He then plotted both sets of data on the same graph. Figure 8-17 shows that, within experimental error, all the data points lie on one straight line, meaning that the termolecular mechanism is correct under all conditions. These data thus reveal that the mechanism represented in Figure 8-16 is incorrect. The actual mechanism is shown in Figure 8-18.

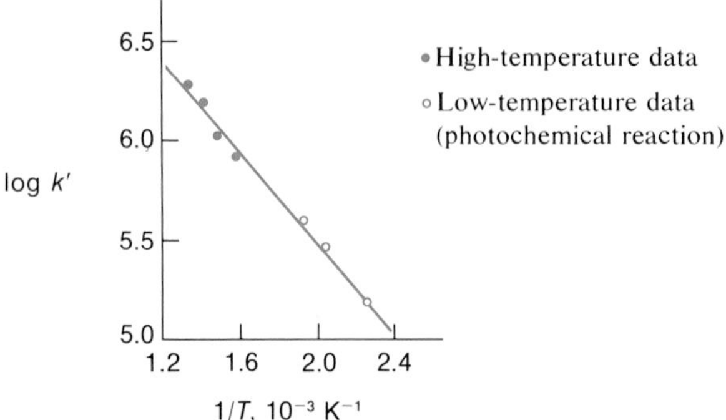

Figure 8-17. Arrhenius plot of Sullivan's data for the $H_2 + I_2$ reaction. These data show that the reaction $H_2 + I_2 \rightarrow 2HI$ does not proceed via a bimolecular step, as had long been assumed. [J. H. Sullivan, *Journal of Chemical Physics* **46**, 73 (1967).]

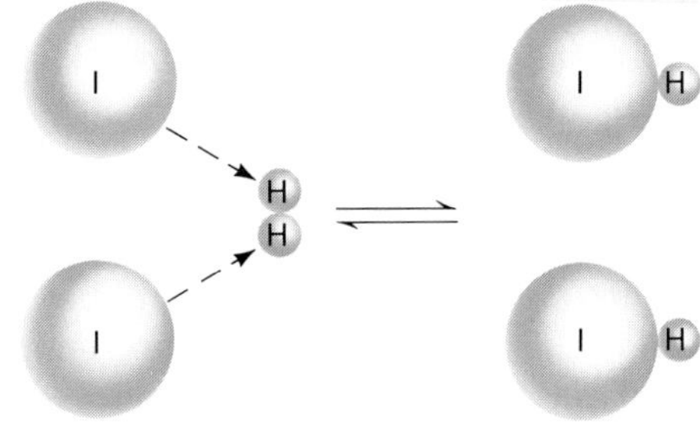

Figure 8-18. Representation of the actual mechanism of the hydrogen-iodine reaction.

The double arrow in Figure 8-18 is meant to emphasize the principle of microscopic reversibility. Thus, the reaction sequence for the disappearance of HI must be

$$2HI \rightarrow H_2 + 2I \rightarrow H_2 + I_2 \qquad (8\text{-}80)$$

The activation energies for the several steps are summarized in Figure 8-19, which shows that a very large activation energy, 43.8 kcal/mole, is required for the reaction between two HI molecules. The activation energy for the reaction between H_2 and I_2, 40.8 kcal/mole, is nearly as large because the change in enthalpy for the overall reaction, $\Delta H = -3$ kcal/mole, is small. Almost all the activation energy for this reaction is required for the dissociation of molecular iodine; the remainder is required to form the activated complex.

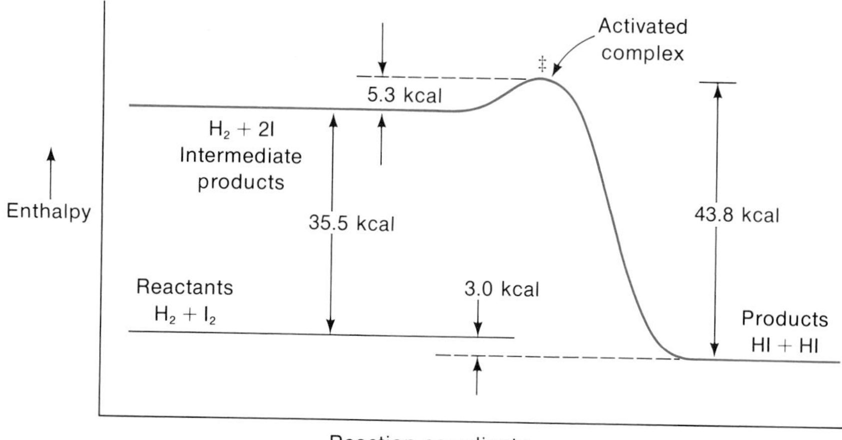

Figure 8-19. Activation-energy diagram for the hydrogen-iodine system. All data are molar quantities.

8.8 THE EFFECT OF CATALYSTS ON RATE

Catalysis is the mechanism by which the rate of a reaction is increased by the presence of a substance—the *catalyst*—that does not undergo any net chemical change in the reaction. A qualitative explanation of how catalysts function is given in Figure 8-20. The catalyst provides a new pathway, of lower activation energy, via which the reaction can proceed. If a pathway of higher activation energy were introduced by the catalyst, the system would simply "ignore" it.

Certain substances, when added to a reaction mixture, can retard the reaction or stop it completely because they block the normal reaction pathway. The terms *inhibitor* and *inhibition* are employed to describe such a substance and its effect. Thus, an inhibitor can change not only the rate of a reaction but also the thermodynamic equilibrium composition of the system. The latter implies, however, that the inhibitor must itself be changed by the net reaction, and it is therefore not a catalyst.

Catalysts cannot affect the equilibrium composition of a reactive system. If a catalyst increased the rate of a hypothetical reaction $A \rightarrow B$ more than it increased the rate of the reverse reaction $B \rightarrow A$, the equilibrium constant $K = k_f/k_r = [B]/[A]$ would change. But, since the equilibrium composition depends only on the change in the state function, free energy, which is independent of the path—and, therefore, of the rate—we conclude that a catalyst must increase the rates of both the forward and reverse reactions by exactly the same factor.

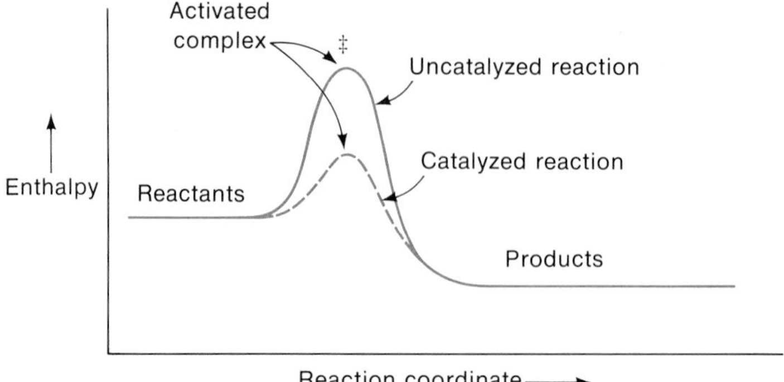

Figure 8-20. Activation-energy diagram for catalyzed and uncatalyzed reactions.

There are two general types of catalysis, heterogeneous and homogeneous. *Heterogeneous catalysis* is catalysis that occurs at interfaces between phases in systems containing more than one phase. The most common example of this phenomenon is a gas-phase reaction that is accelerated at a solid surface. Consider the hydrogenation of an unsaturated organic compound:

$$H_2 + \underset{R}{\overset{R}{\diagdown}}C{=}C\underset{R}{\overset{R}{\diagup}} \rightarrow R{-}\underset{H}{\overset{R}{\underset{|}{\overset{|}{C}}}}{-}\underset{H}{\overset{R}{\underset{|}{\overset{|}{C}}}}{-}R \qquad (8\text{–}81)$$

In the gaseous phase, this reaction occurs immeasurably slowly at ordinary pressures and room temperature. However, the addition of a specific catalyst such as finely divided nickel or platinum causes the reaction to proceed fairly rapidly. The first step in the mechanism by which reaction occurs appears to be the dissociation of molecular hydrogen to hydrogen atoms on the surface of the metal. These atoms are far more reactive than H_2 molecules and rapidly attack the carbon atoms of the double bond. Figure 8-21 illustrates a similar heterogeneous catalytic reaction whose mechanism is fairly well understood.

Heterogeneous catalysts are assuming great importance in our efforts at cleaning the air. The sulfur present in coal and fuel oil is oxidized to SO_2 in the combustion process. One means of controlling the emission of this pollutant is to oxidize it to SO_3 before it leaves the smokestack and dissolve the SO_3 to form sulfuric acid, which can then be sold to partially offset the recovery costs. Vanadium pentoxide, V_2O_5, an orange solid, greatly increases the rate of the oxidation reaction:

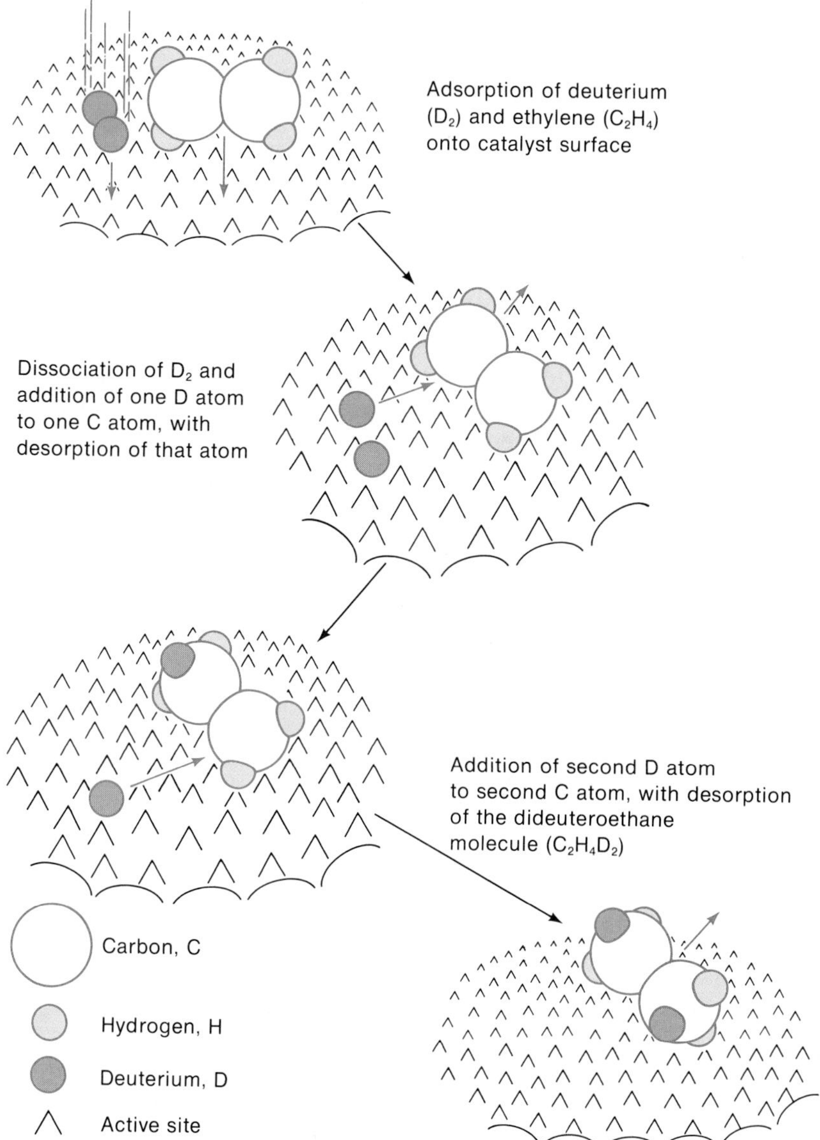

Adsorption of deuterium (D_2) and ethylene (C_2H_4) onto catalyst surface
Dissociation of D_2 and addition of one D atom to one C atom, with desorption of that atom
Addition of second D atom to second C atom, with desorption of the dideuteroethane molecule ($C_2H_4D_2$)
Carbon, C
Hydrogen, H
Deuterium, D
Active site

Figure 8-21. Catalytic deuteration of ethylene on a heterogeneous catalyst. (T. L. Brown, *General Chemistry*, C. E. Merrill, Columbus, Ohio, 1963.)

$$SO_2(g) + \tfrac{1}{2}O_2(g) \xrightarrow{V_2O_5} SO_3(g) \qquad \text{(Fast)} \tag{8-82}$$

This catalytic effect is essential because the SO_2 is a trace component of a high-velocity gas stream as it moves up the stack. If it is going to react at all before escaping to the atmosphere, it must react rapidly. Heterogeneous catalysts are also important in controlling the emission of CO and unburned hydrocarbons from the internal combustion engine. They permit the oxidation of these substances to CO_2 and H_2O (relatively harmless products) to occur at temperatures low enough that N_2 is not oxidized to NO at an appreciable rate.

Homogeneous catalysis is catalysis that occurs in one phase. A wide variety of homogeneous catalysts are known and utilized by chemists, especially in organic syntheses. An example is the acid-catalyzed substitution of a bromine atom for the hydroxyl group of an alcohol. The bromide ion in solution attacks the carbon atom attached to the OH group. This C atom has a positive formal charge because of the electronegativity of the O atom. Thus, it can be thought of as being slightly like a nucleus (i.e., positive). The Br^- ion is said to be *nucleophilic* (nucleus-loving), and the rate-determining step of the mechanism is a bimolecular elementary process. The reaction is therefore called an S_N2 reaction (*s*ubstitution, *n*ucleophilic, *bi*molecular). Hydrogen ions in solution act as homogeneous catalysts by the following mechanism:

$$H^+ + RCH_2{-}O{-}H \rightleftharpoons RCH_2{-}O^+H_2 \qquad \text{(Fast)} \tag{8–83}$$

$$Br^- + RCH_2{-}O^+H_2 \rightarrow \left[Br\cdots CH_2R\cdots OH_2\right] \rightarrow Br{-}CH_2R + H_2O \qquad \text{(Slow)} \tag{8–84}$$

Activated complex

Enzyme Kinetics

Because of their enormous importance and ubiquity in living systems, we will devote this section to the most phenomenal catalysts known, enzymes. All enzymes are proteins (but not vice versa!), large polymers of amino acids whose composition and structure will be discussed in Chapter 11. Until

1968 enzymes were produced only in living cells. Then came the exciting announcement by R. B. Merrifield (b. 1921) of Rockefeller University that chymotrypsin had been synthesized in the laboratory and that it possessed much of the enzymatic activity observed in the natural molecule. (Chymotrypsin is a gastrointestinal-tract enzyme that derives from a precursor produced by the pancreas.)

More than a thousand enzymes have now been identified. Charts showing the majority of the biochemical pathways catalyzed by enzymes in humans simply boggle the mind.* One way to begin to make some sense out of all the compounds listed is to note that the names of most enzymes end in "-ase"; the first part of the name often reveals what the enzyme does. An example is yeast alcohol dehydrogenase. The source of this enzyme is obvious. The last two words indicate that the enzyme catalyzes the dehydrogenation of an alcohol (it also catalyzes the reverse reaction, the hydrogenation of an aldehyde). Another molecule is often required for enzyme catalysis and could be any of a variety of compounds. It is generally a specific type of organic compound called a *coenzyme*, which works in concert with the enzyme to effect the reaction being catalyzed.

Enzymes are large. In contrast to the relatively small molecules we have discussed thus far, these molecules contain from 1000 to 100,000 atoms. As we will see below, only a small portion of the molecule is active in catalysis. Although some quite specific information concerning the nature of this *active site* has been obtained, the role of the large remainder of the molecule and the reasons for the sequence of amino acids in it remain a challenge for a new generation of biochemists.

Enzymes are incredibly efficient catalysts. One measure of the quality of a catalyst is the degree to which it lowers the activation energy of the reaction (see Figure 8-20). For example, the uncatalyzed decomposition of H_2O_2 to water and oxygen has an activation energy of 18 kcal/mole. Finely divided platinum, a very effective artificial catalyst, reduces this value to 12 kcal/mole. The enzyme catalase, by comparison, reduces it to only 0.6 kcal/mole. If we assign a relative reaction rate of 1 to the uncatalyzed reaction, the Pt-catalyzed reaction has a rate of 26,000, whereas the enzyme-catalyzed reaction has a rate greater than 10^{10}!

Another example of enzymatic efficiency is the fixation of nitrogen to ammonia (*nitrogen fixation* is the conversion of nitrogen into stable, biologically assimilable compounds). In the presence of enzymes, this reaction occurs readily at room temperature and atmospheric pressure. However, the most efficient process that chemists have been able to devise is the Haber

---

*A set of four Intermediary Metabolism Charts, by Dr. H. J. Sallach, is available for $3.75 from P-L Biochemicals, Inc., Box 4606, Chicago, Ill. 60680.

Table 8-4. Maximum Turnover Numbers of Some Enzymes.

Enzyme	Turnover Number, s^{-1}
Peroxidase	10
Ribonuclease	10^2
Aldolase	10^2
Chymotrypsin	10^2–10^3
Dehydrogenases	10^3
Catalase	10^7

process, which, even with the best mixed-metal catalyst, requires a temperature of about 500°C and a pressure of about 400 atm.

A useful measure of the efficiency of an enzyme in accelerating a reaction is the *turnover number*. This is the number of reactant molecules converted to product per enzyme molecule per unit time. The turnover numbers given in Table 8-4 have been *normalized* (reduced to a common basis for comparison) by dividing by the number of active sites on the enzyme. Note the enormous magnitude of one of these numbers. Can you easily visualize a process whereby 10 million hydrogen peroxide molecules interact with one catalase molecule and are converted to products—every second?

A crude representation of the mechanism of enzyme catalysis is presented in Figure 8-22. The substance undergoing chemical reaction is called the *substrate* (S). It is bound momentarily at an active site on the enzyme (E), forming the *enzyme-substrate complex* (ES), from which the substrate is then converted to product (P) or released unchanged. This scheme was placed on a firm mathematical basis by the German chemist Leonor

S ⟶ P

S ⟶ P (Uncatalyzed—slow)

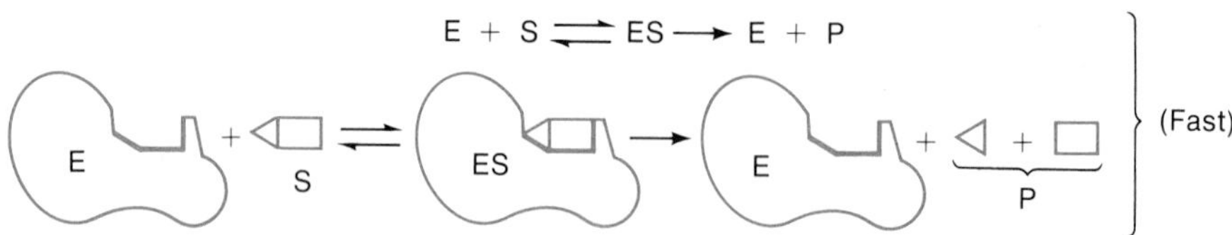

Figure 8-22. Schematic representation of the mechanism of enzyme catalysis. S = substrate, P = product(s), E = enzyme, and ES = enzyme-substrate complex.

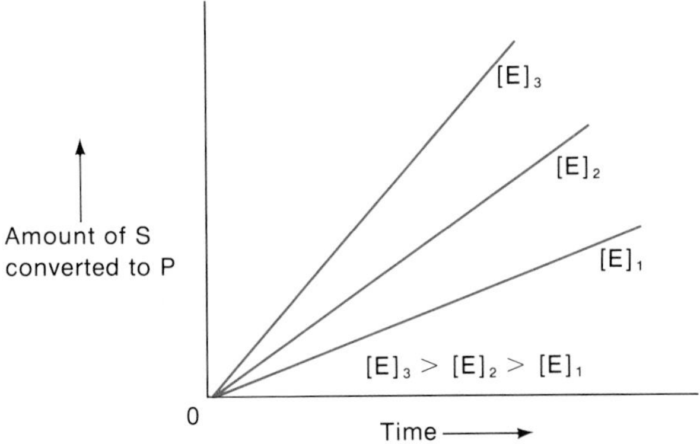

Figure 8-23. Amount of substrate converted to product(s) as a function of time, for several enzyme concentrations.

Michaelis (1875–1949) and his assistant Maud Menten (1879–1960) in 1913, 21 years after it was first proposed. Let us examine the basic experimental facts of enzyme catalysis and then see how well the Michaelis-Menten model accounts for this behavior.

Figure 8-23 displays the effect of increasing the enzyme concentration upon the amount of substrate converted, as a function of time. The slopes of these lines are, of course, the rates of the reaction, and it is of interest to know whether the rate is directly proportional to the enzyme concentration. As can be seen in Figure 8-24, it is. These two plots should seem reasonable in the light of our earlier discussions in this chapter, i.e., the reaction appears to be first-order with respect to the enzyme.

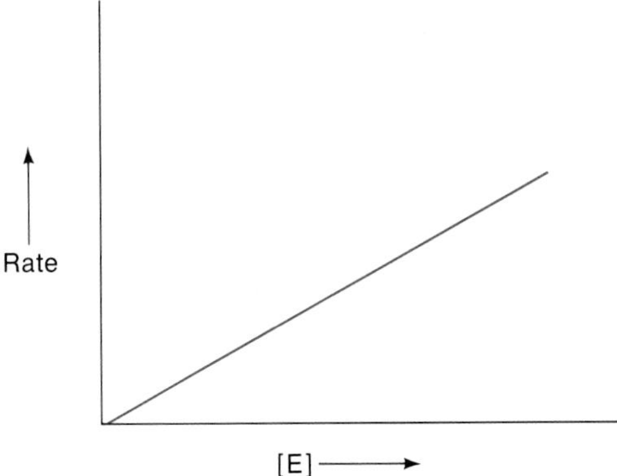

Figure 8-24. Plot of rate versus enzyme concentration for an enzyme-catalyzed reaction.

The catch comes when we look at the next obvious set of data, namely, the variation of the rate with the concentration of the *substrate*. Nothing that we have seen so far in this chapter has prepared us for such peculiar behavior.

Figure 8-25 is a plot of the rate of the enzyme-catalyzed conversion of sucrose to glucose and fructose. The concentration of the enzyme (yeast invertase) is, of course, constant because enzymes are catalysts. The figure tells us that at very low substrate concentrations the rate of conversion depends upon the amount of substrate that is available for interaction with the active sites of the enzyme molecules. At high values of [S], a level is eventually reached at which all the active sites on the enzyme are occupied; hence, further amounts of substrate cannot result in any rate increase. The enzyme molecules are saturated.

We can learn much more about our system by a quantitative examination of some proposed mechanisms. At low values of [S], the rate is directly proportional to [S]. This means that the rate is first-order with respect to S in this region:

$$\text{Rate} = \frac{-d[\text{S}]}{dt} = k[\text{S}] \qquad (\text{Low [S]}) \tag{8-85}$$

At high values of [S], where the rate curve reaches a plateau, the rate is independent of [S] and obeys zero-order kinetics with respect to S:

$$\text{Rate} = \frac{-d[\text{S}]}{dt} = k' \qquad (\text{High [S]}) \tag{8-86}$$

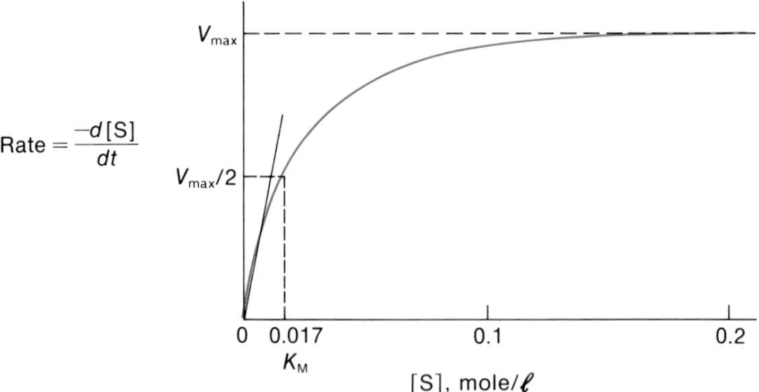

Figure 8-25. Plot of rate versus substrate concentration for an enzyme-catalyzed reaction: the yeast invertase-catalyzed conversion of sucrose to glucose and fructose. K_M is the Michaelis constant in units of moles per liter.

It is clear that a simple mechanism such as

$$E + S \rightarrow E + P \tag{8-87}$$

cannot begin to account for a combination of zero- and first-order kinetic behavior. The great success of the theory of Michaelis and Menten was that it *did* explain this strange behavior. The Michaelis-Menten mechanism is

$$E + S \underset{k_{-1}}{\overset{k_1}{\rightleftharpoons}} ES \xrightarrow{k_2} E + P \tag{8-88}$$

Three reasonable assumptions are required in order to use this mechanism to explain the dependence of the rate upon [S]. The first is that the rate-determining step is the conversion of ES to E + P. This means that the overall rate, or velocity V, is given by

$$\text{Rate} = V = k_2[\text{ES}] \tag{8-89}$$

The second assumption is that [S] $\gg$ [E]. Because of the high turnover numbers of most enzymes, this condition is readily realized. The third assumption is that the reaction goes essentially to completion, as implied by the single arrow on the right side of Reaction 8-88.

We begin by calling the initial concentration of enzyme $[E_0]$. This quantity is always equal to the sum of the concentrations of unbound and bound enzyme, [E] + [ES], i.e., it is the total enzyme concentration. The concentration of unbound enzyme is therefore given by $[E] = [E_0] - [ES]$. We can now express the net rate of production of ES as

$$\frac{d[\text{ES}]}{dt} = \text{rate of formation} - \text{rate of disappearance} \tag{8-90}$$

$$= k_1[\text{E}][\text{S}] - k_{-1}[\text{ES}] - k_2[\text{ES}] \tag{8-91}$$

$$= k_1([\text{E}_0] - [\text{ES}])[\text{S}] - (k_{-1} + k_2)[\text{ES}] \tag{8-92}$$

When the rate of formation of ES equals the rate of disappearance, the net rate of production is zero:

$$\frac{d[\text{ES}]}{dt} = 0 \tag{8-93}$$

This condition is called the *steady state*. The assumption that it exists in a system such as this is called the *steady-state assumption,* and is necessary for the mathematical derivation at hand. Although a number of enzyme-catalyzed reactions exist for which this assumption is not valid, many reactions can be adequately described by the equation that is derived on the basis of it. Let us continue the derivation.

Equations 8-92 and 8-93 can now be combined to give

$$k_1([\mathrm{E}_0] - [\mathrm{ES}])[\mathrm{S}] = (k_{-1} + k_2)[\mathrm{ES}] \qquad (8\text{-}94)$$

which, upon rearrangement, gives

$$\frac{([\mathrm{E}_0] - [\mathrm{ES}])[\mathrm{S}]}{[\mathrm{ES}]} = \frac{k_{-1} + k_2}{k_1} = K_{\mathrm{M}} \qquad (8\text{-}95)$$

where K_{M} is called the *Michaelis constant.* We will see below how it is measured and what it means. For now, note that it has the dimensions of a concentration.

Equation 8-97 can be solved for [ES], which is required in order to solve for the rate:

$$[\mathrm{ES}] = \frac{[\mathrm{E}_0][\mathrm{S}]}{K_{\mathrm{M}} + [\mathrm{S}]} \qquad (8\text{-}96)$$

Substitution of this relation into Equation 8-89 gives

$$\text{Rate} = V = k_2[\mathrm{ES}] = \frac{k_2[\mathrm{E}_0][\mathrm{S}]}{K_{\mathrm{M}} + [\mathrm{S}]} \qquad (8\text{-}97)$$

At high substrate concentrations virtually all the enzyme is present as the complex ES (i.e., $[\mathrm{E}_0] = [\mathrm{ES}]$) and the rate has reached a maximum velocity V_{max}:

$$V_{\mathrm{max}} = k_2[\mathrm{E}_0] \qquad (8\text{-}98)$$

We can now write the *Michaelis-Menten equation* by substituting Equation 8-98 into the above rate equation:

$$V = \frac{V_{\mathrm{max}}[\mathrm{S}]}{K_{\mathrm{M}} + [\mathrm{S}]} \qquad (8\text{-}99)$$

The Michaelis-Menten equation provides the quantitative basis for the shape of Figure 8-25. At very low values of [S], $K_{\mathrm{M}} > [\mathrm{S}]$ and we can make the approximation $K_{\mathrm{M}} + [\mathrm{S}] \cong K_{\mathrm{M}}$. Therefore,

$$V \cong \frac{V_{\mathrm{max}}[\mathrm{S}]}{K_{\mathrm{M}}} = k[\mathrm{S}] \qquad (\text{Low } [\mathrm{S}]) \qquad (8\text{-}100)$$

which corresponds to our earlier observation of first-order kinetics (Equation 8-85). At high values of [S], [S] > K_M and we can make the approximation $K_M + [S] \cong [S]$. Therefore,

$$V \cong \frac{V_{max}[S]}{[S]} = V_{max} = k' \qquad \text{(High [S])} \tag{8-101}$$

which corresponds to our earlier observation of zero-order kinetics (Equation 8-86). At intermediate values of [S], a mixture of zero- and first-order kinetics obtains, and the full Michaelis-Menten equation is required to explain the rate curve in that region.

The Michaelis constant can be measured in a variety of ways. The simplest (and least accurate) way is to note the special relation between K_M and [S] in the Michaelis-Menten equation when $V = V_{max}/2$:

$$\frac{\cancel{V_{max}}}{2} = \frac{\cancel{V_{max}}[S]}{K_M + [S]} \tag{8-102}$$

$$K_M + [S] = 2[S]$$

$$K_M = [S] \tag{8-103}$$

Thus, a rough estimate of K_M can be obtained, as shown in Figure 8-25, by plotting the rate (velocity) versus [S] and measuring the value of [S] graphically when the velocity reaches half its maximum value.

A particularly simple explanation of the meaning of K_M is available for a few enzymes, such as yeast invertase and chymotrypsin. For these enzymes, $k_2 << k_{-1}$ and the expression for K_M becomes

$$K_M = \frac{k_{-1}}{k_1} \tag{8-104}$$

This corresponds to the equilibrium constant K_{eq} for the dissociation of the enzyme-substrate complex ES, in the event that a true equilibrium is established between ES, E, and S:

$$\text{ES} \underset{k_1}{\overset{k_{-1}}{\rightleftharpoons}} \text{E} + \text{S} \qquad \text{(Equilibrium)} \tag{8-105}$$

A large value of K_{eq} would mean a low affinity of the enzyme for the substrate (or vice versa) and a small value of K_{eq} would mean a high affinity between the two. For systems in which Equation 8-105 does represent a true equilibrium, measurements of values of K_M for a variety of substrates thus provide a measure of the strength of the binding between E and S for differing polarities, charges, and shapes. This, in turn, provides valuable information on the chemical nature of the active site(s) on the enzyme. Many systems, however, do not have a true equilibrium between E, S, and ES, as we will see shortly.

The final point we will consider in our discussion of enzymes and their incredible behavior is the reality of the enzyme-substrate complex, for which two kinds of evidence are now available. The existence of the species ES was accepted on faith for 24 years after Michaelis and Menten used it successfully to explain enzyme kinetics. In 1937 the Polish biochemists David Keilin (1887–1963) and Thaddeus Mann (b. 1908) produced definite evidence* of the existence of a complex between peroxidase and its substrate, hydrogen peroxide.

A solution of peroxidase is reddish-brown. When the substrate H_2O_2 is added, a shift in the spectral absorption band occurs immediately. The only logical explanation for this is that the following reaction occurs:

$$\text{Peroxidase} + H_2O_2 \underset{k_{-1}}{\overset{k_1}{\rightleftharpoons}} \text{peroxidase} \cdot H_2O_2 \qquad (8\text{-}106)$$

This reaction corresponds to $E + S \rightleftharpoons ES$ and demonstrates the reality of the enzyme-substrate complex for this system. The remainder of the reaction can be followed spectroscopically if a hydrogen donor, such as an oxidizable dye, H_2A, is available:

$$\text{Peroxidase} \cdot H_2O_2 + H_2A \xrightarrow{k_2} \text{peroxidase} + 2H_2O + A \qquad (8\text{-}107)$$

Both of the above reactions can be monitored with a spectrophotometer and the three rate constants measured. If H_2A is leukomalachite green (an organic dye), the three rate constants have the following values at 25°C: $k_1 = 1.2 \times 10^7$ ℓ/(mole s), $k_{-1} = 0.2$ s^{-1}, and $k_2 = 5.2$ s^{-1}. The value of K_M obtained from a rate-vs-[S] plot at the point $V = V_{max}/2$ is 4.1×10^{-7} mole/ℓ. It is pleasing to compare this with the ratio of specific rate constants:

$$K_M = \frac{k_{-1} + k_2}{k_1} = \frac{0.2 + 5.2}{1.2 \times 10^7} = 4.5 \times 10^{-7} \text{ mole}/\ell \qquad (8\text{-}108)$$

Because $k_2 > k_{-1}$ we do not expect $K_{eq} = k_{-1}/k_1$, the equilibrium constant for the dissociation of ES, to approximate K_M very well. In fact, $K_{eq} = 2 \times 10^{-8}$ mole/ℓ, demonstrating that the simplified picture of K_M as a dissociation constant for ES does not apply to the peroxidase-hydrogen peroxide reaction.

Proceedings of the Royal Society (London)* **B122, 119 (1937).

The second kind of evidence for the existence of enzyme-substrate complexes is more direct and more recent. In 1965 David Phillips (b. 1924) and his coworkers at the Royal Institution in London obtained the three-dimensional structure of the enzyme lysozyme with a molecule of substrate, a complex sugar called a hexasaccharide, attached to its surface. Lysozyme was the first enzyme and only the third protein to yield its intricate structural details to the powerful method of x-ray crystallography. We will learn more about this enzyme in Chapter 11.

8.9 THEORIES OF CHEMICAL KINETICS

Two major theories have been developed for predicting and explaining reaction rates. We will look qualitatively at the principles underlying these theories and compare them briefly. Your detailed understanding of the development and use of these theories must await a later course in physical chemistry or chemical kinetics. However, it is useful, even at an introductory level, to examine the bases of these theories, see the results they provide, and thus gain more insight into the meaning of the Arrhenius pre-exponential factor A, so casually introduced in Equation 8-63:

$$k = Ae^{-E_a/RT}$$

Collision Theory

Collision theory has been worked out in great detail for bimolecular gas-phase reactions. Even though most chemical reactions occur in condensed phases, collision theory is based on the kinetic theory of gases because the gaseous state, being mathematically the most tractable state of matter, is by far the best understood. Yet, for a variety of reasons that are still only partially understood, the collision theory of gases can also be successfully applied to some reactions occurring in the liquid state.

Collision theory is based on the assumptions that, in order for a chemical reaction to occur: (1) the reactant molecules must collide, (2) they must possess sufficient internal energy to react, and (3) they must have approximately the correct orientation with respect to each other. These ideas can be compressed into the following form:

$$\text{Rate} = \text{collision frequency} \times \text{energy requirement} \times \text{steric factor} \qquad (8\text{-}109)$$

The *collision frequency* Z_{AB} of molecules A and B is the number of molecules that collide per unit volume per unit time. It can be derived from the kinetic theory of gases presented in Chapter 6. The result is

$$Z_{AB} = \pi[A][B]d^2\bar{v} \tag{8-110}$$

where [A] and [B] are the concentrations of A and B in, e.g., molecules/ml, d is the *collision diameter*, the internuclear distance for which we can say that a collision has occurred (generally 2–4 Å for atoms), and $\bar{v}$ is the average velocity of the two molecules. The collision frequency for HI molecules at 700 K and 1 atm is about 10^{28} molecules/ml s. Because this factor alone would imply the enormous reaction rate of about 2×10^4 mole/ml s, it is apparent that at least one of the other two factors must be a number much smaller than one.

The *energy requirement* in Equation 8-109 is that postulated by Arrhenius, namely, the fraction of molecules having internal energy equal to or greater than E_a, which is the minimum energy required for the reaction to occur. This fraction is given by the following expression:

$$\text{Energy requirement} = e^{-E_a/RT} \tag{8-111}$$

Clearly, as E_a becomes smaller, the fraction of molecules having at least this much internal energy becomes larger. The energy-requirement term in Equation 8-109 increases and therefore the rate of reaction increases. This explains why a catalyst, by decreasing the activation energy of a reaction, increases the reaction rate (see Figure 8-20).

The inherent vagueness of the *steric factor p*, making it difficult to interpret as well as calculate, is implicit in Figure 8-16. Originally, the steric factor was employed to account statistically for the variety of possible collisions by considering all possible orientations of the reactants, i.e., those that were likely to lead to a reaction as well as those that were not. As we saw with the hydrogen-iodine reaction, a value of p was computed for a reaction that did not even occur by the mechanism on which the calculation was based! Today, the product of the collision frequency and the energy requirement is divided by the experimental rate to give p, which amounts, in truth, to a "fudge factor."

We can now write the rate of reaction according to collision theory as

$$\text{Rate} = p\pi d^2\bar{v}e^{-E_a/RT}[A][B] \tag{8-112}$$

This equation can be compared with the standard rate expression for a bimolecular elementary process:

$$\text{Rate} = k_2[A][B] \tag{8-113}$$

It then becomes obvious that

$$k_2 = p\pi d^2\bar{v}e^{-E_a/RT} \tag{8-114}$$

(In making this substitution it must be borne in mind that the concentration units in Equation 8-112 are molecules/ml, whereas the units in Equation 8-113 are typically moles/ℓ. Hence, the units of k_2 in Equation 8-114 are different from those in Equation 8-113, and the numerical values differ correspondingly.) If we now recall the form of the Arrhenius equation, we see that

$$A = p\pi d^2\bar{v} \tag{8-115}$$

i.e., the pre-exponential factor consists of a "steric" (plus other terms) factor, the collision diameter, and the average molecular velocity; this seems quite reasonable.

Molecular Beams

Before we turn to the other major theory of chemical kinetics, we should consider the relatively new technique of *molecular beams* for studying simple chemical reactions. This technique is based on an apparatus such as that shown schematically in Figure 8-26. The simplicity of this diagram belies the complexity and sophistication of the molecular-beam technique. Great ingenuity (and expense) is required to accomplish a seemingly simple objective: to cause molecules or atoms of two different kinds to collide with each other under precisely controlled conditions. The benefit of the investment is that much of the uncertainty with regard to each of the three terms in the collision theory can be minimized, if not eliminated.

The sources of the particles are ovens with small orifices for the hot gases to diffuse through. The velocity of the potassium atoms is controlled by the selector,

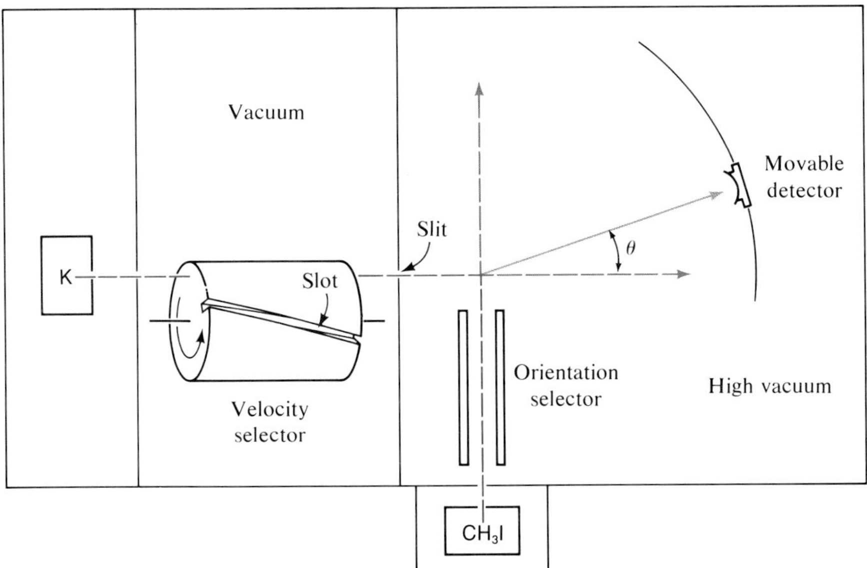

Figure 8-26. Molecular beam apparatus. The reaction is $K + CH_3I \rightarrow KI + CH_3$.

which is a rotating cylinder with a skewed slot cut into it. For a given speed of rotation, only those atoms having the correct velocity can travel down the slot without hitting the sides and being lost. The selector for the methyl iodide is an electrostatic device that controls the orientation of the dipole moment of this molecule. The movable detector records the numbers of reactant and product molecules scattered through a given angle θ.

Such experiments have already yielded much detailed information concerning the mechanisms of bimolecular gas-phase reactions, such as values for the reaction cross section (the area in which reaction can occur) and the partitioning of the internal energy of the products between translational and other components. Much more detailed information is to be expected from this technique with increasing refinements in both experimental methods and theoretical analyses. At present, it is restricted to those particles that can be easily volatilized and that contain relatively few atoms.

Transition-State Theory

The *transition-state theory* (also called the *absolute reaction-rate theory*) focuses upon the properties of the activated complex, which exists at the peaks of the curves (‡) in Figures 8-11, 8-19, and 8-20. The model upon which the theory is based corresponds roughly to the specific model that Michaelis and Menten used successfully for enzyme-catalyzed reactions. The general reaction sequence postulated is

$$\mathrm{A} + \mathrm{B} \rightleftharpoons \mathrm{AB}^{\ddagger} \rightarrow \text{Products} \qquad (8\text{-}116)$$

Transition-state theory, like the collision theory described earlier, is based on the assumption that the rate of a reaction can be predicted from a series of factors whose effects are multiplicative. The rate is taken to be directly proportional to the following three factors: (1) the concentration of the activated complex, (2) the probability that the complex will yield products rather than revert to reactants, and (3) the rate at which the complex decomposes to give products. We will examine each of these factors separately.

The theory assumes a true equilibrium between the reactant molecules and the activated complex. Thus, calculation of the concentration of the complex is straightforward:

$$\mathrm{A} + \mathrm{B} \rightleftharpoons \mathrm{AB}^{\ddagger} \qquad (8\text{-}117)$$

$$K^{\ddagger} = \frac{[\mathrm{AB}^{\ddagger}]}{[\mathrm{A}][\mathrm{B}]} \qquad (8\text{-}118)$$

or

$$[\mathrm{AB}^{\ddagger}] = K^{\ddagger}[\mathrm{A}][\mathrm{B}] \qquad (8\text{-}119)$$

This provides an expression for the first factor mentioned above.

The second factor, called the *transmission coefficient*, is generally given the symbol κ (Greek kappa). The activated complex is at the top of an "energy hill," and κ is the probability that the system will "roll" in the forward direction toward products instead of backward toward reactants. Its value ranges from about 0.5 to 1.0 in most instances.

The final factor is the rate of decomposition of the complex. In a simplified version of how this rate is determined, the activated complex is considered to yield products when one of its chemical bonds is stretched beyond its normal limit during a molecular vibration. The energy of this vibrational motion is given approximately by

$$E_v = \tfrac{1}{2}kT \qquad \text{(8-120)}$$

where k is the Boltzmann constant. This same energy can be approximated by

$$E_v = \tfrac{1}{2}h\nu \qquad \text{(8-121)}$$

where h is Planck's constant and ν is the frequency of vibration of the two atoms in question (or the frequency of a photon of sufficient energy to excite this particular vibration). If the molecule possesses at least $\frac{1}{2}kT$ of energy in this mode of vibration, reaction will occur in the time required for one oscillation. This time is the reciprocal of the frequency, ν^{-1} s. Therefore,

$$\text{Rate of decomposition of complex} = \nu = \frac{kT}{h} \qquad \text{(8-122)}$$

We can now combine these three factors to yield the transition-state prediction of the rate of a reaction:

$$\text{Rate} = K^{\ddagger}\kappa\frac{kT}{h}[\text{A}][\text{B}] \qquad \text{(8-123)}$$

As before, comparison of this rate equation with the standard rate expression for a bimolecular elementary process yields an expression for k_2, the bimolecular rate constant:

$$k_2 = K^{\ddagger}\kappa\frac{kT}{h} \qquad \text{(8-124)}$$

(Remember that the k on the right-hand side is not a rate constant!) Sophisticated methods are available for calculating $K^{\ddagger}$, so rate constants can be computed from transition-state theory as they can from collision theory.

Comparison of the Two Theories

The final results of the two theories, Equations 8-114 and 8-124, do not appear to have much in common. Their similarities and differences, as well as one of the strengths of transition-state theory, become more apparent if we relate the equilibrium constant $K^\ddagger$ to other thermodynamic parameters, via the Gibbs free energy:

$$(\Delta H^\circ)^\ddagger - T(\Delta S^\circ)^\ddagger = (\Delta G^\circ)^\ddagger = -RT \ln K^\ddagger \tag{8-125}$$

or

$$\ln K^\ddagger = \frac{(\Delta S^\circ)^\ddagger}{R} - \frac{(\Delta H^\circ)^\ddagger}{RT} \tag{8-126}$$

$$K^\ddagger = \exp[(\Delta S^\circ)^\ddagger/R] \exp[-(\Delta H^\circ)^\ddagger/RT] \tag{8-127}$$

The standard enthalpy change for the formation of $AB^\ddagger$—$(\Delta H^\circ)^\ddagger$—can often be equated with the activation energy E_a for the formation of this complex. We can thus write

$$K^\ddagger = \exp[(\Delta S^\circ)^\ddagger/R]\exp(-E_a/RT) \tag{8-128}$$

Substitution into Equation 8-124 gives

$$k_2 = \kappa\frac{kT}{h}\exp[(\Delta S^\circ)^\ddagger/R]\exp(-E_a/RT) \tag{8-129}$$

Comparison with Equation 8-114 now shows that both theories entail the energy factor $\exp(-E_a/RT)$ and a somewhat arbitrary probability factor p or κ. They differ in that collision theory requires a collision diameter and an average velocity, and shows a $T^{1/2}$ dependence of the pre-exponential factor (see Equation 6-42), whereas transition-state theory shows a direct proportionality between the pre-exponential factor and the absolute temperature.

The ultimate test of any theory is whether or not it correctly represents nature, or, more accurately, our experimental measurements of natural phenomena. If the rate equation for the bimolecular reaction

$$\text{A} + \text{B} \rightarrow \text{Products} \tag{8-130}$$

is given by

$$\text{Rate} = A\exp(-E_a/RT)[\text{A}][\text{B}] \tag{8-131}$$

where [A] and [B] are expressed as moles/ml, values of A can be obtained experimentally and from each of the two theories. The logarithms of A from each of these three sources for several bimolecular reactions are presented in Table 8-5. The data indicate that, in general, the transition-state theory represents the experimental measurements somewhat better than does the simpler collision theory.

Another advantage of the transition-state theory is that it provides some valuable information regarding the geometry of the activated complex. Rate constants k_2, as well as values of E_a, can be obtained experimentally. If a reasonable estimate for κ is available, an approximate value for $(\Delta S^\circ)^\ddagger$ can be obtained. This is very

Table 8-5. Values of log *A* from Experiment, Collision Theory, and Transition-State Theory for Several Bimolecular Reactions.

Reaction	log *A*		
	Experiment	Collision Theory	Transition-State Theory
$NO + O_3 \rightarrow NO_2 + O_2$	11.9	13.7	11.6
$NO + O_3 \rightarrow NO_3 + O$	12.8	13.8	11.1
$NO_2 + CO \rightarrow NO + CO_2$	13.1	13.6	12.8
$2NO_2 \rightarrow 2NO + O_2$	12.3	13.6	12.7
$NO_2 + F_2 \rightarrow NO_2F + F$	12.2	13.8	11.1
$F_2 + ClO_2 \rightarrow FClO_2 + F$	10.5	13.7	10.9

useful to have because it tells us something about the probable geometry of the activated complex. If $(\Delta S°)^{\ddagger}$ is positive, the complex has less order than the separate reactants. Values of $(\Delta S°)^{\ddagger}$ for bimolecular reactions are almost always negative, indicating the greater order resulting from the coalescence of two molecules into the activated complex.

Bibliography

Sidney W. Benson, **The Foundations of Chemical Kinetics,** McGraw-Hill, New York, 1960. This is a sophisticated, advanced, comprehensive classic in the kinetics literature. We recommend it to the beginner as a reference book, not as a text.

Sidney A. Bernhard, **The Structure and Function of Enzymes,** W. A. Benjamin, New York, 1968. Sidney Bernhard is a leading enzymologist. The first three chapters of this book provide a nice introduction to the forces between molecules and the structure of proteins. The remaining chapters describe in considerable detail the ways in which enzymes function.

Halvor N. Christensen and Graham A. Palmer, **Enzyme Kinetics,** W. B. Saunders, Philadelphia, 1967. Relatively few programmed texts are available in any of the fields of chemistry. This text is one of them. A question is asked that the student should be able to answer, based on his previous learning experience. The answer is given immediately in order to reinforce the learning process. The discussion through the Michaelis-Menten model of enzyme kinetics is particularly appropriate to the treatment of enzyme kinetics in the present chapter.

Arthur A. Frost and Ralph G. Pearson, **Kinetics and Mechanism,** 2nd ed., Wiley, New York, 1961. This is a widely used, upper-division text that is mathematically more sophisticated than the present chapter.

Edward L. King, **How Chemical Reactions Occur,** W. A. Benjamin, New York, 1963. This is an introductory text, written at about the level of the present chapter. We recommend it highly.

Problems

1. The reaction $4PH_3(g) \rightarrow P_4(g) + 6H_2(g)$ occurs in a flask at 600°C in the presence of an inert gas. The reverse reaction is $\frac{3}{2}$-order in the hydrogen pressure and $\frac{1}{4}$-order in the phosphorus pressure. Write the rate expression for the reverse reaction.

2. State whether each of the following assertions is true or false. If it is false, state why it is false.
   a. The addition of a catalyst to a chemically reacting system will increase the concentration of products at equilibrium.
   b. The rate of a reaction depends only on the overall stoichiometry of the reaction.
   c. The rate constant is a proportionality constant between the rate of reaction and the product of concentrations of species that affect the rate.
   d. A synonym for the order of a reaction is the molecularity of the reaction.
   e. The rate-determining step in a mechanism is the slowest step.

3. The kinetics for the hypothetical reaction $3A + 2B \rightarrow 2C + 3D$ was studied and the following data were obtained:

Run No.	[A], Initial Conc., mole/ℓ	[B], Initial Conc., mole/ℓ	Initial Rate, mole/ℓ h
1	6×10^{-3}	1×10^{-3}	36
2	6×10^{-3}	2×10^{-3}	72
3	1×10^{-3}	6×10^{-3}	6
4	2×10^{-3}	6×10^{-3}	24

   What is the rate expression for this reaction?

4. Consider the hypothetical elementary process $aA + bB + cC \rightarrow$ products. Doubling the concentrations of A, B, and C increases the overall rate of reaction by a factor of 64. When [A] and [B] are held constant, doubling [C] quadruples the rate. Quadrupling [A] has the same effect as quadrupling [B]. What are a, b, and c? Is it realistic to expect a reaction to occur by the simultaneous collision of as many particles as are implied in this hypothetical process?

5. The kinetics of the hypothetical reaction $2A + 3B \rightarrow 3C + 2D$ was studied by determining the initial rate of formation of D, which none of the solutions originally contained. The rate expression for this reaction can be written as

$$\frac{d[D]}{dt} = k[A]^a[B]^b[C]^c[D]^d$$

   The following data were obtained:

Run No.	[A], Initial Conc., mole/ℓ	[B], Initial Conc., mole/ℓ	[C], Initial Conc., mole/ℓ	Initial Rate of Formation of D, mole/ℓ h
1	1×10^{-2}	1×10^{-2}	1×10^{-2}	1×10^{-2}
2	2×10^{-2}	1×10^{-2}	1×10^{-2}	4×10^{-2}
3	1×10^{-2}	2×10^{-2}	2×10^{-2}	1×10^{-2}
4	2×10^{-2}	2×10^{-2}	1×10^{-2}	8×10^{-2}

a. Deduce the values of *a*, *b*, *c*, and *d*.
b. Calculate the value of *k*.

6. The kinetics of the reaction

$$CaCO_3(c) + 2HCl(aq) \rightarrow CaCl_2(aq) + CO_2(g) + H_2O(l)$$

was followed at 25°C by measuring the rate of evolution of CO_2. The following data were obtained for 20-mesh $CaCO_3$ in 0.037 *M* HCl with a constant stirring speed:

Time, s	35	49	65	80	93
Volume of CO_2, ml	20	30	40	50	60

What is the rate of the reaction in moles per second? (At 25°C and 1 atm, the molar volume of an ideal gas is 24.5×10^3 ml.)

7. Consider the reaction

$$N_2O_5(g) \rightarrow 2NO_2(g) + \tfrac{1}{2}O_2(g)$$

The data at 328 K for the change in partial pressure of N_2O_5 with time are

Time, s	0	1200	3600	6000	7200
$P_{N_2O_5}$, atm	0.458	0.243	0.076	0.024	0.013

Determine the order of the reaction by using the half-life concept.

8. The decomposition of dimethyl ether,

$$(CH_3)_2O \rightarrow CH_4 + CO + H_2$$

is first-order with respect to $(CH_3)_2O$. The rate constant is 1.35×10^{-3} s^{-1}. Calculate the half-life $\tau_{1/2}$ of this reaction.

9. When the activity of a particular radioactive substance was plotted in the form $\ln(c/c_0)$ vs *t*, a straight line of slope -3.5×10^{-4} s^{-1} was obtained. What is the half-life of the radioactive substance? How long will it take for it to decay to 1/10 of its original amount?

10. The age of whiskey can be determined by measuring its content of tritium, which has a half-life of 12.3 years. Calculate the age of a whiskey sample that is 0.200 times as radioactive as when it was new.

11. The conversion of *tert*-butyl bromide to *tert*-butyl alcohol in a solvent consisting of 90% acetone and 10% water can be written as

$$(CH_3)_3CBr + H_2O \rightarrow (CH_3)_3COH + HBr$$

A plot of the kinetic data shows that the reaction follows first-order kinetics and has a rate constant of $1.4 \times 10^{-5}\ s^{-1}$. How many seconds are required for 3/4 of the original *tert*-butyl bromide to react? How many seconds are required for the *tert*-butyl bromide concentration to be reduced from 1/8 to 1/16 of its original value?

12. The rate of the reaction $A + B \rightarrow C + D$ is first-order in both A and B, and second-order overall. When equal initial concentrations of A and B react, the reaction is 40% complete in 50 s. How long will it take for the reaction to go to 80% completion?

13. What is the activation energy of a reaction that is found to proceed with a rate constant of $2.7\ s^{-1}$ at 25°C and $7.4 \times 10^3\ s^{-1}$ at 60°C?

14. What is the activation energy of a reaction whose rate exactly quadruples when the temperature is increased from 25°C to 57°C?

15. If the activation energy of a reaction is 28 kcal/mole, at what temperature will the rate of the reaction be twice as great as at 400 K?

16. Consider the reaction $N_2O_5 \rightarrow 2NO_2 + \frac{1}{2}O_2$ and the data below, which give the temperature dependence of the rate constant:

T, K	k, s^{-1}
338	4.87×10^{-3}
328	1.50×10^{-3}
318	4.98×10^{-4}
308	1.35×10^{-4}
298	3.46×10^{-5}
273	7.87×10^{-7}

What is the activation energy of this reaction?

17. Why does a pressure cooker cook food faster?

18. The forward rate for the reaction $2NO + O_2 \rightarrow 2NO_2$ obeys the rate expression

$$\frac{-d[O_2]}{dt} = k_f[NO]^2[O_2]$$

What rate expression would you expect for the reverse reaction?

19. The rate of the reaction $Sn^{2+} + 2Ce^{4+} \rightarrow Sn^{4+} + 2Ce^{3+}$ is proportional to $[Sn^{2+}][Ce^{4+}]$. Suggest a possible mechanism to account for this observation.

20. The oxidation of plutonium(III) by nitrous acid in acid solution can be written as

$$Pu^{3+} + HNO_2 + H^+ \rightarrow Pu^{4+} + NO + H_2O$$

The proposed mechanism is

$$HNO_2 + H^+ \underset{k_{-1}}{\overset{k_1}{\rightleftharpoons}} NO^+ + H_2O \qquad \text{(Fast equilibrium)}$$

$$NO^+ + Pu^{3+} \xrightarrow{k_2} NO + Pu^{4+} \qquad \text{(Slow)}$$

Using the steady-state assumption, derive an expression for $[NO^+]$ in terms of the concentrations of the other species.

21. The following mechanism has been proposed for the alkaline hydrolysis of $[CoCl(NH_3)_5]^{2+}$, the chloropentamminecobalt(III) ion:

$$[CoCl(NH_3)_5]^{2+} + OH^- \underset{k_{-1}}{\overset{k_1}{\rightleftharpoons}} [CoCl(NH_2)(NH_3)_4]^+ + H_2O \qquad \text{(Fast equilibrium)}$$

$$[CoCl(NH_2)(NH_3)_4]^+ \xrightarrow{k_2} [Co(NH_2)(NH_3)_4]^{2+} + Cl^- \qquad \text{(Slow)}$$

$$[Co(NH_2)(NH_3)_4]^{2+} + H_2O \xrightarrow{k_3} [Co(OH)(NH_3)_5]^{2+} \qquad \text{(Fast)}$$

(Remember that the formulas of complex ions are always enclosed in brackets, so a second set of brackets may be used to denote "concentration of.")

a. Write the overall chemical reaction.
b. Write the rate expression for this reaction in terms of the product of concentrations $[[CoCl(NH_3)_5]^{2+}][OH^-]$.

22. Consider the gas-phase reaction

$$2NO + Cl_2 \rightarrow 2NOCl$$

for which the following mechanism has been postulated:

$$NO + Cl_2 \underset{k_{-1}}{\overset{k_1}{\rightleftharpoons}} NOCl_2 \qquad \text{(Fast equilibrium)}$$

$$NOCl_2 + NO \xrightarrow{k_2} 2NOCl \qquad \text{(Slow)}$$

Show that the rate expression for this reaction is second-order in $[NO]$ and first-order in $[Cl_2]$.

23. Carbon monoxide reacts with chlorine at high temperatures to give phosgene (all three of these gases happen to be very toxic). The experimental rate expression is

$$\frac{d[COCl_2]}{dt} = k[CO][Cl_2]^{3/2}$$

The following mechanism has been suggested:

$$Cl_2 \underset{k_{-1}}{\overset{k_1}{\rightleftharpoons}} 2Cl \qquad \text{(Fast equilibrium)}$$

$$Cl + CO \underset{k_{-2}}{\overset{k_2}{\rightleftharpoons}} COCl \qquad \text{(Fast equilibrium)}$$

$$COCl + Cl_2 \xrightarrow{k_3} COCl_2 + Cl \qquad \text{(Slow)}$$

a. Show that the suggested mechanism is consistent with the rate expression.
b. What is the algebraic relation between k in the rate expression and the rate constants in the mechanism?

INTERLUDE

"Would you be interested to know that it broke all records for coast-to-coast flight?"

the supersonic transport

In March, 1971, Congress voted to stop its financial support for the development of the supersonic transport. The technical information available to Congress at the time dealt with the anticipated effects of a fleet of about 500 supersonic transports on the oxygen-carbon dioxide balance in the stratosphere and the amounts of water vapor and particulate matter in the stratosphere, and the environmental effects that these changes would be likely to exert in the biosphere. The decision, which was a close one, was based on an uneasiness on the part of many congressmen concerning the probable environmental effects of an SST fleet. There is growing belief that this uneasiness was justified, but apparently for reasons that were not fully understood at the time.

The ultimate environmental effects of about 500 SST's flying about seven hours daily in the stratosphere cannot be predicted with certainty, but chemical theory can lead to predictions that are very probably correct. The position presented here is based largely on the opinions of Professor Harold S. Johnston (b. 1920), a distinguished chemical kineticist at the University of California, Berkeley. He argues that the rate of the photochemical conversion of ozone to oxygen in the stratosphere (see Reaction 8-17) is controlled by the concentration of nitrogen oxides, which are natural trace constituents of the stratosphere but which are also produced in significant amounts by the jet engines of the SST. Because these oxides (NO and NO_2, generally lumped together as NO_x) are believed to be so important, many attempts have been made to quantify their probable effects.

It is interesting to note that, at the time the decision against the SST was made, nitrogen oxides were not considered to be the major environmental threat. There is often no certainty that a scientific problem has been defined correctly. In this instance the congressional technical reports focused mainly upon water vapor and its probable reactions in the stratosphere. It is the opinion of Johnston and many others that water vapor is less significant in the chemistry of the stratosphere than are the nitrogen oxides. This opinion was reported in the press in time to affect the second vote in Congress (in May, 1971), which again went against the SST.

One can think of the Earth's atmosphere as consisting of several concentric, diffusely overlapping layers having characteristic physical and chemical properties. The *troposphere* is the lowest layer, extending from the surface to an altitude of approximately 10 km. It is characterized, *inter alia,* by a decreasing temperature with increasing altitude, a high water content and very low ozone content, and a rapid mixing of its contents in the vertical direction. From 0 km to 10 km the temperature decreases from about 290 K to 220 K, the pressure decreases from about 1 atm to 0.26 atm, and the total concentration of gas molecules decreases from about 2.5×10^{19} molecules/ml to 8.7×10^{18} molecules/ml.

The *stratosphere* extends approximately from 10 km to 50 km. It is characterized by an increasing temperature with increasing altitude (a temperature inversion), a low water content and low but significant ozone content, and slow vertical mixing. Over the range of altitudes from 10 km to 50 km, the temperature increases from about 220 K to 270 K, the pressure decreases from about 0.26 atm to 7.7×10^{-4} atm, and the concentration of molecules decreases from about 8.7×10^{18} molecules/ml to 2.1×10^{16} molecules/ml. This is a concentration range of 35% to 0.084% of the concentration at the Earth's surface.

Ozone, O_3, is a minor but extremely important component of the stratosphere. Its maximum concentration, at 25 km, is about 5×10^{12} molecules/ml, or about 6 parts per million (6 ppm). This figure drops by at least an order of magnitude at 50 km and 10 km. The importance of the ozone is that it constitutes a blanket of chemical protection from the dangerous solar ultraviolet radiation in the wavelength range of about 242 nm to 300 nm.* Without the ozone this radiation would reach the Earth and damage plants and animals so severely that life as we know it could not exist. Even a reduction of 5–10% in the ozone concentration, however, could have serious consequences, such as a sharp increase in the incidence of human skin cancer. The reaction by which ozone is destroyed in the stratosphere is catalyzed by nitrogen oxides.

The main source of the natural nitrogen oxides in the stratosphere is believed to be nitrous oxide, N_2O, which is produced by the bacterial reduction of nitrates in the soil and the oceans. N_2O is chemically rather inert and diffuses unchanged to the stratosphere, where it reacts with free oxygen atoms to form nitric oxide, NO. The NO is readily interconvertible with NO_2 via a number of reactions—some of them photochemical—involving O, O_2, and O_3.

In nature the slow but steady accumulation of NO_x in the stratosphere is offset by its loss at the same rate. This loss can occur only by convection and diffusion into the troposphere or by conversion to nitric acid, which

*Remember that 1 nm = 10 Å and that visible light covers the range 380–780 nm (3800–7800 Å). See Figure 2-8 to refresh your memory of the electromagnetic spectrum.

diffuses into the troposphere and is washed to earth by raindrops. The conversion reaction is brought about by hydroxyl radicals in the lower stratosphere:

$$HO + NO_2 + M \rightarrow HNO_3 + M$$

(M is a *third body,* i.e., any molecule that can dissipate the energy released in the reaction, thereby allowing the newly formed chemical bond to stabilize.) The actual concentration of NO_x in the stratosphere is somewhat uncertain. The concentration of NO has been measured, however, at an altitude of 28 km. The maximum concentration (during daylight) is about 2×10^9 molecules/ml, or about 4 parts per billion (4 ppb).

The balance between the rate of accumulation of NO_x and the rate of loss represents a steady-state equilibrium. Obviously, anything that significantly changes one rate without changing the other by the same amount will upset the equilibrium. It is feared that this is exactly what the SST would do. Table 1 shows the composition of the exhaust from *one* of the jet engines on an SST. It is estimated that a fleet of 500 SST's flying seven hours daily in the Northern Hemisphere would release enough NO to reduce the ozone concentration by about 16% in the Northern Hemisphere and about 8% in the Southern Hemisphere. This would cause an increase in the summertime ultraviolet radiation intensity in the United States equivalent to a latitude shift of about 15° southward. Thus the burning potential of the summer sun in Vermont would be comparable to that in Florida now!

Table 1. Intake and Exhaust Figures for One GE-4 Engine Operating in the Cruise Mode.

Substance	Intake, lb/h	Exhaust, lb/h
Air	1,380,000	
Fuel	33,000	
N_2		1,039,000
O_2		208,000
Ar		19,300
CO_2		103,500
H_2O		41,400
CO		1,400
NO		500 ± 170
SO_2		33
Hydrocarbons		16.5
Soot		5
	1,413,000	$\approx$1,413,000

Above the stratosphere lies the *mesosphere,* extending from about 50 km to 80 km, and the *thermosphere,* extending indefinitely from about 80 km. Ozone is produced by the photolysis* of oxygen, primarily in the upper stratosphere between 40 and 50 km. (A much smaller amount is produced in the 100–120-km region of the thermosphere.) Solar ultraviolet radiation of wavelengths less than 242 nm is absorbed by O_2, giving two O atoms:

$$O_2 \xrightarrow{h\nu(<242\ \text{nm})} 2O$$

The O atoms react with other O_2 molecules to give O_3 (a third body is again needed to dissipate the energy of reaction and allow the O_3 to stabilize rather than instantly fly apart):

$$O + O_2 + M \rightarrow O_3 + M$$

The ozone is eventually destroyed, primarily by the photolysis resulting from its strong absorption in the wavelength range 220–300 nm. Since the ultraviolet radiation up to 242 nm is effectively absorbed by O_2 anyway, ozone's primary protective effect is in the range 242–300 nm. The atmosphere is transparent to ultraviolet radiation in the region from 300 nm to 380 nm (where light begins), but this radiation is relatively harmless.

The initial photolytic reaction

$$O_3 \xrightarrow{h\nu(220\text{–}300\ \text{nm})} O_2 + O$$

is followed by either of two O-atom reactions, one of which consumes another O_3 molecule:

$$O + O_3 \rightarrow 2O_2 \qquad \Delta H° = -93.2\ \text{kcal/mole}$$

$$O + O + M \rightarrow O_2 + M \qquad \Delta H° = -118\ \text{kcal/mole}$$

These exothermic reactions occur primarily in the upper stratosphere and are the chief cause of the temperature inversion in the stratosphere.

Just as with the accumulation and loss of NO_x in the stratosphere, there is a steady-state equilibrium between the rate of formation and the rate of destruction of ozone. Table 2 gives a list of twelve of the important reactions that have a bearing on the ozone balance in the stratosphere. Even this list is not complete, however, because it neglects reactions involving H_2O and OH and perhaps others. The catalytic effect of NO_x on the rate of destruc-

---

***Photolysis* is chemical decomposition induced by visible or uv radiation.

Table 2. Some Elementary Processes Involved in the Ozone Balance in the Stratosphere.

Reaction	Rate Expression
a. $O_2 \xrightarrow{h\nu(<242\text{ nm})} 2O$	$k_a[O_2]$
b. $O + O_2 + M \rightarrow O_3 + M$	$k_b[O][O_2][M]$
c. $O_3 \xrightarrow{h\nu(220\text{–}300\text{ nm})} O_2 + O$	$k_c[O_3]$
d. $O + O + M \rightarrow O_2 + M$	$k_d[O]^2[M]$
e. $O + O_3 \rightarrow 2O_2$	$k_e[O][O_3]$
f. $NO + O_3 \rightarrow NO_2 + O_2$	$k_f[NO][O_3]$
g. $O + NO_2 \rightarrow NO + O_2$	$k_g[O][NO_2]$
h. $NO_2 \xrightarrow{h\nu(300\text{–}400\text{ nm})} NO + O$	$k_h[NO_2]$
i. $NO + NO + O_2 \rightarrow 2NO_2$	$k_i[NO]^2[O_2]$
j. $O + NO + M \rightarrow NO_2 + M$	$k_j[O][NO][M]$
k. $NO_2 + O_3 \rightarrow NO_3 + O_2$	$k_k[NO_2][O_3]$
l. $NO_3 \xrightarrow{h\nu(570\text{–}700\text{ nm})} NO + O_2$	$k_l[NO_3]$

tion of O_3 is implicit in Reactions f and g. Note that the sum of these two reactions is Reaction e (see also Reaction 8-21),

$$O + O_3 \rightarrow 2O_2$$

and that the NO is regenerated, thereby fulfilling the requirement that a catalyst be "unchanged" by the reaction.

The kinetics of this complex set of reactions has been analyzed to compare the rates of Reaction e in the presence and absence of NO_x. The ratio of the rate with NO_x present to the rate with NO_x absent is called the *catalytic ratio, ρ*. Professor Johnston has shown that the catalytic ratio is given by the equation

$$\rho = 1 + \frac{k_g[NO_2]}{k_e[O_3]}$$

Using typical natural conditions in the stratosphere, catalytic ratios of anywhere from 1 to 1000 are estimated. In fact, NO_x appears to be very important in regulating the concentration of ozone in the stratosphere. Figure 1 shows the results of a set of idealized calculations in which the concentration of NO_x was held constant throughout the stratosphere and gas flow

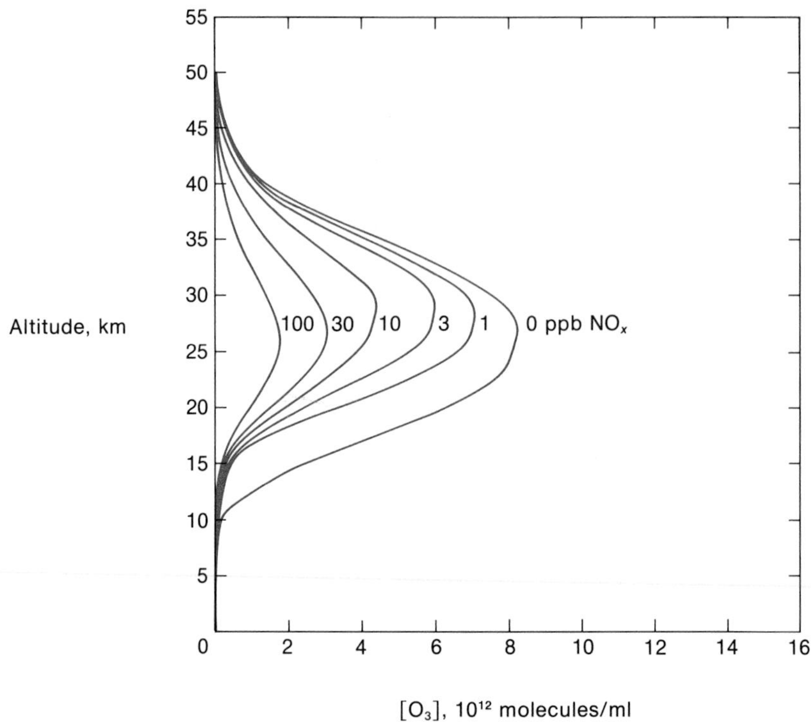

Figure 1. Calculated steady-state concentrations of O_3 in the stratosphere for various concentrations of NO_x (in ppb), under the idealized conditions of constant $[NO_x]$ at all altitudes and neglect of gas flow. (H. S. Johnston, "Catalytic Reduction of Stratospheric Ozone by Nitrogen Oxides," in *Advances in Environmental Science and Technology,* Vol. 4, ed. by J. N. Pitts, Jr., and R. L. Metcalf, Wiley, New York, 1974.)

was neglected. The figure shows a strong dependence of the ozone concentration on the NO_x concentration. In the absence of any NO_x, the ozone concentration at 25 km is calculated to be about 8×10^{12} molecules/ml. The observed maximum value is 5×10^{12} molecules/ml.

The effect of those 500 SST's flying seven hours daily remains impossible to predict with certainty, but there is little doubt that some depletion of the ozone layer would occur. Since the biological implications of such a depletion are not fully understood, caution seems the wisest approach. As is true with so many complex environmental problems, basic science must provide guidelines for action—or nonaction, as in the present example—based on man's always imperfect but steadily growing understanding of nature.

9 chemical equilibrium

We use the term *equilibrium* in our everyday speech to describe a variety of situations. Two children of identical weight, placed at identical distances from the fulcrum of a perfectly balanced see-saw, are said to be in equilibrium. A person is said to have lost his mental equilibrium if he displays abnormal behavioral patterns. Both these usages incorporate part of the fundamental concept of chemical equilibrium. We define *chemical equilibrium* as that state of a system in which no discernible changes in the macroscopic properties of the system occur with time. For example, no changes in the concentration of any component, or the amount of a phase, or the temperature, or the color of the system can be detected if the system is at equilibrium and we continue to monitor its significant properties.

Does this imply that at chemical equilibrium nothing at all is happening within the system? No more so than "mental equilibrium" implies a complete absence of mental activity. Just as any human mind is *always* engaged in processing incoming sensory information and directing bodily functions (e.g., breathing), any system at chemical equilibrium *always* entails a great deal of activity at a microscopic level.

Consider the chemical system in which A + B is converted to C + D:

$$aA + bB \rightleftharpoons cC + dD \qquad \Delta G^\circ < 0 \tag{9-1}$$

We have seen in Chapters 7 and 8 that an equilibrium constant K exists, for which the rigorous thermodynamic definition is

$$K = e^{-\Delta G^\circ/RT} \tag{9-2}$$

An equilibrium constant can always be represented in the form

$$K = \frac{[C]^c[D]^d}{[A]^a[B]^b} \tag{9-3}$$

where the species A, B, C, and D are either solutes in a liquid solution or gases, and their concentrations are those that exist *at equilibrium*, i.e., when the net rate of reaction is zero. The concentrations of pure solids and liquids do not appear in equilibrium-constant expressions because the concentration of a pure solid or liquid is essentially invariant. A gram of NaCl has the same number of moles per liter (moles per cubic decimeter, dm^3) of NaCl as does a kilogram. The concentration of water is 55.5 mole/ℓ, either as pure water or the solvent in an aqueous solution (except in fairly concentrated solutions, which will not be considered here), regardless of the volume. These constant concentrations are "built into" the equilibrium constant.

The usual concentration units are those of molarity (moles per liter) for solutes and partial pressure (atmospheres) for gases. Table 9-1 gives some examples of chemical equations and their corresponding equilibrium-constant expressions, which illustrate the above conventions. (The subscripts on the K's in the right-hand column will be explained in the course of this chapter.)

We have seen that the standard free-energy change $\Delta G°$ for a reaction is related to the equilibrium constant K by the equation

$$\Delta G° = -RT \ln K \tag{9-4}$$

This is an important equation. Surprisingly, it has escaped being given a name. As discussed earlier, a variety of methods are available for obtaining values of $\Delta G°$ from standard tables of thermodynamic data. Thus, equilibrium constants can be computed directly and accurately from such data. This is important because values of K are often difficult to measure. A powerful method for measuring K directly for reactions entailing electron transfer is described in Chapter 10.

Example 9-1. Calculate the equilibrium constant for the following reaction, which is employed by geologists for the detection of limestone in minerals:

$$CaCO_3(c) + 2H^+(aq) \rightarrow Ca^{2+}(aq) + CO_2(g) + H_2O(l)$$

Use the data given in Appendix D.

The standard free-energy change $\Delta G°$ of this reaction is given by

$$\begin{aligned}\Delta G° &= \Sigma\, n_p \Delta G_f° \text{ (Products)} - \Sigma\, n_r \Delta G_f° \text{ (Reactants)}\\ &= \Delta G_f°[Ca^{2+}(aq)] + \Delta G_f°[CO_2(g)] + \Delta G_f°[H_2O(l)]\\ &\quad - \Delta G_f°[CaCO_3(c)] - 2\Delta G_f°[H^+(aq)]\\ &= (-132.18 - 94.26 - 56.69 + 269.78 + 0) \text{ kcal/mole}\\ &= -13{,}350 \text{ cal/mole}\end{aligned}$$

The equilibrium constant at 25°C is obtained from the relation

Table 9-1. **Some Chemical Equations and Their Equilibrium-Constant Expressions.**

Equation	Equilibrium Constant
$H_2O(c) \rightleftharpoons H_2O(g)$	$K = P_{H_2O}$
$H_2O(l) \rightleftharpoons H_2O(g)$	$K = P_{H_2O}$
$H_2O(l) \rightleftharpoons H^+(aq) + OH^-(aq)$	$K_w = [H^+][OH^-]$
$La(IO_3)_3(c) \rightleftharpoons La^{3+}(aq) + 3IO_3^-(aq)$	$K_{sp} = [La^{3+}][IO_3^-]^3$
$2La(IO_3)_3(c) \rightleftharpoons 2La^{3+}(aq) + 6IO_3^-(aq)$	$K = [La^{3+}]^2[IO_3^-]^6 = (K_{sp})^2$
$La^{3+}(aq) + 3IO_3^-(aq) \rightleftharpoons La(IO_3)_3(c)$	$K = \dfrac{1}{[La^{3+}][IO_3^-]^3} = (K_{sp})^{-1}$
$CH_3COOH(aq) \rightleftharpoons H^+(aq) + CH_3COO^-(aq)$	$K_a = \dfrac{[H^+][CH_3COO^-]}{[CH_3COOH]}$
$16H^+(aq) + 2MnO_4^-(aq) + 10Cl^-(aq) \rightleftharpoons 8H_2O(l) + 2Mn^{2+}(aq) + 5Cl_2(g)$	$K = \dfrac{[Mn^{2+}]^2(P_{Cl_2})^5}{[H^+]^{16}[MnO_4^-]^2[Cl^-]^{10}}$
$CuSO_4 \cdot 5H_2O(c) \rightleftharpoons CuSO_4 \cdot 3H_2O(c) + 2H_2O(g)$	$K = (P_{H_2O})^2$
$\frac{1}{2}N_2(g) + \frac{1}{2}O_2(g) \rightleftharpoons NO(g)$	$K_1 = \dfrac{P_{NO}}{(P_{N_2})^{1/2}(P_{O_2})^{1/2}}$
$\frac{1}{2}N_2(g) + O_2(g) \rightleftharpoons NO_2(g)$	$K_2 = \dfrac{P_{NO_2}}{(P_{N_2})^{1/2}P_{O_2}}$
$NO(g) + \frac{1}{2}O_2(g) \rightleftharpoons NO_2(g)$	$K_3 = \dfrac{P_{NO_2}}{P_{NO}(P_{O_2})^{1/2}} = \dfrac{K_2}{K_1}$
$2NO(g) + O_2(g) \rightleftharpoons 2NO_2(g)$	$K = \dfrac{(P_{NO_2})^2}{(P_{NO})^2P_{O_2}} = (K_3)^2$

$$\Delta G° = -RT \ln K$$

$$\log K = -\frac{\Delta G°}{2.303RT} = \frac{13{,}350 \ \cancel{\text{cal/mole}}}{2.303(1.99 \ \cancel{\text{cal/K mole}})(298 \ \cancel{\text{K}})} = 9.78$$

$$K = 6.0 \times 10^9$$

The large negative value of $\Delta G°$ and the large value of K indicate that this reaction proceeds spontaneously at standard conditions. (Note that log K may be given to only three significant figures because both R and T are given to only three significant figures; K itself may be given to only two significant figures because there are only two significant figures in the mantissa of the logarithm.)

Equation 9-4 allows us to examine the meaning of the values of $\Delta G°$ and K. It was shown in Chapter 7 that $\Delta G°$ provides the criterion for the directionality of reactions. If $\Delta G° < 0$ the reaction as written proceeds spontaneously. From Equation 9-4 we see that, for $\Delta G°$ to be less than 0, K must be greater than 1. For K to be significantly greater than 1 in Equation 9-3, there must be higher concentrations of products than reactants at equilibrium (barring possible large effects due to the exponents). If $\Delta G° > 0$ the free energy of the products in their standard state exceeds the free energy of the reactants, K must be less than 1, and the reverse reaction is favored. It is rare that $\Delta G° = 0$, in which case $K = 1$.

9.1 EXTERNAL EFFECTS ON EQUILIBRIUM

It is useful to consider qualitatively what changes are caused on a system at equilibrium by an applied stress, before we examine quantitatively the conditions under which a variety of systems exist at equilibrium. People have been wrestling with this question for some time. The first man to provide a qualitatively correct answer was the French scientist Henri Louis Le Châtelier (1850–1936). His statement,* published in 1884, is a trifle verbose but perhaps it will give you some flavor of the struggles of mid-nineteenth-century scientists as they attempted to describe nature:

> "Every system in stable equilibrium which is submitted to the influence of an exterior cause that tends to produce variations in the temperature, for in the condensation (the pressure, the concentration, the number of molecules per unit volume) either of the whole system, or of some of its parts, can undergo only those interior changes which if they happened alone would bring about a change of temperature, or of condensation, of the opposite sign to that which results from the exterior cause.
>
> "These modifications are generally progressive and partial. They are sudden and complete only when they can happen without changing the individual condensations of the various homogeneous parts of the system in equilibrium, while, however, changing the condensation of the system as a whole.
>
> "They are of no effect when their occurrence cannot bring about changes analogous to that of the exterior cause. Finally, if these modifications are possible, they are not, therefore, necessary. If they do not happen, if the system remains unchanged, the equilibrium, however stable it was, becomes unstable, and can then undergo only those modifications which tend towards the conditions of stability."

*Quoted in M. M. P. Muir, *A History of Chemical Theories and Laws*, John Wiley, New York, 1907.

A modern, considerably shorter statement of Le Châtelier's principle that retains all the generality of his original statement is, *When a stress is applied to a system at equilibrium, the system responds in such a way as to minimize the effect of the stress and restore equilibrium.* Note that the system does not return to the *same* equilibrium state from which it is displaced by the stress, but to a *new* one.

Let us examine the usefulness of Le Châtelier's principle in predicting the effects of various stresses on a system, and compare these predictions with explanations of the observed effects. A system that lends itself well to such an examination is the dimerization of NO_2 to N_2O_4:

$$2NO_2(g) \rightleftharpoons N_2O_4(g) \qquad \Delta H^\circ = -13.9 \text{ kcal/mole} \tag{9-5}$$

What is the Effect of Adding NO_2 at Constant Volume? Le Châtelier's principle predicts that the effect of this stress will be minimized by the conversion of some of the reactant molecules to product molecules. This is, in fact, what happens. The explanation from the kinetic theory of gases is that, since there are more NO_2 molecules in the same volume, there are more NO_2-NO_2 collisions, so more reactions occur and more N_2O_4 is produced. The equilibrium constant $K = P_{N_2O_4}/(P_{NO_2})^2$ clearly shows that, if P_{NO_2} increases, $P_{N_2O_4}$ must also increase in order for equilibrium to be restored.

What is the Effect of Decreasing the Volume or Increasing the Pressure? Le Châtelier's principle predicts the formation of more N_2O_4 because the system can counteract both these stresses—which are equivalent in effect—by the conversion of two molecules (or moles, or volumes) to one. Again, this is what happens, and again, the explanation is that the increased pressure produces more collisions, which produce more dimerization. The equilibrium constant can also be invoked if we express it in concentration units as follows:

$$K = \frac{[N_2O_4]}{[NO_2]^2} = \frac{\dfrac{\text{moles } N_2O_4}{V}}{\left(\dfrac{\text{moles } NO_2}{V}\right)^2} = \frac{\text{moles } N_2O_4}{(\text{moles } NO_2)^2}\, V$$

If V decreases, the ratio of the numbers of moles must increase. This can occur only if the amount of N_2O_4 increases at the expense of the NO_2 molecules.

What is the Effect of Raising the Temperature? Because the forward reaction is exothermic, Le Châtelier's principle predicts that the equilibrium will shift to the left, in order to absorb the heat that is added to the system—and it does. The explanation is a little difficult. As discussed in Chapter 8, increasing the temperature increases the rates of all chemical reactions. Thus, the rates of both the forward and reverse reactions should be increased and we might be tempted to predict that there would be no change in the equilibrium concentrations of NO_2 and N_2O_4. However, it is found experimentally that the rates of endothermic reactions are increased more than the rates of exothermic reactions for a given positive ΔT. It is expected, then, that the rate of the reverse reaction in the present example will be greater, until equilibrium is restored, and more NO_2 will be produced. Finally, the experimental values of the equilibrium constant for Reaction 9-5 provide quantitative substantiation of the above predictions:

T, °C	K_{eq}
0	76
25	8.8
50	7.10

As the temperature is raised, the equilibrium constant, and hence the concentration of N_2O_4, decrease.

9.2 HETEROGENEOUS EQUILIBRIA

This section describes the conditions for equilibrium between the three common states of matter and liquid water. Three other types of heterogeneous systems exist: gas-solid, solid-solid, and gas-liquid-solid. There is no need to be encyclopedic, however. The following discussions will illustrate the basic ideas necessary for dealing with any heterogeneous system and will also give us practice in dealing with systems of very great interest to both the chemist and biologist.

Gas-Liquid Equilibria

Gas-liquid systems are relatively simple. For O_2 gas to be in equilibrium with water, the water must be saturated with oxygen. The partial pressure of the oxygen *over* the water determines the saturation concentration of the oxygen *in* the water, at any given temperature:

$$O_2(g) \rightleftharpoons O_2(aq)_{\text{satd}} \tag{9-6}$$

At moderate pressures of, say, less than 10 atm, this system can be described in the standard way by an equilibrium constant:

$$K_H = \frac{[O_2(aq)]_{\text{satd}}}{P_{O_2}} \tag{9-7}$$

Table 9-2. Henry's-Law Constants for Some Gases Dissolved in H_2O at 38°C.

Gas	K_H, mmole/ℓ Torr
N_2	0.0007
O_2	0.0014
CO_2	0.0325

SOURCE: F. M. Snell, S. Shulman, R. P. Spencer, and C. Moos, *Biophysical Principles of Structure and Function,* Addison-Wesley, Reading, Mass., 1965.

This expression is an illustration of *Henry's law,* discovered by the English chemist William Henry (1775–1836). It states that, at any given temperature, the solubility of a gas in a liquid is directly proportional to the partial pressure of the gas.* The proportionality factor, called the *Henry's-law constant,* is the equilibrium constant K_H in Equation 9-7. This behavior is very different from that of the solubilities of liquids or solids in liquids, which are nearly independent of the pressure on the liquid. With gases a tenfold increase in gas pressure at low-to-moderate pressures results in a tenfold increase in the concentration of the gas dissolved in the liquid.

Values of a few Henry's-law constants for gases dissolved in water are given in Table 9-2. With respect to solubility in water, it is clear that two types of gases exist. Nitrogen and oxygen are typical of gases that are poorly soluble in water. On the other hand, compounds such as CO_2, NH_3, and SO_2 have much higher "solubilities," apparently because they react chemically with water. In using these constants care must be exercised that the units of measure are consistent with those of the other terms that appear in the calculation.

Example 9-2. Calculate the flow of blood that would be required to meet the normal oxygen demand of an adult human if the only mechanism for O_2 transport were the solution of O_2 in the blood plasma. Assume, for simplicity, that blood consists entirely of plasma, that the plasma may be regarded as pure water, and that all O_2 is removed from the plasma by the cells of the body. Some basic data are that the average partial pressure of O_2 in the alveoli of the lungs is $\approx$100 Torr and that an adult human consumes $\approx$300 ml(STP)/min of O_2.

The concentration of O_2 in the plasma is given by Henry's law:

*Note that Raoult's law, which was discussed in Chapter 7, can be regarded as a special case of Henry's law.

$$[O_2] = K_H P_{O_2} = (0.0014 \text{ mmole}/\ell \text{ Torr})(100 \text{ Torr}) = 1.4 \times 10^{-1} \text{ mmole}/\ell$$

$$= 1.4 \times 10^{-4} \text{ mole}/\ell$$

The quantity of oxygen that is required per minute is calculated as

$$\frac{300 \text{ ml(STP)/min}}{22{,}400 \text{ ml(STP)/mole}} = 1.3 \times 10^{-2} \text{ mole/min}$$

Thus, the required blood flow is

$$\frac{1.3 \times 10^{-2} \text{ mole/min}}{1.4 \times 10^{-4} \text{ mole}/\ell} = 93\ \ell/\text{min}$$

Because the normal blood flow in an adult human is only ≈5.4 ℓ/min, it is clear that some mechanism other than the solution of O_2 in the plasma is required to keep us alive. We will examine some aspects of this mechanism, the binding of oxygen by hemoglobin, in Chapter 11.

Liquid-Liquid Equilibria

A common method for the identification of iodide ion in a sample is to oxidize the I^- to I_2 and shake the aqueous solution with carbon tetrachloride, which is immiscible with water. The resulting transfer of I_2 from the H_2O phase to the CCl_4 phase provides a nice illustration of the equilibrium distribution of a solute between two liquid phases:

$$I_2(H_2O) \rightleftharpoons I_2(CCl_4) \tag{9-8}$$

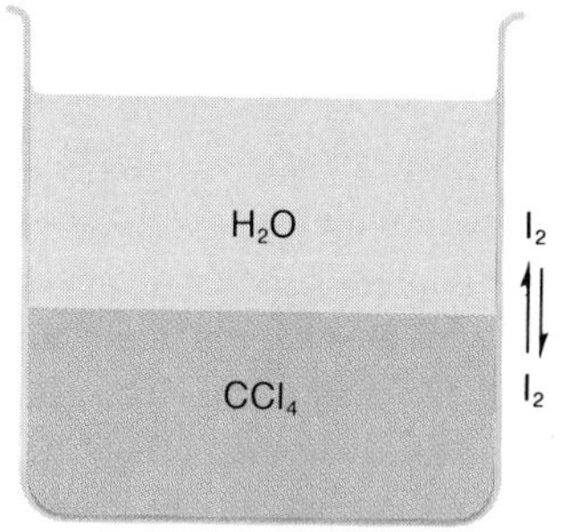

The equilibrium constant for this distribution is

$$K = \frac{[I_2]_{CCl_4}}{[I_2]_{H_2O}} \tag{9-9}$$

i.e., the ratio of the concentrations of the solute in the two liquid phases is a constant at equilibrium at a given temperature. At 25°C the value of K for this system is 87, indicating that I_2 is much more soluble in CCl_4 than in H_2O. This is what we might have predicted, knowing that I_2 and CCl_4 are nonpolar molecules, whereas H_2O is highly polar. Shaking an aqueous solution of I_2 with CCl_4 results in the rapid transfer of most of the iodine (and hence the violet color) to the nonpolar phase. As is true with all extraction processes, especially those for which the values of K do not differ from 1 by more than a few orders of magnitude, it is much better to use several small portions of CCl_4 in succession rather than one large portion because, mathematically, the effects of the individual extractions are multiplicative, not additive. Do you see why?

The technique of liquid-liquid extraction is widely used in organic chemistry. A fairly nonpolar substance, such as an alcohol, can be removed from an aqueous solution by shaking the solution with another organic liquid in which the substance is more soluble than it is in water. The two immiscible phases are separated and the aqueous phase, depleted in solute, can be equilibrated a second time with a fresh portion of the organic liquid, if necessary.

Solid-Liquid Equilibria

Solid-liquid equilibrium phenomena in heterogeneous systems are very important because they form the basis of many of the procedures for separating ions in aqueous solutions. Although wet-chemical precipitations are much less commonly used in industrial and academic analytical laboratories than they used to be, most students of chemistry spend a good deal of time in qualitative analysis laboratories and in doing solubility calculations. This is justified partly by the usefulness of the concepts in question to all scientists and partly by the practice it gives the student in handling stoichiometry and equilibrium-constant expressions.

Some concrete examples may make the best case for studying solid-liquid equilibria in some detail. Most water supplies now contain added fluoride ion, just as they often contain some chlorine, but for a very different purpose. The chlorine is there to kill bacteria, whereas the fluoride is there to inhibit tooth decay, especially in the more vulnerable teeth of children. Tooth enamel consists mainly of hydroxyapatite, $Ca_{10}(PO_4)_6(OH)_2$, a complex mineral that also constitutes the bulk of the inorganic matter in bones. Dental caries (tooth decay) is believed to begin with the attack on this compound by the organic acids produced by oral bacteria. Dental researchers have compared the rate of dissolution of untreated enamel (the "control" specimens) by dilute acids with the rates of dissolution of enamel previously treated with two fluoride compounds dissolved in water. Figure 9-1 shows the percentage reduction in the dissolution rates of the treated teeth with respect to the controls, as a function of the fluoride-ion concentration.

The improvement is dramatic. The reactions thought to occur when the fluoride is applied via dentifrices and when it is administered in drinking water are the following two reactions, respectively:

$$\underset{\text{Hydroxyapatite}}{Ca_{10}(PO_4)_6(OH)_2} + 20F^- + 8H^+ \rightarrow \underset{\text{Calcium fluoride}}{10CaF_2} + 6HPO_4^{2-} + 2H_2O \quad (9\text{-}10)$$

$$\underset{\text{Hydroxyapatite}}{Ca_{10}(PO_4)_6(OH)_2} + 2F^- \rightarrow \underset{\text{Fluoroapatite}}{Ca_{10}(PO_4)_6F_2} + 2OH^- \quad (9\text{-}11)$$

The protection process thus consists of dissolving away a very thin layer of perfectly good enamel and replacing it with either of these two fluoro compounds. Apparently the new compounds are more insoluble than the natural enamel and more resistant to acid attack.

The reason for the large disparity between the two curves in Figure 9-1 is the subject of current investigations. Neither the stannous ion (Sn^{2+}) nor the sodium ion alone has any protective effect. (By "alone" we mean accompanied not by F^- but rather by an inactive anion, such as NO_3^- or Cl^-.) However, the tin compound is known to form hydrated tin(II) oxides, which may have some synergistic effects with the fluoro compounds. Quite clearly, the results discussed above point to the wisdom of brushing with dentifrices containing SnF_2.

The insolubility of certain inorganic compounds is often exploited by the medical profession. For a variety of diagnostic purposes, it is desirable to

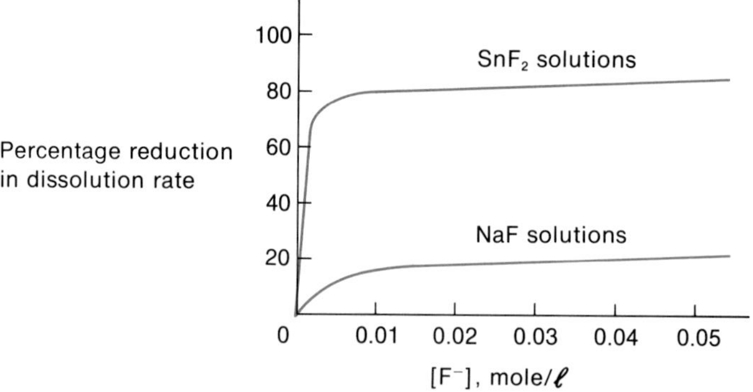

Figure 9-1. Protection of tooth enamel by fluoride treatment. [William E. Cooley, *Journal of Chemical Education* **47**, 177 (1970).]

"see" the physical structure of the upper and lower intestines. Unfortunately, most animal tissues are transparent to x rays. To make them opaque to x rays requires the presence of a heavy, electron-rich element. The Ba^{2+} ion would be ideal if it were not so poisonous. However, because of the extremely low solubility of $BaSO_4$, this white salt can be swallowed as an aqueous slurry, the x-ray photographs recorded, and the salt excreted with no damage to the system.

Small amounts of certain mercury compounds have been found useful by physicians as cathartics, antiseptics, antisyphilitics, and diuretics. However, as the general public has recently become aware, owing to the appearance of high concentrations of mercury in various fish and waterfowl, mercury in large amounts is a deadly poison. (The fanciful phrase, "mad as a hatter," stems from the fact that long ago many hatters did suffer from minor mental disorders such as excessive anxiety, depression, and irritability, not from the tedium of their jobs, but from the chronic effects of the mercuric nitrate, $Hg(NO_3)_2$, that was used in the processing of furs into felt.) Tiny doses of mercury in various forms, including Hg_2Cl_2, can be administered to humans. Mercurous chloride (a formerly popular cathartic called *calomel*) has a very low solubility and provides enough, but not too much, mercury to the system.

Our final example is the solubility of DDT [dichlorodiphenyltrichloroethane, $(ClC_6H_4)_2CHCCl_3$] in water. DDT has been a phenomenal boon to the farmer—especially in underdeveloped countries—in reducing insect populations on agricultural crops. However, in recent years it has become apparent that DDT is also responsible for the decline in the populations of ospreys and eagles, and that it is now found in every corner of the globe—from the penguins of Antarctica to the fatty tissue of all of us. Why is this particular compound so widely distributed in animals and why does it persist for so long? Part of the answer can be seen in the structure of the molecule:

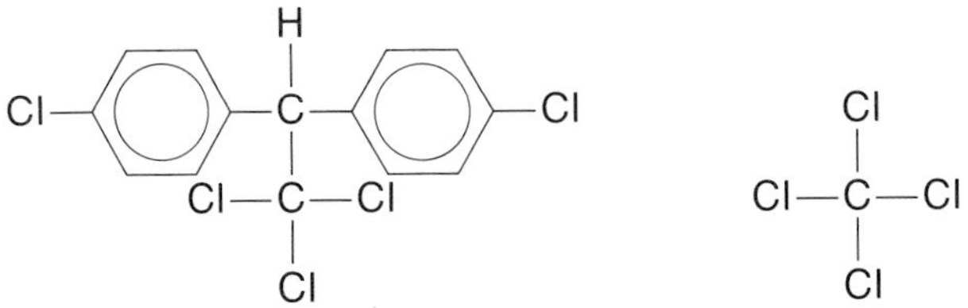

DDT

Carbon tetrachloride

The bonds in DDT are primarily C—C and C—Cl bonds, the strengths of which are typically about 80 kcal/mole (see Table 4-5). These bonds are very stable and are quite resistant to degradation by compounds and conditions normally encountered in our environment. The structure of carbon tetrachloride is shown next to that of DDT to demonstrate the similarity of the two compounds. As we saw in the preceding section, CCl_4 and H_2O are immiscible, and we would expect the solid DDT to be insoluble in H_2O also. Thus, it is not excreted with the aqueous cellular fluids of bacteria and algae or the higher organisms that feed on them, but rather is concentrated in the body fat of the higher organisms because of its relatively nonpolar structure. This is some cause for concern, in that part of the nerve tissue is also fatty. It is clear that either pesticides that are rapidly biodegradable or improved biological controls, such as the selective sterilization of insect populations, will have to be developed if we are to protect our fragile biosphere.

These are just a few examples of the importance of solubility phenomena in our existence. It will be profitable for us to see what kinds of compounds are soluble and insoluble and then do some quantitative work that will enable us to predict the behavior of such systems.

The general rule for liquid-liquid solubility, alluded to above, is that "like dissolves like." Polar liquids are miscible with polar liquids, and nonpolar with nonpolar; most polar liquids do not mix with nonpolar ones. The origin of this behavior lies in the dipole-dipole forces and hydrophilic and hydrophobic interactions between molecules, which were described in Chapter 5. Similar considerations lead us to generally correct predictions of the solubilities of polar and nonpolar organic solids in solvents of the two kinds, because these compounds are rarely ionized in solution. For inorganic solids, which usually *are* ionized in solution, the problem is complicated by solute-solvent interactions, which are often quite strong. Some general rules for the solubilities of ionic compounds in water are presented in Table 9-3.

Equilibria Involving Slightly Soluble Ionic Compounds

There is a large body of experimental information regarding slightly soluble (essentially "insoluble") ionic compounds. For the general solution reaction of a slightly soluble compound,

$$\mathrm{M}_a\mathrm{N}_b(c) \rightleftharpoons a\mathrm{M}(aq) + b\mathrm{N}(aq) \tag{9-12}$$

the equilibrium constant is

$$K = \frac{[\mathrm{M}]^a[\mathrm{N}]^b}{[\mathrm{M}_a\mathrm{N}_b]} \tag{9-13}$$

However, because the concentration of the solid phase is itself a constant (as discussed at the beginning of this chapter), a new equilibrium constant

Table 9-3. General Rules for the Solubilities of Ionic Compounds in Water.

Soluble Compounds	Important Exceptions
All nitrates	
Most acetates	Silver acetate
Most chlorates and perchlorates	
Most chlorides, bromides, and iodides	Halides of Pb(II), Ag(I), and Hg(I)
Many sulfates	Sulfates of heavy divalent metals
Most salts and bases of alkali metals and the ammonium ion	

Insoluble Compounds	Important Exceptions
Most hydroxides	Hydroxides of alkali metals, barium, and the ammonium ion
Most sulfides	Sulfides of metals whose cations have noble-gas structures, and the ammonium ion
Most carbonates, phosphates, arsenates, and sulfites	Compounds of alkali metals and the ammonium ion

called the *solubility-product constant,* or simply the *solubility product,* can be defined as

$$K_{sp} = K[M_aN_b] = [M]^a[N]^b \tag{9-14}$$

This simple equation answers many difficult questions regarding solid-liquid equilibria. It is so simple, in fact, that beginners often take one look at it and assume they know all there is to know about solubility products. The first time they are challenged to use it, however, they generally discover that there are a few fine points that had slipped past them. To reduce stubbed toes to a minimum, we will give some general principles and then a number of examples.

It is useful to define a term called the *ion product,* Q_{ion}:

$$Q_{ion} = [M]^a[N]^b \tag{9-15}$$

This expression is identical in form to that which defines K_{sp}, but quite different in concept. The solubility product is a constant that applies *only* to a saturated solution in equilibrium with the undissolved solid. The ion product, on the other hand, can assume any number of values:

$$K_{sp} = \text{constant at given } T \tag{9-16}$$

$$0 \leq Q_{ion} < \infty \tag{9-17}$$

There are three possible relations between the values of Q_{ion} and K_{sp} in any given solution:

1. $Q_{ion} < K_{sp}$. This case applies to all *unsaturated* solutions, which are at stable equilibrium.
2. $Q_{ion} = K_{sp}$. This is the special case in which a *saturated* solution is in equilibrium with the undissolved solid.
3. $Q_{ion} > K_{sp}$. This case applies to all *supersaturated* solutions, which contain—at least temporarily—more dissolved solute than do the corresponding saturated solutions at the temperature in question.

There are two general kinds of solutions that solubility products describe: (1) solutions in which the only source of ions is the parent solid itself, and (2) solutions obtained by the mixing of other solutions containing the ions in question (plus other ions). Data for solutions in which the ions come only from the parent solid are given either as the solubility s or the solubility product K_{sp} of the salt. The *solubility* of a solid is the number of moles per liter of the solid that are in solution when excess solid is in equilibrium with the solution. Thus, if s is the solubility of the generalized compound M_aN_b in Equation 9-12, we can write

$$K_{sp} = (as)^a(bs)^b = a^ab^bs^{(a+b)} \tag{9-18}$$

If K_{sp} is known, s can be computed; if s is known, K_{sp} can be computed; and if the concentration of either ion is known, either s or K_{sp} can be computed. (A number of K_{sp} values are tabulated in Appendix F.) Two examples follow.

Example 9-3. In order to grow well, pineapples in Hawaii must be sprayed with a solution of $FeSO_4$, even though the plants grow in a bright red soil containing 20% iron. The problem is that much of the iron is in the form of $Fe(OH)_3$. Given that the K_{sp} of $Fe(OH)_3$ is 1.1×10^{-36}, calculate its solubility to explain why the plants receive so little dissolved iron from the soil.

$$Fe(OH)_3(c) \rightleftharpoons Fe^{3+} + 3OH^-$$

$$K_{sp} = [Fe^{3+}][OH^-]^3$$

One mole of $Fe(OH)_3$ yields one mole of Fe^{3+}. Therefore, s = solubility = $[Fe^{3+}]$.

$$[OH^-] = 3[Fe^{3+}] = 3s$$

$$K_{sp} = s(3s)^3 = 27s^4 = 1.1 \times 10^{-36} \text{ mole}^4/\ell^4$$

$$s^4 = \frac{1.1 \times 10^{-36}}{27} = 4.1 \times 10^{-38} \text{ mole}^4/\ell^4$$

$$s = 4.5 \times 10^{-10} \text{ mole}/\ell$$

Example 9-4. The solubility of Ag_2SO_4 is 1.4×10^{-2} mole/ℓ. Calculate K_{sp}.

$$Ag_2SO_4(c) \rightleftharpoons 2Ag^+ + SO_4^{2-}$$

One mole of Ag_2SO_4 yields one mole of SO_4^{2-}. Therefore, s = solubility = $[SO_4^{2-}]$.

$$[Ag^+] = 2[SO_4^{2-}] = 2s$$

$$K_{sp} = [Ag^+]^2[SO_4^{2-}] = (2s)^2 s = 4s^3$$

$$K_{sp} = 4(1.4 \times 10^{-2})^3 = 1.1 \times 10^{-5} \text{ mole}^3/\ell^3$$

Several procedures should be noted in the above calculations and followed in similar ones. The chemical equation and the K_{sp} expression are always written out fully. Care must be exercised in selecting the ion whose concentration is to be set equal to the solubility. Students sometimes think it somewhat "unfair" or "illogical" that one of the ions affects K_{sp} by a factor of 2^2 or 3^3 in comparison with its partner. Because two or three of the ions are required to form the precipitate, however, it is both fair and logical that the calculation reflect this.

The other general kind of solution, the kind resulting from a mixture of other solutions, is often more difficult to handle mathematically, but it is the kind more commonly met in practice. It is often useful to formulate the expression for Q_{ion} from the given concentrations and then compare its value with that of K_{sp}. Three examples will demonstrate some of the kinds of problems that can be handled in this way.

Example 9-5. All introductory qualitative analysis schemes rely heavily on solubility differences for the quantitative separation of ions. A reagent that is particularly useful for this purpose (although exceedingly disagreeable—and poisonous) is H_2S. In aqueous solution it is a *diprotic acid,* which means that it dissociates to give two hydrogen ions. This particular acid is extremely weakly ionized, having an overall dissociation equilibrium constant of only 1.0×10^{-19} mole^2/ℓ^2:

$$H_2S(aq) \rightleftharpoons 2H^+(aq) + S^{2-}(aq)$$

$$K = \frac{[H^+]^2[S^{2-}]}{[H_2S]} = 1.0 \times 10^{-19} \text{ mole}^2/\ell^2$$

In several commonly employed procedures, $[H^+]$ is adjusted to, say, 0.30 mole/ℓ with some strong acid. The solution is then saturated with H_2S, which gives a constant $[H_2S] = 0.10$ mole/ℓ. The concentration of sulfide ion is therefore given by

$$[S^{2-}] = \frac{K[H_2S]}{[H^+]^2} = \frac{1.0 \times 10^{-19} \times 0.10}{0.30^2} = 1.1 \times 10^{-19} \text{ mole}/\ell$$

Given 10 ml of such a solution containing 100 mg each of Cd^{2+}, Ni^{2+}, and Zn^{2+}, can these ions be precipitated quantitatively?

We will answer this question by first calculating the concentration of each metal ion in a solution at equilibrium with its solid sulfide, and then converting these values to milligrams, recognizing that about 0.1 mg is the limit of detection in most wet-chemical elemental analyses.

$$CdS(c) \rightleftharpoons Cd^{2+} + S^{2-}$$

$$K_{sp} = [Cd^{2+}][S^{2-}] = 3.6 \times 10^{-29} \text{ mole}^2/\ell^2$$

$$[Cd^{2+}] = \frac{3.6 \times 10^{-29}}{1.1 \times 10^{-19}} = 3.3 \times 10^{-10} \text{ mole}/\ell$$

$$Cd^{2+} \text{ in 10 ml soln} = (0.010\ \not{\ell})\,(3.3 \times 10^{-10}\ \cancel{\text{mole}}/\not{\ell})\,(112\ \text{g}/\cancel{\text{mole}})$$

$$= 3.7 \times 10^{-10} \text{ g} = 3.7 \times 10^{-7} \text{ mg}$$

Therefore, Cd^{2+} is quantitatively precipitated from the solution. Similarly, using the values $K_{sp} = 1.4 \times 10^{-24}$ and $K_{sp} = 1.2 \times 10^{-23}$ for NiS and ZnS, respectively, we find that

$$Ni^{2+} \text{ in 10 ml soln} = 7.7 \times 10^{-3} \text{ mg}$$

$$Zn^{2+} \text{ in 10 ml soln} = 7.2 \times 10^{-2} \text{ mg}$$

The prediction that the Zn^{2+} will be quantitatively precipitated turns out to be wrong if HCl is the acid used to adjust $[H^+]$ to 0.30 mole/ℓ. This is because Zn^{2+} forms soluble complexes with Cl^-, such as $ZnCl^+$, thus reducing $[Zn^{2+}]$ by a mechanism other than precipitation. Even if HCl is not the acid used, the prediction could be wrong because ZnS solutions supersaturate readily.

The reactions in the above example illustrate the commonly encountered phenomenon of *competing equilibria*, which should be studied closely.

Example 9-6. The bulk of the inorganic matter in bones is hydroxyapatite, $Ca_{10}(PO_4)_6(OH)_2$, for which K_{sp} has the infinitesimal value 10^{-111} mole^{18}/ℓ^{18}.

The concentration of free Ca^{2+} (that which is not bound to protein) in normal blood plasma is about 2.5×10^{-3} mole/ℓ. The concentration of PO_4^{3-} is about 2.2×10^{-3} mole/ℓ, and that of OH^- is 6.3×10^{-7} mole/ℓ. Let us calculate Q_{ion} for this solution, compare it with K_{sp}, and try to explain the results.

The reaction we are considering is

$$Ca_{10}(PO_4)_6(OH)_2(c) \rightleftharpoons 10Ca^{2+} + 6PO_4^{3-} + 2OH^-$$

The value of Q_{ion} is given by

$$\begin{aligned} Q_{ion} &= [Ca^{2+}]^{10}[PO_4^{3-}]^6[OH^-]^2 \\ &= (2.5 \times 10^{-3})^{10}(2.2 \times 10^{-3})^6(6.3 \times 10^{-7})^2 \\ &= (9.5 \times 10^{-27})\,(1.1 \times 10^{-16})\,(40 \times 10^{-14}) \\ &= 420 \times 10^{-57} = 4.2 \times 10^{-55}\ \text{mole}^{18}/\ell^{18} \end{aligned}$$

What in the world is going on? The value of Q_{ion} is $10^{-55}/10^{-111} = 10^{56}$ times larger than K_{sp}! We are forced to conclude that our calculation is wrong. Where should we look to find the source of the error?

In a calculation of this sort, a good place to begin is with the concentrations of the ions. Because their concentrations are raised to such high powers, a slight error in $[Ca^{2+}]$ or $[PO_4^{3-}]$ will give a huge error in the answer. Such infinitesimal values of K_{sp} are often suspect for this reason. Hydroxyapatite solutions may supersaturate readily. Furthermore, thermodynamic considerations indicate that simple concentrations are not exactly correct to use in the K_{sp} expression; they must be corrected for the effects of interionic interactions. The possible applicability of such factors must always be taken into account in calculations of Q_{ion} and K_{sp} for real systems.

Solid-liquid equilibria must obey Le Châtelier's principle and be amenable to the kind of reasoning illustrated above. Thus, the addition to a solution of an ion involved in precipitate formation should cause more precipitate to form and reduce the concentration of the other ion proportionally. This effect is called the *common-ion effect*. The last example of this section provides a quantitative illustration of it.

Example 9-7. At 25°C one liter of water dissolves 2.3 mg $BaSO_4$. Calculate the solubility of $BaSO_4$ in pure water and in a 0.10 *M* Na_2SO_4 solution.

$$BaSO_4(c) \rightleftharpoons Ba^{2+} + SO_4^{2-}$$

$$K_{sp} = [Ba^{2+}][SO_4^{2-}]$$

One mole of $BaSO_4$ yields one mole of Ba^{2+} and one mole of SO_4^{2-}. In pure water the solubility of $BaSO_4$ is therefore

$$s = [Ba^{2+}] = [SO_4^{2-}] = \frac{2.3\ \cancel{mg(BaSO_4)}/\ell}{(10^3\ \cancel{mg}/\cancel{g})[233\ \cancel{g(BaSO_4)}/\text{mole}]} = 1.0 \times 10^{-5}\ \text{mole}/\ell$$

To calculate the solubility in 0.10 M Na_2SO_4, we must know the value of K_{sp}. This is obtained from the solubility value determined above:

$$[Ba^{2+}] = [SO_4^{2-}] = 1.0 \times 10^{-5}\ \text{mole}/\ell \text{ in pure } H_2O$$

$$K_{sp} = (1.0 \times 10^{-5})^2 = 1.0 \times 10^{-10}\ \text{mole}^2/\ell^2$$

In 0.10 M Na_2SO_4, $[SO_4^{2-}] = 0.10$ mole/ℓ. Thus, because a large excess of SO_4^{2-} is already present, we cannot use its concentration as a measure of the solubility of $BaSO_4$. We must define our unknown, the solubility, as $[Ba^{2+}]$, which we will call x:

$$x = [Ba^{2+}]$$

Clearly, x also equals the concentration of the SO_4^{2-} that comes from the dissolved $BaSO_4$, so

$$[SO_4^{2-}] = 0.10 + x\ \text{mole}/\ell$$

The expression for K_{sp} requires that

$$x(0.10 + x) = 1.0 \times 10^{-10}\ \text{mole}^2/\ell^2$$

The solubility in pure water is 1.0×10^{-5} mole/ℓ, and by Le Châtelier's principle we know that the solubility will be even less in this solution, so we can neglect x in the term $0.10 + x$. We then obtain

$$0.10x = 1.0 \times 10^{-10}\ \text{mole}^2/\ell^2$$

and the solubility of $BaSO_4$ in 0.10 M SO_4^{2-} solution is

$$x = 1.0 \times 10^{-9}\ \text{mole}/\ell$$

The important point here is that the solubility has decreased by a factor of 10,000 owing to the presence of a moderate concentration of one of the ions. This is the common-ion effect. It can be used to good advantage in many chemical separations involving ionic equilibria.

9.3 HOMOGENEOUS EQUILIBRIA

Gas-Phase Equilibria

Several of the concepts relating to gas-phase equilibria have been discussed earlier with regard to the $2NO_2 \rightleftharpoons N_2O_4$ reaction. The following example demonstrates the power of the equilibrium-constant concept to describe such systems.

Example 9-8. The equilibrium constant for the nitrogen dioxide–dinitrogen tetroxide reaction at 0°C is 76 atm^{-1}. Calculate the equilibrium concentrations of both species if 1.00 mole of N_2O_4 is introduced into a 22.4-ℓ container. Assume that both gases behave ideally.

$$2NO_2(g) \rightleftharpoons N_2O_4(g) \qquad K = \frac{P_{N_2O_4}}{(P_{NO_2})^2}$$

$$PV = nRT$$

$$P_{\text{initial}} = \frac{nRT}{V} = \frac{(1.00\ \text{mole})(0.0821\ \ell\ \text{atm/K mole})(273\ \text{K})}{22.4\ \ell} = 1.00\ \text{atm}$$

Let x = the loss in $P_{N_2O_4}$ resulting from the dissociation. The partial pressure of N_2O_4 falls from 1.00 to $1.00 - x$ and the partial pressure of NO_2 increases from 0 to $2x$. Therefore,

$$K = \frac{1.00 - x}{(2x)^2} = 76\ \text{atm}^{-1}$$

$$1.00 - x = 300x^2$$

$$300x^2 + x - 1.00 = 0$$

The best way to solve for x is by the *quadratic formula*,

$$ax^2 + bx + c = 0 \tag{9-19}$$

for which the exact solution is

$$x = \frac{-b \pm \sqrt{b^2 - 4ac}}{2a} \tag{9-20}$$

The plus-or-minus sign in front of the square-root sign can always be correctly interpreted on the basis of the physical situation, i.e., one choice will lead to a physical impossibility and the other will not. We can now solve for x as follows:

$$x = \frac{-1 + \sqrt{1 + 1200}}{600} = \frac{-1 + 35}{600} = \frac{34}{600} = 0.057 \text{ atm}$$

Thus, the equilibrium partial pressures are $P_{N_2O_4} = 0.94$ atm and $P_{NO_2} = 0.11$ atm, giving a total equilibrium pressure of 1.05 atm.

The above example demonstrates that calculating gas-phase equilibria entails primarily the counting of particles, both before and after reaction occurs. If this can be done and the stoichiometry of the reaction is known, it is a straightforward process to calculate either K or the equilibrium concentrations, depending on what data are given. The following example illustrates the simplicity of such calculations nicely.

Example 9-9. A snowbound skier has 100 walnuts in his backpack. To pass the time, he carefully cracks some of the nuts into two half-shells and the intact kernel. Night falls and he cannot see how many of the nuts he has cracked. Because of the intense cold, he must put his ski mittens on, so he can only count the number of pieces he has, which turns out to be 130. How can he calculate the degree of dissociation of the walnuts?

We can represent the half-shells by the symbol S, the kernels by K, and hence the complete nut by KS_2. The "reaction" is thus

	$KS_2 \rightarrow$	$K +$	$2S$
Initial numbers:	100	0	0
Final numbers:	$100 - x$	x	$2x$

where x = the number of nuts that were cracked. The total number of pieces after cracking is

$$130 = 100 - x + x + 2x = 100 + 2x$$

so $x = 15$, and the percent dissociation is $15/100 = 15\%$.

Liquid-Phase Equilibria

Earlier in our discussion of the states of matter, we referred to the ubiquity and importance of water. It is indeed the "universal solvent." Because of these factors and because it is the principal solvent used in general chemistry laboratories, we will restrict our discussion solely to aqueous solutions. We will examine only acid-base and complex-ion equilibria, devoting the major amount of our time to the former.

Acid-Base Behavior

Definitions

Acids and bases constitute two of the largest classes of compounds that the chemist and biologist encounter. They are interesting for the chemist to study and useful in the control of the acidity of reaction systems. Most biological compounds can function as acids *or* bases. The control of the acidity of blood by proteins is of vital importance to our well-being.

There are numerous ways to define acids and bases. Arrhenius viewed acids as those compounds that liberate hydrogen ions in solution and bases as those compounds that liberate hydroxide ions in solution. Lewis defined acids as electron-pair acceptors and bases as electron-pair donors. We will find it most useful to adopt the Brønsted-Lowry definition, which asserts that an acid is a compound (neutral or ionic) that donates protons and a base is a compound (neutral or ionic) that accepts protons. This conception, according to which all acids and bases function as conjugate pairs, was proposed simultaneously and independently in 1923 by the Danish chemist Johannes Brønsted (1879–1947) and the English chemist Thomas Lowry (1874–1936).

We have glibly used the term *protons* without considering the unusual properties of this species. A proton is a subatomic particle with a diameter about 100,000 times smaller than that of the hydrogen atom. Thus, with its +1 charge, it has an enormous charge density. There is definite evidence that this particle has no independent existence in an aqueous solution. It is always associated with at least one H_2O molecule as the oxonium ion, H_3O^+, and more probably with a cluster of four H_2O molecules as the $H_9O_4^+$ ion. [When a proton is hydrated to an unspecified degree, it is a hydronium ion, $H^+(H_2O)_n$.*] Because symbols are written not only to represent reality but also to be convenient and simple to use, we will use the symbol H^+ to represent any hydrated proton.

Examples

It is useful to consider a variety of acids and bases and their relative strengths before we begin a more quantitative discussion of their properties. Some typical acids and bases are listed in Table 9-4. The adjective *strong* is reserved for those compounds that appear to be totally ionized in solution.

---

*See footnote on p. 111.

Table 9-4. Some Typical Acids and Bases.

Acids			
Strong		Weak	
Name	Formula	Name	Formula
Hydrobromic	HBr	Bisulfate ion	HSO_4^-
Hydrochloric	HCl	Nitrous	HNO_2
Nitric	HNO_3		
Perchloric	$HClO_4$	Benzoic	—COOH
Sulfuric	H_2SO_4 (first proton)		
		Acetic	CH_3COOH
		Hydrocyanic	HCN
Bases			
Strong		Weak	
Name	Formula	Name	Formula
Potassium hydroxide	KOH		
Sodium hydroxide	$NaOH$	Pyridine	N
		Ammonia	NH_3
		Methylamine	CH_3NH_2
		Dimethylamine	$(CH_3)_2NH$
		Trimethylamine	$(CH_3)_3N$

Note that H_2SO_4 is a strong acid with respect to the first proton to dissociate from it, but that the resulting HSO_4^- ion is a weak acid. This means that, to some extent, HSO_4^- is a stable species in solution.

Strong Acids and Bases

Because strong acids and bases are totally dissociated in water, we know the exact concentrations of all species in such solutions when the molarities of the solutions are known. For instance, a 1.00 *M* HCl solution is 1.00 *M* in H^+ and 1.00 *M* in Cl^-. Both ions are hydrated, but, as we have already mentioned, we will simply represent them as shown here. In dealing with strong acids and bases, the only interesting chemistry occurs when acids are added to basic solutions, and vice versa. The calculations of the resulting concentrations of species are examples of *stoichiometric problems*. Some aspects of stoichiometry were discussed in Chapter 2. A few comments and examples here should suffice to enable you to handle these problems with ease.

The reaction

$$H^+ + OH^- \rightleftharpoons H_2O \qquad (9\text{-}21)$$

has an equilibrium constant of $1.0 \times 10^{14}\ \ell^2/\text{mole}^2$ at 25°C. We can safely conclude that, whenever a strong acid and a strong base are mixed, the reaction will go completely to the right until one or the other is consumed.

The concentrations of H^+ and OH^- are not independent of each other in an aqueous system, as is evident from Reaction 9-21 and its reverse reaction:

$$H_2O \rightleftharpoons H^+ + OH^- \qquad (9\text{-}22)$$

The apparent equilibrium constant for this reaction, $K = [H^+][OH^-]/[H_2O]$, is simplified to give the *ion product of water,*

$$K_w = K[H_2O] = [H^+][OH^-]$$

in accord with the conventions described at the beginning of this chapter. The value of K_w is $1.0 \times 10^{-14}\ \text{mole}^2/\ell^2$ at 25°C; it increases to 9.6×10^{-14} at 60°C. This temperature dependence of the equilibrium constant must be remembered when dealing with aqueous acid-base systems at temperatures much above or below 25°C. We will generally assume room temperature in our calculations.

Example 9-10. (a) Calculate $[H^+]$ and $[OH^-]$ in pure water. (b) Calculate $[H^+]$ if $[OH^-]$ is 8.3×10^{-3} mole/ℓ. (c) Calculate $[OH^-]$ in concentrated HCl, which is 12 *M* in H^+.

(a) In pure water the only source of H^+ and OH^- is the water itself, and the ion concentrations must be equal, according to Reaction 9-22. Therefore we may let $x = [H^+] = [OH^-]$ and write

$$x^2 = 1.0 \times 10^{-14}\ \text{mole}^2/\ell^2$$

$$x = 1.0 \times 10^{-7}\ \text{mole}/\ell$$

(b) If $[OH^-] = 8.3 \times 10^{-3}$ mole/ℓ,

$$[H^+] = \frac{10 \times 10^{-15}}{8.3 \times 10^{-3}} = 1.2 \times 10^{-12}\ \text{mole}/\ell$$

(c) If $[H^+] = 12$ mole/ℓ,

$$[OH^-] = \frac{100 \times 10^{-16}}{12} = 8.3 \times 10^{-16}\ \text{mole}/\ell$$

It is apparent from the calculations above that the range of concentrations of H^+ and OH^- in aqueous solutions is enormous, and very impractical to deal with on a routine basis. If you have ever tried plotting a series of numbers such as 1×10^{-1}, 1×10^{-2}, 1×10^{-3}, and 1×10^{-4} on graph paper, you will understand why the concept of pH was introduced. In this expression the H stands for the hydrogen ion concentration and the p is a mathematical operator (hence, not italicized) that means "the negative logarithm of." The quantity pH is thus defined as the negative logarithm of the hydrogen ion concentration:

$$\mathrm{pH} = -\log[\mathrm{H}^+] \qquad (9\text{-}23)$$

In exactly the same way, pOH is defined as the negative logarithm of the hydroxide ion concentration,

$$\mathrm{pOH} = -\log[\mathrm{OH}^-] \qquad (9\text{-}24)$$

and pK is defined as the negative logarithm of the equilibrium constant:

$$\mathrm{p}K = -\log K \qquad (9\text{-}25)$$

The following examples illustrate the usefulness of the pH concept in describing and calculating the acidity (or basicity) of solutions.

Example 9-11. Calculate the pH of the three solutions described in Example 9-10.

(a) $[\mathrm{H}^+] = 1.0 \times 10^{-7}$ mole/ℓ.

$$\mathrm{pH} = -\log(1.0 \times 10^{-7}) = -(\log 1.0 + \log 10^{-7}) = -\log 10^{-7}$$
$$= -(-7.00) = 7.00$$

The pH of water (at 25°C) is 7.00. Because $[\mathrm{OH}^-]$ is also 1.0×10^{-7} mole/ℓ, the pOH is also 7.00. It should be evident that

$$\mathrm{pH} + \mathrm{pOH} = 14.00 = \mathrm{p}K_w \qquad (9\text{-}26)$$

(b) $[\mathrm{H}^+] = 1.2 \times 10^{-12}$ mole/ℓ.

$$\mathrm{pH} = -\log(1.2 \times 10^{-12}) = -(\log 1.2 + \log 10^{-12})$$
$$= -(0.08 - 12.00) = -(-11.92) = 11.92$$

Note that the answer appears to have four significant figures, which would be unjustified in this instance. However, it is a logarithm, and the characteristic of a logarithm serves only to specify the *position* of the decimal point in the number, not the number itself. (For example, the negative logarithm of 1.2×10^{-1} is just 0.92.) Hence, only the mantissa—two digits in this instance—is "significant." You can verify for yourself that the same is true of the answers to parts a and c of this example.

(c) $[H^+] = 12$ mole/ℓ.

$$\text{pH} = -\log 12 = -\log(1.2 \times 10) = -(\log 1.2 + \log 10)$$
$$= -(0.08 + 1.00) = -1.08$$

Note that, when $[H^+]$ is equal to or greater than 1.0 mole/ℓ, the pH is equal to or less than 0, respectively. The limits of attainable pH values are about -1.5 to $+15.5$, representing a difference of 17 orders of magnitude in the hydrogen ion concentration.

The reverse of the above kind of calculation is also an important skill to be acquired, so that $[H^+]$-to-pH and pH-to-$[H^+]$ calculations can be done quickly and surely.

Example 9-12. Calculate $[H^+]$ for the following solutions: (a) pH = 9.47; (b) pH = 0; (c) pOH = 10.31.

(a) $\text{pH} = 9.47 = -\log[H^+]$.

$$\log[H^+] = -9.47 = -10.00 + 0.53$$
$$[H^+] = 3.4 \times 10^{-10} \text{ mole}/\ell$$

(b) $\text{pH} = 0 = -\log[H^+]$.

$$\log[H^+] = 0$$
$$[H^+] = 1.0 \text{ mole}/\ell$$

(c) $\text{pOH} = 10.31 = -\log[OH^-]$.

$$\text{pH} = 14.00 - 10.31 = 3.69 = -\log[H^+]$$
$$\log[H^+] = -3.69 = -4.00 + 0.31$$
$$[H^+] = 2.0 \times 10^{-4} \text{ mole}/\ell$$

Two further kinds of calculations need to be mastered before we take on the more difficult task of handling weak-acid (or weak-base) calculations. The stoichiometries can best be illustrated by a few examples.

Example 9-13. Every day a large industrial plant discharges 10,000 gallons of water having a pH of 1.26 into a nearby river. If NaOH (technical grade, flake) costs \$0.25/lb, what will be the daily cost of neutralizing the effluent so that the water returned to the river has a pH of 7.00?

$$\mathrm{pH} = 1.26 = -\log[\mathrm{H^+}]$$

$$\log[\mathrm{H^+}] = -1.26 = -2.00 + 0.74$$

$$[\mathrm{H^+}] = 5.5 \times 10^{-2}\ \mathrm{mole}/\ell$$

The amount of H^+ discharged per day is

$$(1.0 \times 10^4\ \cancel{\mathrm{gal}})\ (4\ \cancel{\mathrm{qt}}/\cancel{\mathrm{gal}})\ (0.95\ \cancel{\ell}/\cancel{\mathrm{qt}})\ (5.5 \times 10^{-2}\ \mathrm{mole}/\cancel{\ell})$$

$$= 4 \times 0.95 \times 5.5 \times 10^2 = 2.1 \times 10^3\ \mathrm{moles}$$

Since $H^+ + OH^- \rightarrow H_2O$, the amount of OH^- required per day must equal the amount of H^+ discharged per day. And, since moles OH^- = moles NaOH, the daily cost of neutralization must therefore be

$$(2.1 \times 10^3\ \cancel{\mathrm{mole}})\ (40\ \cancel{\mathrm{g}}/\cancel{\mathrm{mole}})\ (2.2 \times 10^{-3}\ \cancel{\mathrm{lb}}/\cancel{\mathrm{g}})\ (\$0.25/\cancel{\mathrm{lb}}) = \$46$$

Example 9-14. An HCl solution of unknown concentration is standardized by titration against a standard solution (a solution of known concentration) of NaOH. If 32.8 ml of acid is required to neutralize 26.9 ml of the NaOH solution, which is known to be 0.106 *M*, what is the concentration of the HCl solution?

Let x be the HCl concentration in mmole/ml.

$$\mathrm{mmoles\ H^+} = \mathrm{mmoles\ OH^-}$$

$$(32.8\ \cancel{\mathrm{ml}})\ (x\ \mathrm{mmole}/\cancel{\mathrm{ml}}) = (26.9\ \cancel{\mathrm{ml}})\ (0.106\ \mathrm{mmole}/\cancel{\mathrm{ml}})$$

$$x = \frac{26.9 \times 0.106}{32.8} = 0.0869\ \mathrm{mmole/ml}$$

The last problem we might attack is the calculation of the titration curve for a strong acid titrated with a strong base. This calculation is of less intrinsic interest than that of the titration curve for a weak acid, but it will provide us with some needed practice. We seek the pH of a solution of a strong acid after incremental portions of a strong base have been added.

Example 9-15. Calculate and plot the titration curve for the addition of 0.100 *M* KOH to 50.0 ml of 0.100 *M* HNO_3.

(a) At the beginning of the titration: no KOH added.

$$[H^+] = 0.100 \text{ mole}/\ell \qquad pH = 1.00$$

(b) After addition of 10.0 ml KOH. This amount of base corresponds to (10.0 ~~ml~~) (0.100 mmole/~~ml~~) = 1.00 mmole OH^-, which would neutralize 1.00 mmole of the original 5.00 mmoles H^+ to H_2O, leaving 4.00 mmoles in 50.0 + 10.0 = 60.0 ml solution.

$$[H^+] = \frac{4.00 \text{ mmole}}{60.0 \text{ ml}} = 6.67 \times 10^{-2} \text{ mole}/\ell \qquad pH = 1.18$$

The pH has increased only very slightly.

(c) After 20.0 ml KOH. Now 2.00 mmoles of the H^+ have been consumed, leaving 3.00 mmoles H^+ in 70.0 ml solution.

$$[H^+] = \frac{3.00 \text{ mmole}}{70.0 \text{ ml}} = 4.29 \times 10^{-2} \text{ mole}/\ell \qquad pH = 1.37$$

(d) After 30.0 and 40.0 ml KOH. Similar reasoning shows that the pH values have now climbed to 1.60 and 1.96. The solution is still strongly acidic.

(e) After 50.0 ml KOH. At this point an amount of OH^- exactly equal to that of the original H^+ has been added to the solution. This point is called the *equivalence point* of the titration. The solution is now water with a little KNO_3 dissolved in it, which does not affect the pH in any way. Thus, the pH at the equivalence point is the same as that in pure water, 7.00.

(f) After 60.0 ml KOH. The first 50.0 ml base have been totally consumed by the acid originally present. The next 10.0 ml base (1.00 mmole OH^-) are simply diluted by the water already present.

$$[OH^-] = \frac{1.00 \text{ mmole}}{110 \text{ ml}} = 9.09 \times 10^{-3} \text{ mole}/\ell \qquad pOH = 2.04$$

$$pH = 14.00 - 2.04 = 11.96$$

The solution is now strongly basic.

The data from the above example are plotted in Figure 9-2. A smooth curve has been drawn to show the overall shape of the function. The pH increases slowly and almost linearly from the beginning of the titration until about 90% of the equivalent amount of base has been added. Then the pH begins to increase rapidly, culminating in the vertical rise of the curve at

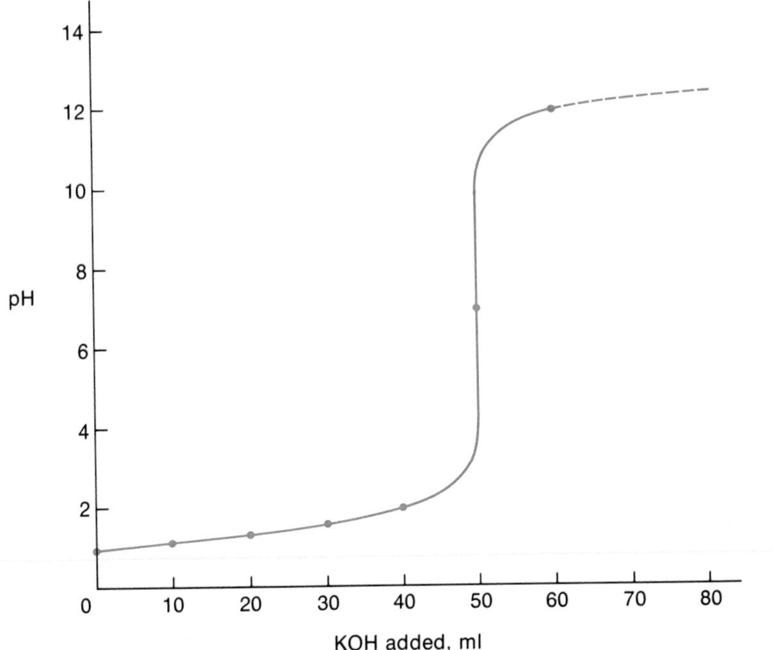

Figure 9-2. Titration of 50.0 ml of 0.100 *M* HNO_3 with 0.100 *M* KOH. The change in pH as a function of base added reveals a striking effect at the equivalence point.

the equivalence point, the point at which the number of moles of OH^- added exactly equals the number of moles of H^+ originally present. The pH then tapers off, asymptotically approaching the value of the titrant, 13.0, as the contribution of the original acid solution to the total volume becomes increasingly insignificant.

The one feature of the curve that we will refer to again is the vertical rise at the equivalence point. The value of $[H^+]$ decreases by six orders of magnitude upon the addition of only 0.20 ml base spanning the equivalence point! This phenomenal decrease provides the "handle" by which we can trigger a visible change in the solution to indicate when to stop the addition of base—namely, a change in the color of an *indicator*. The functioning of indicators will be discussed in the sections on weak acids, since that is what they are.

The corresponding changes in pH at the equivalence points for the titrations of 0.01 *M* acid with 0.01 *M* base and 0.001 *M* acid with 0.001 *M* base are only four and two pH units, respectively. Since a change of at least two pH units is required to trigger most indicators, the 0.001 *M* solutions are the most dilute that can be titrated using indicators to determine the equivalence points.

Equilibria Involving Weak Acids and Weak Bases

There is a large class of soluble, ionizable compounds that do not ionize completely in aqueous solution. These include many acids, bases, and complex ions, and a few inorganic salts. Mercuric chloride, $HgCl_2$, is an example of the last category. We will not discuss these compounds further except to note that the concentration of $HgCl_2(aq)$ is not negligible, as it is with most salts. Complex ions will be treated at the end of this chapter.

In our discussion of acids and bases that do not ionize completely and are hence termed *weak,* we will focus almost exclusively on acids. A weak acid can be represented by HA, its *conjugate base* by A^-, and its dissociation reaction by

$$HA \rightleftharpoons H^+ + A^- \tag{9-27}$$

$$K_a = \frac{[H^+][A^-]}{[HA]} \tag{9-28}$$

This is consistent with the Brønsted-Lowry definition of an acid as a proton donor. The subscript "a" means that the equilibrium constant is an *acid* dissociation constant.

Similarly, a weak base can be represented by B, its *conjugate acid* by BH^+, and its protonation reaction with water by

$$B + H_2O \rightleftharpoons BH^+ + OH^- \tag{9-29}$$

$$K_b = \frac{[BH^+][OH^-]}{[B]} \tag{9-30}$$

This is consistent with the Brønsted-Lowry definition of a base as a proton acceptor. The subscript "b" means that the equilibrium constant is a *base* protonation constant.* Of course, the proton does not have to come from

*Note that, whereas the Brønsted-Lowry and Arrhenius definitions of acids are essentially the same, this is not true of the definitions of bases. According to the Arrhenius definition, a base can be represented by BOH and its dissociation reaction by

$$BOH \rightarrow B^+ + OH^- \qquad K_b = \frac{[B^+][OH^-]}{[BOH]}$$

The constant K_b is now a base *dissociation* constant for a hydroxide. This illustrates the importance of always keeping the relevant definitions clearly in mind.

a water molecule, but for the present purposes it is convenient to describe both acids and bases in terms of their interactions with water.

We choose to discuss weak-acid equilibria almost exclusively, for two reasons: (1) The principles learned can be applied just as well to weak-base equilibria, so by not discussing bases we can cut our work in half. (2) The protonation of weak Brønsted-Lowry bases is often represented by the dissociation of their conjugate acids. For instance, the protonation of the weak base ammonia in aqueous solution can be represented by

Protonation of the base (Base; Conjugate acid)

$$\underset{\text{Base}}{NH_3} + H_2O \rightleftharpoons \underset{\text{Conjugate acid}}{NH_4^+} + OH^- \qquad K_b = 1.77 \times 10^{-5}\ \text{mole}/\ell \tag{9-31}$$

but also by

Dissociation of the acid (Acid; Conjugate base)

$$\underset{\text{Acid}}{NH_4^+} \rightleftharpoons \underset{\text{Conjugate base}}{NH_3} + H^+ \qquad K_a = 5.65 \times 10^{-10}\ \text{mole}/\ell \tag{9-32}$$

Indeed, the equilibrium properties of the organic bases within the biologically important nucleic acid molecules are always described in terms of the dissociation constants K_a of their conjugate acids.

Exercise 9-1. Demonstrate that the analytical expressions for K_a and K_b for the conjugate acid-base pair in Equations 9-31 and 9-32 are related by the expression $K_aK_b = K_w$. Verify that, in analogy with Equation 9-26, $pK_a + pK_b = 14.00 = pK_w$.

Table 9-5 contains the dissociation constants of a number of common weak acids and of the conjugate acids of a number of common weak bases, in aqueous solution. Additional data are given in Appendix G.

Three aspects of weak-acid equilibria will be of interest to us in the ensuing discussions. The first comprises the simplest properties and phenomena characterizing a system containing a weak acid and its conjugate base: dissociation, fractional distribution curves, buffers, hydrolysis, and indicators. These are the fundamental concepts underlying the whole of our discussion. They are not difficult concepts to grasp, and the algebraic skills required are minimal. We will get our introduction to them by examining a titration curve in detail. Titrations are important in their own right. In addition to their usefulness for illustrating the above-mentioned concepts, they are used to determine the number of equivalents of acid or base present, to determine concentrations, and to measure dissociation constants.

Table 9-5. Dissociation Constants of Some Common Weak Acids and Conjugate Acids of Common Weak Bases.

Name	Formula	K_a	pK_a
Bisulfate ion	HSO_4^-	1.20×10^{-2}	1.92
Nitrous	HNO_2	4.6×10^{-4}	3.34
Benzoic	C_6H_5—COOH	6.46×10^{-5}	4.19
Acetic	CH_3COOH	1.76×10^{-5}	4.75
Pyridinium ion	$C_5H_5NH^+$	5.62×10^{-6}	5.25
Ammonium ion	NH_4^+	5.65×10^{-10}	9.25
Hydrocyanic	HCN	4.93×10^{-10}	9.31
Trimethylammonium ion	$(CH_3)_3NH^+$	1.55×10^{-10}	9.81
Methylammonium ion	$CH_3NH_3^+$	2.70×10^{-11}	10.57
Dimethylammonium ion	$(CH_3)_2NH_2^+$	1.85×10^{-11}	10.73

At the end of the discussion on titration, we will come back to look at buffers more thoroughly.

The second aspect of weak-acid equilibria that we will discuss is *polyprotic acids,* i.e., acids that have two or more dissociable protons. Their elucidation requires only slight extensions of the fundamental concepts mentioned above. The biologically important amino acids will serve as our models.

Finally, we will discuss in detail the exact calculation of concentrations in dilute solutions of moderately strong acids and in solutions of salts of weak acids and weak bases.

Titration of a Weak Acid. We will calculate the titration curve for 1.00 ℓ of 0.100 *M* acetic acid. The structure of acetic acid and several representations of its formula are as follows:

H—C(H)(H)—C(=O)—O—H

Acetic acid

$C_2H_4O_2$

$HC_2H_3O_2$

CH_3COOH

$CH_3C(=O)$—OH

Ac—OH

HOAc

For convenience acetic acid is often abbreviated HOAc. As shown above, the H and O represent the OH group and the symbol Ac represents the *acetyl* group $CH_3C\overset{O}{\underset{\diagdown}{\diagup\!\!\diagup}}$, i.e., the rest of the molecule.* Note that, despite the appearance of the HO in the abbreviation, HOAc is not an alcohol. It is a *carboxylic acid,* comprising the acetyl group and the OH group. The latter is the site of the dissociation that makes this species an acid. Because of the strong electronegativity of oxygen, the H atom attached to the O atom dissociates fairly readily in aqueous solution, with the dissociation constant $K_a = 1.76 \times 10^{-5}$ mole/ℓ at 25°C. The H atoms on the remaining anion, OAc^- (i.e., $C_2H_3O_2^-$, CH_3COO^-), have values of $K_a < 10^{-20}$ and therefore need not be considered in any ionization equilibria.

We will titrate this liter of HOAc with 10.0 *M* NaOH, a very concentrated solution of a strong base. Thus, only 10.0 ml of titrant (that solution of known concentration that is added from the burette) will be required in order to reach the equivalence point. To within a maximum error of $(10/1010) \times 100\% = 0.99\%$, therefore, we may assume the volume of the system to remain constant at 1.00 ℓ; this will simplify the calculations.

The titration divides naturally into three stages that require different conceptual approaches and somewhat different methods of calculation.

No NaOH Added. Dissociation of Weak Acids in Water. These two phrases are equivalent. The calculation to be performed here will provide the first point on the titration curve; it is also a model of the techniques to be used whenever the concentrations of the species resulting from the dissociation of a weak acid in water are to be calculated.

The dissociation reaction in the present example is

$$HOAc \rightleftharpoons H^+ + OAc^- \tag{9-33}$$

One approach to calculating the concentrations of all species present in this solution is to write down two sets of concentrations. The first set applies to the system before any dissociation occurs, and is hypothetical only. The second set is the real equilibrium description. In order to be able to calculate these equilibrium concentrations, an unknown variable must be introduced. It is generally best to let the unknown equal the *concentration* of weak acid that dissociates, and it is a good habit always to state the meaning of the unknown explicitly.

---

*In organic chemistry the symbol Ac generally represents the generic *acyl* group $RC\overset{O}{\underset{\diagdown}{\diagup\!\!\diagup}}$,of which the acetyl group is one example (some others are formyl, propionyl, butenyl, etc.). The acyl group was introduced in connection with PAN on p. 131.

Let x be the concentration of the HOAc that dissociates. Because each molecule of HOAc that dissociates gives one cation and one anion, we obtain the following relations:

	$HOAc \rightleftharpoons$	$H^+ +$	OAc^-
Initial concentrations:	0.100	0	0
Equilibrium concentrations:	$0.100 - x$	x	x

The initial concentration of H^+ is not really zero, of course, because of the autoionization of water: it equals 10^{-7} mole/ℓ. However, experience shows that this is almost always negligible in comparison with the concentration of H^+ released by the acid. Only when the solution is extremely dilute (less than about 0.001 M) or when K_a is of the same order of magnitude as K_w is it wrong to neglect the initial concentration of H^+. (When you examine the solution for x given below, you will see why this is so.) The more refined calculations described at the end of this chapter must then be employed.

The expression for the equilibrium constant can now be formulated. Because we have eliminated any competing effects due to the water, we have one equation in one unknown. With a modicum of algebraic skill, the solution is guaranteed.

$$K_a = \frac{[H^+][OAc^-]}{[HOAc]} = \frac{x^2}{0.100 - x} = 1.76 \times 10^{-5} \text{ mole}/\ell \qquad (9\text{-}34)$$

The general equation

$$\frac{x^2}{c_0 - x} = \text{constant}$$

where c_0 is the initial concentration of the weak acid (i.e., before dissociation), will be encountered repeatedly. The exact solution of this equation requires the application of the quadratic formula or the method of successive approximations. However, an exact solution is often not necessary, or even meaningful, given the limited number of significant figures in the data. If $c_0 > K_a$ by a factor of 1000 or more, x can generally be disregarded in comparison with c_0, so the equation becomes

$$\frac{x^2}{c_0} = \text{constant}$$

or, in the present example,

$$\frac{x^2}{0.100} = 1.76 \times 10^{-5} \text{ mole}/\ell$$

$$x = 1.33 \times 10^{-3} \text{ mole}/\ell$$

We find that $x \cong 0.001$, so our disregarding it in comparison with 0.100 was accurate to within 1%, which is generally good enough.

We have made two approximations in the above calculation: one based on the relative magnitudes of the initial and equilibrium concentrations of H^+, and the other on the relative magnitudes of the initial concentration of HOAc and the equilibrium concentration of H^+. Both illustrate the maxim that

$$\text{LARGE} \pm \text{small} \doteq \text{LARGE}$$

where the symbol $\doteq$ means "is almost exactly equal to." (Note that this maxim does *not* apply if the numbers in question are multiplied or divided.)

The value for x is simply the solution to an algebraic equation. We must always bear this in mind and remember what relation the unknown x bears to physical reality. It was defined in this example as the concentration of the HOAc that dissociated, and hence equals $[H^+]$:

$$[H^+] = 1.33 \times 10^{-3} \text{ mole}/\ell \qquad \text{pH} = 2.88$$

This point (0 ml NaOH, pH 2.88) can now be plotted as the first point on the titration curve, which is shown in Figure 9-3.

We will also begin making a second plot, one that demonstrates clearly the relation between the concentrations of the acid and its conjugate base, as a function of the pH. Such plots are called *fractional distribution curves.* In the present example they represent the mole fraction of acetic acid, X_{HOAc}, and the mole fraction of the acetate ion, X_{OAc^-}, both versus pH. These mole fractions are defined as*

$$X_{HOAc} = \frac{[HOAc]}{[HOAc] + [OAc^-]}$$

$$X_{OAc^-} = \frac{[OAc^-]}{[HOAc] + [OAc^-]}$$

Clearly, $X_{HOAc} + X_{OAc^-} \equiv 1$.

Earlier we found that $x = [H^+] = 1.33 \times 10^{-3}$ mole/ℓ. By the stoichiometry of the dissociation reaction and the definition of x, it is obvious that

*In these mole-fraction expressions the concentration of the solvent, which is a constant ($[H_2O] = 55.5$ mole/ℓ), can be ignored because it is "built into" the solute concentration terms.

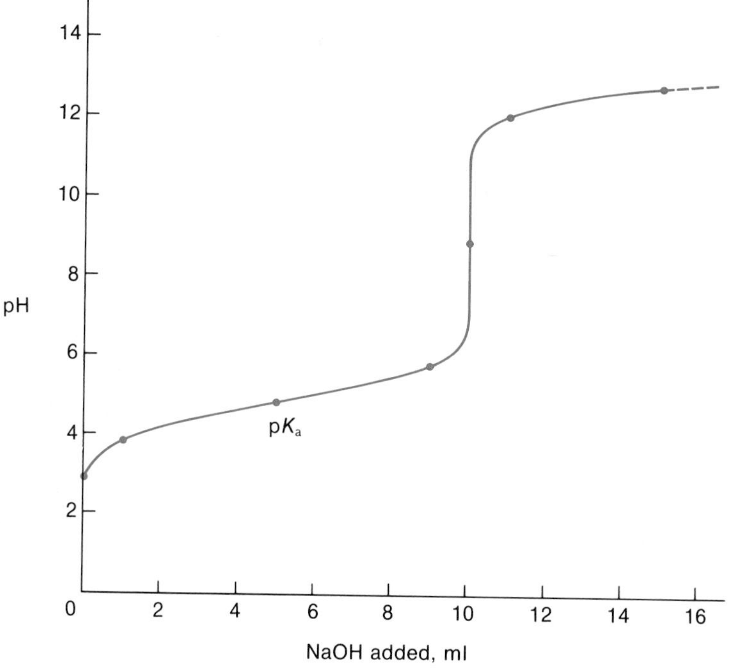

Figure 9-3. Titration of 1.00 ℓ of 0.100 M HOAc with 10.0 M NaOH. The change in pH at the equivalence point is not quite as pronounced as in the titration of a strong acid.

$x = [\text{OAc}^-] = 1.33 \times 10^{-3}$ mole/ℓ also. Therefore, at equilibrium,

$$[\text{HOAc}] = 0.100 - x = 0.099 \text{ mole}/\ell$$

From this it follows that

$$X_{\text{HOAc}} = 0.99$$

$$X_{\text{OAc}^-} = 0.0133$$

(Note that the values of [HOAc] and, therefore, X_{HOAc} may be given to only two significant figures. Within this limit of accuracy, the sum of X_{HOAc} and X_{OAc^-} is 1.00, which is correct.) We can now enter the first pair of points on the second plot, which is shown in Figure 9-4. As incremental amounts of base are added, the additional points shown in Figures 9-3 and 9-4 will be calculated.

After Addition of 0.0100 to 0.0900 Mole NaOH. Buffers. Let us calculate the pH after 1.00, 5.00, and 9.00 ml of 10.0 M NaOH have been added.

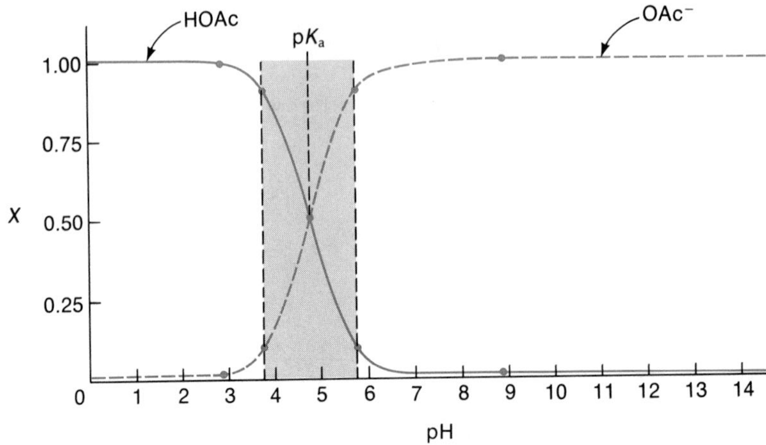

Figure 9-4. Fractional distribution curves for the $HOAc/OAc^-$ system. The buffer region is shaded. At the crossover point, $pH = pK_a = 4.75$. Note that this is the midpoint of the titration, not the equivalence point. The latter is at pH 8.88.

We began the titration with $(1.00\ \ell)(0.100\ \text{mole}/\ell) = 0.100$ mole $= 100$ mmoles of HOAc. The addition of 1.00 ml of 10.0 *M* NaOH provides 10.0 mmoles of OH^-. We would expect this amount of base to titrate 10.0 mmoles of the HOAc quantitatively, producing 10.0 mmoles of OAc^- and leaving 90.0 mmoles of HOAc. This assumption is readily checked by calculating the appropriate equilibrium constant:

$$\text{HOAc} + \text{OH}^- \rightleftharpoons \text{OAc}^- + \text{H}_2\text{O} \tag{9-35}$$

$$K = \frac{[\text{OAc}^-]}{[\text{HOAc}][\text{OH}^-]} = \frac{\cancel{[\text{H}^+]}[\text{OAc}^-]}{[\text{HOAc}]} \times \frac{1}{\cancel{[\text{H}^+]}[\text{OH}^-]} = \frac{K_a}{K_w}$$

$$= \frac{1.76 \times 10^{-5}}{1.00 \times 10^{-14}} = 1.76 \times 10^{9}\ \ell/\text{mole}$$

Indeed, the reaction as written goes virtually to completion.

Thus, we now know the concentrations of the two species HOAc and OAc^-, which together determine $[H^+]$ and therefore the pH. We could insert these concentrations into the expression for K_a and compute $[H^+]$ directly. However, there is a very useful equation by which the pH of a solution in which the concentrations of a weak acid and its conjugate base are unequal can be calculated directly. Again using HA to denote any weak acid, we can derive this equation from the expression for the acid dissociation constant as follows:

$$K_a = \frac{[H^+][A^-]}{[HA]}$$

$$[H^+] = K_a \frac{[HA]}{[A^-]}$$

$$\log[H^+] = \log K_a + \log \frac{[HA]}{[A^-]}$$

$$-\log[H^+] = -\log K_a - \log \frac{[HA]}{[A^-]}$$

$$-\log[H^+] = -\log K_a + \log \frac{[A^-]}{[HA]}$$

$$pH = pK_a + \log \frac{[A^-]}{[HA]} \tag{9-36}$$

Equation 9-36 is called the *Henderson-Hasselbalch equation*, after the American biochemist Lawrence Henderson (1878–1942) and the Danish physiologist Karl Hasselbalch (1874–1962). It permits the simple and accurate calculation of pH throughout the relatively flat region of the titration curve for a weak acid, i.e., the region in which neither of the two concentration terms in the equation is overwhelmingly preponderant.

In the present example of the titration of acetic acid, we have

$$K_a = 1.76 \times 10^{-5} \text{ mole}/\ell$$

$$pK_a = -\log K_a = -(0.25 - 5.00) = 4.75$$

Because the log term in the Henderson-Hasselbalch equation is a ratio of two concentrations, we can insert the *amounts* of each species rather than the concentrations—the volumes would cancel anyway:

$$pH = 4.75 + \log \frac{10.0}{90.0} = 4.75 + \log(1.11 \times 10^{-1}) = 4.75 - 0.95 = 3.80$$

This point can be entered on the titration curve in Figure 9-3 and the points computed for the fractional distribution curves in Figure 9-4. At pH 3.80 the mole fractions are

$$X_{HOAc} = \frac{90.0}{100} = 0.900$$

$$X_{OAc^-} = \frac{10.0}{100} = 0.100$$

With the addition of 5.00 ml NaOH, exactly half the acetic acid has been neutralized, and $[HOAc] = [OAc^-]$. This is an especially important situation because it provides a rapid means of determining the pK_a of a weak acid:

$$pH = pK_a + \log \frac{[A^-]}{[HA]}$$

$$pH = pK_a + \log 1$$

$$pH = pK_a$$

Thus, if a complete titration curve is measured, the pH at the midpoint (also called the half-equivalence point) is equal to the pK_a (4.75 in the present example). Alternatively, a strong base of known concentration can be added to the weak acid until the equivalence point is reached. The subsequent addition of exactly half that volume of a strong acid of exactly the same concentration will yield the point at which $pH = pK_a$. It is nice to have a five-minute technique for determining a pK_a value—assuming that the proper solutions of NaOH and HCl and a suitable means of detecting the equivalence point are available.

The mole fractions X_{OAc^-} and X_{HOAc} are clearly both equal to 0.50 at the midpoint of the titration. This point represents the crossover point of the two curves in Figure 9-4. It is already apparent that the curves are very steep in the pH region near the pK_a. Let us calculate one further point in the central region of the titration curve before examining the significance of this steepness.

With the addition of 9.00 ml of NaOH, 90.0 mmoles of HOAc have been titrated to 90.0 mmoles of OAc^-, leaving 10.0 mmoles of HOAc. This is all the information we need to calculate the pH:

$$pH = pK_a + \log \frac{[OAc^-]}{[HOAc]} = 4.75 + \log \frac{90.0}{10.0} = 5.70$$

where we again use the amounts of the solutes rather than their concentrations. The mole fractions are $X_{OAc^-} = 0.90$ and $X_{HOAc} = 0.10$.

These points are sufficient to permit us to sketch in the first portions of the titration curve and the fractional distribution curves. Both plots tell us the same thing, but in somewhat different ways. Figure 9-3 shows us that, between about 1 and 9 ml of NaOH added, the titration curve is quite flat compared with the region between 0 and 1 ml. There is very little change in pH upon the addition of considerable amounts of base. Figure 9-4 shows why this is so. Over the pH range covered by the addition of these amounts of base, the mole fraction of acetic acid drops from 0.9 to 0.1 and the mole fraction of acetate ion does the opposite. The added base results in a very small change in pH because virtually all of it is consumed in this reaction (Reaction 9-35).

A little thought should convince you that, if we begin with a 0.100 *M* acetate solution at pH 5.70 and add 10.0 *M* HCl, we exactly retrace both curves to the starting points. The reaction of interest is now

$$H^+ + OAc^- \rightarrow HOAc \qquad (9\text{-}37)$$

and it also goes virtually to completion.

Exercise 9-2. Verify this last statement by calculating the equilibrium constant for Reaction 9-37.

From Figure 9-4 you can see that solutions containing appreciable concentrations of both acetate ion and acetic acid can absorb moderate amounts of either acid or base with very little resultant change in the pH. This is due to mutually compensating changes in the equilibrium distributions of species concentrations in the following two reactions:

$$HOAc \rightleftharpoons H^+ + OAc^-$$

$$OAc^- + H_2O \rightleftharpoons HOAc + OH^-$$

Note that both these reactions go strongly to the left, and their sum is

$$H_2O \rightleftharpoons H^+ + OH^-$$

Le Châtelier's principle tells us what happens if a moderate amount of acid or base is added to such a system. The addition of acid drives the first reaction to the left, consuming most of the added H^+ and an equal amount of OAc^-. Both the decrease in $[OAc^-]$ and the corresponding increase in $[HOAc]$ cause the second reaction to be driven to the left also. This tends to restore the original concentrations of these two species in the reaction system. Analogously, the addition of base drives the second reaction to the left, consuming most of the added OH^- and an equal amount of HOAc. Both the decrease in $[HOAc]$ and the corresponding increase in $[OAc^-]$ cause the first reaction to be driven to the left also, with the same result as before. The net result in each case is that the system tends to minimize the change in the ratio of the concentrations of the weak acid and its conjugate base. This result is manifested as a relatively small change in the pH, as we would expect from a consideration of the Henderson-Hasselbalch equation.

The property just described is characteristic of all solutions containing appreciable concentrations of both a weak acid and its salt (or a weak base and its salt) and constitutes the definition of a *buffer*. Buffers are of great importance in many systems. We will give a more detailed discussion of their preparation and properties after we complete our calculations for Figures 9-3 and 9-4.

After Addition of 0.100 Mole NaOH. Hydrolysis. The chemical reaction that has occurred at this point is formally given as

$$100 \text{ mmoles HOAc} + 100 \text{ mmoles OH}^- \rightarrow 100 \text{ mmoles OAc}^- + 100 \text{ mmoles H}_2\text{O} \quad (9\text{-}38)$$

Application of the Henderson-Hasselbalch equation to this relation results in catastrophe because Reaction 9-38 implies that [HOAc] = 0. This would place a zero in the denominator of the fraction in Equation 9-36, which results in a mathematically undefined quantity. Whenever the mathematics goes haywire it is a good sign that our physical description is probably in error. Both the math and our physical intuition tell us that the concentration of a species in solution never goes to zero.

Exercise 9-3. Thermodynamic calculations show that the concentration of an ion in a solution at equilibrium is 10^{-36} mole/ℓ. Calculate this concentration on an atomic basis and think through the implications of the resulting number.

The error in our initial analysis of the concentrations of species present when 0.100 mole of base has been added stems from the fact that, although a reaction may go *virtually* to completion, no reaction ever goes *entirely* to completion. The value of K for Reaction 9-35 is so high, as calculated earlier, that the reaction as written (i.e., the forward reaction) is indeed the predominant one. Since an amount of base exactly equivalent to the amount of acid originally present has been added, this point on the titration curve is the equivalence point. However, now that no appreciable excess of HOAc remains, we must consider the magnitude of the reverse reaction, that of the acetate ion with water.

A chemical reaction in which water decomposes and reacts with another substance is called *hydrolysis*:

$$\text{OAc}^- + \text{H}_2\text{O} \rightleftharpoons \text{HOAc} + \text{OH}^- \quad (9\text{-}39)$$

Note that this reaction is simply the reverse of the titration reaction. Its equilibrium constant K_b is therefore just the reciprocal of the equilibrium constant for Reaction 9-35. It should also be apparent that K_b is the protonation constant of the conjugate base of HOAc, i.e., OAc^-. The value of K_b for the acetate ion is

$$K_b = \frac{[\text{HOAc}][\text{OH}^-]}{[\text{OAc}^-]} = \cancel{[\text{H}^+]}[\text{OH}^-] \times \frac{1}{\cancel{[\text{H}^+]}[\text{OAc}^-]/[\text{HOAc}]} = \frac{K_w}{K_a}$$

$$= \frac{1.00 \times 10^{-14}}{1.76 \times 10^{-5}} = 5.68 \times 10^{-10} \text{ mole}/\ell$$

We saw earlier (in Exercise 9-1) that, for aqueous solutions of Brønsted-Lowry acids and bases, K_a and K_b are always related in this manner.

The concentration of OAc^- is very nearly 0.100 mole/ℓ. (Remember our assumption that the volume of the system remains constant at 1.00 ℓ throughout the titration.) The solution at this point in the titration is identical to that obtained by dissolving 0.100 mole NaOAc in water and diluting to 1.00 ℓ. Indeed, you should now sense that the calculation we are engaged in applies to an aqueous solution of the salt of *any* weak acid. Because a small amount of OH^- is produced by Reaction 9-39, we predict that all such solutions should be slightly basic. Thus we expect that the pH at the equivalence point in this titration will not be 7.00, as we found for strong acids, but about 8 or 9. Because $K_b \cong 10^{-10}$ we do not expect the solution to be strongly basic.

Hydrolysis problems can be solved by means of the standard techniques we used earlier. Let x equal the concentration of OAc^- that hydrolyzes according to Reaction 9-39. This reaction produces equal concentrations of HOAc and OH^-, of magnitude x:

$$K_b = \frac{x^2}{0.100 - x} = 5.68 \times 10^{-10} \text{ mole}/\ell$$

Because $c_0 >> K_b$ we can surely neglect the x in the denominator.

$$x^2 = 5.68 \times 10^{-11} \text{ mole}^2/\ell^2$$

$$x = [\text{HOAc}] = [\text{OH}^-] = 7.54 \times 10^{-6} \text{ mole}/\ell$$

$$[\text{H}^+] = 1.33 \times 10^{-9} \text{ mole}/\ell \qquad \text{pH} = 8.88$$

Our prediction that solutions of salts of weak acids are slightly basic is borne out. This point can now be placed on the titration curve in Figure 9-3.

Exercise 9-4. Demonstrate that the same concentration of HOAc as was calculated above is obtained from the equation for K_a for HOAc, using the above-determined concentrations of H^+ and OAc^-. This is an important demonstration that *all* equilibrium relations that can be legitimately written for a solution must be satisfied simultaneously.

The mole fraction of OAc^- is almost exactly 1.00. The concentration of OAc^- used in the calculation was $0.100 - x$, where x turned out to be of the order of 10^{-5} mole/ℓ. To three significant figures, therefore, $[OAc^-] = 0.100$ mole/ℓ and hence

$$X_{OAc^-} = \frac{[OAc^-]}{[HOAc] + [OAc^-]} = \frac{0.100}{0.100} = 1.00$$

Similarly,

$$X_{HOAc} = \frac{[HOAc]}{[HOAc] + [OAc^-]} = \frac{7.54 \times 10^{-6}}{0.100} = 7.54 \times 10^{-5} \cong 0$$

We see that virtually all the acetate is present in the deprotonated form at pH 8.88. This leads us to conclude that the two curves in Figure 9-4 will be virtually horizontal from this point out to the highest pH that we plot. Similarly, there are two virtually horizontal lines from a pH of about 2.5 down to the lowest pH.

Before we can contrast the characteristics of these two figures with regard to any properties that might be useful in detecting the equivalence point, we need to add a few more points to complete the titration curve.

After Addition of Excess NaOH. Dilution. The addition of 11.0 ml of 10.0 *M* NaOH provides 1.00 ml more NaOH than is required to titrate the HOAc completely. An examination of the possible reactions provides us with the clue we need to find that property of the solution that is the most important in determining the pH.

In descending order the magnitudes of the various values of K are

1. $HOAc \rightleftharpoons H^+ + OAc^-$ $\qquad K_a = 1.76 \times 10^{-5}$ mole/ℓ
2. $OAc^- + H_2O \rightleftharpoons HOAc + OH^-$ $\qquad K_b = 5.68 \times 10^{-10}$ mole/ℓ
3. $H_2O \rightleftharpoons H^+ + OH^-$ $\qquad K_w = 1.00 \times 10^{-14}$ mole^2/ℓ^2

The excess milliliter of NaOH provides an OH^- concentration of (1.00 ml)(10.0 mmole/ml)/(1000 ml) = 0.0100 mole/ℓ. What effect does this concentration have on the above three equilibria?

The first reaction above is of no consequence in determining the pH because, as we have already seen, the concentration of HOAc at the equivalence point is only about 10^{-5} mole/ℓ. Thus the amount of H^+ liberated by the HOAc (about 10^{-9} mole/ℓ) can be neutralized by the OH^- without appreciably changing the OH^- concentration, which is 10^7 times greater. The next two reactions are also of no consequence in determining the pH of the solution because their equilibrium constants are very small and they both produce OH^- as a product. Therefore, by Le Châtelier's principle, the presence of 0.0100 mole of OH^- from the excess base in the solution shifts these equilibria even further to the left.

With respect to the addition of excess NaOH, the system should thus be regarded as chemically passive in terms of the weak acid and its conjugate base, the anion. For all practical purposes it can be regarded simply as pure water, which dilutes the NaOH. The added OH^- represses the dissociation of H_2O, decreasing the H^+ concentration and increasing the pH:

$$[OH^-] = 1.00 \times 10^{-2} \text{ mole}/\ell$$

$$[H^+] = 1.00 \times 10^{-12} \text{ mole}/\ell$$

$$pH = 12.00$$

Similar calculations show that, with the addition of 15.0 ml NaOH, the $pH = 12.70$. If the dilution with 10.0 *M* NaOH is continued indefinitely, the pH will approach the value 15 asymptotically.

Equivalence-Point Detection. Indicators. Now that both plots are complete, we see that the fractional distribution curves in Figure 9-4 contain no feature that could give us a clue as to when the equivalence point of the titration has been reached. However, the titration curve in Figure 9-3 possesses a striking feature at the point at which 10.0 ml of base have been added: The curve becomes nearly vertical, as in the strong-acid curve shown in Figure 9-2. This change of several pH units upon the addition of only a few hundredths of a milliliter of titrant gives us the means of detecting the equivalence point.

Trace amounts of compounds called *indicators* are added to the unknown acid solutions. These organic compounds, which we represent by HIn, are typically weak acids with the useful property of having strikingly different colors for the acid form HIn and the conjugate base form In^-. A typical indicator is *phenolphthalein,* a diprotic acid. The structural formulas and colors of the two forms of this compound are given below.

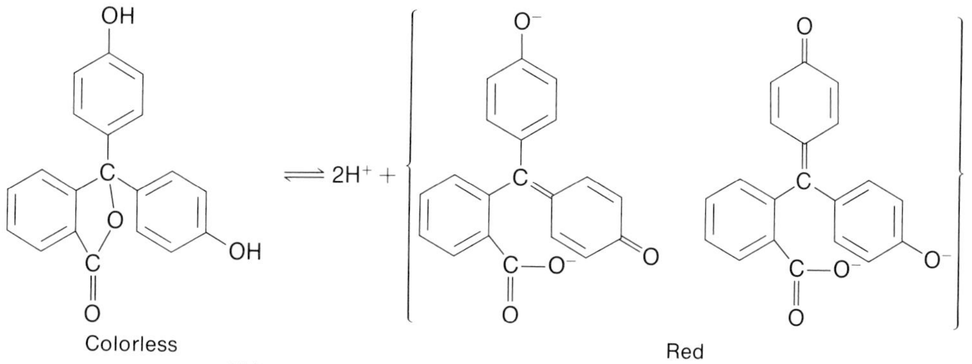

Colorless $\rightleftharpoons 2H^+ +$ Red

$$H_2In \rightleftharpoons 2H^+ + In^{2-}$$

Note the two resonance structures of In^{2-}, the red (basic) form of phenolphthalein, and the extensive system of conjugated π electrons in the red form, which is not a property of the colorless (acidic) form.

A solution containing predominantly the acidic form of phenolphthalein is colorless. If the basic form predominates, the solution is pink to red. A useful generalization for all indicators is that, when the concentration of one form of the indicator is ten times greater than that of the other form, the solution will have the color of the first form. For phenolphthalein this implies the following relations:*

$$\frac{[HIn]}{[In^-]} = 10 \qquad \text{Solution is colorless}$$

$$\frac{[In^-]}{[HIn]} = 10 \qquad \text{Solution is red}$$

How can we adjust these concentration ratios, and can they be made to work to our advantage in signaling an abrupt change in pH? The answers quickly become apparent when we recall that indicators are typically weak acids (some are weak bases). The ratios given above can be accurately calculated from the acid dissociation constant K_{HIn} and are found to depend solely on the overall H^+ concentration in the solution:

$$K_{HIn} = \frac{[H^+][In^-]}{[HIn]}$$

$$\frac{[HIn]}{[In^-]} = \frac{[H^+]}{K_{HIn}} = 10$$

$$\frac{[In^-]}{[HIn]} = \frac{K_{HIn}}{[H^+]} = 10$$

It is clear that, when $[H^+]$ is ten times greater than K_{HIn} for phenolphthalein, the upper ratio obtains and the solution is colorless. When $K_{HIn} = 10[H^+]$, the solution is red.

These calculations can be cast into the more useful form of the Henderson-Hasselbalch equation. For example, when $[HIn] = 10[In^-]$,

$$pH = pK_{HIn} + \log\frac{[In^-]}{[HIn]} = pK_{HIn} + \log\frac{1}{10} = pK_{HIn} - 1.00$$

When $[In^-] = 10[HIn]$,

*Note that we here revert to the generic formula HIn for convenience, even though phenolphthalein is a diprotic acid.

$$pH = pK_{HIn} + \log 10 = pK_{HIn} + 1.00$$

Thus, a solution containing phenolphthalein is colorless at any pH that is less than pK_{HIn} by one pH unit or more, and red at any pH that is greater than pK_{HIn} by one pH unit or more. We see that the indicator will surely change color over a pH range of two units—usually less.

In order to signal the equivalence point in the acetic acid titration, the indicator should begin turning color after pH 7 and complete the change before pH 11. Phenolphthalein has a pK_{HIn} of about 9.0 and thus appears to be ideally suited for this titration.

The above calculation demonstrates the general principle that, for a given titration, an indicator with a pK_{HIn} as close as possible to the pH at the equivalence point should be selected. Surprisingly, perhaps, if you go to look up these pK_{HIn} values, you will probably not be able to find them. The data on indicators are given as that range of pH values over which the indicator changes from one color to the other. This is more useful information because it tells us what we will actually see. The intensities of the colors of the two forms may be different and the sensitivity of the human eye to the two colors may be different. Some typical indicators, spanning most of the pH range, are listed in Table 9-6.

We see from Table 9-6 that most indicators change color within a range of less than two pH units. Let us now calculate the pH change brought about by a volume of titrant that is within an acceptable range of experimental error. The maximum error in most titrations is about ±0.2%. In the 10.0-ml titration discussed above, this error might correspond to going 0.02 ml beyond the equivalence point. Then $[OH^-] = (0.02 \text{ ml})(10.0 \text{ mmole/ml})/(1000 \text{ ml}) = 2 \times 10^{-4}$ mole/ℓ, and the pH = 10.3. Thus we have

Table 9-6. Some Acid-Base Indicators.

Common Name	Transition pH Range	Color of Acidic Form	Color of Basic Form
Thymol blue	1.2–2.8	Red	Yellow
Methyl orange	3.1–4.4	Red	Orange
Bromcresol green	4.0–5.6	Yellow	Blue
Methyl red	4.4–6.2	Red	Yellow
Bromthymol blue	6.2–7.6	Yellow	Blue
Cresol red	7.2–8.8	Yellow	Red
Phenolphthalein	8.0–10.0	Colorless	Red
Alizarin yellow	10.0–12.0	Yellow	Lilac

SOURCE: Lange's *Handbook of Chemistry*, revised 11th ed., McGraw-Hill Book Company, New York, 1973.

a comfortable margin of 10.3 − 8.9 = 1.4 pH units beyond the equivalence point in which to observe the indicator change color. The point at which the color change of the indicator first becomes visible is called the *end point* of the titration. For the reasons just discussed, it may be somewhat different from the equivalence point but still within the acceptable error range; it will be *very* different if the indicator is poorly chosen.

Buffers in More Detail. We have seen in a general way how buffers maintain a pH that is nearly constant despite the addition of moderate amounts of acid or base. We will now examine some of their properties in more detail.

To recognize the importance of buffers, we need only look inside us. The pH of normal human blood is between 7.35 and 7.45. If it should drop to 7.2 or less, the body would go into severe *acidosis,* which results in a coma. This pH change corresponds to a change in $[H^+]$ from 0.00000004 mole/ℓ to 0.00000006 mole/ℓ. Most people have no concept whatever of the exceedingly fine chemical line between a healthy human and one near death! Similarly, if the pH should creep up to 7.6, severe *alkalosis* would occur, resulting in tetany, a condition characterized by rigid muscles.

Clearly, our bodies need a sure-fire mechanism to prevent such changes from occurring. The buffer systems that accomplish this are derived from the ionizable groups on proteins and the equilibria existing among several of the forms of phosphoric acid and carbonic acid and their salts. Hemoglobin provides about 60%, the serum albumins and globulins about 20%, and the inorganic systems the remaining 20% of the buffer capacity of our blood. All these systems entail molecules that can liberate more than one proton. We will defer our discussion of their buffer properties until the next section. For now, we will concentrate on the buffer properties of the commonly encountered acetic acid-acetate system.

We will first calculate how to prepare a buffer solution based on this system. There are three distinct ways to prepare a buffer solution for a monoprotic acid and its anion. The first method is to measure the correct amounts of each species, dissolve them in water, and dilute to the appropriate volume. This technique always entails two calculations. First, for a given pH, the ratio of the two concentrations is determined from the expression for K_a or the Henderson-Hasselbalch equation. The second relation between the two concentrations is obtained by following a standard convention in describing buffer solutions: A solution that is x molar in buffer is one in which the sum of the concentrations of the two or more conjugate acid-base species equals x. An example will make these concepts more concrete.

Example 9-16. Describe how to prepare 500 ml of 0.10 *M* acetate buffer of pH 4.90 at 25°C, using NaOAc as the salt. For HOAc, $pK_a = 4.75$.

Because the data are given in logarithmic form, the Henderson-Hasselbalch equation proves most convenient to use:

$$\text{pH} = \text{p}K_a + \log\frac{[\text{OAc}^-]}{[\text{HOAc}]}$$

$$\log\frac{[\text{OAc}^-]}{[\text{HOAc}]} = \text{pH} - \text{p}K_a = 4.90 - 4.75 = 0.15$$

$$\frac{[\text{OAc}^-]}{[\text{HOAc}]} = 1.4$$

$$[\text{OAc}^-] = 1.4[\text{HOAc}]$$

The convention regarding concentrations gives us

$$[\text{HOAc}] + [\text{OAc}^-] = 0.10 \text{ mole}/\ell$$

Substituting the above relation for $[\text{OAc}^-]$ into this equation gives

$$[\text{HOAc}] + 1.4[\text{HOAc}] = 0.10 \text{ mole}/\ell$$

$$[\text{HOAc}] = \frac{0.10}{2.4} = 0.042 \text{ mole}/\ell$$

$$[\text{OAc}^-] = 0.058 \text{ mole}/\ell$$

The amounts represented by these equilibrium concentrations can be used to make the buffer solution because they are significant amounts with respect to each other: the OAc^- present will repress the ionization of the HOAc and the HOAc will repress the hydrolysis of the OAc^-. The concentrations are converted to amounts by using the standard procedure of multiplying the concentration by the volume and the molecular weight (for the OAc^- calculation the molecular weight used must be that of NaOAc because that is what is actually added to the water):

$$(0.50\ \cancel{\ell})\ (0.042\ \cancel{\text{mole}/\ell})\ (60\ \text{g}/\cancel{\text{mole}}) = 1.3\ \text{g HOAc}$$

$$(0.50\ \cancel{\ell})\ (0.058\ \cancel{\text{mole}/\ell})\ (82\ \text{g}/\cancel{\text{mole}}) = 2.4\ \text{g NaOAc}$$

Dissolve these quantities in water and dilute to 500 ml.

A moment's reflection at one step in the above calculation can prevent it from going awry. The desired pH is greater than the $\text{p}K_a$. This should tell you immediately that $[\text{OAc}^-]$ must exceed $[\text{HOAc}]$. Because the numbers calculated satisfy this requirement, we can be reasonably sure that we have correctly negotiated the sign and antilogarithm conversions.

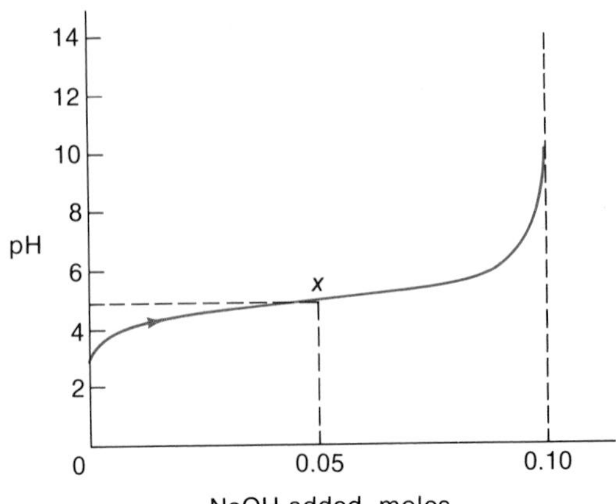

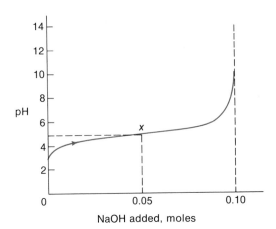

Figure 9-5. Preparation of an acetate buffer by titration of HOAc with NaOH.

Let us now examine two other methods of preparing exactly the same solution. These methods are shown schematically in Figures 9-5 and 9-6. Figure 9-5 displays the titration of HOAc with NaOH. If 0.050 mole of HOAc were dissolved in water, sufficient NaOH added until the pH was 4.90, and this solution diluted to 500 ml, the resulting solution would be identical to that prepared by the method described in Example 9-16.*

The amount of NaOH to add need not be computed if a pH meter is used while the NaOH is being added. If a pH meter is not available, the amount of NaOH needed can be readily determined from the calculation already done in Example 9-16. The total amount x of acetate buffer in question here is 0.050 mole. Example 9-16 shows that, of this total amount, the amount of OAc^- in the solution must be $(0.50\ \ell)(0.058\ \text{mole}/\ell) = 0.029$ mole. This requires that

$$0.029\ \text{mole HOAc} + 0.029\ \text{mole OH}^- \rightarrow 0.029\ \text{mole OAc}^- + 0.029\ \text{mole H}_2\text{O}$$

Thus, $(0.029\ \cancel{\text{mole}})(40\ \text{g}/\cancel{\text{mole}}) = 1.2$ g NaOH must be added to 0.050 mole HOAc and diluted to 500 ml to give a 0.10 *M* acetate buffer of pH 4.90.

Exercise 9-5. Describe qualitatively how to prepare 500 ml of 0.10 *M* acetate buffer of pH 4.90 if the only substances available are NaOAc, HCl, and H_2O. Would the resulting solution be identical to the one prepared by the two methods described above?

Describe quantitatively how to carry out this preparation. Reference to Figure 9-6, which displays the titration of NaOAc with HCl, may be helpful.

---

*As the solution is diluted, the ratio $[OAc^-]/[HOAc]$ remains constant. Therefore, by the Henderson-Hasselbalch equation, the pH remains constant also.

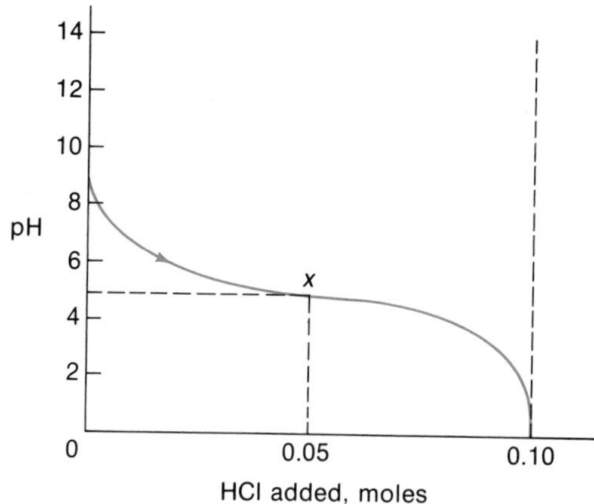

Figure 9-6. Preparation of an acetate buffer by titration of NaOAc with HCl.

The term *buffer capacity* was used earlier. Qualitatively, we can think of buffer capacity in terms of the amount of H^+ or OH^- that the system can absorb without a significant change in the pH. The larger the volume of the solution and the more concentrated it is in the two conjugate forms of the buffer substance, the greater is its buffer capacity.

A clue to the quantitative definition of buffer capacity is found in the fractional distribution curves for a solution of a weak acid. The pH range in which the HOAc/OAc^- system serves as a good buffer is shaded in Figure 9-4. Let us focus on the X_{HOAc} curve. The buffer capacity is maximal where this curve has the greatest slope and minimal where it has the smallest slope. The slope of the curve is greatest at the point $X_{HOAc} = X_{OAc^-} = 0.50$; it approaches zero asymptotically on either side of this point and is always negative. This behavior suggests that the slope would serve as a useful measure of the buffer capacity. The latter, however, is always defined as a positive quantity, so it is the negative of this slope (i.e., a positive slope) to which the buffer capacity is directly proportional. The slope itself is given by the derivative of X_{HOAc} with respect to the pH, so we obtain the mathematical expression

$$\text{Buffer capacity} \propto \frac{-dX_{HOAc}}{d\text{pH}}$$

If the log term in the Henderson-Hasselbalch equation is rewritten in terms of X_{HOAc} alone, an analytical expression for this derivative can be obtained and converted back to an expression in terms of [HOAc] and [OAc^-]. Multiplication of both sides of this expression by the constant ([HOAc] + [OAc^-]), which represents the total concentration of acetate species in the system, yields an equation for the buffer capacity:

$$\text{Buffer capacity} = \frac{-d[\text{HOAc}]}{d\text{pH}} = 2.303\frac{[\text{HOAc}][\text{OAc}^-]}{[\text{HOAc}] + [\text{OAc}^-]} \qquad (9\text{-}40)$$

As with every new physical or chemical relation that we obtain, we should examine the chemical sense of Equation 9-40. At the limits [HOAc] or $[OAc^-] = 0$, the buffer capacity is zero, as we would expect. If [HOAc] $= [OAc^-]$ the buffer capacity is a maximum. The number of moles per liter of HOAc that must be consumed in order to increase the pH by one unit is equal to 2.303 times the product of the concentrations of the acid and base forms of the buffer, divided by their sum. It should be apparent that the number of moles per liter of HOAc consumed is equal to the number of moles per liter of OH^- added. For simplicity, therefore, the buffer capacity for any system is usually defined as the number of moles per liter of base that must be added in order to increase the pH by one unit. The slope of this function is always positive, so we use the positive form of the derivative:

$$\text{Buffer capacity} = \frac{d[\text{Base}]}{d\text{pH}} \qquad (9\text{-}41)$$

The greater the amount of base needed to increase the pH by one unit, the greater the buffer capacity.

Titration of Polyprotic Acids. As implied by their name, *polyprotic acids* contain more than one proton that can be released into solution. Some examples of these acids are H_2S, H_2CO_3, H_3PO_4, and amino acids, some of which can be symbolized as H_2aa^+ (a diprotic amino acid cation). The two or three protons are generally released with very different dissociation constants. The oxygen acids, such as H_2SO_3 and H_3AsO_4, have successive values of K_a typically differing by a factor of about 10^5. Other kinds of polyprotic (also called *polybasic*) acids have K_a values that are closer together (e.g., citric, tartaric, and succinic acids) or further apart (e.g., hydrosulfuric acid).

The dissociation equilibria for hydrogen sulfide are given by*

$$H_2S \rightleftharpoons H^+ + HS^- \qquad K_1 = \frac{[H^+][HS^-]}{[H_2S]} = 9.1 \times 10^{-8}\ \text{mole}/\ell$$

$$HS^- \rightleftharpoons H^+ + S^{2-} \qquad K_2 = \frac{[H^+][S^{2-}]}{[HS^-]} = 1.1 \times 10^{-12}\ \text{mole}/\ell$$

*Successive acid dissociation constants are labeled K_1, K_2, K_3, . . . , beginning with the first proton to be released. Note that the product K_1K_2 for H_2S equals the overall dissociation equilibrium constant K given in Example 9-5.

We stated earlier that all possible equilibria in a given solution must be satisfied simultaneously. This is especially important to remember here. Any aqueous solution containing H_2S, HS^-, or S^{2-} in any combination must satisfy both K_1 and K_2 as well as K_w, which is always valid in aqueous solutions.

As with weak monoprotic acids, most of the important equilibrium properties of polyprotic acids can be discovered by studying their titration curves and fractional distribution curves. We will look at the acid-base properties of amino acids as an example. All the methods to be applied here are applicable to any polyprotic acid. Only the values of the individual acid dissociation constants are different.

Amino acids are the monomers (building blocks) from which proteins, one of the vital macromolecules present in all organisms, are made. The reaction of various proteins with certain enzymes (which are also proteins) or with concentrated acids at elevated temperature for a day or so yields a mixture consisting mainly of 20 different amino acids. Amino acids derive their name from the amino ($—NH_2$) and carboxylic acid ($—COOH$) groups, both of which are present in all amino acids. With the wondrous variety of proteins that are known (enzymes, hemoglobin, keratin in hair and nails, actomyosin and myoglobin in muscle, etc.), it may seem surprising that 19 of the 20 common amino acids have the same basic structural formula:

$$
\begin{array}{c}
\mathrm{R} \\
| \\
\mathrm{H_2N—C—COOH} \\
| \\
\mathrm{H}
\end{array}
$$

But, like the letters of the alphabet, they can be combined in an almost infinite variety of ways. Except for proline, which differs somewhat, they all contain the $—CHNH_2COOH$ portion of the molecule. The remaining portion, designated R, provides the diversity: R can be H, CH_3, CH_2COOH, etc.

Several lines of evidence indicate that this structure is not exactly correct. Amino acids are fairly soluble in water, whereas similar neutral organic amines and acids are not. Amino acids melt at high temperatures (typically 200–300°C), many with decomposition, whereas most other organic amines and acids melt smoothly at relatively low temperatures. The most puzzling data in light of the above structure, however, are the titration data. If this neutral species, with R being CH_3 (L-alanine), were titrated separately with acid and base, the following results would be obtained:

$$\mathrm{H_2N{-}\underset{H}{\overset{CH_3}{C}}{-}COOH} \;(1\text{ mole}) \xrightarrow[pK_1 = 2.3]{1\text{ mole HCl}} \mathrm{H_3\overset{+}{N}{-}\underset{H}{\overset{CH_3}{C}}{-}COOH + Cl^-}$$

$$\mathrm{H_2N{-}\underset{H}{\overset{CH_3}{C}}{-}COOH} \;(1\text{ mole}) \xrightarrow[pK_2 = 9.7]{1\text{ mole NaOH}} \mathrm{H_2N{-}\underset{H}{\overset{CH_3}{C}}{-}COO^- + Na^+ + H_2O}$$

Here pK_1 is not the pK of the reaction as written, but the pK_a of the protonated form of the amino acid at the upper right. Similarly, pK_2 is not the pK of the reaction as written, but the pK_a of the neutral form of the amino acid.

The result of adding base to neutral L-alanine is indeed a negative ion, and adding acid produces a positive ion. However, the pK_a value of 9.7 for the dissociation of a COOH group is inconsistent with all known carboxylic acid data. The same argument holds for the pK_a value of 2.3 for the dissociation of an NH_3^+ group (the conjugate-acid form of an amine, which is a base). The fact that the two values look as though they were interchanged (most organic acids have pK_a values of 3 or 4 and the conjugate-acid forms of most organic amines have pK_a values of 9 or 10) suggests the answer: they *are* interchanged.

Amino acids exist both in the solid state and in solution as *zwitterions* (from the German *Zwitter,* hybrid), which are neutral molecules that have both positively and negatively charged groups in them. Neutral amino acids do possess both positive and negative charges and can therefore behave like ions. This explains the anomalous solubility and melting-point data and suggests the correct titration reactions:

$$\mathrm{H_3\overset{+}{N}{-}\underset{H}{\overset{CH_3}{C}}{-}COO^-} \;(\text{Zwitterion, 1 mole}) \xrightarrow[pK_1 = 2.3]{1\text{ mole HCl}} \mathrm{H_3\overset{+}{N}{-}\underset{H}{\overset{CH_3}{C}}{-}COOH + Cl^-}$$

$$\mathrm{H_3\overset{+}{N}{-}\underset{H}{\overset{CH_3}{C}}{-}COO^-} \;(\text{Zwitterion, 1 mole}) \xrightarrow[pK_2 = 9.7]{1\text{ mole NaOH}} \mathrm{H_2N{-}\underset{H}{\overset{CH_3}{C}}{-}COO^- + Na^+ + H_2O}$$

Table 9-7. pK_a Values of Several Polyprotic Acids at 18–25°C.

Acid	pK_1	pK_2	pK_3
Glycine	2.34	9.60	
Alanine	2.34	9.69	
Aspartic acid	1.88	3.65	9.60
Lysine	2.18	8.95	10.53
Histidine	1.82	6.00	9.17
H_3PO_4	2.12	7.21	12.67
H_2CO_3	6.37	10.25	
Citric acid	3.14	4.77	6.39
Tartaric acid	2.98	4.34	

SOURCE (amino acid data): A. White, P. Handler, and E. L. Smith, *Principles of Biochemistry*, 5th ed., McGraw-Hill Book Company, New York, 1973.

The same products are formed as above, but now the observed pK_a values make sense when compared with those of other compounds. Table 9-7 lists the pK_a values for several amino acids as well as for some typical inorganic and organic acids.

We will select histidine as our model polyprotic acid, in part because the second ionization equilibrium, the dissociation of a proton from the imidazole ring, is primarily responsible for the buffering action of proteins—particularly hemoglobin—in blood. Histidine has the general structure of amino acids given above, with the imidazole ring connected via a methylene group, —CH_2—, to the base unit of the amino acid:

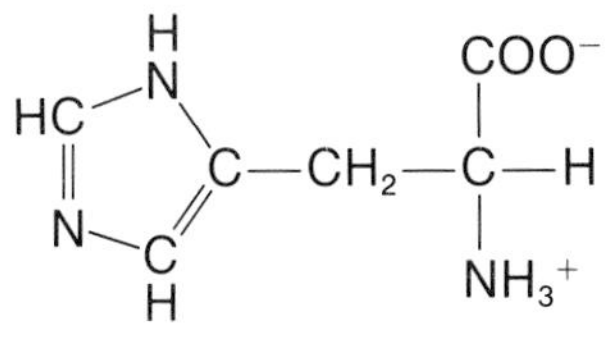

H
N
HC
C—CH_2—C—H
N
C
H
COO^-
NH_3^+

The four different forms of histidine and the reversible reactions by which they are interconverted are shown in Figure 9-7. Note that the pK values are not those of the reactions as written (in either direction); each is the pK_a value of the protonated (acidic) form of the histidine molecule on its left.

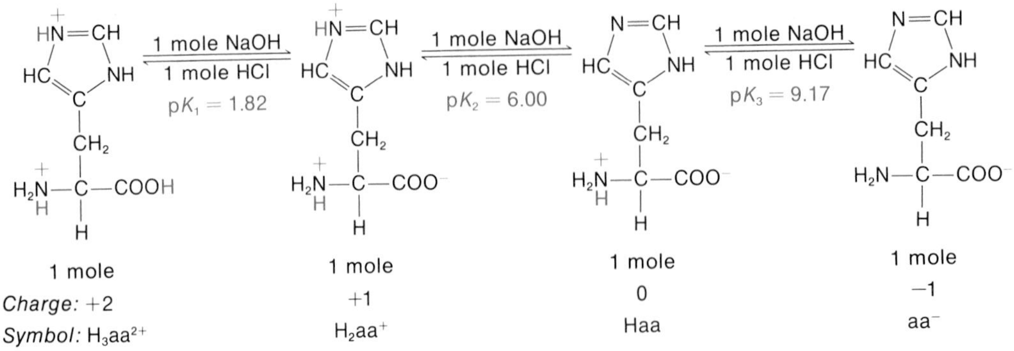

$$\underset{\substack{\text{1 mole}\\ \text{Charge: } +2\\ \text{Symbol: } H_3aa^{2+}}}{\text{(imidazolium)}-CH_2-C(H)(\overset{+}{N}H_3)-COOH} \underset{\text{1 mole HCl}}{\overset{\text{1 mole NaOH}}{\rightleftharpoons}}_{pK_1 = 1.82} \underset{\substack{\text{1 mole}\\ +1\\ H_2aa^+}}{\text{(imidazolium)}-CH_2-C(H)(\overset{+}{N}H_3)-COO^-} \underset{\text{1 mole HCl}}{\overset{\text{1 mole NaOH}}{\rightleftharpoons}}_{pK_2 = 6.00} \underset{\substack{\text{1 mole}\\ 0\\ Haa}}{\text{(imidazole)}-CH_2-C(H)(\overset{+}{N}H_3)-COO^-} \underset{\text{1 mole HCl}}{\overset{\text{1 mole NaOH}}{\rightleftharpoons}}_{pK_3 = 9.17} \underset{\substack{\text{1 mole}\\ -1\\ aa^-}}{\text{(imidazole)}-CH_2-C(H)(NH_2)-COO^-}$$

$$K_1 = \frac{[H^+][H_2aa^+]}{[H_3aa^{2+}]} \qquad K_2 = \frac{[H^+][Haa]}{[H_2aa^+]} \qquad K_3 = \frac{[H^+][aa^-]}{[Haa]}$$

Figure 9-7. Four forms of the amino acid histidine and their interconversions by acid-base reactions. Note the three acidic protons and the sequence of their release. The entire molecule, minus these protons, is abbreviated aa (technically, aa^-), for amino acid.

Now we must try our hand at constructing the titration curve for histidine. We choose to start with the fully protonated substance H_3aa^{2+}. We will examine the addition of 0.100 *M* NaOH to 50 ml of 0.100 *M* H_3aa^{2+}. Because we already know the general shape of monoprotic weak-acid titration curves, we will only calculate the pH values when certain key amounts of base have been added. Recall that there were three especially important points on the titration curve of 1.00 ℓ of 0.100 *M* HOAc, namely, when 0, 50, and 100 mmoles of base had been added. The analogous points, plus even multiples of them, will be the central part of our discussion here as well.

Point a. No NaOH Added. The pH at the start of the titration is just that of the solution of fully protonated acid, which is determined by the degree of dissociation of the first proton (the carboxyl-group proton). We can neglect the further dissociation of H_2aa^+ to Haa because $[H_2aa^+]$ is so small and because $[H^+]$ is large enough to repress this reaction.

Let x be the concentration of H_3aa^{2+} that dissociates:

	H_3aa^{2+}	$\rightleftharpoons H^+$	$+\ H_2aa^+$
Initial concentrations:	0.100	0	0
Equilibrium concentrations:	$0.100 - x$	x	x

$$K_1 = \frac{[H^+][H_2aa^+]}{[H_3aa^{2+}]} = \frac{x^2}{0.100 - x}$$

$$= \text{antilog}\,(-1.82) = 1.5 \times 10^{-2}\ \text{mole}/\ell \tag{9-42}$$

The initial concentration of acid and the value of K_1 differ by only a factor of seven. Hence, x cannot be neglected with respect to 0.100. Two recourses are available. The exact solution by the quadratic formula is usually the fastest. Thus we can solve Equation 9-42 as follows:

$$\frac{x^2}{0.100 - x} = 1.5 \times 10^{-2}$$

$$x^2 = 1.5 \times 10^{-3} - (1.5 \times 10^{-2})x$$

$$x^2 + (1.5 \times 10^{-2})x - 1.5 \times 10^{-3} = 0$$

$$x = \frac{-1.5 \times 10^{-2} + \sqrt{2.3 \times 10^{-4} + 6.0 \times 10^{-3}}}{2}$$

$$x = [H^+] = [H_2aa^+] = 3.2 \times 10^{-2} \text{ mole}/\ell$$

$$\text{pH} = 1.49$$

The other way to proceed is by the *method of successive approximations.* Although this takes a little longer, it is a good idea to learn its use for the solution of cubic or quartic equations. This method consists of approximating one of the concentrations, using K_a to calculate the other concentrations, using those concentrations to improve on the first approximation, and continuing until the concentrations calculated do not change significantly.

The calculation begins with the approximation that the equilibrium concentration of H_3aa^{2+} is 0.100 mole/ℓ. Substitution in Equation 9-42 gives

$$\frac{x^2}{0.100} = 1.5 \times 10^{-2}$$

$$x^2 = 1.5 \times 10^{-3}$$

$$x = [H^+] = [H_2aa^+] = 3.9 \times 10^{-2} \text{ mole}/\ell$$

These concentrations are now used to refine our guess for $[H_3aa^{2+}]$:

$$[H_3aa^{2+}] = 0.100 - x = 0.100 - 0.039 = 0.061$$

$$\frac{x^2}{0.061} = 1.5 \times 10^{-2}$$

$$x = 3.0 \times 10^{-2} \text{ mole}/\ell$$

One more substitution yields almost the same answer as that given by the exact calculation above:

$$[H_3aa^{2+}] = 0.100 - x = 0.100 - 0.030 = 0.070$$

$$\frac{x^2}{0.070} = 1.5 \times 10^{-2}$$

$$x = 3.3 \times 10^{-2} \text{ mole}/\ell$$

(Note that only two significant figures are justified in any product or quotient here.) We leave it as an exercise for you to perform the final step, leading to the value 0.032. The oscillating sequence of the results—0.039, 0.030, 0.033, and 0.032—is characteristic of this method. One point on the titration curve of fully protonated histidine is now available and is plotted in Figure 9-8. Several additional points can be accurately located with considerably less effort.

Point b. After Addition of 25 ml NaOH. At this point (25 ml)(0.100 mmole/ml) = 2.5 mmoles of base have been added. [The original amount of H_3aa^{2+} was (50 ml) (0.100 mmole/ml) = 5.0 mmoles.] We must decide which of the three protons on the histidine will react with this base. The added hydroxide titrates the proton that is released most readily, i.e., the proton related to pK_1. As noted earlier, this is the carboxyl-group proton.

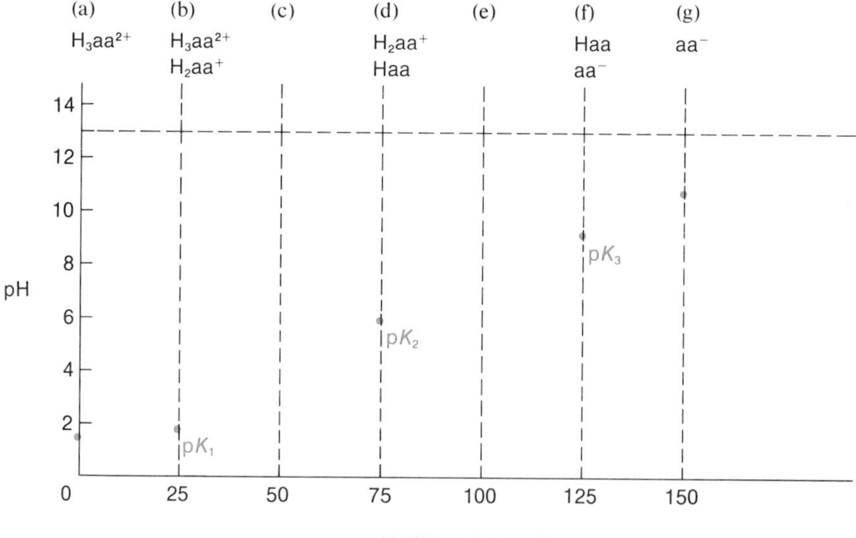

Figure 9-8. Titration of 50 ml of 0.100 *M* histidine with 0.100 *M* NaOH. The first five points on the curve are plotted.

$$2.5 \text{ mmoles } H_3aa^{2+} + 2.5 \text{ mmoles } OH^- \rightarrow$$

$$2.5 \text{ mmoles } H_2aa^+ + 2.5 \text{ mmoles } H_2O$$

At this point pH = pK_1 because 2.5 mmoles of H_2aa^+ have been produced and 2.5 mmoles of H_3aa^{2+} remain. By the Henderson-Hasselbalch equation,

$$\text{pH} = \text{p}K_1 + \log\frac{\cancel{[H_2aa^+]}}{\cancel{[H_3aa^{2+}]}} = \text{p}K_1 = 1.82$$

and we can add this point to Figure 9-8.

Point c. After Addition of 50 ml NaOH. This point in the titration corresponds to the equivalence point in the acetic acid titration, where we found the pH to be 8.88. We will skip this point for now and return to it shortly.

Point d. After Addition of 75 ml NaOH. Half (2.5 mmoles) of the second group of acidic protons on the molecules have now been neutralized by the hydroxide. These protons, which are related to pK_2, are released by the cationic imidazole ring in H_2aa^+. In analogy with point b,

$$\text{pH} = \text{p}K_2 + \log\frac{\cancel{[Haa]}}{\cancel{[H_2aa^+]}} = \text{p}K_2 = 6.00$$

and we have another point for Figure 9-8.

Points e and f. Reasoning similar to that above suggests that we postpone treatment of point e for now and that the pH at point f equals pK_3 = 9.17. This last group of protons is released by the protonated amine group of the zwitterionic form of histidine.

Point g. After Addition of 150 ml NaOH. The predominant species is now aa^-. Because only one reaction of aa^-, the acquisition of a proton from water by hydrolysis, is now possible, the calculation of the pH here is analogous to that in the hydrolysis of a solution of NaOAc at the equivalence point of the acetic acid titration.

$$aa^- + H_2O \rightleftharpoons Haa + OH^-$$

$$K_{b_3} = \frac{[Haa][OH^-]}{[aa^-]} = \frac{K_w}{K_3} = 1.5 \times 10^{-5} \text{ mole}/\ell$$

Since the final amount of aa^- (neglecting its hydrolysis) is essentially the same as the initial amount of H_3aa^{2+}, we can estimate its concentration as $[aa^-] \cong (50 \cancel{\text{ml}})(0.100 \text{ mmole/ml})/(200 \cancel{\text{ml}}) = 0.025 \text{ mole}/\ell$. Now letting x be the concentration of aa^- that hydrolyzes, we obtain

$$\frac{x^2}{0.025 - x} = 1.5 \times 10^{-5}$$

Neglecting the x in the denominator appears to be marginally justified, and the calculation yields

$$x = [\text{Haa}] = [\text{OH}^-] = 6.2 \times 10^{-4} \text{ mole}/\ell$$

$$\text{pH} = 10.79$$

This is about 2% of the value of $[\text{aa}^-]$ given above, and we will accept this amount of error in that calculation. With increasing dilution with 0.100 *M* NaOH, the pH approaches the value 13 asymptotically.

From the HOAc titration curve calculated earlier, we can surmise that the histidine titration curve should consist of three segments of approximately the same shape, each segment corresponding to the titration of a monoprotic acid. Before we calculate the pH values at points c and e, let us examine these segments. In the vicinity of each pK_a value, the curves are quite flat as the two species buffer against the effect of additional NaOH. In the vicinity of each equivalence point, the curves rise steeply. This information, coupled with the five points accurately determined above, enable us to plot Figure 9-9.

The curves in Figure 9-9 give us approximate pH values of 3–5 and 7–8 for points c and e, respectively. The low value of the pH at point c especially suggests that hydrolysis of the H_2aa^+ is not the only reaction of importance, because the solution would be basic if that were true. We must recognize that this species is both an acid *and* a base, and that an alternative reaction is possible, namely, the dissociation of a proton.

When several competing reactions are possible, a certain simplifying assumption can usually be made. This assumption is that the reaction having the largest equilibrium constant is the sole determinant of the equilibrium concentrations in the solution. The two reactions mentioned above and their equilibrium constants are

1. $H_2aa^+ + H_2O \rightleftharpoons H_3aa^{2+} + OH^-$ $\quad K_{b_1} = \frac{K_w}{K_1} = 6.7 \times 10^{-13} \text{ mole}/\ell$

2. $H_2aa^+ \rightleftharpoons H^+ + Haa$ $\quad K_2 = 1.0 \times 10^{-6} \text{ mole}/\ell$

Clearly, the second of these reactions should predominate because of its much larger equilibrium constant. However, there is still something wrong here. Note that the two equations suggest the implausible result that, in the same solution, one reaction produces H^+ while another produces OH^-. When such a contradiction occurs, it implies that yet another reaction is the predominant one. This reaction, number 3 below, is called *autoprotolysis.*

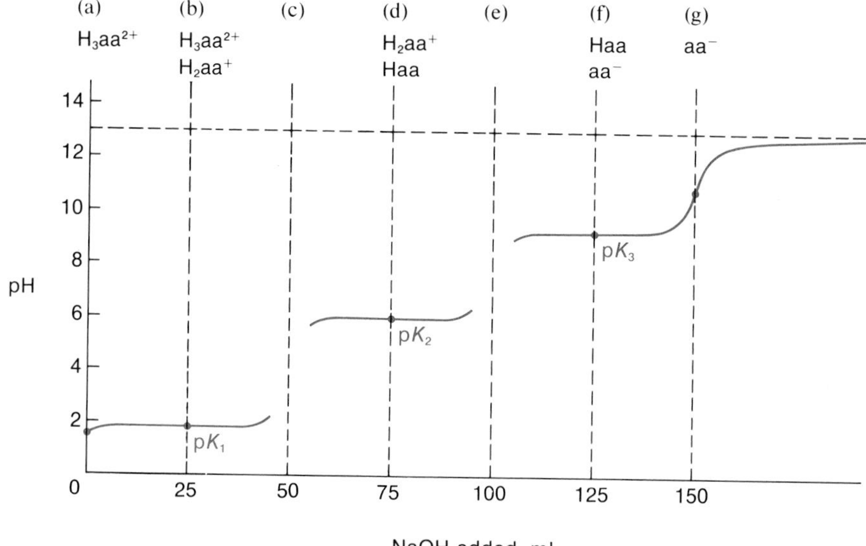

Figure 9-9. Partly complete histidine titration curve.

It represents the transfer of a proton from one H_2aa^+ molecule (acting as an acid) to another (acting as a base):

3. $2H_2aa^+ \rightarrow H_3aa^{2+} + Haa \qquad K_{auto} = \dfrac{[H_3aa^{2+}][Haa]}{[H_2aa^+]^2} = \dfrac{K_2}{K_1} = 6.7 \times 10^{-5}$

Because this equilibrium constant (which is dimensionless) is 67 times larger than K_2, we need consider only this reaction in our calculation of the pH. Since H_3aa^{2+} and Haa are produced in equal amounts, we can write

$$K_{auto} = \frac{[H_3aa^{2+}]^2}{[H_2aa^+]^2} = \frac{K_2}{K_1}$$

Taking square roots gives us

$$\frac{[H_3aa^{2+}]}{[H_2aa^+]} = \sqrt{\frac{K_2}{K_1}} \tag{9-43}$$

The left-hand side of this equation is the reciprocal of a ratio that occurs in the equation for K_1:

$$\frac{[H_3aa^{2+}]}{[H_2aa^+]} = \frac{[H^+]}{K_1} \tag{9-44}$$

Combining these two equations gives

$$\frac{[H^+]}{K_1} = \sqrt{\frac{K_2}{K_1}}$$

whence

$$[H^+] = \sqrt{K_1 K_2} \tag{9-45}$$

Taking logarithms and multiplying through by -1 yields a useful and surprisingly simple relation that is valid in all solutions of compounds that can undergo autoprotolysis:

$$\text{pH} = \frac{\text{p}K_1 + \text{p}K_2}{2} \tag{9-46}$$

Since all the terms on the right-hand side of this equation are constants, the equation must define a unique value of the pH. This value is that at the equivalence point for the first stage of the titration.

Exercise 9-6. Verify that Equation 9-46 is derived from Equation 9-45 by following the steps outlined above.

The pH at point c, the first equivalence point, is thus given by

$$\text{pH} = \frac{\text{p}K_1 + \text{p}K_2}{2} = \frac{1.82 + 6.00}{2} = 3.91$$

Similarly, the pH at point e, the second equivalence point, is

$$\text{pH} = \frac{\text{p}K_2 + \text{p}K_3}{2} = \frac{6.00 + 9.17}{2} = 7.59$$

These points can now be added to the titration curve for histidine and the remainder of the curve sketched in. Figure 9-10 presents the complete curve.

The above calculations demonstrate that the titration curve for any polyprotic acid can be sketched rather quickly from a few points that are obtained with little or no calculation. The titration curves for triprotic acids

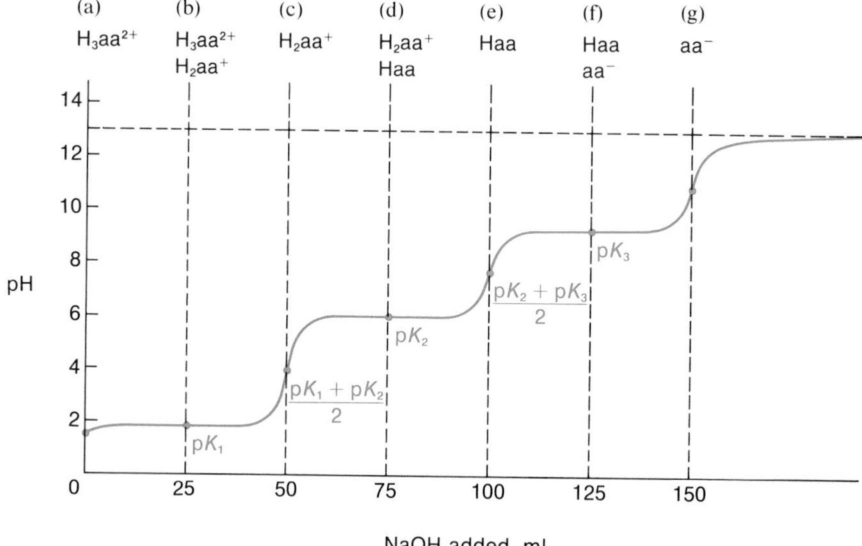

Figure 9-10. Complete histidine titration curve.

such as H_3PO_4 or histidine can be constructed by (1) directly plotting one of the three pK_a values at the midway point in the titration of each species, (2) taking the arithmetic average of two pK_a values to obtain the pH of the equivalence point that lies between them on the curve, and (3) employing standard calculations for weak-acid dissociation and hydrolysis at the beginning and end of the titration. Of course, as many intermediate points as desired can be added in the flat regions of the curve (but not the steep regions) by application of the Henderson-Hasselbalch equation, using whatever pK_a value is appropriate for the segment in question. After the final equivalence point g, standard strong-base dilution calculations should be used.

You should cultivate the facility to construct and interpret titration curves for both mono- and polyprotic acids. They are, in themselves, interesting displays of the chemical behavior of the several acid-base forms. They suggest what indicators should prove useful in detecting any of the several equivalence points that may exist, and show whether the rise in pH in the vicinity of the equivalence point is sufficient to cause the indicator to change color. They are the basis for the measurement of most pK_a values. Construction of titration curves as described above provides an excellent means of becoming proficient in the four principal computational techniques encountered in acid-base equilibria: dissociation of weak acids, buffers, autoprotolysis, and hydrolysis. Finally, they are useful in determining which are the predominant species in a solution at a given value of the pH. Before we turn to the techniques of making exact calculations for the above problems, we should construct and study the fractional distribution curves for histidine because these make it even clearer which species predominate at any given value of the pH.

The basic principles underlying the construction of these curves were given earlier for the HOAc/OAc$^-$ system. The key points to locate first on the diagram are the three pK_a values. At these points the concentrations of the conjugate acid-base species involved in the acid dissociation constant in question are equal. Thus, two of the curves intersect there at $X = 0.50$. The rest of the diagram can be constructed as a series of symmetric, sigmoidal curves, as shown in Figure 9-11. The buffer regions are denoted by the three shaded areas, which extend one pH unit to either side of the pK_a values. Histidine is an effective buffer in each of these regions.

We recall that histidine, one of the amino-acid constituents of hemoglobin, is primarily responsible for the buffering action of hemoglobin in blood. But the pH of normal blood is about 7.4, which is almost dead center *between* two of the buffer regions shown in Figure 9-11, i.e., at a position where we would expect the buffer capacity to be minimal. The explanation is that our data are all for free histidine rather than histidine bound in hemoglobin, and

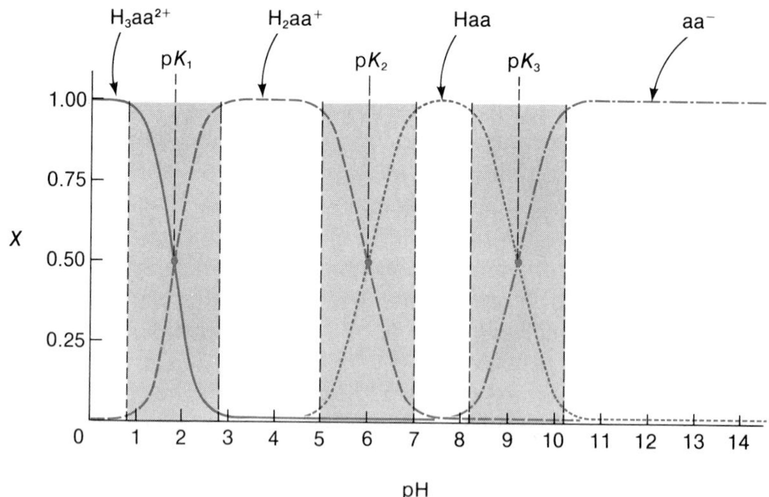

Figure 9-11. Fractional distribution curves for the histidine acid-base system. The three buffer regions are shaded.

the pK_a values are not the same. There are 36 histidines in each hemoglobin molecule, and their pK_a values vary considerably, depending on the immediate chemical environment. The values of pK_2 (dissociation of a proton from the imidazole ring) range from about 6.7 to 7.9, making histidine a very good buffer for blood.

Citric acid represents an even more unusual buffer system. Citric acid gives citrus fruits their tangy taste. It has the formula

```
        CH2—COOH
        |
   HO—C—COOH
        |
        CH2—COOH
```

In contrast to histidine, the three pK_a values of citric acid are rather close: pK_1 = 3.14, pK_2 = 4.77, and pK_3 = 6.39. This means that the end points in the titration of citric acid are visually indistinguishable and the titration behavior is not easy to interpret. Because the pK_a values are so close, the concentrations of at least two species are significant over a wide range of the pH, as is evident from the fractional distribution curves in Figure 9-12. Citric acid then has the useful property of being an effective buffer over the wide pH range of about 2.1 to 7.4. [Note that, because $(\mathrm{p}K_3 - \mathrm{p}K_2) \doteq (\mathrm{p}K_2 - \mathrm{p}K_1)$, the plot of the fractional distribution curves is almost perfectly symmetric.]

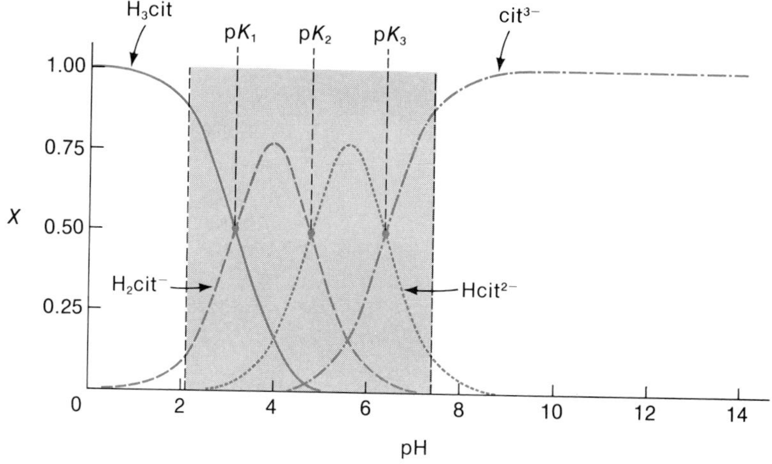

Figure 9-12. Fractional distribution curves for the citric acid system. The buffer region is shaded.

The reasoning and the approximate calculations employed in the examples given above are valid in the great majority of acid-base problems. But occasionally we must re-examine our simplifying assumptions and apply a more rigorous treatment. We turn our attention now to these somewhat more sophisticated calculations.

Exact Calculations in Acid-Base Systems. It is not often that exact calculations must be made, but perhaps you will feel more confident of your ability to handle any acid-base calculation if you have at least seen how to go about it. Side benefits of this rigorous excursion are that you will be better able to judge when some of the approximations invoked earlier may be invalid and that we can introduce the two conservation relations that must hold at all times. We will tackle two general problems.

The first problem is represented by a very dilute solution of a weak acid, say, 1.00×10^{-5} *M* acetic acid. What are the concentrations of all species in such a solution? Our previous approach was to let x be the concentration of the HOAc that dissociates and fill in the table below the dissociation equation:

	HOAc	$\rightleftharpoons$ H$^+$	+ OAc$^-$
Initial concentrations:	1.00×10^{-5}	0	0
Equilibrium concentrations:	$(1.00 \times 10^{-5}) - x$	x	x

This approach is now incorrect, primarily because the initial concentration of H^+ is not 0: it is actually 10^{-7} mole/ℓ. The extent of the dissociation of HOAc will be of comparable magnitude and this will have two effects: (1) the H^+ originally present will repress the ionization of HOAc and (2) the ionization of HOAc will repress the ionization of H_2O so that it will no longer contribute 10^{-7} mole/ℓ of H^+ to the solution. Our approximate picture has become so slippery as to be totally untenable. We must now seek rigorously correct expressions.

As we have previously stated, two such expressions that must always be valid simultaneously are

$$K_a = \frac{[H^+][OAc^-]}{[HOAc]} = 1.76 \times 10^{-5} \text{ mole}/\ell \tag{9-47}$$

$$K_w = [H^+][OH^-] = 1.00 \times 10^{-14} \text{ mole}^2/\ell^2 \tag{9-48}$$

But this is not enough, because there are four unknowns in only two equations. The two additional relations are derived from universally valid conservation laws. The first relation states that the sum of the concentrations of all the positive species in the solution must be equal to the sum of the concentrations of all the negative species. This is a consequence of the *law of the conservation of charge*. It requires that

$$[H^+] = [OH^-] + [OAc^-] \tag{9-49}$$

because no other cation or anion has been added with the HOAc. The second relation states that the total concentration of acetate must be a constant (which we designate c_0), whether or not it is protonated. This is a consequence of the *law of the conservation of mass*. It requires that

$$[HOAc] + [OAc^-] = c_0 \tag{9-50}$$

We now have four equations in four unknowns. It only remains to solve them.

We might start with Equation 9-49 and see whether we can express $[OH^-]$ and $[OAc^-]$ in terms of the known quantities K_a, K_w, and c_0, in which case we can solve for $[H^+]$. If we can manage this, everything else ought to fall into place.

From Equation 9-48, we see that $[OH^-]$ can be replaced by $K_w/[H^+]$:

$$[H^+] = \frac{K_w}{[H^+]} + [OAc^-] \tag{9-51}$$

Two of our other equations contain $[OAc^-]$. Because they both also contain [HOAc], we must be able to combine them in such a way as to eliminate [HOAc] and express $[OAc^-]$ only in terms of the constants. Equation 9-50 yields

$$[HOAc] = c_0 - [OAc^-] \tag{9-52}$$

Substitution of this expression in Equation 9-47 then gives

$$K_a = \frac{[H^+][OAc^-]}{c_0 - [OAc^-]}$$

$$K_a c_0 - K_a[OAc^-] = [H^+][OAc^-]$$

$$[H^+][OAc^-] + K_a[OAc^-] = K_a c_0$$

$$[OAc^-] = \frac{K_a c_0}{[H^+] + K_a} \tag{9-53}$$

Substitution of Equation 9-53 in Equation 9-51 gives us a rigorously correct expression for $[H^+]$ in terms of the three constants:

$$[H^+] = \frac{K_w}{[H^+]} + \frac{K_a c_0}{[H^+] + K_a} \tag{9-54}$$

The only catch is that the simple, first-order equation we hoped for has turned out to be a cubic equation, for which no simple methods of solution are available. Multiplication of Equation 9-54 by the term $[H^+]([H^+]+K_a)$ gives this cubic equation as

$$[H^+]^3 + K_a[H^+]^2 - (K_w + K_a c_0)[H^+] - K_w K_a = 0 \tag{9-55}$$

The method of successive approximations is the most satisfactory way to solve Equation 9-55. First we guess a value for $[H^+]$. If all the HOAc dissociated, we would have $[H^+] = 1.00 \times 10^{-5}$ mole/ℓ. This would be the *upper* limit for $[H^+]$. When it is substituted in Equation 9-55 and the left-hand side evaluated, the sign of the result is positive. The *lower* limit is provided by the autoionization of H_2O. When the value 1.00×10^{-7} mole/ℓ is substituted in Equation 9-55, the sign of the result is negative. Now the criterion for deciding whether a guess is too large or too small is obvious, namely, is the left-hand side of Equation 9-55 positive or negative? The guesses for $[H^+]$ are refined accordingly until the result of substituting a guessed value into the cubic equation is a number sufficiently close to zero.

Exercise 9-7. Use the method of successive approximations to solve Equation 9-55 for $[H^+]$ in an acetic acid solution for which the initial concentration $c_0 = 1.00 \times 10^{-5}$ mole/ℓ.

It is interesting to play a little with Equation 9-55 and see whether we can extract any familiar relations from it. It turns out that this equation is directly related to two equations we have already used. It can be solved for K_a by transposing the K_a terms and dividing by the resulting coefficient of K_a:

$$K_a = \frac{[H^+]([H^+]^2 - K_w)}{c_0[H^+] - ([H^+]^2 - K_w)} \tag{9-56}$$

The terms have been grouped as shown here because we wish to compare K_w and $[H^+]^2$. If the pH = 6.0, $[H^+] = 1.0 \times 10^{-6}$ mole/ℓ and $[H^+]^2 = 1.0 \times 10^{-12}$ mole²/ℓ². This is 100 times larger than K_w. Thus, if 1% accuracy will suffice, $[H^+]^2 - K_w \doteq [H^+]^2$ when pH ≤ 6.0. Equation 9-56 immediately reduces to a familiar equation if we first make this substitution and then divide the numerator and denominator by $[H^+]$:

$$K_a = \frac{[H^+][H^+]^2}{c_0[H^+] - [H^+]^2} = \frac{[H^+]^2}{c_0 - [H^+]} \tag{9-57}$$

Equation 9-57 is identical in form to Equation 9-29. We see that the basic requirement for the two approximations that are used in calculating most weak-acid dissociative equilibria (see the discussion surrounding Equation 9-29) to be valid is that the acid be sufficiently strong that the pH is 6.0 or less. When this requirement *is* met, the valid approximations are that $[H^+]_{equil} \gg [H^+]_{initial}$, and $c_0 \gg [H^+]$ (because $c_0 \gg K_a$). Because of the latter approximation, Equation 9-57 reduces to the simple expression $K_a = [H^+]^2/c_0$.

Exercise 9-8. Write the four basic equilibrium relations that apply to a very dilute solution of NaOAc. Combine these into an exact expression for $[OH^-]$ in terms of c_0, K_w, and K_b. *Hint:* For the charge-conservation expression, remember that $[Na^+] = c_0$.

A second problem in which mathematical rigor is required is the study of a solution of the salt of a weak acid and a weak base. A common example is a solution of ammonium acetate, NH_4OAc. The two ions react with water as follows:*

$$NH_4^+ \rightleftharpoons H^+ + NH_3 \qquad K_a' = 5.65 \times 10^{-10} \text{ mole}/\ell \tag{9-58}$$

$$OAc^- + H_2O \rightleftharpoons HOAc + OH^- \qquad K_b = 5.68 \times 10^{-10} \text{ mole}/\ell \tag{9-59}$$

Things look tough—six unknowns in two equations. Fortunately, we can find four more equations:

$$H_2O \rightleftharpoons H^+ + OH^- \qquad K_w = 1.00 \times 10^{-14} \text{ mole}^2/\ell^2 \tag{9-60}$$

$$\text{Charge conservation:} \quad [NH_4^+] + [H^+] = [OAc^-] + [OH^-] \tag{9-61}$$

$$\text{Mass conservation:} \begin{cases} [NH_4^+] + [NH_3] = c_0 & (9\text{-}62) \\ [OAc^-] + [HOAc] = c_0 & (9\text{-}63) \end{cases}$$

Again we begin with the charge-conservation relation and seek $[H^+]$ as a function of K_a', K_b, K_w, and c_0. From Equation 9-61,

$$[H^+] = [OH^-] + [OAc^-] - [NH_4^+]$$

The task before us is to express each of the terms on the right in terms of these four constants. From the equation for K_w,

$$[OH^-] = \frac{K_w}{[H^+]}$$

From the equation for K_b and Equation 9-63, we can solve for $[OAc^-]$ to obtain

$$[OAc^-] = \frac{[HOAc][OH^-]}{K_b} = \frac{(c_0 - [OAc^-])[OH^-]}{K_b}$$

If we now collect terms, make the substitution $[OH^-] = K_w/[H^+]$, and solve for $[OAc^-]$, we obtain

*We designate the acid dissociation constant of NH_4^+ as K_a' in order to distinguish it from that of HOAc, which we will call K_a when it appears later in the calculation. Note also that, *by coincidence*, $pK_a' = pK_b = 9.25$ in this example. Therefore, $pK_a = pK_b' = 4.75$ because the sum of the pK values for any conjugate acid-base pair in aqueous solution must equal the constant $pK_w = 14.00$ (at 25°C).

$$[\mathrm{OAc^-}] = \frac{c_0 K_w/[\mathrm{H^+}]}{K_b + (K_w/[\mathrm{H^+}])} = \frac{c_0 K_w}{K_b[\mathrm{H^+}] + K_w}$$

By an analogous calculation, starting with the equation for K'_a and Equation 9-62, we can solve for $[NH_4^+]$ to obtain

$$[\mathrm{NH_4^+}] = \frac{c_0[\mathrm{H^+}]}{K'_a + [\mathrm{H^+}]}$$

Combining the results of these three calculations gives us

$$[\mathrm{H^+}] = \frac{K_w}{[\mathrm{H^+}]} + \frac{c_0 K_w}{K_b[\mathrm{H^+}] + K_w} - \frac{c_0[\mathrm{H^+}]}{K'_a + [\mathrm{H^+}]} \tag{9-64}$$

Unfortunately, this turns out to be a formidable, quartic equation and even the method of successive approximations becomes a bit tedious. Rather than solve Equation 9-64 explicitly, let us see—as in the previous example—whether we can simplify it to obtain any familiar relations. We can accomplish this by going back to the charge-conservation relation, Equation 9-61, and combining it with the two mass-conservation relations before making the necessary substitutions.

The sums of the concentrations of ammonia species and acetate species must be equal because the original compound was NH_4OAc, a 1:1 ionic solid:

$$[\mathrm{NH_4^+}] + [\mathrm{NH_3}] = [\mathrm{OAc^-}] + [\mathrm{HOAc}] \tag{9-65}$$

If this equation is subtracted from Equation 9-61, a new identity results:

$$[\mathrm{H^+}] + [\mathrm{HOAc}] = [\mathrm{NH_3}] + [\mathrm{OH^-}] \tag{9-66}$$

Now we can substitute for all the terms except $[H^+]$, yielding

$$[\mathrm{H^+}] + \frac{K_b[\mathrm{OAc^-}]}{[\mathrm{OH^-}]} = \frac{K'_a[\mathrm{NH_4^+}]}{[\mathrm{H^+}]} + \frac{K_w}{[\mathrm{H^+}]}$$

The second term on the left requires some further attention. Using the substitutions $K_b = K_w/K_a$ (note that K_a is the acid dissociation constant of HOAc) and $[OH^-] = K_w/[H^+]$, we obtain

$$\frac{K_b[\mathrm{OAc^-}]}{[\mathrm{OH^-}]} = \frac{(\cancel{K_w}/K_a)[\mathrm{OAc^-}]}{\cancel{K_w}/[\mathrm{H^+}]} = \frac{[\mathrm{H^+}][\mathrm{OAc^-}]}{K_a}$$

which is just the expression we would have obtained had we derived [HOAc] from the dissociation of HOAc instead of the hydrolysis of OAc^-. We now have

$$[\mathrm{H^+}] + \frac{[\mathrm{H^+}][\mathrm{OAc^-}]}{K_a} = \frac{K'_a[\mathrm{NH_4^+}]}{[\mathrm{H^+}]} + \frac{K_w}{[\mathrm{H^+}]}$$

Multiplication by $[H^+]$ gives the rigorously correct equation

$$[H^+]^2\left(1 + \frac{[OAc^-]}{K_a}\right) = K_a'[NH_4^+] + K_w \tag{9-67}$$

Now we must introduce our first approximations. We know that $K_a \cong 10^{-5}$ mole/ℓ and $K_a' \cong 10^{-9}$ mole/ℓ. Therefore, if both OAc^- and NH_4^+ are present in appreciable concentrations, say, greater than 10^{-2} mole/ℓ, the inequalities $1 << [OAc^-]/K_a$ and $K_a'[NH_4^+] >> K_w$ are valid and a considerable simplification results:

$$\frac{[H^+]^2[OAc^-]}{K_a} = K_a'[NH_4^+]$$

It follows that

$$[H^+]^2 = K_a K_a' \frac{[NH_4^+]}{[OAc^-]} \tag{9-68}$$

Before any reaction occurs in the solution, $[NH_4^+] \equiv [OAc^-]$. If $K_a' \cong K_b$, as happens to be true in the present example, both reactions occur to approximately the same extent. Therefore, $[NH_4^+] \cong [OAc^-]$ at equilibrium, and Equation 9-68 reduces to a form that should look familiar:

$$[H^+] = \sqrt{K_a K_a'} \tag{9-69}$$

Equation 9-69 is identical in form to Equation 9-45, which gives the equilibrium concentration of H^+ for the autoprotolysis of a species such as H_2aa^+. We now recognize that, in a sense, Reactions 9-58 and 9-59 constitute an "autoprotolysis" because of the near-identity of their equilibrium constants. Adding these two reactions gives

$$NH_4^+ + OAc^- + H_2O \rightleftharpoons NH_3 + HOAc + H^+ + OH^-$$

Subtracting the reaction for the autoionization of water leads to a straightforward proton-transfer reaction,

$$NH_4^+ + OAc^- \rightleftharpoons NH_3 + HOAc$$

for which the equilibrium constant is given by $K_a'/K_a = 3.21 \times 10^{-5}$. We would expect a reaction such as this to result in a neutral solution. Sure enough, solving Equation 9-69 gives us $[H^+] = 9.97 \times 10^{-8}$ mole/ℓ, or pH $= 7.00$.

The detailed analyses presented in this section provide examples of the general procedures that must be followed in making exact calculations in acid-base systems. They make clear the conditions under which approximate calculations are valid in such systems.

Complex-Ion Equilibria

Examples of complex ions abound. Hemoglobin, the vital, oxygen-carrying protein of the blood, can be regarded as a giant complex of Fe^{2+}. There are many examples of smaller and simpler coordination compounds, the existence of which is often indicated by color changes in solution. Dissolving a little $CuSO_4$ in water produces a light blue color. We can describe this reaction as

$$CuSO_4 \xrightarrow{H_2O} Cu^{2+} + SO_4^{2-} \qquad (9\text{-}70)$$

The fact that solutions of a variety of other sulfate salts are colorless suggests that the color of the cupric sulfate solution is due to the cupric ion. The representation of this ion in Equation 9-70 is incorrect, however, in the same way that representing a proton in solution as H^+ instead of H_3O^+ or $H_9O_4^+$ is incorrect. In fact, Cu^{2+} is hydrated in aqueous solution to give what is probably the tetraaquocopper(II) ion, $Cu(H_2O)_4^{2+}$:*

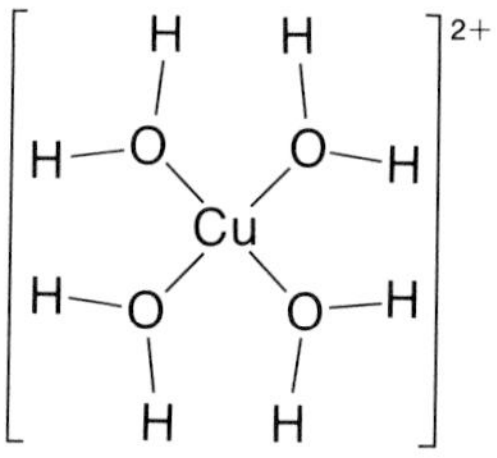

$$\left[Cu(OH_2)_4\right]^{2+}$$

The addition of ammonia (which hydrolyzes to give OH^-) to a cupric-ion solution produces a surprising result. A light blue precipitate forms, which, upon further addition of NH_3, dissolves to give a deep blue solution. The reactions that occur are

$$\underset{\text{Light blue}}{Cu(H_2O)_4^{2+}} + 2OH^- \rightleftharpoons \underset{\text{Light blue}}{Cu(OH)_2\downarrow} + 4H_2O$$

$$\upharpoonleft\downharpoonright\ 4NH_3 \qquad (9\text{-}71)$$

$$\underset{\text{Deep blue}}{Cu(NH_3)_4^{2+}} + 2OH^-$$

*In this section we will omit the brackets that are customarily used in written formulas to denote complex ions (recall Chapter 4), so as to avoid confusion with the use of brackets to denote species concentrations in solution.

The addition of NH_3 to the cupric ion in Reaction 9-71 is known to occur in steps, and it is possible to identify all the intermediate species between $Cu(H_2O)_4^{2+}$ and $Cu(NH_3)_4^{2+}$. Because these species are in equilibrium with each other, it should not be surprising that the equilibrium constants for all the steps have been measured and can be used to predict equilibrium concentrations.

A common laboratory experiment is the separation of two metal ions, cupric and cobaltous, from each other by *ion-exchange chromatography,* a technique discussed in the Interlude following this chapter. The separation is effected by anion-exchange chromatography because the metal ions are present as the chloro anions $CuCl_4^{2-}$ and $CoCl_4^{2-}$ in 6 *M* HCl:

$$Cu(H_2O)_4^{2+} + 4Cl^- \rightarrow \underset{\text{Deep green}}{CuCl_4^{2-}} + 4H_2O$$

$$Co(H_2O)_4^{2+} + 4Cl^- \rightarrow \underset{\text{Deep blue}}{CoCl_4^{2-}} + 4H_2O$$

It is possible that we would have no beautiful silver plate if it were not for a knowledge of complex ions. As discussed in the next chapter, the tendency of silver ion to be reduced ($Ag^+ + e^- \rightarrow Ag$) is so great that the silver from a $AgNO_3$ solution plates out too rapidly, forming large silver crystals that do not adhere well to the metal surface being plated. A way had to be found to decrease the concentration of free Ag^+ to slow the reaction and allow an adherent, fine-grained deposit to form. The problem was solved by the addition of cyanide ion, CN^-, which forms a very stable complex with Ag^+:

$$Ag^+ + 2CN^- \rightarrow \underset{\text{Colorless}}{Ag(CN)_2^-}$$

An excess of CN^- must be present or insoluble AgCN will form.

Quantitative Calculations

Chemists have found stable complexes of a wide variety of metals—mostly transition metals—and ligands ranging from small anions and neutral molecules to chelating agents such as EDTA. Because these complexes are in equilibrium (in aqueous solution) with other complexes or the hydrated metal ion, it is not surprising that there is a large number of equilibrium constants governing the allowed concentration of each species. We will begin with a brief discussion of how equilibrium constants for complex-ion reactions are determined, especially because the problems encountered here are more difficult than those involving solubility phenomena and weak-acid dissociation.

Ferric ion forms very stable complexes with the thiocyanate ion:

$$Fe(H_2O)_6{}^{3+} + SCN^- \rightleftharpoons Fe(H_2O)_5(SCN)^{2+} + H_2O$$

Because the water molecules make the writing and reading of complex-ion formation unnecessarily difficult, we will generally write such reactions as

$$Fe^{3+} + SCN^- \rightleftharpoons Fe(SCN)^{2+}$$

Our present theoretical knowledge is insufficient to allow us to calculate the concentration of each species in a mixture of ferric and thiocyanate ions, so experiments must be done to determine the equilibrium constant of this reaction.

It would be simple to add Ag^+, precipitate AgSCN quantitatively, and weigh the precipitate as a measure of $[SCN^-]$. However, as the free SCN^- was removed from solution, more of the complex would dissociate, giving more free SCN^-, etc., so the result would be meaningless. Fortunately, the separate ions in this example are colorless but the complex has a deep red color. Thus, if solutions of known concentrations of Fe^{3+} and SCN^- are prepared and mixed, the absorbance of the resulting red solution can be measured with a spectrophotometer and compared with the absorbances of solutions of known $[Fe(SCN)^{2+}]$ to obtain the unknown $[Fe(SCN)^{2+}]$. The equilibrium concentrations of Fe^{3+} and SCN^- are then calculated by subtracting the amounts bound in the complex from the initial concentrations of these ions in the mixed solution. The three concentrations are substituted into the expression for the equilibrium constant of the reaction to obtain its numerical value:

$$K = K_{stab} = \frac{[Fe(SCN)^{2+}]}{[Fe^{3+}][SCN^-]} = 91 \ \ell/\text{mole}$$

Note that this form of the equilibrium constant, called a *stability constant,* is opposite to that of the dissociation constant of a weak acid. Complex-ion equilibria are generally expressed in terms of stability constants. The larger the value of K_{stab} for a complex ion, the more stable that ion is.

Another example of how values of K_{stab} are determined is provided by the work of the brilliant Danish chemist Jannik Bjerrum (b. 1909). He used a glass electrode to measure $[H^+]$ in aqueous Ag^+ and NH_3 solutions in order to determine the stability constants of the two silver-ammonia complexes. In other words, he measured the concentration of an ion that is not even directly involved in the equilibria, as a means of determining the equilibrium concentrations of those species that are. This technique is valid because $[H^+]$ and $[NH_3]$ are directly related, as we know from our discussion of weak-acid equilibria:

$$NH_4^+ \rightleftharpoons H^+ + NH_3$$

$$K_a = \frac{[H^+][NH_3]}{[NH_4^+]} = 5.65 \times 10^{-10}\ \text{mole}/\ell$$

$$[NH_3] = \frac{K_a[NH_4^+]}{[H^+]}$$

If the pH of the solution of an ammonium salt is kept below 7, the concentration of NH_4^+ can be regarded as almost exactly equal to the molarity of the added ammonium salt. The product $K_a[NH_4^+]$ can then be regarded as a constant, and we obtain

$$[NH_3] = \frac{K'}{[H^+]} \tag{9-72}$$

The pH values of 2.0 *M* NH_4NO_3 and 0.020 *M* $AgNO_3$ solutions were measured, Equation 9-72 and the appropriate stoichiometric relations were employed, and values of n, the average number of NH_3 molecules complexed to the Ag^+ ion, were determined. The results are shown in Figure 9-13.

The curve in Figure 9-13 shows that, in a solution initially 0.020 *M* in Ag^+, the $Ag(NH_3)^+$ species exists predominantly between $[NH_3]$ values of 10^{-3} and 10^{-4} mole/ℓ and the diamminesilver(I) complex is found predominantly at $[NH_3]$ values greater than 10^{-2} mole/ℓ. This suggests that complex-ion equilibria are similar to polyprotic-acid equilibria in that they are characterized by a succession of equilibrium constants. Each one repre-

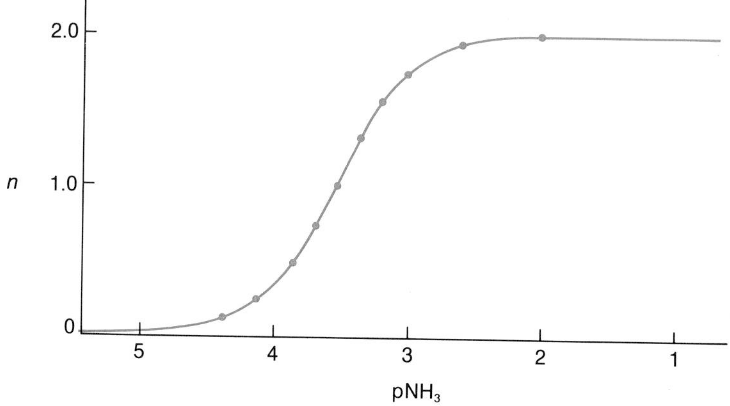

Figure 9-13. The average number n of ammonia molecules bound to a silver ion in a solution initially 0.020 *M* in Ag^+, as a function of $pNH_3 = -\log[NH_3]$. (F. Basolo and R. C. Johnson, *Coordination Chemistry*, W. A. Benjamin, New York, 1964.)

Table 9-8. Logarithms of the Stability Constants for Several Complex Ions.

Metal Ion	Ligand	log K_1	log K_2	log K_3	log K_4	log K_5	log K_6
Ag^+	NH_3	3.1	3.8				
Ag^+	CN^-	5.3	5.0	1.0	−0.6		
Cu^{2+}	NH_3	4.5	3.8	3.2	2.2		
Fe^{3+}	SCN^-	1.96	2.02				
Co^{3+}	NH_3	7.3	6.7	6.1	5.6	5.1	4.4

sents the binding of a certain number of ligands to the central atom. The two stability constants obtained by Bjerrum for this system are

$$Ag^+ + NH_3 \rightleftharpoons Ag(NH_3)^+ \qquad K_1 = \frac{[Ag(NH_3)^+]}{[Ag^+][NH_3]} = 1.3 \times 10^3 \ \ell/\text{mole}$$

$$Ag(NH_3)^+ + NH_3 \rightleftharpoons Ag(NH_3)_2{}^+ \qquad K_2 = \frac{[Ag(NH_3)_2{}^+]}{[Ag(NH_3)^+][NH_3]} = 6.3 \times 10^3 \ \ell/\text{mole}$$

The similarity is further revealed by considering the overall reaction

$$Ag^+ + 2NH_3 \rightleftharpoons Ag(NH_3)_2{}^+$$

for which the stability constant is

$$K_{1,2} = \frac{[Ag(NH_3)_2{}^+]}{[Ag^+][NH_3]^2} = K_1K_2 = 8.2 \times 10^6 \ \ell^2/\text{mole}^2$$

The values of K_1 and K_2 in the present example illustrate a phenomenon sometimes encountered in complex-ion chemistry: the second value is somewhat larger than the first. The binding of one ammonia molecule causes the second one to be bound even more tightly. The stability constants of several of the complex ions formed by Ag^+, Cu^{2+}, Fe^{3+}, and Co^{3+} are given in Table 9-8.*

We conclude our discussion of the quantitative treatment of complex ions with several examples.

Example 9-17. One liter of solution is prepared to be 5.0×10^{-2} M in diethylenetriamine (dien) and 2.0×10^{-1} M in $Ni(NO_3)_2$. The stability constants for the

*A more extensive table of stability constants of complex ions is given in Appendix H.

addition of one and two dien molecules to the Ni^{2+} ion are $K_1 = 5.0 \times 10^{10}$ and $K_2 = 1.6 \times 10^8$, respectively. From the initial concentrations and the relative magnitudes of the two constants, we can assume that, at equilibrium, $[Ni^{2+}] > [Ni(dien)^{2+}] \gg [Ni(dien)_2{}^{2+}]$. Calculate the concentrations of these three species in this solution.

The reaction for the addition of the first dien molecule is

$$Ni^{2+} + dien \rightleftharpoons Ni(dien)^{2+} \qquad K_1 = \frac{[Ni(dien)^{2+}]}{[Ni^{2+}][dien]} = 5.0 \times 10^{10}\ \ell/\text{mole}$$

Because the stability constant is so large and because the dien is present in a limiting amount, we can approximate the concentration of each species by first assuming complete reaction of the dien:

$$0.050 \text{ mole dien} + 0.050 \text{ mole } Ni^{2+} \rightarrow 0.050 \text{ mole } Ni(dien)^{2+}$$

Thus, $[Ni(dien)^{2+}] = 0.050$ mole/ℓ. This leaves $[Ni^{2+}] = 1.5 \times 10^{-1}$ mole/ℓ (by subtraction from 0.20). We can now calculate [dien]:

$$[dien] = \frac{[Ni(dien)^{2+}]}{K_1[Ni^{2+}]} = \frac{5.0 \times 10^{-2}}{5.0 \times 10^{10} \times 1.5 \times 10^{-1}} = 6.7 \times 10^{-12}\ \text{mole}/\ell$$

The $Ni(dien)_2{}^{2+}$ concentration can now be determined from the concentrations of $Ni(dien)^{2+}$ and dien derived above.

$$Ni(dien)^{2+} + dien \rightleftharpoons Ni(dien)_2{}^{2+} \qquad K_2 = \frac{[Ni(dien)_2{}^{2+}]}{[Ni(dien)^{2+}][dien]} = 1.6 \times 10^8\ \ell/\text{mole}$$

$$[Ni(dien)_2{}^{2+}] = K_2[Ni(dien)^{2+}][dien]$$

$$= 1.6 \times 10^8 \times 5.0 \times 10^{-2} \times 6.7 \times 10^{-12} = 5.4 \times 10^{-5}\ \text{mole}/\ell$$

Complex-ion formation is a useful tool in controlling the formation of precipitates. Several qualitative analysis schemes employ ammoniacal solutions to dissolve AgCl precipitates. Similarly, EDTA can be used to control the precipitation of insoluble sulfides. The next example will provide some practice with the use of competing equilibria involving complex ions and precipitate formation.

Example 9-18. A solution is prepared to be 2.0 *M* in NH_3, 5.0×10^{-2} *M* in $AgNO_3$, and 5.0×10^{-2} *M* in NaCl. Determine whether AgCl(*c*) is present in this solution at equilibrium.

$$AgCl(c) \rightleftharpoons Ag^+ + Cl^- \qquad K_{sp} = [Ag^+][Cl^-] = 1.6 \times 10^{-10}\ mole^2/\ell^2$$

$$Ag^+ + NH_3 \rightleftharpoons Ag(NH_3)^+ \qquad K_1 = \frac{[Ag(NH_3)^+]}{[Ag^+][NH_3]} = 1.3 \times 10^3\ \ell/mole$$

$$Ag(NH_3)^+ + NH_3 \rightleftharpoons Ag(NH_3)_2^+ \qquad K_2 = \frac{[Ag(NH_3)_2^+]}{[Ag(NH_3)^+][NH_3]} = 6.3 \times 10^3\ \ell/mole$$

We begin our consideration of the problem by examining the relative concentrations of the three ions of interest and the relative magnitudes of the three equilibrium constants. Because $[NH_3]$ is initially 40 times greater than $[Ag^+]$ and because the stability constants of the two complexes are larger than 10^3, we will make the trial assumptions that: (1) AgCl does not precipitate and (2) virtually all the Ag^+ is bound as the diammine complex. This implies that

$$[Ag(NH_3)_2^+] = 5.0 \times 10^{-2}\ mole/\ell$$

and therefore

$$[NH_3] = 2.0 - 2(0.05) = 1.9\ mole/\ell$$

The actual Ag^+ concentration can be determined from the overall stability constant:

$$Ag^+ + 2NH_3 \rightleftharpoons Ag(NH_3)_2^+ \qquad K_{1,2} = \frac{[Ag(NH_3)_2^+]}{[Ag^+][NH_3]^2} = K_1K_2 = 8.2 \times 10^6\ \ell^2/mole^2$$

$$[Ag^+] = \frac{[Ag(NH_3)_2^+]}{K_{1,2}[NH_3]^2} = \frac{5.0 \times 10^{-2}}{8.2 \times 10^6 \times 1.9^2} = 1.7 \times 10^{-9}\ mole/\ell$$

The effects of our trial assumptions are to maximize both the calculated value of $[Ag^+]$ and the assumed value of $[Cl^-]$ at equilibrium, so their product Q_{ion} must also be the maximum possible in this system. The criterion for precipitate formation is whether $Q_{ion} > K_{sp}$:

$$Q_{ion} = [Ag^+][Cl^-] = 1.7 \times 10^{-9} \times 5.0 \times 10^{-2} = 8.5 \times 10^{-11}\ mole^2/\ell^2 < K_{sp}$$

Because $Q_{ion} < K_{sp}$, the solution is unsaturated and no AgCl will precipitate.

A wide variety of complex ions are found in living systems. The Fe^{2+} ion in heme was discussed in Chapters 4 and 5. The Cu^{2+} ion forms complexes with amino acids and polypeptide chains. The antibiotic Aureomycin and the vitamin riboflavin (called vitamin B_2 or vitamin G) form complexes with divalent cations such as Fe^{2+} and Ni^{2+}. The Mg^{2+} ion forms a wide variety of complex ions of great importance in living systems, such as chlorophyll and the complex with adenosine triphosphate, ATP. The EDTA ligand can be used as a competitor for these complexes, as the last example in this chapter demonstrates.

Example 9-19. How many moles of EDTA must be added to one liter of a solution that is 0.10 *M* in $Ni(riboflavin)^{2+}$ in order to free all of the vitamin?

1. Ni^{2+} + riboflavin $\rightleftharpoons$ $Ni(riboflavin)^{2+}$ $\qquad K_1 = 1.3 \times 10^4$ ℓ/mole

2. Ni^{2+} + EDTA $\rightleftharpoons$ $Ni(EDTA)^{2+}$ $\qquad K_2 = 4.0 \times 10^{18}$ ℓ/mole

Because the stability constant of the EDTA complex is about 10^{14} times greater than that of the vitamin complex, we may assume that all that is required is to add an amount of EDTA equivalent to that of the vitamin complex already present. This assumption is justified by subtracting Reaction 1 from Reaction 2:

$$\mathrm{Ni(riboflavin)^{2+} + EDTA \rightleftharpoons Ni(EDTA)^{2+} + riboflavin} \qquad K = \frac{K_2}{K_1} = 3.1 \times 10^{14}$$

Because this reaction is quantitative, we need add only 0.10 mole EDTA to free all of the vitamin from the nickel complex.

Bibliography

Adrien Albert and E. P. Serjeant, **The Determination of Ionization Constants**, 2nd ed., Wiley, New York, 1971. This brief text provides a nice extension of the treatment of the dissociation of weak acids given in the present chapter. The treatment is extended to include a discussion of activity coefficients.

Allen J. Bard, **Chemical Equilibrium**, Harper & Row, New York, 1966. The entire spectrum of chemical equilibrium concepts and calculations is discussed in this book. These include acid-base, solubility, complex-ion, and oxidation-reduction equilibria. A useful chapter on numerical and computer methods concludes the volume.

James N. Butler, **Solubility and pH Calculations**, Addison-Wesley, Reading, Mass., 1964. This book lives up to its title. Elementary but sound discussions of solubility calculations and pH calculations are given. The logarithmic concentration plots introduced on page 48 are particularly useful once their significance is grasped.

Paul H. Carnell and Rosetta N. Reusch, **Molecular Equilibrium**, W. B. Saunders, Philadelphia, 1963. This is a programmed guide to the study of chemical equilibrium. Your understanding of a wide variety of chemical equilibrium concepts is certain to be improved if you work your way through this material.

Emil J. Margolis, **Chemical Principles in Calculations of Ionic Equilibria**, Macmillan, New York, 1966. Professor Margolis has been teaching students of college chemistry for four decades. This text is a distillation of his approach to the calculations involved in problems dealing with chemical equilibria.

Problems

1. Write the equilibrium-constant expression for each of the following reactions and include the appropriate units (atm for pressure and mole/ℓ for concentration).

a. $Zn(c) + 2H^+(aq) \rightleftharpoons Zn^{2+}(aq) + H_2(g)$

b. $NH_4SH(c) \rightleftharpoons NH_3(g) + H_2S(g)$

c. $Ag_2CrO_4(c) \rightleftharpoons 2Ag^+(aq) + CrO_4^{2-}(aq)$

d. $Co(NH_3)_6^{3+}(aq) + 6H^+(aq) \rightleftharpoons Co^{3+}(aq) + 6NH_4^+(aq)$

2. Consider the following reactions and their equilibrium constants:

$$HCN \rightleftharpoons H^+ + CN^- \qquad K_1 = 4.9 \times 10^{-10}$$

$$NH_3 + H_2O \rightleftharpoons NH_4^+ + OH^- \qquad K_2 = 1.8 \times 10^{-5}$$

$$H_2O \rightleftharpoons H^+ + OH^- \qquad K_w = 1.0 \times 10^{-14}$$

Calculate the value of the equilibrium constant for the reaction

$$NH_3 + HCN \rightleftharpoons NH_4^+ + CN^-$$

3. At 25°C nitrogen reacts with hydrogen to give ammonia:

$$N_2(g) + 3H_2(g) \rightleftharpoons 2NH_3(g)$$

a. Write the equilibrium-constant expression.

b. If the partial pressures of the three gases in a closed container are $P_{N_2} = 40$ Torr, $P_{H_2} = 25$ Torr, and $P_{NH_3} = 600$ Torr, what is the equilibrium constant for the reaction?

4. The reaction

$$N_2(g) + O_2(g) \rightleftharpoons 2NO(g)$$

is endothermic in the forward direction. What effects will (a) an increase in temperature, (b) an increase in pressure, and (c) the addition of a catalyst have on the equilibrium constant? Give reasons for each of your answers.

5. The reaction

$$4HCl(g) + O_2(g) \rightleftharpoons 2Cl_2(g) + 2H_2O(g) \qquad \Delta H° = -27 \text{ kcal/mole}$$

occurs in a closed vessel. Suggest four different ways in which the equilibrium concentration of $H_2O(g)$ can be increased.

6. The reaction

$$CO(g) + H_2O(g) \rightleftharpoons CO_2(g) + H_2(g)$$

occurs in a 6.0-ℓ vessel at a temperature for which the equilibrium constant is 1. If 0.30 atm CO, 2.0 atm H_2O, 2.0 atm CO_2, and 2.0 atm H_2 are introduced into the vessel at this temperature, will any net reaction occur? Will CO_2 be formed or consumed at this temperature?

7. In a hypothetical reaction, 8 mmoles of A_2O_3, 25 mmoles of MO_2, and 40 mmoles of H_2O are mixed and reacted according to the following equation:

$$3A_2O_3 + 4MO_2 + 9H_2O \rightleftharpoons 6H_3AO_4 + 2M_2O$$

Let x be the number of millimoles of A_2O_3 existing at equilibrium. Express the concentration of each of the other four species at equilibrium in terms of x.

8. Consider the hypothetical reaction $A + B \rightleftharpoons 2C$ occurring in a closed vessel, which is heated to 425°C and held there until equilibrium is reached. If the initial mixture contained 0.20 g of A and 0.30 g of B, and it is found that 0.16 g of A is left unreacted, what is the equilibrium constant for the reaction? Assume that the molecular weights of A and B are 6.0 g/mole and 8.0 g/mole, respectively.

9. At 690 K the equilibrium constant for the reaction

$$CO_2(g) + H_2(g) \rightleftharpoons CO(g) + H_2O(g)$$

is 0.10. If 4.00 moles of CO_2 and 3.00 moles of H_2 react at a total pressure of 1 atm, what are the partial pressures of CO_2, H_2, CO, and H_2O in the equilibrium mixture?

10. At 699 K the equilibrium constant for the reaction

$$H_2(g) + I_2(g) \rightleftharpoons 2HI(g)$$

is 55.3. If 2.00 moles of H_2 and 2.00 moles of I_2 are placed in a 4.00-ℓ vessel and allowed to react, what mass of hydrogen iodide will be present at equilibrium?

11. At 725°C the equilibrium constant for the reaction

$$2BaO_2(c) \rightleftharpoons 2BaO(c) + O_2(g)$$

is 0.497 atm. If this reaction is initiated in a closed, evacuated, 10-ℓ vessel, how many grams of barium peroxide will decompose to barium oxide?

12. The solubility of oxygen in water is 3.93×10^{-3} g/100 g H_2O at 25°C and $P_{O_2} = 1.00$ atm. Calculate the concentration of oxygen (moles/ml) in an air-saturated solution at 25°C and $P_{O_2} = 735$ Torr. (The density of water at 25°C is 0.997 g/ml.)

13. A scuba diver who comes up too rapidly from depths below 100 feet is liable to suffer from the bends, a painful condition caused by bubbles of N_2 forming in the blood vessels, with distention of the tissues. If the diver breathes air, which is 78 mole % N_2, how many milliliters of N_2 (at 38°C and 1 atm) could be released as bubbles in 1 ℓ of his blood when he surfaces quickly from a depth of 500 ft? (34 ft of H_2O = 1 atm.)

14. The distribution constant for iodine between water and carbon tetrachloride at 25°C is 87:

$$I_2(H_2O) \rightleftharpoons I_2(CCl_4) \qquad K = 87$$

If 35 ml of 0.050 M aqueous I_2 is extracted with 50 ml of CCl_4, what is the number of grams of I_2 in each layer at equilibrium?

15. A distribution constant shows that the ratio of formaldehyde concentrations in diethyl ether and water at 25°C is 1/9. How many liters of ether are needed to extract 95% of the formaldehyde from 500 ml of a 0.020 M aqueous solution? Would it take less ether to extract 95% of the formaldehyde if five successive, equal-volume extractions were performed? Why?

16. Calculate the solubility product for each of the following:
    a. A saturated solution containing 3.99×10^{-5} mole of $BaCO_3$ per liter.
    b. Lead phosphate, $Pb_3(PO_4)_2$, with a solubility of 1.27×10^{-4} g/ℓ.
    c. A saturated solution made by dissolving 0.060 g BaF_2 in 45 ml H_2O.

17. The solubility product of $Mg(OH)_2$ is 1.2×10^{-11}. How many grams of $Mg(OH)_2$ will dissolve in 250 ml of water?

18. How many grams of Na_2SO_4 must be dissolved in 10 ℓ of 0.01 M $AgNO_3$ to saturate the solution with Ag_2SO_4? The solubility product of Ag_2SO_4 is 1.4×10^{-5}.

19. An ex-chemistry student decides to test the water in his home for the presence of chloride ions. First, using a volumetric flask, he dissolves 2.5 g of $AgNO_3$ with acidified distilled water to make a solution of 250 ml. He then adds ten drops (≈0.50 ml) of this solution to 10 ml of the water to be tested. The solubility product of AgCl is 1.56×10^{-10}. What is the lowest concentration of chloride ion that he can detect by the formation of a precipitate?

20. The solubility product of calcium sulfate, $CaSO_4$, is 2.45×10^{-5}. Calculate the solubility of $CaSO_4$ in (a) pure water, (b) a 0.50 M $Ca(NO_3)_2$ solution, and (c) a 0.80 M Na_2SO_4 solution.

21. The solubility product of lead sulfate, $PbSO_4$, is 1.06×10^{-8}. Calculate the number of grams of $PbSO_4$ that will dissolve in 2 ℓ of a 0.200 M Na_2SO_4 solution.

22. 80 ml of 0.50 *M* $BaCl_2$ is mixed with 80 ml of 0.15 *M* Na_2CrO_4. The solubility product of $BaCrO_4$ is 2.4×10^{-10}. Calculate the concentrations of Ba^{2+} and CrO_4^{2-} left in the solution. What mass of precipitate is formed?

23. The solubility product of silver bromate, $AgBrO_3$, is 5.8×10^{-5}. 15 g each of $AgNO_3$ and $NaBrO_3$ are mixed in a volumetric flask and distilled water is added to the 250-ml mark. Calculate the values of $[Ag^+]$, $[NO_3^-]$, $[Na^+]$, and $[BrO_3^-]$ in the resulting solution.

24. A solution is 0.0500 *M* in Fe^{2+} and 0.200 *M* in NH_4Cl. The solubility product of $Fe(OH)_2$ is 1.64×10^{-14} and the value of K_b for NH_3 is 1.77×10^{-5}. Calculate the concentration of ammonia required to form a saturated solution of $Fe(OH)_2$.

25. Calculate the pH of the following solutions:
    a. 200 ml of 3.65×10^{-5} *M* $HClO_4$
    b. 300 ml of 0.0055 *M* KOH
    c. A solution prepared by mixing 18.3 ml of 0.300 *M* $HClO_4$ with 38.0 ml of 0.104 *M* NaOH
    d. A solution prepared by mixing 250 ml of a solution in which pH = 2.40 with 425 ml of a solution in which pH = 3.70 (assume that the source of H^+ in the two original solutions is HNO_3).

26. Calculate $[H^+]$ for solutions having (a) a pH of 3.56, (b) a pH of 8.69, and (c) a pOH of 12.25.

27. Imagine that you have 100 ml of pure water (pH = 7.00) and you add enough HCl to make the solution 1.0×10^{-7} *M* in HCl. Assuming the HCl dissociates completely, what is the pH of the solution?

28. The acid HA is 5% dissociated in a 0.30 *M* solution. Calculate the acid dissociation constant and the pH of the solution.

29. Calculate the pH of a 0.300 *M* solution of NH_4Cl. (K_b for NH_3 is 1.77×10^{-5}.)

30. Calculate the pH of the solution prepared by diluting 150 ml of 0.435 *M* hydroxylamine ($K_b = 1.07 \times 10^{-8}$) to 250 ml in a volumetric flask with distilled water.

31. Calculate the pH of the solution prepared by diluting 150 ml of 0.295 *M* acetic acid ($K_a = 1.76 \times 10^{-5}$) to 500 ml in a volumetric flask with distilled water.

32. Calculate the pH of the solution prepared by diluting a mixture of 10.0 g of sodium formate and 200 ml of 1.50 *M* formic acid ($K_a = 1.77 \times 10^{-4}$) to 1 ℓ in a volumetric flask with distilled water.

33. How many milliliters each of 1.0 *M* HOAc ($K_a = 1.76 \times 10^{-5}$) and 1.0 *M* NaOAc should be mixed to prepare 1 ℓ of a buffer of pH 4.0?

34. How many milliliters of 1.0 *M* NaOH should be added to 1.0 ℓ of 0.50 *M* HClO ($K_a = 2.95 \times 10^{-8}$) to give a buffer of pH 8.0?

35. What is the pH of the mixture that results when 0.40 ml of 1.0 *M* HCl is added to 150 ml of 0.20 *M* benzoic acid ($K_a = 6.46 \times 10^{-5}$)?

36. How many grams of sodium acetate must be added to 300 ml of 0.200 M acetic acid ($K_a = 1.76 \times 10^{-5}$) if the pH of the resulting solution is to be 6.0?

37. To 1 ℓ of a buffer consisting of 0.50 M $NaNO_2$ and 0.60 M HNO_2 ($K_a = 4.6 \times 10^{-4}$) is added 4.0 ml of 1.0 M NaOH. Calculate the change in the pH of the solution.

38. To 1 ℓ of a buffer consisting of 0.50 M $NaNO_2$ and 0.60 M HNO_2 ($K_a = 4.6 \times 10^{-4}$) is added 2.0 ml of 1.5 M HCl. Calculate the change in the pH of the solution.

39. Calculate the pH of the solutions obtained by adding the following volumes of 0.100 M HCl to 25.0 ml of 0.200 M ethylamine ($K_b = 6.41 \times 10^{-4}$):

a. 0 ml	e. 49.9 ml
b. 10.0 ml	f. 50.0 ml
c. 25.0 ml	g. 50.1 ml
d. 45.0 ml	h. 60.0 ml

40. Sketch the titration curve and the fractional distribution curves for ethylamine from the previous problem. Select an indicator suitable for this titration from Table 9-6.

41. One drop of indicator, HIn (yellow in the ionized form and red in the molecular form, $K_{HIn} = 10^{-5}$), is added to 200 ml of 0.100 M HOAc ($K_a = 1.76 \times 10^{-5}$). How many grams of NaOAc must be added to this solution to reach the point at which the solution begins to look yellow to the human eye?

42. 50 ml of 0.0050 M H_2SO_4 is titrated with 0.10 M NaOH. Assume that an inappropriate indicator is chosen—one that changes color at pH = 4.00. By how many milliliters of NaOH solution will the titration be in error?

43. After eating too much Mexican food, an amateur chemist gets an upset stomach with an accompanying sour taste in his mouth. He decides to mix a potion of a quarter-teaspoon ($\approx$0.65 g) of baking soda, $NaHCO_3$, in a 300-ml glass of water. After drinking his potion, the amateur chemist burps.
a. Why does the baking soda relieve the sour taste?
b. Calculate the pH of the potion. (For carbonic acid, $K_1 = 4.30 \times 10^{-7}$ and $K_2 = 5.61 \times 10^{-11}$.)
c. What causes the burp?

44. Sulfuric acid, H_2SO_4, in aqueous solution behaves like a strong acid; however, HSO_4^- is a relatively weak acid ($K_2 = 1.20 \times 10^{-2}$). What are the concentrations of HSO_4^-, H^+, and SO_4^{2-} in 0.50 M H_2SO_4?

45. Calculate the pH of a solution that is 0.0300 M in Na_2HPO_4 and 0.0400 M in Na_3PO_4. (For phosphoric acid, $K_1 = 7.52 \times 10^{-3}$, $K_2 = 6.23 \times 10^{-8}$, and $K_3 = 2.2 \times 10^{-13}$.)

46. Describe in detail how to prepare 100 ml of a 0.10 M buffer, pH 6.34, if all you have available is KH_2PO_4, K_2HPO_4, and water. (For phosphoric acid, $pK_2 = 7.21$.)

47. The bicarbonate buffer system is biologically important. To study this system, let us titrate 100 ml of 0.200 M Na_2CO_3 with 0.200 M HCl. (For carbonic acid, $K_1 = 4.30 \times 10^{-7}$ and $K_2 = 5.61 \times 10^{-11}$.)

a. Calculate the pH after the addition of 0, 50.0, 100, 150, and 200 ml of 0.200 M HCl.

b. Sketch the titration curve.

c. Judging from Table 9-6, what indicator would be suitable for the second equivalence point?

48. Lysine is one of the amino acids that make up proteins. Its fully protonated structure is

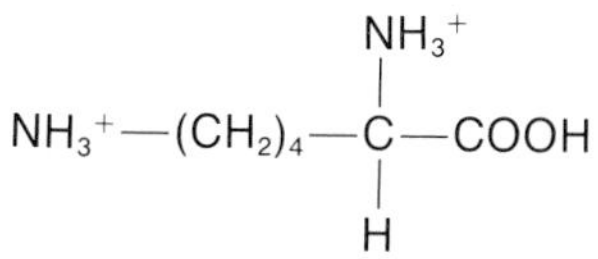

$$\begin{array}{c} \quad\quad\quad\quad\quad\quad\quad NH_3^+ \\ \quad\quad\quad\quad\quad\quad\quad | \\ NH_3^+\text{—}(CH_2)_4\text{—}C\text{—}COOH \\ \quad\quad\quad\quad\quad\quad\quad | \\ \quad\quad\quad\quad\quad\quad\quad H \end{array}$$

Three of its protons are dissociable; in analogy with the histidine example discussed in this chapter, therefore, the molecule as shown above can be represented as H_3aa^{2+} (the neutral form is Haa). The three protons dissociate with the values $pK_1 = 2.18$, $pK_2 = 8.95$, and $pK_3 = 10.53$. Sketch the titration curve of 50.0 ml of 0.100 M H_3aa^{2+} with 0.100 M NaOH. Accurately plot the points corresponding to the addition of 0, 25.0, 50.0, 75.0, 100, and 125 ml of 0.100 M NaOH. Indicate on the curve the regions in which lysine would serve as an effective buffer.

49. For the titration of lysine in the previous problem, sketch the fractional distribution curves of H_3aa^{2+}, H_2aa^+, Haa, and aa^-, as a function of pH. Label the axes and curves carefully and indicate by shading where lysine is a good buffer.

50. The compound o-phthalic acid is a diprotic acid with the following structure:

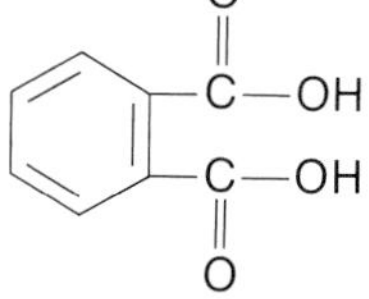

$$C_6H_4(\text{—}\overset{O}{\overset{\|}{C}}\text{—}OH)_2$$

and the acid dissociation constants $K_1 = 1.3 \times 10^{-3}$ and $K_2 = 3.9 \times 10^{-6}$.

a. Representing o-phthalic acid by the symbol H_2A, write the acid dissociation reactions of H_2A and HA^- in water, showing H_2O as a reactant in each case. Identify the two Brønsted-Lowry conjugate acid-base pairs in each reaction.

b. Write the mass-balance expression for all the phthalate species, letting c represent the total concentration.

c. Write the electroneutrality expression (charge balance) for an aqueous solution of H_2A.

51. A solution that is 0.15 M in $[Zn(NH_3)_4]^{2+}$ is prepared. What are the concentrations of Zn^{2+} when the ammonia concentration is 0.10, 0.20, and 0.30 mole/ℓ? The overall stability constant for $[Zn(NH_3)_4]^{2+}$ (i.e., the product of the four individual stability constants) is $K_{stab} = 3.2 \times 10^9$.

52. The stability constants of $[Co(OH)]^{2+}$ and $[Ni(OH)]^+$ are 1.0×10^{12} and 5.0×10^4, respectively. Calculate the pH of solutions that are prepared by dissolving nitrate salts of the two metals in water until the solutions are 0.15 M in Co^{3+} and Ni^{2+}. Correlate the charge on the metal ion with the acidity of the solution.

INTERLUDE

"Now, when we take three hundred millilitres of a compound containing hydrogen and oxygen in a ratio of two to one and add three millilitres of an eight-hundredths-per-cent chlorine solution, one millilitre of a three-ten-thousandths-per-cent stannous-fluoride solution, and fifty millilitres of treated industrial wastes and solids, we get drinking water."

making hard water soft

The adjective *hard* for describing water seems rather strange. It is difficult to think of a fountain, a placid trout stream, or a morning mist as being hard. Perhaps this term is more easily understood by a parent scouring out the tub after the kids' bath or by a homeowner discovering that he needs a new water heater because the old one is clogged with scale (a precipitate of oxides and carbonates). The structure of a soap micelle, which can lead to the infamous bathtub ring, is shown in Figure 5-7.

Hard water is characterized principally by the presence of appreciable amounts of Ca^{2+} and Mg^{2+}—usually as the bicarbonates or sulfates—and, to a lesser extent, Fe^{2+}, Fe^{3+}, Mn^{2+}, and silica. Such water is "softened" by removing these metal ions. This eliminates bathtub rings and boiler scale. Even though we Americans are particularly fortunate among the world's peoples to have copious amounts of reasonably safe water delivered to our taps, we have come to expect that, in addition, our water should be *soft*.

There are two categories of hard water. The type of hardness is determined by the anions that accompany the metal cations. (For convenience from now on, we will consider only the calcium ion as the source of hardness.) If the primary contaminant is calcium bicarbonate, $Ca(HCO_3)_2$ (as in the limestone regions of Pennsylvania), the water is said to have *temporary hardness*. This hardness can be eliminated by boiling the water, which converts the bicarbonate to carbonate, carbon dioxide, and water. The offending calcium ions are removed from solution by precipitation as calcium carbonate:

$$Ca^{2+}(aq) + 2HCO_3^-(aq) \xrightarrow{\text{Heat}} CaCO_3(c) + CO_2(g) + H_2O(l) \quad (1)$$

This is a rather expensive procedure to apply on a large scale because of the high fuel costs. A cheaper way to effect the softening is to add slaked lime, $Ca(OH)_2$:

$$Ca^{2+}(aq) + 2HCO_3^-(aq) + Ca(OH)_2(c) \rightarrow 2CaCO_3(c) + 2H_2O(l) \quad (2)$$

Water that cannot be softened by boiling is said to have *permanent hardness*. This hardness is attributable primarily to calcium sulfate and calcium chloride. One way to soften such water is to add a soluble carbonate, such as Na_2CO_3. As in Reactions 1 and 2 above, this precipitates the Ca^{2+} as $CaCO_3$, which can be removed from the system:

$$Ca^{2+}(aq) + CO_3^{2-}(aq) \rightarrow CaCO_3(c) \quad (3)$$

All three of the above reactions require the periodic removal of a precipitate, which necessitates some machinery and a certain amount of effort. Furthermore, none of the reactions removes all the ions from the water.

The simplest and most elegant method of water softening is that of *ion exchange*. Many homes in regions where the water is very hard have individual ion-exchange units to replace the Ca^{2+} ions by Na^+ ions. (Note that this is not purification of the water; it is merely the substitution of one impurity—a more tolerable one—for another.) Most chemical and industrial laboratories requiring especially pure water, called *deionized water*, use ion-exchange units that substitute H^+ and OH^- for the traces of all other cations and anions that remain in the water even after it has been distilled. Because of the importance of ion exchange, we will devote the remainder of this Interlude to it.

Ion exchange has been occurring in the clays of the Earth's crust for a very long time. Clays are hydrous aluminosilicate compounds (complex molecules of hydrogen, aluminum, silicon, and oxygen), many of which have variable amounts of other cations, such as Ca^{2+}, Mg^{2+}, Na^+, K^+, Fe^{2+}, and Fe^{3+}. They are able to selectively exchange ions in the groundwater flowing through the clay with ions contained in the clay itself. It is only in the last three decades that we have mimicked nature successfully in this regard. In fact, as with many other chemical syntheses and processes, we have developed ion exchangers that are much better than any found in nature.

Ion exchangers are used routinely in modern chemical and biochemical laboratories to perform two important tasks: *separation* and *total exchange*. The former is just that: the separation of ions that are quite similar in composition and charge (e.g., rare-earth ions) under very mild conditions. Total exchange is the replacement of all the cations or all the anions in the solution with an equivalent number of H^+ or OH^- ions, respectively. After discussing the principles of ion exchange, we will look at total exchange in more detail. Separation methods are not discussed here, but are discussed in numerous quantitative analysis texts.

ION-EXCHANGE RESINS

Ion-exchange resins are specially designed, synthetic organic polymers that are generally produced in the form of small beads. A tube packed with these beads and equipped with hose or pipe fittings at each end is called an *ion-exchange column* and is the basic unit of an ion exchanger. Because of the very large volumes of dilute aqueous solutions that flow through typical ion-exchange columns, the resins must have extremely low solubility in water.*

Typical ion-exchange resins have polymeric structures such as

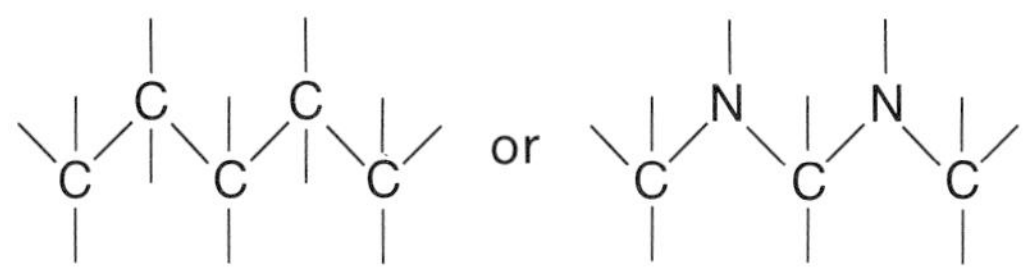

C C C C C or C N C N C (zig-zag chains)

Such zig-zag chains of atoms are called the *backbone* of the molecule. Although it is stereochemically less accurate, it is generally more convenient to represent the backbone as a straight chain:

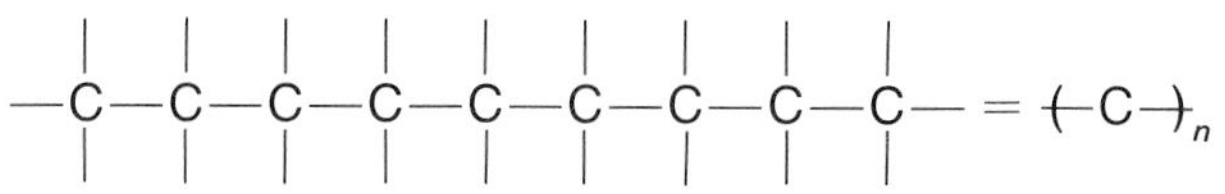

$$-\mathrm{C}-\mathrm{C}-\mathrm{C}-\mathrm{C}-\mathrm{C}-\mathrm{C}-\mathrm{C}-\mathrm{C}-\mathrm{C}- = (\!-\mathrm{C}-\!)_n$$

where n may be in the thousands.

An example of an industrially important polymer that is also an ion-exchange resin (albeit a very poor one) is Bakelite, a plastic that is produced by reacting phenol with formaldehyde:

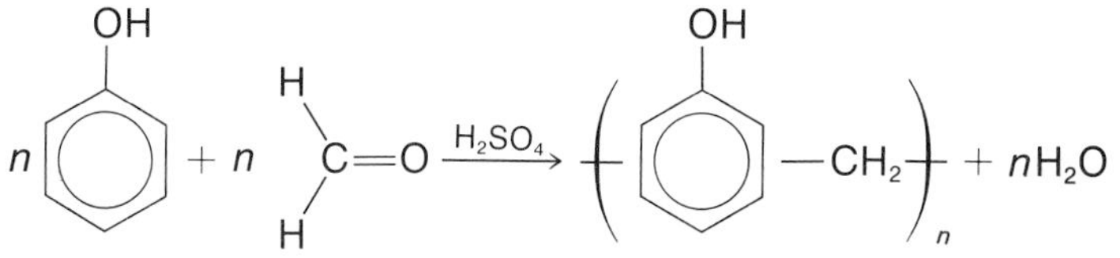

$$n\,\mathrm{C_6H_5OH} + n\,\mathrm{H_2C{=}O} \xrightarrow{\mathrm{H_2SO_4}} (\!-\mathrm{C_6H_3(OH)}-\mathrm{CH_2}-\!)_n + n\mathrm{H_2O}$$

*There is no such thing as perfect insolubility, of course, and all deionized water is contaminated by traces of organic matter from the resin used.

(The points of attachment of the CH_2 groups to the ring are unspecified here because they may be any of the C atoms except the one to which the OH group is attached.) Because a water molecule is eliminated when each pair of reactant molecules is joined, this type of reaction is called *condensation polymerization.* It is very common in the production of industrial and biological polymers.

In order to ascertain a polymer's ion-exchange properties, if any, we must consider all the side groups attached to the backbone. There are three basic possibilities. The first, of little interest to us here, is chemically unreactive groups such as H, CH_3, etc., which are an unintended result of the particular monomer(s) chosen to make the polymer.

The second possibility is a *cross-link,* i.e., a chemical link between two polymer chains. A cross-link may be anything from a single chemical bond to a segment of the polymer backbone itself. (Bakelite, for example, is highly cross-linked via CH_2 groups.) The amount of cross-linking in ion-exchange resins is important because it tends to minimize the solubility of the resin and it determines the extent to which the resin will swell in use. When ions from the aqueous solution flowing through the column are attracted into the pores of the resin beads (we will soon see what it is that attracts them), the ion concentration inside the pores becomes very high. This can cause osmotic pressures of as much as 1000 between the solution inside the pores and the solution outside, among the beads. The effect of this pressure is to force water into the pores, causing the resin to swell. The swelling is minimized by cross-linking in the polymer, however, as shown in Figure 1. Thus, most ion-exchange resins are highly cross-linked

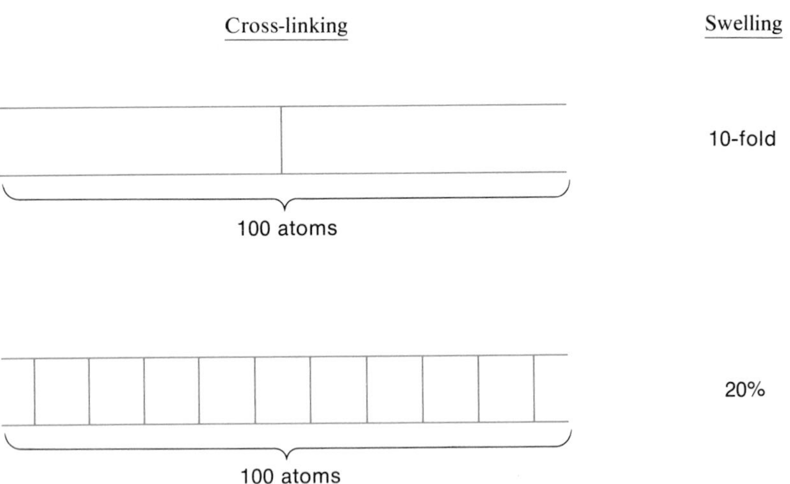

Figure 1. The tendency of ion exchange resins to swell owing to osmotic pressure is minimized by extensive cross-linking between the polymer chains.

Table 1. Some Ion-Exchanging Groups.

	Molecular Form	Ionic Form	Name	
Cation exchangers (acids)	R—COOH	R—COO^-	Carboxylic acid (carboxylate)	
	R—SO_3H	R—SO_3^-	Sulfonic acid (sulfonate)	
	R—OH	R—O^-	Phenol (phenolate)	
	R—SH	R—S^-	Mercaptan (mercaptide)	
Anion exchangers (bases)	R—NH_2	R—NH_3^+	Primary amine	(Substituted ammonium ions)
	R_1R_2—NH	R_1R_2—NH_2^+	Secondary amine	
	$R_1R_2R_3$—N	$R_1R_2R_3$—NH^+	Tertiary amine	

to prevent extreme swelling upon contact with the solution. Even then, the resins must always be pre-soaked in a beaker before they are used in a column.

The third possibility for a side group attached to the backbone is an ion-exchanging group, of which there are two general types. *Acidic* groups, —AH, are negatively charged because they dissociate in H_2O to give —A^- (bound) and $H^+(aq)$. They therefore attract and exchange positive ions (cations). *Basic* groups, —B, are positively charged because they are protonated in H_2O to give —BH^+ (bound) and $OH^-(aq)$. They therefore attract and exchange negative ions (anions). Some examples of cation exchangers and anion exchangers are given in Table 1. In each formula there, R denotes the remainder of the molecule.

The functions of the cation exchangers and anion exchangers are illustrated by the following two typical reactions:

$$R\text{—}SO_3^-H^+ + Na^+ \rightleftharpoons R\text{—}SO_3^-Na^+ + H^+ \tag{4}$$

$$R\text{—}NH_3^+OH^- + Cl^- \rightleftharpoons R\text{—}NH_3^+Cl^- + OH^- \tag{5}$$

It is useful to see how equilibrium is established between the resin and the ions in solution. We will calculate a hypothetical equilibrium ion distribution for Reaction 4.

Example 1. Calculate the concentration of Na^+ remaining in solution after a 1.0 *M* solution of Na^+ is equilibrated in a beaker with a cation-exchange resin of the H^+-donating type. Assume that the initial amount of H^+ on the resin is equivalent to 1 mole/ℓ (i.e., the concentration would be 1 mole/ℓ if all the H^+ were released into solution) and that the equilibrium constant for the exchange reaction is 10.

Let x be the concentration of Na^+ exchanged by the resin and let (r) denote species bound to the resin. Thus, we can write Reaction 4 as

$$H^+(r) + Na^+ \rightleftharpoons Na^+(r) + H^+ \qquad (6)$$

Initial concentrations:	1.0	1.0	0	0
Equilibrium concentrations:	$1.0 - x$	$1.0 - x$	x	x

$$K_{eq} = \frac{[Na^+]_r[H^+]}{[H^+]_r[Na^+]} = \frac{x^2}{(1.0 - x)^2} = 10$$

$$\frac{x}{1.0 - x} = \sqrt{10}$$

$$x = 0.76$$

Therefore the concentration of Na^+ remaining in solution after one exchange is $1.00 - 0.76 = 0.24$ mole/ℓ.

The result in the above example may be surprising if you expected total exchange between the cations. The reason for the relatively high concentration of unexchanged cation is obviously the small equilibrium constant. Values of the equilibrium constant between 1 and 10 are typical for such systems, however.

To achieve a quantitative exchange (i.e., one that is "total" within the normally acceptable error limit of about $\pm 0.2\%$) requires further steps. If the equilibrated solution from the above example is transferred to a second beaker containing fresh cation-exchange resin and swirled until a new equilibrium is established, the concentration of Na^+ remaining in solution is reduced to 0.03 mole/ℓ. In such repeated equilibrations, the sum of the concentrations of H^+ and Na^+ in solution always equals the initial concentration of Na^+. In other words, one H^+ ion is released for each Na^+ ion bound. If the metal ion to be exchanged were a divalent ion such as Ca^{2+}, two H^+ ions would be released for each metal ion bound. Similarly, one Al^{3+} ion would release three H^+ ions.

The equilibrium calculations show that several successive exchanges are required to achieve quantitative accuracy. This is both inefficient and tedious. A far more efficient method is to pass the solution through a column. A typical experimental setup is shown in Figure 2.

To do the experiment, the ion-exchange resin of choice is primed by equilibrating it with a large excess of the ion to be released—in this case, H^+. It is then rinsed thoroughly with distilled water and placed in the buret as an aqueous slurry. The amount of available H^+ can be calculated from the known number of acidic side groups per unit volume of the resin. This

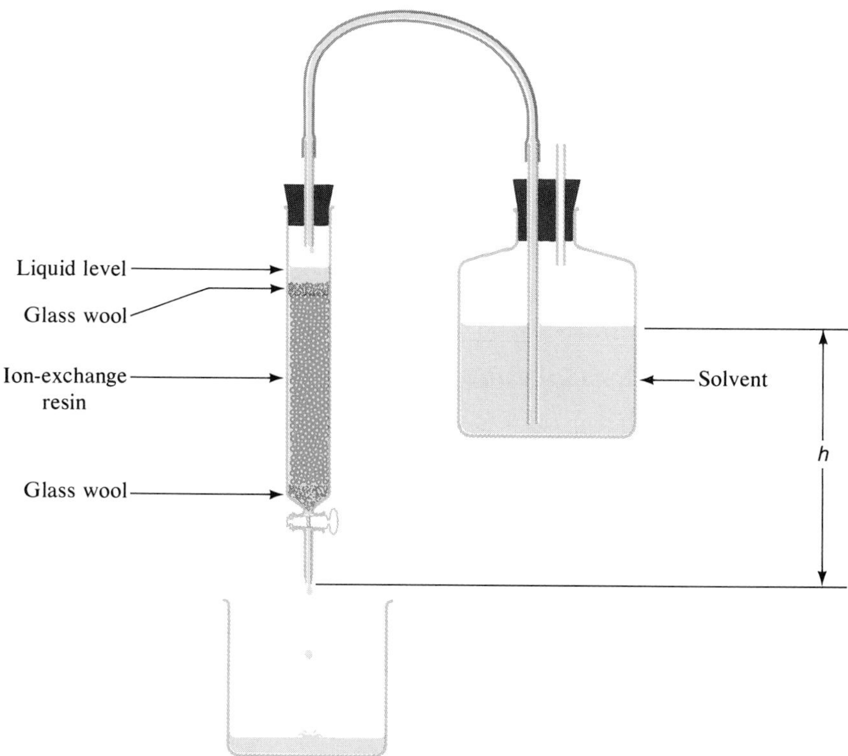

Figure 2. An ion-exchange column for achieving total exchange in an experimental solution.

amount should represent at least a 2-to-3-fold excess over the amount of the ion to be removed from the experimental solution—in this case, Na^+. With the stopcock still closed, the experimental solution is pipetted into the buret (the glass-wool plug prevents the top layer of resin beads from being churned up). The solution is now a "slug" of liquid just above the resin. It is essential that no air bubbles be entrained in the column during the experiment, so a backup flow of pure solvent (water) must be provided to keep the liquid level in the buret above the resin at all times. This is done by attaching a reservoir of the solvent to the buret, as shown in the figure. The hydrostatic head represented by the height h ensures that, as liquid is allowed to drip from the buret, atmospheric pressure on the reservoir forces an equal volume of solvent into the buret. Thus, the liquid level is held constant.

Now the stopcock is opened slightly to allow the slug of solution, preceded by the water in the column and followed by the water from the reservoir, to move slowly down the column. If the flow rate is slow enough, ion-exchange equilibrium is established repeatedly as the slug descends. In effect, as many as several dozen successive equilibrations may be achieved, which is more than enough to guarantee an exchange that is asymptotically total to an extent far beyond our ability to detect. How nice it is to have such a simple, efficient technique!

USES OF ION EXCHANGE

The purpose of the total-exchange technique described above is to replace all the cations or anions in a solution with ions of another kind, generally H^+ or OH^-, with which the resin was originally charged. This enables the total cation or anion concentration of a solution to be accurately determined by titrating the effluent from the column with a standard solution of base or acid, respectively. (The effluent is first diluted to a precise volume in a volumetric flask, of course.) This is often much easier and more accurate than titrating the original ion.

Total exchange is widely used today to provide deionized water for industry and for scientific laboratories. Tap water—better yet, distilled water—first enters a cation-exchange column, where virtually all the cations in solution are replaced by H^+ ions. It then passes through an anion-exchange column, where the anions are exchanged for OH^- ions. Because the solution is electrically neutral to begin with, exactly equal numbers of H^+ and OH^- ions are released. Most of these combine, in accord with the familiar equilibrium constant $K_w = 10^{-14}$, to produce a little more of what was desired in the first place: pure water. Figure 3 illustrates this process.

Now we are ready to return to the homeowner and the problem of how to make hard water soft. Some municipalities in regions that have unusually hard water provide ion-exchange treatment at their water plants. Others use the slaked-lime method. But many have no such facilities, so individual homeowners have their own ion-exchange units if they choose to soften their water. Few such units provide deionization, however. The amounts of sulfate, chloride, and other anions found in most water do not significantly interfere with normal household functions, so anion exchange is unnecessary. A cation exchanger alone is used to provide total exchange of the cations. Since the addition of H^+ to the water would make it somewhat acidic, the ion-exchange resin is designed to release harmless Na^+ ions instead. Thus, the water is softened but not deionized.

Because Ca^{2+} is divalent and Na^+ is monovalent, one Ca^{2+} ion displaces two Na^+ ions and occupies two negatively charged sites. If these sites are

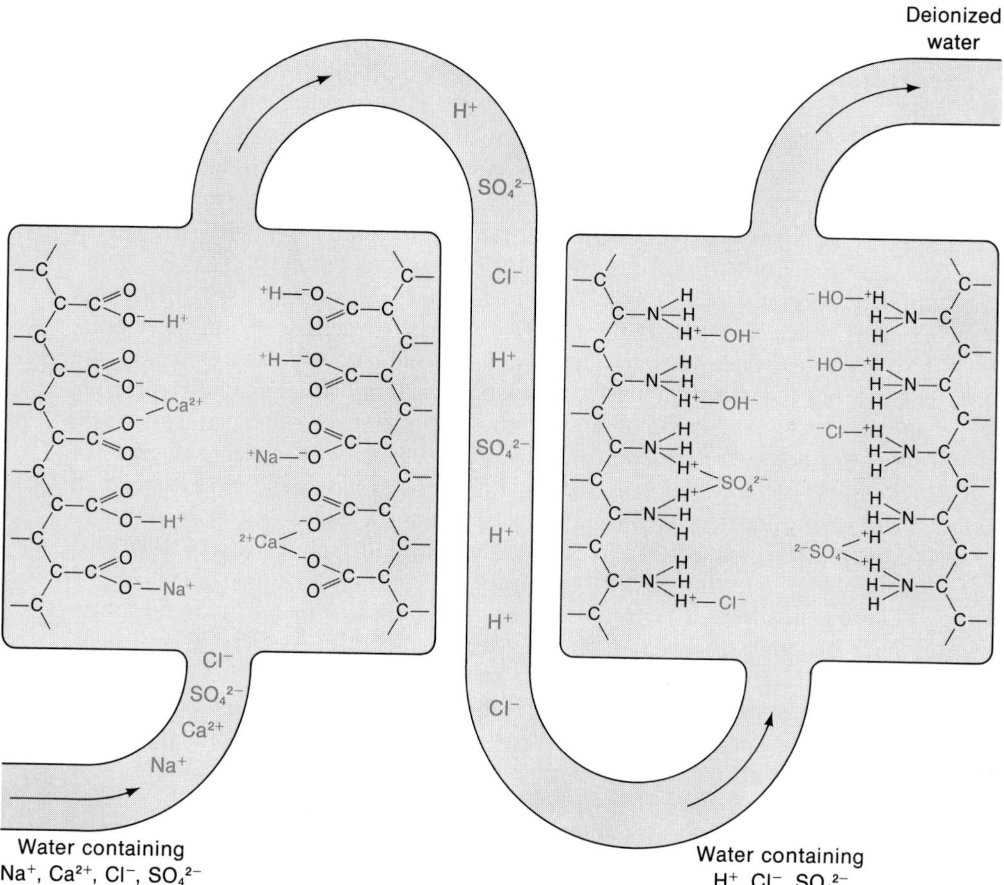

Figure 3. A cation exchanger (acidic side groups) and anion exchanger (basic side groups) in series, for producing deionized water.

carboxylate groups, as in Figure 3, the exchange reaction can be written as

$$2\text{R—CO}_2{}^-\text{Na}^+ + \text{Ca}^{2+} \rightleftharpoons (\text{R—CO}_2{}^-)_2\text{Ca}^{2+} + 2\text{Na}^+ \tag{7}$$

After a certain amount of water of a given hardness passes through a cation exchange column, most of the negative sites on the resin are occupied by Ca^{2+} ions, and those rings begin to appear around the bathtub again. The column can be regenerated by filling it with brine, a concentrated solution of NaCl. The concentration of Na^+ is so high that the equilibrium in Reaction 7 is driven strongly to the left, releasing the Ca^{2+} ions into solution. The brine, which now contains $CaCl_2$ as well as NaCl, is flushed out, and the column is ready to go again. This cycle can be repeated hundreds of times before the ion-exchange resin must be replaced.

There is only one problem with this system: the brine. It is "flushed out," we said, but where does it go? If the homeowner does the regeneration himself, it goes down the sewer, like most other domestic waste water. This encourages the homeowner (and most other people) to indulge subconsciously in the attitude, "Out of sight, out of mind." Unfortunately, this attitude is becoming increasingly untenable in modern society because of the environmental degradation that often results from it. For example, municipal sewage plants have no capability for removing Ca^{2+}, Na^+, and Cl^- ions. These ions, and others, therefore pass directly into our waterways. This may not be a serious problem if the effluent is discharged into the oceans, but it can significantly increase the salinity of lakes and rivers. We tend to forget that most of us are downstream from others and that still others are downstream from us. The same water, with varying degrees of processing, is used over and over as it moves toward the sea. As it picks up an increasing burden of soluble salts, it becomes less and less desirable to the people who must drink it and the farmers and manufacturers who must use it, not to mention the fish who must live in it.

There is an alternative to the homeowner's flushing brine down the sewer. It may be more expensive for him but cheaper for society as a whole, because it is environmentally more sound. He can subscribe to a water-softening service that delivers a fresh ion-exchange column periodically and takes the spent column to a central regeneration plant. But that plant produces waste brine and it too must go somewhere. The catch is that more and more municipalities are forbidding such plants to discharge their brine into the sewer system. Instead, the brine is placed in evaporation vats or ponds. The recovered salt can be recycled for other purposes and helps to offset the operating costs.

Our desire for the better life often leads us into environmental difficulties that make life worse in other ways. Only as we are willing to recognize the environmental consequences of our actions and to modify our actions accordingly, even if the cost is greater, will we be able to prevent the harm that we are inflicting upon our fragile ecosystem and preserve the world as a viable home for our descendants.

10 oxidation-reduction reactions

We have just completed our study of aqueous acid-base reactions, which, in the Brønsted-Lowry sense, involve the transfer of protons between chemical species. Oxidation-reduction reactions, on the other hand, involve the transfer of electrons between chemical species. When such a transfer occurs, one species is *oxidized* (loses one or more electrons) and another species is *reduced* (gains one or more electrons). Such reactions are often called *redox reactions*. In every redox reaction there is a change in the oxidation state of at least two atoms.

Chemists have established a set of rules for assigning an oxidation state to any atom in a compound or ion. The primary purpose in doing this is to provide a convenient bookkeeping method for electrons. It does not necessarily mean that the number of electrons "belonging" to each atom in a chemical species is determined by the oxidation state of the atom in question. We have already seen from quantum mechanics that the electronic charge in a molecule is a property of the entire molecule and that in covalent structures the charge imbalances are usually rather small. As a first example of oxidation-reduction, however, it is easiest to consider a reaction involving monatomic ions, in which oxidation state and ionic charge are the same:

$$Zn(c) + H^+(aq) \rightarrow Zn^{2+}(aq) + H_2(g) \qquad (10\text{-}1)$$

Balancing this reaction is trivially easy because we know that, like atoms, electric charge is neither created nor destroyed. As long as nothing else enters into this reaction, we see that two H^+ ions must react with each Zn atom to produce one Zn^{2+} ion and one H_2 molecule:

$$Zn(c) + 2H^+(aq) \rightarrow Zn^{2+}(aq) + H_2(g) \qquad (10\text{-}2)$$

The oxidation state of all elements is 0 because they are neutral. The oxidation state of H^+ is +1 and that of Zn^{2+} is +2. Thus the sum of the oxidation states on the left side of this reaction equals the sum of the oxidation states on the right. The reaction actually entails the transfer of two electrons from each atom of zinc metal to two hydrogen ions. We can represent this transfer in terms of two half-reactions that can be added algebraically, with cancellation of the electrons:

$$\textbf{Oxidation:} \quad \overset{0}{Zn} \rightarrow \overset{+2}{Zn^{2+}} + 2e^-$$

$$\textbf{Reduction:} \quad 2\overset{+1}{H^+} + 2e^- \rightarrow \overset{0}{H_2}$$

$$Zn + 2H^+ \rightarrow Zn^{2+} + H_2$$

Zinc is oxidized in this reaction because its oxidation state has increased from 0 to +2. Hydrogen is reduced in this reaction because its oxidation state has decreased from +1 to 0. A chemical species is oxidized if it loses electrons. A chemical species is reduced if it gains electrons.

Whenever one species is oxidized, another species is reduced, and vice versa. Elemental zinc is called a *reducing agent* in the above reaction because it donates the electrons that reduce the hydrogen ions. The latter are the *oxidizing agents* because, by accepting these electrons and being reduced, they oxidize the zinc.

Assigning oxidation states to atoms in chemical species is done by the rules that were presented in Chapter 3. They are repeated here for your convenience.

1. The oxidation state of all elements is 0.
2. The oxidation state of a monatomic ion equals the charge on the ion.
3. Oxygen has an oxidation state of −2 in all compounds except those designated as *peroxides*, in which its oxidation state is −1, and those designated as *superoxides*, in which its oxidation state is effectively $-\frac{1}{2}$.
4. Hydrogen has an oxidation state of +1 in all compounds except those designated as *metal hydrides*, in which its oxidation state is −1.
5. The sum of the oxidation states of the elements in a compound or ion equals the charge on that species.

6. The most electronegative element in a complex ion is assigned the negative oxidation state.

Another set of rules consistent with those just stated is:

1. Electrons shared by unlike atoms are counted with the more electronegative atom.
2. Electrons shared by like atoms are divided equally.

We discussed the relation between the observed oxidation states of the elements and their position in the periodic table in Chapter 3. We suggest that you reread the section on oxidation states in that chapter.

10.1 BALANCING REDOX REACTIONS

Balancing redox reactions requires that an accounting be made of the electrons transferred in the reaction. Thus the number of electrons donated by the reducing agent must be equal to the number of electrons accepted by the oxidizing agent. The relative amounts of oxidizer and reducer can therefore be determined by the *half-reaction method*, in which each process (oxidation and reduction) is considered separately and the two half-reactions are added algebraically, with cancellation of the electrons.

For example, consider the following redox reaction:

$$Cr_2O_7^{2-} + H_2S \rightarrow Cr^{3+} + S(c) \qquad \text{(Acidic solution)}$$

This reaction occurs only in acidic solutions because neither H_2S nor dichromate ion is stable except under acidic conditions. Dichromate ion is a very strong oxidizing agent because of the high oxidation state (high tendency to accept electrons and be reduced) of the Cr atom in this ion.* In balancing this reaction we treat the reduction and oxidation processes separately as follows, beginning with the reduction half-reaction:

$$\textbf{Reduction:} \quad Cr_2O_7^{2-} \rightarrow Cr^{3+}$$

1. First balance the principal atom on both sides of the half-reaction:

$$Cr_2O_7^{2-} \rightarrow 2Cr^{3+}$$

*The dark brown "cleaning solution" found in many laboratories is a saturated solution of $K_2Cr_2O_7$ in H_2SO_4. It is so powerful an oxidizing agent that few substances can withstand its effect.

2. Then balance oxygen and hydrogen atoms on both sides. Because the solution is acidic, we may use only H^+ and H_2O for this purpose (the OH^- concentration is negligible):

$$Cr_2O_7^{2-} + 14H^+ \rightarrow 2Cr^{3+} + 7H_2O$$

3. When all the atoms are balanced, balance the charges by adding electrons to the appropriate side of the half-reaction. The net charge on the left side is +12 and on the right side, +6. Therefore we must add six electrons to the left side to balance the half-reaction:

$$Cr_2O_7^{2-} + 14H^+ + 6e^- \rightarrow 2Cr^{3+} + 7H_2O$$

4. Check the final, balanced half-reaction by calculating the oxidation state of the principal atom on both sides, to be sure the correct number of electrons is being transferred:

$$\overset{2(+6)}{Cr_2}O_7^{2-} + 6e^- \rightarrow \overset{2(+3)}{2Cr^{3+}} \quad (+6e^-)$$

We now treat the oxidation half-reaction analogously:

Oxidation: 1. $H_2S \rightarrow S(c)$
2. $H_2S \rightarrow S(c) + 2H^+$
3. $H_2S \rightarrow S(c) + 2H^+ + 2e^-$
4. *Check*: $H_2\overset{-2}{S} \rightarrow \overset{0}{S}(c) + 2e^- \quad (-2e^-)$

Now the reduction and oxidation half-reactions must be added so that the number of electrons on the left and right sides cancel. In the present example this requires that the oxidation half-reaction be multiplied by the factor 3.

Reduction: $Cr_2O_7^{2-} + 14H^+ + 6e^- \rightarrow 2Cr^{3+} + 7H_2O$

Oxidation: $3H_2S \rightarrow 3S(c) + 6H^+ + 6e^-$

$$Cr_2O_7^{2-} + 3H_2S + 14H^+ \rightarrow 2Cr^{3+} + 3S(c) + 7H_2O + 6H^+$$

Finally, any component appearing on both sides of the reaction should be canceled. In the present example 14 protons on the left and 6 protons on the right gives 8 protons on the left:

$$Cr_2O_7^{2-} + 3H_2S + 8H^+ \rightarrow 2Cr^{3+} + 3S(c) + 7H_2O \qquad (10\text{-}3)$$

Another example is given below, using another common oxidizing agent, permanganate ion, in basic solution:

$$MnO_4^- + NO_2^- \rightarrow MnO_2(c) + NO_3^- \qquad \text{(Basic solution)}$$

The reduction and oxidation half-reactions are developed as follows:

Reduction:
1. $MnO_4^- \rightarrow MnO_2(c)$
2. $MnO_4^- + 2H_2O \rightarrow MnO_2(c) + 4OH^-$
3. $MnO_4^- + 2H_2O + 3e^- \rightarrow MnO_2(c) + 4OH^-$

   $+7 \xrightarrow{+3e^-} +4$

4. *Check*: $MnO_4^- + 3e^- \rightarrow MnO_2(c)$

Oxidation:
1. $NO_2^- \rightarrow NO_3^-$
2. $NO_2^- + 2OH^- \rightarrow NO_3^- + H_2O$
3. $NO_2^- + 2OH^- \rightarrow NO_3^- + H_2O + 2e^-$

   $+3 \xrightarrow{-2e^-} +5$

4. *Check*: $NO_2^- \rightarrow NO_3^- + 2e^-$

We now multiply the reduction and oxidation half-reactions by 2 and 3, respectively, in order to be able to cancel the electrons in the final addition:

$$\textbf{Reduction:}\quad 2MnO_4^- + \overset{1}{\cancel{4}}H_2O + \cancel{6e^-} \rightarrow 2MnO_2(c) + \overset{2}{\cancel{8}}OH^-$$

$$\textbf{Oxidation:}\quad 3NO_2^- + \cancel{6OH^-} \rightarrow 3NO_3^- + \cancel{3H_2O} + \cancel{6e^-}$$

$$2MnO_4^- + 3NO_2^- + H_2O \rightarrow 2MnO_2(c) + 3NO_3^- + 2OH^- \qquad (10\text{-}4)$$

Note that, in balancing half-reactions that occur in acidic solutions, the side that contains an excess of O atoms must have two H^+ ions added to it for each excess O atom, producing one molecule of H_2O on the opposite side. In basic solutions, one H_2O molecule must be added for each excess O atom, producing two OH^- ions on the opposite side.

The half-reaction method of balancing redox reactions is particularly useful in balancing organic reactions, as we will see later in this chapter. There is an alternate method that is useful in balancing inorganic reactions such as the dichromate/hydrogen sulfide reaction discussed above. This method is called *balancing by oxidation states*. Its name implies a heavy reliance upon the last step of the half-reaction method.

The method begins with the requirement of our knowing (or correctly guessing) the two principal reactants and the two principal products. The principal reactants are the atoms, molecules, or ions that undergo oxidation and reduction, respectively.* Thus, in the present instance, we begin with the same information as before:

$$Cr_2O_7^{2-} + H_2S \rightarrow Cr^{3+} + S(c) \qquad \text{(Acidic solution)}$$

Now the specific atoms undergoing oxidation and reduction are identified and oxidation states are assigned to them, both as reactants and as products:

$$\overset{+6}{Cr_2}O_7^{2-} + H_2\overset{-2}{S} \rightarrow \overset{+3}{Cr^{3+}} + \overset{0}{S}(c)$$

If there is more than one such atom in any of the species involved in the reaction, the numbers of that atom are balanced on both sides of the reaction. At the same time, the oxidation states for that atom are multiplied by the number of atoms of that kind in question, i.e., by the number of atoms of that kind in the species and/or by the stoichiometric coefficient of the species in the reaction as it is now written:

$$\overset{2(+6)}{Cr_2}O_7^{2-} + H_2\overset{-2}{S} \rightarrow \overset{2(+3)}{2Cr^{3+}} + \overset{0}{S}(c)$$

The total numbers of electrons transferred are now indicated by arrows connecting the atoms undergoing redox:

$$\overbrace{\overset{2(+6)}{Cr_2}O_7^{2-} + H_2\overset{-2}{S} \rightarrow \overset{2(+3)}{2Cr^{3+}}}^{+6e^-} + \overset{0}{S}(c)$$
$$H_2S \xrightarrow{-2e^-} S(c)$$

These numbers of electrons are balanced by multiplying one (or both) of them by the appropriate integer(s). In the present example we multiply the $-2e^-$ by the factor 3. This provides for six electrons lost, to equal the six electrons gained:

$$\overbrace{\overset{2(+6)}{Cr_2}O_7^{2-} + H_2\overset{-2}{S} \rightarrow \overset{2(+3)}{2Cr^{3+}}}^{+6e^-} + \overset{0}{S}(c)$$
$$H_2S \xrightarrow{3(-2e^-)} S(c)$$

---

*It is common to speak of a molecule or polyatomic ion as being oxidized or reduced, even though it is only one atom within such a species that actually undergoes a change in oxidation state during a reaction. Thus we say that the $Cr_2O_7^{2-}$ ion is reduced when, during a reaction, the oxidation state of the Cr atom decreases from +6 to +3.

Whatever integers are required as multipliers are now inserted as stoichiometric factors before the appropriate species. If one atom of sulfur loses two electrons, there must be three atoms undergoing oxidation in order for there to be a loss of six electrons:

$$Cr_2O_7^{2-} + 3H_2S \rightarrow 2Cr^{3+} + 3S(c)$$

Here the arrows and oxidation numbers have been dropped, having served their purpose. The tricky job of balancing electron transfer is now complete. The only job remaining is to balance the oxygen and hydrogen atoms. If we have done our electron counting properly, we are guaranteed that the charge will be equal on both sides of the reaction when this is done.

There are two almost equally desirable ways to proceed. First, note that the solution under consideration is acidic. Therefore the only species to be added to either side of the reaction are H^+ and H_2O because the OH^- concentration is so low. We can finish balancing the reaction by noting that there are seven O atoms on the left and none on the right. The only oxygen-containing species that we can add to the right is H_2O. Adding $7H_2O$ to the right balances the O atoms. A count of the H atoms on both sides now reveals that $8H^+$ must be added to the left, which leads to both mass and charge balance.

Alternatively, we can finish balancing the reaction by noting that the charge on the left side is -2 and on the right side, $+6$. Because H^+ is the only charged species that we can add, charge balance requires that $8H^+$ be added to the left side of the reaction. Atom balance of either O or H atoms then requires the addition of $7H_2O$ to the right side. The final result is identical to Reaction 10-3:

$$Cr_2O_7^{2-} + 3H_2S + 8H^+ \rightarrow 2Cr^{3+} + 3S(c) + 7H_2O$$

10.2 ELECTROCHEMICAL CELLS

Electrochemical cells are devices that convert electric energy directly to chemical energy or vice versa. Their basic components are a vessel containing a dissolved or molten electrolyte, a pair of electrodes immersed in the electrolyte, and an external circuit. Electrochemical cells can be classified in two groups. The *electrolytic cell* is a cell that uses external electric energy to cause a chemical reaction to occur. This process is called *electrolysis*. It is used industrially for winning some metals, e.g., aluminum, magnesium, sodium, and potassium, from their ores and for purifying most metals. The *galvanic cell* (sometimes called voltaic cell) is a cell that uses the free energy of a chemical reaction to produce an electromotive force (emf). A typical example is a flashlight battery. Thus, the electrolytic cell uses electric energy to drive a chemical reaction that does not occur spontaneously ($\Delta G > 0$), whereas the galvanic cell exploits the spontaneity of a chemical reaction ($\Delta G < 0$) to produce electric energy.

Electrolytic Cells

The simplest kind of electrolytic cell, in principle, is that used to electrolyze molten salts such as NaCl. A diagram of such a cell is shown in Figure 10-1. The external dc source is, in effect, an electron pump that drives electrons to the cathode of the electrolytic cell, giving it a negative charge, and withdrawing electrons from the anode, giving it a positive charge. In any electrochemical cell (electrolytic or galvanic), the *cathode*

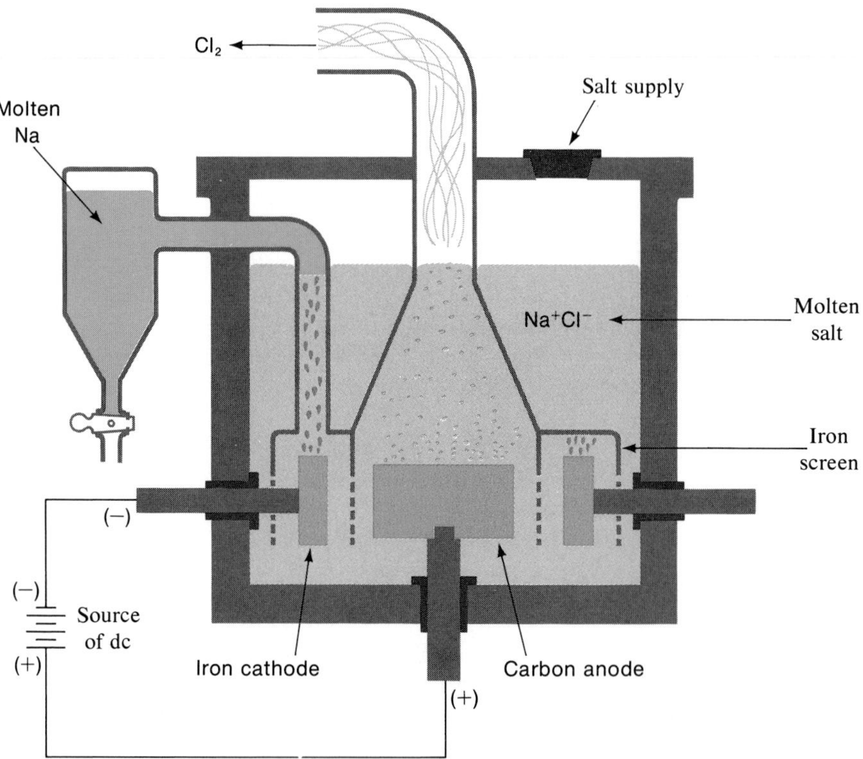

Figure 10-1. The electrolysis of molten sodium chloride, showing the manner in which the two products are collected. (L. Pauling, *General Chemistry*, 3rd ed., W. H. Freeman and Company. Copyright © 1970.)

is the electrode *to* which electrons in the external circuit flow, and the *anode* is the electrode *from* which electrons in the external circuit flow. Thus, in the external circuit of any electrochemical cell, electrons always flow from the anode to the cathode. This statement is worth memorizing.

In the cell the cathode donates electrons to positive ions (*cations*), thereby reducing these ions. The anode accepts electrons from negative ions (*anions*), thereby oxidizing these ions. In this particular cell the only reactive species are Na^+ and Cl^-. The only possible reactions are therefore

Reduction (cathode): $2Na^+ + 2e^- \rightarrow 2Na(l)$

Oxidation (anode): $2Cl^- \rightarrow Cl_2(g) + 2e^-$

$$2Na^+ + 2Cl^- \rightarrow 2Na(l) + Cl_2(g) \qquad (10\text{-}5)$$

This is not a spontaneous reaction because, even if we heat sodium chloride enough to vaporize it, the reaction does not occur to an appreciable extent. In the standard state (all substances in their most stable physical form at 1 atm and 298 K) the reaction is written

$$2NaCl(c) \rightarrow 2Na(c) + Cl_2(g)$$

From Appendix D we find that the standard free-energy change for the reaction is

$$\begin{aligned}\Delta G^\circ_{\text{reaction}} &= 2\Delta G^\circ_f[Na(c)] + \Delta G^\circ_f[Cl_2(g)] - 2\Delta G^\circ_f[NaCl(c)] \\ &= 2(0) + 0 - 2(-91.785) = 183.570 \text{ kcal/mole}\end{aligned}$$

or -2 times the standard free energy of formation of NaCl.

To obtain the value of ΔG for the actual reaction in the electrolytic cell, however, we would have to use the values of ΔG_f for the three species in their actual (nonstandard) states in the cell. The reaction is now written

$$2NaCl(l) \rightarrow 2Na(l) + Cl_2(g)$$

and the equation for the free-energy change is

$$\Delta G_{\text{reaction},T} = 2\Delta G_{f,T}[Na(l)] + \Delta G_{f,T}[Cl_2(g)] - 2\Delta G_{f,T}[NaCl(l)]$$

where T is the temperature of the molten salt. Thermodynamic data over a very wide temperature range are readily available for a great many substances. Using such data for $T = 1100$ K (NaCl melts at 1074 K), we find that $\Delta G = 148.5$ kcal/mole for this reaction. For $T = 1600$ K (NaCl boils at 1686 K), $\Delta G = 135.0$ kcal/mole. Clearly the reaction is still thermodynamically impossible by a wide margin. The point of this discussion, however, is that the reaction can be *made* to occur by driving it with an amount of external electric energy equivalent to at least the free-energy change of the reaction. Recalling the principles of thermodynamics, what do you think would happen if the reverse reaction were attempted, i.e., if liquid Na and gaseous Cl_2 were mixed?

The conduction of electricity in electrolytic cells is due to the migration of ions under the influence of the externally applied emf. In the present example, Na^+ ions migrate toward the cathode and Cl^- ions migrate toward the anode. The net effect of these migrations and the reduction and oxidation reactions that ensue is electrically equivalent to the conduction of electrons from the cathode to the anode inside the cell. In this way the circuit is completed when the external dc source is turned on. Note that, for every two Na atoms liberated at the cathode, two electrons must pass through the external circuit and one Cl_2 molecule is evolved at the anode.

Two other examples of electrolysis follow. The electrolysis of seawater can be described by the following scheme:

Reduction (cathode): $2H_2O + 2e^- \rightarrow H_2(g) + 2OH^-$

Oxidation (anode): $2Cl^- \rightarrow Cl_2(g) + 2e^-$

$$2H_2O + 2Cl^- \rightarrow H_2(g) + Cl_2(g) + 2OH^- \qquad (10\text{-}6)$$

Since NaCl is the principal electrolyte in seawater, you may wonder why H_2O is reduced at the cathode rather than Na^+. The reason is simply that H_2O molecules are more easily reduced than Na^+ ions, i.e., H_2O has a greater tendency to accept electrons. Actually, both these species are difficult to reduce, but, when forced by an external emf, H_2O yields first. We will re-examine this aspect of the electrolysis of seawater later in this chapter.

Exercise 10-1. Calculate the value of $\Delta G°$ for Reaction 10-6. What does the answer tell you about the stability of seawater? Are you surprised?

The second example of electrolysis is that of water slightly acidified with H_2SO_4 (the electrolyte):

Reduction (cathode): $4H^+ + 4e^- \rightarrow 2H_2(g)$

Oxidation (anode): $2H_2O \rightarrow O_2(g) + 4H^+ + 4e^-$

$$2H_2O \rightarrow 2H_2(g) + O_2(g) \qquad (10\text{-}7)$$

This time you may wonder (you *should* wonder) why SO_4^{2-} does not appear. Here H_2O is not reduced because H^+ is more easily reduced. Instead, H_2O is oxidized, because it is more easily oxidized than SO_4^{2-}. In analogy with the seawater example, both species are difficult to oxidize, but, when forced by an external emf, H_2O yields first.

Exercise 10-2. Write the redox reaction scheme that you think would describe the electrolysis of a dilute solution of Na_2SO_4. Calculate the values of $\Delta G°$ for this reaction and for Reaction 10-7.

The Quantitative Basis of Electrolysis

Since there is an obvious relation between the number of electrons passing through the external circuit of an electrolytic cell and the number of molecules of chemical product obtained, it is useful to calculate the charge of one mole of electrons.

We already know that the electron charge $e = 4.80 \times 10^{-10}$ esu. A more convenient unit of charge in electrochemistry, however, is the *coulomb* (abbreviation C), which is defined in terms of the fundamental units ampere and second:

$$1 \text{ coulomb} = 1 \text{ ampere-second}$$

In other words, one coulomb is the quantity of electric charge transferred in one second by a steady current of one ampere.

In terms of coulombs, the electron charge $e = 1.6022 \times 10^{-19}$ C. The charge of one mole of electrons is therefore

$$eN_A = (1.6022 \times 10^{-19} \text{ C})(6.0222 \times 10^{23} \text{ mole}^{-1}) = 96{,}487 \text{ C/mole}$$

This quantity is called the *Faraday constant* (symbol $\mathscr{F}$ or F) after the British physicist and chemist Michael Faraday (1791–1867), and the quantity 96,487 coulombs is called one *faraday* (no abbreviation).* Thus we have

$$\mathscr{F} = eN_A = 96{,}487 \text{ C/mole}$$

$$1 \text{ faraday} = 96{,}487 \text{ C}$$

*Neither of these should be confused with the *farad* (abbreviation F), which is the unit of electric capacitance.

Clearly the number of moles of product obtained in an electrolytic cell is related to the number of faradays that pass through the circuit. The relation is summarized in *Faraday's laws of electrolysis:*

1. The mass of substance produced at the cathode or anode in an electrolytic reaction is directly proportional to the quantity of electricity that passes through the circuit.
2. The masses of different substances produced by the same quantity of electricity are directly proportional to the molecular weights of the substances and inversely proportional to the numbers of electrons transferred per molecule in their production.

Let us use Reaction 10-7 as an example of the application of these laws. Faraday's first law says that the masses of H_2 and O_2 produced at the electrodes are directly proportional to the number of electrons passing through the circuit. That this is true is obvious from the stoichiometries of the half-reactions. We see that the transfer of 4 electrons produces 2 molecules of H_2 and 1 molecule of O_2. Therefore a flow of 4 moles of electrons (4 faradays) produces 2 moles (4 g) of H_2 and 1 mole (32 g) of O_2; a flow of 8 faradays produces 4 moles (8 g) of H_2 and 2 moles (64 g) of O_2; etc.

Faraday's second law says that the masses of H_2 and O_2 produced are directly proportional to their molecular weights (2 and 32 g/mole, respectively) and inversely proportional to the numbers of electrons transferred per molecule in their production (2 and 4, respectively). Therefore the ratio of the masses is given by

$$\frac{\text{Mass H}_2}{\text{Mass O}_2} = \frac{2}{32} \times \left(\frac{2}{4}\right)^{-1} = \frac{4}{32} = \frac{1}{8}$$

For any given quantity of electricity that passes through the circuit, the mass of O_2 produced at the anode is 8 times the mass of H_2 produced at the cathode.

That the inverse proportionality specified in Faraday's second law *must* be true can be seen by recognizing that the number of electrons accepted by the anode in the oxidation half-reaction must equal the number of electrons donated by the cathode in the reduction half-reaction. If twice as many electrons must be given up for the formation of one O_2 molecule as must be received for the formation of one H_2 molecule, then only half as many O_2 molecules as H_2 molecules will be formed by the passage of a given number of electrons. This is an illustration of a general principle that applies to any electrolysis reaction and to comparisons of the numbers of atoms or molecules produced in different electrolysis reactions, for a given quantity of electricity.

Example 10-1. What is the current in amperes that flows through an electrolytic cell containing dilute H_2SO_4 if 200 ml(STP) of H_2 is generated in 10 minutes?

The current is calculated by dividing the total number of coulombs by the time elapsed in seconds. The number of coulombs is calculated from the number of faradays required to produce the number of moles of product in question. From Reaction 10-7 we see that the transfer of 4 electrons yields 2 molecules of H_2, hence that a flow of 4 faradays yields 2 moles of H_2. Writing this as 2 faradays/mole, we now obtain the expression for the current:

$$\text{Current} = \frac{[200\ \cancel{\text{ml(STP)}}](2\ \cancel{\text{faradays}}/\cancel{\text{mole}})(9.65 \times 10^4\ \text{C}/\cancel{\text{faraday}})}{[22.4 \times 10^3\ \cancel{\text{ml(STP)}}/\cancel{\text{mole}}](10\ \cancel{\text{min}})(60\ \text{s}/\cancel{\text{min}})}$$

$$= 2.88\ \text{C/s} = 2.88\ \text{A}$$

Galvanic Cells

A galvanic cell is the opposite of an electrolytic cell in that it converts the free energy of a spontaneous chemical reaction to electric energy. An external circuit connecting the electrodes is required, but no external dc source. As an example, consider the redox reaction of zinc metal with a solution of cupric ion:

$$\text{Zn}(c) + \text{Cu}^{2+} \rightarrow \text{Zn}^{2+} + \text{Cu}(c) \tag{10-8}$$

From Appendix D we find that the standard free-energy change for this reaction is

$$\Delta G^\circ_{\text{reaction}} = \Delta G^\circ_f[\text{Zn}^{2+}(aq)] + \Delta G^\circ_f[\text{Cu}(c)] - \Delta G^\circ_f[\text{Zn}(c)] - \Delta G^\circ_f[\text{Cu}^{2+}(aq)]$$
$$= -35.18 + 0 - 0 - 15.53 = -50.71\ \text{kcal/mole}$$

The reaction is spontaneous, a fact readily demonstrated by placing a piece of zinc in a solution containing cupric ion. The reaction begins immediately and cannot be stopped unless the zinc is removed. This is surely an impractical way to make a cell, especially since the chemical energy is converted to heat, not electricity.

To make a practical galvanic cell, the cell must be designed in such a way that no reaction occurs unless current can flow in an external circuit. The zinc metal and cupric ion must therefore be separated! This is not as ridiculous as it sounds. Consider two beakers, one containing a 1 *M* $ZnSO_4$ solution and the other containing a 1 *M* $CuSO_4$ solution, with electrodes of the corresponding pure metals. This is shown in Figure 10-2.

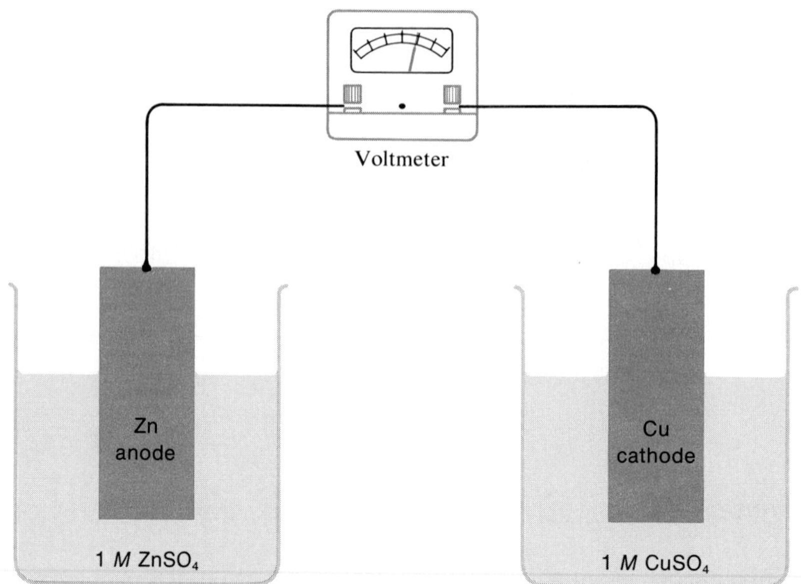

Figure 10-2. An incomplete galvanic cell. There is an electric potential difference, but no current can flow.

The proposed reaction is

Reduction (cathode): $Cu^{2+}(1\ M) + 2e^- \rightarrow Cu(c)$

Oxidation (anode): $Zn(c) \rightarrow Zn^{2+}(1\ M) + 2e^-$

$$Zn(c) + Cu^{2+}(1\ M) \rightarrow Zn^{2+}(1\ M) + Cu(c) \qquad (10\text{-}9)$$

In the setup shown in Figure 10-2, there is an external circuit but no internal circuit, because the reactants are now separated. Therefore no current can flow, even though there is an electric potential difference between the two electrodes. In order to get electric energy from this cell, the circuit can be completed by means of a *salt bridge*. This is a tube containing a saturated KCl solution or an agar (a gelatinous substance extracted from certain marine algae) saturated with KCl or some other strong electrolyte. A typical salt bridge is shown in Figure 10-3.

The salt bridge allows ionic conduction to occur (via the opposite migrations of K^+ and Cl^- ions) between the two solutions without allowing the solutions themselves to mix. This prevents the "short circuit" that would result if the complete redox reaction occurred at the zinc surface, as described earlier, and forces electrons to flow from the anode to the cathode through the external circuit. The resulting electric current can be used to

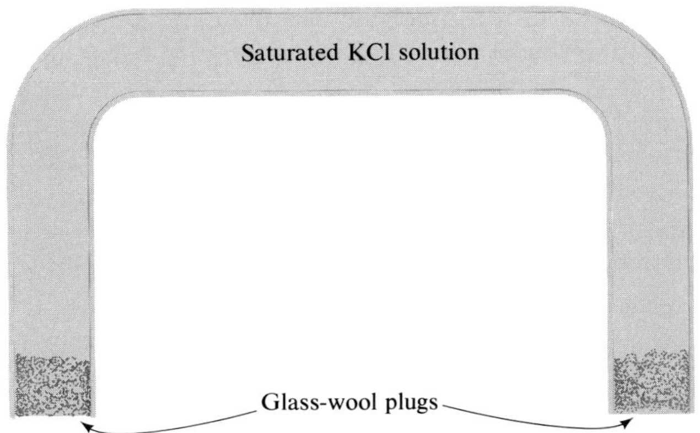

Figure 10-3. A KCl salt bridge. The glass wool inhibits mixing of the salt solution with those of the cell.

light a bulb, ring a bell, drive an electrolytic cell, etc. In effect, the galvanic cell consists of two half-cells, in each of which a half-reaction occurs. The complete cell is shown in Figure 10-4.

This particular cell is often called the *Daniell cell* after the British chemist John Daniell (1790–1845), who invented it. It was the first really reliable source of steady electric current ever devised. This cell could be used as a

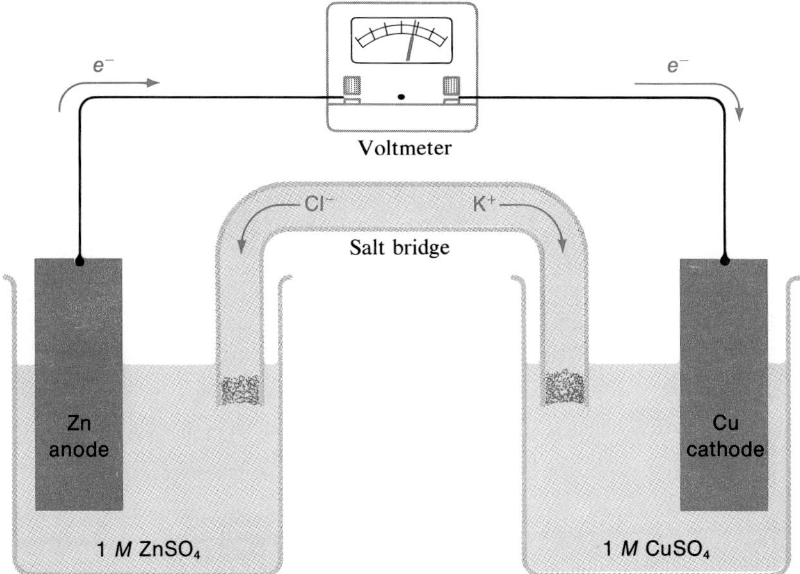

Figure 10-4. A complete galvanic cell—here, the Daniell cell.

constant-voltage source of direct current until the zinc anode was completely dissolved, provided that the concentrations of Zn^{2+} and Cu^{2+} in the two solutions could be maintained at 1 mole/ℓ. Under these conditions the voltmeter would read a steady 1.100 V, regardless of the sizes of the electrodes or the volumes of the solutions. The voltage does depend on the concentrations, however, and also on the temperature. We will hear more about this shortly.

A shorthand notation for the Daniell cell is

$$Zn(c) \mid Zn^{2+}(1\ M) \parallel Cu^{2+}(1\ M) \mid Cu(c)$$

The double line represents a salt bridge or liquid junction between the two half-cells and the single lines denote phase boundaries within the half-cells (in this example, the boundaries between the solid and aqueous phases of the metals).

The same kind of notation, called a *cell diagram*, is used for all galvanic cells. Starting with the overall reaction written in either the forward or reverse direction (the forward direction is customary, of course), it is conventional in writing the cell diagram always to write the oxidation half-reaction on the left and the reduction half-reaction on the right. The two half-reactions are based on the overall reaction *as written*.* This means that the left side always represents the anode of the cell so specified, and the right side always represents the cathode. By the definitions of anode and cathode, electrons always flow from anode to cathode in the external circuit, so the cell diagram corresponds to an electron flow from left to right in the external circuit of the cell as specified.

If the cell diagram is based on a *spontaneous* reaction (i.e., one written in the forward direction), this left-to-right flow of electrons corresponds to a *positive* electric potential difference (voltage) between the electrodes. If the cell diagram is based on a *nonspontaneous* reaction (i.e., one written in the reverse direction), the left-to-right flow of electrons corresponds to a *negative* voltage. This is because the actual flow of electrons in a cell so specified would be from right to left. An actual cell does not "know" how we specified it in writing. It follows the laws of chemistry, not the conventions of chemists. But these conventions are very important, as we will see.

---

*If Reaction 10-9 were written in the reverse (nonspontaneous) direction,

$$Cu(c) + Zn^{2+}(1\ M) \rightarrow Cu^{2+}(1\ M) + Zn(c)$$

the oxidation and reduction half-reactions would be reversed and the cell diagram for the galvanic cell so specified would be

$$Cu(c) \mid Cu^{2+}(1\ M) \parallel Zn^{2+}(1\ M) \mid Zn(c)$$

Standard Cell Potentials

The fact that the voltage generated by a galvanic cell is independent of the volumes of the solutions indicates that it is an intensive property of the system. The limiting value of this voltage, when there is zero current flowing in the cell, is called the electromotive force of the cell, or the *cell potential* (symbol $\mathscr{E}_{\text{cell}}$ or E_{cell}). As with various thermodynamic state functions with which we are familiar, it is most convenient to specify the cell potential for systems that are in the standard state (each solution 1 *M* in the solute in question, all gases at 1 atm, and $T = 298$ K). It is then called the *standard cell potential* $\mathscr{E}^\circ_{\text{cell}}$.

For any given cell diagram written in the conventional manner (oxidation on the left, reduction on the right), the standard cell potential is defined as the standard potential of the electrode on the right (the cathode) minus the standard potential of the electrode on the left (the anode):

$$\mathscr{E}^\circ_{\text{cell}} = \mathscr{E}^\circ_{\text{cathode}} - \mathscr{E}^\circ_{\text{anode}} \tag{10-10}$$

The two terms on the right side of this equation are called *standard electrode potentials*. Each corresponds to the potential difference between the electrode in question and the solution with which it is in contact, i.e., it is the voltage generated by a given half-reaction occurring in a half-cell at standard conditions. For the Daniell cell these half-reactions and half-cells are those shown above Reaction 10-9 and in Figure 10-4, respectively.

Note that $\mathscr{E}^\circ_{\text{cell}}$ is actually a "delta" quantity ($\Delta\mathscr{E}^\circ_{\text{cell}}$), but it is conventional in electrochemistry to omit the Δ. It is also conventional to omit the subscripts "cell," "cathode," and "anode" except where they are necessary, as in Equation 10-10. It is almost always clear from the context whether $\mathscr{E}^\circ$ denotes a standard electrode potential or a standard cell potential. In analogy with thermodynamic state functions such as ΔH and ΔG, the value of $\mathscr{E}$ for a galvanic cell is always obtained by applying the "right-side-minus-left-side" arithmetic to the process *as written*. But note that the "process as written" referred to here is not the overall reaction itself, but the cell diagram written on the basis of that reaction. In other words, the cell potential is obtained not by taking "products minus reactants" but by taking "cathode half-cell minus anode half-cell."

It would be nice if we could measure the absolute values of individual electrode potentials (half-cell potentials) experimentally. This is impossible, however, because it would require another metallic electrode to complete the measuring circuit. The half-cell would then no longer be a half-cell, and the measurement would be meaningless. The situation is analogous to that concerning enthalpies and free energies, for which absolute values cannot be measured either. The problem is resolved in the same way: define an arbitrary standard and express all measured values relative to it.

What is needed is a reference electrode to use as a standard for comparison. The one universally agreed upon is the *standard hydrogen electrode*, which is defined as a platinum electrode that is in contact with H_2 gas at 1 atm and an aqueous solution that is 1 *M* in H^+. The potential of this electrode, whether it be considered as a cathode (reduction of H^+ to H_2) or an anode (oxidation of H_2 to H^+), is defined to be exactly zero at all temperatures. Thus we can write

$$2H^+(1\ M) + 2e^- \rightarrow H_2(1\ \text{atm}) \qquad \mathscr{E}^\circ = 0.0000\ \text{V}$$

$$H_2(1\ \text{atm}) \rightarrow 2H^+(1\ M) + 2e^- \qquad \mathscr{E}^\circ = 0.0000\ \text{V}$$

We can now define the standard electrode potential more precisely: The standard electrode potential for any half-reaction is defined as the cell potential of the whole cell in which the electrode on the left (the anode) is the standard hydrogen electrode and the electrode on the right (the cathode) is the electrode in question, operating at standard conditions. We know that the cathode is always the site of reduction, so it is evident that all standard electrode potentials are defined as standard *reduction* potentials, i.e., they all denote the potential of a half-reaction written as a reduction half-reaction. This convention is now well established, although it was common until fairly recently to give many electrode potentials as oxidation potentials (one must therefore be especially careful in reading the older literature on this subject). The standard oxidation potential for any half-reaction, now an obsolete concept, is equal in magnitude but opposite in sign to the standard reduction potential. An extensive table of standard reduction potentials is given in Appendix I.

Let us see how all these conventions and definitions work out in practice. The value of $\mathscr{E}^\circ$ for the reduction of Cu^{2+} to Cu is obtained by measuring the potential of the following cell:

$$\text{Pt} \mid H_2(1\ \text{atm}) \mid H^+(1\ M) \parallel Cu^{2+}(1\ M) \mid Cu(c)$$

(The symbol Pt on the left denotes the inert electrode used to register the potential—defined as zero—of the standard hydrogen electrode.) The overall reaction for this cell is written

$$Cu^{2+}(1\ M) + H_2(1\ \text{atm}) \rightarrow Cu(c) + 2H^+(1\ M) \qquad (10\text{-}11)$$

and an illustration of the cell set up in accord with this reaction as written is shown in Figure 10-5.

By definition, the measured cell potential in Figure 10-5 is the standard reduction potential for the following half-reaction:

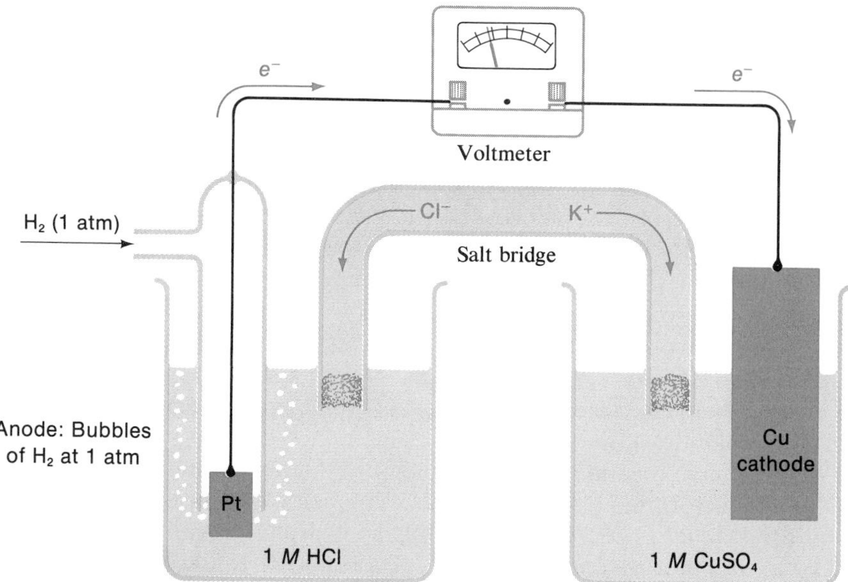

Figure 10-5. A galvanic cell. Using the standard hydrogen electrode as a reference electrode, the value of the standard reduction potential for the half-reaction $Cu^{2+} + 2e^- \rightarrow Cu(c)$ can be obtained.

$$Cu^{2+} + 2e^- \rightarrow Cu(c) \qquad \mathscr{E}^\circ = +0.337 \text{ V}$$

How do we know that the sign is positive? It is an experimental fact. Reaction 10-11 is spontaneous in the direction written, so, in a cell set up accordingly, the standard hydrogen electrode *is* the anode, the copper electrode *is* the cathode, and the electrons *do* flow from left to right in the external circuit. With the voltmeter leads hooked up according to the reaction as written, therefore, the measured potential is positive. What this means is that Cu^{2+} is easier to reduce than H^+, i.e., that Cu^{2+} is a better oxidizing agent than H^+, at standard conditions.

Now suppose we apply the same method to the reduction of Zn^{2+} to Zn. We measure the potential of the following cell:

$$Pt \mid H_2(1 \text{ atm}) \mid H^+(1\ M) \parallel Zn^{2+}(1\ M) \mid Zn(c)$$

for which the overall reaction is written

$$Zn^{2+}(1\ M) + H_2(1 \text{ atm}) \rightarrow Zn(c) + 2H^+(1\ M) \qquad (10\text{-}12)$$

The measured cell potential is the standard reduction potential for the following half-reaction:

$$Zn^{2+} + 2e^- \rightarrow Zn(c) \qquad \mathscr{E}^\circ = -0.7628\ V$$

Here the sign is negative because Reaction 10-12 is *not* spontaneous in the direction written. In the actual cell the zinc electrode is not the cathode, but the anode (Zn is oxidized to Zn^{2+}), and the standard hydrogen electrode is the cathode (H^+ is reduced to H_2). Therefore, if the cell and voltmeter are set up according to the reaction *as written* (as they must be, by convention), the electrons flow from right to left in the external circuit and the measured potential is negative. What this means is that Zn^{2+} is more difficult to reduce than H^+, i.e., that Zn^{2+} is a poorer oxidizing agent than H^+, at standard conditions.

We now have the values of the standard reduction potentials for the two half-reactions of the Daniell cell, each value being expressed relative to an arbitrary standard (the standard hydrogen electrode). Neither value is useful by itself, but if we combine them by Equation 10-10 to obtain a standard cell potential, the hidden factor introduced by the arbitrary standard drops out. The result is the true cell potential, correct in both magnitude and sign. Note that $\mathscr{E}^\circ_{\text{anode}}$ in Equation 10-10, although it denotes the potential associated with the *oxidation* half-reaction, is nevertheless given as a standard *reduction* potential, i.e., the potential associated with the reverse of the oxidation half-reaction. In writing out the reaction scheme, therefore, we add the half-reactions to obtain the overall reaction, as usual, but we subtract the standard reduction potentials, in accord with Equation 10-10:

	Add	*Subtract*
Reduction (cathode):	$Cu^{2+} + 2e^- \rightarrow Cu(c)$	$\mathscr{E}^\circ = +0.337\ V$
Oxidation (anode):	$Zn(c) \rightarrow Zn^{2+} + 2e^-$	$-(\mathscr{E}^\circ = -0.763\ V)$
	$Zn(c) + Cu^{2+} \rightarrow Zn^{2+} + Cu(c)$	$\mathscr{E}^\circ = +1.100\ V$

The experimentally measured value is, indeed, +1.100 V.

Clearly it is unnecessary to set up and measure the potential of a galvanic cell if the values of the standard reduction potentials for the two half-reactions are known. The standard cell potential can be obtained in seconds, using only the simplest arithmetic. Thus, a moderate number of known standard reduction potentials can be combined in a great many ways to obtain the standard cell potentials for a great many reactions, with minimal effort. This delightful situation is analogous to that in thermochemistry, where standard enthalpy and free-energy changes for innumerable reactions can be calculated very easily from the data in tables of standard enthalpies and free energies of formation.

From the discussion of sign conventions in this section, you have probably already guessed that the sign of a standard cell potential is an indicator of the spontaneity of the overall reaction, at standard conditions. You are right. For Reaction 10-8 we found that $\Delta G° = -50.71$ kcal/mole and $\mathscr{E}° = +1.100$ V. If $\Delta G°$ is negative, $\mathscr{E}°$ is positive and the reaction as written is spontaneous. If $\Delta G°$ is positive, $\mathscr{E}°$ is negative and the reaction as written is not spontaneous. We will hear more about the relation between $\Delta G°$ and $\mathscr{E}°$ shortly.

Electrolysis Re-Examined

Now we have all the information necessary for a quantitative re-examination of the electrolysis of seawater, which we discussed earlier. We repeat the reaction scheme, this time including the values of the standard reduction potentials, from which we obtain the standard cell potential:

	Add	*Subtract*
Reduction (cathode):	$2H_2O + 2e^- \rightarrow H_2(g) + 2OH^-$	$\mathscr{E}° = -0.8277$ V
Oxidation (anode):	$2Cl^- \rightarrow Cl_2(g) + 2e^-$	$-(\mathscr{E}° = +1.3595$ V)
	$2H_2O + 2Cl^- \rightarrow H_2(g) + Cl_2(g) + 2OH^-$	$\mathscr{E}° = -2.1872$ V

Recall that, by convention, the anode potential $\mathscr{E}° = +1.3595$ V is the potential corresponding to the *reduction* half-reaction $Cl_2(g) + 2e^- \rightarrow 2Cl^-$, which is the *reverse* of the actual oxidation half-reaction. That is why this potential is *subtracted* from the cathode potential rather than added to it. If it were still conventional to use standard *oxidation* potentials for anode half-reactions, the value would be $\mathscr{E}° = -1.3595$ V and it would be *added* to the cathode potential, giving exactly the same result for the cell potential. The negative value of the standard cell potential shows that the reaction does not proceed spontaneously in the direction written, at standard conditions. This is hardly surprising, because we know that the reaction as written is that of an electrolytic cell, not a galvanic cell. It is therefore necessary to apply an electric potential of at least 2.1872 V to this cell in order to drive the reaction, if all species are at standard conditions:

$$[Cl^-] = [OH^-] = 1 \text{ mole}/\ell$$

$$P_{H_2} = P_{Cl_2} = 1 \text{ atm}$$

$$T = 298 \text{ K}$$

Now we return to the question, raised earlier, about why H_2O is reduced in this reaction and not Na^+. From Appendix I we see that

$$Na^+ + e^- \rightarrow Na\,(c) \qquad \mathscr{E}^\circ = -2.714 \text{ V}$$

The standard reduction potential of Na^+ has a much greater negative value than that of H_2O, which means that Na^+ is much more difficult to reduce than H_2O, at standard conditions. If Na^+ *were* reduced in the electrolysis of seawater, the standard cell potential would be $\mathscr{E}^\circ = -2.714 - 1.360 = -4.074$ V. This reaction would be much more difficult to drive than the actual one, and therefore does not occur.

Exercise 10-3. Using the values of the standard reduction potentials of O_2 and $S_2O_8^{2-}$ (which are the products of the oxidation of H_2O and SO_4^{2-}, respectively), explain why, in the electrolysis of dilute H_2SO_4, it is H_2O rather than SO_4^{2-} that is oxidized at the anode.

The sign conventions for the electrodes in electrochemical cells are somewhat confusing but very important because we have to know how to hook up our voltmeter leads. Unfortunately the conventions are opposite for the two kinds of cells. In electrolytic cells the *cathode* is called the negative pole. In galvanic cells the *anode* is called the negative pole. When the two kinds of cells are connected, or when a cell is connected to a voltmeter or generator, the rule for connecting the poles is

$$(+) \text{ to } (+) \qquad (-) \text{ to } (-)$$

Thus the polarity labels on a storage battery (e.g., the battery in a car) are always valid, whether the battery is being charged or discharged.

It is important to remember that, in *all* cells, reduction occurs at the cathode and oxidation at the anode. Knowing this, one can always deduce the direction of electron flow in the circuit, once the cathode and anode are identified. Figure 10-6 should help to clarify the conventions discussed above.

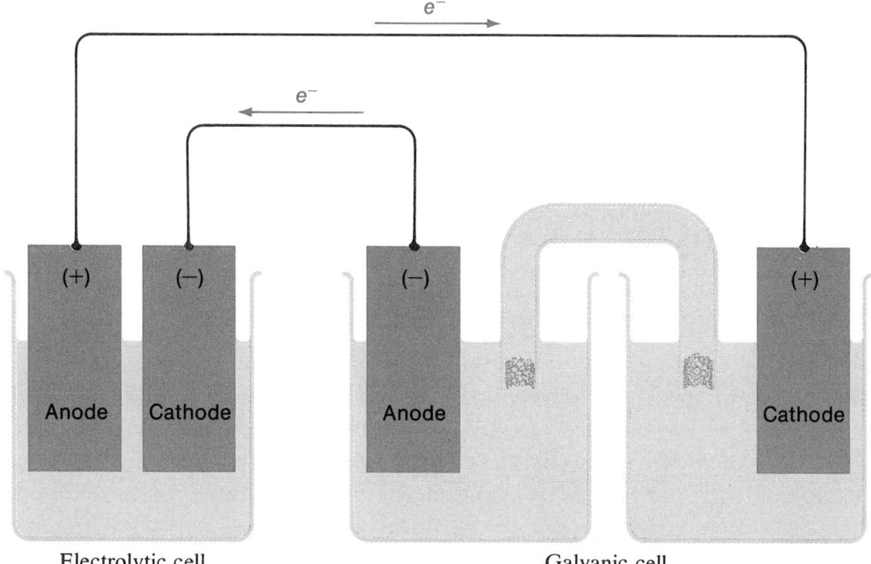

Figure 10-6. Electrode sign conventions for electrolytic and galvanic cells. Note the directions of flow of the electrons: electrons always flow from the anode to the cathode in the external circuit. Therefore the anode is always the site of oxidation, and the cathode, of reduction, regardless of sign conventions.

10.3 THE CONCENTRATION DEPENDENCE OF CELL POTENTIALS

In Chapter 7 we derived the equation (Equation 7-29) for the free-energy change in an ideal-gas reaction in terms of the partial pressures of the reactant and product gases. We saw that, *at equilibrium*, $\Delta G = 0$ and therefore the quotient in the logarithmic term must be a constant—the equilibrium constant K—because $\Delta G°$ is a constant for any reaction. The expression for K can also be given as a quotient of species concentrations in solution, as we have seen many times in Chapter 9. If the system is *not* at equilibrium, however, this quotient is called the *reaction quotient* and given the symbol Q. Thus, for the generalized chemical reaction (not at equilibrium)

$$a\mathrm{A} + b\mathrm{B} \rightleftharpoons c\mathrm{C} + d\mathrm{D}$$

we can write

$$Q = \frac{[C]^c[D]^d}{[A]^a[B]^b} \tag{10-13}$$

Note that Q can have any value between 0 and ∞, depending on the relative concentrations of species at the moment in question. As equilibrium is approached, the value of Q approaches that of K, and at equilibrium $Q = K$. This relation is analogous to the relation between Q_{ion} and K_{sp}, which we saw in Chapter 9.

We can now rewrite Equation 7-29 as

$$\Delta G = \Delta G^\circ + RT \ln Q \tag{10-14}$$

An important result of thermodynamics is that, in a spontaneous reaction carried out in a reversible manner at constant temperature and pressure, the decrease in the free energy of the system is equivalent to the work (excluding pressure-volume work) done by the system on its surroundings. In other words, the free energy of chemical substances is converted to work. Since ΔG is a negative quantity and the work done is a positive quantity, we can say that ΔG is equal to the negative of the work done. (For a nonspontaneous reaction, ΔG is positive and the work done by the system on its surroundings is negative, so the relation is exactly the same.)

The work done by an electrochemical cell on its surroundings (the external circuit) is electric work, which is equal to the amount of charge transferred (coulombs) times the potential drop (volts). Like all work, it has the units of energy because

$$1 \text{ volt} = 1 \text{ joule/ampere-second} = 1 \text{ joule/coulomb}$$

and therefore

$$1 \text{ coulomb-volt} = 1 \text{ joule}$$

For an electrochemical cell, therefore, we obtain the following expression for ΔG in units of joules per mole:

$$\Delta G = -n\mathscr{F}\mathscr{E} \tag{10-15}$$

where n (dimensionless) equals the number of electrons transferred in the redox reaction, $\mathscr{F}$ is the Faraday constant, and $\mathscr{E}$ is the cell potential measured at the existing conditions. Note that the Faraday constant converts the *intensive* property $\mathscr{E}$ to the *extensive* property ΔG. The standard-state analog of Equation 10-15 is, of course,

$$\Delta G^\circ = -n\mathscr{F}\mathscr{E}^\circ \tag{10-16}$$

The free-energy change and the cell potential always have opposite signs, as we saw earlier, so the sign of a measured or calculated cell potential is just as conclusive an indicator of the spontaneity of a reaction as the sign of the free-energy change.

If we now substitute Equations 10-15 and 10-16 into Equation 10-14, we obtain

$$\mathscr{E} = \mathscr{E}^\circ - \frac{RT}{n\mathscr{F}} \ln Q = \mathscr{E}^\circ - 2.303\frac{RT}{n\mathscr{F}} \log Q \tag{10-17}$$

This important equation relating the cell potential to the species concentrations is called the *Nernst equation* after the German physical chemist H. Walther Nernst (1864–1941).

The quantity 2.303 $RT/n\mathscr{F}$ occurs so often in electrochemistry that it is useful to evaluate it at 298 K. First we convert the Faraday constant to a different set of units. Recognizing that 1 coulomb = 1 joule/volt, we have

$$\mathscr{F} = \frac{96{,}487 \text{ J/V mole}}{4.1840 \text{ J/cal}} = 23{,}061 \text{ cal/V mole}$$

Therefore

$$2.303\frac{RT}{n\mathscr{F}} = \frac{2.303(1.987 \text{ cal/K mole})(298 \text{ K})}{n(2.306 \times 10^4 \text{ cal/V mole})} = \frac{0.0592}{n} \text{ V}$$

and the Nernst equation now becomes

$$\mathscr{E} = \mathscr{E}^\circ - \frac{0.0592}{n} \log Q \tag{10-18}$$

We will see how this equation is used in the following section.

Sign Conventions and the Nernst Equation

We saw earlier that in galvanic cells the cathode is the positive pole and should be connected to the positive pole on the voltmeter. Let us use the Daniell cell as an example of the effect of an *incorrect* connection on the cell potential as calculated by the Nernst equation.

We begin by considering the *correct* reaction scheme, given in the section on standard cell potentials. Copper is the cathode and therefore the positive pole. For this reaction as written, the Nernst equation is

$$\mathscr{E} = 1.100 - \frac{0.0592}{2} \log \frac{[Zn^{2+}]}{[Cu^{2+}]}$$

and the standard free-energy change is*

$$\Delta G^\circ = -n\mathscr{F}\mathscr{E}^\circ = -2\,(23.06\ \text{kcal/V mole})\,(1.100\ \text{V})$$
$$= -50.73\ \text{kcal/mole}$$

However, we could just as well divide the entire reaction scheme by 2 and write, just as correctly,

	Add	*Subtract*
Reduction (cathode):	$\frac{1}{2}Cu^{2+} + e^- \rightarrow \frac{1}{2}Cu(c)$	$\mathscr{E}^\circ = +0.337$ V
Oxidation (anode):	$\frac{1}{2}Zn(c) \rightarrow \frac{1}{2}Zn^{2+} + e^-$	$-(\mathscr{E}^\circ = -0.763$ V)
	$\frac{1}{2}Zn(c) + \frac{1}{2}Cu^{2+} \rightarrow \frac{1}{2}Zn^{2+} + \frac{1}{2}Cu(c)$	$\mathscr{E}^\circ = +1.100$ V

As written here, $n = 1$ and the Nernst equation becomes

$$\mathscr{E} = 1.100 - \frac{0.0592}{1} \log \frac{[Zn^{2+}]^{1/2}}{[Cu^{2+}]^{1/2}} = 1.100 - \frac{0.0592}{2} \log \frac{[Zn^{2+}]}{[Cu^{2+}]} \qquad (10\text{-}19)$$

because, for any quantities x and n, $\log x^n = n \log x$. Thus, the Nernst equation yields a cell potential that is invariant with the stoichiometric coefficients used to balance the reaction because the cell potential is an intensive property of the system. It does not depend on the size (number of moles) of the system.

For this same reaction as written here (with $n = 1$), the standard free-energy change is

$$\Delta G^\circ = -n\mathscr{F}\mathscr{E}^\circ = -(23.06\ \text{kcal/V mole})\,(1.100\ \text{V})$$
$$= -25.37\ \text{kcal/mole}$$

*The slight discrepancy between this value and that calculated earlier from the standard free energies of formation is due to a slight *junction potential* introduced into the system by the salt bridge and not taken into account here.

which is 1/2 of that calculated earlier. The free-energy change is an extensive property of the system. Therefore its value does depend on the stoichiometric coefficients used to balance the reaction, i.e., it depends on the number of moles considered to be involved in the reaction as written.

Now suppose we are totally ignorant of what is the cathode and what is the anode and we hook the positive pole of the voltmeter to the zinc electrode instead of the copper electrode. We would then write the reaction scheme—incorrectly—as

	Add	*Subtract*
Reduction (cathode):	$Zn^{2+} + 2e^- \rightarrow Zn(c)$	$\mathscr{E}° = -0.763$ V
Oxidation (anode):	$Cu(c) \rightarrow Cu^{2+} + 2e^-$	$-(\mathscr{E}° = +0.337$ V)
	$Cu(c) + Zn^{2+} \rightarrow Cu^{2+} + Zn(c)$	$\mathscr{E}° = -1.100$ V

The voltmeter would show a negative cell potential, indicating that the true spontaneous reaction is the reverse of the reaction as written here. We can still write the Nernst equation for this reaction, as follows:

$$\mathscr{E} = -1.100 - \frac{0.0592}{2} \log \frac{[Cu^{2+}]}{[Zn^{2+}]} = -\left(1.100 - \frac{0.0592}{2} \log \frac{[Zn^{2+}]}{[Cu^{2+}]}\right)$$

so the cell potential is -1 times the cell potential for the reverse reaction, just as we would expect. The standard free-energy change is

$$\Delta G° = -n\mathscr{F}\mathscr{E}° = -2\,(23.06 \text{ kcal/V mole})\,(-1.100 \text{ V})$$
$$= +50.73 \text{ kcal/mole}$$

again indicating that the true spontaneous reaction is the reverse of the reaction as written here, and that the two values of $\Delta G°$ in question have equal magnitudes but opposite signs.

10.4 CONCENTRATION CELLS

If a galvanic cell is constructed from two half-cells in which the half-reactions are identical, the result is called a *concentration cell*. Such a cell is shown in Figure 10-7. The reaction scheme for this cell is

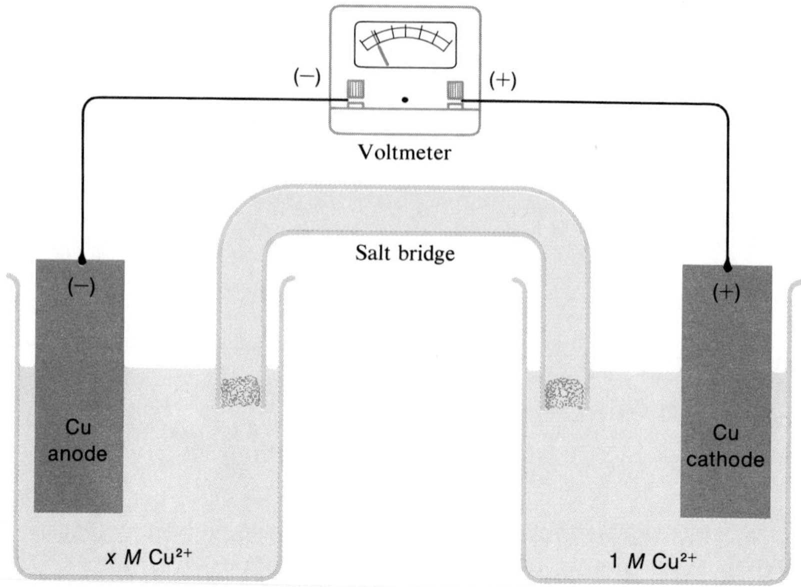

Figure 10-7. A cupric ion/copper concentration cell. The electrode labels shown here are correct if $x < 1$. If $x > 1$ the labels must be reversed.

	Add	*Subtract*	
Reduction (cathode):	$Cu^{2+} + 2e^- \rightarrow Cu(c)$	$\mathscr{E}° = +0.337$ V	
Oxidation (anode):	$Cu(c) \rightarrow Cu^{2+} + 2e^-$	$-(\mathscr{E}° = +0.337$ V$)$	
	Cu^{2+} (*cathode*) $\rightarrow Cu^{2+}$ (*anode*)	$\mathscr{E}° = 0.000$ V	(10-20)

The Nernst equation for this reaction is

$$\mathscr{E} = \mathscr{E}° - \frac{0.0592}{2}\log\frac{[Cu^{2+}]_{anode}}{[Cu^{2+}]_{cathode}} = 0 - \frac{0.0592}{2}\log\frac{x}{1}$$

$$= -0.0296 \log x$$

This equation says that if $x = 1$, i.e., if the Cu^{2+} concentration in the anode half-cell is equal to that in the cathode half-cell, then $\mathscr{E} = 0$ because there is no driving force for the reaction. If $x = [Cu^{2+}]_{anode}$ is less than 1, then $\mathscr{E}$ will be positive and the reaction will be spontaneous. Electrons will flow from left to right in the external circuit of Figure 10-7. Copper will dissolve from the anode, thereby increasing $[Cu^{2+}]_{anode}$, and copper will plate onto the cathode, thereby decreasing $[Cu^{2+}]_{cathode}$, until the two concentrations are equal and the reaction stops.

On the other hand, if $x = [Cu^{2+}]_{anode}$ is greater than 1, then $\mathscr{E}$ will be negative and there will be no reaction, right? Right that $\mathscr{E}$ will be negative for the reaction based on this assumption, but wrong that there will be no reaction. The cell in Figure 10-7 does not "know" that we called the left electrode the anode and the right electrode the cathode, so the reaction will occur, but in the opposite direction: electrons will flow from right to left in the external circuit. Without advance knowledge of the value of x, our assignment of the "anode" and "cathode" labels was arbitrary—a shot in the dark that was right for $x < 1$ but wrong for $x > 1$. In the latter case we must reverse the labels (and the voltmeter leads). Now x denotes $[Cu^{2+}]_{cathode}$, the log term in the Nernst equation is $\log(1/x)$, and, because $x > 1$, $\mathscr{E}$ is again positive. This illustrates that the correct assignment of electrode labels is always determined by *what actually happens.* Note that, because Reaction 10-20 is independent of the assignment of labels in Figure 10-7, it is always valid. The expression for Q is therefore invariant and the log term in the Nernst equation for this system is always $\log([Cu^{2+}]_{anode}/[Cu^{2+}]_{cathode})$.

We see that the tendency in a concentration cell is always for the two concentrations to equalize and that the electrode representing the higher initial concentration is always the cathode. The same Gibbs free-energy change that drives the process of diffusion from regions of high concentration to regions of low concentration when two solutions are mixed also provides a potential difference in the concentration cell. This is the driving force for the reaction. If $x = [Cu^{2+}]_{anode} = 0.1$ mole/ℓ we have

$$\mathscr{E} = -0.0296 \log 10^{-1} = +0.0296 \text{ V}$$

For every factor of 10 by which the solution in the anode half-cell is less concentrated than that in the cathode half-cell, there is an additional potential difference of 0.0296 V. The concentration cell represents an especially simple means of determining the ion concentration in a solution of unknown concentration, provided that the ion in question is the only reactive species present.

10.5 THE RELATION BETWEEN STANDARD CELL POTENTIAL AND EQUILIBRIUM CONSTANT

When the chemical reaction driving a galvanic cell is at equilibrium, the free-energy change for the reaction must equal zero and therefore the cell potential must also be zero. At chemical equilibrium the cell can no longer be used as a power source. Under these conditions, $\mathscr{E} = 0$ and $Q = K$, so the Nernst equation becomes

$$\mathscr{E}° = \frac{0.0592}{n} \log K \qquad (10\text{-}21)$$

or

$$\log K = 16.9n\mathscr{E}° \qquad (10\text{-}22)$$

where the factor 16.9 has the unit V^{-1}. Standard cell potentials are the most accurate sources of equilibrium constants for many ionic systems.

Example 10-2. One of the most widely used galvanic cells is the lead storage battery, which provides the current to start automobile engines. The battery consists of a series of lead-alloy plates that are designed as lattices and filled alternatingly with spongy metallic lead and lead dioxide. The electrolyte is a fairly concentrated solution of H_2SO_4. The two half-reactions (both given as reduction half-reactions) in this cell are

$$PbSO_4(c) + 2e^- \rightarrow Pb(c) + SO_4^{2-} \qquad \mathscr{E}° = -0.355\ V$$

$$PbO_2(c) + SO_4^{2-} + 4H^+ + 2e^- \rightarrow PbSO_4(c) + 2H_2O \qquad \mathscr{E}° = +1.685\ V$$

(a) Write the balanced, overall, spontaneous reaction and obtain the value of $\mathscr{E}°$ for this reaction. (b) Calculate the value of $\Delta G°$ for this reaction. (c) Calculate the equilibrium constant for this reaction.

(a) We know that the standard cell potential must be positive, so it cannot be obtained by subtracting +1.685 V from −0.355 V; it must be the other way around. Therefore the first half-reaction above must represent the anode and must be written as an oxidation half-reaction. (Remember, however, that the electrode potential is still written as a reduction potential, by convention.) We can now write the reaction scheme in the usual way:

	Add	*Subtract*
Reduction (cathode):		
	$PbO_2(c) + SO_4^{2-} + 4H^+ + 2e^- \rightarrow PbSO_4(c) + 2H_2O$	$\mathscr{E}° = +1.685\ V$
Oxidation (anode):		
	$Pb(c) + SO_4^{2-} \rightarrow PbSO_4(c) + 2e^-$	$-(\mathscr{E}° = -0.355\ V)$
	$Pb(c) + PbO_2(c) + 2SO_4^{2-} + 4H^+ \rightarrow 2PbSO_4(c) + 2H_2O$	$\mathscr{E}° = +2.040\ V$

(b) The standard free-energy change is obtained directly from the standard cell potential:

$$\Delta G^\circ = -n\mathscr{F}\mathscr{E}^\circ = -2(23.06 \text{ kcal/V mole})(2.040 \text{ V}) = -94.08 \text{ kcal/mole}$$

(c) The equilibrium constant (at 298 K) is obtained from the equilibrium Nernst equation:

$$\log K = 16.9n\mathscr{E}^\circ = (16.9 \text{ V}^{-1})(2)(2.04 \text{ V}) = 69.0$$

$$K = 10^{69.0}$$

It should be noted that the equilibrium constants for many redox reactions are extraordinarily large.

10.6 OXIDATION-REDUCTION IN LIVING SYSTEMS

Living cells require chemical energy to synthesize the molecules necessary for their growth. Since cell division is the process necessary for the propagation of life, the synthesis of the molecules of life is a continuous process. Life, when viewed as a vast and fabulously intricate system of ongoing chemical reactions, is a thermodynamically spontaneous process, so the net free-energy change associated with cell growth and division must be negative. Common sources of chemical energy for sustaining life are the carbohydrates. Glucose, $C_6H_{12}O_6$, is a typical example. Cells ranging from simple bacteria to human cells utilize glucose as a source of carbon and a source of energy.

The common oxygen-containing organic compounds can be interconverted by oxidation or reduction. Their relations to each other are demonstrated by the sequential oxidation of the simplest hydrocarbon, methane, to carbon dioxide:

$$\xrightleftharpoons[\text{Reduction}]{\text{Oxidation}}$$

$$CH_4 \xrightleftharpoons[+2e^-]{-2e^-} H{-}CH_2{-}OH \xrightleftharpoons[+2e^-]{-2e^-} H{-}C({=}O){-}H \xrightleftharpoons[+2e^-]{-2e^-} H{-}C({=}O){-}OH \xrightleftharpoons[+2e^-]{-2e^-} O{=}C{=}O$$

Methane	Methanol	Formaldehyde	Formic acid	Carbon dioxide
CH_4	CH_4O	CH_2O	CH_2O_2	CO_2
−4	−2	0	+2	+4

The number below each formula in this reaction sequence is the oxidation state of the carbon atom in that compound.

In organic compounds containing more than one carbon atom, the oxidation state calculated for carbon is an average value for all the carbon atoms. The individual carbon atoms within a given molecule often have very different effective oxidation states, depending on their locations in the molecule and the other atoms to which they are bound, so the average oxidation state generally has little physical significance. Nevertheless, it is a useful device in deciding whether or not a given organic reaction is a redox reaction. To do this it is not necessary to know the structural formulas of the compounds in question, just the empirical formulas. For each carbon-containing reactant and product, the average oxidation state of the carbon is multiplied by the number of carbon atoms present, i.e., by the number of carbon atoms in the molecule and/or by the stoichiometric coefficient of the molecule in the balanced reaction. The resulting numbers are added algebraically for the carbon-containing reactants and, separately, the carbon-containing products. If the two totals are equal, the reaction is not a redox reaction. If they are unequal, it will be by an integer representing the number of electrons transferred in the reaction. The total for the products will be greater than that for the reactants if the carbon compound is oxidized, and smaller if it is reduced.

Living cells are able to utilize the free energy released by organic reactions in enzyme-catalyzed processes called *fermentations.* An example of a fermentation reaction is

$$\underset{\text{Glucose}}{C_6H_{12}O_6} \xrightarrow{\text{Enzymes}} \underset{\text{Lactic acid}}{2CH_3\text{—}\overset{\displaystyle H}{\underset{\displaystyle OH}{\overset{|}{\underset{|}{C}}}}\text{—}COOH} \qquad \Delta G^{\circ\prime} = -47 \text{ kcal/mole} \qquad (10\text{-}23)$$

More precisely, it is an example of *glycolysis*, which is the metabolic breakdown of sugars—especially glucose—in living organisms. Reaction 10-23 is actually a complex sequence of ten enzyme-catalyzed reactions. In one of these the carbon compound is oxidized and in another it is reduced by the same amount. The net effect for the process is no oxidation-reduction (note that the average oxidation state of the carbon in both the glucose and the lactic acid is 0). The sum of the ten free-energy changes (not all of which are negative) is -47 kcal/mole. The prime sign in the $\Delta G^{\circ\prime}$ denotes a special condition—one that is particularly useful for biologists—imposed on the usual definition of the standard state. This condition is that pH = 7.0, i.e., $[H^+] = 1.0 \times 10^{-7}$ mole/ℓ rather than 1.0 mole/ℓ (pH = 0).

Some metabolic processes, e.g., glycolysis, can be either *aerobic* (occurring in the presence of oxygen) or *anaerobic* (occurring in the absence of oxygen). Reaction 10-23 is an example of anaerobic glycolysis. In the presence of oxygen, aerobic cells (those requiring oxygen for life) are able to utilize a much larger portion of the original free energy of the glucose by metabolizing the lactic acid in a process called *respiration*:

$$\underset{\text{Lactic acid}}{CH_3\text{—}\overset{\displaystyle H}{\overset{|}{\underset{\underset{\displaystyle OH}{|}}{C}}}\text{—}COOH} + 3O_2 \xrightarrow{\text{Respiration}} 3CO_2 + 3H_2O \qquad \Delta G^{\circ\prime} = -320 \text{ kcal/mole} \tag{10-24}$$

For each molecule—or mole—of glucose, there is more than 13 times more free energy available from the respiration of lactic acid than from anaerobic glycolysis, which precedes it:

$$\underset{\text{Glucose}}{C_6H_{12}O_6} \xrightarrow[\Delta G^{\circ\prime} = -47 \text{ kcal/mole}]{\text{Anaerobic glycolysis}} \underset{\text{Lactic acid}}{2C_3H_6O_3} \xrightarrow[\Delta G^{\circ\prime} = -639 \text{ kcal/mole}]{+6O_2 \text{ (Respiration)}} 6CO_2 + 6H_2O \tag{10-25}$$

Respiration is an oxidation process and must therefore have an oxidizing agent associated with it. Obviously that would just be oxygen, right? Wrong—it is not that simple. There are many intermediate agents involved in the process. We will learn more about this soon, but first we will discuss the general problem of balancing organic redox reactions.

The principles underlying the balancing of organic redox reactions are the same as those for inorganic reactions. To begin with, we forget about the rather complicated structural formulas of the compounds and write just the empirical formulas. For example, consider the oxidation of lactic acid. Since the only carbon-containing product is CO_2, we begin by writing the oxidation half-reaction as

$$\underset{\text{Lactic acid}}{C_3H_6O_3} \rightarrow 3CO_2$$

Now, using only H_2O and H^+ as sources of O and H atoms, we balance the O atoms by adding 3 water molecules to the left side, and then the H atoms by adding 12 protons to the right side:

$$\underset{\text{Lactic acid}}{C_3H_6O_3} + 3H_2O \rightarrow 3CO_2 + 12H^+ + 12e^- \qquad (10\text{-}26)$$

The 12 electrons are added to the right side to achieve charge balance. We now have a balanced oxidation half-reaction.

When lactic acid is oxidized, oxygen is reduced, and we know that water is a product of the overall reaction. To obtain the reduction half-reaction, therefore, we begin by writing

$$O_2 \rightarrow 2H_2O$$

Proceeding as before, we add 4 protons and 4 electrons to the left side:

$$O_2 + 4H^+ + 4e^- \rightarrow 2H_2O \qquad (10\text{-}27)$$

We now have a balanced reduction half-reaction. To obtain the overall reaction, we multiply by 3 and add the two half-reactions:

$$\begin{array}{ll} \textbf{Reduction:} & 3O_2 + 12H^+ + 12e^- \rightarrow 6H_2O \\ \textbf{Oxidation:} & \underline{C_3H_6O_3 + 3H_2O \rightarrow 3CO_2 + 12H^+ + 12e^-} \\ & C_3H_6O_3 + 3O_2 \rightarrow 3CO_2 + 3H_2O \qquad (10\text{-}28) \end{array}$$

Note that 12 electrons are transferred in the oxidation of lactic acid. This oxidation occurs not in one step but in six steps, two electrons at a time. Part of the larger sequence of reactions (not all of which are oxidation reactions) resulting in this stepwise oxidation is called the *tricarboxylic acid cycle* or the *citric acid cycle* or, popularly, the *Krebs cycle*, after its discoverer, the German-British biochemist Sir Hans Adolf Krebs (b. 1900). This is one of the most important metabolic pathways in living organisms. It is a cyclic series of fermentation reactions that occur in *mitochondria*, the tiny "power plants" found in almost all cells. The Krebs cycle represents the final, critical phase in the oxidation of nutrients (the organism's "fuel") to CO_2 and H_2O, with the release of large amounts of energy. A diagram of the Krebs cycle is shown in Figure 10-8.

All the major nutrients of cells, notably carbohydrates, fats, and proteins, ultimately pass through the Krebs cycle. First, however, they must be degraded by complex series of enzymatic reactions to a particular two-carbon species, the acetyl group, $CH_3{-}C{\overset{O}{\underset{\diagdown}{/\!\!/}}}$, which is derived from the acetate

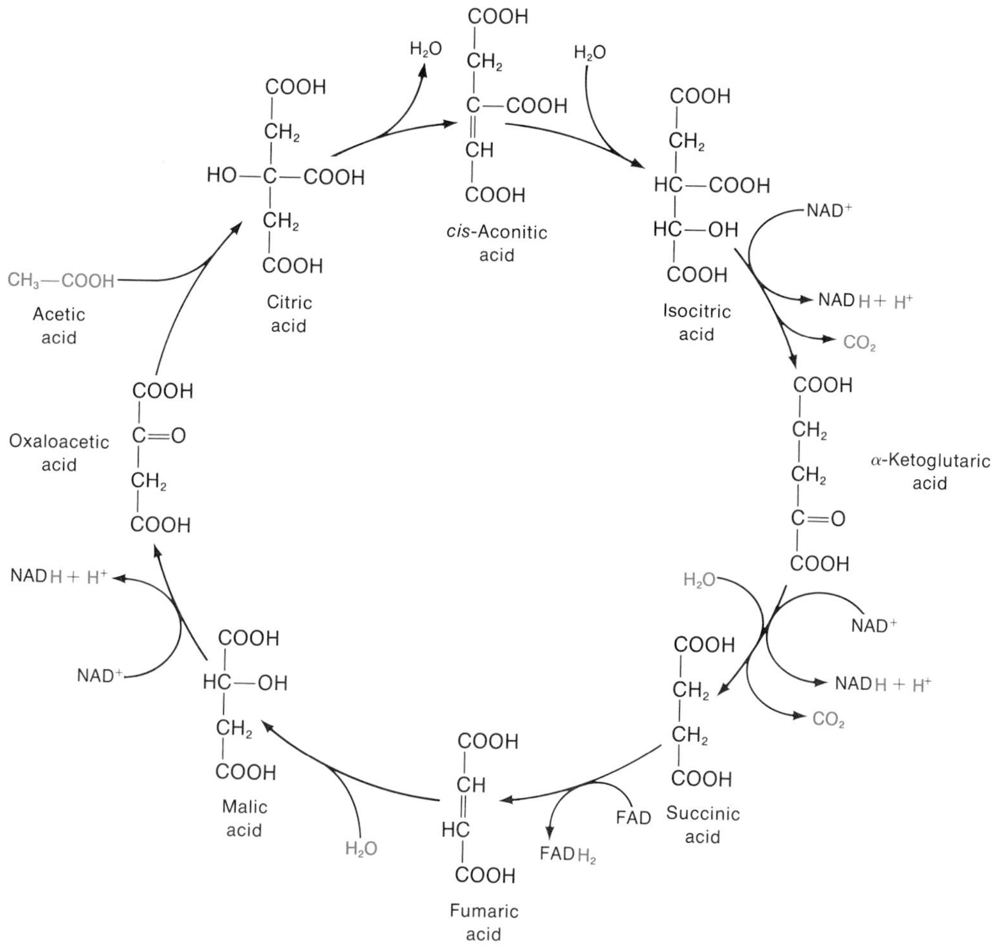

Figure 10-8. The Krebs cycle, the major energy producer in living organisms.

ion or acetic acid.* This species is the only one that can enter the Krebs cycle. The entry is effected by an enzyme called *coenzyme A*, to which the acetyl group is attached to form *acetylcoenzyme A*. As the acetyl group reacts with oxaloacetic acid (a four-carbon, dicarboxylic acid) to give citric acid (a six-carbon, tricarboxylic acid), the coenzyme A departs and the Krebs cycle for that acetyl group has begun.

---

*Since most carboxylic acids are weak acids with $pK_a \cong 5$, they are largely dissociated in the cell, which is at $pH \cong 7$. For convenience, however, we will write them all as undissociated acids in the discussion to follow.

The many metabolic routes from nutrient to acetic acid are beyond the scope of this book. Let us just consider the fate of lactic acid, the final product of anaerobic glycolysis. Lactic acid is a metabolic "dead end." For it to undergo respiration in an aerobic cell, it must be converted *back* to *pyruvic acid*, which is the next-to-final product of anaerobic glycolysis (and the final product of aerobic glycolysis). In aerobic cells pyruvic acid enters the mitochondria and is converted to acetic acid, which then enters the Krebs cycle in the form of acetyl coenzyme A. Pyruvic acid is thus the one link between glycolysis of either kind and the Krebs cycle. The two oxidation half-reactions in the conversion of lactic acid to acetic acid can be written as follows:

$$\underset{\text{Lactic acid}}{CH_3{-}\overset{\displaystyle H}{\underset{\displaystyle OH}{\overset{|}{\underset{|}{C}}}}{-}COOH} \rightarrow \underset{\text{Pyruvic acid}}{CH_3{-}\underset{\displaystyle O}{\underset{\|}{C}}{-}COOH} + 2H^+ + 2e^- \tag{10-29}$$

$$\underset{\text{Pyruvic acid}}{CH_3{-}\underset{\displaystyle O}{\underset{\|}{C}}{-}COOH} + H_2O \rightarrow \underset{\text{Acetic acid}}{CH_3COOH} + CO_2 + 2H^+ + 2e^- \tag{10-30}$$

Note that two electrons are transferred in each half-reaction. These half-reactions constitute the first two steps (mentioned earlier) in the complete metabolic oxidation of lactic acid to CO_2 and H_2O. The remaining four steps occur in the Krebs cycle.

Much of the free energy released in these oxidation reactions is stored (temporarily) in the reduced forms of the oxidizing agents that bring them about. The most important biochemical oxidizing agent is *nicotinamide adenine dinucleotide* (NAD^+). The half-reaction for the reduction of this oxidizing agent is

$$\underset{\text{Oxidized form}}{NAD^+} + H^+ + 2e^- \rightarrow \underset{\text{Reduced form}}{NADH} \tag{10-31}$$

Note that NAD^+ accepts two electrons. The structure of NAD^+ is shown below, with labels indicating the major components of this molecule.

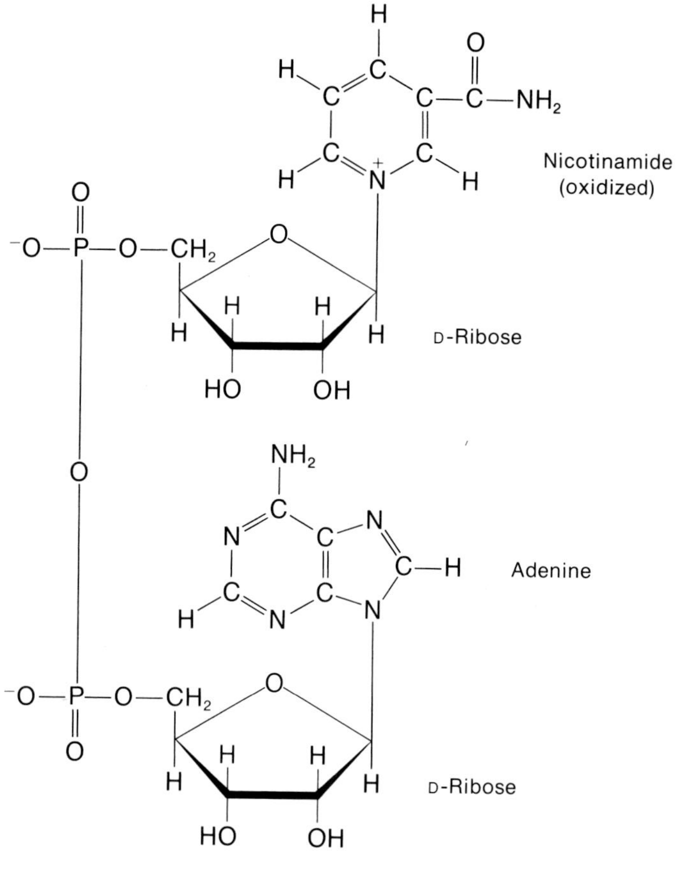

Nicotinamide (oxidized)

D-Ribose

Adenine

D-Ribose

NAD^+

In considering the reduction half-reaction, we can ignore all the atoms below the nicotinamide portion of the molecule because they remain unaffected. If we collectively call those atoms R, we can write the half-reaction as follows:

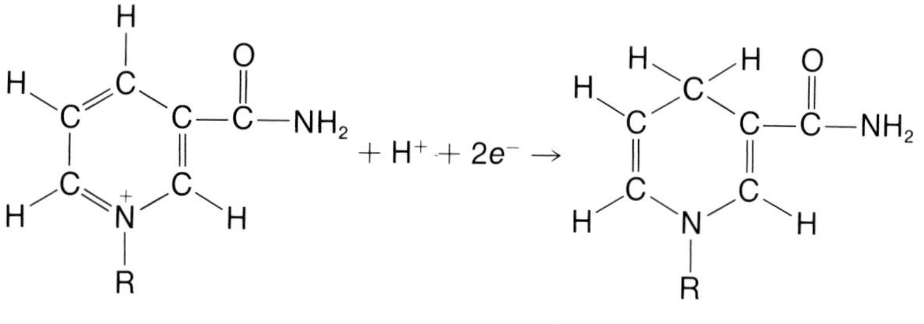

$$RC_6H_6ON_2^+ + H^+ + 2e^- \rightarrow RC_6H_7ON_2$$

NAD^+ NADH

Another biochemically important oxidizing agent is *flavin adenine dinucleotide* (FAD). The half-reaction for the reduction of this oxidizing agent is

$$\underset{\text{Oxidized form}}{\text{FAD}} + 2\text{H}^+ + 2e^- \rightarrow \underset{\text{Reduced form}}{\text{FADH}_2} \tag{10-32}$$

Note that FAD also accepts two electrons. The structure of FAD is shown below.

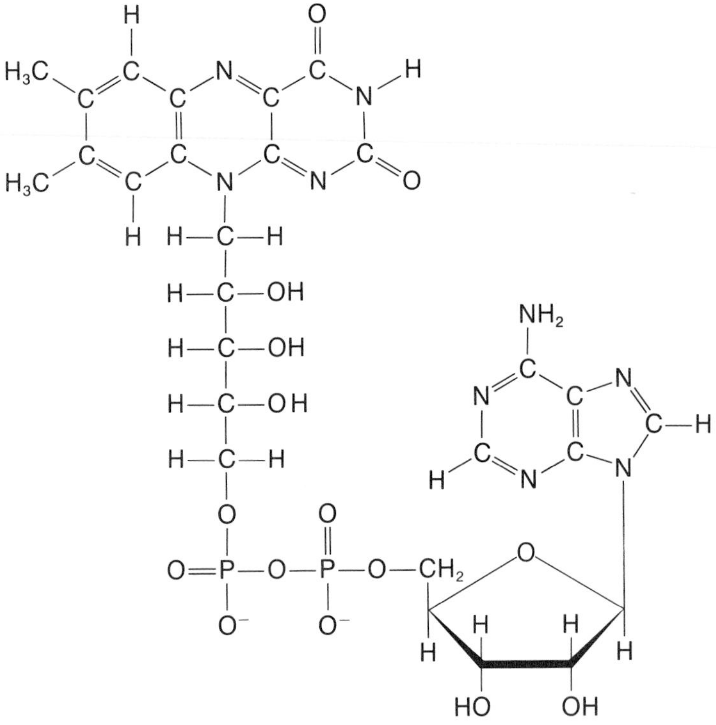

FAD

We can use the letter R as before and write the half-reaction as follows:

$$\underset{\text{FAD}}{RC_{12}H_9O_2N_4} + 2H^+ + 2e^- \rightarrow \underset{\text{FADH}_2}{RC_{12}H_{11}O_2N_4}$$

Let us now return to the Krebs cycle. We see that there are four oxidation steps; three of them produce NADH and the fourth produces $FADH_2$. In each of these steps, two electrons are transferred and two H atoms leave the cycle. In two of them, one CO_2 molecule leaves the cycle. Thus, the entire Krebs cycle can be summarized in one very simple reaction:

$$CH_3COOH + 2H_2O \rightarrow 2CO_2 + 8H \qquad (10\text{-}33)$$

The free energy stored in the reduced compounds NADH and $FADH_2$ is used by the cell in the formation of *adenosine triphosphate* (ATP) from *adenosine diphosphate* (ADP), a process called *oxidative phosphorylation.* ATP is a high-energy compound that plays a key role in many metabolic processes. The details of these processes are found in most biology and biochemistry texts.

Now let us apply some of what we have learned about oxidation-reduction to the Krebs cycle. First we will examine one of the oxidation steps in the cycle, for example, the one from α-ketoglutaric acid to succinic acid:

$$\underset{\alpha\text{-Ketoglutaric acid}}{HOOC{-}CH_2{-}CH_2{-}\overset{\overset{\displaystyle O}{\|}}{C}{-}COOH} + H_2O \rightarrow \underset{\text{Succinic acid}}{HOOC{-}CH_2{-}CH_2{-}COOH} + CO_2$$

Writing this half-reaction as

$$C_5H_6O_5 + H_2O \rightarrow C_4H_6O_4 + CO_2 \qquad (10\text{-}34)$$

we see that two protons and two electrons must be added to the right side to balance it. We then add the oxidation and reduction half-reactions to get the overall reaction:

Oxidation: $C_5H_6O_5 + H_2O \rightarrow C_4H_6O_4 + CO_2 + 2H^+ + 2e^-$

Reduction: $NAD^+ + H^+ + 2e^- \rightarrow NADH$

$$C_5H_6O_5 + NAD^+ + H_2O \rightarrow C_4H_6O_4 + NADH + CO_2 + H^+ \qquad (10\text{-}35)$$

Note how the reactants and products are represented in the Krebs cycle. If we look at another step, for example, the one from fumaric acid to malic acid, and we are uncertain as to whether it is an oxidation step, we can find the answer by checking the balance of the reaction:

$$\underset{\text{Fumaric acid}}{HOOC{-}CH{=}CH{-}COOH} + H_2O \rightarrow \underset{\text{Malic acid}}{HOOC{-}CH_2{-}\overset{\overset{\displaystyle OH}{|}}{C}H{-}COOH}$$

$$C_4H_4O_4 + H_2O \rightarrow C_4H_6O_5 \qquad (10\text{-}36)$$

Obviously this reaction is balanced. It is not necessary to add any electrons, so this step in the Krebs cycle is not a redox reaction. Of course, the same method also works for more complicated reactions, such as the reaction of oxaloacetic acid with acetic acid:

$$\underset{\text{Oxaloacetic acid}}{HOOC{-}CH_2{-}\overset{\overset{\displaystyle O}{\|}}{C}{-}COOH} + \underset{\text{Acetic acid}}{CH_3COOH} \rightarrow \underset{\text{Citric acid}}{HOOC{-}CH_2{-}\overset{\overset{\displaystyle COOH}{|}}{\underset{\underset{\displaystyle OH}{|}}{C}}{-}CH_2{-}COOH}$$

$$C_4H_4O_5 + C_2H_4O_2 \rightarrow C_6H_8O_7 \qquad (10\text{-}37)$$

The reaction is balanced; it is not a redox reaction.

The Discovery of Dynamic Electricity

At the beginning of the section on electrochemical cells, we mentioned that the galvanic cell is sometimes called a voltaic cell. Therein lies an interesting story. In the late eighteenth century, static electricity was all the rage in scientific circles. Few laboratories were without devices such as Leyden jars, in which a large static charge could be built up and then released as a jolt of electricity. In 1780 the Italian anatomist Luigi Galvani (1737–1798) observed that the muscles of dissected frog legs twitched when struck by an electric spark. This was not too surprising. In 1786, however, he attempted to confirm Benjamin Franklin's discovery of the electrical nature of lightning by seeing whether a nerve-muscle preparation of frog legs would twitch when exposed to the outdoor atmosphere during a thunderstorm. The muscles were hung from an iron grill by copper hooks, and rested against the grill. To his surprise, Galvani found that the muscles twitched *regardless* of the weather. In fact, they twitched whenever two different metals in contact with each other were applied to the muscle and nerve!

Galvani had discovered *dynamic electricity*—the generation and flow of electric current—which was soon to revolutionize both physics and chemistry. Being a physician, Galvani took the view that the effect originated not in the metal junction but in the muscle, owing to an intrinsic "animal electricity." He clung to this theory even after 1794, when he was proved wrong by his friend, the renowned physicist Count Alessandro Volta (1745–1827). Volta showed that an electric current could be generated without any animal tissue, merely by putting together pieces of two different metals with some saline liquid or a damp cloth between them. In the ensuing scientific controversy, Galvani rapidly lost ground to Volta's ever more conclusive resu ts, and he died in great disappointment. Volta went on to construct the first crude batteries—initially "wet cells" (galvanic cells), and subsequently "dry cells" (voltaic piles). Despite being rather unwieldy and unreliable (Daniell had not yet invented the Daniell cell), they had an enormous impact on the development of physical science, both experimental and theoretical.

Today we know that Galvani was not entirely wrong after all. It is true that animal electricity in the form in which he envisioned it does not exist (although certain species, notably the electric eel and the torpedo fish, can deliver astonishingly powerful electric shocks). In a more subtle yet very real sense, however, it does exist. All living organisms—and, for that matter, dead ones that are decaying—are "alive" with electrochemical reactions of the most fundamental importance. Although he could not prove it, Galvani correctly sensed that, in chancing to touch a frog's leg in a special way, he had cast the first light on one of life's deepest mysteries.

Bibliography

Irving M. Klotz, **Energy Changes in Biochemical Reactions,** Academic Press, New York, 1967. Professor Klotz is a noted thermodynamicist. He has written a wide-ranging survey of the molecular basis for the energy changes on which life depends. It will be useful to you in your study of Chapter 11 as well as the present chapter. It is listed here because it gives a rather thorough discussion of the mechanisms by which electrons are transferred in biological systems.

Albert L. Lehninger, **Bioenergetics,** 2nd ed., W. A. Benjamin, Menlo Park, Calif., 1971. The content of Professor Lehninger's text is very similar to that of the one listed above, and Lehninger's writing style is superb.

Problems

1. For each of the following reactions, identify the reducing agent and then balance the reaction by oxidation states.
   a. $Fe^{2+} + Cr_2O_7^{2-} \rightarrow Fe^{3+} + Cr^{3+}$ (Acidic solution)
   b. $Bi(OH)_3 + Sn(OH)_3^- \rightarrow Bi + Sn(OH)_6^{2-}$ (Basic solution)
   c. $Zn + HAsO_2 \rightarrow Zn^{2+} + AsH_3$ (Acidic solution)
   d. $MnO_2 \rightarrow MnO_4^- + MnO_3^{3-}$ (Basic solution)

2. For each of the following reactions, identify the oxidizing agent and then balance the reaction by the half-reaction method.
   a. $Cu + NO_3^- \rightarrow Cu^{2+} + NO$ (Acidic solution)
   b. $CrO_2^- + ClO^- \rightarrow CrO_4^{2-} + Cl^-$ (Basic solution)
   c. $MnO_4^- + H_2C_2O_4 \rightarrow Mn^{2+} + CO_2$ (Acidic solution)
   d. $Ag_2O + CH_2O \rightarrow Ag + HCOO^-$ (Basic solution)

3. Balance the following reactions by any method you choose.
   a. $MnO_4^- + Bi^{3+} \rightarrow Mn^{2+} + HBiO_3$ (Acidic solution)
   b. $PbO_2 + Cl^- \rightarrow Pb(OH)_3^- + ClO^-$ (Basic solution)
   c. $ICl_4^- \rightarrow I_2 + IO_3^- + Cl^-$ (Acidic solution)
   d. $ClO_2 \rightarrow ClO_2^- + ClO_3^-$ (Basic solution)
   e. $Cu_2S + NO_3^- \rightarrow Cu^{2+} + S_8 + NO_2$ (Acidic solution)
   f. $CrI_3 + Cl_2 \rightarrow CrO_4^{2-} + IO_3^- + Cl^-$ (Basic solution)

4. Look up the standard reduction potentials for the following half-reactions:

$$MnO_4^- + 4H^+ + 3e^- \rightarrow MnO_2(c) + 2H_2O$$

$$Ce^{4+} + e^- \rightarrow Ce^{3+}$$

$$Fe^{2+} + 2e^- \rightarrow Fe(c)$$

$$Ag^+ + e^- \rightarrow Ag(c)$$

   a. Which of the above ions is the strongest reducing agent? Oxidizing agent?
   b. Which of the above ions can reduce Fe^{2+} to Fe?
   c. Which of the above ions can oxidize Ag to Ag^+?

5. Aluminum is produced by the electrolysis of molten salts of Al(III). The annual world production of aluminum is about 14×10^6 metric tons. Calculate the current required for this production, assuming that all of it occurs in just one plant, with round-the-clock operation.

6. The volume of Lake Cayuga in New York State is estimated to be 8.2×10^{12} ℓ. A power station not far above Cayuga's waters generates electricity at the rate of 1.5×10^6 C/s (at an appropriate voltage). At this rate how many hours would it take to electrolyze all the water in Lake Cayuga?

7. A fork can be silver-plated by immersing it and a piece of pure Ag in a solution of silver cyanide, AgCN. The fork is made the cathode, where the half-reaction is $Ag^+ + e^- \rightarrow Ag(c)$. If the surface area of the fork is 35.0 cm² and a current of 1.00 A is passed through the cell for 2 h, how thick a coating will be deposited on the fork? The density of Ag is 10.5 g/cm³.

8. A given current liberates 3.6 g of O_2 from water in 25 min. How long will it take for the same current to deposit 1.0 mole of Zn from a $ZnSO_4$ solution?

9. A precise tool for the measurement of dissolved oxygen is the Clark oxygen electrode. If this electrode is immersed in 10.0 ml of air-saturated water and the current is 2.00 μA, what fraction of the oxygen in solution will be reduced at the electrode in 2 h? The reduction half-reaction is $O_2 + 2H_2O + 4e^- \rightarrow 4OH^-$ and the concentration of O_2 in an air-saturated solution at 25°C is 2.59×10^{-7} mole/ml.

10. Write the cell diagram for each of the following reactions:

a. $2Ag + Hg_2Cl_2 \rightleftharpoons 2AgCl + 2Hg$
b. $Ni + Sn^{2+} \rightleftharpoons Ni^{2+} + Sn$
c. $4Ag + O_2 + 4H^+ \rightleftharpoons 4Ag^+ + 2H_2O$
d. $5Cl_2 + I_2 + 6H_2O \rightleftharpoons 10Cl^- + 2IO_3^- + 12H^+$

11. State whether each of the following reactions will proceed spontaneously in the forward direction, and in which direction the reaction will proceed if all dissolved species are initially at unit molarity.

a. $Ba + 2Na^+ \rightleftharpoons Ba^{2+} + 2Na$
b. $Fe^{3+} + Ag \rightleftharpoons Fe^{2+} + Ag^+$
c. $Cu^{2+} + H_2 \rightleftharpoons Cu + 2H^+$
d. $2Cr^{3+} + 6Fe^{3+} + 7H_2O \rightleftharpoons Cr_2O_7^{2-} + 6Fe^{2+} + 14H^+$

12. An electrode that is used for the measurement of $[H^+]$ is the *quinhydrone electrode,* which is made by dipping a Pt electrode into a saturated solution of an equimolar mixture of the organic compounds *p*-quinone (Q) and hydroquinone (H_2Q). The Pt should make contact with the undissolved solids at the bottom of the half-cell. Consider the following reduction half-reactions:

$$Q + 2H^+ + 2e^- \rightarrow H_2Q \qquad \mathscr{E}° = +0.70 \text{ V}$$

$$H_2O_2 + 2H^+ + 2e^- \rightarrow 2H_2O \qquad \mathscr{E}° = +1.77 \text{ V}$$

a. Write the cell diagram for the following reaction:

$$H_2O_2 + H_2Q \rightarrow 2H_2O + Q$$

b. Sketch a galvanic cell based on the cell diagram you have just written. Label the anode, the cathode, the direction of electron flow in the external circuit, the positive electrode, the negative electrode, and the directions in which cations and anions will tend to migrate in the cell.

13. The standard hydrogen electrode ($\mathscr{E}° = 0.0000$ V) is the reference electrode for all systems (note that the value of $\mathscr{E}°'$ for this electrode is -0.421 V). If the quinhydrone electrode ($\mathscr{E}°' = +0.28$ V) were selected as the reference electrode for biological systems instead of the standard hydrogen electrode, i.e., if the quinhydrone electrode were given the value $\mathscr{E}°' = 0.0000$ V, what would be the standard reduction potential of each of the following half-reactions:

a. Cu^{2+}/Cu^+ Hemocyanin
b. Fe^{3+}/Fe^{2+} Cytochrome a
c. Oxaloacetate/Malate
d. $NAD^+/NADH$

14. For each of the following cells, write the overall reaction, compute the actual cell potential, and state which electrode bears the negative charge.

a. $Ag \mid Ag^+(0.05\ M) \parallel Fe^{3+}(0.3\ M),\ Fe^{2+}(0.02\ M) \mid Pt$

b. $Zn \mid Zn^{2+}(0.1\ M) \parallel Ag^+(0.2\ M) \mid Ag$

c. $Ag \mid AgBr(c),\ Br^-(0.05\ M) \parallel H^+(6.0\ M) \mid H_2(1\ \text{atm}) \mid Pt$

d. $Pt \mid H_2(1\ \text{atm}) \mid H^+(10^{-4}\ M) \parallel Cl^-(0.5\ M),\ AgCl(c) \mid Ag$

15. Consider a cell consisting of an iron rod immersed in a 0.010 *M* solution of $FeSO_4$ and a manganese rod immersed in a 0.10 *M* solution of $MnSO_4$, the solutions being connected by a salt bridge and the metal rods by a voltmeter.

a. Write the spontaneous chemical reaction and its cell potential.

b. If you wanted to increase the cell potential by 0.02 V, which solution should be diluted, and with how much water?

16. Consider the following reduction half-reactions:

$$\underset{\text{Pyruvate}}{CH_3-\overset{\overset{\large O}{\|}}{C}-COO^-} + 2H^+ + 2e^- \rightarrow \underset{\text{Lactate}}{CH_3-\overset{\overset{\large OH}{|}}{C}H-COO^-} \qquad \mathscr{E}^{\circ\prime} = -0.185\ \text{V}$$

$$Br_2(l) + 2e^- \rightarrow 2Br^- \qquad \mathscr{E}^\circ = +1.065\ \text{V}$$

a. Write the half-reactions and the balanced, overall reaction for the oxidation of lactate to pyruvate by bromine.

b. Sketch the cell based on this reaction. Label the cathode, the anode, and the direction of electron flow in the external circuit.

c. Calculate the equilibrium constant for the spontaneous reaction at 25°C.

d. If precisely 1 mole of bromine and 1 mole of lactate ion were mixed at pH 7.0 and 25°C, what would be the concentrations of lactate, pyruvate, and bromide ions at equilibrium? Assume a total aqueous volume of 1 ℓ.

17. Calculate the solubility product of Ag_2S, using the standard reduction potentials in Appendix I.

18. The half-reaction $Mg(OH)_2 + 2e^- \rightarrow Mg + 2OH^-$ has the standard reduction potential $\mathscr{E}^\circ = -2.67$ V. The solubility of $Mg(OH)_2$ is 1.45×10^{-4} mole/ℓ. Calculate $\mathscr{E}^\circ$ for the half-reaction $Mg^{2+} + 2e^- \rightarrow Mg(c)$.

19. Consider the hypothetical reaction

$$2M(c) + 3P^{2+} \rightarrow 2M^{3+} + 3P(c) \qquad \mathscr{E}^\circ = -0.01\ \text{V}$$

When an excess of M is shaken in a solution of P^{2+}, the equilibrium concentration of P^{2+} is found to be 10^{-1} mole/ℓ. Calculate the initial concentration of P^{2+}.

INTERLUDE

Richter

fuel cells

As we have seen before, the generation of electricity by burning fossil fuels and using the steam to drive turbines causes serious environmental problems. For example, the combustion of coal releases large amounts of SO_2 and particulate matter into the atmosphere. Eventually, there will be no more coal left to burn. In addition, the second law of thermodynamics imposes a fundamental limitation on the operating efficiency of all heat engines (a *heat engine* is any device for transforming heat into mechanical work). All heat engines operate between two temperatures: the higher temperature T_1 is that of the heat *source*, and the lower temperature T_2 is that of the heat *sink*. Only a limited proportion of the heat input from the source can be converted to work, however. The rest flows to the heat sink and is lost as waste heat. (Heat always flows from a hotter body to a colder body, never the reverse.) This inevitable consequence of the second law can be expressed as

$$\text{Efficiency} = \frac{T_1 - T_2}{T_1}$$

where T_1 and T_2 are absolute temperatures. Clearly, the efficiency of a heat engine can be 100% only if the temperature T_2 is absolute zero, which is impossible. The maximum efficiency of most modern steam turbines is about 40%, but most other heat engines, such as reciprocating steam engines and internal combustion engines, are fortunate to achieve a 30% efficiency.

It is small wonder, then, that chemists and chemical engineers have been extremely active in seeking better ways to extract the free energy contained in the bonds of molecules. One of the most significant results of their efforts is the *fuel cell.* A fuel cell is a device in which the energy of a fuel is converted directly to electricity by an electrochemical reaction, rather than to heat by a combustion reaction. More precisely, it is an electrochemical cell in which, at a uniform temperature, the free-energy change of a fuel-oxidizer reaction is converted directly and continuously to electric energy for as long as the reactants are supplied and the products removed at an appropriate rate. Thus, the only basic difference between a fuel cell and a galvanic cell or any other battery is that, in a fuel cell, the chemicals that generate the electron flow are stored separately and supplied to the electrodes on demand. In principle, therefore, a fuel cell can generate electricity indefinitely, i.e., for as long as there is fuel to "burn."

A fuel cell is not a heat engine, because it operates at a uniform temperature, which may be anywhere from room temperature to about 1000°C. The maximum theoretical efficiency is therefore not limited by the available working temperatures, as it is in heat engines, and can approach 100%. Inevitably, of course, there is some waste heat in real fuel cells and some loss of efficiency due to undesirable side reactions and degradation of the electrodes and electrolyte. Nevertheless, overall efficiencies up to 75%—occasionally even 95%—can be realized. This is what makes fuel cells so attractive as potential sources of electric energy.

Like galvanic cells, all fuel cells consist of a pair of electrodes and an electrolyte, plus an external circuit for electron flow and an internal mechanism for allowing ion migration to complete the circuit. In addition, they must have the associated hardware necessary for the controlled delivery of fuel (gases or liquids such as hydrogen, hydrocarbons, methanol, hydrazine, etc.) and oxidizer (air or oxygen) to the electrodes, and for the controlled removal of the reaction products and waste heat. Depending on the operating conditions, they may also require temperature and pressure control systems. The output of electricity, which is always low-voltage dc, is controlled by voltage or current regulators. The basic idea of a fuel cell can easily be shown in "black-box" style, as in Figure 1.

The fuel cell is nothing new. The first one was invented in 1839 by Sir William Grove (1811–1896), an English physicist and lawyer who was an ardent advocate of energy conservation. It was based on the combination of gaseous hydrogen and oxygen to form water, a reaction that was familiar to scientists even then. In effect, Grove achieved the reverse of the electrolysis of water: the spontaneous, controlled reaction of H_2 and O_2 at platinum electrodes in a dilute solution of H_2SO_4, with the generation of a steady electric current. At the anode, the H_2 was oxidized to H^+ ions, which migrated to the cathode. There they reacted with the O_2, which was reduced, giving H_2O.

One hundred eighteen years passed—and then there was Sputnik.* Suddenly the idea of generating electricity from a very light, very-high-energy fuel, with high thermodynamic efficiency, and obtaining only pure water as a "waste" product, took on a whole new meaning. Several types of pure H_2/O_2 fuel cells were rapidly developed to a high degree of refine-

*Actually, fuel-cell research was not entirely dormant until Sputnik appeared. Several types of fuel cells were developed in the first half of this century. The work picked up noticeably after World War II, when fuel cells were found useful for powering portable communications and radar systems, and for special applications in submarines.

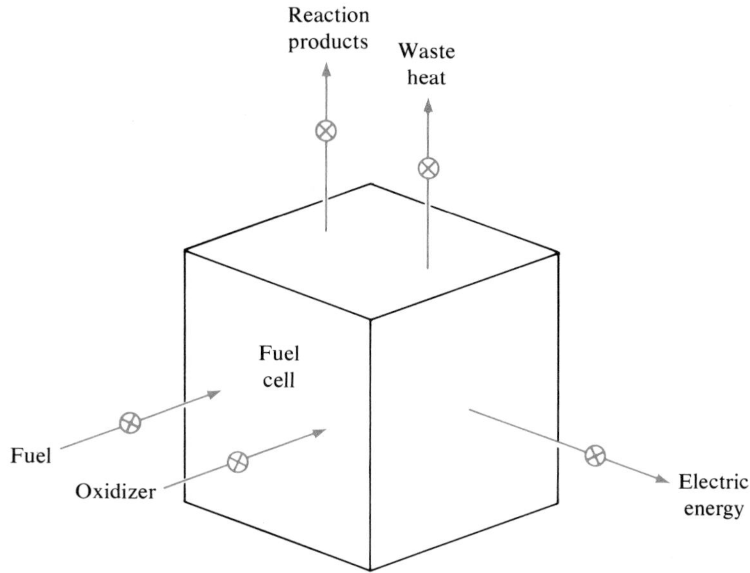

Figure 1. A fuel cell shown in "black-box" style. Each input and output is controlled in a suitable manner.

ment and used in the Soviet and American manned spacecraft. With hardly any moving parts, they generated electricity (and drinking water) at power levels up to 3 kW, while producing no pollution, no noise, and almost no waste heat. Surely they are an environmentalist's (as well as astronaut's) dream come true. Let us look at one in more detail.

Alkaline electrolytes are generally less corrosive than acidic ones, and permit higher power output at ambient temperature and pressure. They are therefore preferred for use in fuel cells, even though they are highly susceptible to degradation by CO_2 (a normal component of air and a combustion product of all hydrocarbons), which must therefore be rigorously excluded from the system. The H_2/O_2 fuel cell operating with an alkaline electrolyte, usually aqueous KOH, is the most highly developed and reliable of all fuel cells. It is based on the following reaction scheme:

	Add	*Subtract*
Reduction (cathode):	$O_2(g) + 2H_2O + 4e^- \rightarrow 4OH^-$	$\mathscr{E}° = +0.401$ V
Oxidation (anode):	$2H_2(g) + 4OH^- \rightarrow 4H_2O + 4e^-$	$-(\mathscr{E}° = -0.828$ V$)$
	$2H_2(g) + O_2(g) \rightarrow 2H_2O$	$\mathscr{E}° = +1.229$ V

The mechanism is somewhat different from that in the acidic H_2/O_2 fuel cell described above, but the results are the same. At the cathode, the O_2 is reduced to OH^- ions, which migrate to the anode. There they react with the H_2, which is oxidized, giving H_2O.

A diagram of such a fuel cell is shown in Figure 2. The electrodes are made of porous carbon in order to provide the gas-liquid-solid interface that is essential for the reactions to be able to occur. Because these reactions are normally very slow, the electrodes are impregnated with catalysts to speed them up enough to make the fuel cell practical. Typically, finely divided Pt, Pd, or Ni is used in the anode and Ag, Pt, or certain transition-metal oxides and organic complexes in the cathode. Note that the electrolyte is left unchanged by the overall reaction. For each OH^- ion produced at the cathode, one is consumed at the anode. For each four H_2O molecules produced at the anode, two are consumed at the cathode and, in a well-designed system operating at the proper conditions, the other two are carried away with the excess H_2. The K^+ ions, although slightly attracted

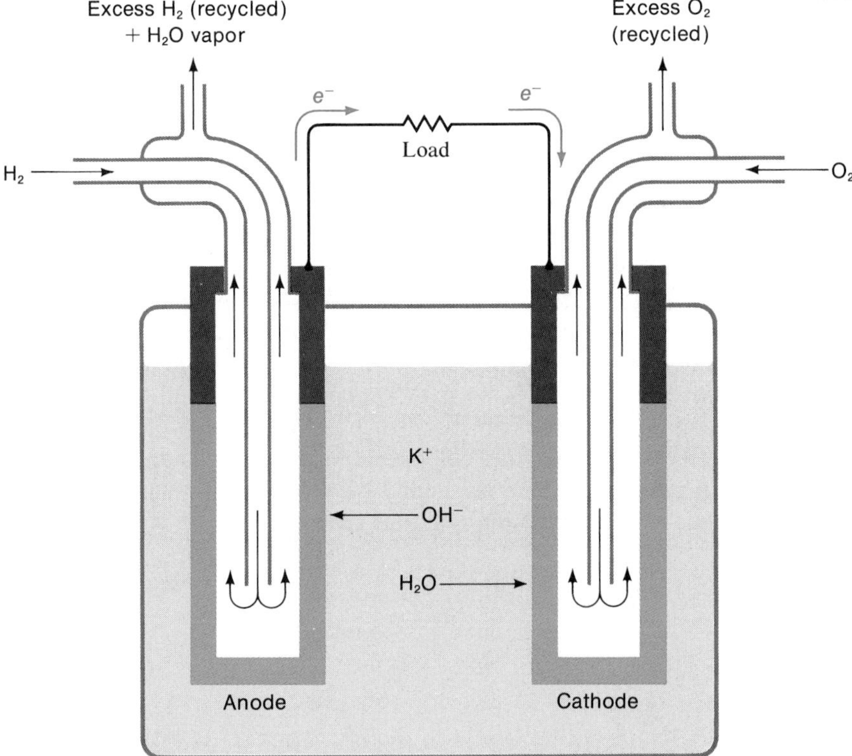

Figure 2. Diagram of an alkaline fuel cell using the reaction $2H_2 + O_2 \rightarrow 2H_2O$.

to the cathode while current is flowing, are not reduced there because K^+ is very difficult to reduce ($\mathscr{E}° = -2.925$ V). By contrast, O_2 is easy to reduce ($\mathscr{E}° = +0.401$ V).

You should be feeling quite suspicious by now. If fuel cells are so efficient and environmentally desirable, why can't you drive down to your friendly electrical contractor and buy one? The obvious answer is that he has none to sell. A clue to the primary reason for this is in the preceding paragraph. Catalysts such as those mentioned there are so expensive that electricity from most fuel cells still costs about a thousand times more than the same amount derived from conventional sources, despite the higher energy-conversion efficiency of fuel cells. Until ways can be found to obtain the same catalytic action with much less catalyst, or until catalysts can be found that are at least as efficient as Pt, Pd, and Ag but much cheaper, there appears to be no chance of making fuel cells economically competitive. They would remain useful only for specialized military and aerospace applications in which cost is much less important than performance.

Furthermore, there are still formidable scientific and engineering problems to be overcome. Chief among these is the design of catalytic electrode-electrolyte systems that are not significantly degraded over long periods of continuous, hard use with commercial-quality (i.e., relatively crude) fuels, at high power levels. This problem is most severe in the high-temperature fuel cells, in which, on the other hand, the best success has been found in developing cheap catalysts. Ironically, the fuel cells using expensive fuels, such as hydrazine, are generally simpler and cheaper to construct than those that use cheap fuels, such as crude hydrogen from hydrocarbons. There is thus a trade-off between the cost of the system and the cost of the energy it produces. Despite these problems, and many more, there is hope. More important, there is hard work, and it may yet pay off.

Another example of a fuel cell that has been made to work is the methane-oxygen cell. Instead of burning natural gas and using the released thermal energy to boil water for driving steam turbines, the reaction can be used directly to provide electric power. With an acid electrolyte such as concentrated H_3PO_4 or H_2SO_4, the reaction scheme can be written as

Add *Subtract*

Reduction (cathode):

$$2O_2(g) + 8H^+ + 8e^- \rightarrow 4H_2O \qquad \mathscr{E}° = +1.229 \text{ V}$$

Oxidation (anode):

$$CH_4(g) + 2H_2O \rightarrow CO_2(g) + 8H^+ + 8e^- \qquad -(\mathscr{E}° = +0.169 \text{ V})$$

$$CH_4(g) + 2O_2(g) \rightarrow CO_2(g) + 2H_2O \qquad \mathscr{E}° = +1.060 \text{ V}$$

This fuel cell is quite weak and inefficient, however. Much better performance is obtained by converting the natural gas to crude H_2 and CO_2

(the industrial processes for doing this are called *cracking* and *reforming*), removing the bulk of the CO_2 with an alkaline scrubber, and using the crude H_2 alone as fuel. The cost of this conversion is more than compensated for by the increased power output of the fuel cell. In fact, an acidic H_2/air fuel cell based on just this process is presently the only one that appears to have any chance of commercial success in the foreseeable future.

Although alkaline fuel cells are usually preferred for technical and economic reasons, natural gas cannot be used directly as a fuel with an alkaline electrolyte because the latter is rapidly degraded by the CO_2 produced in the reaction (even the CO_2 in the air stream cannot be tolerated). Of course, the natural gas can be converted to crude H_2, as described above. However, the CO_2 (as well as CO, residual CH_4, and various other impurities) must now be removed so thoroughly, to protect the sensitive electrolyte, that the process becomes economically self-defeating.

One of the most attractive fuels for fuel cells is the rocket propellant, hydrazine, N_2H_4.* (See Table 3-7 for a review of the oxidation states of nitrogen.) Hydrazine is a colorless, fuming, corrosive, toxic liquid that reacts violently with many substances—which does not sound very attractive, to be sure. But it has a very high energy yield (exothermicity) in relation to its mass and volume, a very high electrochemical reactivity (it is a powerful reducing agent in alkaline media), and its oxidation products (N_2 and H_2O) do not contaminate alkaline electrolytes—or the atmosphere. Furthermore, its rapid oxidation in fuel cells does not, in certain circumstances, require the use of Pt or Pd catalysts, a most attractive feature.

Hydrazine is supplied to the fuel cell not as a gas or pure liquid, but as a dilute solution in concentrated aqueous KOH. The oxidizer can be air, O_2, or H_2O_2. Air is the cheapest, of course, and therefore the most desirable. The reaction scheme can be written as follows:

	Add	*Subtract*
Reduction (cathode):		
	$O_2(g) + 2H_2O + 4e^- \rightarrow 4OH^-$	$\mathscr{E}° = +0.40$ V
Oxidation (anode):		
	$N_2H_4(aq) + 4OH^- \rightarrow N_2(g) + 4H_2O + 4e^-$	$-(\mathscr{E}° = -1.16$ V$)$
	$N_2H_4(aq) + O_2(g) \rightarrow N_2(g) + 2H_2O$	$\mathscr{E}° = +1.56$ V

A diagram of an alkaline N_2H_4/air fuel cell is shown in Figure 3. The liquid fuel stream, called the *anolyte*, is recycled, with the continuous addition of N_2H_4 and removal of N_2 and excess H_2O. The electrodes are porous to permit diffusion of the reactants and products as required, and the

*An excellent account of this fuel cell is given by G. E. Evans and K. V. Kordesch in *Science* **158,** 1148 (1967).

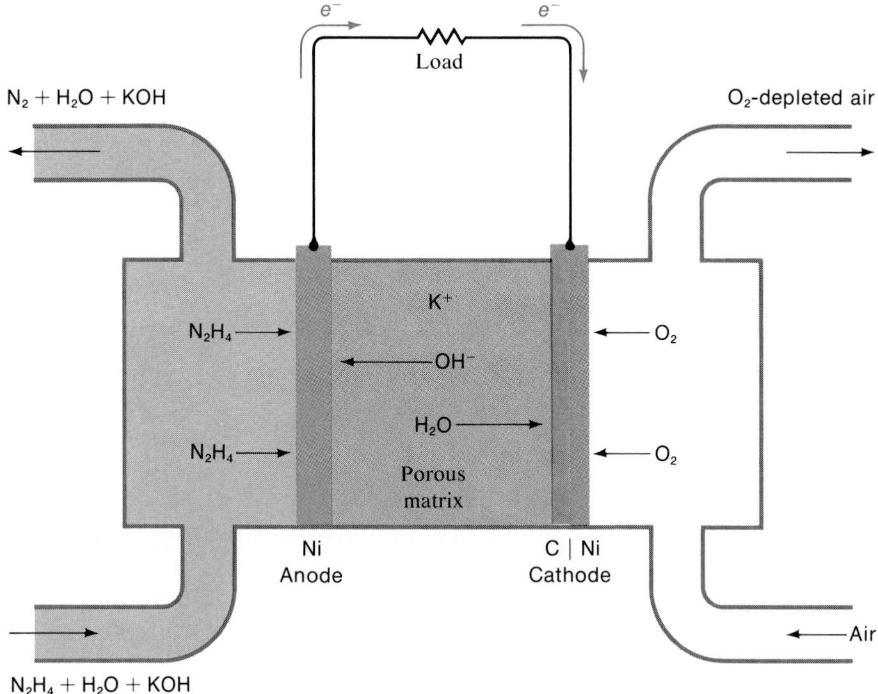

Figure 3. Diagram of a hydrazine-air fuel cell. [Adapted from G. E. Evans and K. V. Kordesch, *Science* **158**, 1148 (1967). Copyright © 1967 by the American Association for the Advancement of Science.]

electrolyte is a porous matrix of an inert substance (e.g., asbestos) that is saturated with aqueous KOH. The operating temperature of this fuel cell is typically 40–80°C.

In reality, neither of the half-reactions shown above for the hydrazine fuel cell is quite that simple, and the actual electrode potentials deviate considerably from the predicted thermodynamic equilibrium potentials. The result is a cell potential of only about 1 V at best. The fuel cell is so efficient (up to 95%), powerful (up to 5 kW), and reliable (several thousand hours), however, that this is not a serious problem. What *is* a serious problem is the high cost of hydrazine, making the electricity from N_2H_4/air fuel cells up to several hundred times as expensive as that from conventional sources.

It is clear that significant commercial power generation by fuel cells is a distant goal at best, but a very worthwhile one. Human perseverance and ingenuity being what they are, there is good reason to believe that the difficult problems discussed above will eventually be solved. It will take the combined talent and effort of many technically trained people to do it. Will you accept the challenge?

11 the chemistry of life

The last chapter of this text is devoted to a discussion of the chemistry of molecules originally derived from biological sources. Of course, virtually all the molecules we will study, except for the nucleic acids, have been synthesized by scientists *in vitro,* which means "in glass," i.e., without the aid of any living organism. We will still refer to these molecules as *biomolecules,* however, because they were first produced *in vivo,* within living organisms, and are almost always associated with biological functions.

Because many of you may be interested in careers related to biology in some form, an introduction to the chemistry of life is appropriate at this point. The discovery of the *genetic code* and many of its far-reaching implications have been the subjects of much writing in the popular press in recent years. We hope you will find it interesting to learn some of the details of the actual molecular structures and interactions that constitute the genetic code. Furthermore, we hope that this chapter will serve as a capstone for much of the material you have learned up to now. It is useful to see the applications of many of the concepts of chemistry to problems of biology. For example, hydrogen bonds ultimately determine the three-dimensional structures of proteins and nucleic acids. Weaker intermolecular forces, such as London and dipole-dipole forces, play significant roles in determining protein structure. Thermodynamics is at the heart of many physicochemical techniques that are of paramount importance in learning the sizes and shapes of biomolecules. In order to begin to understand those incredible catalysts, the enzymes, we must look at the structure of proteins in some detail. The properties of acids and bases will be very useful in understanding the chemistry of the amino acids and nucleotides, and of their polymers.

Organic chemistry and biochemistry are so closely interrelated that it is often advantageous to consider them together, as a conceptual whole. In the following pages we will examine some of the classes of organic compounds, the ways to name them, and a few typical reactions. However, we will do this in the context of small and large biomolecules.

A useful procedure for naming organic compounds is to classify them according to the presence of functional groups. For example, in Chapter 5, the names and structures of some hydrocarbons, alcohols, ethers, aldehydes, ketones, and carboxylic acids were presented. In Chapter 4 we learned of the chelating properties of ethylenediamine, $H_2NCH_2CH_2NH_2$, a simple amine. Chapter 9 contains an extensive discussion of the titration of the weak carboxylic acid, acetic acid. In addition, Chapters 5 and 9 both contain discussions of amino acids. Table 11-1 lists some of the major classes of

Table 11-1. **Some Major Classes of Organic Compounds.**

	Functional Group		Example	
Class	Structure	Name	Structure	Name
Hydrocarbon	–	–	$CH_3CH_2CH(C_2H_5)CH(CH_3)CH_3$	2-Methyl-3-ethyl-pentane
Carboxylic acid	$-C(=O)-OH$	Carboxyl	$CH_3CH_2-C(=O)-OH$	Propanoic acid
Ester	$-C(=O)-O-$	–	$H-C(=O)-O-CH_2CH_3$	Ethyl formate
Amine	$-NH_2$	Amino	CH_3-NH_2	Methylamine
Amide	$-C(=O)-NH_2$	Amido	$CH_3CH_2-C(=O)-NH_2$	Propionamide
Alcohol	$-OH$	Hydroxyl	$CH_3CH(OH)CH_2CH_3$	2-Butanol
Mercaptan	$-SH$	Sulfhydryl	$CH_3CH_2CH_2-SH$	Propanethiol
Ether	$-O-$	–	CH_3-O-CH_3	Dimethyl ether
Aldehyde	$-C(=O)-H$	Carbonyl	$CH_3-C(=O)-H$	Acetaldehyde
Ketone	$-C(=O)-$	Carbonyl	$CH_3-C(=O)-C_6H_5$	Methylphenyl ketone

organic compounds. You should study and memorize them. Only the esters, amides, and mercaptans have not been discussed before in this book.

Our goal in this chapter is an understanding of the basic principles of the genetic code and a feel for some of the excitement of current research in this field. Our study begins with an examination of the chemistry and structure of small biological molecules.

small biological molecules—the monomers

11.1 THE AMINO ACIDS

Small quantities of amino acids are found as discrete molecules in cells. Most of them, however, are contained in large molecules called *proteins*. A mixture of most of the amino acids is produced by heating proteins in concentrated inorganic acid, a process called *acid hydrolysis*. All the biologically important amino acids possess the following general structural formula:

$$
\begin{array}{c}
\mathrm{R} \\
| \\
\mathrm{H{-}C{-}COOH} \\
| \\
\mathrm{NH_2}
\end{array}
$$

As we saw in Chapter 9, however, a more realistic representation of this formula is the zwitterionic form:

$$
\begin{array}{c}
\mathrm{R} \\
| \\
\mathrm{H{-}C{-}COO^-} \\
| \\
\mathrm{NH_3^+}
\end{array}
$$

Such compounds are called *α-amino acids* because the α-carbon atom, shown at the center of the above two structures, is the one to which the amino group is attached. The carbon atoms in the side chain extending from the α-carbon are given the sequential labels β, γ, δ, ϵ, etc.

Physical Properties

Several of the properties of the amino acids are of great importance in their biological functioning and in our handling of proteins. We will briefly

Table 11-2. Values of pK_a for Some Side Chains of Amino Acids.

Amino Acid	Ionizable Side Chain	pK_a
Aspartic acid Glutamic acid	Carboxyl	4
Histidine	Imidazolyl	6
Cysteine	Sulfhydryl	9
Lysine	Amino	9.5
Tyrosine	Phenolic hydroxyl	10
Arginine	Guanidino	13

examine three of these properties. The first is optical activity. All but one of the amino acids are optically active and occur in nature in the L-configuration. (See Chapter 5 for a discussion of optical activity and Figure 5-13 for a representation of the absolute configuration of amino acids.)

The second property we will consider is acid-base behavior. As we will see later, the α-NH_3^+ and α-COO^- groups in amino acids lose their ability to function as weak acids when they are polymerized to proteins. These groups are linked together to form the *amide group*, (H)N—C(=O), which does not dissociate or bind H^+. However, the acid-base properties of proteins are very important in the functioning of proteins. These properties are contained in the side chains, which will be discussed in more detail shortly. It is useful, in considering the properties of proteins in aqueous environments, to keep in mind the pK_a values shown in Table 11-2. These values are approximate, and fluctuations of ± 1 are to be expected for a given side chain in different proteins.

The third property we will consider is the ability of three of the side chains to absorb light. As we saw in our discussion of conjugated double-bond systems in organic molecules, such systems absorb light in the visible and ultraviolet regions of the spectrum. Long-chain conjugated systems such as β-carotene absorb light in the visible (lower-energy) region of the spectrum and short-chain conjugated systems absorb light in the ultraviolet (higher-energy) region. Benzene, for instance, has an absorption maximum at 243 nm. The amino acids phenylalanine, tyrosine, and tryptophan all contain at least one benzene ring. Their absorption spectra, which are highly pH-dependent, are shown in Figure 11-1 for the value pH = 6.

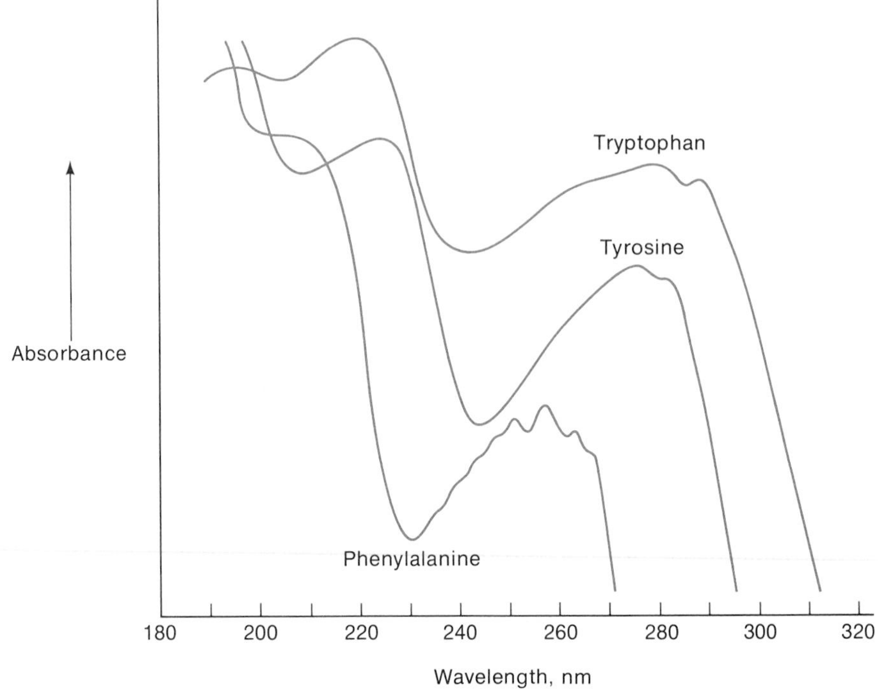

Figure 11-1. Ultraviolet absorption spectra of the three aromatic amino acids at pH 6. [D. B. Wetlaufer, *Advances in Protein Chemistry* **17**, 303 (1962).]

Side-Chain Structures

There are about 75 naturally occurring compounds with the general structure of an amino acid. We will restrict ourselves to just 20, the most common ones resulting from the acid hydrolysis of a wide variety of naturally occurring proteins. In the following discussion the amino acid side chains are grouped in a way that makes them easiest to learn. A slightly modified grouping will be found useful when we consider the locations and functions of these side chains in proteins.

Hydrocarbon Side Chains

The names, abbreviations, and structures of the amino acids containing hydrocarbon side chains are given in Table 11-3. Hydrocarbon side chains

Table 11-3. Hydrocarbon Side Chains of Amino Acids.

Amino Acid	Abbreviation	Structure	Side Chain
Glycine	gly	$^{-}OOC{-}C(NH_3^+)(H){-}H$	Hydrogen
Alanine	ala	$^{-}OOC{-}C(NH_3^+)(H){-}CH_3$	Methyl
Valine	val	$^{-}OOC{-}C(NH_3^+)(H){-}CH(CH_3)_2$	Isopropyl
Leucine	leu	$^{-}OOC{-}C(NH_3^+)(H){-}CH_2CH(CH_3)_2$	Isobutyl
Isoleucine	ile	$^{-}OOC{-}C(NH_3^+)(H){-}CH(CH_3)CH_2CH_3$	*sec*-Butyl
Phenylalanine	phe	$^{-}OOC{-}C(NH_3^+)(H){-}CH_2{-}C_6H_5$	Benzyl
Proline	pro	$^{-}OOC{-}C(H)$ in ring $H_2C{-}CH_2{-}CH_2{-}\overset{+}{N}H_2$	3-Carbon (cyclic)

are those containing only C and H atoms. The names of the amino acids are all "trivial," i.e., they are coined names that are derived from a source of the compound (asparagine from asparagus, for instance) or on the basis of some other arbitrary criterion. Organic chemists have adopted a more rational way of naming most carbon compounds. An example is the column of names of the side chains in Table 11-3. The method consists of numbering the carbon skeleton so as to give the longest possible chain and to give substituents the smallest possible position-identifying numbers. The side chains are named according to the nomenclature of the hydrocarbon groups given in Table 11-4.

Example 11-1. The *octane rating* of gasoline in an internal combustion engine is based on the performance of the fuel compared with that of the compound iso-octane. Iso-octane is a trivial name for the following compound:

$$\overset{1}{CH_3}-\overset{CH_3\ \ 2}{\underset{CH_3}{C}}-\overset{3}{CH_2}-\overset{CH_3\ \ 4}{CH}-\overset{5}{CH_3}$$

Give the systematic name of this compound.

The longest identifiable chain of carbon atoms has five carbons. There are three side chains, all of them methyl groups. We therefore name the compound as a tri-substituted pentane: 2,2,4-trimethylpentane.

Table 11-4. Some Hydrocarbon Groups.

Alkyl Groups	
Name	Formula
Methyl	CH_3-
Ethyl	CH_3CH_2-
n-Propyl	$CH_3CH_2CH_2-$
Isopropyl	$(CH_3)_2CH-$
n-Butyl	$CH_3CH_2CH_2CH_2-$
sec-Butyl	$CH_3\underset{\vert}{C}HCH_2CH_3$
Isobutyl	$(CH_3)_2CHCH_2-$
tert-Butyl	$(CH_3)_3C-$
Aryl Groups	
Name	Formula
Phenyl	C_6H_5- (benzene ring)—
Benzyl	(benzene ring)$-CH_2-$

The above example illustrates the importance of learning the names and formulas of the first half-dozen or so linear hydrocarbons (see Table 5-1). Table 11-4 contains several hydrocarbon groups that are not shown in Table 11-3. It has been suggested that nature did not select these groups for amino acids because the side chains on other amino acids have approximately the same sizes and shapes. For instance, the side chains of serine and cysteine have about the same size and shape as the ethyl group, and the side chain of methionine is quite similar to the *n*-butyl group.

Several of the hydrocarbon groups have properties of particular importance in protein structures. The glycine side chain is just a hydrogen atom. This is the smallest possible group, so it represents the smallest volume of any group with regard to packing arrangements. This is significant in the structure of silk, which has a high content of glycine. The H atom also means that glycine is not optically active, because it has two H atoms attached to the α-carbon. The α-carbon is therefore not an asymmetric center. The side chain of phenylalanine is significant in that it is a completely nonpolar group with a π-electron system that interacts strongly with other similar groups. Proline is the one real maverick among the amino acids because, in either the zwitterionic or nonzwitterionic forms, it has one less H atom attached to the N atom than do the other amino acids. This is due to the three-carbon side chain attached to the nitrogen atom. The side chain is also attached to the α-carbon, forming a five-membered ring containing one nitrogen atom. This cyclic structure has profound structural consequences for proteins that contain proline.

Hydrocarbons, and hence hydrocarbon side chains, participate in virtually no low-energy, "clean" chemical reactions. They can be oxidized by powerful oxidizing agents and they can be halogenated in the presence of light. In both of these rather drastic chemical processes, a complex mixture of products results. Except in regard to one property, hydrocarbon groups can be considered to be inert in amino acids. This one property is the ability to form *hydrophobic bonds,* an important factor in protein structure. The concept of hydrophobic bonds, formulated by Walter Kauzmann (b. 1916) of Princeton University in 1957, is that a more stable aqueous system results if nonpolar, hydrophobic groups are clustered together with other hydrophobic groups and are not individually dispersed throughout the solvent. The driving force for this clustering is not that bonds are formed between the hydrophobic groups. Rather, it is entropy. When the hydrophobic group is in an aqueous environment, it forces the water molecules around it to adopt a rigid, cage-like structure called a *clathrate* structure. An example is shown in Figure 11-2. It is apparent from this figure that the water molecules are constrained to fixed positions, with each water molecule tetrahedrally hydrogen-bonded to four other water molecules.

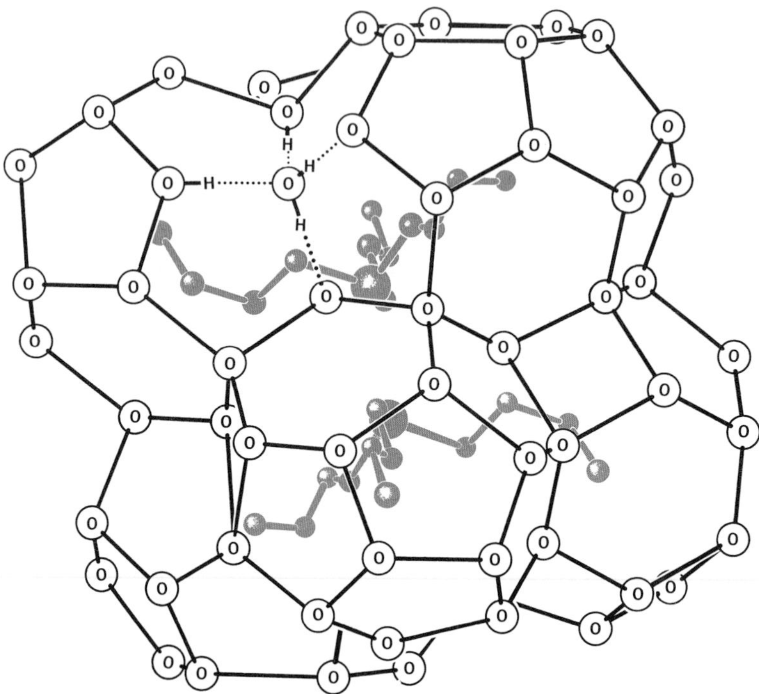

Figure 11-2. Water clathrate structure. The "caged" species are $(n\text{-}C_4H_9)_3S^+$ ions. (R. E. Dickerson and I. Geis, *The Structure and Action of Proteins,* W. A. Benjamin, Menlo Park, Cal., 1969. Copyright © 1969 by R. E. Dickerson and I. Geis.)

If the nonpolar groups cluster together, however, the water molecules are released to assume more random structures, causing the entropy of the system to increase. This leads to a decrease in the Gibbs free energy, or a free-energy stabilization, in accord with the all-important Gibbs equation:

$$\Delta G = \Delta H - T\Delta S \tag{11-1}$$

Kauzmann has calculated that, for every side chain listed above (except hydrogen) that is removed from an aqueous to a nonpolar environment, the free energy of the system decreases by 4 kcal/mole. Although this is small in comparison with the roughly 100-kcal/mole energies of covalent bonds, it does not mean that these hydrophobic bonds are unimportant in determining structures in aqueous solutions. There are large numbers of nonpolar groups in a protein, and large negative values of ΔG can result from their withdrawal from the aqueous exterior of the molecule and clustering together in the interior. This is called the *oil-drop* model of a protein, for obvious reasons. Note the similarity of this model and the model for the action of soaps and detergents shown in Figure 5-7.

The other important feature of hydrophobic bonds is that they are weak enough that they can be readily broken. This is a desirable feature if the structure of the protein must change slightly as it performs its functions of binding a substrate, a molecule of oxygen, etc.

Acidic Residues

Two amino acids—aspartic acid and glutamic acid—possess a carboxyl group in the side chain, in addition to the α-COO^- group. Such side chains are called *acidic residues,* and are shown in Table 11-5. The names of the parent compounds of these side chains represent a trivial name and a systematic name. Most carboxylic acids derived from hydrocarbons are named by adding the suffix "-oic acid" to the name of the hydrocarbon. Thus, the systematic name for acetic acid is ethanoic acid.

Carboxyl groups are quite reactive, as we know from Chapter 9. Because its pK_a value is generally about 4, the carboxyl group is essentially completely ionized in biological systems, in which the pH is about 7. The carboxylate ion can therefore form ionic bonds with groups such as $—NH_3^+$:

$$—COO^- \text{----} {}^+H_3N—$$

We know that such ionic bonds are quite strong and should therefore make significant contributions to the stability of protein molecules. In addition, the carboxylate ion can act as an electron-pair donor in the formation of hydrogen bonds. Because it is negatively charged, however, it cannot act

Table 11-5. Carboxylic Acid Side Chains of Amino Acids.

Amino Acid	Abbreviation	Structure	Parent Compound of Side Chain
Aspartic acid	asp	$^-OOC—C(NH_3^+)(H)—CH_2COOH$	Acetic acid
Glutamic acid	glu	$^-OOC—C(NH_3^+)(H)—CH_2CH_2COOH$	Propanoic acid

as an electron-pair acceptor. The resonance structures of this ion are shown on page 158.

Carboxylic acids undergo a wide variety of reactions, of which we will mention only two. One is the reaction with an alcohol to give an ester. Using the symbols R_1 and R_2 to denote any two organic groups, we can write the acid-catalyzed reaction of a carboxylic acid with an alcohol as follows:

$$\underset{\text{Acid}}{R_1-\overset{\overset{\displaystyle O}{\|}}{C}-OH} + \underset{\text{Alcohol}}{R_2-OH} \overset{H^+}{\rightleftharpoons} \underset{\text{Ester}}{R_1-\overset{\overset{\displaystyle O}{\|}}{C}-O-R_2} + H_3O^+ \tag{11-2}$$

If the acid were propanoic acid and the alcohol were ethanol, the resulting ester would be ethyl propanoate. Such esterification reactions are important in many metabolic pathways, especially when the acid is the inorganic acid H_3PO_4. As we will see, this reaction produces the nucleotides.

The second reaction we will mention is the reaction of a carboxylic acid with ammonia to form an amide. This reaction can be written as follows:

$$R-\overset{\overset{\displaystyle O}{\|}}{C}-OH + NH_3 \rightarrow R-\overset{\overset{\displaystyle O}{\|}}{C}-O^- + NH_4^+ \xrightarrow{\text{Heat}} R-\overset{\overset{\displaystyle O}{\|}}{C}-NH_2 + H_2O \tag{11-3}$$

Thus, the two amino acids with acidic residues can be converted to their respective amides, and vice versa. This interchange confers quite different properties on the protein because the amide is incapable of ionization.

Basic Residues

The three amino acids that have basic residues as side chains are histidine, lysine, and arginine. They are shown in Table 11-6. We will consider the chemistry only of lysine, because it represents a large class of organic compounds, the amines. All amines can be thought of as derivatives of ammonia, NH_3. They are usually named, however, as derivatives of the parent organic compound with "-amine" as a suffix. Thus, $CH_3CH_2NH_2$ is ethylamine.

The amines are classified according to the number of substituent groups on the nitrogen atom:

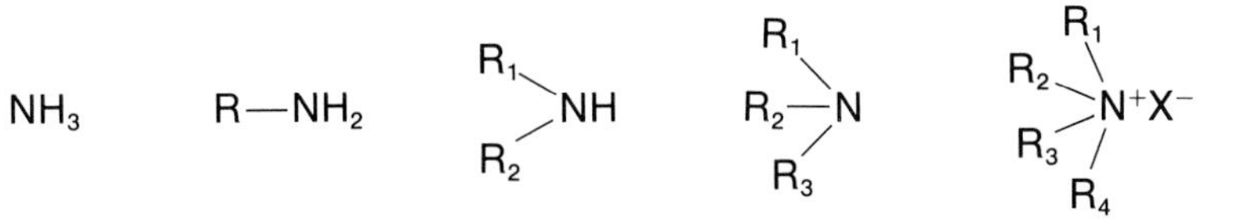

NH_3	$R—NH_2$	R_1R_2NH	$R_1R_2R_3N$	$R_1R_2R_3R_4N^+X^-$
Ammonia	Primary amine	Secondary amine	Tertiary amine	Quaternary ammonium salt

This classification is significant because the different amines have very different chemical reactivities. The ε-NH_2 group of lysine is a primary amine. The α-nitrogen of proline can be considered to be a secondary amine because only one hydrogen is bonded to it.

The acid-base properties of histidine were discussed in Chapter 9. The pK_a values of the three basic amino acids given in Table 11-2 suggest that lysine and arginine will be protonated under most physiological conditions. The amino and guanidino residues therefore carry positive charges and are important in forming ionic bonds with carboxylate ions in proteins and in determining the *isoelectric point* of the protein. This is the pH at which the protein molecule does not move when an electric field is placed across the solution, because there is no net charge on the protein. Histidine, on the other hand, has a pK_a value close to the physiological pH and is therefore important in determining the buffering properties of proteins.

Table 11-6. Basic Side Chains of Amino Acids.

Amino Acid	Abbreviation	Structure	Side Chain
Histidine	his	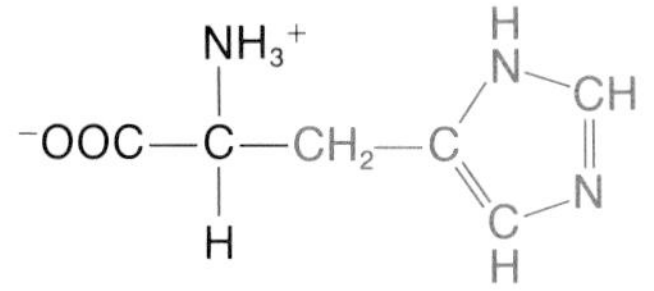 $^-OOC—C(NH_3^+)(H)—CH_2—C$ (ring: C—N(H)—CH=N—C(H)=)	Imidazolyl
Lysine	lys	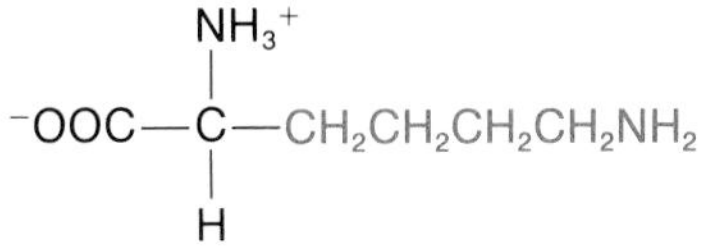 $^-OOC—C(NH_3^+)(H)—CH_2CH_2CH_2CH_2NH_2$	*n*-Butylamino
Arginine	arg	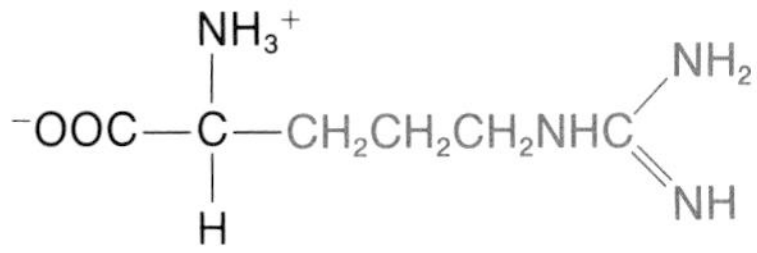 $^-OOC—C(NH_3^+)(H)—CH_2CH_2CH_2NHC(NH_2)=NH$	Guanidino

Amines undergo a wide variety of chemical reactions, including alkylation, arylation, and oxidation. All α-amino acids undergo reaction with ninhydrin in aqueous solution to give a blue pigment, except for proline, which gives a yellow pigment.

$$2\ \text{(Ninhydrin)} + H{-}\overset{\displaystyle R}{\underset{\displaystyle NH_3^+}{C}}{-}COO^- \longrightarrow$$

Ninhydrin α-Amino acid

$$\text{(Blue pigment)} + RCHO + CO_2 + 3H_2O \qquad (11\text{-}4)$$

Blue pigment Aldehyde

This reaction, which is specific for α-amino acids, is very useful for the quantitative measurement of amino acids. For instance, it is used to measure the amount of each amino acid as it emerges from an ion-exchange column. Carbon dioxide is released quantitatively and could be used to monitor the reaction, but the spectroscopic measurement of the pigment is simpler.

Like the carboxylate ion, amino groups can act as electron-pair donors in the formation of hydrogen bonds, but they can also act as electron-pair acceptors.

Polar but Neutral Side Chains

The amino acids in this group are the most heterogeneous we will consider. Their side-chain structures are shown in Table 11-7. The fundamental similarity between these six side chains is that they are neutral at physiological pH, yet they all possess oxygen and/or nitrogen atoms. These electronegative atoms can form hydrogen bonds or dipole-dipole bonds with similar groups.

We will restrict our attention primarily to the properties of alcohols. Alcohols are usually named by adding the suffix "-ol" to the parent compound. Two common examples are ethanol and isopropanol. Ethanol is

Table 11-7. **Polar but Neutral Side Chains of Amino Acids.**

Amino Acid	Abbreviation	Structure	Parent Compound of Side Chain
Serine	ser	$^{-}OOC—C(NH_3^+)(H)—CH_2OH$	Methanol
Threonine	thr	$^{-}OOC—C(NH_3^+)(H)—CH(OH)CH_3$	Ethanol
Asparagine	asn	$^{-}OOC—C(NH_3^+)(H)—CH_2—C(=O)—NH_2$	Acetamide
Glutamine	gln	$^{-}OOC—C(NH_3^+)(H)—CH_2CH_2—C(=O)—NH_2$	Propionamide
Tryptophan	trp	$^{-}OOC—C(NH_3^+)(H)—CH_2$—(indol-3-yl, N—H)	Indole
Tyrosine	tyr	$^{-}OOC—C(NH_3^+)(H)—CH_2—C_6H_4—OH$	*p*-Hydroxy-toluene

produced by the millions of gallons, both industrially and in the fermentation of a variety of sugars and polysaccharides. (See the Interlude on Wine-making at the end of this chapter.) Its properties are well-known to everyone. Isopropanol is used in a great variety of industrial processes and products, including rubbing alcohol.

Like the amines, alcohols are classified as primary, secondary, and tertiary, depending on the number of substituents attached to the carbon atom bearing the OH group. These three classes of alcohols display unique reactivities in the large class of organic reactions called *nucleophilic substitutions,* which we encountered briefly in Chapter 8. Nucleophilic substitution reactions are of great importance in organic chemistry because of their tremendous diversity. Bonds between a C atom and N, O, P, S, or any halogen atom can be produced by these reactions. An example is the substitution of a Cl atom for the OH group in methanol, giving methyl chloride:

$$CH_3OH + HCl \rightarrow CH_3Cl + H_2O \tag{11-5}$$

The second large class of reactions in which alcohols participate is *elimination* reactions, an example of which is the acid-catalyzed dehydration of ethanol, giving ethylene:

$$CH_3CH_2OH \xrightarrow[\text{Heat}]{H_2SO_4} CH_2{=}CH_2 + H_2O \tag{11-6}$$

Aliphatic alcohols (those derived from alkanes) have pK_a values less than the pH of water and hence do not ionize in water. On the other hand, aromatic alcohols (those derived from aromatic hydrocarbons) have pK_a values of about 10. Tyrosine has a pK_3 of 9.5, corresponding to the ionization of its OH group, which is on the benzene ring. Although this value is so high that tyrosine does not ordinarily ionize in biological systems, the acid-base properties of tyrosine have been investigated extensively to determine its location in proteins and to establish whether or not the H atom of the OH group participates in hydrogen bonding. (It does in some proteins.)

The parent compounds of the other polar but neutral side chains are indole and two amides. The special properties of tryptophan are its strong ultraviolet absorption at 280 nm and the fact that it is destroyed by acid hydrolysis. Other methods than the usual HCl hydrolysis employed in ion-exchange chromatography must be used to detect tryptophan. Asparagine and glutamine are probably produced after aspartic acid and glutamic acid are incorporated into proteins during protein synthesis *in vivo.* They are important hydrogen-bonding sites because of the unshared electron pairs on both the O and N atoms of the amide group and because of the hydrogens on the N atom.

Sulfur-Containing Side Chains

Amino acids and, hence, proteins consist primarily of C, H, O, and N. But this is also true of the nucleotides and nucleic acids. The one unique element found in proteins is sulfur. It is found in two amino acids, cysteine and methionine, which are shown in Table 11-8.

Organic sulfur compounds are often named as derivatives of the corresponding oxygen compounds, with the term "thio" used to denote the sulfur

Table 11-8. Sulfur-Containing Side Chains of Amino Acids.

Amino Acid	Abbreviation	Structure	Parent Compound of Side Chain
Cysteine	cys	$^{-}OOC{-}C(NH_3^+)(H){-}CH_2SH$	Methyl mercaptan
Methionine	met	$^{-}OOC{-}C(NH_3^+)(H){-}CH_2CH_2SCH_3$	Methylethyl sulfide

atom. For example, mercaptans are often called *thiols,* meaning thioalcohols. Methionine is a thio analog of an ether. Ethers and thioethers are quite unreactive, so methionine does not contribute any distinctive properties to proteins except by functioning primarily as a hydrophobic group.

Cysteine, on the other hand, is very reactive and is one of the most important amino acids in determining the structures of protein molecules. The SH group can be fairly easily oxidized. If oxidation occurs in the presence of another SH group (typically, one from another cysteine molecule), an S—S bond can result:

$$2R{-}SH \xrightarrow{\text{Oxidation}} R{-}S{-}S{-}R + 2H \qquad (11\text{-}7)$$

The importance of this reaction in proteins is that it is responsible for the formation of covalent cross-links between two adjacent chains. When this happens, the resulting linked pair of cysteine molecules is called cystine (note the subtle difference in spelling):

$$\underset{\text{Cysteine}}{\overset{\text{Cysteine}}{\begin{array}{c} | \\ CH_2 \\ | \\ SH \\ \\ SH \\ | \\ CH_2 \\ | \end{array}}} \xrightarrow{\text{Oxidation}} \underset{\text{Cystine}}{\begin{array}{c} | \\ CH_2 \\ | \\ S \\ | \\ S \\ | \\ CH_2 \\ | \end{array}} + 2H \qquad (11\text{-}8)$$

This places a severe constraint on the three-dimensional structure of the protein. It is the only known covalent bond between chains in proteins.

11.2 THE MONOSACCHARIDES

Monosaccharides are the monomers from which *polysaccharides*, or carbohydrates, are formed. Carbohydrates are the most abundant organic compounds on earth. It is estimated that the annual production of carbohydrates in nature is 10^{11} tons. Green plants are responsible for virtually all of this production, which occurs by the process of *photosynthesis*. The thermodynamics of this process is somewhat surprising:

$$x\mathrm{CO_2} + x\mathrm{H_2O} \xrightarrow[\text{Green plants}]{\text{Light}} \underset{\text{Carbohydrate}}{(\mathrm{CH_2O})_x} + x\mathrm{O_2} \qquad (11\text{-}9)$$

$$\Delta G° = +686 \text{ kcal/mole } \mathrm{C_6H_{12}O_6}$$

$$\Delta H° = +673 \text{ kcal/mole } \mathrm{C_6H_{12}O_6}$$

$$\Delta S° = -43.6 \text{ cal/mole } \mathrm{C_6H_{12}O_6}$$

The wavelength of light required to produce this large an increase in $\Delta G°$ would have to be 42 nm.

Exercise 11-1. Calculate the energy of one mole of photons (one einstein) of wavelength 42 nm. Use the equation $E = N_A h\nu$ and express your answer in kcal/einstein.

Because virtually no radiation of this very short wavelength reaches the surface of the Earth, it is apparent that the energy must arrive in smaller packets (i.e., as longer-wavelength photons) and be stored in a sequence of interrelated chemical reactions. Scientists today are actively working at unraveling the last details of this sequence. The first steps appear to be the reduction of CO_2 to a compound similar to D-glyceric acid, the oxidized form of D-glyceraldehyde. The latter is the parent compound of the largest class of monosaccharides, the *aldoses*. Dihydroxyacetone is the parent compound of the *ketoses*. These names are derived from the fact that the compounds are polyhydroxyaldehydes and polyhydroxyketones, respectively. The structural formulas of the parent compounds are shown below:

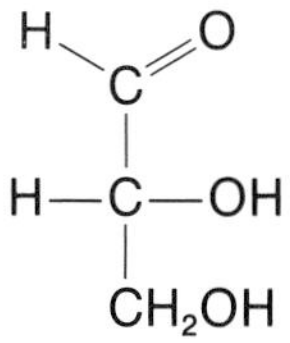

```
H     O
 \   //
   C
   |
H—C—OH
   |
  CH2OH
```
D-Glyceraldehyde
(an aldose)

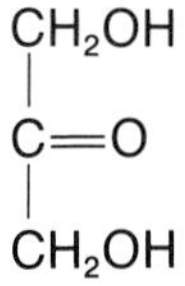

```
CH2OH
 |
C=O
 |
CH2OH
```
Dihydroxyacetone
(a ketose)

Of these two compounds, both of which are trioses (having three C atoms), only D-glyceraldehyde is optically active.

In contrast to the naturally occurring amino acids in proteins, all of which have the L-configuration, most naturally occurring monosaccharides are structurally related to D-glyceraldehyde and are hence called D-monosaccharides. The absolute configuration of D-glyceraldehyde is shown below:

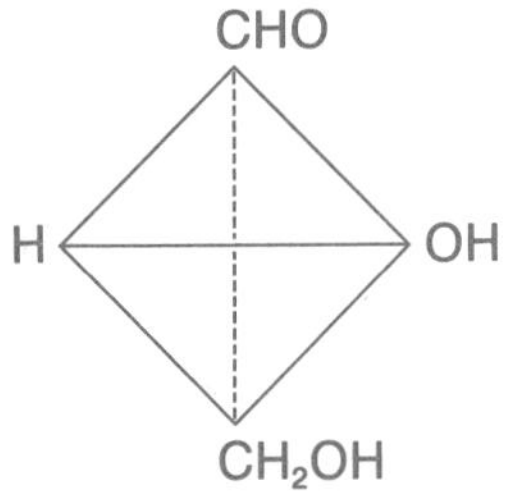

CHO

H OH

CH_2OH

D-(+)-Glyceraldehyde

D-Glyceraldehyde is only one of dozens of known monosaccharides. Additional members of this class of compounds can be thought of as derivatives of the trioses D-glyceraldehyde and dihydroxyacetone by the insertion of additional H—C—OH groups (with vertical bonds above and below the C). Higher members of this series are called tetroses, pentoses, hexoses, and heptoses. We will look at the structures and properties of four representative pentoses and hexoses.

A number of questions relating to the structure and nomenclature of the monosaccharides remain to be examined. It will be helpful to look at one compound, glucose, to demonstrate certain conventions. Glucose is the monomer from which cellulose, the principal constituent of all plants, is constructed, and it represents over one-half of all the organic matter on earth. D-Glucose is an aldohexose, a six-carbon derivative of D-glyceraldehyde. The configuration of the hydrogen and hydroxyl groups about each carbon atom and the numbering of the carbon chain are shown in the Fischer projection formulas below (the asymmetric centers are denoted by asterisks):

```
   H   1  O              H     O
     \ C//                 \ C//
       |                     |
 H —2C*— OH            HO — C* — H
       |                     |
HO —3C*— H              H — C* — OH
       |                     |
 H —4C*— OH            HO — C* — H
       |                     |
 H —5C*— OH            HO — C* — H
       |                     |
     6CH2OH                CH2OH
   D-Glucose             L-Glucose
```

Since there are two possible configurations about each of the four asymmetric centers, there are $2^4 = 16$ optical isomers of D-glucose. One of them is L-glucose, the mirror image of D-glucose. These two compounds form an enantiomorphic pair and are identical in chemical and most physical properties except for the opposite rotations of plane-polarized light. There are eight such enantiomorphic pairs in all, each pair having quite different chemical and physical properties.

The properties and some reactions of the carbonyl group found in aldehydes and ketones were discussed in Chapter 5. The aldehyde portion of D-glucose can be oxidized to a carboxylic acid or reduced to an alcohol. However, it does not give a bisulfite addition compound (see page 207) because D-glucose exists in solution almost exclusively in the form of two cyclic hemiacetals. The spontaneous interconversion of these two cyclic forms of D-glucose via the open-chain form is shown by the following Fischer projection formulas:

```
     ┌──────┐               H     O               ┌──────┐
 H —1C— OH  │                 \ C//          HO — C — H  │
     |      │                   |                 |      │
 H —2C— OH  │               H — C — OH        H — C — OH │
     |      O                   |                 |      O
HO —3C— H   │   ⇌          HO — C — H    ⇌   HO — C — H  │
     |      │                   |                 |      │
 H —4C— OH  │               H — C — OH        H — C — OH │
     |      │                   |                 |      │
 H —5C──────┘               H — C — OH        H — C──────┘
     |                          |                 |
   6CH2OH                     CH2OH             CH2OH
 α-D-Glucose                D-Glucose         β-D-Glucose
 [α]20D = +112°                               [α]20D = +19°
```

The symbol $[\alpha]_D^{20}$ denotes the *specific rotation* of the compound (i.e., the optical rotation under certain standard conditions) at 20°C and for radiation of wavelength corresponding to the D-line of the sodium-vapor emission spectrum. (The sodium D-line is actually a pair of very closely spaced lines in the yellow region of the spectrum.)

Because of the interconversion (or *mutarotation*) shown above, a small fraction of D-glucose does exist in the open-chain form in solution, giving rise to some of the typical reactions of aldehydes. Most of the chemistry of D-glucose, however, is determined by the predominance of the ring structures. When the ring closes between carbon atoms 1 and 5, a new asymmetric center is created at carbon atom 1. If the OH group is on the right side of this atom in the Fischer formula, the compound is said to have the α-configuration; if it is on the left side, the compound has the β-configuration. Because all the aldohexoses undergo this cyclization reaction, there are always five asymmetric centers, one of which is capable of rearrangement. The total number of optical isomers of each aldohexose therefore becomes $2^5 = 32$. If pure α-D-glucose is dissolved in water, its specific rotation is +112°. However, this value changes slowly until the value +52.5° is reached, indicating that the equilibrium mixture consists of 36% of the α-isomer and 64% of the β-isomer.

The Fischer projection formulas for the cyclic forms of monosaccharides are easy to draw but admittedly unrealistic. A much more realistic representation, which still gives absolute configurations, is that of the *Haworth projection formulas,* named for the English chemist Sir Walter Haworth (1883–1950). The Haworth projection formulas for the cyclic forms of D-glucose are shown below:

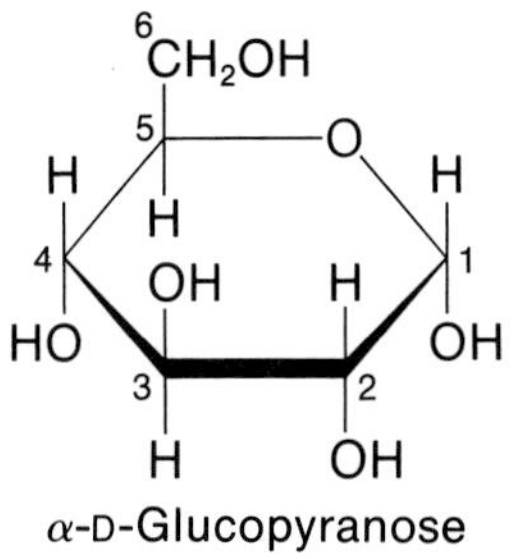

α-D-Glucopyranose
(α-D-glucose)

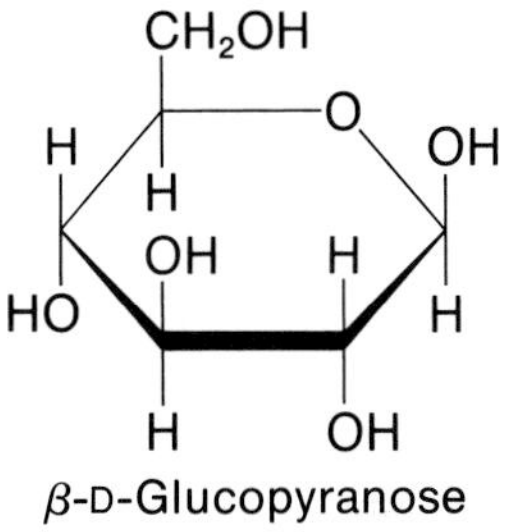

β-D-Glucopyranose
(β-D-glucose)

Note that we have introduced a new name here: glucopyranose. This indicates that the compounds in question can be considered to be derivatives of the six-membered heterocyclic compound* pyran. Many other

*A *heterocyclic compound* is an organic ring compound in which at least one of the atoms in the ring is not a C atom. Typically, it is N, O, or S.

cyclic monosaccharides are called furanoses because they can be considered to be derivatives of the five-membered heterocyclic compound furan. The structures of pyran and furan are shown below, and the structures of three furanoses (one hexose and two pentoses) are shown in Table 11-9.

Pyran Furan

Table 11-9. The Structures of Three Common Monosaccharides.

Name	Open-Chain Structure	Fischer Structure	Haworth Structure
D-Fructose	CH_2OH C=O HO—C—H H—C—OH H—C—OH CH_2OH	CH_2OH HO—C— HO—C—H H—C—OH O H—C— CH_2OH	6 $HOCH_2$ O OH 5 H HO 2 H CH_2OH 4 3 1 HO H β-D-Fructofuranose
D-Ribose	H O C H—C—OH H—C—OH H—C—OH CH_2OH	HO—C—H H—C—OH O H—C—OH H—C— CH_2OH	5 $HOCH_2$ O OH 4 H H 1 H H 3 2 HO OH β-D-Ribofuranose
D-2-Deoxyribose	H O C H—C—H H—C—OH H—C—OH CH_2OH	HO—C—H H—C—H O H—C—OH H—C— CH_2OH	5 $HOCH_2$ O OH 4 H H 1 H H 3 2 HO H β-D-2-Deoxyribofuranose

D-Fructose is found in honey and fruits. Because of the sweet taste of many monosaccharides, they and compounds that contain them are often called *sugars*. The best-known sugar, sucrose, is a compound consisting of one glucose molecule and one fructose molecule joined by a covalent bond. Sucrose is an outstanding source of metabolic "quick energy."

The two pentoses shown in Table 11-9 are of great biological importance, and we will see them again shortly. They are an integral part of the backbone structures of the two kinds of nucleic acids. In fact, the two kinds of nucleic acids are distinguished on the basis of which kind of sugar—ribose or deoxyribose—they contain. Note the numbering of the ring system and the subtle difference between the two compounds. They differ only at the C-2 position, the one compound having H instead of OH (hence, the prefix 2-deoxy).

The cyclic forms of the monosaccharides show the R—O—R structure characteristic of ethers. Here, of course, the two R groups are actually joined to form a single ring structure. Ethers were introduced along with the other major classes of organic compounds in Chapter 5. They are not very reactive and are used in the laboratory mainly as solvents.

Because of the aldehyde and ketone moieties in the open-chain forms of the monosaccharides, plus the large numbers of polar hydroxyl groups, these molecules undergo a wide variety of chemical reactions. We will examine two reactions of particular importance. One is the reaction of an alcohol with the hydroxyl group of the hemiacetal carbon (C-1 in the ribofuranoses) in dilute acid, giving a *glycoside*:

$$\underset{\text{Hemiacetal}}{\text{R—OH} + \text{HO—C—H} \;\; (\text{—C—} \;\; \text{O ring})} \xrightarrow{H^+} \underset{\beta\text{-O-Glycoside}}{\text{R—O—C—H} \;\; (\text{—C—} \;\; \text{O ring})} + H_2O \quad (11\text{-}10)$$

The reaction leads to an α- or β-glycoside, which cannot mutarotate. This is an important fact, the consequences of which we will see later on. If the original sugar is glucose, the resulting molecule is a glucoside. If the R—OH compound happens to be another sugar, the resulting molecule is a disaccharide, e.g., sucrose.

Glycosides such as that shown above are sometimes called O-glycosides to emphasize that the linkage between the compounds is through an oxygen atom. The important class of compounds to be discussed shortly, the nucleosides, are N-glycosides. The hydroxyl group attached to C-1 of a ribofuranose or deoxyribofuranose reacts with a heterocyclic organic base to form an N-glycosidic linkage:

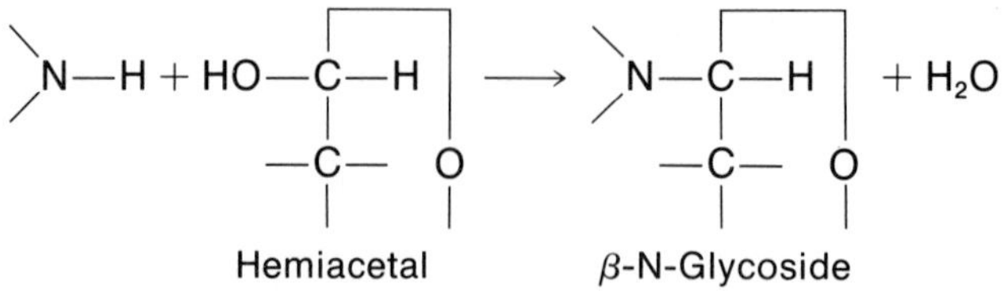

$$\begin{array}{c} \backslash \\ \text{N—H} + \text{HO—C—H} \\ / \quad\quad\quad\quad | \\ \quad\quad\quad \text{—C—}\quad \text{O} \end{array} \longrightarrow \begin{array}{c} \backslash \\ \text{N—C—H} \\ / \quad | \\ \text{—C—}\quad \text{O} \end{array} + H_2O \qquad (11\text{-}11)$$

Hemiacetal $\qquad$ β-N-Glycoside

The other reaction of biological significance in which monosaccharides participate is ester formation. The most important esters are those formed with phosphoric acid, H_3PO_4, which has pK_a values of 2, 7, and 13. Thus the predominant form of this acid at pH 7 is HPO_4^{2-}, the monohydrogen-phosphate ion:

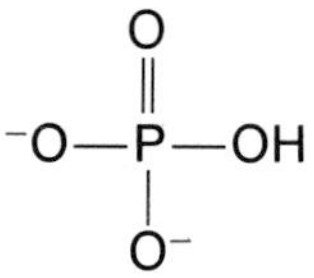

$$^{-}O-\overset{\overset{\displaystyle O}{\|}}{\underset{\underset{\displaystyle O^{-}}{|}}{P}}-OH$$

The formation of the phosphate ester of a monosaccharide is illustrated by the following reaction:

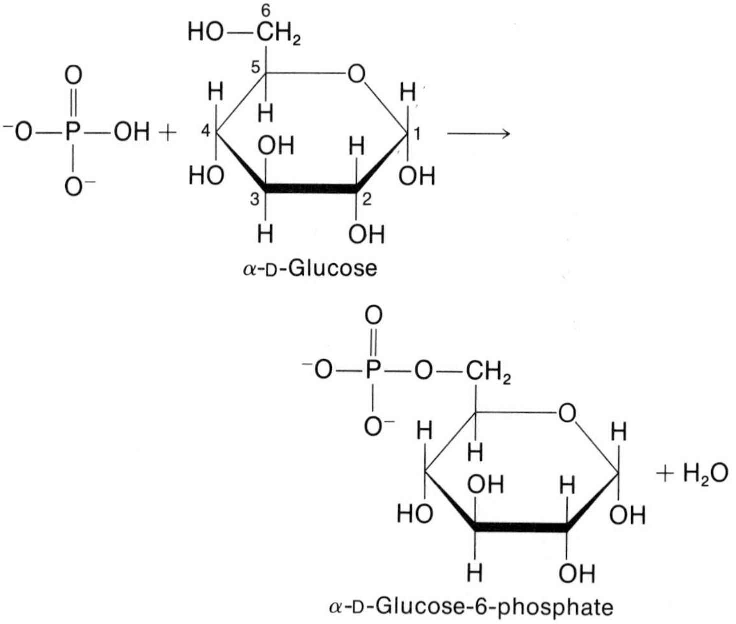

$$^{-}O-\overset{\overset{\displaystyle O}{\|}}{\underset{\underset{\displaystyle O^{-}}{|}}{P}}-OH + \alpha\text{-D-Glucose} \longrightarrow \alpha\text{-D-Glucose-6-phosphate} + H_2O \qquad (11\text{-}12)$$

α-D-Glucose

α-D-Glucose-6-phosphate

The phosphate esterification of nucleosides gives nucleotides, which are the monomers of nucleic acids.

11.3 THE NUCLEOTIDES

Nucleotides are found in nature primarily in the giant polymers called *nucleic acids*. There are two kinds of nucleic acids found in all cells: *ribonucleic acid* (RNA) and *deoxyribonucleic acid* (DNA). The bulk of the DNA is found in the nucleus of the cell. However, significant amounts (of the order of several percent) have been found in other cellular organelles, such as the mitochondria. The RNA is synthesized in the nucleus but performs its function outside the nucleus, in the cytoplasm. It is found in and on small globular particles called *ribosomes*, which are attached to the endoplasmic reticulum. The structure of a typical cell is shown in Figure 11-3.

The term *nucleic acid* derives from the original discovery of DNA in the nuclei of cells and the fact that both DNA and RNA are rather strong acids because of the presence of the phosphate group. The nucleotide monomers of nucleic acids consist of the phosphate ester of a pentose sugar (ribose or deoxyribose) that is bound through an N-glycosidic linkage to one of a number of heterocyclic, nitrogen-containing organic bases. These bases are all derived from the parent compounds purine and pyrimidine and are hence referred to collectively as purines and pyrimidines. The relations between these several classes of compounds are shown below:

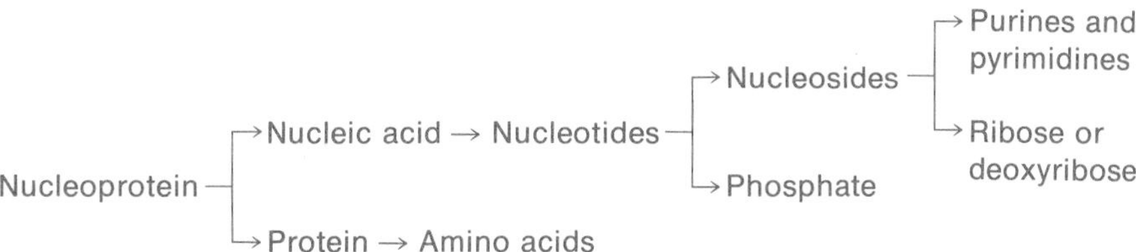

Let us now look at the five purines and pyrimidines that constitute the great majority of the bases found in nucleic acids. Their structures, plus those of the two parent compounds, are shown in Table 11-10. The H atoms shown in color denote the positions of the N-glycosidic linkage to the pentose: at the N-9 atom in the purines and the N-1 atom in the pyrimidines. Take a close look at the groups to the left of the colored diagonal lines. We will soon see how the particular properties of those groups are responsible in a fundamental way for the geometry of the master molecule of life, DNA.

The bases are fairly insoluble in aqueous solutions, probably because of the rather large amount of aromatic character of the purine and pyrimidine rings. The bases can assume charges, however, because they possess several weak-acid and weak-base groups. The protonated form of the amino groups of adenine, guanine, and cytosine has pK_a values of about 4, so those groups are neutral at physiological pH. The NH groups at N-1 of guanine and at N-3 of uracil and thymine have pK_a values of about 9.5.

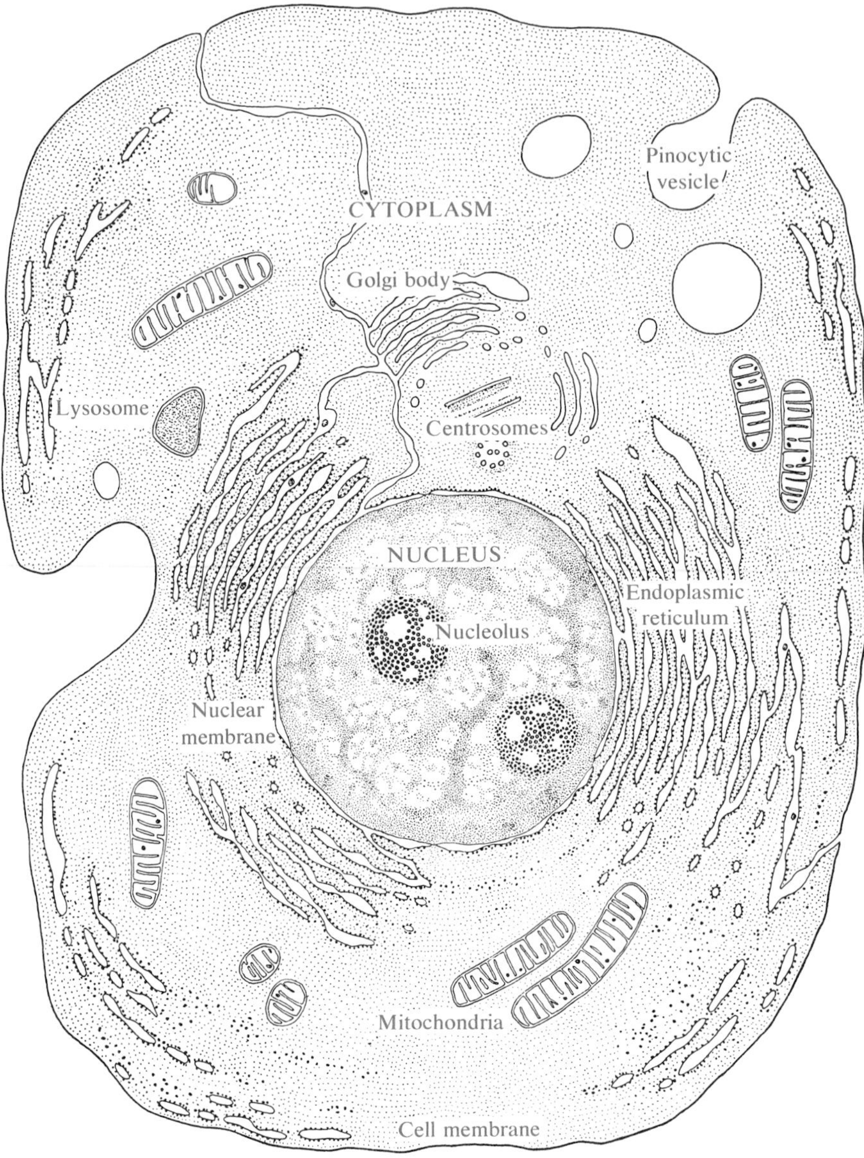

Figure 11-3. Diagram of a typical cell. The mitochondria are the sites of the oxidative reactions that provide the cell with energy. The dots that line the endoplasmic reticulum are ribosomes, the sites of protein synthesis. (J. Brachet, "The Living Cell." Copyright © 1961 by Scientific American, Inc. All rights reserved.)

Table 11-10. **The Organic Bases Most Commonly Found in Nucleic Acids.**

Parent Compound	Base		
Purine	Adenine (A)	Guanine (G)	
Pyrimidine	Cytosine (C)	Uracil (U)	Thymine (T)

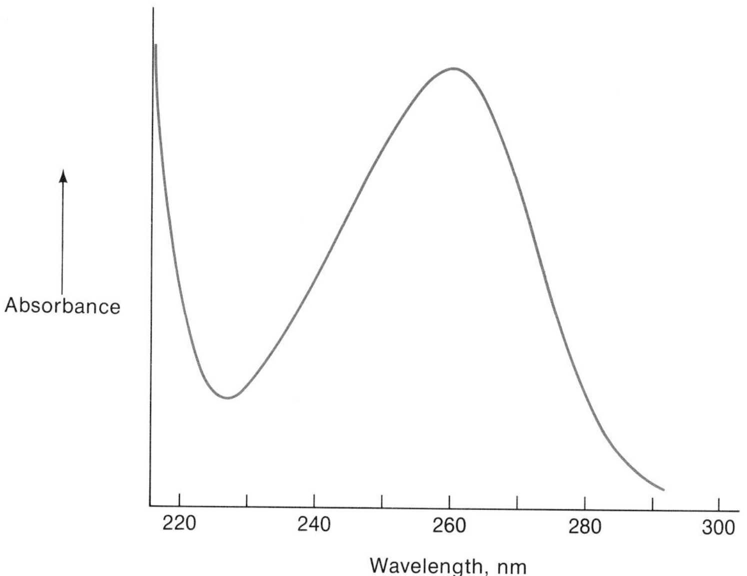

Figure 11-4. The ultraviolet absorption spectrum of adenosine 5′-monophosphate.

The bases also display characteristic absorption spectra in the ultraviolet. Figure 11-4 shows the spectrum of adenosine 5′-monophosphate, a nucleotide containing adenine. Most nucleotides have spectra similar to this, with a maximum at about 260 nm. This contrasts with two of the three aromatic amino acids (see Figure 11-1), which have maxima at about 280 nm, a difference that is used to good advantage for analytical purposes. The absorption spectra of the nucleotides are also highly pH-dependent, indicating the existence of ionizable groups.

Let us now examine the structures of some nucleotides. We begin with the nucleosides, which, as we saw earlier (Reaction 11-11), result from the formation of an N-glycosidic linkage between an organic base and a sugar, e.g., adenine and deoxyribose:

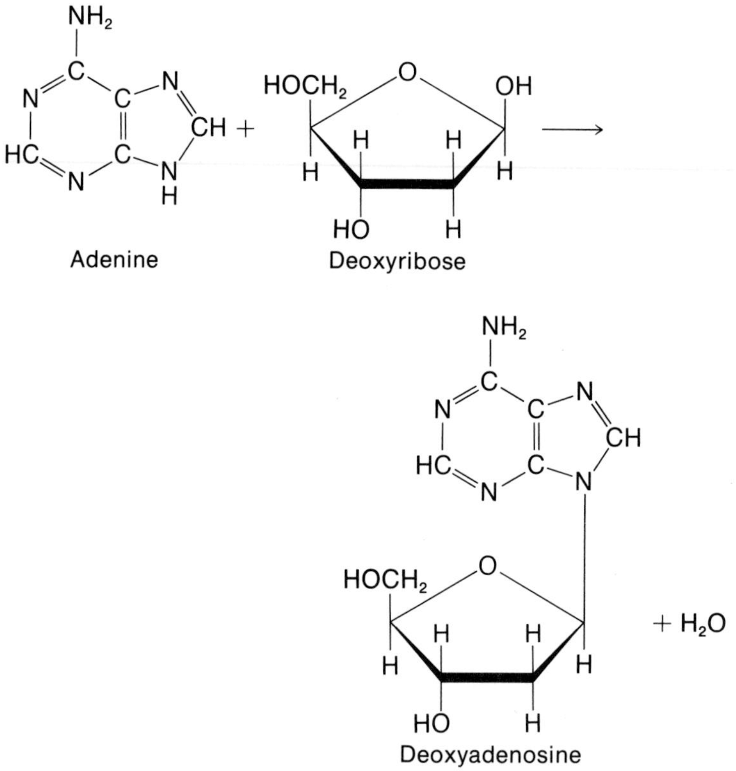

Adenine + Deoxyribose ⟶

Deoxyadenosine $+ H_2O$ (11-13)

Note that we labeled the sugar deoxyribose instead of β-D-2-deoxyribofuranose. This is because the sugars in all the naturally occurring nucleosides and nucleotides have the properties β, D, and furanose in common, and the property deoxy, when it applies, always applies to the C-2 position on the ring. The structures of four more nucleosides are shown below:

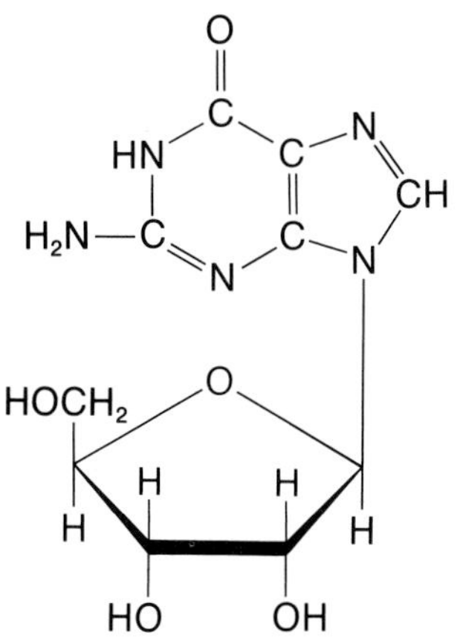

Guanosine

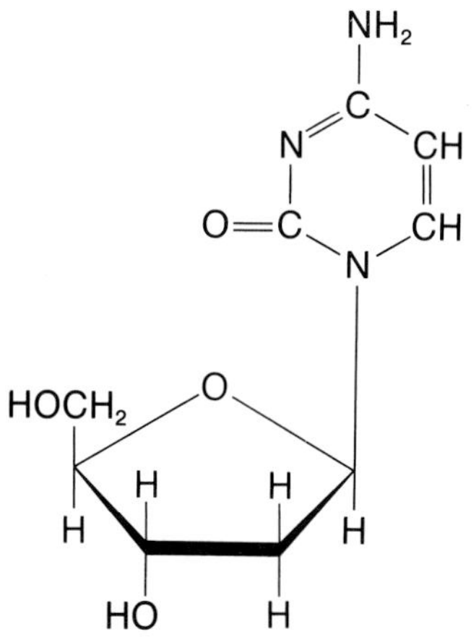

Deoxycytidine

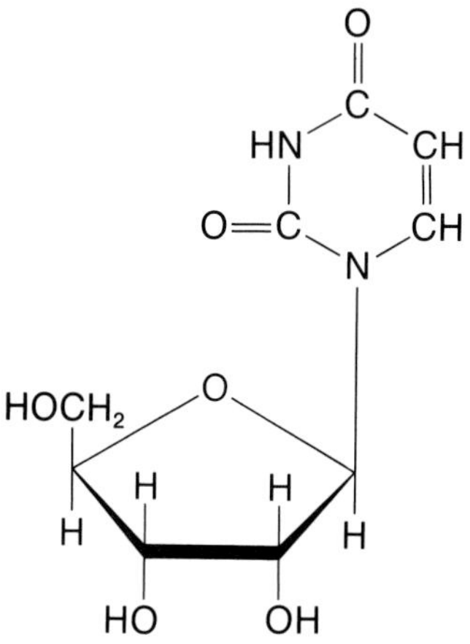

Uridine

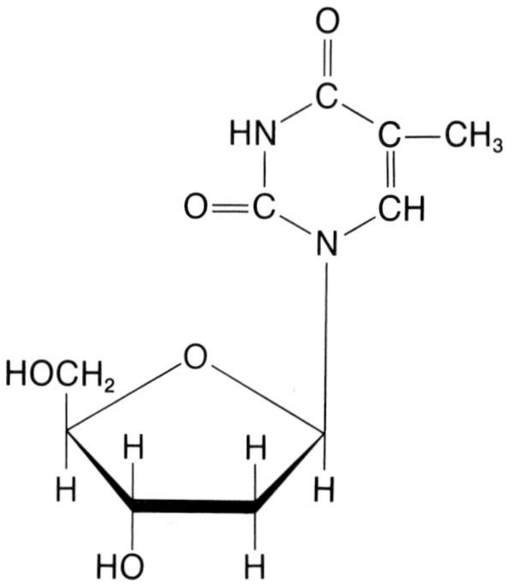

Deoxythymidine

The names of the nucleosides are seen to be derivatives of the base name. It is not important that you memorize such names, but it is useful to be familiar with all the basic constituents of the nucleosides. There are only seven: the five bases and the two sugars.

Only one further step is now required to produce a nucleotide. This is to esterify the sugar with phosphoric acid. Three sites are available on the ribofuranose ring and two on the deoxyribofuranose ring. Because there are two separate ring systems in a nucleoside, it is customary to add prime signs to the numbers of the carbon atoms of the sugar ring, to avoid confusion. The formation of deoxyadenosine 5′-monophosphate (dAMP) can be written as follows:

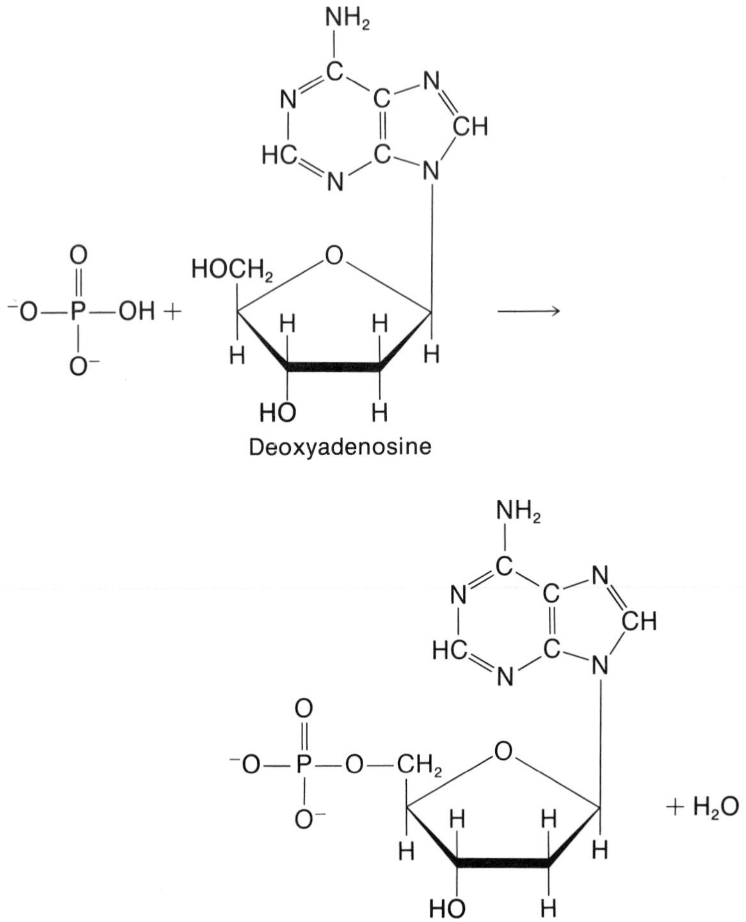

$$^{-}O{-}\overset{\overset{\displaystyle O}{\|}}{\underset{\underset{\displaystyle O^-}{|}}{P}}{-}OH + \text{Deoxyadenosine} \longrightarrow \text{Deoxyadenosine 5'-monophosphate (dAMP)} + H_2O \qquad (11\text{-}14)$$

Deoxyadenosine

Deoxyadenosine
5′-monophosphate
(dAMP)

Three other types of structures commonly found in nucleotides are shown on the opposite page. (In addition to monophosphates and diphosphates, triphosphates also exist. We will encounter one such triphosphate shortly.) Note that the phosphate group is usually attached at either the 3′- or 5′-position in naturally occurring nucleic acids, but it can also be attached at the 2′-position, either in the normal way or as the interesting 2′,3′-cyclic monophosphate.

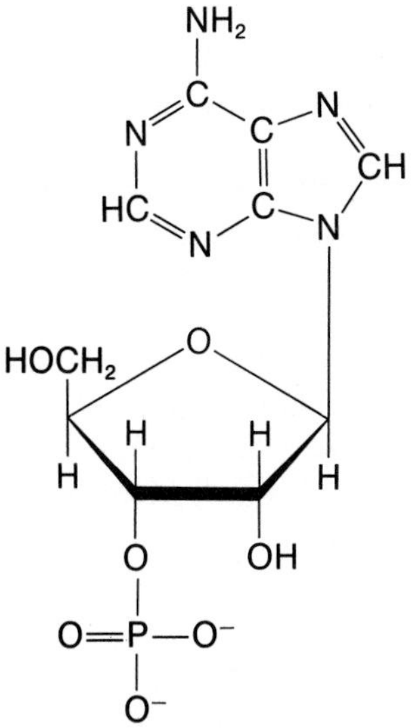

Adenosine 3′-
monophosphate
(AMP)

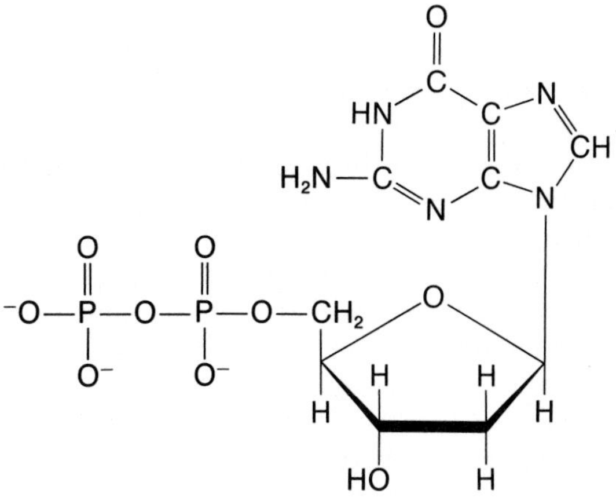

Deoxyguanosine
5′-diphosphate
(dGDP)

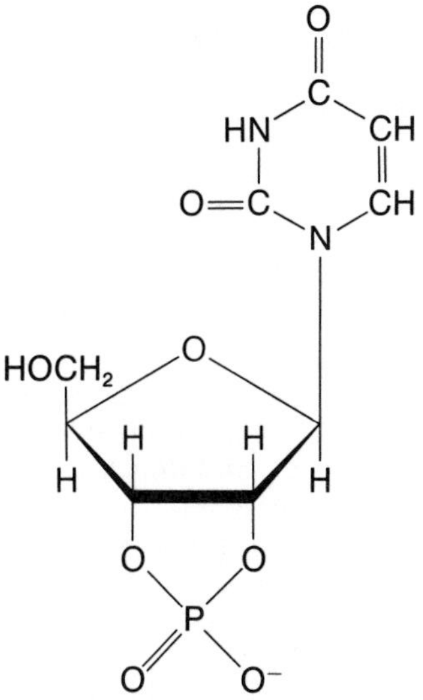

Uridine 2′, 3′-
cyclic monophosphate
(cyclic UMP)

The compound dGDP, shown above, is representative of a large class of extremely important compounds, the 5′-diphosphate and 5′-triphosphate nucleotides. They are called *high-energy compounds* because of the great decrease in Gibbs free energy that occurs upon cleavage of one or two phosphate groups from the chain. Because of the cumulative charges on the phosphate groups, it is not surprising that the molecule becomes more stable when such cleavage occurs. The importance of these compounds and their free energies of reaction is that biological systems can store energy in the phosphate bonds when an energy source is available, and later retrieve this energy upon demand. The most important high-energy compound is adenosine 5′-triphosphate (ATP), which was mentioned in Chapter 10 in connection with the Krebs cycle. Its structure is shown below, with the high-energy bonds shown as wavy lines:*

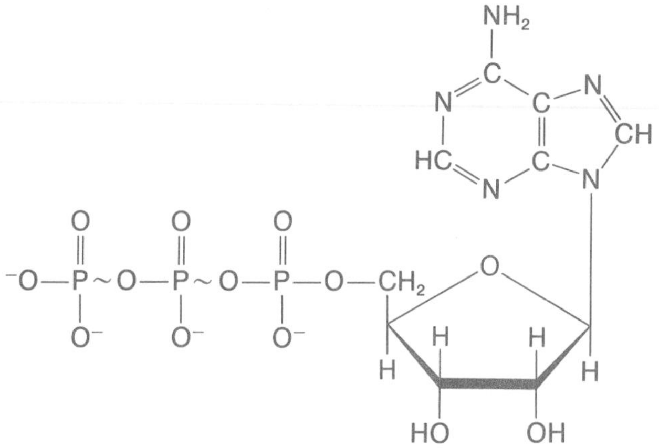

Adenosine 5′-triphosphate
(ATP)

Example 11-2. Show that it is thermodynamically possible for the esterification of an acid ($\Delta G^\circ = +4$ kcal/mole) to occur through a coupled mechanism with the hydrolysis of ATP ($\Delta G^\circ = -6$ kcal/mole).

The two reactions and their ΔG° values can be added (the symbol PP_i denotes the inorganic diphosphate group):

*It is an oddity of the history of science that the mere introduction of this vivid symbol for a high-energy bond, and especially the term itself, did much to stimulate the interest of scientists in the relatively new field of *bioenergetics*. The credit for this belongs to the great German-American biochemist Fritz Lipmann (b. 1899), who shared the Nobel Prize with Krebs in 1953.

$$\begin{array}{lr} \text{R—OH} + \text{HOOC—R}' \longrightarrow \text{R—O—}\overset{\displaystyle \text{O}}{\overset{\|}{\text{C}}}\text{—R}' + \text{H}_2\text{O} & \Delta G^\circ = +4\ \text{kcal/mole} \\ \text{ATP} + \text{H}_2\text{O} \longrightarrow \text{AMP} + \text{PP}_i & \Delta G^\circ = -6\ \text{kcal/mole} \\ \hline \text{R—OH} + \text{HOOC—R}' + \text{ATP} \longrightarrow \text{R—O—}\overset{\displaystyle \text{O}}{\overset{\|}{\text{C}}}\text{—R}' + \text{AMP} + \text{PP}_i & \Delta G^\circ = -2\ \text{kcal/mole} \end{array}$$

Reactions for which ΔG° is negative are thermodynamically possible. Of course, we cannot predict whether they will proceed at a useful rate.

The above example illustrates that thermodynamically unfavorable reactions can proceed if coupled to more favorable reactions. But it overlooks an equally important requirement: that there must be something to do the coupling, some common intermediate. The two reactions cannot simply proceed side-by-side with the free energy floating mysteriously from one set of reactants to the other. In the above example this function is probably served by an acyl adenylate, which is an acyl derivative of AMP:

$$\text{R—}\overset{\displaystyle \text{O}}{\overset{\|}{\text{C}}}\text{—O—}\underset{\displaystyle \text{O}^-}{\underset{|}{\overset{\displaystyle \text{O}}{\overset{\|}{\text{P}}}}}\text{—O—Ribose—Adenine}$$

The distinctive acid-base and spectral absorption characteristics of nucleotides are determined essentially by the bases and the phosphate groups. We will soon examine the role of each of the three parts of the nucleotides in determining the structure of the nucleic acid molecule. But first let us look briefly at some of the general aspects of polymerization.

general aspects of polymerization

The task before us now is to join the three classes of monomers that we have discussed—amino acids, monosaccharides, and nucleotides—to form the corresponding polymers: proteins, polysaccharides, and nucleic acids. All three classes of compounds, indeed, all polymerization processes, share some common features, which we will explore now in a general way.

11.4 MODES OF POLYMERIZATION

A useful analogy for the assembling of monomers into polymers is the stringing of beads. The beads are sometimes joined with a hook —⊃ and eye —o mechanism. This model can serve as a fairly decent representation for a wide variety of polymerization processes.

We could represent a monomer, then, as A—o. It might also be o—A—o or ⊂—A—o or ⊂—A—⊃. For the purposes of our simple model, we will assume that no bond can result between two eyes —oo— or two hooks —⊃⊂— and that only hook-and-eye combinations result in chemical bonds. This model will serve us well in the representation of bonding in proteins and nucleic acids. A slight modification will make it applicable to polysaccharides also. We now proceed to look at some mixtures of monomers and see what polymers result.

Amino acids and nucleotides can be represented by the monomers o—A—⊃. This means that any two such monomers of the same kind can be joined as follows:

o—A—⊃ + o—A—⊃ → o—A—⊃o—A—⊃ (11-15)

The important point here is that, after the first bond is formed, exactly one free hook and one free eye remain per molecule of A_2. This means that the dimer can undergo a reaction with two more monomers to form a tetramer:

o—A—⊃ + o—A—⊃o—A—⊃ + o—A—⊃ →

o—A—⊃o—A—⊃o—A—⊃o—A—⊃ (11-16)

The resulting molecule A_4 again retains one free hook and one free eye. Thus it is not difficult to imagine the chain's continuing to grow to a great length, with repeating units of o—A—⊃. We can abbreviate this polymer as (o—A—⊃)$_n$, where n can be any number from 2 up into the millions. Other ways to write the formula of the polymer are

—A—A—A—A— or . . . AAAA . . . or A_n

Several points about our model should be considered. Even after the chain has stopped growing, exactly one free hook and one free eye remain. Furthermore, only straight chains are produced from these kinds of monomers. No branching of any kind is possible. And finally, because of our initial assumption about the bonding, we have produced a perfectly regular polymer with every monomer exactly in place and with identical bonds connecting them. There is no possibility of the following structure:

○—A—⊃⊂—∀—○○—A—⊃

What is the result of stringing equal numbers of beads of two different colors (monomers of two different kinds)? As long as the two beads, say, red ones ○—R—⊃ and yellow ones ○—Y—⊃, possess the same bonding properties as described above, the same linear structure will result, except that now there will be a random sequence of red and yellow beads:

$$. . . \text{RYYRRYRYYRRRYYRY} . . . \quad \text{or} \quad (\text{RY})_n$$

The structure is completely random because we assume that the hooks and eyes of the R and Y beads are identical. Thus there is an equal chance that the hook of the R bead will latch onto the eye of an R bead or a Y bead. If the ratio of red to yellow were 3 : 1 instead of 1 : 1, we would expect a chain something like the following to result:

$$. . . \text{RRYRRRRYRYRRYRRR} . . . \quad \text{or} \quad (\text{R}_3\text{Y})_n$$

If it were necessary to construct a red-yellow polymer of a definitely repeating sequence, such as . . . RYRYRY . . . . it would require monomers with special properties, e.g., ⊂—R—⊃ and ○—Y—○. Maintaining our assumption that only hook-and-eye bonds are possible, the resulting chain would look like this:

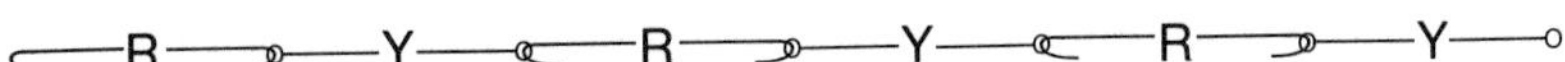

Reactions of this kind are of great importance in the production of synthetic polymers such as those used as ion-exchange resins. However, they do not occur in living systems, so we will not consider them further.

Obviously, we are not restricted to beads of only two different colors. We must use beads of four different colors to produce a chain similar to a nucleic acid, and of 20 different colors to produce a chain similar to a protein. We will consider the implications of the many different ways those beads can be strung together, but first we must examine the special types of beads required to produce a polysaccharide.

Most polysaccharides consist of repeating pyranose units. This means that our beads should have five hooks, because there are five carbon atoms in the pyranose ring. We can make our point clearly, however, with beads having only three hooks. Note that our beads now have hooks only and that the hooks can bond to each other. One of the results of mixing three such beads and allowing the hooks to bond randomly is shown below:

(11-17)

In our previous examples the result of polymerization was to give a total number of free hooks and eyes equal to the number possessed by each monomer before it reacted. Now, however, the total number is greater, no matter how the monomers are joined. Also, because there are now three hooks per monomer, the possibility of *branching* has been introduced, leading to structures such as the following:

This variation in the bonding pattern produces some major differences in the properties of the polysaccharides. Although the beads representing the monosaccharides have five hooks, there are a number of examples of chains in which only two of the hooks are used. This naturally gives rise to linear polymers.

11.5 DISTRIBUTION OF MONOMERS IN POLYMERS

The properties of proteins and nucleic acids are critically dependent on precisely how the monomers are arranged in the polymer chain. The sequence of monomers totally determines the physical, chemical, and biological properties of these giant molecules. Let us consider the number of different possibilities of linking a finite number of monomers, as a prelude to the genetic code.

Two different monomers, A and B, can be linked in two, not one, different ways. If we retain our hook-and-eye model, the following two structures show why the dimers AB and BA are different:

o—A—⊃o—B—⊃ ≠ o—B—⊃o—A—⊃

They differ in which of the two connectors on a given monomer, a hook or an eye, is used to form the bond. Extending this model, we see that three different monomers can be linked in six ways to form a trimer:

ABC ACB BAC BCA CAB CBA

Exercise 11-2. Determine the number of ways in which four different monomers, A, B, C, and D, can be linked to form a tetramer. *Answer:* 24.

Perhaps by now you have discovered the pattern in the number of isomers of the polymer arising from a given number of different monomers. For two different monomers, there were $2 = 2 \cdot 1$ isomers; for three different monomers, $6 = 3 \cdot 2 \cdot 1$ isomers; and for four different monomers, $24 = 4 \cdot 3 \cdot 2 \cdot 1$ isomers. Clearly the number of isomers of the polymer is n factorial ($n!$), where n is the number of different monomers. Thus $n! = n(n-1)(n-2) \cdots 3 \cdot 2 \cdot 1$. The number of possible isomers (remember that each isomer is a unique species, different from all the rest) mounts spectacularly as n increases. If we double n from 4 to 8, the number of isomers becomes 40,320. If $n = 20$, as it does for the common amino acids, the number is 2.4×10^{18}, or 2.4 quintillion. That is an unimaginably large number of possible protein structures, but it is infinitesimal in comparison with the *real* number of possibilities. Real proteins contain far more than 20 monomer units—typically, hundreds, but it can be tens of thousands. Naturally there is a great deal of duplication of individual amino acid monomers in any given protein, and most proteins do not contain all 20 of the amino acids anyway. These factors must be accounted for in the calculation of the number of possible isomers.

One rather small protein has a molecular weight of 34,000. It contains 288 amino acid monomers of only 12 different kinds. If we were able to make just *one molecule* of each of the possible isomers of this protein, the resulting mass would be 10^{280} g. Compare this number with that of the estimated total mass of the known universe: 10^{55} g. It seems that there are a great many possible proteins that do not exist.

It should be apparent that, if the sequence of monomers is of any importance in determining the properties of a polymer, there must be some way in which a particular sequence can be selected and constructed. Otherwise, the resulting plethora of different molecules could not possibly be expected to display any useful functions. "Like mother, like son" would have no meaning. The set of directions for creating the proper sequences of amino acids in proteins is called the *genetic code.* The existence of such a code has long been obvious, but its basic nature and some of the details of its operation have been discovered only in the last two decades. These discoveries represent one of the greatest triumphs in the history of science. We will discuss the genetic code and its cracking in some detail at the end of this chapter.

biopolymers

We now have enough background information to consider the structures and properties of the three major classes of naturally occurring polymers—the proteins, polysaccharides, and nucleic acids. In each case we will see how the monomers coalesce to form the polymers, examine the structures of the polymers, and then see how these structures give rise to the particular biological functions of the polymers. The elucidation of such structure-function relations is at the very heart of man's attempt to gain a fundamental understanding of living organisms at the molecular level.

11.6 PROTEINS

Proteins consist primarily of amino acids. The amino acids are linked in the manner shown in Reaction 11-15, with the carboxyl group of one amino acid and the amino group of the other acting as the hook and eye. Thus glycylvaline can be formed as follows:

$$\underset{\text{Glycine}}{^{+}H_3N\text{—}CH_2\text{—}COO^-} + \underset{\text{Valine}}{^{+}H_3N\text{—}\underset{\underset{CH_3\ \ CH_3}{CH}}{\overset{|}{CH}}\text{—}COO^-} \rightarrow$$

$$\underset{\text{Glycylvaline}}{^{+}H_3N\text{—}CH_2\text{—}\overset{\overset{O}{\|}}{C}\text{—}NH\text{—}\underset{\underset{CH_3\ \ CH_3}{CH}}{\overset{|}{CH}}\text{—}COO^-} + H_2O \qquad (11\text{-}18)$$

Because one free NH_2 group and one free COOH group still exist in the molecule, it is easy to see how very long chains can result from reactions with additional amino acids. The backbone of such a structure is called a *polypeptide chain.* It is shown below, with $R_1, R_2, R_3, \ldots, R_n$ denoting the side chains that define the identities of the various amino acids that constitute the molecule:

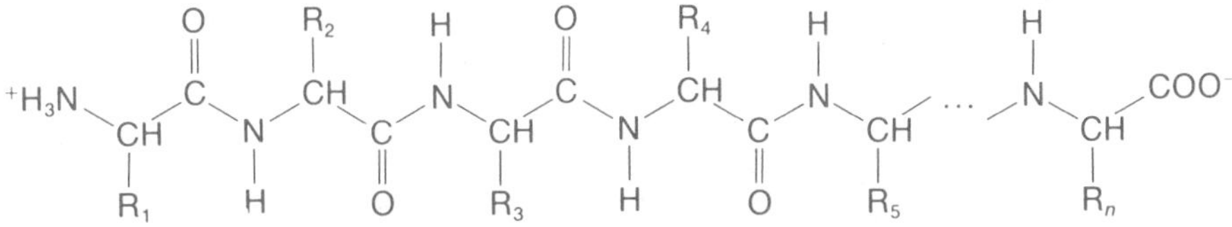

```
          O      R2      H      O      R4      H             H
          ‖      |       |      ‖      |       |             |
+H3N     C      CH      N      C      CH      N      ...    N      COO-
     CH     N       C      CH     N       C      CH             CH
     |      |       ‖      |      |       ‖      |              |
     R1     H       O      R3     H       O      R5             Rn
```

Some proteins consist of a single polypeptide chain, whereas others have two or more such chains that are linked in various ways.

The specific sequence of amino acids in the chain is called the *primary structure* of the protein. Each —NH—CHR—CO— segment in the chain is called a *residue* because it represents the residual amino acid left after the elimination of the H_2O molecule in the formation of the bond between one amino acid and the next. This carbon-nitrogen bond, $\mathrm{{-}C({=}O){-}N(H){-}}$, is called the *peptide bond* (or amide bond, because the group in question is an amide group). The properties of the peptide bond have been studied in great detail by Linus Pauling and Robert Corey (1897–1971), who used x-ray diffraction to determine the precise values of bond angles and bond lengths in many small polypeptide chains. This information was crucial in their proposal of the existence and structures of particular kinds of helices in protein molecules. The dimensions of the peptide bond are shown in Figure 11-5. Note that two of the bond lengths are known to within one-thousandth of an angstrom, a tribute to the skill of the crystallographers. The six atoms covered by the shaded area all lie in the same plane.

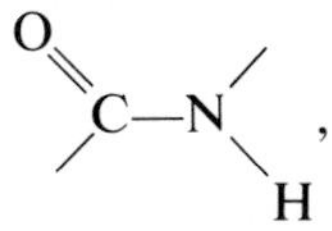

Exercise 11-3. Draw atomic orbital diagrams for the six coplanar atoms in Figure 11-5.

The C—N bond length of 1.325 Å is considerably shorter than the normal length of a C—N single bond (about 1.47 Å), indicating a substantial amount of double-bond character. Recalling the principles of chemical bonding that we learned in Chapter 4, we can show this partial double-bond character as follows:

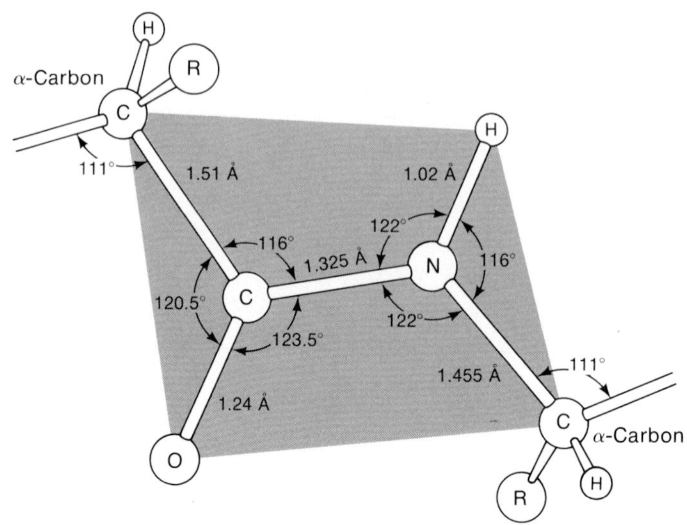

Figure 11-5. The dimensions of the peptide bond. (Adapted from R. E. Dickerson and I. Geis, *The Structure and Action of Proteins,* W. A. Benjamin, Menlo Park, Cal., 1969. Copyright © 1969 by R. E. Dickerson and I. Geis.)

$$\left\{ \begin{array}{c} \ddot{\text{O}} \\ \| \\ -\text{C}-\ddot{\text{N}}- \\ | \\ \text{H} \end{array} \qquad \begin{array}{c} \ominus \\ :\ddot{\text{O}}: \\ | \\ -\text{C}=\overset{\oplus}{\text{N}}- \\ | \\ \text{H} \end{array} \right\} \quad \text{or} \quad \begin{array}{c} \text{O} \\ \vdots\| \\ -\text{C}\overset{\cdots}{=}\text{N}- \\ | \\ \text{H} \end{array}$$

It is this feature of the peptide bond that prevents free rotation about the C—N axis and forces the structure to remain planar.

Knowledge of the primary structure of a protein is essential for an understanding of the more subtle details of the structure and of the biological function of the molecule. The first step in determining a primary structure is to determine the amino acid composition. This is now done routinely, using one of several commercially available amino acid analyzers. The protein is hydrolyzed in the absence of air in 6 *M* HCl at 110°C for about one day. This decomposes the protein to its constituent amino acids:

$$\mathrm{^{+}H_3N{-}CH(R_1){-}C({=}O){-}NH{-}CH(R_2){-}C({=}O){-}NH{-}CH(R_3){-}C({=}O){-}} \xrightarrow[\text{24 h at 110°C}]{6\,M\ \text{HCl}}$$

$$\mathrm{^{+}H_3N{-}CHR_1{-}COO^- +}$$

$$\mathrm{^{+}H_3N{-}CHR_2{-}COO^- +}$$

$$\mathrm{^{+}H_3N{-}CHR_3{-}COO^- + \cdots} \qquad (11\text{-}19)$$

The resulting mixture of amino acids is separated in an ion-exchange column and the individual amino acids measured quantitatively. As each "slug" of amino acid emerges from the column, it is mixed with ninhydrin (see Reaction 11-4). The resulting blue solution is passed through a spectrophotometer, and the absorbance, which is directly proportional to the concentration, is measured and plotted on a strip-chart recorder. The results of such analyses are often given as the number of moles of each amino acid per 100,000 g of protein.

The determination of the sequence of residues in the protein requires a great deal more effort. Specific reagents are available for hydrolyzing the N-terminal amino acid (the amino acid at the $^{+}H_3N$— end of the molecule) and the C-terminal amino acid (the one at the —COO^- end), and for identifying them. Either of these procedures produces a new polypeptide chain that is one unit shorter, and exposes a new amino acid at the terminus in question. The procedure can then be repeated. Unfortunately, experimental difficulties after four or five such cleavages make further sequence analysis more difficult. The problem is solved by introducing another protein, an enzyme, which preferentially cleaves the polypeptide chain at particular amino acid residues. This yields a series of smaller chains whose sequences can be determined fairly readily. If several enzymes with different specificities are employed, overlapping sequences can be obtained. You can appreciate some of the difficulty but also some of the fun of pursuing such studies by working through the following exercise.

Exercise 11-4. Gramicidin-S is an important antibiotic. Total hydrolysis of this short polypeptide chain (too short to be considered a true protein) yields L-leu, L-orn, D-phe, L-pro, and L-val. (Note the unusual appearance of a D-amino acid.) Random partial hydrolysis yields the following dipeptides: L-leu-D-phe, L-val-L-orn,

L-pro-L-val, L-orn-L-leu, and D-phe-L-pro. No free α-NH_3^+ or α-COO^- groups are found, and the molecular weight is found to be 1200. Write the primary structure of Gramicidin-S.

Complete primary structures are now known for hundreds of proteins. Because we will look at the three-dimensional structure and the function of the protein lysozyme in this chapter, its primary structure, consisting of one polypeptide chain, is given in Figure 11-6. Note that there are eight cysteine residues. They can cross-link the polypeptide chain with itself in the form of four cystine residues (see Reaction 11-8), thereby constraining it to adopt only certain three-dimensional structures.

Enough protein structures have now been determined for scientists to have searched for patterns or regularities that may exist. The following list summarizes some of their findings to date.

1. Each protein has a unique structure.
2. No apparent regularity or periodicity exists in amino acid sequences in most proteins. The only known exceptions are collagen, elastin, silk fibroin, and homologous proteins (i.e., the same protein, such as hemoglobin, from a variety of related species).
3. With respect to disulfide bridges (cross-links), all possible types of proteins have been found. Some proteins have none, some have intrachain bridges, and some have interchain bridges.
4. The distributions of glutamine and asparagine are independent of the distributions of the related amino acids glutamic acid and aspartic acid.
5. Excluding glycine, which is not optically active, only L-amino acids have been found in proteins.
6. No covalent bonds other than the peptide and disulfide bonds have been found between amino acid residues in globular proteins (proteins in which the folding and cross-linking of polypeptide chains leads to an overall globular structure). Despite the wide range of differences in size, composition, and function of such proteins, these two bonds are the only ones found. Other types of covalent bonds between amino acid residues have been found, however, in fibrous proteins.

Item number 2 in this list is of particular interest to scientists in the field of molecular evolution. The primary structures of hemoglobin, for example, have been determined for a wide variety of living organisms. Although each one is unique, the similarities are striking. The number and nature of amino acid differences in the sequences from two related species can be taken as clues to the time and circumstances of the mutational change in a

LYSOZYME—CHICKEN

```
    1 2 3 4 5 6 7 8 9 0 1 2 3 4 5 6 7 8 9 0 1 2 3 4 5 6 7 8 9 0
  1 K-V-F-G-R-C-E-L-A-A-A-M-K-R-H-G-L-D-N-O-R-G-O-S-L-G-N-W-V-C-
 31 A-A-K-F-E-S-N-F-N-T-Q-A-T-N-R-N-T-D-G-S-T-D-O-G-I-L-Q-I-N-S-
 61 R-W-W-C-N-D-G-R-T-P-G-S-R-N-L-C-N-I-P-C-S-A-L-L-S-S-D-I-T-A-
 91 S-V-N-C-A-K-K-I-V-S-D-G-D-G-M-N-A-W-V-A-W-R-N-R-C-K-G-T-D-V-
121 Q-A-W-I-R-G-C-R-L
```

```
     1   2   3   4   5   6   7   8   9   10  11  12  13  14  15
  1 Lys-Val-Phe-Gly-Arg-Cys-Glu-Leu-Ala-Ala-Ala-Met-Lys-Arg-His-
 16 Gly-Leu-Asp-Asn-Tyr-Arg-Gly-Tyr-Ser-Leu-Gly-Asn-Trp-Val-Cys-
 31 Ala-Ala-Lys-Phe-Glu-Ser-Asn-Phe-Asn-Thr-Gln-Ala-Thr-Asn-Arg-
 46 Asn-Thr-Asp-Gly-Ser-Thr-Asp-Tyr-Gly-Ile-Leu-Gln-Ile-Asn-Ser-
 61 Arg-Trp-Trp-Cys-Asn-Asp-Gly-Arg-Thr-Pro-Gly-Ser-Arg-Asn-Leu-
 76 Cys-Asn-Ile-Pro-Cys-Ser-Ala-Leu-Leu-Ser-Ser-Asp-Ile-Thr-Ala-
 91 Ser-Val-Asn-Cys-Ala-Lys-Lys-Ile-Val-Ser-Asp-Gly-Asp-Gly-Met-
106 Asn-Ala-Trp-Val-Ala-Trp-Arg-Asn-Arg-Cys-Lys-Gly-Thr-Asp-Val-
121 Gln-Ala-Trp-Ile-Arg-Gly-Cys-Arg-Leu
```

Composition

12	Ala	A	8	Cys	C	8	Asp	D	2	Glu	E
3	Phe	F	12	Gly	G	1	His	H	6	Ile	I
6	Lys	K	8	Leu	L	2	Met	M	13	Asn	N
3	Tyr	O	2	Pro	P	3	Gln	Q	11	Arg	R
10	Ser	S	7	Thr	T	6	Val	V	6	Trp	W

Total number of amino acids = 129

Figure 11-6. The primary structure of the protein lysozyme from the chicken. (M. O. Dayhoff, ed., *Atlas of Protein Sequence and Structure 1972,* Vol. 5, National Biomedical Research Foundation, Washington, D.C., 1972.)

common ancestor that led to species differentiation. Such molecular evidence is perhaps the most compelling of all in favor of the fundamental correctness of the theory of evolution, and may prove to be the most accurate source of information in tracing evolutionary history.

At the beginning of this section, we said that proteins consist primarily of amino acids. Other groups are also present in many proteins, however, and sometimes constitute the site of biological activity in the molecule. These proteins are called *conjugated proteins* and the non-amino acid group is called the *prosthetic group*. Some such proteins and their prosthetic groups are: mucoproteins (carbohydrates), nucleoproteins (nucleic acids), and phosphoproteins (phosphoric acid). Myoglobin, one of the two proteins whose structure we will examine in some detail, is an example of a hemoprotein, which contains a porphyrin ring structure (see Figure 5-21).

If the description of protein structure ended here, proteins would be characterized simply by a statement of the sequence of amino acids in the polymer. They would be regarded as long, floppy, somewhat amorphous chains of amino acids. In reality, proteins contain several higher levels of structural organization. The protein molecule, of course, knows nothing about this; it simply exists. The concept of levels of structure is a matter of convenience in describing very complicated systems.

The *secondary structure* of a protein is concerned with the configuration of the polypeptide chain(s). There are three principal ways in which such chains assume ordered configurations. The most common one is the *α helix,* which was postulated by Pauling and his coworkers* in 1951 and subsequently confirmed by others in numerous x-ray crystallographic studies of proteins. The proposal of Pauling et al. was based on precise x-ray studies of amino acids and small polypeptides. The assumptions they invoked in arriving at plausible configurations of large polypeptide chains were the following: (1) the amide group that constitutes the peptide bond is planar because of the large amount of double-bond character in the C—N bond; (2) the bond lengths and bond angles in this group are those determined by Pauling and Corey for the small polypeptides; (3) the N atom of each amide group forms a hydrogen bond with the O atom of another amide group,

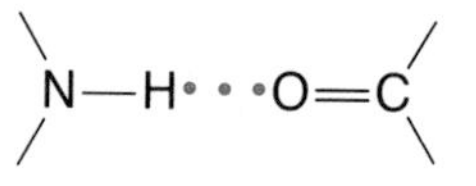

$$\rangle N{-}H\cdots O{=}C\langle$$

(4) the N-O distance defined by this hydrogen bond is 2.72 Å; (5) the N, H, and O atoms are collinear to within 30°, i.e.,

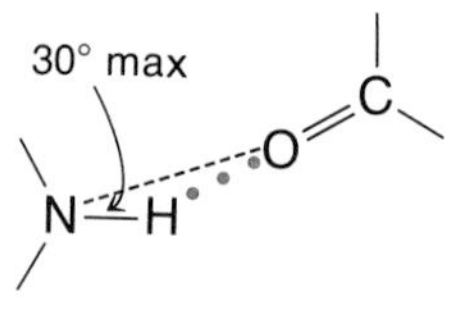

30° max

N—H···O=C

(6) the energy of the hydrogen bond is about 8 kcal/mole.

---

*L. Pauling, R. B. Corey, and H. R. Branson, *Proceedings of the National Academy of Sciences of the USA* **37**, 205 (1951). This short article is one of the epochal works of modern chemistry, and is not difficult reading. The same volume of this journal contains seven further articles on protein structure by Pauling and Corey.

The fundamental role of the hydrogen bond in this proposal is obvious. Two kinds of helices were found that satisfied all of the above criteria reasonably well. One of them, the α helix, has been found to be the predominant one in proteins and in synthetic polypeptides as well. The α helix is shown in Figure 11-7. The ball-and-stick model shows the structure of the

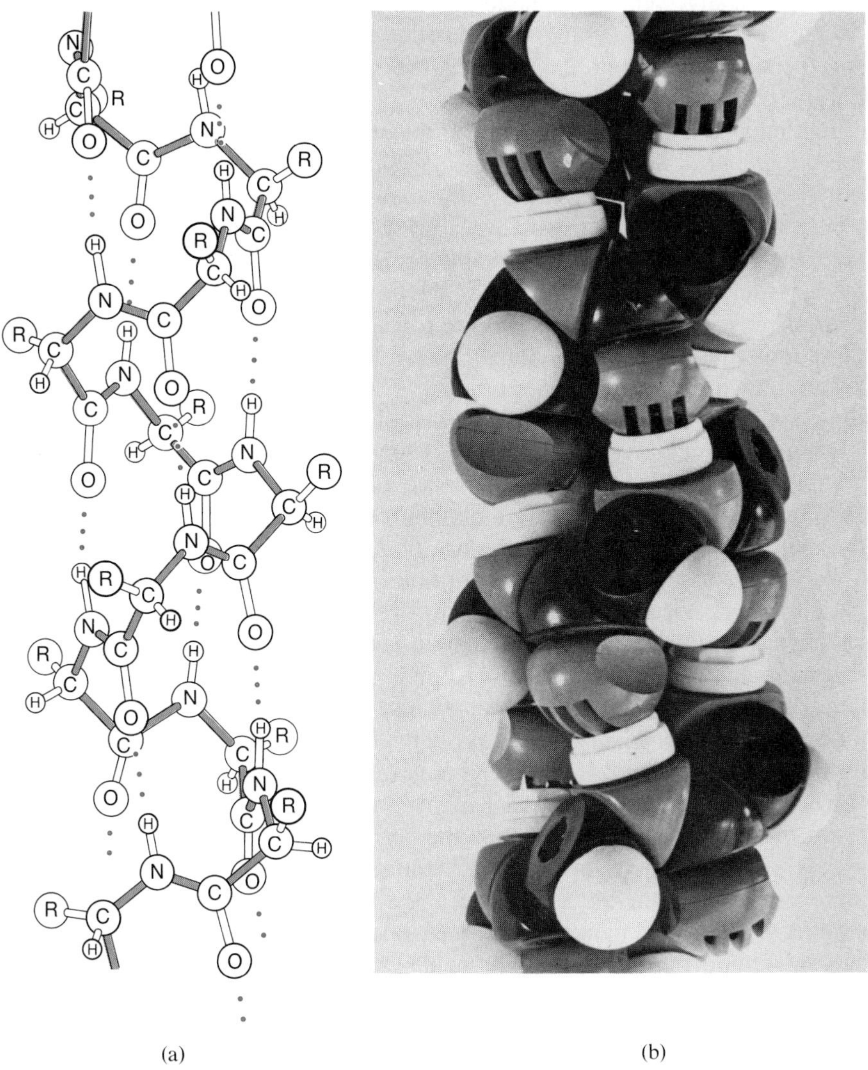

Figure 11-7. The α helix. (a) Ball-and-stick model. (b) Space-filling model, with the side chains (R) removed. [Part (a) from L. Pauling, *The Nature of the Chemical Bond,* 3rd ed., Cornell University Press, Ithaca, N.Y., 1960.]

helical backbone, which has 3.6 amino acid residues per turn. The space-filling model shows how "solid" the core of the helix is, denying access even to small molecules and ions. Both models show that the hydrogen bonds are parallel to the axis of the helix, that the helical structure is determined solely by the polypeptide bonds and hydrogen bonds, and that the side chains of the residues project into the surrounding space.

Theoretically, both left-handed and right-handed helices are possible. Calculations have shown the right-handed form to be somewhat more stable for L-amino acids, and this form is the only one that has been found in natural proteins.

Pauling and Corey also discovered the second major type of configuration for polypeptide chains. Called the *β pleated sheet,* it consists of rows of polypeptide chains lying side-by-side and hydrogen-bonded to each other. The resulting sheet-like structure appears pleated when viewed from the side, with the amino acid side chains projecting up and down. With respect to the amino and carboxyl ends of the polypeptide chains, the chains may be lined up in the same direction (the parallel configuration) or in alternating directions (the antiparallel configuration). Both are found in nature, but the latter, shown in Figure 11-8, is the more common. It is found primarily as the class of proteins called silk fibroins, of which all silk is made.

The third major type of protein configuration is the *collagen helix,* a complex, triple-helical structure that we will not discuss in detail. Collagen is a tough, fibrous protein, the most abundant protein in all mammals. It is the major fibrous constituent of skin, bones, tendons, cartilage, and teeth.

Again, we cannot stop our description of protein structure here. If there were nothing more to say at this point, we would conclude that, except for collagen and some related proteins, all proteins were either long, narrow, helical rods with intramolecular H bonds parallel to the molecular axis, or large, pleated sheets with intermolecular H bonds perpendicular to the molecular axis. In reality, many proteins are globular as a result of folding in the polypeptide chains. This is a manifestation of *tertiary structure.*

There are several reasons that many proteins are not straight α helices. First, proline does not fit well into the helix because it does not contain an α-amino group. Wherever a proline residue appears in the chain, therefore, there is a kink in the helix. Second, the disulfide bridges between different regions of the same chain require at least one bend in the backbone. Finally, and most important of all, there is a wide variety of attractive forces between the side chains: hydrophobic bonds, ion-ion interactions, hydrogen bonds, and the various van der Waals forces. It appears that that particular conformation of the protein that allows the maximum number of these attractive forces to come into play is the most stable configuration.

Long before today's extremely accurate x-ray methods were available, chemists had devised means of determining the overall sizes and shapes of

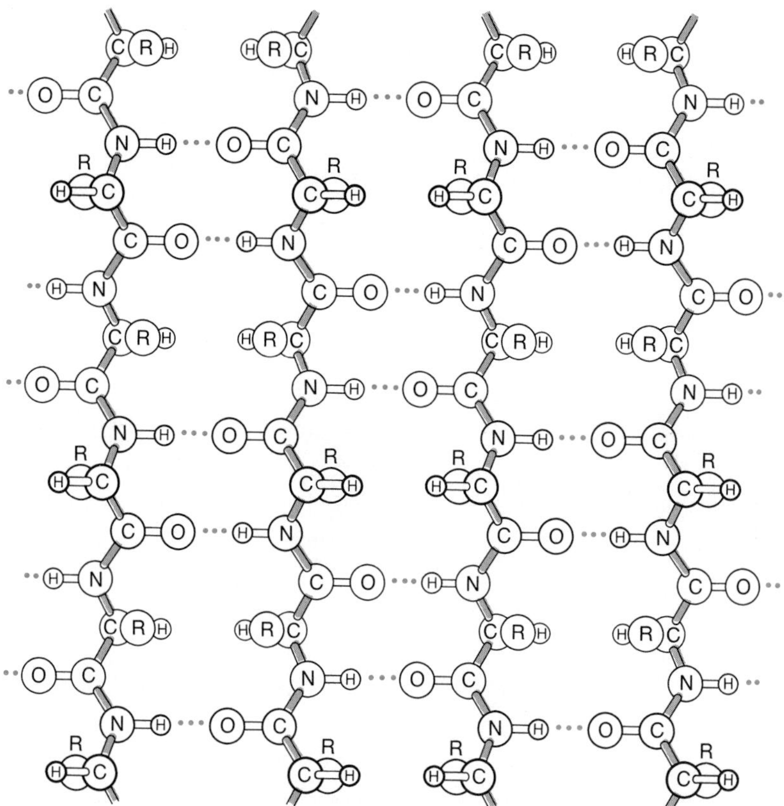

Figure 11-8. The β pleated sheet, antiparallel configuration. (L. Pauling, *The Nature of the Chemical Bond,* 3rd ed., Cornell University Press, Ithaca, N.Y., 1960.)

proteins. Some examples from the classic work of the American biophysicist John L. Oncley (b. 1910) are shown in Figure 11-9. Although the folding of the polypeptide chains cannot be seen here, the general tertiary structures are quite apparent.

Many proteins consist of two or more discrete polypeptide chains, called *subunits.* The last level of protein structure, the *quaternary structure,* is concerned with the spatial relations between the subunits, i.e., the manner in which they are connected and packed. For example, hemoglobin, with a molecular weight of 64,500, contains four subunits. Accurate x-ray analyses are required to locate their positions. The bonds holding them together are weak and easily ruptured, which permits the isolation of the subunits. As detailed physicochemical and x-ray data on enzymes become available, more and more of them are found to possess subunits. The study of the

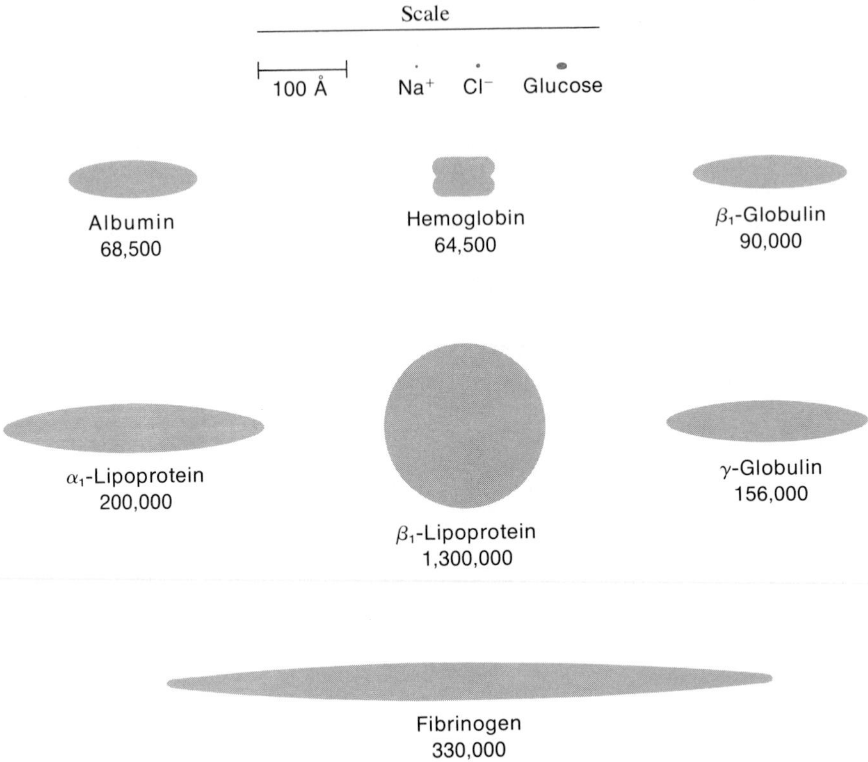

Figure 11-9. Relative dimensions of various proteins. The numbers are molecular weights. (J. L. Oncley, *Conference on the Preservation of the Cellular and Protein Components of Blood,* American National Red Cross, Washington, 1949.)

structure-function relations in proteins and their subunits is an area of active research today, and surely the most exciting of all in protein chemistry.

For over 70 years scientists have known the functions of certain proteins, such as the oxygen-binding capacity of hemoglobin. But until very recently, the question of how the structure of the molecule gave rise to this function was a complete mystery. The problem was that no one knew the structure of any protein. Now the detailed structures of more than 20 proteins are known, and many other structure determinations are in progress. We will look at structure-function relations in two of these proteins.

At the end of the section on enzyme kinetics in Chapter 8, we mentioned that lysozyme was the first enzyme for which the detailed three-dimensional structure was determined. Lysozyme is a small, globular protein found in many body tissues and secretions (especially tears), and in great quantities in chicken egg whites. Its function is to catalyze the lysis, or decomposition, of the cell walls of certain bacteria, a reaction that would otherwise be imperceptibly slow. These cell walls contain linear polysaccharide chains consisting of alternating units of the monosaccharides N-acetylglucosamine (NAG) and N-acetylmuramic acid (NAM). The polysaccharide chains are extensively cross-linked by short peptide chains, resulting in an enormous biopolymer that assumes a bag-like shape. A hexasaccharide segment of the polysaccharide chain acts as the substrate for lysozyme, and is cleaved between two of the monosaccharide units when the enzyme-substrate complex forms. When this occurs enough times, the integrity of the bacterial cell wall is destroyed and the bacterium dies. (It seems hardly coincidental that tears are so rich in this protective enzyme.)

In 1965 Phillips and his coworkers announced the results of their brilliant x-ray crystallographic analyses of chicken lysozyme and the lysozyme-substrate complex formed with the trisaccharide $(NAG)_3$, an unreactive species that remains bound to the active site long enough for the x-ray photographs to be obtained. Based on the results of the latter structure determination, they were able to construct a detailed model of the enzyme with the complete, reactive hexasaccharide segment $(NAG—NAM)_3$ attached to the active site. A simplified ball-and-stick model of this complex is shown in Figure 11-10.*

The enzyme is essentially globular, as are all other known enzymes. An interesting feature of the polypeptide chain is that three separate portions of it have the α helix configuration, yet in another region, where the chain doubles back upon itself several times, it has the antiparallel β pleated sheet configuration. Almost all the polar, hydrophilic groups are on the surface, and most of the nonpolar, hydrophobic groups are in the interior. This common characteristic of globular protein structures has led to their being likened to oil drops, as mentioned earlier.

The most important feature of lysozyme is a crevice that constitutes the active site. Phillips et al. were able to show exactly how the hexasaccharide segment in the bacterial cell wall can slip into this crevice. This can be inferred from the face-on view of the enzyme-substrate complex shown in Figure 11-10, but is revealed more clearly by the side views of the space-filling models in Figure 11-11.

Detailed analyses of the possible and probable atomic interactions between amino acid residues in the polypeptide chain of the enzyme and the monosaccharide units in the substrate enabled Phillips et al. to postulate

---

*Magnificent, detailed, full-color drawings of the lysozyme structure by the noted scientific illustrator Irving Geis appear in Phillips's article, "The Three-Dimensional Structure of an Enzyme Molecule," published in *Scientific American,* November 1966, page 78. Both the article and the drawings are masterly contributions to the scientific literature and are well worth your close study.

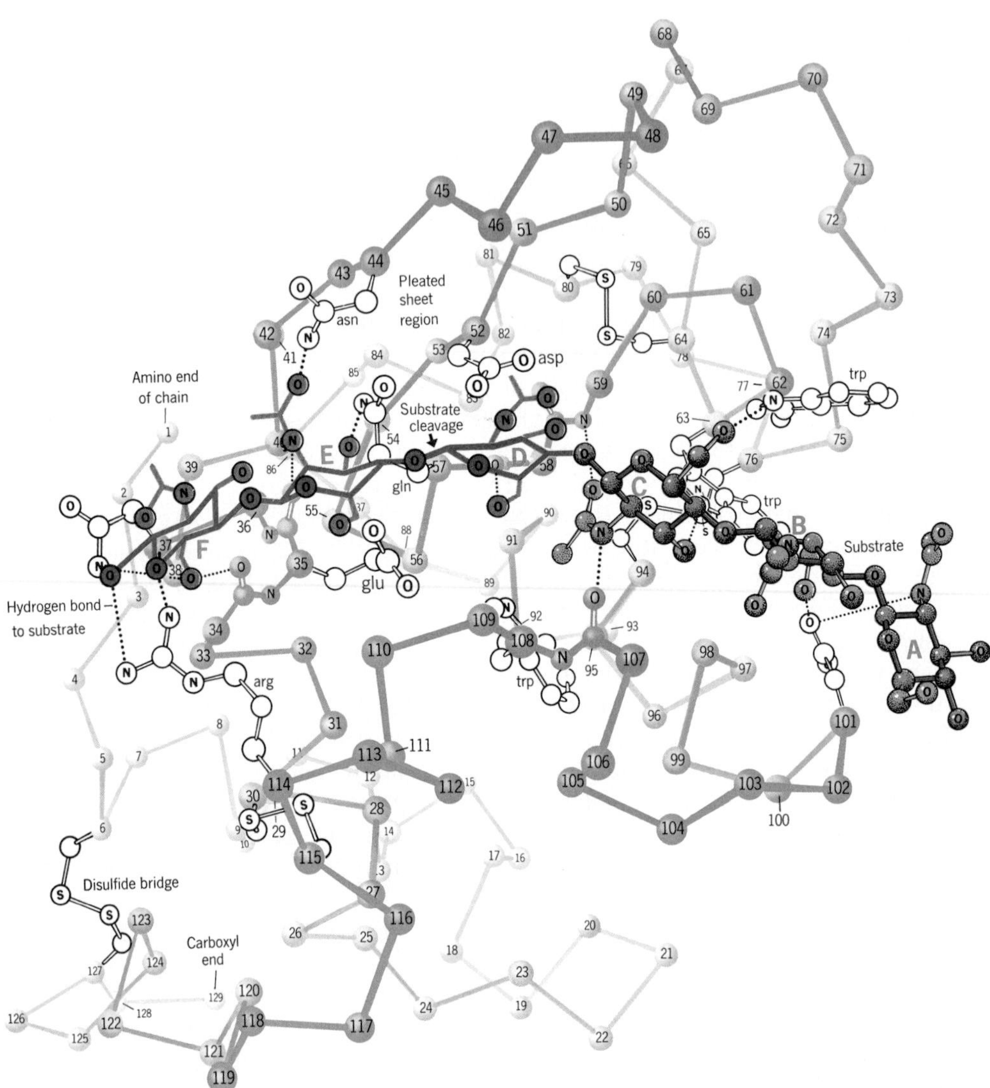

Figure 11-10. Ball-and-stick model of the polypeptide chain of chicken lysozyme, showing the 129 amino acid residues. The crevice that forms the active site runs horizontally across the molecule and is occupied, in this drawing, by the hexasaccharide substrate (shown in darker color, with the six rings labeled A through F). The amino acid side chains that are believed to interact with the substrate are shown as unshaded line drawings. Note the point of cleavage of the substrate, between rings D and E. (R. E. Dickerson and I. Geis, *The Structure and Action of Proteins,* W. A. Benjamin, Menlo Park, Cal., 1969. Copyright © 1969 by R. E. Dickerson and I. Geis.)

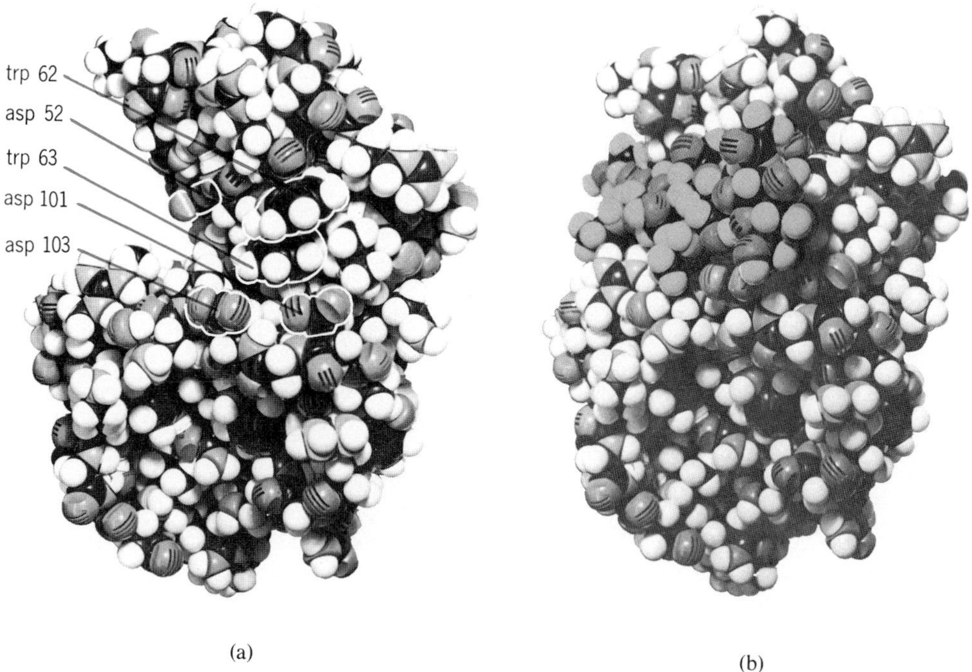

Figure 11-11. Space-filling models of lysozyme. (a) The enzyme alone, showing the crevice that forms the active site. (b) The enzyme-substrate complex, showing how neatly the substrate fits into the active site. (R. E. Dickerson and I. Geis, *The Structure and Action of Proteins,* W. A. Benjamin, Menlo Park, Cal., 1969. Copyright © 1969 by R. E. Dickerson and I. Geis.)

an exact mechanism by which the substrate is cleaved. All available physicochemical and biochemical evidence indicates that this mechanism is correct. Thus, at long last, the venerable Michaelis-Menten mechanism (Reaction 8-88) has been verified at the atomic level, an accomplishment that ranks among the most outstanding in the history of enzymology.

Two other proteins for which the structure-function relations are understood are the oxygen-binding proteins hemoglobin and myoglobin. The evolution of the higher animals would not have been possible without hemoglobin (in the vertebrates) and various other oxygen-binding proteins (in many invertebrates). This is because the only other mechanism—molecular diffusion—for transporting O_2 to the tissues of the body works only for a limited distance, usually not more than a few centimeters. All insects rely on diffusion for O_2 transport, and the longest known insect, the tropical stick insect *Pharnacia serratipes,* measures 33 cm. It seems unlikely that insects much longer than this could possibly exist, Hollywood films notwithstanding.

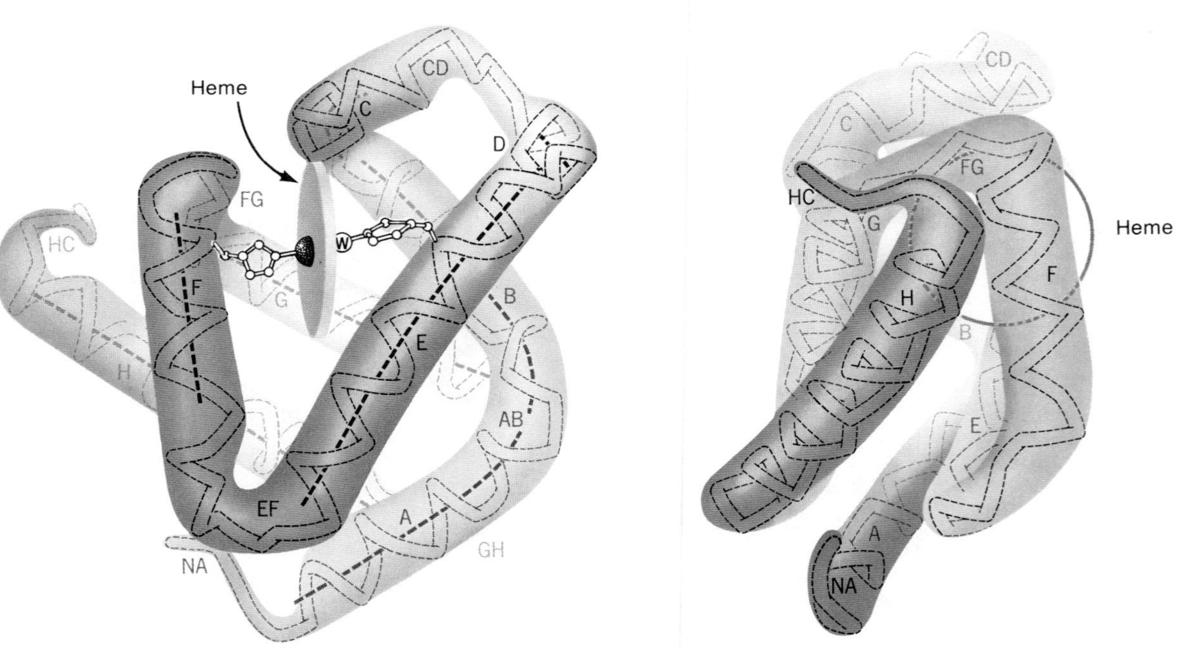

Figure 11-12. The myoglobin molecule. (a) Front view. Histidine residues interact with the heme to the left and right, and the O_2 molecule is labeled W. (b) Side view. (R. E. Dickerson and I. Geis, *The Structure and Action of Proteins,* W. A. Benjamin, Menlo Park, Cal., 1969. Copyright © 1969 by R. E. Dickerson and I. Geis.)

Hemoglobin and myoglobin are a molecular team. Hemoglobin binds O_2 at the lungs and transports it via the bloodstream to cells throughout the body. Most cells are muscle cells, all of which contain myoglobin. The hemoglobin transfers the O_2 to the myoglobin, which stores it until it is needed, and then releases it. The hemoglobin picks up the waste product CO_2 and transports it back to the lungs, where it is expelled. A rather clever, not to mention indispensable, pair of molecules.

Myoglobin and hemoglobin were the first two proteins for which the detailed structures were determined, outstanding pieces of research that earned the 1962 Nobel Prize in chemistry for John Kendrew (b. 1917) and Max Perutz (b. 1914), both of Cambridge University. Myoglobin is the smaller and simpler of the two molecules (it is, in fact, essentially one-quarter of the hemoglobin molecule), so we will limit ourselves to a brief discussion of myoglobin.

Myoglobin is a *heme* protein, which means that it contains a heme group (see Figure 5-21) as its prosthetic group. Almost identical groups are found in several other proteins, and in the chlorophylls. The overall structure of myoglobin is shown in Figure 11-12. The polypeptide, or *globin*, part of the molecule consists of eight segments of α helix (labeled A through H) connected by short nonhelical segments. The eight helices are arranged roughly in the shape of a shallow box that almost entirely surrounds the heme. Only the two propanoic acid side chains of the heme project outside this box. Since the heme group is essentially nonpolar, it is of interest to see whether the amino acid residues lining the "pocket" that contains it are hydrophobic, as one would expect. The complete x-ray analysis of the myoglobin structure showed that these residues are, indeed, hydrophobic and that the hydrophilic residues are spread uniformly across the outer surface of the molecule.

A more detailed diagram of myoglobin is shown in Figure 11-13.* The heme group is an elegant example of a coordination compound (see Chapter 4). In the porphyrin ring the iron atom is coordinated to four coplanar nitrogen atoms, each of which donates an electron pair to the bonding orbitals of the iron. Iron is usually octahedrally coordinated, however, and it is here also. A fifth electron pair is donated by a nitrogen atom of the imidazole ring of histidine F8 (the eighth residue of the helical segment labeled F). The sixth coordination site is the one on the opposite side of the porphyrin ring. This is the most important one, where, in oxymyoglobin, the oxygen molecule is bound through one of its lone pairs to the iron atom. Another histidine residue, E7, interacts weakly with the iron atom through the oxygen molecule. The heme group is covalently bound to the globin only via this coordination complex. The rest of the bonding appears to

*For another spectacular illustration of a protein molecule by Geis, see Kendrew's excellent article, "The Three-Dimensional Structure of a Protein Molecule," published in *Scientific American,* December 1961, page 96.

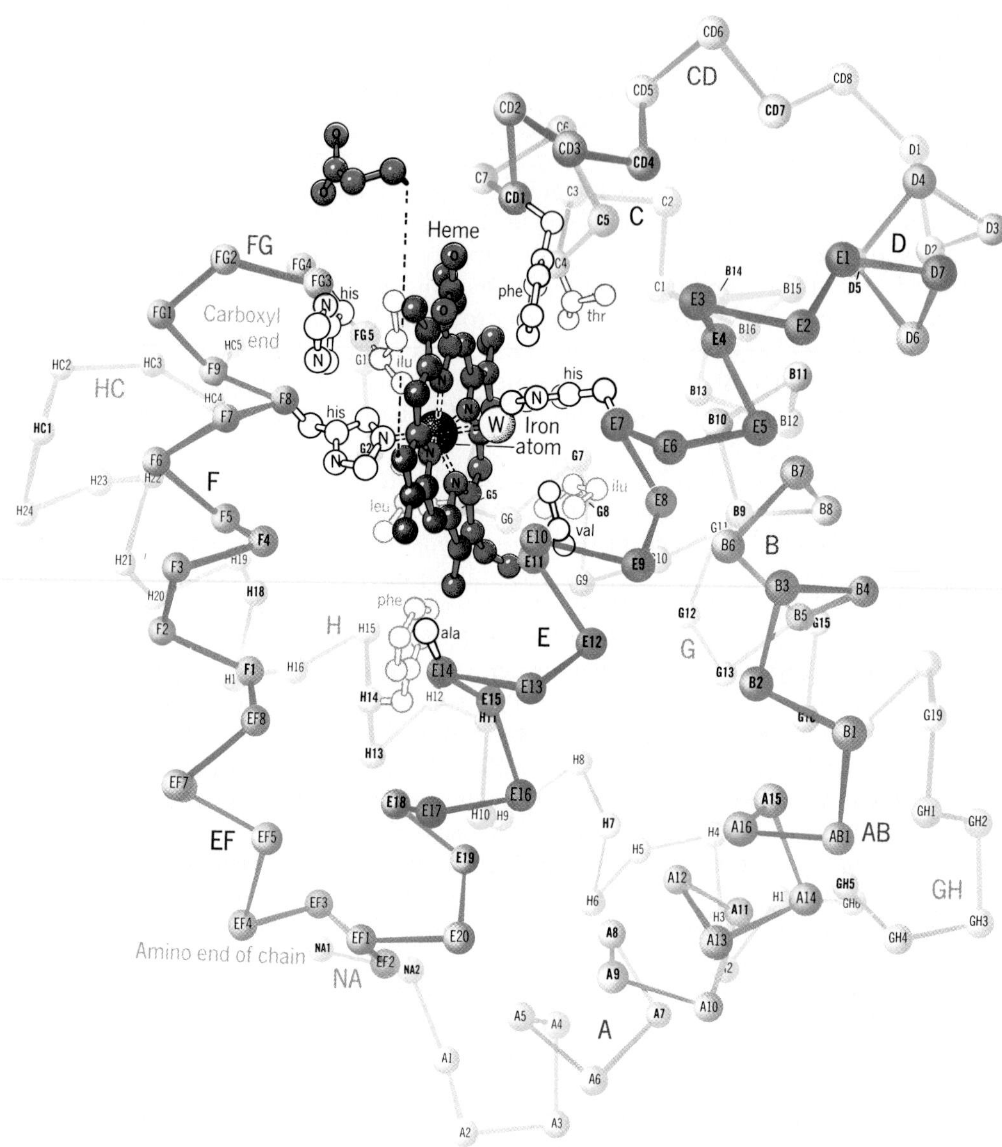

Figure 11-13. Ball-and-stick model of the myoglobin molecule. One propanoic acid side chain of the heme has been displaced for clarity. The O_2 molecule is labeled W. The amino acid side chains that interact with the heme are shown as unshaded line drawings. (R. E. Dickerson and I. Geis, *The Structure and Action of Proteins,* W. A. Benjamin, Menlo Park, Cal., 1969. Copyright © 1969 by R. E. Dickerson and I. Geis.)

consist primarily of hydrophobic interactions with the amino acid side chains, and two hydrogen bonds from the heme's propanoic acid side chains.

The oxidation state of the iron atom in myoglobin is +2, and the main purpose of the protective heme and globin structures seems to be to keep it that way. Oxidation to the +3 state does sometimes occur, however. The resulting compound, called metmyoglobin, is unable to bind oxygen, and binds a water molecule instead. Strangely enough, in deoxymyoglobin (normal myoglobin without the oxygen) the sixth coordination site is vacant rather than occupied by a water molecule. Because myoglobin, like hemoglobin, binds carbon monoxide preferentially to oxygen at this site, the breathing of carbon monoxide causes death by oxygen starvation.

11.7 POLYSACCHARIDES

Polysaccharides are an important energy source in plants and animals, serving as a metabolic reserve of monosaccharides. They also serve as the structural elements of plants, providing the skeleton for the organism. They are composed of enormous numbers of repeating units of identical monosaccharides, or repeating units of disaccharides composed of two different monosaccharides. We will discuss only the former type, which includes several of the most important polysaccharides. The molecular weights of these biopolymers are thought to be between about 10^4 and 10^8.

The biological roles of the polysaccharides appear rather mundane by comparison with those of the proteins and nucleic acids, but the polysaccharides are nonetheless of fundamental importance in many life processes and of considerable intrinsic interest as chemical compounds. Recently several groups of scientists have begun to look at their structures as a challenging problem in the prediction of the geometric properties of long-chain polymers, based on calculations of the properties of the monomers. It has also recently been discovered that the immune response and blood grouping in humans result from specific polysaccharide sequences in cell membranes.

The reaction of two D-glucose molecules yields three kinds of disaccharides, each of which is known to occur in polysaccharides. They are maltose, isomaltose, and cellobiose. Depending on whether the glucose molecules both have the α-configuration or the β-configuration or whether there is one of each, there are certain combinations of them that can result in various forms of the three kinds of disaccharides. Of the various possibilities, certain ones are found to be predominant in naturally occurring polysaccharides.

The distinguishing characteristic in the three kinds of glucose dimers is the nature of the glycosidic linkage between them. In the maltoses it is

α-1,4, meaning that C-1 of an α-D-glucose molecule is linked to C-4 of the other glucose molecule, which may have either the α-configuration or the β-configuration. The former structure, shown below, is the repeating unit in the polysaccharide amylose, or starch:

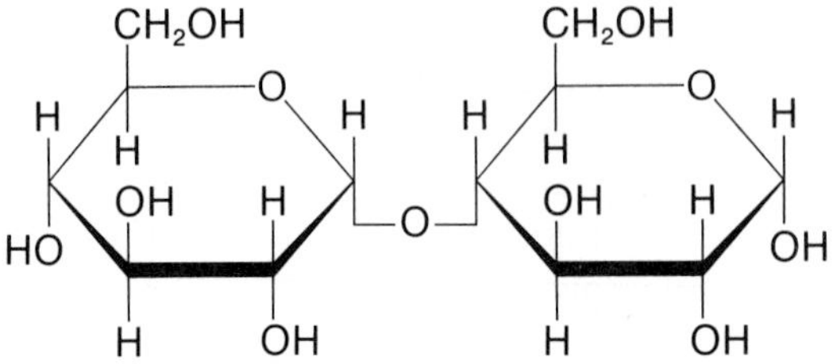

α-Maltose
(4-O-α-D-glucopyranosyl-α-D-glucopyranose)

The polysaccharides amylopectin and glycogen are branched-chain polymers in which the branching is due to the disaccharide unit isomaltose. Here the glycosidic linkage is α-1,6:

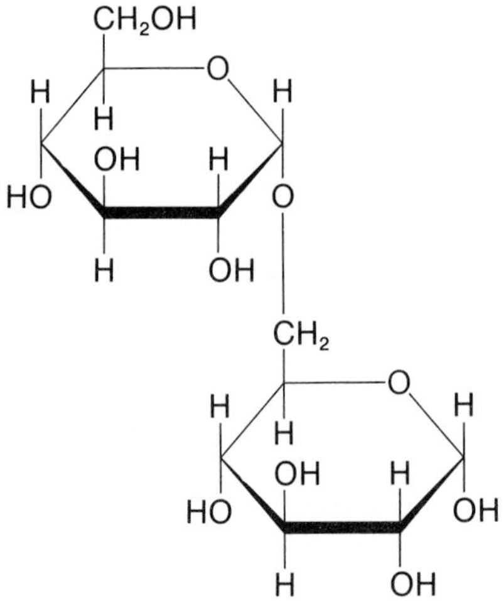

α-Isomaltose
(6-O-α-D-glucopyranosyl-α-D-glucopyranose)

Finally, the polysaccharide cellulose consists of repeating units of the disaccharide β-cellobiose, in which the glycosidic linkage is β-1,4 (see top of opposite page). The seemingly minor difference between the α-1,4 linkage in starch and the β-1,4 linkage in cellulose can actually be one of life and death. Most mammals can digest starch but not cellulose, because they lack the specific enzyme required to hydrolyze the β-1,4 linkage. A cow can survive on grass (the traditional kind), but a human being cannot.

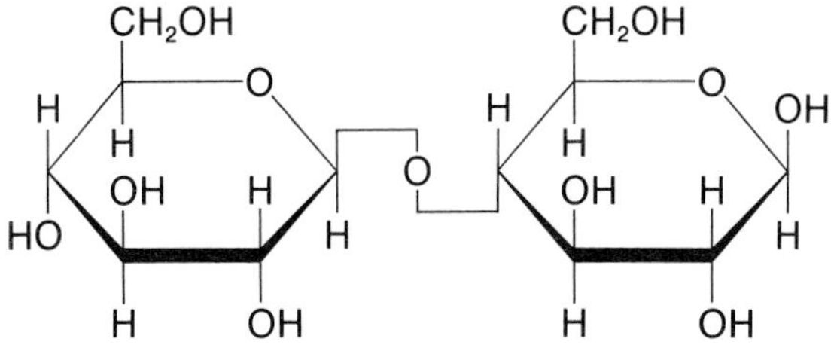

β-Cellobiose
(4-O-β-D-glucopyranosyl-β-D-glucopyranose)

Now let us look at the polysaccharides themselves. The starches amylose and amylopectin are found in small (5–100 μm diam) granules in plants. As the fruits ripen, these starches are converted to sugars. Amylose consists of long, unbranched chains of α-D-glucose monomers (or α-maltose dimers) linked to form a polymer with a molecular weight of as much as 500,000:

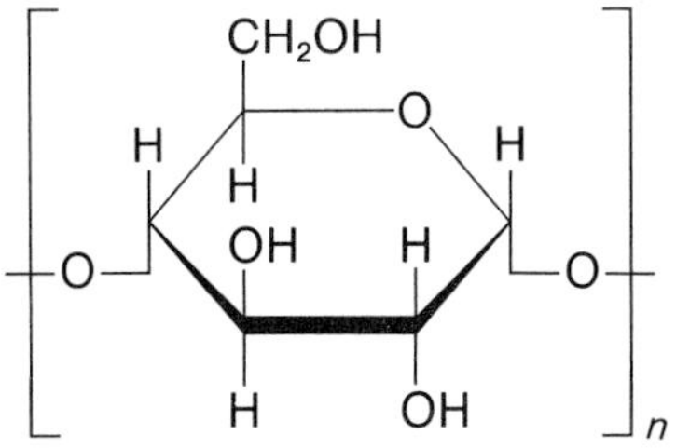

Physicochemical studies have shown this polymer to have a flexible, helical structure in solution. On page 384 we saw that the color of the complex formed between this helix and the tri-iodide ion I_3^- is used as a test for the presence of free I_2.

Amylopectin is a branched-chain polymer containing both α-1,4 and α-1,6 glycosidic linkages, the latter acting as branching points on amylose chains. The structure is analogous to the generalized branched-chain structure shown on page 632, with one of every 24–30 glucose units being at the end of a chain. The molecular weight may be as high as several million. The structure of glycogen is identical to that of amylopectin except that it is much more highly branched, with one of every 8–12 glucose units being at the end of a chain. Glycogen is the animal counterpart of starch in that it is the principal source of reserve carbohydrate (ultimately, glucose) in animal cells.

The action of two different enzymes on starches is interesting. One of them is α-amylase, a digestive enzyme found in saliva and pancreatic juice. It catalyzes the hydrolysis of the α-1,4 linkages randomly throughout the amylose molecule, ultimately yielding a mixture of glucose and maltose. The other enzyme, β-amylase, is found in malt (barley grain that has been allowed to germinate, then heated and dried). It catalyzes the same reaction, but in an entirely different way. Here the amylose molecule is systematically cleaved into maltose units, one by one, from one end of the chain to the other.

Both enzymes also catalyze the hydrolysis of amylopectin, but, because neither has any effect on an α-1,6 linkage, the results are different. The α-amylase yields glucose, maltose, and numerous small polysaccharide fragments called *dextrins,* which contain α-1,6 linkages. The β-amylase works its way down each chain until it is stopped by an α-1,6 linkage. The result is numerous maltose units plus one highly-branched "core" of the amylopectin molecule, called a *limit dextrin.*

Dextrins should not be confused with *dextrans,* which are also branched polysaccharides of D-glucose. Here the principal linkage is α-1,6, with numerous branch points formed by α-1,2, α-1,3, and α-1,4 linkages. The extent of the branching can be controlled in artificial syntheses of dextrans. A tightly branched network having holes of a specified size can be produced. One of the types of dextrans available commercially is called Sephadex. It is used in biochemical laboratories throughout the world to fractionate mixtures of polymers gently and cleanly.

Our last polysaccharide is cellulose, the main structural component of the cell walls of plants:

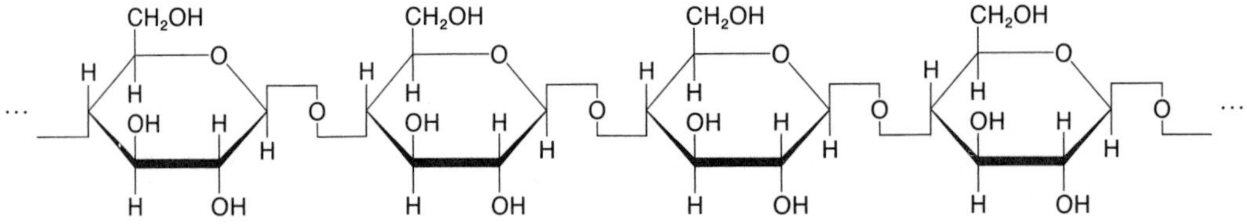

The cellulose chains are unbranched. The β-1,4 linkage permits the chains to be quite close to each other, with extensive interchain hydrogen bonding. This results in long microfibrils consisting of bundles of parallel cellulose chains and possessing great mechanical strength.

11.8 NUCLEIC ACIDS

Nucleic acids share some of the properties of the proteins and polysaccharides. They are built up by the successive elimination of a molecule of water from two monomers. They are intermediate between the compositional and structural complexity of the proteins and the simplicity of many polysaccharides. In contrast to the proteins, a few of which have been synthesized *in vitro,* no viable DNA has been synthesized without using some material that was obtained from a living cell. DNA shares the property of helicity with proteins, although there are two strands in the DNA helix versus one in the α helix. Nucleic acids are distinctive in having large amounts of phosphorus. Because the nucleic acids direct the synthesis of proteins, they can be regarded as the master molecules of the cell. When we complete our discussion of the structures of the several types of DNA and RNA, we will examine this fascinating function of the nucleic acids.

The Primary Structure of the Nucleic Acids

Both types of nucleic acids have the same kind of primary structure, i.e., they both have the same kind of backbone. Hydrolysis of nucleic acids with acid, base, or specific enzymes has shown that their backbones consist of alternating molecules of phosphate and ribose or deoxyribose, with attached purines or pyrimidines. The phosphodiester linkages occur at the 3′ and 5′ positions on the sugar rings. Thus, two ribonucleotides are linked as follows:

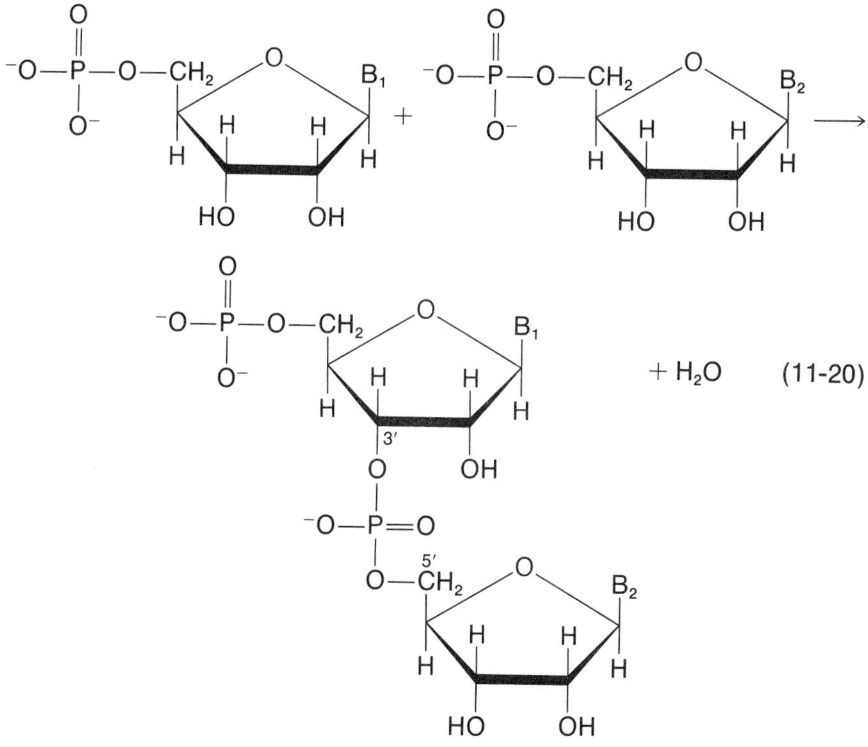

$^{-}O-P(=O)(O^{-})-O-CH_2$ (ribose, B_1; HO, OH) + $^{-}O-P(=O)(O^{-})-O-CH_2$ (ribose, B_2; HO, OH) $\longrightarrow$

$^{-}O-P(=O)(O^{-})-O-CH_2$ (ribose, B_1; 3′–O, OH) — $^{-}O-P=O$ — O–5′CH_2 (ribose, B_2; HO, OH) $+\ H_2O$ (11-20)

The bases B_1 and B_2 could be adenine (A), cytosine (C), guanine (G), thymine (T), or uracil (U). The dinucleotide shown here could be considered to be part of an RNA but not a DNA molecule because the sugars are ribose, not deoxyribose.

It is a somewhat tiring but useful exercise to write the complete structure of a small segment of DNA or RNA. Figure 11-14 shows the complete structure of a tetranucleotide segment of a DNA molecule. Note that all the sugars are deoxyribose. A much simpler diagram of the same segment is shown below:

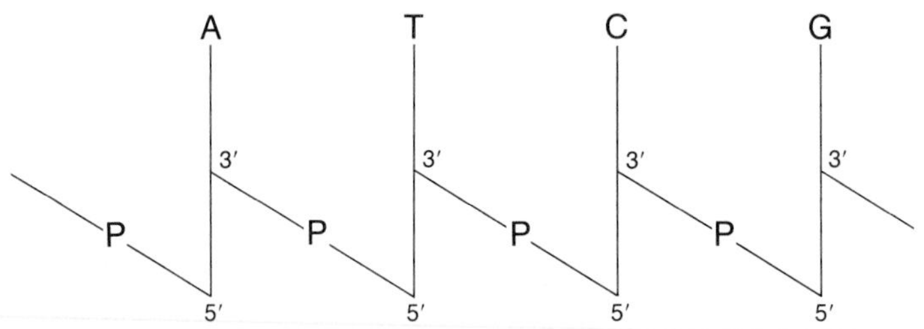

Here the vertical lines represent the nucleosides (sugar plus base) as labeled, and the diagonal lines with the letter P represent the phosphodiester linkages. Still simpler is the following abbreviation:

pApTpCpG

for which it is understood that, by convention, a letter p to the left of a base symbol denotes a 5′-phosphate linkage, and to the right, a 3′-linkage. Thus, by convention, the base sequence in a polynucleotide is always written in the $5' \rightarrow 3'$ direction.

Exercise 11-5. Draw the detailed structure of the RNA pentanucleotide segment UpGpUpCpAp.

Both DNAs and RNAs have the primary structure shown in Figure 11-14, the only difference being in the sugar. DNA molecules are among the largest known, with as many as tens of millions of nucleotides and molecular weights up to tens of billions. One striking feature that we should not overlook is the charge on the molecule. Nucleic acids are polyanions bearing enormous negative charges, owing to the ionized phosphate groups. This is of importance in considering the structure of the molecule in solution and how it might be crystallized from a solution.

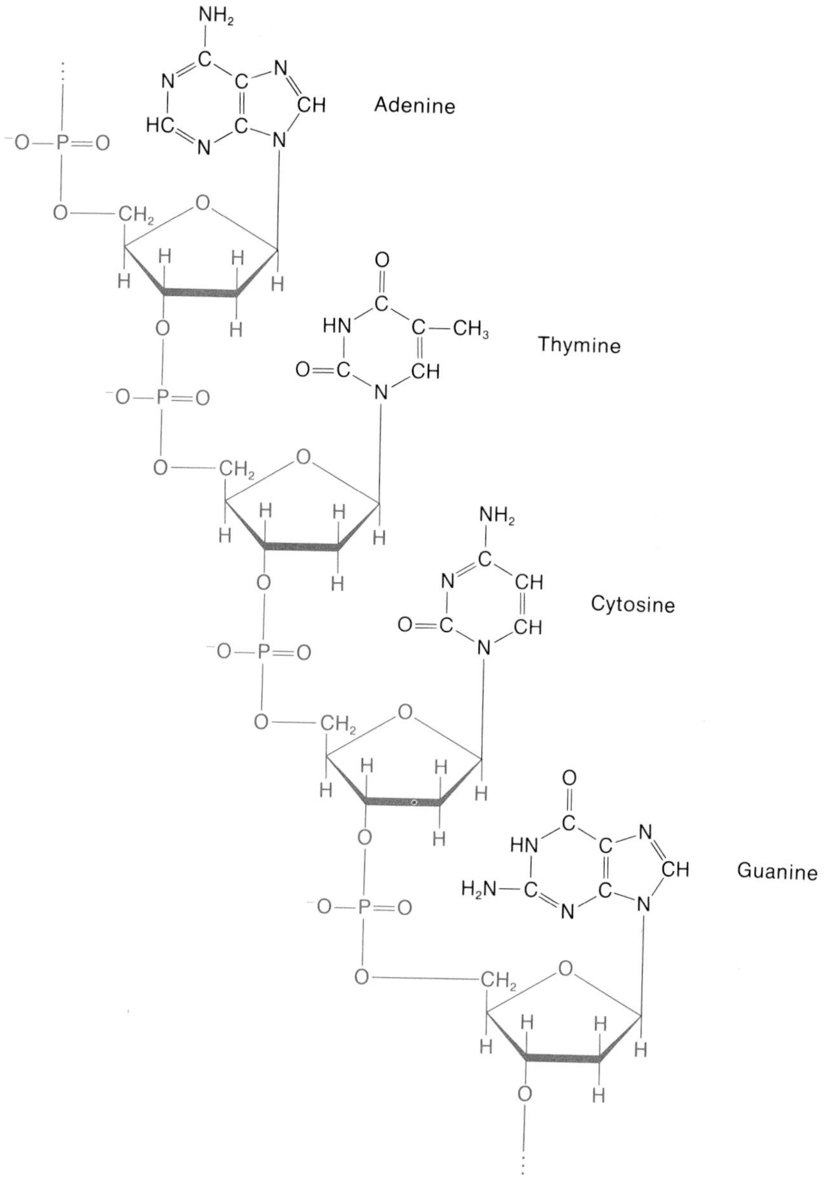

Figure 11-14. A tetranucleotide segment of a DNA molecule.

This completes our discussion of the primary structure of the nucleic acids. Unlike the proteins, for a number of which the exact amino acid sequences are known, no complete base sequence is known for any DNA molecule, and only a few sequences for some of the smaller RNAs. The reasons for this are the huge sizes of DNA molecules, the presence of only four different nucleotides (which makes the sequential analysis more difficult rather than simpler), and the very few specific enzymes that are available for cleaving the nucleotide chain at a particular location. Much information is known, however, about overall base composition, and it played a vital role in the determination of the secondary structure of DNA. It will be profitable now to discuss the structures of DNA and the several RNAs separately.

The Structure of DNA

In the late 1940s, the Austrian-American biochemist Erwin Chargaff (b. 1905) studied the base compositions of a wide variety of DNAs. With few exceptions he found the following to be true: (1) all cells within an organism have the same DNA, (2) the base composition is characteristic of the organism, (3) the base composition, as measured by, say, mole% (G + C), varies widely, and (4) the numbers of moles of the four bases per mole of DNA obey the equations

$$A = T \qquad \text{or} \quad A/T = 1$$

$$G = C \qquad \text{or} \quad G/C = 1$$

$$A + G = T + C \qquad \text{or} \quad \text{purines} = \text{pyrimidines}$$

$$A + C = T + G$$

These equations have come to be known as *Chargaff's rules*. Whereas the above ratios were found to be quite constant (except in a few viral DNAs), the ratio (A + T)/(G + C) varied widely. In higher animals it ranged up to 1.9, and in microorganisms, down to 0.4.

Another interesting fact about DNA was known at that time. You may recall that one of the characteristic features of the nucleotides is their absorption maximum at about 260 nm. A reasonable assumption might be that the absorbance of a DNA molecule at 260 nm would be the sum of the absorbances of its constituent nucleotides. The absorbance of DNA, however, is found to be as much as 40% *less* than that of its constituents. This effect is known as *hypochromism*. Another strange effect occurs as a solution of DNA is heated. Figure 11-15 shows the effect of heating on the absorbance of DNA.

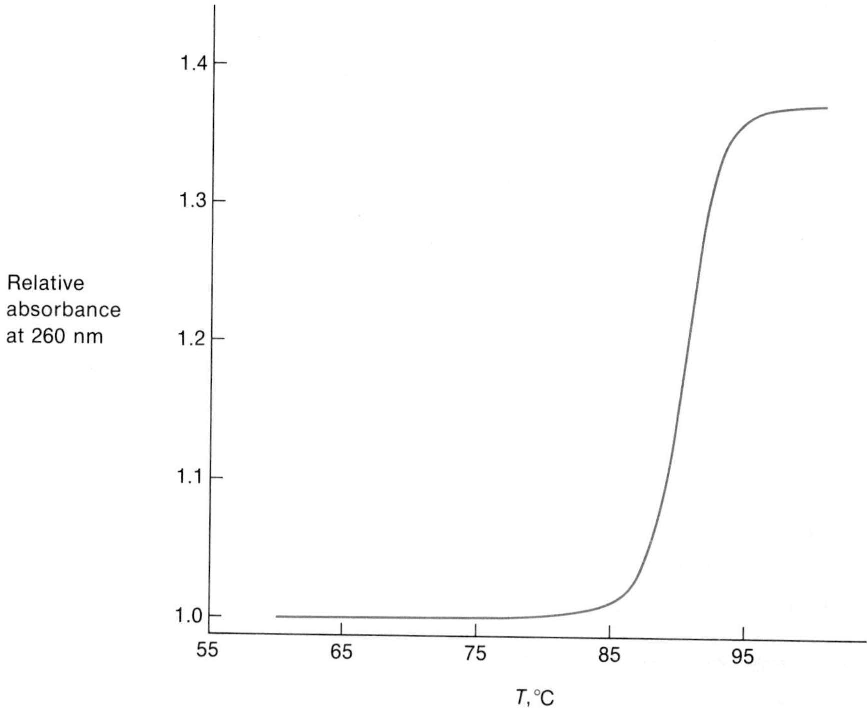

Figure 11-15. The relative absorbance of DNA at 260 nm as a function of temperature.

The information outlined above, on the base composition and the hypochromicity of DNA, set the stage for one of the great scientific adventures of all time. In his well-known book, *The Double Helix,* James Watson (b. 1928) has described the human as well as scientific elements in his work with Francis Crick (b. 1916) that led to the correct formulation of the secondary structure of DNA in 1953.* We recommend this book highly.

Watson and Crick provided almost none of the experimental data on which their proposed structure depended. Maurice Wilkins (b. 1916) and Rosalind Franklin (1920–1958) at King's College, London, provided the necessary x-ray data, Jerry Donohue (b. 1920), an American crystallographer, helped formulate the hydrogen-bonding scheme, and Chargaff provided the base composition data. The 1962 Nobel Prize in medicine and physiology was awarded to Wilkins for his x-ray work and to Watson and Crick for their brilliant synthesis of the data to formulate the correct structure.

Their basic hypothesis was that DNA consists of two polynucleotide strands, wound helically around each other with appropriate base pairs

*J. D. Watson, *The Double Helix,* Atheneum, New York, 1968.

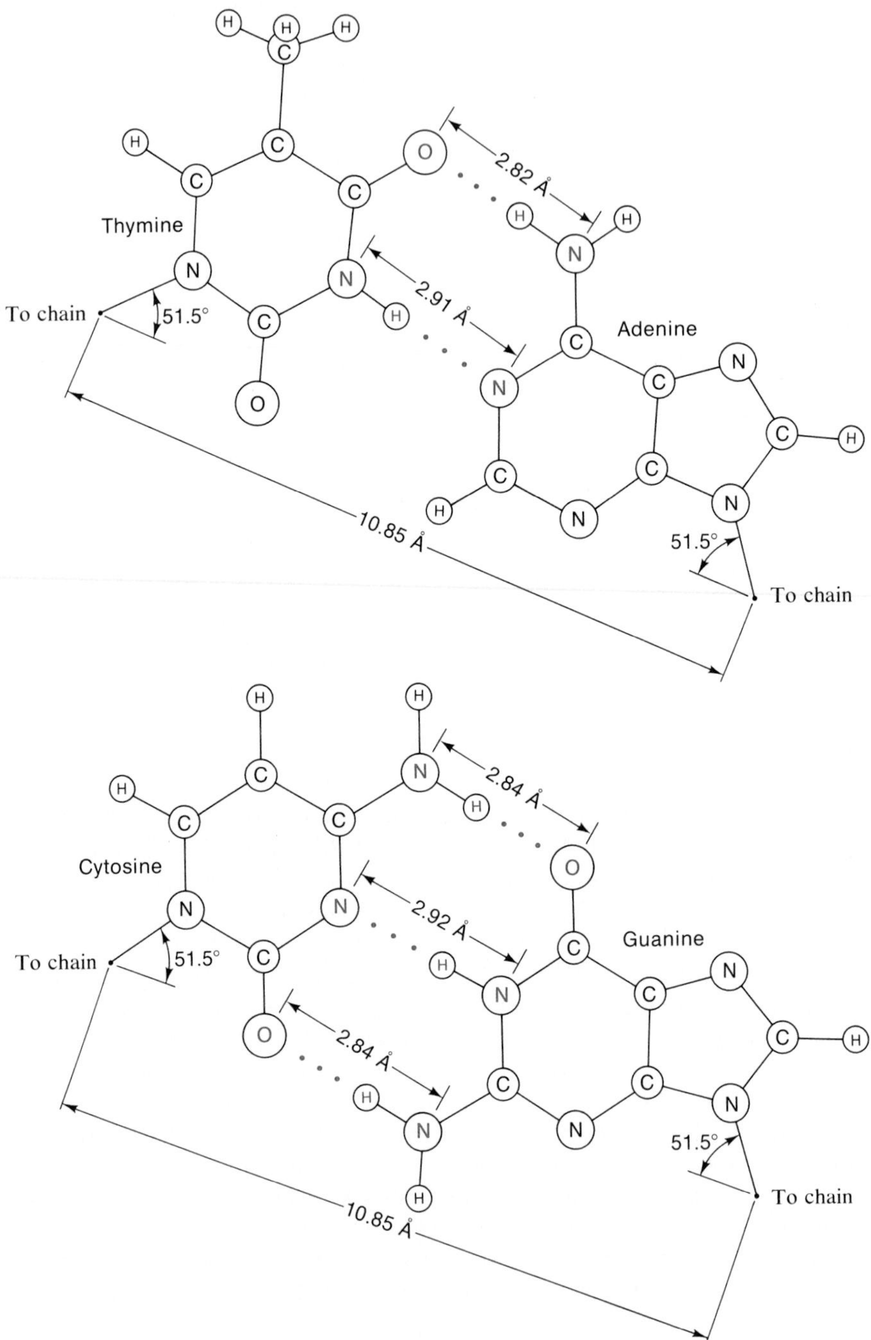

Figure 11-16. The base pairs found in the Watson-Crick double helix.

linking the two strands through hydrogen bonds. The key to their success was determining which bases were paired. They postulated that A always paired with T, and G always with C. The former pair, now called a *base pair,* was held together by two hydrogen bonds, and the latter base pair by three hydrogen bonds. The presently accepted base-pairing schemes, along with certain critical dimensions, are shown in Figure 11-16. The significance of the colored diagonal lines in Table 11-10 is now apparent.

Several features of these base pairs should be noted. If the double helix is to be at all regular, two requirements must be (and are) met: The distances between the C-1 atoms of the two ribose rings must be essentially the same for the two pairs, and so must the angles that the base pairs make with the two chains. For a long time it was believed that the hydrogen bonds in the base pairs held the two chains together. More recently it has been realized that hydrogen bonds of virtually the same strength could be formed with water molecules, and thus no decrease in the energy of the system would result from hydrogen bonding between the bases.

The primary forces now believed to be responsible for the stability of the Watson-Crick double helix are the van der Waals interactions between the planar base pairs, which are "stacked" on top of each other in the molecule. Figure 11-17 shows a schematic and a space-filling model of the DNA double helix. The sugar-phosphate groups form the backbone of each helix and the bases point inward. The π-electron systems of the base pairs virtually fill the core of the helix and produce strong positive interactions. These interactions are similar to those holding together the planes of conjugated carbon atoms in graphite. The hydrogen bonds between bases provide the correct orientation of the base pairs, but their role in the energetics of helix formation appears to be little more than compensating for the energy lost in breaking the base-water hydrogen bonds.

The double helix accounts nicely for the two sets of information we outlined at the beginning of this section. The base-pairing scheme requires that $A = T$, $G = C$, and $A + G = T + C$, i.e., the number of purines equals the number of pyrimidines. The stacking of the bases is known from model-compound studies to lead to a reduction in the absorbance. As a solution of DNA is heated, we would expect that some of the AT and GC hydrogen bonds would break and the chains would start to unwind. This would eliminate the base stacking and cause the absorbance to increase, as is observed in Figure 11-15. Such curves are called *melting curves.* The melting temperature T_m corresponding to the inflection point of the curve depends on the base composition of the DNA. Because three hydrogen bonds are formed between G and C, we would expect that, as the mole% (G + C) increased, so would T_m. Figure 11-18 illustrates the linear relation between these two variables, which has been observed in over 40 DNAs.

The explanations of the hypochromicity of DNA and the data in Figures 11-15 and 11-18 are of great significance. Because the two strands of DNA

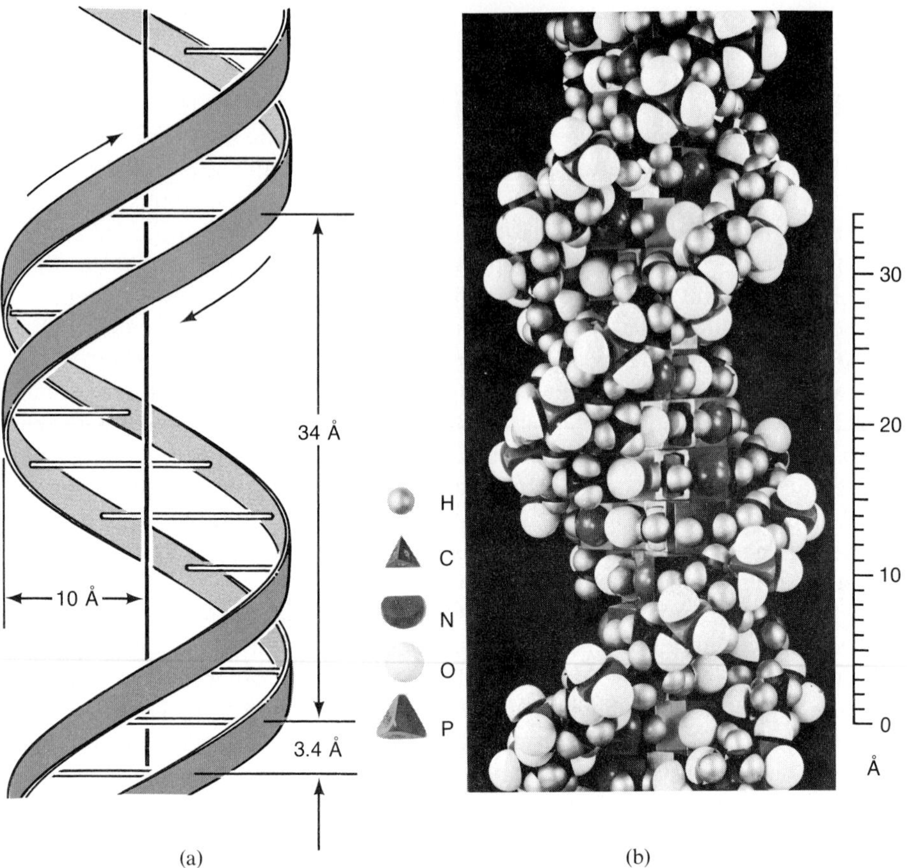

Figure 11-17. The DNA double helix. (a) Schematic. (b) Space-filling model. [Part (b) courtesy of Prof. M. H. F. Wilkins, King's College, London.]

can unwind and because a given base can hydrogen-bond to only one other base, we have the first glimmering of how one DNA molecule can give rise to two exact copies of itself. We will postpone a detailed look at this process, called *replication,* until we consider the other half of the nucleic acid team, the ribonucleic acids.

The Structure of the RNAs

We have been referring to ribonucleic acid in the plural. There are three principal kinds of RNA, which differ in size, function, and location in the cell. The two things they have in common are the sugar-phosphate backbone and predominantly the same four bases, A, G, C, and U. Their structure is that shown in Figure 11-14, except that the sugar is ribose and uracil replaces thymine.

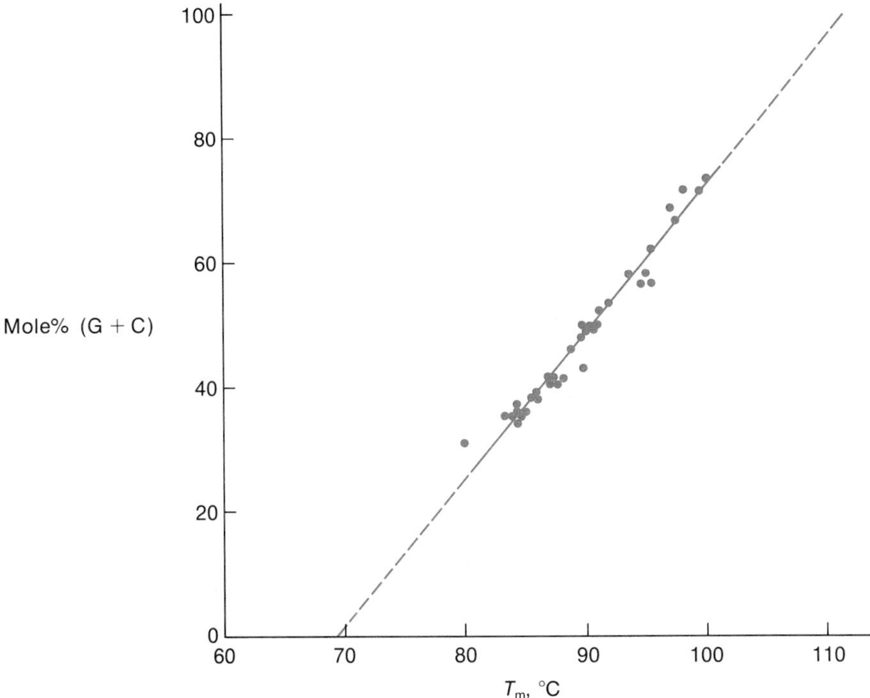

Figure 11-18. The melting temperature of a variety of DNAs as a function of the mole percent of guanine plus cytosine. (J. Marmur, C. Schildkraut, and P. Doty, in *The Molecular Basis of Neoplasia,* University of Texas Press, Austin, Texas, 1962.)

In contrast to DNA, an RNA molecule shows no regular relations among the four bases. This and other evidence rule out the Watson-Crick structure. In fact, all RNAs consist of single strands. As we will see, however, this does not preclude some forms of secondary structure. Generally speaking, RNA molecules are long, floppy, highly charged polynucleotide strands. We will discuss the structures of the three principal kinds of RNA very briefly here and consider their roles in the section on the genetic code.

Ribosomal RNA, or rRNA, constitutes 75–80% of the RNA of a cell. It is found in the ribosomes, which are small globular particles with the general shape shown below.

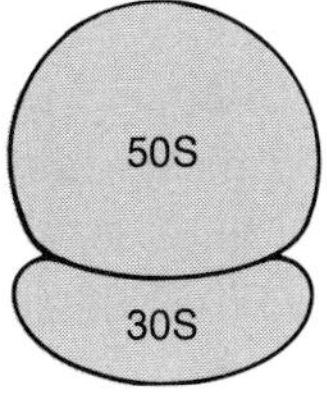

The symbol S will be explained later. For now, it can be used as a measure of size. The 50S part of the ribosome contains two molecules of RNA, of

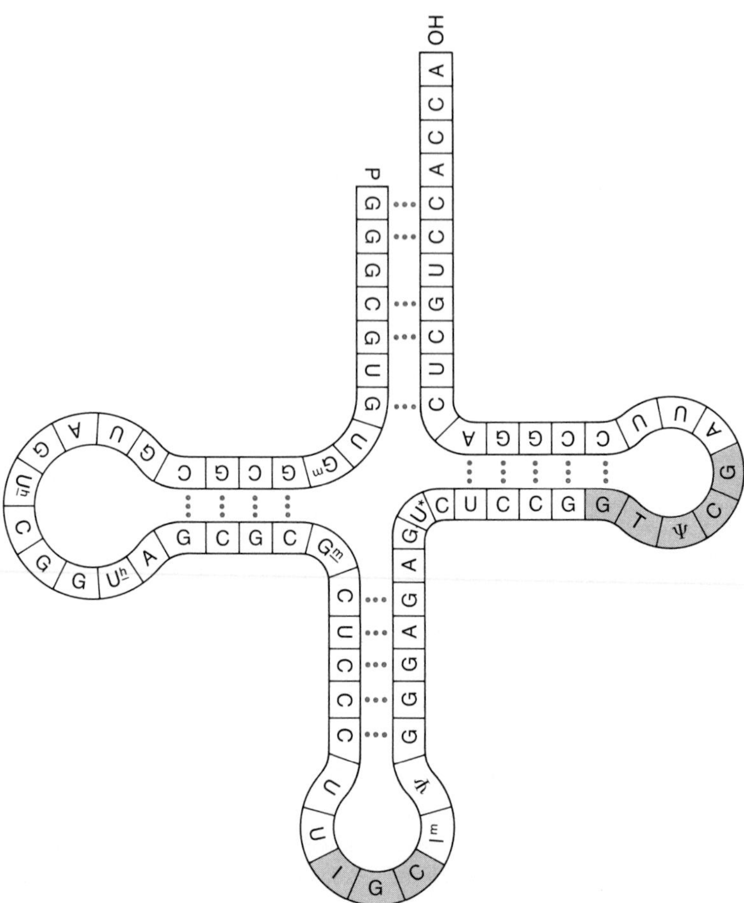

Figure 11-19. A proposed secondary structure for alanine tRNA, showing hydrogen bonds between various nucleotides in the chain. (R. W. Holley, "The Nucleotide Sequence of a Nucleic Acid." Copyright © 1966 by Scientific American, Inc. All rights reserved.)

sizes 23S and 5S, and about 30 proteins. The 30S part of the ribosome contains one molecule of 16S RNA, which has a molecular weight of about one million, and about 20 proteins.

Messenger RNA, or mRNA, is made by DNA in the cell nucleus and diffuses to the site of protein synthesis, the ribosomes. Its size is 30S and its molecular weight is also about one million. The base composition of mRNA is found to be exactly complementary to that of the DNA of the nucleus.

Transfer RNA, or tRNA, is a class of RNAs constituting 10–15% of the RNA of a cell. The tRNAs are the smallest (molecular weight about 25,000) of the ribonucleic acids. Not surprisingly, their structures are known in more detail than those of the other kinds. The primary structures of several tRNAs have been determined. In addition, chemists have predicted the existence of some helical regions owing to the possibilities of base pairing when the chain loops back on itself. Figure 11-19 shows a possible secondary structure for alanine tRNA based on the formation of the maximum possible number of base-pair hydrogen bonds. Interestingly, all tRNAs are believed to have the same type of cloverleaf structure, with three major loops and perhaps a smaller fourth one. Note the characteristic presence of a number of bases other than the usual A, G, C, and U. These bases are also capable of participating in hydrogen bonds. The sequence of bases shaded in the right-hand loop is invariant in all the tRNAs investigated to date. The significance of the base sequence shaded in the bottom loop will be explained later.

Some of the tRNAs have been crystallized and their tertiary (three-dimensional) structures are being investigated by means of x-ray diffraction. Recently some new classes of RNAs have been reported. These include the supposed precursors to mRNA, which are called *heterogeneous nuclear RNA,* or hnRNA, and many different RNAs in viruses.

the genetic code

Biochemists have known for a long time that DNA is the genetic material. Chemical mutagens acting only on DNA produce nearly 100% mutagenesis, or changes in the organism's characteristic structure. The concentration of DNA is constant in all cells of the same species, even under famine conditions. The characteristics of a cell can be entirely changed by the injection of DNA into the cell by a virus. This function of DNA was shown most conclusively by the Canadian physician Oswald Avery (1877–1955) and his colleagues in 1944. They showed that the ability of an extract from cells to transform other cells is totally lost upon the addition of an enzyme, deoxyribonuclease, which degrades DNA but not RNA or proteins.

If DNA is the central repository of genetic information, what is the role of RNA in synthesizing proteins? How are the amino acids directed to assemble themselves in one particular sequence out of the nearly infinite number possible? The answers to these questions and a host of others are central to an understanding of the chemistry of life. We can merely scratch the surface here, but we hope you will be stimulated to want to read further in such books as Watson's *Molecular Biology of the Gene* or Lehninger's *Short Course in Biochemistry.**

*See the Bibliography for the references to these books.

Francis Crick was the first to formulate the *central dogma of molecular biology*. This set of rules states that information flows from DNA to RNA to protein. Although some recent results suggest that information can also flow in the reverse direction, this pattern appears to be correct in most normal cells. Three processes are involved in this scheme: *replication, transcription,* and *translation*. We will consider them in turn.

11.9 REPLICATION

The double-helix structure of DNA immediately suggests a model for replication, based on the complementarity of the purine and pyrimidine bases: A bonds only to T and G bonds only to C. The bonds are hydrogen bonds, which, as we have already seen, can be broken by the heat denaturation of DNA. This allows the two strands to separate and uncoil. Each strand could then serve as a template for a new DNA molecule. A diagram of such a scheme is shown in Figure 11-20. For the scheme to work, it is essential that the template (single strand) be stable enough to survive the replication process. Since this process consists of the breaking and making of hydrogen bonds, which are rather weak, the requirement is easily fulfilled.

The mechanism shown in Figure 11-20 is called *semiconservative replication* because each double-stranded daughter molecule contains one strand from the parent molecule. This is shown more clearly in Figure 11-21. Upon continued replication, the two original parent strands remain intact, so there will always be found two DNA molecules having one parent strand each.

The experiment proving that the replication of DNA is indeed semiconservative was performed by Matthew Meselson (b. 1930) and Franklin Stahl (b. 1929) in 1958. Before we describe this famous experiment, however, let us examine the technique that made it possible—*analytical ultracentrifugation*.

The first analytical ultracentrifuge was built in 1924 by the Swedish physical chemist The Svedberg (1884–1971).* This instrument is designed to whirl a sample of a solution so rapidly (up to 67,000 rpm) that a centrifugal force equivalent to hundreds of thousands of times greater than that of normal gravity is developed. (The chamber must be kept under high vacuum to minimize frictional air drag and the severe mechanical stresses that it would cause at such enormous angular velocities of the rotor.) The rate of settling (or sedimentation) of suspended biopolymers such as proteins and nucleic acids can then be observed and interpreted in terms of the size and

---

*Svedberg's first name (short for Theodor) is pronounced *tay*.

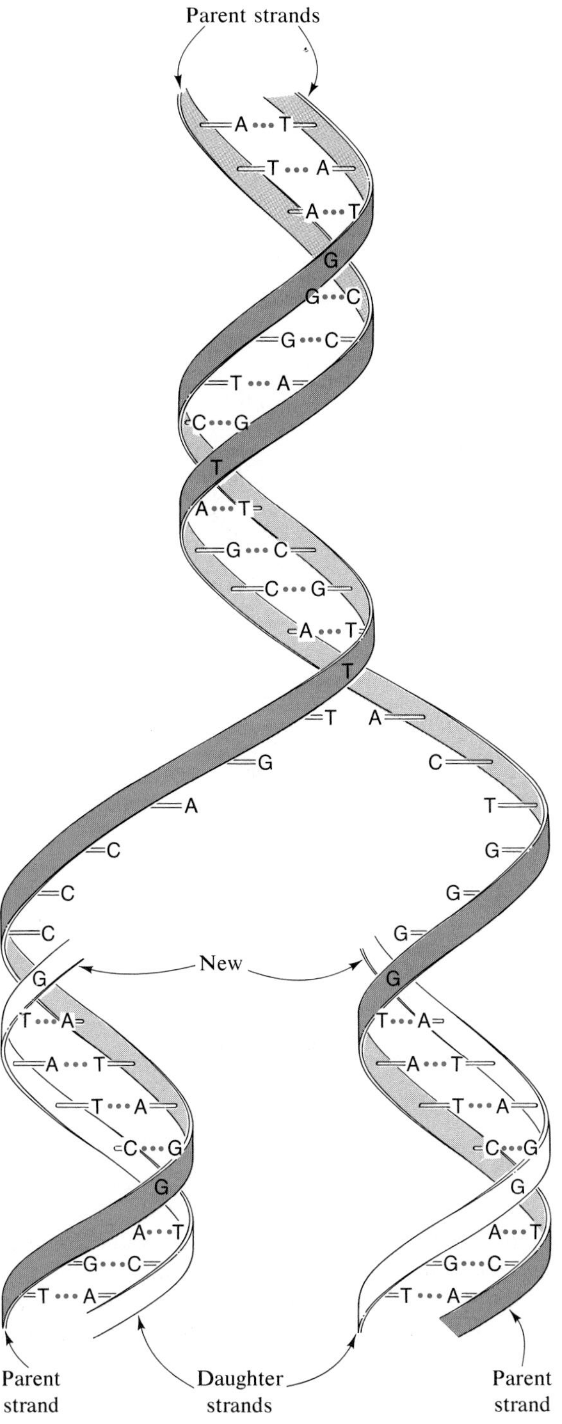

Figure 11-20. The presently accepted model for DNA replication.

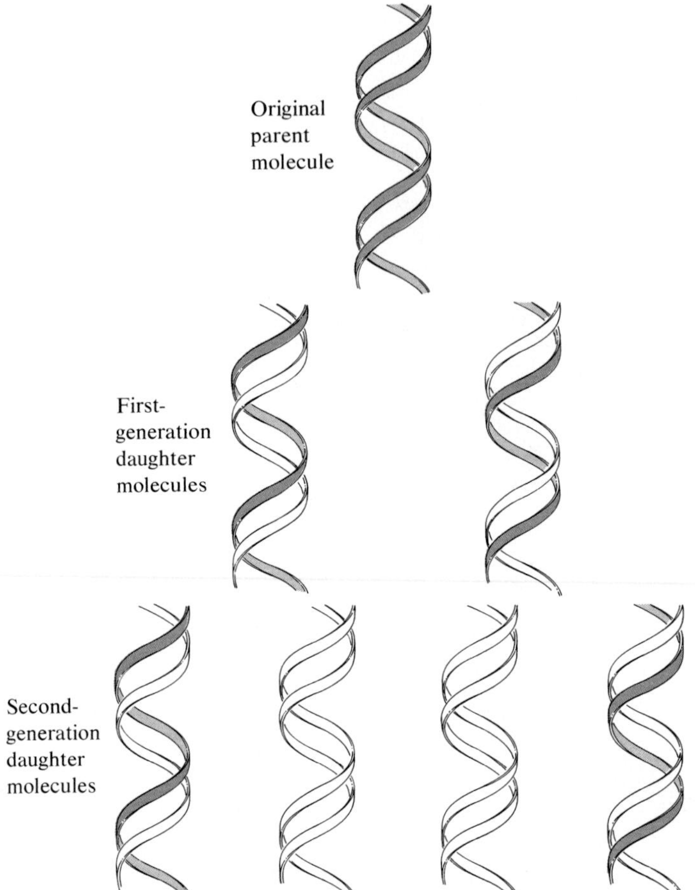

Figure 11-21. The mechanism of DNA replication proposed by Watson and Crick.

shape of the molecule. Figure 11-22 is a schematic of the three optical systems of the analytical ultracentrifuge, which make possible the continuous observation of the polymer in the cell. The optical system most commonly used for DNA is the ultraviolet absorption system, since DNA absorbs in the ultraviolet.

The sedimentation velocity dr/dt of the polymer molecule is directly proportional to the centrifugal acceleration $\omega^2 r$, where ω is the angular velocity of the ultracentrifuge rotor in radians per second and r is the distance from the center of rotation to the point of observation in the cell. The proportionality constant is called the *sedimentation coefficient s*. It is related to the size and shape of the polymer, and is given by the following equation:

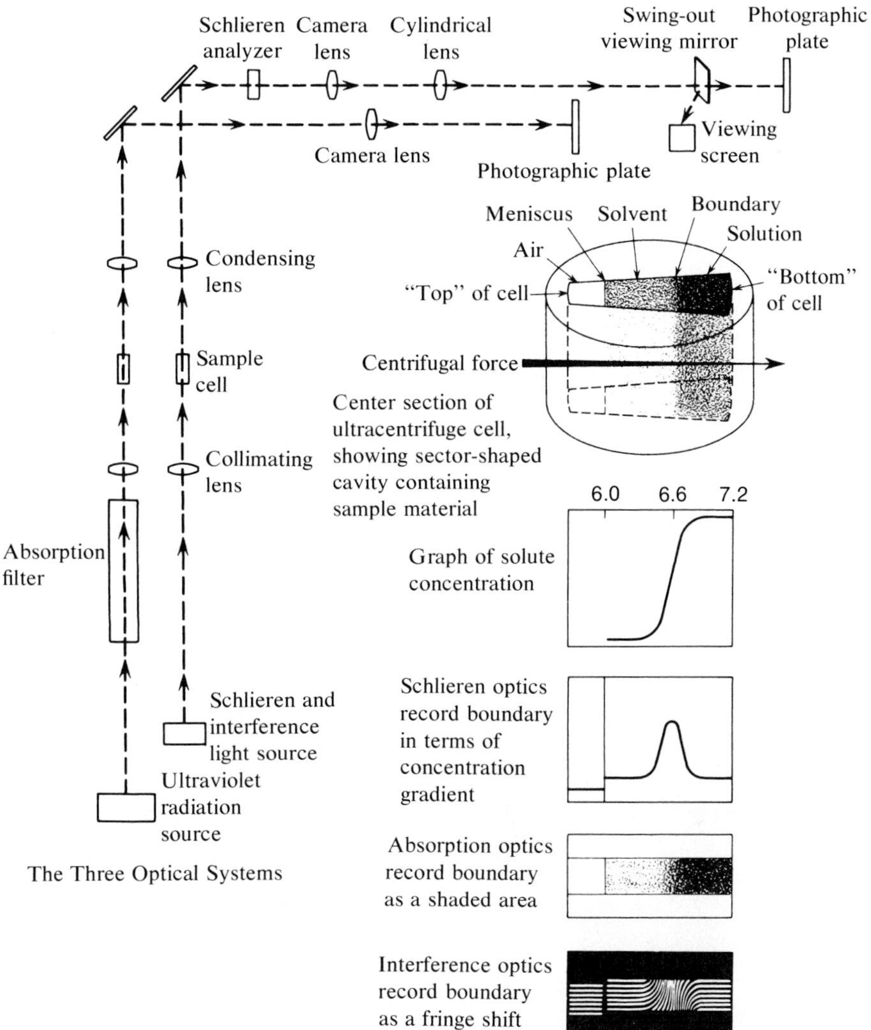

Figure 11-22. The three optical systems of the Spinco Model E analytical ultracentrifuge. (Courtesy of Spinco Division, Beckman Instruments, Inc.)

$$s = \frac{1}{\omega^2 r}\frac{dr}{dt} \tag{11-21}$$

The unit of the sedimentation coefficient is the svedberg (abbreviation S), which is 10^{-13} s. Typical values of s range from 1×10^{-13} s to 200×10^{-13} s, or from 1 S to 200 S. (Note the distinction between the three forms of the letter s used here.) This explains the previous discussion of 16S RNA, 30S ribosome, etc. A moving sedimentation boundary of a DNA sample is shown in Figure 11-23.

The experimental test of the semiconservative replication mechanism requires a physical separation of parent and daughter molecules by some means. The method used by Meselson and Stahl was to grow *E. coli* bacteria, the source of the DNA, in a ^{15}N-containing medium. The ^{15}N-labeled DNA has a slightly higher density than normal (^{14}N) DNA. Upon transfer of the ^{15}N bacteria to a ^{14}N medium for further growth, the new synthesis makes daughter DNA with a density slightly less than that of the parent DNA. The method used for separating these DNAs was sedimentation equilibrium in a density gradient. The density gradient was created by an increasing concentration of a CsCl solution from top to bottom in the ultracentrifuge cell. The DNA undergoes sedimentation in this gradient and forms a band where its density is equal to that of the CsCl solution. Figure 11-24 shows an artificially created density gradient in a mixture of benzene and bromobenzene. The four sets of balls in the gradient have slightly different densities and are therefore separated.

The results of the Meselson-Stahl experiment are shown in Figure 11-25.

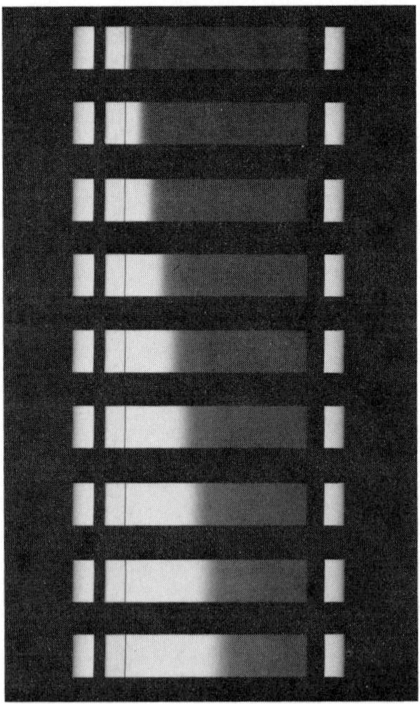

Figure 11-23. Ultraviolet absorption photographs of a DNA sample, obtained at 9945 rpm.

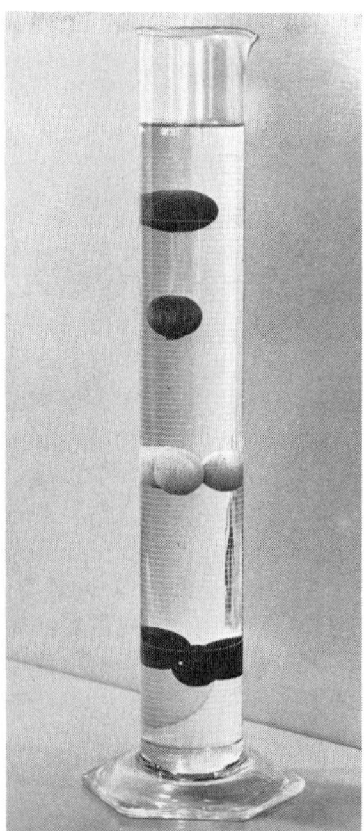

Figure 11-24. Balls of four different densities banded in a linear density gradient.

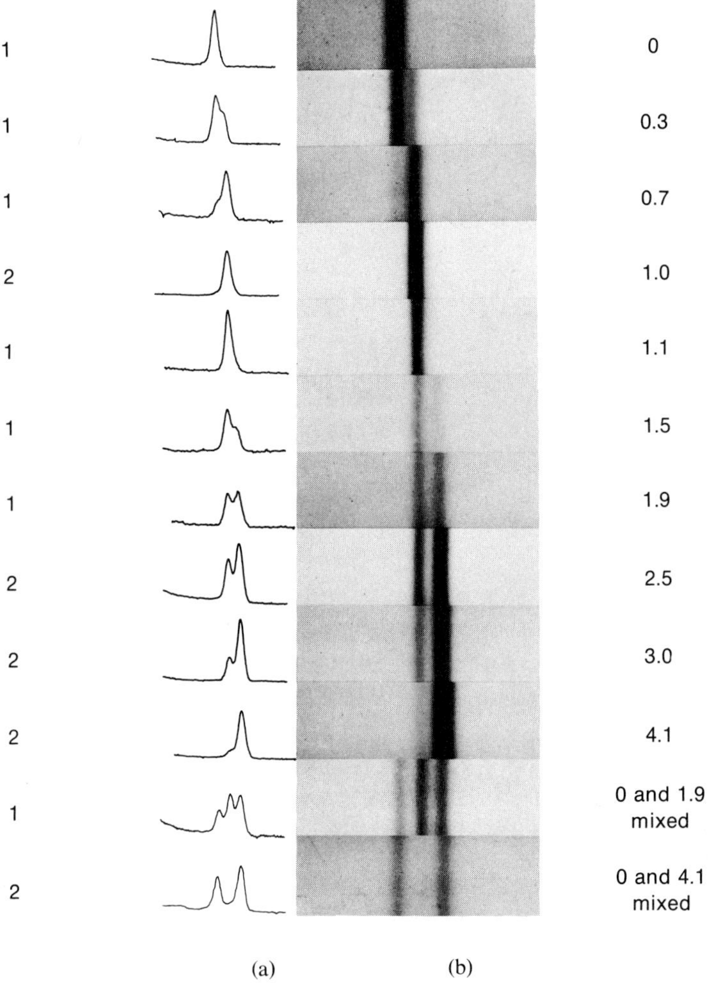

Figure 11-25. The results of the Meselson-Stahl experiment, which proved that DNA replication is semiconservative. (a) Ultraviolet absorption photographs showing DNA bands resulting from density-gradient centrifugation of lysates of bacteria sampled at various times after the addition of an excess of ^{14}N substrates to a growing ^{15}N-labeled culture. Each photograph was taken after 20 hours of centrifugation at 44,770 rpm under the conditions described in the text. The density of the CsCl solution increases to the right. Regions of equal density occupy the same horizontal position on each photograph. The time of sampling is measured from the time of the addition of ^{14}N in units of the generation time. The generation times were estimated from measurements of bacterial growth. (b) Microdensitometer tracings of the DNA bands shown in the adjacent photographs. The microdensitometer pen displacement above the base line is directly proportional to the concentration of DNA. The degree of labeling of a species of DNA corresponds to the relative position of its band between the bands of fully labeled and unlabeled DNA shown in the lowermost frame, which serves as a density reference. A test of the conclusion that the DNA in the band of intermediate density is just half-labeled is provided by the frame showing the mixture of generations 0 and 1.9. When allowance is made for the relative amounts of DNA in the three peaks, the peak of intermediate density is found to be centered at $50 \pm 2\%$ of the distance between the ^{14}N and ^{15}N peaks. [M. Meselson and F. W. Stahl, *Proceedings of the National Academy of Sciences of the USA* **44**, 671 (1958).]

We see that, after one generation time, or one round of DNA replication, a hybrid band (^{14}N-^{15}N) of DNA slightly less dense than the parent (^{15}N) DNA is formed. After several generations, a still lighter product of pure ^{14}N DNA is found. Since only these three products are ever found and their appearance corresponds to the known generation time of the bacterium, the semiconservative replication mechanism is proved.

11.10 TRANSCRIPTION

DNA produces the several RNAs by essentially the same mechanism by which it produces new DNA molecules. In addition to a strand of template DNA, the ribonucleoside triphosphates, a specific enzyme (RNA polymerase), and Mg^{2+} are required to effect the synthesis. This synthesis can be represented by the following reaction (recall that PP_i denotes the inorganic diphosphate group):

$$\begin{array}{c} \text{DNA template} \\ \begin{array}{|c|} \text{—A}\cdots\text{T—} \\ \text{—C}\cdots\text{G—} \\ \text{—T}\cdots\text{A—} \\ \text{—G}\cdots\text{C—} \end{array} \end{array} + \begin{array}{c} \text{ATP} \\ \text{GTP} \\ \text{CTP} \\ \text{UTP} \end{array} \xrightarrow[\text{RNA polymerase}]{Mg^{2+}} \begin{array}{c} \text{DNA template} \\ \begin{array}{|c|} \text{—A}\cdots\text{T—} \\ \text{—C}\cdots\text{G—} \\ \text{—T}\cdots\text{A—} \\ \text{—G}\cdots\text{C—} \end{array} \end{array} + \begin{array}{c} \text{RNA} \\ \begin{array}{c|} \text{U—} \\ \text{G—} \\ \text{A—} \\ \text{C—} \end{array} \end{array} + 4PP_i \qquad (11\text{-}22)$$

The sequence of bases in the RNA molecule is complementary to that in one of the DNA strands. We will see that this aspect of transcription is essential for the synthesis of viable proteins by the RNA.

11.11 TRANSLATION

Translation, the final process in protein synthesis, is by far the most complicated. Early in the investigation of the details of this process, some scientists explored the possibility of direct bonds between the amino acids and the mRNA molecule. We now know, however, that this does not occur. Instead, the bonding occurs indirectly via adaptor molecules. These adaptors are tRNA molecules that are specific for each amino acid. The bonding of the amino acid to the tRNA can be considered to be the first step of translation, called *amino acid activation*. The sequence of reactions is shown in Figure 11-26. The amino acid is bound to the end of the tRNA molecule, which always terminates with the sequence —C—C—A.

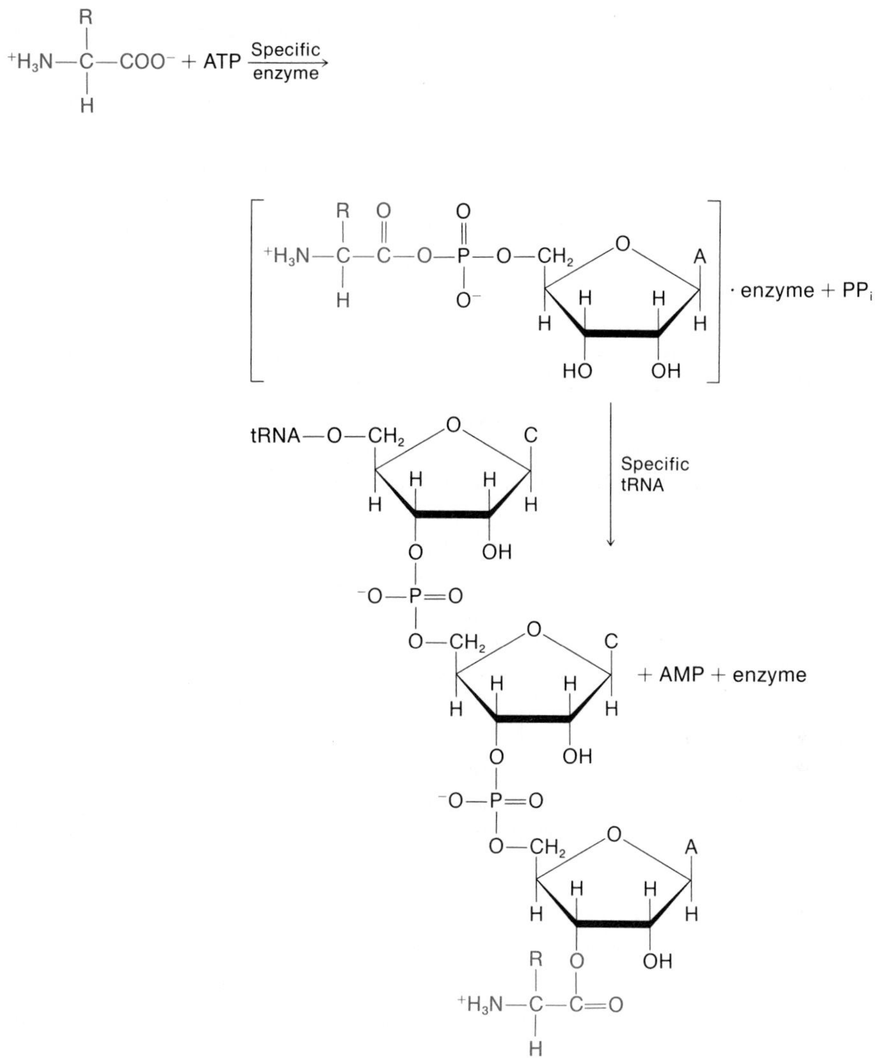

$$
{}^{+}H_3N-\underset{\substack{| \\ H}}{\overset{\substack{R \\ |}}{C}}-COO^- + ATP \xrightarrow[\text{enzyme}]{\text{Specific}}
$$

$$
\left[{}^{+}H_3N-\underset{\substack{| \\ H}}{\overset{\substack{R \\ |}}{C}}-\overset{\substack{O \\ \|}}{C}-O-\underset{\substack{| \\ O^-}}{\overset{\substack{O \\ \|}}{P}}-O-CH_2-\text{ribose(A)} \right] \cdot \text{enzyme} + PP_i
$$

$$
\xrightarrow{\text{Specific tRNA}}
$$

tRNA—O—CH_2—ribose(C)—O—PO_2^-—O—CH_2—ribose(C)—O—PO_2^-—O—CH_2—ribose(A)—O—C(=O)—C(R)(H)—NH_3^+ + AMP + enzyme

Figure 11-26. The reaction sequence of amino acid activation, in which an amino acid is bound to a specific tRNA molecule.

The next step of translation is *chain initiation.* All polypeptides are formed on the surfaces of ribosomes. The process is initiated by the formation of a complex between a molecule of mRNA, the smaller (30S) part of the ribosome, and a modified form of an amino acid, N-formyl methionine, bound to its tRNA. Next, the larger (50S) part of the ribosome is attached and protein synthesis can begin. It would appear that the first amino acid in every polypeptide chain ought to be N-formyl methionine. However, an enzyme removes it soon after the chain begins to grow.

The mRNA molecule has a directional sense, either $5' \rightarrow 3'$ or $3' \rightarrow 5'$. Chain reading always begins at the $5'$ end. Similarly, polypeptide chains always begin growing at the N-terminal end. Three enzymes, called F1, F2, and F3, as well as GTP and Mg^{2+}, are required to carry out the above initiation steps.

The final steps of translation are *chain elongation* and *chain termination.* The manner in which the ribosomes interact with the mRNA molecule and the tRNA-amino acid complexes is shown in Figure 11-27.

The function of rRNA molecules is to maintain the structure of the ribosome. The tRNA molecules transfer the activated amino acids to the mRNA molecule. The mRNA carries the genetic message from the DNA. But the crucial question is: How do the tRNAs know when to transfer which amino acid to the growing polypeptide chain at the correct point? The answer lies in the genetic code, which was first postulated by the Russian-American physicist George Gamow (1904–1968) in 1954. The details of the code were worked out primarily (and largely independently) by the American biochemist Marshall Nirenberg (b. 1927) and the Indian-American chemist Har Gobind Khorana (b. 1922) in the early 1960s, a feat for which they shared the Nobel Prize in 1968.

There are twenty amino acids and only four nucleotide bases. If one base coded for one amino acid, we would have only a four-letter code, and only four amino acids could be directed to positions in the chain. If a pair of

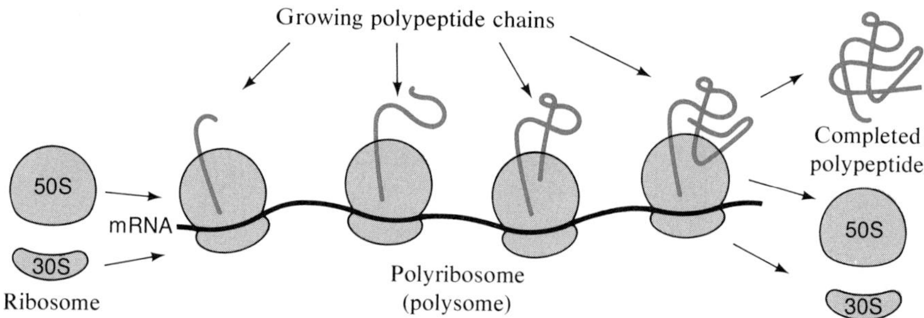

Figure 11-27. Protein synthesis by an mRNA molecule at the surface of a polyribosome.

adjacent bases coded for one amino acid, there would be $4^2 = 16$ possibilities for coding because AG, for example, represents a different code word than GA. But 16 code words are still not enough; we need at least 20. If a triplet of adjacent bases coded for one amino acid, there would be $4^3 = 64$ possibilities, e.g., AUG or UUU or CAU. This would be an extremely redundant code, with about three times as many code words as are needed. It turns out that nature, in this instance, is extremely redundant.

The code was cracked using synthetic polyribonucleotides and the cell's own synthesizing machinery. It was first found that pure polyuridylic acid, which contains only uracil as a base, generated pure polyphenylalanine. When the triplet nature of the code was established, it became clear that the code word, now called a *codon,* for phenylalanine was UUU. Similar experiments with the other three pure polyribonucleotides yielded three more pure polypeptides and three more codons. Then mixed polyribonucleotides were used. Statistical analysis of the frequencies of particular base sequences compared with the amino acid compositions of the resulting polypeptides led to the correct assignment of most of the codons. The last ambiguous cases, as well as the exact sequence of each triplet, were resolved by using trinucleotides of known base sequence, binding them to ribosomes, adding a mixture of radioactively labeled amino acids, and determining which tRNA-amino acid complex was bound to the ribosome. The ingenious simplicity of this technique was that the ribosomes, with their telltale polypeptides attached, could be separated from the unbound tRNA-amino acid complexes by simple filtration. The complete genetic code is shown in Figure 11-28.

Now the reason for the tRNA as an intermediary between the amino acid and the mRNA should be clear. The amino acid obviously has no base sequence with which to recognize a part of the mRNA sequence, but the tRNA does. The base sequence IGC at the bottom of the alanine tRNA molecule shown in Figure 11-19 is called the *anticodon.* It is complementary to the alanine codon on the mRNA strand, so the required hydrogen bonds are formed and the alanine is added to the growing polypeptide chain.

The process of protein synthesis is summarized in Figure 11-29. In transcription, one of the strands of the DNA double helix serves as a template for the production of mRNA, with uracil replacing thymine. In translation, the mRNA is then "read" at the surfaces of the ribosomes, and a specific protein emerges, one amino acid residue at a time.

The detailed events at the ribosome are shown in Figure 11-30. The process begins with the codon AUG or GUG of the mRNA at the peptide site on the ribosome. The tRNA for N-formyl methionine ($tRNA_F$) goes to the peptide site on the ribosome, initiating the protein synthesis. The tRNA for normal methionine ($tRNA_M$) goes to the amino acid site. A peptide bond

FIRST LETTER	SECOND LETTER: U		C		A		G		THIRD LETTER
U	UUU	Phe	UCU	Ser	UAU	Tyr	UGU	Cys	U
	UUC	Phe	UCC	Ser	UAC	Tyr	UGC	Cys	C
	UUA	Leu	UCA	Ser	UAA	*Stop*	UGA	*Stop*	A
	UUG	Leu	UCG	Ser	UAG	*Stop*	UGG	Trp	G
C	CUU	Leu	CCU	Pro	CAU	His	CGU	Arg	U
	CUC	Leu	CCC	Pro	CAC	His	CGC	Arg	C
	CUA	Leu	CCA	Pro	CAA	Gln	CGA	Arg	A
	CUG	Leu	CCG	Pro	CAG	Gln	CGG	Arg	G
A	AUU	Ile	ACU	Thr	AAU	Asn	AGU	Ser	U
	AUC	Ile	ACC	Thr	AAC	Asn	AGC	Ser	C
	AUA	Ile	ACA	Thr	AAA	Lys	AGA	Arg	A
	AUG	Met	ACG	Thr	AAG	Lys	AGG	Arg	G
G	GUU	Val	GCU	Ala	GAU	Asp	GGU	Gly	U
	GUC	Val	GCC	Ala	GAC	Asp	GGC	Gly	C
	GUA	Val	GCA	Ala	GAA	Glu	GGA	Gly	A
	GUG	Val	GCG	Ala	GAG	Glu	GGG	Gly	G

Figure 11-28. The genetic code. The three codons marked *Stop* do not code for any amino acids, but rather for termination of the polypeptide chain.

is formed between the N-formyl methionine and the methionine. The ribosome then moves along the mRNA, placing the methionine at the peptide site. This allows the $tRNA_F$ to be released to the solution, and frees the amino acid site for the next tRNA, the one for valine ($tRNA_{Val}$). The second peptide bond is formed between methionine and valine, and so the polypeptide chain grows. This process continues until a termination codon reaches the amino acid site, at which point the completed protein is released to the solution.

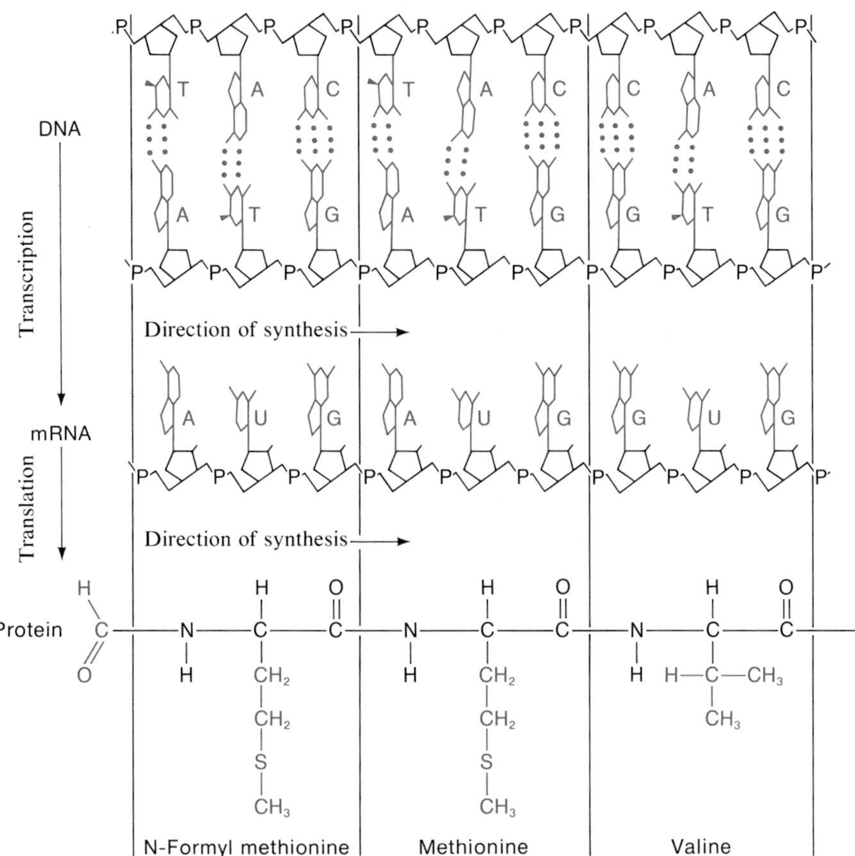

Figure 11-29. The transmission of genetic information from DNA to mRNA and from mRNA to protein. The HCO group at the beginning of the protein molecule is the formyl group. (B. F. C. Clark and K. A. Marcker, "How Proteins Start." Copyright © 1968 by Scientific American, Inc. All rights reserved.)

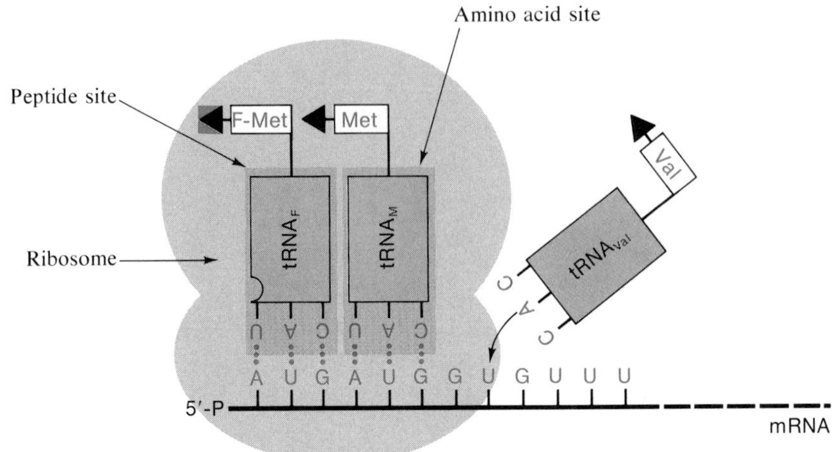

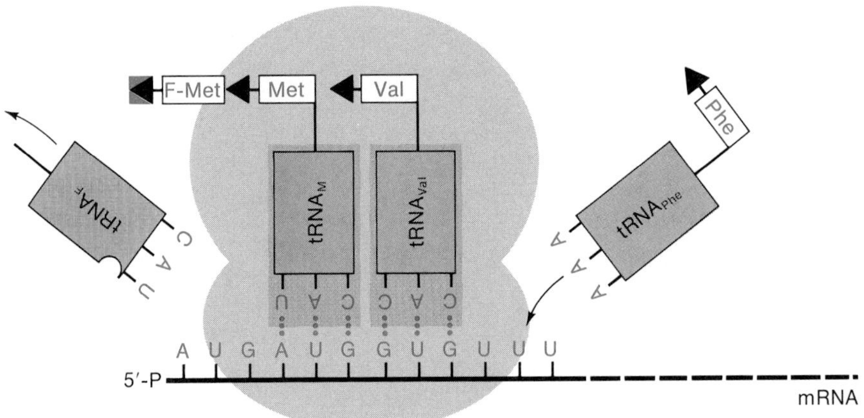

Figure 11-30. The details of protein synthesis at the ribosome. (B. F. C. Clark and K. A. Marcker, "How Proteins Start." Copyright © 1968 by Scientific American, Inc.)

Bibliography

J. M. Barry and E. M. Barry, **An Introduction to the Structure of Biological Molecules,** Prentice-Hall, Englewood Cliffs, N.J., 1969. The content of this book closely parallels that of the present chapter, at a nearly identical level. Following an introductory discussion of structural principles and molecular asymmetry, the structures of carbohydrates, proteins, and nucleic acids are presented. The book concludes with a good introduction to the chemistry of lipids.

Thomas P. Bennett and Earl Frieden, **Modern Topics in Biochemistry,** Macmillan, New York, 1966. The order and level of presentation of the material in this book are nearly identical to those of the present chapter. The structures of the monomers of proteins, carbohydrates, and nucleic acids, as well as of the polymers themselves, are discussed in turn. This is followed by a discussion of macromolecular biosynthesis.

Paul L. Cook and John W. Crump, **Organic Chemistry: A Contemporary View,** D. C. Heath, Lexington, Mass., 1969. A concise but readable survey of organic chemistry is presented. The structural theory of organic chemistry, methods for structural determination, and organic reactions and syntheses are covered. The book concludes with a short discussion of natural and synthetic polymers.

M. O. Dayhoff, ed., **Atlas of Protein Sequence and Structure 1972**, Vol. 5, National Biomedical Research Foundation, Washington, D.C., 1972. This is a compilation of most of the known peptide and protein sequences. Many interspecies sequence comparisons are made. The tables are prefaced by a short course on selected topics of biochemistry, emphasizing evolutionary protein studies and the genetic code.

R. E. Dickerson and I. Geis, **The Structure and Action of Proteins,** W. A. Benjamin, Menlo Park, Cal., 1969. This is probably the finest introductory-level book on proteins. It combines the talents of a gifted scientist-writer with those of a noted scientific illustrator. Professor Dickerson begins with atomic orbitals and brings the reader all the way to a detailed discussion of the structure-function relations of seven proteins. Mr. Geis's drawings illuminate almost every page. We recommend this book highly.

Werner Herz, **The Shape of Carbon Compounds,** W. A. Benjamin, New York, 1963. A somewhat classical presentation of organic chemistry is given in this book. It is a good source of information on the molecular shapes of optical isomers.

Kenneth D. Kopple, **Peptides and Amino Acids,** W. A. Benjamin, New York, 1966. This book gives a more detailed discussion of amino acids and peptides than that in the present chapter. Methods of peptide synthesis in common use up to 1964 are particularly well described.

Albert L. Lehninger, **Short Course in Biochemistry,** Worth, New York, 1973. Professor Lehninger has written a shorter, simpler version of his well-known text, *Biochemistry.* The sections on biomolecules and the transfer of genetic information are particularly relevant to the present chapter. There are also two large sections dealing with metabolism and phosphate-bond energy.

Robley J. Light, **A Brief Introduction to Biochemistry,** W. A. Benjamin, New York, 1968. A brief overview of biochemistry through 1967 is presented. Topics of particular interest, such as aging and cancer, are discussed in the last chapter.

James D. Watson, **Molecular Biology of the Gene**, 2nd ed., W. A. Benjamin, New York, 1970. This is a classic work—a wide-ranging and zesty account of molecular biology by one of its most outstanding practitioners. The material in the first four chapters is particularly relevant to the present chapter. The level of difficulty is not far above that of the present text. There is a considerable amount of biology, and the treatment is almost completely nonmathematical.

James D. Watson, **The Double Helix,** Atheneum, New York, 1968. This is an extremely readable and entertaining account of the events of the hectic months in 1952–53 that culminated in one of the greatest triumphs of modern science, the correct formulation of the three-dimensional structure of DNA by Watson and Crick. Although there has been criticism of Watson's highly personal views of some of the people who figured prominently in the story, one can only admire his candor in revealing the human foibles (his own included) that are as much at home in the ivory towers of learning as anywhere else.

Problems

1. Draw the structural formulas of the following compounds:
   a. 2-Methyl-3,5-diethylheptane
   b. 1-Chloro-2,2-dimethylpropane
   c. 5-Methyl-2-hexene
   d. 3-Methyl-2-butanol
   e. Methylethyl ether
   f. 3-Chloro-3-ethyl-5-methylheptanoic acid
   g. 2-Pentanone
   h. 5-Methyl-2-hexenal

2. Give the systematic name for each of the following compounds:

```
a. CH3—CH2—CH—CH—CH2—CH3
            |   |
           CH3  CH—CH2—CH3
                |
                CH3

         CH3       CH3
         |         |
b. CH3—C—CH2—CH—CH=CH2
         |
         CH3
```

c. $\overset{CH_3}{\underset{H}{}}\!\!>C{=}C<\!\!\overset{H}{\underset{CH_3}{}}$

d. $CH_3{-}CH_2{-}\underset{OH}{\overset{CH_3-CH-CH_2-CH_3}{\overset{|}{\underset{|}{C}}}}{-}CH_2{-}CH_3$

e. $CH_3{-}CH_2{-}O{-}CH_2{-}CH_3$

f. $CH_3{-}\underset{CH_3}{\overset{CH_3}{C}}{-}CH_2{-}\underset{CH_3}{\overset{CH_3}{C}}{-}C\overset{O}{\underset{H}{\lessgtr}}$

g. $CH_3{-}\underset{\|\atop O}{C}{-}CH_2{-}\underset{CH_3}{CH}{-}CH_3$

3. Draw the structural formulas of three isomeric tertiary alkyl chlorides with the molecular formula $C_6H_{13}Cl$, and of three isomeric secondary alkyl chlorides with the molecular formula $C_5H_{11}Cl$.

4. Draw the structural formulas of the isomers of each of the following molecular formulas:

a. C_5H_{12}
b. $C_4H_{10}O$
c. $C_6H_4Cl_2$ (benzene-ring isomers)

5. Draw the resonance structures of phenylalanine, tryptophan, and arginine.

6. Complete the following reactions, indicating the reaction conditions where appropriate:

a. $C_6H_5{-}CH_2CH_2CH_2C(=O)OH + C_2H_5OH \xrightarrow{H_2SO_4}$ _____ + _____

b. $C_6H_5{-}C(=O)OH$ + _____ $\longrightarrow C_6H_5{-}C(=O)O^-NH_4^+ \longrightarrow$ _____ $+ H_2O$

c. $C_6H_5{-}CH_2NH_2$ + _____ $\longrightarrow C_6H_5{-}CH_2OH$ + _____

d. _____ $+ 2C_2H_5OH \longrightarrow C_2H_5OOC(CH_2)_4COOC_2H_5$ + _____

7. Complete the following table:

Name	Open-Chain Structure	Fischer Structure	Haworth Structure
D-δ-Glucono-lactone		C=O H—C—OH HO—C—H H—C—OH H—C CH_2OH (ring O linking C=O and C-5)	
D-Galactose	H—C=O H—C—OH HO—C—H HO—C—H H—C—OH CH_2OH		
2,3,4,6-Tetra-O-methyl-D-glucose			CH_2OCH_3 O H H H OCH_3 H H_3CO OH H OCH_3

8. Decide whether each of the following structures is named correctly. Make whatever changes in the name are necessary.

CH_2OH
O
H OCH_3
H
OH H
HO H
H OH

Ethyl-α-D-glucose

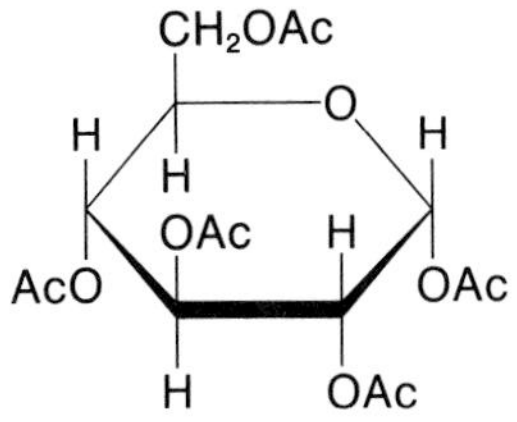

Penta-O-acetyl-β-L-fructose

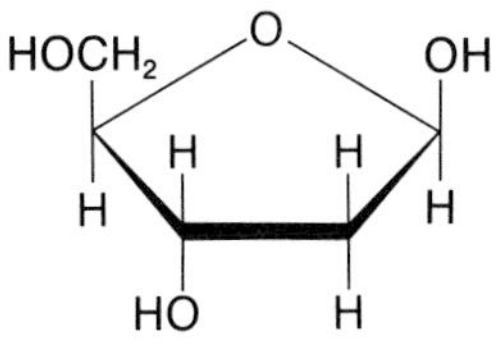

α-D-Ribofuranose

9. The specific rotations of α-D-lactose and β-D-lactose are $+90°$ and $+35°$, respectively. When either of these isomers is dissolved in water and the solution allowed to reach equilibrium, the specific rotation is $+55°$. Calculate the ratio of the two isomers in the equilibrium mixture.

10. The equilibrium constant for the isomerization reaction

$$\alpha\text{-D-Glucose} \rightleftharpoons \alpha\text{-D-Galactose}$$

is 12.3 at 30°C. Calculate the free-energy change for this reaction when the concentrations of α-D-glucose and α-D-galactose are maintained at 1.0×10^{-2} mole/ℓ and 1.0×10^{-1} mole/ℓ, respectively.

11. The nuclear zone of an *E. coli* cell is about 3000 Å in diameter. The DNA molecule that makes up the bulk of this zone consists of about 3.5 million nucleotide pairs spaced 3.4 Å apart. Calculate the ratio of the length of the DNA molecule to the diameter of the nuclear zone.

12. Through controlled polymerization, a polypeptide containing 2 units of alanine, 1 unit of tyrosine, 3 units of serine, and 2 units of leucine is formed, in the sequence stated. Draw the formula of this polypeptide.

13. A biochemist could not convince a skeptic that the 20 amino acids are indeed capable of forming all the different proteins, so he decided to prove it mathematically. He assumed that there were approximately 10^6 species of living organisms, ranging in complexity from *E. coli* to human beings, with an approximate total of 10^{12} different kinds of protein molecules. He also assumed that each protein consisted of only 20 amino acids (one of each), a drastic underestimate, and then calculated the ratio of the number of possible protein molecules to the total of the different kinds of protein molecules in living organisms. What was the answer, and did he make his point?

14. The hemoglobin in the red corpuscles of most mammals contains 0.335% iron by weight. Calculate its minimum molecular weight.

15. Ribonuclease is 1.65% leucine and 2.48% isoleucine by weight. Bearing in mind that a water molecule is lost in forming a peptide bond, calculate the minimum molecular weight of ribonuclease.

16. The distance between the two asymmetric centers in the peptide bond is 3.6 Å. What is the length of a polypeptide chain containing 1000 amino acid residues if (a) it exists only in the extended form, or (b) it exists only in the α-helical form?

17. The peptide group forms the backbone structure of proteins.
a. Draw the atomic structure of this group.
b. What is the 3-dimensional structure of this group?
c. Comment on the nature of the C—N bond.
d. Identify the atoms of this group that participate in bonding in the α helix, and the kind of bonding. How many residues farther along the chain are these atoms bonded?
e. What is the orientation of this group with respect to the axis of the α helix?

18. Define the following terms and give an example of each, using amino acids whenever possible.
a. Hydrogen bond
b. Ionic bond
c. Disulfide bond
d. Primary structure
e. Secondary structure
f. Tertiary structure
g. Quaternary structure
h. Hydrophobic bond

19. Nylon-66 is made from the monomers adipic acid, $HOOC—CH_2—CH_2—CH_2—CH_2—COOH$, and 1,6-diaminohexane.
a. Draw the structure of the product of the reaction of one molecule of each monomer.
b. Where does the 66 come from in the name of this polymer?
c. The side chains can bond side-by-side to form a structure similar to the β pleated sheet in proteins. Draw two short, schematic chains of the polymer side-by-side and indicate how this bonding can occur.

20. State three similarities and three differences between the Watson-Crick DNA helix and the Pauling-Corey α helix for proteins.

21. It is now known that adenine hydrogen bonds to thymine and that guanine hydrogen bonds to cytosine in DNA. However, this was not obvious to Watson and Crick. Draw reasonable structures for a cytosine-adenine base pair and a cytosine-thymine base pair.

22. The electron microscope is a powerful tool for many jobs, one of which is determining molecular weights. Electron microscopes can accurately measure particle lengths. Thus, if we know the linear density ρ of a molecule, the electron microscope can be used to determine its molecular weight. Calculate ρ in units of atomic mass units per angstrom for DNA from the following data: radius of helix to phosphate = 9 Å; spacing between adjacent base pairs = 3.4 Å; distance between phosphates along helix = 7 Å; average molecular weight of a nucleotide = 325. From this density, determine the molecular weight of a DNA molecule that is found to be 6.8×10^{-3} mm long.

23. A short segment of a DNA molecule is represented as

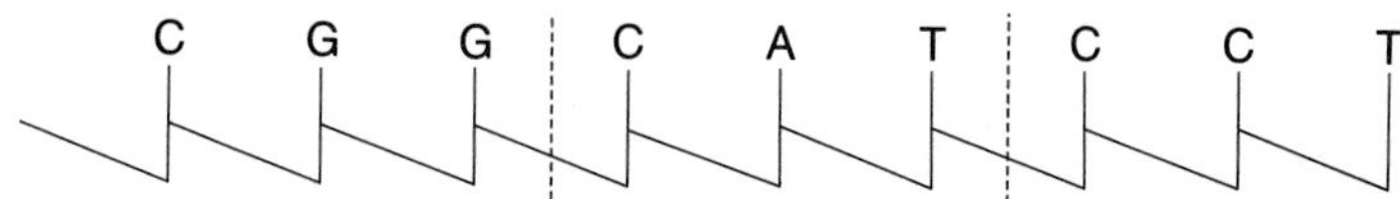

a. Draw the detailed atomic structure of the segment between the dashed lines.

b. The genetic information contained in this polynucleotide is transcribed to an mRNA molecule, beginning at the left. Would the short polypeptide sequence resulting from translation of this message most likely be found in the interior or on the surface of a protein molecule? Support your answer.

"I guess we should have written it down, but I never dreamed we'd all forget the secret."

Drawing by Al Ross

INTERLUDE

winemaking

Man has delighted in wine for a long time. Archaeologists believe that wine was already well-known in the Neolithic Age, about twelve thousand years ago. The fermentation of grain and honey to make beer and mead may have begun even earlier. When Noah left the ark, he promptly planted a vineyard (and subsequently got drunk). Winemakers have been at work ever since, and today the annual world production is about eight billion gallons, or about two gallons for every human being on earth.

There is a boom in wine production in America today. Thousands of acres of new vineyards are being planted in California. Wine has become not only a big business, but a popular do-it-yourself home project as well. Many people are interested in "natural foods." Wine is (or could be) a totally natural beverage. A few handfuls of grapes can be crushed and allowed to stand for about a week, and a cup of rough wine will be ready to drink. The sugars to produce the ethanol, the yeasts to catalyze the fermentation, the pigments to give color, and the dozens of other substances that add up to the subtly complex mixture called wine are all tidily packaged in the natural grape. Of course, to produce a really fine wine, let alone a great one, a good deal of effort—and often a good deal of luck—are required.

Countless volumes have been written on the wines of the world: how to taste, judge, select, and store wine, and how to serve gourmet wines with gourmet foods. Our aim in this Interlude is to understand some of the chemistry of winemaking. The discussion will illustrate many of the things you have learned from this book: stoichiometry in Chapter 2, organic molecules in Chapters 4 and 5, thermochemistry in Chapter 7, enzyme kinetics in Chapter 8, the pH concept in Chapter 9, redox in Chapter 10, and some of the biological monomers in Chapter 11.

We begin with a brief look at the production of wine. Figure 1, from an excellent article in *Scientific American,* shows the basic steps and intro-

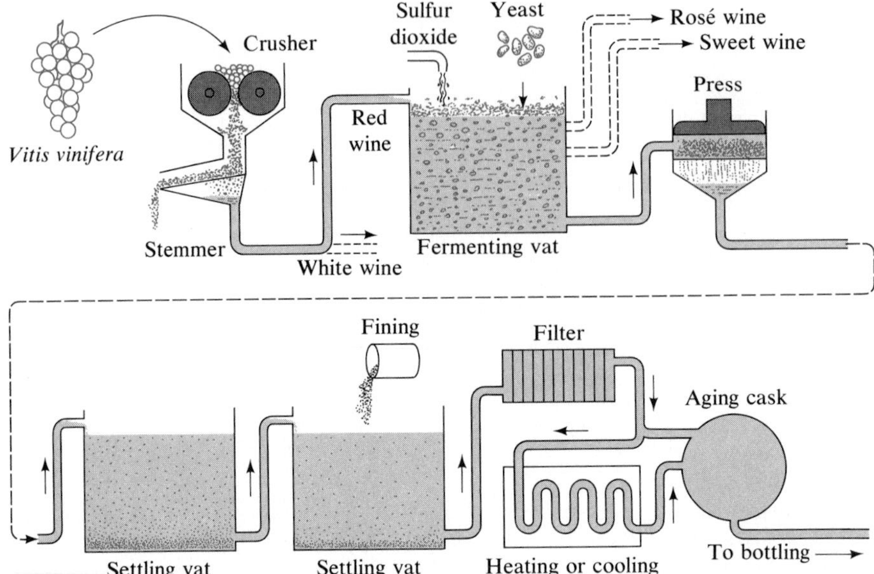

Figure 1 The western U.S. method of producing red wine duplicates the European method. The grapes are crushed between rollers, forming an intermediate product known as *must*. The must is piped to a fermenting vat, where yeasts speed the transformation of sugars to alcohol, and then to a press, where skin, seeds, and pulp are removed. The juice proceeds through two settling vats, where the "fining" process removes impurities. It is then filtered, sometimes heated or cooled, and aged in casks prior to bottling. (M. A. Amerine, "Wine." Copyright © 1964 by Scientific American, Inc. All rights reserved.)

duces a number of terms. The stemmer removes the stems from the bunches of grapes. The press squeezes the juice from the *must* (which is the pulp of the grapes plus the juice), and retains the pulp, skins, and seeds. The process of removing the wine from one settling vat to another, leaving the dead or dormant yeast cells and other detritus behind, is called *racking*.

About 15% by weight of a typical grape is stem, skin, and seeds. The remaining 85% is pulp and juice. The waxy film on the surface of a grape (called the *bloom* of the grape) collects cells of various molds, about 100,000 cells of wine yeasts, and perhaps 10 million cells of other yeasts called wild yeasts. These fungal organisms drift in the air of the vineyard and either settle naturally on the grapes or are placed there by insects such as the ubiquitous fruit fly *Drosophila melanogaster*. If grapes are crushed and the must is untreated, the wild yeasts begin the fermentation process and produce up to 4% ethanol. During this period the wine yeasts gradually take over and then complete the fermentation, raising the ethanol concentration to 12–14%. This purely natural process is seldom relied upon, however, because the results tend to be haphazard and quality control is difficult. In modern winemaking a small amount of SO_2 is added to the must to inhibit most of the wild yeasts. Then a pure wine yeast is added to effect the fermentation. The preferred strain is *Saccharomyces cerevisiae*.* Like most other yeasts, it reproduces by simple, asexual budding.

The grape skins themselves contain a variety of pigments, and are left with the must for the making of red wine, providing the beautiful colors of the clarets and burgundies. They are removed immediately after crushing for the making of white wine, and after about one day of fermentation for the making of rosé wine. The skins also contain tannins, which are complex mixtures of esters and ethers of various carbohydrates. Tannins from the skin and seeds give red wines their characteristic astringency.

The grape seeds contain tannins and have a high content of oils derived from solid fatty acids (palmitic and stearic) and liquid fatty acids (oleic and linoleic). Minute amounts of these oils are extracted from the seeds during the fermentation of red wine, and contribute to the flavor.

The must of the grape (the pulp and juice) contains most of the necessary ingredients for a good wine. The must is essentially an 18–25% (by weight) sugar solution. The principal sugars are dextrose and levulose, which are the common names for D-glucose and D-fructose:

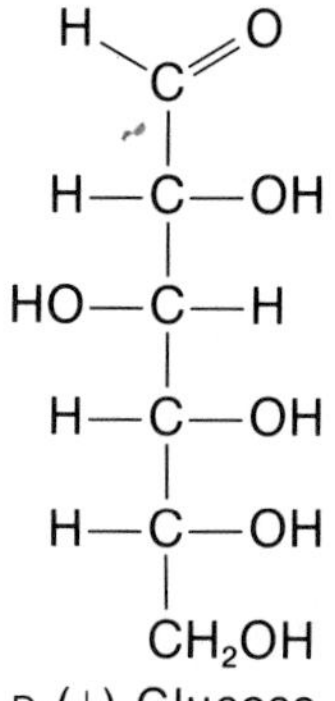

```
 H    O                         CH2OH
  \  //                           |
   C                              C=O
   |                              |
H—C—OH                      HO—C—H
   |                              |
HO—C—H                       H—C—OH
   |                              |
H—C—OH                       H—C—OH
   |                              |
H—C—OH                         CH2OH
   |
 CH2OH
```

D-(+)-Glucose (dextrose) D-(−)-Fructose (levulose)

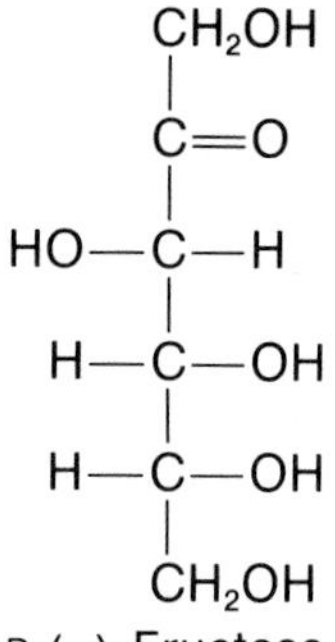

They are present in about equal concentrations in the grape. Note that dextrose is detrorotatory and levulose is levorotatory (hence their names), and that both have the formula $C_6H_{12}O_6$. For convenience we will regard them as the same compound.

Several organic acids are also present, of which the two principal ones are L-malic and L-tartaric:

---

*The genus name comes from the Sanskrit (via Greek and Latin) word for sugar, plus the Greek word for fungus. The species name comes from the Latin word for beer.

$$\begin{array}{c}COOH\\ |\\ HO{-}C{-}H\\ |\\ H{-}C{-}H\\ |\\ COOH\end{array}\qquad\qquad\begin{array}{c}COOH\\ |\\ HO{-}C{-}H\\ |\\ H{-}C{-}OH\\ |\\ COOH\end{array}$$

L-(−)-Malic acid   L-(+)-Tartaric acid

The total acid concentration in the must is between 0.3 and 1.5 wt%.

Minor components abound in the must, e.g., the 20 amino acids (both free and in proteins). There are also numerous pigments, some tannins, many minerals (e.g., Fe^{2+}, Fe^{3+}, Na^{+}, K^{+}, Mg^{2+}, Ca^{2+}, SO_4^{2-}, and phosphates), at least half a dozen vitamins (e.g., ascorbic acid, thiamine, and riboflavin), and a variety of odoriferous compounds. The structures of two common grapes are shown in Figure 2.

We now turn to the process by which this complex sugar solution is turned into wine. We know that a sugar solution that has been boiled can be stored in a refrigerator indefinitely. Let us examine the thermodynamics of the reaction for the conversion of sugar, $C_6H_{12}O_6$, to ethanol to see whether there is anything peculiar in the fact that grapes ferment to give ethanol but a sterile sugar solution does not. The overall reaction by which fermentation proceeds was discovered by Gay-Lussac in 1810:

$$C_6H_{12}O_6(c) \rightarrow 2C_2H_5OH(l) + 2CO_2(g) \qquad \Delta G^{\circ\prime} = -54 \text{ kcal/mole} \qquad (1)$$

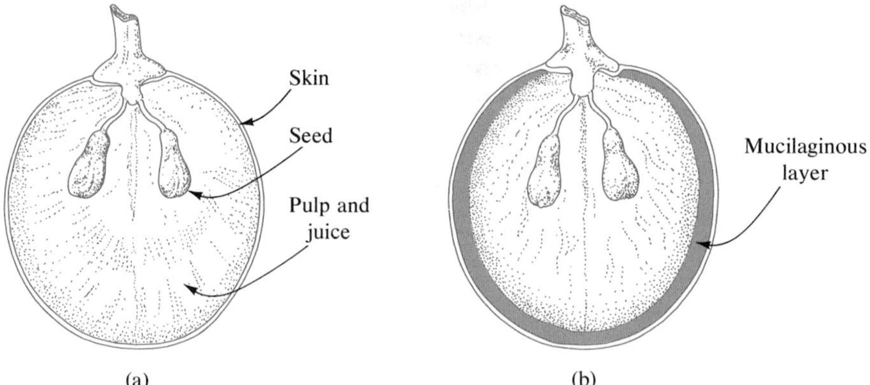

Figure 2. Cross-sectional views of two common grapes. (a) Cabernet franc (a variety of *Vitis vinifera*) is an Old World grape of relatively low acidity that flourishes in California. (b) The Concord grape (a variety of *Vitis labrusca*) of the northeastern U.S. has a mucilaginous layer separating the skin and pulp, hence its "slipskin" classification. This grape is generally lower in sugar and higher in acid than the varieties of Old World grapes. (M. A. Amerine, "Wine." Copyright © 1964 by Scientific American, Inc. All rights reserved.)

The value of $\Delta G^{\circ\prime}$ indicates that crystalline sugar ought to decompose spontaneously to ethanol and carbon dioxide. Because we know that it does not, we conclude that there is a very large activation energy for the reaction as written, making the rate imperceptibly slow. The same is true of this reaction in aqueous solution. As we saw in Chapter 8, the way to make the reaction go is to provide a catalyst that reduces the value of E_a. The enzymes of the yeast perform this function admirably, although they would probably prefer not to. Let us see why.

The energy derived from Reaction 1 or any other reaction that the enzymes catalyze can be used by the yeast cells themselves. Another possible reaction of the sugar, respiration, can provide them with much *more* energy. The reaction is

$$C_6H_{12}O_6(c) + 6O_2(g) \rightarrow 6CO_2(g) + 6H_2O(l)$$

$$\Delta G^{\circ\prime} = -688 \text{ kcal/mole} \quad (2)$$

Given a choice between Reactions 1 and 2, the yeast cells would certainly prefer to meet their metabolic energy requirements by catalyzing the respiration rather than the fermentation. This would be bad news for the winemaker. How does he prevent it? Simply by denying the yeasts access to much oxygen. The initial fermentation of the must is so vigorous that the frothing caused by the generation of CO_2 excludes the ambient O_2 from the vat. From then on the winemaker must control the rate and amount of oxidation of the developing wine very carefully, because oxidation is one of the most crucial factors in determining the ultimate quality of the product. Most important is the length of time the wine is allowed to age and mellow in wooden casks before bottling. Oxygen diffuses slowly through the wood and into the wine, where it undergoes numerous reactions of which we still know very little.

The thermodynamics and kinetics of the initial fermentation present a practical problem for the winemaker. The standard enthalpy change for Reaction 1 is −20 kcal/mole, which means that 20,000 calories are released to the must per mole of sugar fermented. As the system heats up, the rate of reaction increases, causing heat to be released even faster, until . . . ? Until a temperature of about 32°C is reached, at which point the must "sticks," or stops fermenting. To avoid this problem and keep the reaction under control, refrigerated coils are used in the large fermentation vats.

The enzyme-catalyzed conversion of glucose to ethanol is called *alcoholic fermentation*. Reaction 1 obscures the fact that it is actually a complex, 12-step process. Nine of these steps are shown in Figure 3. Note that there are two redox reactions: the oxidation of glyceraldehyde 3-phosphate by NAD^+ and the reduction of acetaldehyde by NADH. Thus, there is no net oxidation-reduction in alcoholic fermentation.

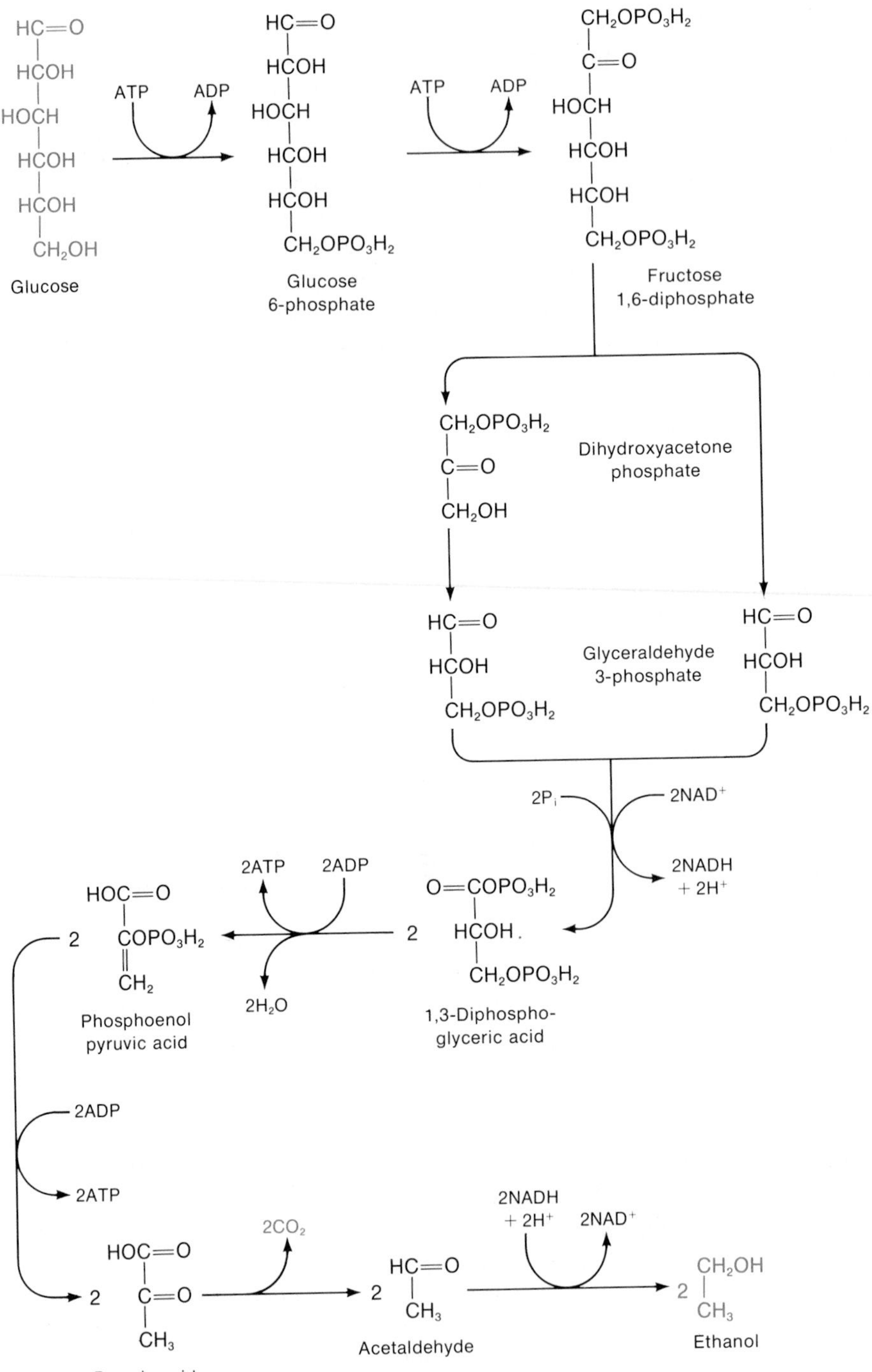

Glucose
ATP
ADP
Glucose 6-phosphate
ATP
ADP
Fructose 1,6-diphosphate
Dihydroxyacetone phosphate
Glyceraldehyde 3-phosphate
2P$_i$
2NAD$^+$
2NADH + 2H$^+$
1,3-Diphospho-glyceric acid
2ATP
2ADP
$2H_2O$
Phosphoenol pyruvic acid
2ADP
2ATP
Pyruvic acid
$2CO_2$
Acetaldehyde
2NADH + 2H$^+$
2NAD$^+$
Ethanol

The entire sequence of reactions through the formation of pyruvic acid is identical to that of aerobic glycolysis, which we discussed briefly in Chapter 10. As we saw there, pyruvic acid is the final product of aerobic glycolysis and the next-to-final product of anaerobic glycolysis (the final product being lactic acid). By its reaction in the mitochondria to give acetyl coenzyme A, pyruvic acid acts as the precursor to the Krebs cycle, which occurs in most living organisms. Alcoholic fermentation, however, occurs in many microorganisms (notably yeasts), but not in animals. This is because animals lack the enzyme pyruvate decarboxylase, which is required for the conversion of pyruvic acid (actually, pyruvate ion *in vivo*), to acetaldehyde. The latter is the precursor to ethanol.

The maximum possible alcohol content of a table wine is set by the amount of sugar in the grapes or by the yeast cells, most of which are inhibited at about 15 vol% ethanol.* The actual, permissible alcohol content is set by national or state regulations. In California, white wines must contain 10.0–14.0 vol% alcohol and red wines must contain 10.5–14.0 vol%. German wines, especially from the Moselle and northern Rhine River valleys, are generally at the lower ends of these ranges. Because of the relatively short growing seasons in those northern latitudes, the grapes often do not develop enough sugar to produce more than about 9% ethanol.

Care must be taken at all steps in the winemaking to exclude vinegar bacteria of the genus *Acetobacter,* which convert the glucose to acetic acid. Barring any such disaster, subsequent changes in the wine during aging are subtle but all-important. In some sherries, for example, a small amount of ethanol is converted back to acetaldehyde as the aging progresses. This is generally desirable for sherries because the acetaldehyde imparts a pungent aroma to the wine. It is not desirable for table wines.

The acid content of wine is also important. It is expressed in terms of *fixed-acid* and *volatile-acid* content. The former is due mainly to malic and tartaric acids. For convenience, the entire fixed-acid content is calculated

*Table wines are "natural" wines, i.e., those to which no alcohol has been added. Most table wines are *dry* (very little residual sugar). Apéritif and dessert wines, most of which are *sweet* (moderate residual sugar), are those to which brandy made from distilled wine has been added to raise the alcohol content to about 20%. This is done during the fermentation, so the latter is inhibited and some of the sugar remains unconverted to alcohol—hence the sweetness.

Figure 3. Alcoholic fermentation is the metabolic breakdown of the six-carbon sugar glucose to the two-carbon product ethanol. The carbon backbone of the intermediate product fructose 1,6-diphosphate is cleaved to give two molecules of glyceraldehyde 3-phosphate (an aldose), one of which is obtained via its isomer, dihydroxyacetone phosphate (a ketose). The enzymes, coenzymes, and inorganic ions required for each step in the sequence have been identified by biochemists.

as tartaric acid. If this acidity is greater than about 0.8 wt%, the wine will tend to be too tart. If it is less than about 0.4 wt%, it will probably be insipid and flat. In general, the wines from northern climates have too much acidity and those from southern climates, too little acidity. One way to increase the acidity is to include in the must some grapes that are picked early. Such grapes have a higher malic acid content than ripe grapes because malic acid is metabolized by the grapes during the ripening.

One way to decrease the acidity of a wine is to let nature herself take care of it. Enzymes in several species of *Lactobacillus* bacteria, which grow readily in wines of high acidity, catalyze the following reaction, called the *malo-lactic fermentation*:

$$\underset{\text{L-(}-\text{)-Malic acid}}{\mathrm{HO{-}C(COOH)(CH_2COOH){-}H}} \xrightarrow[\mathrm{NAD}\ \to\ \mathrm{NADH} + \mathrm{H^+}]{\mathrm{Mn^{2+}}} \underset{\text{L-(+)-Lactic acid}}{\mathrm{HO{-}C(COOH)(CH_3){-}H}} + \mathrm{CO_2} \qquad (3)$$

The value of $\mathrm{p}K_a$ for lactic acid at 25°C is 3.86, whereas those for malic acid are $\mathrm{p}K_1 = 3.40$ and $\mathrm{p}K_2 = 5.11$. Thus, lactic acid is slightly the weaker of the two. Furthermore, because one of the carboxyl groups of malic acid is lost as CO_2 in this reaction, the net effect is to reduce the acidity of the wine. Without this vital mellowing process, the otherwise highly acid wines from northern climates would be undrinkable. The reaction is generally allowed to occur before bottling in order to assure microbiological stability of the wine and to prevent bacterial clouding and CO_2 formation in the bottle.

Chemists, of course, use pH as their quantitative measure of acidity. The pH of most wines is in the range 3 to 4, like that of the grapes themselves. Table wines should have a pH of about 3.4 to 3.6, to ensure a fresh, fruity taste and resistance to spoilage.

The volatile-acid content of a wine is due mainly to acetic acid. One type of reaction we have seen earlier in this book is esterification. With large quantities of ethanol present, it is not surprising that much of the acetic acid reacts to give ethyl acetate:

$$\mathrm{CH_3COOH + C_2H_5OH \rightarrow CH_3COOC_2H_5 + H_2O} \qquad (4)$$

We have seen that wine is an extraordinarily complex mixture of chemicals that, in its infinite variety, has given nourishment and pleasure to mankind since before the dawn of history. In the words of one of America's leading enologists, Maynard Amerine (b. 1911), "Wine is a chemical symphony composed of ethyl alcohol, several other alcohols, sugars, other carbohydrates, polyphenols, aldehydes, ketones, enzymes, pigments, at least half a dozen vitamins, 15 to 20 minerals, more than 22 organic acids, and other grace notes that have not yet been identified." We have discussed a few of the chords in the symphony of wine. To discuss them all would require a sizable book. And then there are all those unknown grace notes.

We hope that you have enjoyed this book. These last pages should have shown that, once you have begun to master the principles of chemistry, you will have a better understanding of, and hence a deeper appreciation for, the world around you. Cheers!

general bibliography

General Chemistry Problem Books

The following books have been selected to provide you with additional exercise in solving quantitative problems in chemistry. All the books provide answers to the questions posed and have numerous solved examples, and several provide worked solutions to every question posed. Several of the titles indicate a specific field of chemistry.

Sidney W. Benson, **Chemical Calculations,** 3rd ed., Wiley, New York, 1971.

Ian S. Butler and Arthur E. Grosser, **Relevant Problems for Chemical Principles,** W. A. Benjamin, New York, 1970.

J. S. Finlayson, **Basic Biochemical Calculations,** Addison-Wesley, Reading, Mass., 1969.

Arthur E. Grosser and Ian S. Butler, **Relevance in Chemical Science,** W. A. Benjamin, Menlo Park, Cal., 1971.

Rex Montgomery and Charles A. Swenson, **Quantitative Problems in the Biochemical Sciences,** 2nd ed., W. H. Freeman, San Francisco, 1976.

Conway Pierce and R. Nelson Smith, **General Chemistry Workbook,** 4th ed., W. H. Freeman, San Francisco, 1971.

William M. Risen, Jr., and George P. Flynn, **Problems for General and Environmental Chemistry,** Appleton-Century-Crofts, New York, 1972.

Michell J. Sienko, **Chemistry Problems,** 2nd ed., W. A. Benjamin, Menlo Park, Cal., 1972.

Programmed Text

Lassila et al. have written a very useful self-study supplement to the well-known text by Dickerson, Gray, and Haight. If you are having trouble understanding a particular aspect of chemistry, we recommend this volume to you. The section on the structure of molecules is particularly well done.

Jean D. Lassila, Gordon M. Barrow, Malcolm E. Kenney, Robert L. Litle, and Warren E. Thompson, **Programmed Reviews of Chemical Principles,** 2nd ed., W. A. Benjamin, Menlo Park, Cal., 1974.

The History of Chemistry

The Houghton Mifflin Company has published a series of paperbacks entitled "Classic Researches in General Chemistry." These volumes explore the historical background of a particular aspect of chemistry. They are interesting particularly because they give long quotes from some of the people who were instrumental in developing the science of chemistry.

O. Theodor Benfey, **From Vital Force to Structural Formulas,** Houghton Mifflin, Boston, 1964.

J. J. Lagowski, **The Chemical Bond,** Houghton Mifflin, Boston, 1966.

J. J. Lagowski, **The Structure of Atoms,** Houghton Mifflin, Boston, 1964.

Mathematical Aids

You may find the following volumes helpful in building the mathematical skills necessary to deal with the types of problems found in this text and in following the discussion when calculus is employed.

Eugene Roberts, **A Programmed Sequence on Exponential Notation,** W. H. Freeman, San Francisco, 1962.

Eugene Roberts, **A Programmed Sequence on the Slide Rule,** W. H. Freeman, San Francisco, 1962.

James N. Butler and Daniel G. Bobrow, **The Calculus of Chemistry,** W. A. Benjamin, New York, 1965.

Chemistry and the Environment

The following two volumes present useful information for an understanding of the impact of man on his environment.

Cleaning Our Environment: The Chemical Basis for Action, The American Chemical Society, Washington, D.C., 1969.

Committee on Resources and Man, National Academy of Sciences–National Research Council, **Resources and Man,** W. H. Freeman, San Francisco, 1969.

Miscellaneous

Because nuclear chemistry is important to both the chemist and biochemist, you may wish to explore this interesting field further by reading Choppin's book. The *Scientific American* Reader contains 36 articles illustrating many of the concepts discussed in this book, in a manner similar to that of the Interludes. Particular emphasis is given to topics in environmental chemistry and molecular biology. There is an epilogue on the future of chemistry by Linus Pauling.

Gregory R. Choppin, **Nuclei and Radioactivity,** W. A. Benjamin, New York, 1964.

James B. Ifft and John E. Hearst, **General Chemistry** (Readings from *Scientific American*), W. H. Freeman, San Francisco, 1974.

appendix A physical constants, units, and conversion factors

Physical Constants

Mass of the neutron	$m_n = 1.6749 \times 10^{-24}$ g
Mass of the proton	$m_p = 1.6726 \times 10^{-24}$ g
Mass of the electron	$m_e = 9.1096 \times 10^{-28}$ g
Electron charge	$e = 1.6022 \times 10^{-19}$ C
	$= 4.80 \times 10^{-10}$ esu
Gravitational constant	$G = 6.67 \times 10^{-8}$ dyn cm^2/g^2
Acceleration due to gravity at sea level	$g = 980.665$ cm/s^2
Speed of light in vacuum	$c = 3.00 \times 10^{10}$ cm/s
Planck's constant	$h = 6.63 \times 10^{-27}$ erg s
Avogadro's number	$N_A = 6.022169 \times 10^{23}$ mole^{-1}
Gas constant	$R = 0.08206$ ℓ atm/K mole
	$= 1.987$ cal/K mole
	$= 8.314$ J/K mole
Boltzmann constant	$k = R/N_A = 1.381 \times 10^{-16}$ erg/K
Faraday constant	$\mathcal{F} = eN_A = 96{,}487$ C/mole
	$= 23{,}061$ cal/V mole
Base of natural logarithms	$e = 2.718 \ldots$

$$\ln x = 2.303 \log x$$

Units and Conversion Factors

Mass (m)

1 gram (g) $= 10^{-3}$ kilogram
1 kilogram (kg) $= 2.205$ pounds
1 pound (lb) $= 0.4536$ kg

Length (l)

1 meter (m)	$= 10^{-3}$ kilometer
1 kilometer (km)	$= 0.6214$ mile
1 mile (mile)	$= 1.609$ km
1 centimeter (cm)	$= 0.3937$ inch
1 inch (in.)	$= 2.540$ cm
1 millimeter (mm)	$= 10^{-3}$ m $= 10^{-1}$ cm
1 micrometer (μm)	$= 10^{-6}$ m $= 10^{-4}$ cm
1 nanometer (nm)	$= 10^{-9}$ m $= 10^{-7}$ cm
1 angstrom (Å)	$= 10^{-10}$ m
	$= 10^{-8}$ cm
	$= 10^{-7}$ mm
	$= 10^{-4}$ μm
	$= 10^{-1}$ nm

Time (t)

1 second (s) = the fundamental unit

Volume (l^3)

1 liter (ℓ)	$= 10^3$ milliliters
1 milliliter (ml)	$= 1$ cm^3
1 gallon (gal)	$= 3.785$ ℓ
1 quart (qt)	$= 0.946$ ℓ

Force (mlt^{-2})

1 dyne (dyn)	$= 1$ g cm/s^2
1 newton (N)	$= 1$ kg m/s^2 $= 10^5$ dyn

Energy (ml^2t^{-2})

1 erg (erg)	$= 1$ dyn cm $= 1$ g cm^2/s^2
1 joule (J)	$= 1$ N m $= 1$ kg m^2/s^2 $= 10^7$ erg
1 calorie (cal)	$= 4.184$ J $= 10^{-3}$ kilocalorie
1 kilocalorie (kcal)	$= 41.29$ liter atmospheres
	$= 1.162 \times 10^{-3}$ kilowatt-hour
	$= 3.966$ British thermal units
	$= 2.611 \times 10^{22}$ electron volts
1 liter atmosphere (ℓ atm)	$= 1.013 \times 10^2$ J
1 kilowatt-hour (kWh)	$= 3.600 \times 10^6$ J
1 British thermal unit (Btu)	$= 1.055 \times 10^4$ J
1 electron volt (eV)	$= 1.602 \times 10^{-19}$ J
1 eV/molecule	$= 23.061$ kcal/mole
1 kcal/mole	$= 4.3363 \times 10^{-2}$ eV/molecule

Power (ml^2t^{-3})

1 watt (W) = 1 J/s = 10^{-3} kilowatt
1 kilowatt (kW) = 0.2390 kcal/s
= 0.9478 Btu/s

Pressure $(ml^{-1}t^{-2})$

1 atmosphere (atm) = 1.013×10^6 dyn/cm^2
= 1.013×10^5 N/m^2
= 14.70 lb/in.^{2}
= 760 mmHg = 760 torr
1 torr (Torr) = 1.316×10^{-3} atm
= 1.333×10^3 dyn/cm^2

Electric Current (I)

1 ampere (A) = the fundamental unit

Electric Charge (tI)

1 coulomb (C) = 1 A s
= 6.241×10^{18} electron charges
= 2.998×10^9 electrostatic units
1 electrostatic unit (esu) = 3.336×10^{-10} C
= 2.082×10^9 electron charges
1 faraday (faraday) = 9.649×10^4 C
= 2.893×10^{14} esu

Temperature

1 kelvin (K) = the fundamental unit
K = °C + 273.15
°C = 0.56(°F − 32)
°F = 1.8(°C) + 32

appendix B exponentials and logarithms

Exponentials with Base 10

The use of exponential notation is very important in chemistry because numbers from 10^{-28} to 10^{23} are commonly encountered. We will briefly review the meaning and use of these exponential numbers. In exponential notation, 1,000,000 is written 10^6 because it equals $10 \times 10 \times 10 \times 10 \times 10 \times 10$. Similarly, 10,000 is written 10^4 and 1000 is written 10^3. In dealing with numbers less than one, negative exponentials are employed. For example, 1/10 is written 10^{-1} and 1/1,000,000 is written 10^{-6}.

A number written in exponential notation is in the form

$$A \times 10^n$$

where A is usually a number between 1 and 10 and n (the *exponent*) is almost always an integer, either positive or negative. For example, one mile equals 5280 feet, which is written

$$5.280 \times 1000 = 5.280 \times 10^3$$

if the 0 is a significant figure, or 5.28×10^3 if it is not. (Here, of course, it is.) Similarly, one foot is 1/5280 of a mile, which equals

$$0.0001894 = 1.894 \times 10^{-4}$$

Addition and subtraction of exponential numbers require that the exponent be the same in all the numbers. For example, $482 - 31 = 451$. In exponential notation, the subtraction $(4.82 \times 10^2) - (3.1 \times 10^1)$ cannot be made directly. We must write both numbers with the same exponent:

$$\begin{array}{r} 4.82 \times 10^2 \\ -0.31 \times 10^2 \\ \hline 4.51 \times 10^2 \end{array}$$

In addition and subtraction, the number of significant figures must always be considered. For example, in adding 1.098 to 23,200, where the latter number is known to only three significant figures, we write

$$(2.32 \times 10^4) + (0.0001098 \times 10^4) = 2.32 \times 10^4$$

Since the larger number has only three significant figures, addition of the smaller number has no effect. (Even if all five digits in the larger number were significant, we could only add 1 to it, not 1.098, because no digits beyond the decimal point of the larger number are specified.)

Multiplication of exponential numbers follows the rule

$$10^a \times 10^b = 10^{a+b}$$

For example,

$$(2 \times 10^3)\,(8 \times 10^{-7}) = 2 \times 8 \times 10^3 \times 10^{-7} = 16 \times 10^{-4} = 1.6 \times 10^{-3}$$

Another example:

$$(3.76 \times 10^7)\,(9.61 \times 10^9) = (3.76 \times 9.61)\,(10^7 \times 10^9) = 3.61 \times 10^{17}$$

Note that, on multiplying numbers, the product has as many significant figures as the numbers being multiplied. If the multipliers have different numbers of significant figures, the product is accurate only to the number of figures of the *least* significantly known multiplier. The same principle applies to division.

Division of exponential numbers follows the rule

$$\frac{10^a}{10^b} = 10^{a-b}$$

For example,

$$\frac{2 \times 10^4}{5 \times 10^6} = \frac{2}{5} \times \frac{10^4}{10^6} = 0.4 \times 10^{-2} = 4 \times 10^{-3}$$

Another example:

$$\frac{7.943 \times 10^{-9}}{1.2 \times 10^4} = \frac{7.943}{1.2} \times 10^{-9-4} = 6.6 \times 10^{-13}$$

The answer is given to only two significant figures because 1.2×10^4 has only two significant figures.

Raising exponential numbers to powers follows the rule

$$(10^a)^b = 10^{ab}$$

An example is the square of 3×10^{-7}:

$$(3 \times 10^{-7})^2 = 3^2 + 10^{-7\times 2} = 9 \times 10^{-14}$$

A more complex example is the cube of 7.61×10^{-2}:

$$(7.61 \times 10^{-2})^3 = 7.61^3 \times 10^{-2\times 3} = 441 \times 10^{-6} = 4.41 \times 10^{-4}$$

Taking roots of exponential numbers follows the rule

$$(10^a)^{1/b} = 10^{a/b}$$

Taking the square root of a number is equivalent to raising it to the 1/2 power. Taking the cube root is equivalent to raising to the 1/3 power, and so forth. The square root of 4.9×10^{21} is

$$(4.9 \times 10^{21})^{1/2} = (49 \times 10^{20})^{1/2} = 49^{1/2} \times (10^{20})^{1/2} = 7.0 \times 10^{20/2} = 7.0 \times 10^{10}$$

Note that the first step in taking a root is writing the exponential number so that the exponent is divisible by the root. Another example is the cube root of 6.28×10^{-16}:

$$(6.28 \times 10^{-16})^{1/3} = (628 \times 10^{-18})^{1/3} = 628^{1/3} \times 10^{-18/3} = 8.56 \times 10^{-6}$$

Exponentials with Base e

Many exponential relations in chemistry are written using the natural base e instead of 10. The number e equals 2.718. . . . All the above rules of addition, subtraction, multiplication, division, raising to powers, and taking roots apply to the base e as well. Any exponential to the base e can be converted to an exponential to the base 10 as follows:

$$e^x = 10^{x/2.303}$$

If $x = 1$,

$$e = 10^{1/2.303} = 10^{0.4342}$$

and if $x = 2.303$,

$$e^{2.303} = 10$$

The *Handbook of Chemistry and Physics* contains a table of values of e^x, and many slide rules and electronic calculators can be used to obtain this information as well. In the following example, we make use of the value $e^{1.5} = 4.5$:

$$3e^{13} = 3e^{1.5} \times e^{11.5} = 3(4.5) \times 10^{11.5/2.3} = 13.5 \times 10^5 = 1.35 \times 10^6$$

Another example:

$$2 \times 10^{-3} = 2e^{-2.3 \times 3} = 2e^{-6.9} = 2e^{-0.9} \times e^{-6} = 2(0.41) \times e^{-6} = 0.82e^{-6}$$

Logarithms with Base 10

The logarithm to the base 10 is defined by the equation

$$x = 10^{\log x}$$

A partial list of logarithms to the base 10 is given below:

x	$\log x$
10^{-13}	-13
10^{-8}	-8
10^{-3}	-3
10^{-1}	-1
1	0
10	1
10^3	3
10^{23}	23

From the properties of exponentials, it is easy to show that, if $ab = c$, then

$$\log ab = \log a + \log b = \log c$$

Tables of logarithms can be found in the *Handbook of Chemistry and Physics*, and again, many slide rules and electronic calculators can be used to obtain logarithms. Example:

$$x = 3.14 \times 7.68$$

$$\log x = \log(3.14 \times 7.68) = \log 3.14 + \log 7.68 = 0.4969 + 0.8854 = 1.3823$$

$$x = \text{antilog } 1.3823 = \text{the number whose log equals } 1.3823$$

$$x = \text{antilog } 0.3823 \times \text{antilog } 1.0000 = 2.41 \times 10^1 = 24.1$$

Example:

$$3000 = 3 \times 10^3$$

$$\log 3000 = \log 3 + \log 10^3 = 0.4771 + 3 = 3.4771$$

Example:

$$0.0008 = 8 \times 10^{-4}$$

$$\log 0.0008 = \log 8 + \log 10^{-4} = 0.9031 + (-4) = -3.0969$$

Example:

$$\mathrm{pH} \equiv -\log[\mathrm{H}^+] = 7.82$$

$$\log[\mathrm{H}^+] = -7.82 = 0.18 - 8.00$$

$$[\mathrm{H}^+] = 1.5 \times 10^{-8}$$

To take the logarithm of a number x, first write the number in the exponential form $A \times 10^n$, where A is a number between 1 and 10 and n is an integer (either positive or negative):

$$x = A \times 10^n$$

$$\log x = n + \log A$$

To take the antilogarithm of a number x, first write the number in the form of an integer (either positive or negative) plus a *positive* decimal between 0 and 1. For example,

$$x = -23.82 = -24.00 + 0.18 = 0.18 - 24.00$$

$$\text{antilog } x = \text{antilog } 0.18 \times \text{antilog}(-24) = 1.5 \times 10^{-24}$$

Mathematical operations with logarithms are summarized below:

Multiplication:	$\log ab = \log a + \log b$
Division:	$\log(a/b) = \log a - \log b$
Raising to a power:	$\log a^n = n \log a$
Taking a root:	$\log a^{1/n} = (1/n) \log a$

Logarithms with Base e

A logarithm to the base e is given the symbol ln and is called a *natural logarithm*. It is defined by the equation

$$x = e^{\ln x}$$

The rules for mathematical operations with natural logarithms are identical to those for logarithms to the base 10. Example:

$$E_a = Ae^{-\Delta H^{\ddagger}/RT}$$

$$\ln E_a = \ln A - \frac{\Delta H^{\ddagger}}{RT}$$

Example:

$$x = 2.8e^{15}$$

$$\ln x = \ln 2.8 + 15 = 1.03 + 15 = 16.03$$

Example:

$$x = 10^{-23}$$

$$\ln x = \ln 10^{-23} = -23 \ln 10 = -23(2.30) = -52.9$$

The two kinds of logarithms are easily interconverted:

$$\ln x = 2.303 \log x$$

$$\log x = \frac{\ln x}{2.303}$$

appendix C nomenclature of inorganic compounds

Some elements of the nomenclature of inorganic complex ions are discussed in Chapter 4, and of organic compounds in Chapters 4, 5, and 11. This appendix is designed to aid in naming inorganic compounds. Mastering chemical nomenclature is little different from learning a new language, such as German. In order to understand the German scientific literature, you must, e.g., learn that the compound H_2 is called *Wasserstoff*. English-speaking chemists call it *hydrogen*. Your task now is to memorize the names of enough compounds and become sufficiently familiar with the several systems of naming compounds that chemistry ceases to be a "foreign language."

The first thing to learn about naming chemical compounds is that there is usually more than one way to do it. We begin with the simplest system, in which a trivial name, i.e., one that has no sensible origin, is assigned to a compound. Some examples are

H_2O	Water
NH_3	Ammonia
Hg_2Cl_2	Calomel

Some names, such as *quicklime* for CaO, derive from the origin of the compound—in this case, limestone, $CaCO_3$. Such word origins are often remembered only by etymologists, but the names have persisted for so long that they are an established part of the language. Can you imagine anyone seriously asking for a drink of dihydrogen oxide? The word *water* serves the purpose much better.

As we come to less common and more complex compounds, the use of trivial names gives way to a more systematic approach. If there are only two elements in the compound, it is customary to name the more metallic element first and the less metallic, or more electronegative, element second, with the suffix "-ide." Some examples are

KCl	Potassium chloride
NaBr	Sodium bromide
CaO	Calcium oxide
HI	Hydrogen iodide
BaS	Barium sulfide

For compounds containing still only two elements but more than two atoms, the prefixes "mono-," "di-," "tri-," etc., become necessary. Some examples of such compounds are the oxides of nitrogen, given in Table 3-7. Another such series is that of the oxides of chlorine. Because chlorine, like nitrogen, is slightly less electronegative than oxygen, the word chlorine comes first:

Cl_2O	Dichlorine monoxide
ClO	Chlorine monoxide
ClO_2	Chlorine dioxide
ClO_3	Chlorine trioxide
Cl_2O_7	Dichlorine heptoxide
ClO_4	Chlorine tetroxide

If no confusion can result, the prefixes "mono-" and "di-" are sometimes dropped.

A class of compounds in which such prefixes are seldom used is that in which the metal atom usually exhibits only one oxidation state. Depending on the oxidation state of the other element, the number of anions per cation is then fixed. Some examples are

$ZnBr_2$	Zinc bromide
CaH_2	Calcium hydride
Na_2O	Sodium oxide
Al_2S_3	Aluminum sulfide

The next level of complexity in naming inorganic compounds arises when there are three elements present. Very often, one of these elements is oxygen. Such compounds are named by combining the suffix "-ate" with the name of the less electronegative of the two nonmetallic elements. For example, $NaNO_3$ is sodium nitrate. The problem with this is that there is a similar compound with nitrogen in the +3 oxidation state, $NaNO_2$. Such compounds with the element in a lower oxidation state use the suffix "-ite," so $NaNO_2$ is sodium nitrite. But the number of chemical compounds is not bounded by the chemists' vocabulary, and there are several such examples entailing more than two oxidation states. To solve this problem, the prefix "hypo-" (meaning "below") is used in the name of the compound in which the less electronegative element is in the lowest oxidation state, and the prefix "per-" (meaning "highest") is used when it is in the highest oxidation state. Some examples of the use of this system are shown in the following table.

Formula	Oxidation State of Less Electro-negative Atom	Name of Salt	Formula and Name of Corresponding Acid
KNO_2	+3	Potassium nitrite	HNO_2 Nitrous acid
KNO_3	+5	Potassium nitrate	HNO_3 Nitric acid
Rb_2SO_3	+4	Rubidium sulfite	H_2SO_3 Sulfurous acid
Rb_2SO_4	+6	Rubidium sulfate	H_2SO_4 Sulfuric acid
$CsClO$	+1	Cesium hypochlorite	$HClO$ Hypochlorous acid
$CsClO_2$	+3	Cesium chlorite	$HClO_2$ Chlorous acid
$CsClO_3$	+5	Cesium chlorate	$HClO_3$ Chloric acid
$CsClO_4$	+7	Cesium perchlorate	$HClO_4$ Perchloric acid

In the inorganic acids, the suffixes "-ous" and "-ic" are used to denote the lower and higher oxidation states, respectively. These same suffixes are also used with the names of a number of metals, namely, those that usually exhibit more than one oxidation state. Some examples are cobaltous and cobaltic, and mercurous and mercuric. The nomenclature is complicated slightly by the fact that, for a few such metals, these terms are derived from the Latin name of the element rather than the English name.

All but eleven of the elements are given a symbol corresponding to one or two letters in the English name of the compound. (The first letter is always capitalized and the second letter is never capitalized.) One of these exceptions is tungsten, whose symbol (W) is derived from the German name of the element, *Wolfram*. The other ten have symbols derived from their Latin names. These are: *stibium* (Sb) for antimony, *cuprum* (Cu) for copper, *aurum* (Au) for gold, *ferrum* (Fe) for iron, *plumbum* (Pb) for lead, *hydrargyrum* (Hg) for mercury, *kalium* (K) for potassium, *argentum* (Ag) for silver, *natrium* (Na) for sodium, and *stannum* (Sn) for tin. The use of the suffixes "-ous" and "-ic" with three of these metals is illustrated below.

CuI	Cuprous iodide
CuI_2	Cupric iodide
$FeBr_2$	Ferrous bromide
$FeBr_3$	Ferric bromide

$SnCl_2$	Stannous chloride
$SnCl_4$	Stannic chloride

The system works well as long as there are only two major oxidation states of the metal atom, as in these examples.

The most rational and self-consistent system of nomenclature of inorganic compounds is that adopted in 1957 by the ultimate authority in such matters, the International Union of Pure and Applied Chemistry. These rules, popularly called the IUPAC Rules, are the model for chemists throughout the world to follow, and are becoming ever more dominant in the chemical literature. A brief summary of a few of these rules, as they pertain to coordination compounds, is given in Chapter 4 (pp. 183–184). Note that the oxidation state of the metal atom is specified by a Roman numeral whenever there could be some doubt about it, but not otherwise. Let us see how the examples shown above are named according to this system.

CuI	Copper(I) iodide
CuI_2	Copper(II) iodide
$FeBr_2$	Iron(II) bromide
$FeBr_3$	Iron(III) bromide
$SnCl_2$	Tin(II) chloride
$SnCl_4$	Tin(IV) chloride

The best way for you to become familiar with the various systems of nomenclature is to practice. Each time you see the name of a compound, try to envision the formula and structure that it designates. Each time you see a new chemical formula, try to name it, using one or more of the principles described above.

appendix D thermodynamic properties

This table gives the standard enthalpy of formation (ΔH_f°), the standard Gibbs free energy of formation (ΔG_f°), the standard absolute entropy (S°), and the standard heat capacity at constant pressure (C_P) for a number of substances. All the data are for 298.15 K and 1 atm. The units of ΔH_f° and ΔG_f° are kcal/mole and the units of S° and C_P are cal/K mole. The physical state is specified for each substance: (*c*) denotes a crystalline solid, (*l*) denotes a liquid, (*g*) denotes a gas, and (*aq*) denotes a 1 molal solution.

These data were abstracted from the classic volume by F. D. Rossini et al., *Selected Values of Chemical Thermodynamic Properties,* National Bureau of Standards Circular 500, U.S. Government Printing Office, Washington, D.C., 1952.

The Representative Elements

	Substance	ΔH_f°	ΔG_f°	S°	C_P
Group IA	H(*g*)	52.089	48.575	27.3927	4.9680
	H^+(*aq*)	0	0	0	0
	H_2(*g*)	0	0	31.211	6.892
	Li(*g*)	37.07	29.19	33.143	4.9680
	Li(*c*)	0	0	6.70	5.65
	Li^+(*aq*)	−66.554	−70.22	3.4	—
	LiH(*c*)	−21.61	−16.72	5.9	8.3
	LiF(*c*)	−146.3	−139.6	8.57	10.04
	Li_2CO_3(*c*)	−290.54	−270.66	21.60	23.28
	Na(*g*)	25.98	18.67	36.715	4.9680
	Na(*c*)	0	0	12.2	6.79
	Na^+(*aq*)	−57.279	−62.589	14.4	—
	Na_2O(*c*)	−99.4	−90.0	17.4	16.3
	NaOH(*aq*)	−112.236	−100.184	11.9	—
	NaF(*c*)	−136.0	−129.3	14.0	11.0

The Representative Elements *(continued)*

	Substance	ΔH_f°	ΔG_f°	S°	C_P
	NaCl(*c*)	−98.232	−91.785	17.30	11.88
	Na_2SO_3(*c*)	−260.6	−239.5	34.9	28.7
	Na_2SO_4(*c*)	−330.90	−302.78	35.73	30.50
	$NaNO_3$(*c*)	−111.54	−87.45	27.8	22.24
	Na_2CO_3(*c*)	−270.3	−250.4	32.5	26.41
	$NaHCO_3$(*c*)	−226.5	−203.6	24.4	20.94
	K(*g*)	21.51	14.62	38.296	4.968
	K(*c*)	0	0	15.2	6.97
	K^+(*aq*)	−60.04	−67.466	24.5	–
	KF(*c*)	−134.46	−127.42	15.91	11.73
	KCl(*c*)	−104.175	−97.592	19.76	12.31
	$KClO_3$(*c*)	−93.50	−69.29	34.17	23.96
	$KClO_4$(*c*)	−103.6	−72.7	36.1	26.33
	KBr(*c*)	−93.73	−90.63	23.05	12.82
	$KBrO_3$(*c*)	−79.4	−58.2	35.65	25.07
	KI(*c*)	−78.31	−77.03	24.94	13.16
	KIO_3(*c*)	−121.5	−101.7	36.20	25.42
	K_2SO_4(*c*)	−342.66	−314.62	42.0	31.1
	KNO_3(*c*)	−117.76	−93.96	31.77	23.01
	$KMnO_4$(*c*)	−194.4	−170.6	41.04	28.5
	Rb(*g*)	20.51	13.35	40.628	4.9680
	Rb(*c*)	0	0	16.6	7.27
	Rb^+(*aq*)	−58.9	−67.45	29.7	–
	RbBr(*c*)	−93.03	−90.38	25.88	12.68
	RbI(*c*)	−78.5	−77.8	28.21	12.50
	Cs(*g*)	18.83	12.24	41.944	4.9680
	Cs(*c*)	0	0	19.8	7.42
	Cs^+(*aq*)	−59.2	−67.41	31.8	–
	CsBr(*c*)	−94.3	−91.6	29	12.4
	CsI(*c*)	80.4	−79.7	31	12.4
Group IIA	Be(*c*)	0	0	2.28	4.26
	BeO(*c*)	−146.0	−139.0	3.37	6.07
	Mg(*c*)	0	0	7.77	5.71
	Mg^{2+}(*aq*)	−110.41	−108.99	−28.2	–
	MgO(*c*)	−143.84	−136.13	6.4	8.94

The Representative Elements *(continued)*

	Substance	ΔH_f°	ΔG_f°	S°	C_P
	$Mg(OH)_2(c)$	−221.00	−199.27	15.09	18.41
	$MgF_2(c)$	−263.5	−250.8	13.68	14.72
	$MgCl_2(c)$	−153.40	−141.57	21.4	17.04
	$MgSO_4(c)$	−305.5	−280.5	21.9	23.01
	$Mg(NO_3)_2(c)$	−188.72	−140.63	39.2	33.94
	$MgCO_3(c)$	−266	−246	15.7	18.05
	$Ca(c)$	0	0	9.95	6.28
	$Ca^{2+}(aq)$	−129.77	−132.18	−13.2	–
	$CaO(c)$	−151.9	−144.4	9.5	10.23
	$Ca(OH)_2(c)$	−235.80	−214.33	18.2	20.2
	$CaF_2(c)$	−290.3	−277.7	16.46	16.02
	$CaCl_2(c)$	−190.0	−179.3	27.2	17.36
	$CaBr_2(c)$	−161.3	−156.8	31	–
	$CaI_2(c)$	−127.8	−126.6	34	–
	$Ca_3(PO_4)_2(c, \alpha)$	−986.2	−929.7	57.6	55.35
	$CaCO_3$(*c*, calcite)	−288.45	−269.78	22.2	19.57
	$Sr(c)$	0	0	13.0	6.0
	$Sr^{2+}(aq)$	−130.38	−133.2	−9.4	–
	$SrO(c)$	−141.1	−133.8	13.0	10.76
	$SrCl_2(c)$	−198.0	−186.7	28	18.9
	$Ba(c)$	0	0	16	6.30
	$Ba^{2+}(aq)$	−128.67	−134.0	3	–
	$BaO(c)$	−133.4	−126.3	16.8	11.34
	$BaCl_2(c)$	−205.56	−193.8	30	18.0
	$Ra(c)$	0	0	17	–
	$Ra^{2+}(aq)$	−126	−134.5	13	–
Group IIIA	$B(c)$	0	0	1.56	2.86
	$B_2O_3(c)$	−302.0	−283.0	12.91	14.88
	$B_2H_6(g)$	7.5	19.8	55.66	13.48
	$H_3BO_3(c)$	−260.2	−230.2	21.41	19.61
	$BF_3(g)$	−265.4	−261.3	60.70	12.06
	$Al(c)$	0	0	6.769	5.817
	$Al^{3+}(aq)$	−125.4	−115.0	−74.9	–
	$Al_2O_3(c, \alpha)$	−399.09	−376.77	12.186	18.88
	$AlCl_3(c)$	−166.2	−152.2	40	21.3

The Representative Elements (*continued*)

	Substance	ΔH_f°	ΔG_f°	S°	C_P
Group IVA	$C(g)$	171.698	160.845	37.7611	4.9803
	$C(c$, diamond)	0.4532	0.6850	0.5829	1.449
	$C(c$, graphite)	0	0	1.3609	2.066
	$CO(g)$	−26.4157	−32.8079	47.301	6.965
	$CO_2(g)$	−94.0518	−94.2598	51.061	8.874
	$CO_3^{2-}(aq)$	−161.63	−126.22	−12.7	–
	$CH_4(g)$	−17.889	−12.140	44.50	8.536
	$HCO_3^-(aq)$	−165.18	−140.31	22.7	–
	$HCHO(g)$	−27.7	−26.3	52.26	8.45
	$HCOOH(g)$	−86.67	−80.24	60.0	–
	$HCOOH(l)$	−97.8	−82.7	30.82	23.67
	$HCOOH(aq)$	−98.0	−85.1	39.1	–
	$H_2CO_3(aq)$	−167.0	−149.00	45.7	–
	$CH_3OH(g)$	−48.08	−38.69	56.8	–
	$CH_3OH(l)$	−57.02	−39.73	30.3	19.5
	$CH_3OH(aq)$	−58.77	−41.88	31.6	–
	$CCl_4(l)$	−33.3	−16.4	51.25	31.49
	$CN^-(aq)$	36.1	39.6	28.2	–
	$HCN(aq)$	25.2	26.8	30.8	–
	$C_2O_4^{2-}(aq)$	−197.0	−161.3	12.2	–
	$C_2H_2(g)$	54.194	50.000	47.997	10.499
	$C_2H_4(g)$	12.496	16.282	52.45	10.41
	$C_2H_6(g)$	−20.236	−7.860	54.85	12.585
	$C_2H_2O_4(c)$	−197.6	−166.8	28.7	26
	$CH_3CHO(g)$	−39.76	−31.96	63.5	15.0
	$CH_3COOH(l)$	−116.4	−93.8	38.2	29.5
	$CH_3CH_2OH(l)$	−66.356	−41.77	38.4	26.64
	$CH_3OCH_3(g)$	−44.3	−27.3	63.72	15.76
	$HOCH_2CH_2OH(l)$	−108.58	−77.12	39.9	–
	$Si(c)$	0	0	4.47	4.75
	$SiO_2(c$, quartz)	−205.4	−192.4	10.00	10.62
	$SiH_4(g)$	−14.8	−9.4	48.7	10.24
	$SiC(c)$	−26.7	−26.1	3.935	6.37
	$Ge(c)$	0	0	10.14	6.24
	$Sn(c$, gray)	0.6	1.1	10.7	6.16

The Representative Elements *(continued)*

	Substance	ΔH_f°	ΔG_f°	S°	C_P
	Sn(*c*, white)	0	0	12.3	6.30
	$SnO_2(c)$	−138.8	−124.2	12.5	12.57
	Pb(*c*)	0	0	15.51	6.41
	$Pb^{2+}(aq)$	0.39	−5.81	5.1	–
	PbO(*c*, red)	−52.40	−45.25	16.2	–
	$PbO_2(c)$	−66.12	−52.34	18.3	15.4
Group VA	N(*g*)	112.965	–	–	–
	$N_2(g)$	0	0	45.767	6.960
	NO(*g*)	21.600	20.719	50.339	7.137
	$NO_2(g)$	8.091	12.390	57.47	9.06
	$NO_3^-(aq)$	−49.372	−26.41	35.0	–
	$N_2O_4(g)$	2.309	23.491	72.73	18.90
	$NH_3(g)$	−11.04	−3.976	46.01	8.523
	$NH_3(aq)$	−19.32	−6.37	26.3	–
	$NH_4^+(aq)$	−31.74	−19.00	26.97	–
	$NH_4Cl(c)$	−75.38	−48.73	22.6	20.1
	$(NH_4)_2SO_4(c)$	−281.86	−215.19	52.65	44.81
	P(*c*, white)	0	0	10.6	5.55
	$PO_4^{3-}(aq)$	−306.9	−245.1	−52	–
	$HPO_4^{2-}(aq)$	−310.4	−261.5	−8.6	–
	$H_2PO_4^-(aq)$	−311.3	−271.3	21.3	–
	As(*c*, gray)	0	0	8.4	5.97
	$AsO_4^{3-}(aq)$	−208	−152	−34.6	–
	$As_2O_5(c)$	−218.6	−184.6	25.2	27.85
	Sb(*c*)	0	0	10.5	6.08
	Bi(*c*)	0	0	13.6	6.1
Group VIA	O(*g*)	59.159	54.994	38.4689	5.2364
	$O_2(g)$	0	0	49.003	7.017
	$O_3(g)$	34.0	39.06	56.8	9.12
	$OH^-(aq)$	−54.957	−37.595	−2.519	−32.0
	$H_2O(g)$	−57.7979	−54.6357	45.106	8.025
	$H_2O(l)$	−68.3174	−56.6902	16.716	17.996
	S(*c*, rhombic)	0	0	7.62	5.40
	S(*c*, monoclinic)	0.071	0.023	7.78	5.65
	$SO_2(g)$	−70.96	−71.79	59.40	9.51
	$SO_3(g)$	−94.45	−88.52	61.24	12.10

The Representative Elements *(continued)*

	Substance	ΔH_f°	ΔG_f°	S°	C_P
	$SO_4^{2-}(aq)$	−216.90	−177.34	4.1	4.0
	$H_2S(g)$	−4.815	−7.892	49.15	8.12
	$H_2S(aq)$	−9.4	−6.54	29.2	—
	Se(*c*, gray)	0	0	10.0	5.95
	Te(*c*)	0	0	11.88	6.15
Group VIIA	F(*g*)	18.3	14.2	37.917	5.436
	$F^-(aq)$	−78.66	−66.08	−2.3	−29.5
	$F_2(g)$	0	0	48.6	7.52
	HF(*g*)	−64.2	−64.7	41.47	6.95
	$Cl^-(aq)$	−40.023	−31.350	13.17	−30.0
	$Cl_2(g)$	0	0	53.286	8.11
	$ClO_2(g)$	24.7	29.5	59.6	—
	$ClO_2^-(aq)$	−16.5	3.5	24.0	—
	$ClO_3^-(aq)$	−23.50	−0.62	39.0	−18
	$ClO_4^-(aq)$	−31.41	−2.57	43.5	—
	HCl(*g*)	−22.063	−22.769	44.617	6.96
	Br(*g*)	26.71	19.69	41.8052	4.9680
	$Br^-(aq)$	−28.90	−24.574	19.29	−30.7
	$Br_2(g)$	7.34	0.751	58.639	8.60
	$Br_2(l)$	0	0	36.4	—
	$BrO_3^-(aq)$	−9.6	10.9	38.9	−19
	HBr(*g*)	−8.66	−12.72	47.437	6.96
	I(*g*)	25.482	16.766	43.1841	4.9680
	$I^-(aq)$	−13.37	−12.35	26.14	−31.0
	$I_2(g)$	14.876	4.63	62.280	8.81
	$I_2(c)$	0	0	27.9	13.14
	$I_3^-(aq)$	−12.4	−12.31	41.5	—
	$IO_3^-(aq)$	−55.0	−32.4	27.7	−19
	HI(*g*)	6.20	0.31	49.314	6.97
Group 0	He(*g*)	0	0	30.126	4.9680
	Ne(*g*)	0	0	34.948	4.9680
	Ar(*g*)	0	0	36.983	4.9680
	Kr(*g*)	0	0	39.19	4.9680
	Xe(*g*)	0	0	40.53	4.9680
	Rn(*g*)	0	0	42.10	4.9680

The Transition Elements

	Substance	ΔH_f°	ΔG_f°	S°	C_P
Group IVB	Ti(*c*)	0	0	7.24	6.010
	TiO_2(*c*, rutile)	−218.0	−203.8	12.01	13.16
	Zr(*c*)	0	0	9.18	—
Group VB	V(*c*)	0	0	7.05	5.85
	V_2O_3(*c*)	−290	−271	23.58	24.83
	V_2O_5(*c*)	−373	−344	31.3	31.00
	Nb(*c*)	0	0	8.3	—
	Ta(*c*)	0	0	9.9	6.05
Group VIB	Cr(*c*)	0	0	5.68	5.58
	CrO_4^{2-}(*aq*)	−206.3	−168.8	9.2	—
	$Cr_2O_7^{2-}$(*aq*)	−349.1	−300.5	51.1	—
	Mo(*c*)	0	0	6.83	5.61
	MoO_3(*c*)	−180.33	−161.95	18.68	17.59
	W(*c*)	0	0	8.0	5.97
Group VIIB	Mn(*c*)	0	0	7.59	6.29
	Mn^{2+}(*aq*)	−52.3	−53.4	−20	—
	MnO_2(*c*)	−124.5	−111.4	12.7	12.91
	MnO_4^-(*aq*)	−123.9	−101.6	45.4	—
Group VIIIB	Fe(*c*)	0	0	6.49	6.03
	Fe^{2+}(*aq*)	−21.0	−20.30	−27.1	—
	Fe^{3+}(*aq*)	−11.4	−2.52	−70.1	—
	Fe_2O_3(*c*)	−196.5	−177.1	21.5	25.0
	Os(*c*)	0	0	7.8	5.9
	Co(*c*)	0	0	6.8	6.11
	$CoCl_2$(*c*)	−77.8	−67.5	25.4	18.8
	Rh(*c*)	0	0	7.6	6.1
	Ir(*c*)	0	0	8.7	5.9
	Ni(*c*)	0	0	7.20	6.21
	Ni^{2+}(*aq*)	−15.3	−11.1	−38.1	—
	NiO(*c*)	−58.4	−51.7	9.22	10.60
	Pd(*c*)	0	0	8.9	6.3
	Pt(*c*)	0	0	10.0	6.35

The Transition Elements (*continued*)

	Substance	ΔH_f°	ΔG_f°	S°	C_P
Group IB	Cu(*c*)	0	0	7.96	5.848
	Cu^{2+}(*aq*)	15.39	15.53	−23.6	—
	CuO(*c*)	−37.1	−30.4	10.4	10.6
	$CuSO_4$(*c*)	−184.00	−158.2	27.1	24.1
	Ag(*c*)	0	0	10.206	6.092
	Ag^+(*aq*)	25.31	18.430	17.67	9
	Ag_2O(*c*)	−7.306	−2.586	29.09	15.67
	AgCl(*c*)	−30.362	−26.224	22.97	12.14
	Au(*c*)	0	0	11.4	6.03
Group IIB	Zn(*c*)	0	0	9.95	5.99
	Zn^{2+}(*aq*)	−36.43	−35.184	−25.45	—
	ZnO(*c*)	−83.17	−76.05	10.5	9.62
	Cd(*c*, α)	0	0	12.3	6.19
	Hg(*l*)	0	0	18.5	6.65
	Hg_2Cl_2(*c*)	−63.32	−50.350	46.8	24.3

appendix E application of calculus to rate equations

The methods of calculus are powerful tools in solving numerous equations in chemistry. The rate laws of Chapter 8 are examples of this. These equations are *differential equations*. They contain the differential expression for the rate of a chemical reaction, dc/dt, on one side of the equation. The other side of the equation is either a constant or some function of the variable c (the concentration of the species in question). The following sections illustrate how these differential equations are transformed into simple algebraic equations relating concentration to time.

Zero-Order Reactions

The rate law for a zero-order reaction is

$$-\frac{dc}{dt} = k \tag{1}$$

If we multiply through by dt, which is an infinitesimal increment in time, we obtain

$$-dc = k\,dt \tag{2}$$

The process of summing up these small increments in concentration and time is known as *integration* and is represented by the symbol

$$\int_{x_1}^{x_2}$$

The x's represent the limits of integration, the values at which the summation begins and ends. For our limits of integration, we choose $c = c_0$ at $t = 0$ and $c = c$ at $t = t$. Integration then yields

$$-\int_{c_0}^{c} dc = k \int_0^t dt \tag{3}$$

$$-(c - c_0) = kt \tag{4}$$

or

$$c = -kt + c_0 \tag{5}$$

First-Order Reactions

The rate law for a first-order reaction is

$$-\frac{dc}{dt} = kc \tag{6}$$

Multiplication by dt/c yields

$$-\frac{dc}{c} = k\,dt \tag{7}$$

$$-\int_{c_0}^{c} \frac{dc}{c} = k \int_0^t dt \tag{8}$$

The integral on the left is of the general form $\int dx/x$, for which the general solution is ln x. Thus, Equation 8 yields

$$-\ln\frac{c}{c_0} = kt \tag{9}$$

Equation 9 is easily rearranged to give Equations 8-34 and 8-35 (see Appendix B on Exponentials and Logarithms).

Second-Order Reactions

The rate law for a second-order reaction is

$$-\frac{dc}{dt} = kc^2 \tag{10}$$

Multiplication by dt/c^2 yields

$$-\int_{c_0}^{c} \frac{dc}{c^2} = k \int_0^t dt \tag{11}$$

The left-hand integral has the following general form and general solution:

$$\int \frac{dx}{x^n} = \frac{x^{-n+1}}{-n+1} + \text{constant} \tag{12}$$

Thus, Equation 11 yields

$$-\left[-\frac{1}{c} - \left(-\frac{1}{c_0}\right)\right] = kt \tag{13}$$

or

$$\frac{1}{c} - \frac{1}{c_0} = kt \tag{14}$$

If you have taken or are taking a course in calculus, the methods outlined above should be quite familiar to you. We hope that this appendix has reinforced your appreciation of the usefulness and power of calculus.

appendix F solubility products

The following data are from the *Handbook of Chemistry and Physics*, 54th edition, published by the Chemical Rubber Company. Additional data can be obtained there. The values were measured at temperatures from 12 to 28°C, with most in the range 18–25°C.

Compound	Formula	K_{sp}
Aluminum hydroxide	$Al(OH)_3$	3.7×10^{-15}
Barium carbonate	$BaCO_3$	8.1×10^{-9}
Barium chromate	$BaCrO_4$	2.4×10^{-10}
Barium fluoride	BaF_2	1.73×10^{-6}
Barium sulfate	$BaSO_4$	1.08×10^{-10}
Cadmium sulfide	CdS	3.6×10^{-29}
Calcium carbonate	$CaCO_3$	8.7×10^{-9}
Calcium fluoride	CaF_2	3.95×10^{-11}
Calcium sulfate	$CaSO_4$	2.45×10^{-5}
Cobalt sulfide	CoS	3×10^{-26}
Cupric sulfide	CuS	8.5×10^{-45}
Cuprous sulfide	Cu_2S	2×10^{-47}
Ferric hydroxide	$Fe(OH)_3$	1.1×10^{-36}
Lead carbonate	$PbCO_3$	3.3×10^{-14}
Lead chromate	$PbCrO_4$	1.77×10^{-14}
Lead fluoride	PbF_2	3.7×10^{-8}
Lead sulfate	$PbSO_4$	1.06×10^{-8}
Lead sulfide	PbS	3.4×10^{-28}
Lithium carbonate	Li_2CO_3	1.7×10^{-3}
Magnesium ammonium phosphate	$MgNH_4PO_4$	2.5×10^{-13}

Compound	Formula	K_{sp}
Magnesium carbonate	$MgCO_3$	2.6×10^{-5}
Magnesium hydroxide	$Mg(OH)_2$	1.2×10^{-11}
Manganese hydroxide	$Mn(OH)_2$	4×10^{-14}
Manganese sulfide	MnS	1.4×10^{-15}
Mercuric sulfide	HgS	4×10^{-53}
Mercurous chloride	Hg_2Cl_2	2×10^{-18}
Nickel sulfide	NiS	1.4×10^{-24}
Silver bromide	AgBr	7.7×10^{-13}
Silver carbonate	Ag_2CO_3	6.15×10^{-12}
Silver chloride	AgCl	1.56×10^{-10}
Silver chromate	Ag_2CrO_4	9×10^{-12}
Silver dichromate	$Ag_2Cr_2O_7$	2×10^{-7}
Silver hydroxide	AgOH	1.52×10^{-8}
Silver iodide	AgI	1.5×10^{-16}
Silver sulfide	Ag_2S	1.6×10^{-49}
Silver thiocyanate	AgSCN	1.16×10^{-12}
Strontium carbonate	$SrCO_3$	1.6×10^{-9}
Zinc hydroxide	$Zn(OH)_2$	1.8×10^{-14}
Zinc sulfide	ZnS	1.2×10^{-23}

appendix G dissociation constants of acids in aqueous solution

Most of the following data are from the *Handbook of Chemistry and Physics,* 54th edition, published by the Chemical Rubber Company. Additional data for acids and data for bases and amino acids can be obtained there. The values were measured at temperatures from 12.5 to 25°C, with most in the range 18–25°C.

Acid	Step	K_a	pK_a
Acetic		1.76×10^{-5}	4.75
Ascorbic	1	7.94×10^{-5}	4.10
	2	1.62×10^{-12}	11.79
Benzoic		6.46×10^{-5}	4.19
Boric	1	7.3×10^{-10}	9.14
	2	1.8×10^{-13}	12.74
	3	1.6×10^{-14}	13.80
Carbonic	1	4.30×10^{-7}	6.37
	2	5.61×10^{-11}	10.25
Chloroacetic		1.40×10^{-3}	2.85
Citric	1	7.10×10^{-4}	3.14
	2	1.68×10^{-5}	4.77
	3	6.4×10^{-6}	6.39
Formic		1.77×10^{-4}	3.75
Fumaric (*trans*)	1	9.30×10^{-4}	3.03
	2	3.62×10^{-5}	4.44

Acid	Step	K_a	pK_a
Hydrocyanic		4.93×10^{-10}	9.31
Hydrogen sulfide	1	9.1×10^{-8}	7.04
	2	1.1×10^{-12}	11.96
Lactic		1.39×10^{-4}	3.86
Malic	1	3.9×10^{-4}	3.40
	2	7.8×10^{-6}	5.11
Nitrous		4.6×10^{-4}	3.37
Oxalic	1	5.90×10^{-2}	1.23
	2	6.40×10^{-5}	4.19
Phenol		1.28×10^{-10}	9.89
Phosphoric	1	7.52×10^{-3}	2.12
	2	6.23×10^{-8}	7.21
	3	2.2×10^{-13}	12.67
o-Phthalic	1	1.3×10^{-3}	2.89
	2	3.9×10^{-6}	5.51
Propionic		1.34×10^{-5}	4.87
Succinic	1	6.89×10^{-5}	4.16
	2	2.47×10^{-6}	5.61
Sulfuric	2	1.20×10^{-2}	1.92
Sulfurous	1	1.54×10^{-2}	1.81
	2	1.02×10^{-7}	6.91

appendix H stability constants of complex ions

The following tables contain the logarithms to the base 10 of the stability constants for the reactions of ligand L with metal ion M. The reactions associated with each stability constant are

K_1	$M + L \rightarrow ML$
K_2	$ML + L \rightarrow ML_2$
K_3	$ML_2 + L \rightarrow ML_3$
etc.	

If the stability constant for the reaction

$$M + 4L \rightarrow ML_4$$

is needed, it can be obtained by multiplying the stability constants $K_1K_2K_3K_4$ or adding their logarithms. This stability constant is given the symbol β_4. The tables give constants for six ligands: OH^-, NH_3, Cl^-, CN^-, ethylenediamine (en), and ethylenediamine tetraacetate (EDTA).

The data in these tables are from L. G. Sillen and A. E. Martell, *Stability Constants of Metal-Ion Complexes,* The Chemical Society, London, 1964.

Central Atom	log K_1	log K_2	log K_3	log K_4	log K_5	log K_6
			Ligand: OH^-			
Cu(II)	6.6					
Ag(I)	2.3	1.9				
Co(II)	4.0					
Hg(II)	11.8	10.3				
Zn(II)	3.8			0.1		
Cd(II)	4.3	3.4	2.6	1.7		
Ni(II)	4.7					

Central Atom	log K_1	log K_2	log K_3	log K_4	log K_5	log K_6
			Ligand: NH_3			
Cu(II)	4.5	3.8	3.2	2.2		
Ag(I)	3.1	3.8				
Co(II)	2.1	1.6	1.1	0.8	0.2	−0.6
Hg(II)	8.8	8.7	1.0	0.8		
Zn(II)	2.4	2.4	2.5	2.2		
Cd(II)	2.7	2.1	1.4	0.9		
Ni(II)	2.8	2.2	1.7	1.2	0.8	0.03
			Ligand: Cl^-			
Cu(II)	≈0	−0.7	−1.5	−2.2		
Ag(I)	3.5	1.9	0.2	−0.4		
Co(II)	−2.4	−7.9	4.4	3.1		
Hg(II)	6.7	6.5	0.9	1.0		
Zn(II)	−0.3	0.3	−0.3	0.2		
Cd(II)	2.0	0.6	0.1	0.3		
Ni(II)	−0.3	0.2				
			Ligand: CN^-			
Cu(II)				($\beta_4 = 25$)		
Ag(I)	5.3	5.0	1.0	−0.6		
Co(II)						($\beta_6 = 19.1$)
Hg(II)	18.0	16.7	3.8	3.0		
Zn(II)				2.7		
Cd(II)	5.5	5.1	4.6	3.6		
Ni(II)				($\beta_4 = 13.8$)	−0.6	−1.0
			Ligand: $H_2N—CH_2—CH_2—NH_2$ en			
Cu(II)	10.7	9.3	1.0			
Ag(I)	4.7	3.0				
Co(II)	5.9	4.8	3.1			
Hg(II)	14.3	9.0				
Zn(II)	5.7	4.7	1.7			
Cd(II)	5.6	4.6	2.1			
Ni(II)	7.7	6.4	4.6			

Central Atom	log K_1	log K_2	log K_3	log K_4	log K_5	log K_6
Ligand: $(^-OOC-CH_2)_2N-CH_2-CH_2-N(CH_2-COO^-)_2$ EDTA						
Cu(II)	18.7					
Ag(I)	7.3					
Co(II)	16.2					
Hg(II)	22.1					
Zn(II)	16.4					
Cd(II)	16.6					
Ni(II)	18.6					

appendix I standard reduction potentials

Inorganic Systems

Most of the following data are from *Lange's Handbook of Chemistry*, 11th edition, McGraw-Hill, New York, 1973. All the potentials listed are based on the convention that the half-reaction in question constitutes the right side (the cathode) of the following cell diagram:

$$\mathrm{Pt} \mid \mathrm{H_2}(1\ \mathrm{atm}) \mid \mathrm{H^+}(1\ M) \parallel \mathrm{X^+}(1\ M) \mid \mathrm{X}$$

where X^+ is the substance being reduced, at standard conditions. The standard hydrogen electrode on the left side is arbitrarily assigned the value 0.0000 V. All the data are for the standard state except where otherwise indicated.

Reduction Half-Reaction	$\mathscr{E}°$, V
$F_2(g) + 2e^- \rightarrow 2F^-$	+2.87
$O_3(g) + 2H^+ + 2e^- \rightarrow O_2(g) + H_2O$	+2.07
$S_2O_8^{2-} + 2e^- \rightarrow 2SO_4^{2-}$	+2.01
$Ag^{2+} + e^- \rightarrow Ag^+$	+2.000 (4 M $HClO_4$)
$Pb^{4+} + 2e^- \rightarrow Pb^{2+}$	+1.8
$H_2O_2 + 2H^+ + 2e^- \rightarrow 2H_2O$	+1.77
$MnO_4^- + 4H^+ + 3e^- \rightarrow MnO_2(c) + 2H_2O$	+1.695
$PbO_2(c) + SO_4^{2-} + 4H^+ + 2e^- \rightarrow PbSO_4(c) + 2H_2O$	+1.685
$HClO_2 + 2H^+ + 2e^- \rightarrow HClO + H_2O$	+1.64
$2HClO + 2H^+ + 2e^- \rightarrow Cl_2(g) + 2H_2O$	+1.63

Reduction Half-Reaction	$\mathscr{E}°$, V
$Ce^{4+} + e^- \rightarrow Ce^{3+}$	+1.61 (1 *M* HNO_3)
$H_5IO_6 + H^+ + 2e^- \rightarrow IO_3^- + 3H_2O$	+1.6
$2HBrO + 2H^+ + 2e^- \rightarrow Br_2(l) + 2H_2O$	+1.59
$2BrO_3^- + 12H^+ + 10e^- \rightarrow Br_2(l) + 6H_2O$	+1.52
$MnO_4^- + 8H^+ + 5e^- \rightarrow Mn^{2+} + 4H_2O$	+1.51
$Au(OH)_3(c) + 3H^+ + 3e^- \rightarrow Au(c) + 3H_2O$	+1.36
$Cl_2(g) + 2e^- \rightarrow 2Cl^-$	+1.3595
$Cr_2O_7^{2-} + 14H^+ + 6e^- \rightarrow 2Cr^{3+} + 7H_2O$	+1.33
$ClO_2(g) + H^+ + e^- \rightarrow HClO_2$	+1.27
$MnO_2(c) + 4H^+ + 2e^- \rightarrow Mn^{2+} + 2H_2O$	+1.23
$O_2(g) + 4H^+ + 4e^- \rightarrow 2H_2O$	+1.229
$2IO_3^- + 12H^+ + 10e^- \rightarrow I_2(c) + 6H_2O$	+1.20
$Pt^{2+} + 2e^- \rightarrow Pt(c)$	≈+1.2
$HCrO_4^- + 7H^+ + 3e^- \rightarrow Cr^{3+} + 4H_2O$	+1.195
$ClO_4^- + 2H^+ + 2e^- \rightarrow ClO_3^- + H_2O$	+1.19
$AuCl_2^- + e^- \rightarrow Au(c) + 2Cl^-$	+1.15 (1 *M* Cl^-)
$ClO_3^- + 2H^+ + e^- \rightarrow ClO_2(g) + H_2O$	+1.15
$N_2O_4(g) + 2H^+ + 2e^- \rightarrow 2HNO_2$	+1.07
$Br_2(l) + 2e^- \rightarrow 2Br^-$	+1.065
$VO_2^+ + 2H^+ + e^- \rightarrow VO^{2+} + H_2O$	+1.000
$HNO_2 + H^+ + e^- \rightarrow NO(g) + H_2O$	+1.00
$HIO + H^+ + 2e^- \rightarrow I^- + H_2O$	+0.99
$Pd^{2+} + 2e^- \rightarrow Pd(c)$	+0.987 (4 *M* $HClO_4$)
$AuBr_2^- + e^- \rightarrow Au(c) + 2Br^-$	+0.963
$NO_3^- + 3H^+ + 2e^- \rightarrow HNO_2 + H_2O$	+0.94
$2Hg^{2+} + 2e^- \rightarrow Hg_2^{2+}$	+0.920 (1 *M* $HClO_4$)
$2NO_3^- + 4H^+ + 2e^- \rightarrow N_2O_4(g) + 2H_2O$	+0.80
$Ag^+ + e^- \rightarrow Ag(c)$	+0.7995
$Hg_2^{2+} + 2e^- \rightarrow 2Hg(l)$	+0.793
$Fe^{3+} + e^- \rightarrow Fe^{2+}$	+0.771
$H_2SeO_3 + 4H^+ + 4e^- \rightarrow Se(c) + 3H_2O$	+0.740
$PtCl_4^{2-} + 2e^- \rightarrow Pt(c) + 4Cl^-$	+0.73
$O_2(g) + 2H^+ + 2e^- \rightarrow H_2O_2$	+0.682
$Au(CN)_2^- + e^- \rightarrow Au(c) + 2CN^-$	+0.611
$2AgO(c) + H_2O + 2e^- \rightarrow Ag_2O(c) + 2OH^-$	+0.599
$H_3AsO_4 + 2H^+ + 2e^- \rightarrow HAsO_2 + 2H_2O$	+0.559
$UO_2^+ + 4H^+ + e^- \rightarrow U^{4+} + 2H_2O$	+0.55
$I_3^- + 2e^- \rightarrow 3I^-$	+0.5446 (0.5 *M* H_2SO_4)
$I_2(c) + 2e^- \rightarrow 2I^-$	+0.5345
$Cu^+ + e^- \rightarrow Cu(c)$	+0.52

Reduction Half-Reaction	$\mathscr{E}°$, V
$S_2O_3^{2-} + 6H^+ + 4e^- \rightarrow 2S(c) + 3H_2O$	+0.5
$O_2(g) + 2H_2O + 4e^- \rightarrow 4OH^-$	+0.401
$2H_2SO_3 + 2H^+ + 4e^- \rightarrow S_2O_3^{2-} + 3H_2O$	+0.40
$Cu^{2+} + 2e^- \rightarrow Cu(c)$	+0.337
$VO^{2+} + 2H^+ + e^- \rightarrow V^{3+} + H_2O$	+0.337
$UO_2^{2+} + 4H^+ + 2e^- \rightarrow U^{4+} + 2H_2O$	+0.334
$Re^{3+} + 3e^- \rightarrow Re(c)$	≈+0.3
$BiO(c) + 2H^+ + 2e^- \rightarrow Bi(c) + H_2O$	+0.28
$PbO_2(c) + H_2O + 2e^- \rightarrow PbO(c) + 2OH^-$	+0.28
$Hg_2Cl_2(c) + 2e^- \rightarrow 2Hg(l) + 2Cl^-$	+0.2676
$HAsO_2 + 3H^+ + 3e^- \rightarrow As(c) + 2H_2O$	+0.248
$As_2O_3(c) + 6H^+ + 6e^- \rightarrow 2As(c) + 3H_2O$	+0.234
$Ge^{2+} + 2e^- \rightarrow Ge(c)$	+0.23
$AgCl(c) + e^- \rightarrow Ag(c) + Cl^-$	+0.2223
$SO_4^{2-} + 4H^+ + 2e^- \rightarrow H_2SO_3 + H_2O$	+0.17
$Cu^{2+} + e^- \rightarrow Cu^+$	+0.159
$Sb_2O_3(c) + 6H^+ + 6e^- \rightarrow 2Sb(c) + 3H_2O$	+0.152
$S(c) + 2H^+ + 2e^- \rightarrow H_2S(g)$	+0.141
$HgO(c) + H_2O + 2e^- \rightarrow Hg(l) + 2OH^-$	+0.098
$AgBr(c) + e^- \rightarrow Ag(c) + Br^-$	+0.0713
$2H^+ + 2e^- \rightarrow H_2(g)$	0.0000
$P(c, \text{white}) + 3H^+ + 3e^- \rightarrow PH_3(g)$	−0.04
$CO_2(g) + 2H^+ + 2e^- \rightarrow CO(g) + H_2O$	−0.12
$WO_2(c) + 4H^+ + 4e^- \rightarrow W(c) + 2H_2O$	−0.12
$Pb^{2+} + 2e^- \rightarrow Pb(c)$	−0.126
$Sn^{2+} + 2e^- \rightarrow Sn(c)$	−0.136
$AgI(c) + e^- \rightarrow Ag(c) + I^-$	−0.152
$CO_2(g) + 2H^+ + 2e^- \rightarrow HCOOH$	−0.2
$Mo^{3+} + 3e^- \rightarrow Mo(c)$	≈−0.2
$Ni^{2+} + 2e^- \rightarrow Ni(c)$	−0.246
$V(OH)_4^+ + 4H^+ + 5e^- \rightarrow V(c) + 4H_2O$	−0.25
$V^{3+} + e^- \rightarrow V^{2+}$	−0.255
$H_3PO_4 + 2H^+ + 2e^- \rightarrow H_3PO_3 + H_2O$	−0.276
$Co^{2+} + 2e^- \rightarrow Co(c)$	−0.277
$Tl^+ + e^- \rightarrow Tl(c)$	−0.3360
$PbSO_4(c) + 2e^- \rightarrow Pb(c) + SO_4^{2-}$	−0.3553
$Cu_2O(c) + H_2O + 2e^- \rightarrow 2Cu(c) + 2OH^-$	−0.361
$Ti^{3+} + e^- \rightarrow Ti^{2+}$	−0.37
$As(c) + 3H^+ + 3e^- \rightarrow AsH_3(g)$	−0.38
$Se(c) + 2H^+ + 2e^- \rightarrow H_2Se(g)$	−0.40

Reduction Half-Reaction	$\mathscr{E}°$, V
$Fe^{2+} + 2e^- \rightarrow Fe(c)$	−0.440
$Bi_2O_3(c) + 3H_2O + 6e^- \rightarrow 2Bi(c) + 6OH^-$	−0.46
$2CO_2(g) + 2H^+ + 2e^- \rightarrow H_2C_2O_4$	−0.49
$H_3PO_2 + H^+ + e^- \rightarrow P(c) + 2H_2O$	−0.51
$Sb(c) + 3H^+ + 3e^- \rightarrow SbH_3(g)$	−0.51
$2SO_3^{2-} + 3H_2O + 4e^- \rightarrow S_2O_3^{2-} + 6OH^-$	−0.58
$U^{4+} + e^- \rightarrow U^{3+}$	−0.61
$Ni(OH)_2(c) + 2e^- \rightarrow Ni(c) + 2OH^-$	−0.66
$Ag_2S(c) + 2e^- \rightarrow 2Ag(c) + S^{2-}$	−0.69
$Te(c) + 2H^+ + 2e^- \rightarrow H_2Te(g)$	−0.72
$Cr^{3+} + 3e^- \rightarrow Cr(c)$	−0.74
$Zn^{2+} + 2e^- \rightarrow Zn(c)$	−0.7628
$2H_2O + 2e^- \rightarrow H_2(g) + 2OH^-$	−0.8277
$SiO_2(c) + 4H^+ + 4e^- \rightarrow Si(c) + 2H_2O$	−0.86
$H_3BO_3 + 3H^+ + 3e^- \rightarrow B(c) + 3H_2O$	−0.87
$SO_4^{2-} + H_2O + 2e^- \rightarrow SO_3^{2-} + 2OH^-$	−0.93
$V^{2+} + 2e^- \rightarrow V(c)$	−1.18
$Mn^{2+} + 2e^- \rightarrow Mn(c)$	−1.182
$SiF_6^{2-} + 4e^- \rightarrow Si(c) + 6F^-$	−1.2
$ZrO_2(c) + 4H^+ + 4e^- \rightarrow Zr(c) + 2H_2O$	−1.43
$Al^{3+} + 3e^- \rightarrow Al(c)$	−1.66
$U^{3+} + 3e^- \rightarrow U(c)$	−1.80
$Be^{2+} + 2e^- \rightarrow Be(c)$	−1.85
$Pu^{3+} + 3e^- \rightarrow Pu(c)$	−2.02
$Sc^{3+} + 3e^- \rightarrow Sc(c)$	−2.08
$H_2 + 2e^- \rightarrow 2H^-$	−2.25
$Mg^{2+} + 2e^- \rightarrow Mg(c)$	−2.37
$Am^{3+} + 3e^- \rightarrow Am(c)$	−2.38
$Sm^{3+} + 3e^- \rightarrow Sm(c)$	−2.41
$Na^+ + e^- \rightarrow Na(c)$	−2.714
$Ca^{2+} + 2e^- \rightarrow Ca(c)$	−2.87
$Sr^{2+} + 2e^- \rightarrow Sr(c)$	−2.89
$Ba^{2+} + 2e^- \rightarrow Ba(c)$	−2.90
$Ra^{2+} + 2e^- \rightarrow Ra(c)$	−2.92
$Cs^+ + e^- \rightarrow Cs(c)$	−2.923
$K^+ + e^- \rightarrow K(c)$	−2.925
$Rb^+ + e^- \rightarrow Rb(c)$	−2.925
$Li^+ + e^- \rightarrow Li(c)$	−3.045

(continued)

Systems of Biological Importance

The following data are from the *Handbook of Biochemistry: Selected Data for Molecular Biology,* 2nd edition, published by the Chemical Rubber Company, 1970. The quantity $\mathscr{E}^{\circ\prime}$ is the reduction potential of the substance in the biological standard state, for which it is specified that pH = 7.0, i.e., $[H^+] = 1.0 \times 10^{-7}$ mole/ℓ rather than 1.0 mole/ℓ.

The oxidized form of the substance in question is on the left side of the slash, and the reduced form on the right. The abbreviation "ox/red" is used for some substances that do not have different names for the oxidized and reduced forms. Where an oxidized/reduced pair of metal ions is followed by the name of an organic substance, it is understood that the ions are the central metal ions in the coordination compound known by that name.

Reduction Half-Reaction	$\mathscr{E}^{\circ\prime}$, V
O_2/H_2O	+0.816
Cu^{2+}/Cu^+ Hemocyanin	+0.540
NO_3^-/NO_2^-	+0.421
Epinephrine, ox/red	+0.380
O_2/H_2O_2	+0.295
Fe^{3+}/Fe^{2+} Cytochrome a	+0.29
p-Quinone/Hydroquinone	+0.28
Fe^{3+}/Fe^{2+} Cytochrome c	+0.254
Fe^{3+}/Fe^{2+} Hemoglobin	+0.17
Dehydroascorbic acid/Ascorbic acid	+0.058
Fe^{3+}/Fe^{2+} Myoglobin	+0.046
Fumarate/Succinate	+0.031
Luciferin, ox/red	−0.05
Hydroxypyruvate/Glycerate	−0.158
Oxaloacetate/Malate	−0.166
Pyruvate/Lactate	−0.185
Dihydroxyacetone phosphate/α-Glycero-phosphate	−0.192
Acetaldehyde/Ethanol	−0.197
$FAD/FADH_2$	−0.219
$FMN/FMNH_2$ (pH 7.09)	−0.219
Glutathione, ox/red	−0.23
Fe^{3+}/Fe^{2+} Peroxidase (horseradish)	−0.271
Fructose/Sorbitol	−0.272
Acetone/Isopropanol	−0.281
Lipoic acid, ox/red	−0.29

Reduction Half-Reaction	$\mathscr{E}^{\circ\prime}$, V
$NAD^+/NADH$	−0.320
Cystine/Cysteine	−0.340
Gluconolactone/Glucose	−0.364
Fe^{3+}/Fe^{2+} Ferredoxin (*Clostridium*)	−0.413
CO_2/Formate	−0.42
H^+/H_2	−0.421
SO_4^{2-}/SO_3^{2-}	−0.454
Oxalate/Glyoxalate	−0.50
Acetate/Acetaldehyde	−0.581

index

References to material found in footnotes are designated by the letter n following the page number.

540 34714
H436c

Hearst, John E.

Contemporary chemistry

THE ATOMIC WEIGHTS OF THE ELEMENTS

Element	Symbol	Atomic Number	Atomic Weight	Element	Symbol	Atomic Number	Atomic Weight
Actinium	Ac	89	[227]	Molybdenum	Mo	42	95.94
Aluminum	Al	13	26.9815	Neodymium	Nd	60	144.24
Americium	Am	95	[243]	Neon	Ne	10	20.183
Antimony	Sb	51	121.75	Neptunium	Np	93	[237]
Argon	Ar	18	39.948	Nickel	Ni	28	58.71
Arsenic	As	33	74.9216	Niobium	Nb	41	92.906
Astatine	At	85	[210]	Nitrogen	N	7	14.0067
Barium	Ba	56	137.34	Nobelium	No	102	[254]
Berkelium	Bk	97	[247]	Osmium	Os	76	190.2
Beryllium	Be	4	9.0122	Oxygen	O	8	15.9994
Bismuth	Bi	83	208.980	Palladium	Pd	46	106.4
Boron	B	5	10.811	Phosphorus	P	15	30.9738
Bromine	Br	35	79.909	Platinum	Pt	78	195.09
Cadmium	Cd	48	112.40	Plutonium	Pu	94	[244]
Calcium	Ca	20	40.08	Polonium	Po	84	[210]
Californium	Cf	98	[251]	Potassium	K	19	39.102
Carbon	C	6	12.01115	Praseodymium	Pr	59	140.907
Cerium	Ce	58	140.12	Promethium	Pm	61	[145]
Cesium	Cs	55	132.905	Protactinium	Pa	91	[231]
Chlorine	Cl	17	35.453	Radium	Ra	88	[226]
Chromium	Cr	24	51.996	Radon	Rn	86	[222]
Cobalt	Co	27	58.9332	Rhenium	Re	75	186.2
Copper	Cu	29	63.54	Rhodium	Rh	45	102.905
Curium	Cm	96	[247]	Rubidium	Rb	37	85.47